Aeronomy of the Earth's Atmosphere and Ionosphere

IAGA Special Sopron Book Series

Volume 2

Series Editor

Bengt Hultqvist
The Swedish Institute of Space Physics, Kiruna, Sweden

The International Association of Geomagnetism and Aeronomy is one of the eight Associations of the International Union of Geodesy and Geophysics (IUGG).

IAGA's Mission

The overall purpose of IAGA is set out in the first statute of the Association:

- to promote studies of magnetism and aeronomy of the Earth and other bodies of the solar system, and of the interplanetary medium and its interaction with these bodies, where such studies have international interest;
- to encourage research in these subjects by individual countries, institutions or persons and to facilitate its international coordination;
- to provide an opportunity on an international basis for discussion and publication of the results of the researches; and
- to promote appropriate standardizations of observational programs, data acquisition systems, data analysis and publication.

Volumes in this series:

The Earth's Magnetic Interior
Edited by E. Petrovský, D. Ivers, T. Harinarayana and E. Herrero-Bervera

Aeronomy of the Earth's Atmosphere and Ionosphere
Edited by M.A. Abdu, D. Pancheva and A. Bhattacharyya

The Dynamic Magnetosphere
Edited by W. Liu and M. Fujimoto

The Sun, the Solar Wind, and the Heliosphere
Edited by M.P. Miralles and J. Sánchez Almeida

Geomagnetic Observations and Models
Edited by M. Mandea and M. Korte

For titles published in this series, go to
http://www.springer.com/series/8636

Aeronomy of the Earth's Atmosphere and Ionosphere

Editors

Mangalathayil Ali Abdu
National Institute for Space Research,
Sao Jose dos Campos, SP, Brazil

Dora Pancheva
Bulgarian Academy of Sciences, Sofia, Bulgaria

Co-editor

Archana Bhattacharyya
Indian Institute of Geomagnetism,
Navi Mumbai, India

Springer

planetearth
Earth Sciences for Society

IAGA

Editors
Dr. Mangalathayil Ali Abdu
National Institute for Space Research
Sao Jose dos Campos
SP, Brazil
maabdu@dae.inpe.br

Dr. Dora Pancheva
Geophysical Institute
Bulgarian Academy of Sciences
Sofia, Bulgaria
dpancheva@geophys.bas.bg

Co-editor
Dr. Archana Bhattacharyya
Indian Institute of Geomagnetism
Navi Mumbai, India
abh@iigs.iigm.res.in

ISBN 978-94-017-7782-7 ISBN 978-94-007-0326-1 (eBook)
DOI 10.1007/978-94-007-0326-1
Springer Dordrecht Heidelberg London New York

Cover illustration: Background photo taken at sunset by Skylab. Credit NASA Earth Observatory.

Printed on acid-free paper

Springer is part of Springer Science+Business Media (www.springer.com)

Foreword by the Series Editor

The IAGA Executive Committee decided in 2008, at the invitation of Springer, to publish a series of books, which should present the status of the IAGA sciences at the time of the IAGA 2009 Scientific Assembly in Sopron, Hungary, the "IAGA Special Sopron Series". It consists of five books, one for each of the IAGA Divisions, which together cover the IAGA sciences:

Division I – Internal Magnetic Field
Division II – Aeronomic Phenomena
Division III– Magnetospheric Phenomena
Division IV– Solar Wind and Interplanetary Field
Division V – Geomagnetic Observatories, Surveys and Analyses.

The groups of Editors of the books contain members of the IAGA Executive Committee and of the leadership of the respective Division, with, for some of the books, one or a few additional leading scientists in the respective fields.

The IAGA Special Sopron Series of books are the first ever (or at least in many decades) with the ambition to present a full coverage of the present status of all the IAGA fields of the geophysical sciences. In order to achieve this goal each book contains a few "overview papers", which together summarize the knowledge of all parts of the respective field. These major review papers are complemented with invited reviews of special questions presented in Sopron. Finally, in some of the books a few short "contributed" papers of special interest are included. Thus, we hope the books will be of interest to both those who want a relatively concise presentation of the status of the sciences and to those who seek the most recent achievements.

I want to express my thanks to the editors and authors who have prepared the content of the books and to Petra van Steenbergen at Springer for good cooperation.

Kiruna, Sweden
October 2010

Bengt Hultqvist

Preface

This book presents a comprehensive set of articles on a series of wide ranging topics in the broad area of the Aeronomy of the Earth's Atmosphere and Ionosphere. Important progress achieved in recent years in our understanding of the field from the lower atmosphere to higher domains of the thermosphere and ionosphere, and extending from polar to equatorial latitudes, is discussed in this book. Overview papers on broader areas are complemented by review papers on topics of specific interest as well as shorter papers of special interest. Together they cover the current status of our understanding of the field in terms of the dynamics, chemistry, energetics and electrodynamics of the atmosphere–ionosphere system and the coupling processes that control the wide ranging characteristics of the system behaviour, and its spatial and temporal variabilities. The scientific results presented at a good number of well focussed Division II and interdivisional symposia held at the 2009 IAGA Sopron Assembly represented the rapid advances in our knowledge of the field achieved during the last few years through observational/experimental studies, sophisticated data analysis techniques and global general circulation model (GCM) and other simulation studies. In an attempt to consolidate these outstanding results in a coherent way in one place, and to mark the progress achieved in recent years in our knowledge of aeronomy, articles from leading scientists in the field who did not attend the Sopron Assembly are also included in the Book. This book is designed to be a useful reference source for graduate students as well as experienced researchers.

The book's content is structured in five sections as follows:

Part I deals with some outstanding problems of the mesosphere and lower thermosphere related to the chemistry and dynamics of the coupling processes, and vertical coupling through tides and planetary waves and climatology of mesospheric temperature, that are addressed in two overview articles. The subsequent articles are concerned with variabilities in ozone, atomic oxygen, aerosols and other minor constituent, and gravity wave effects on noctiluscent clouds.

The coupling processes that involve upward propagating gravity waves and their consequences and manifestations in the mesosphere, thermosphere and ionosphere are discussed in *Part II*. Here the focus is on questions related to upward propagation of gravity waves from sources of their generation in tropospheric convective regions of tropical latitudes, the dissipation of these waves in thermosphere and ionospheric F-region, and their role in providing seed perturbations for instability growth that leads to irregularity development in the post sunset equatorial ionosphere. Some questions related to atmosphere–ionosphere interaction through upward propagating tides and planetary waves are also discussed.

Part III deals with the electrodynamics and structuring of the ionosphere–thermosphere system. It starts with an overview article on the electrodynamics of the ionosphere–thermosphere coupling and is followed by a series of articles on equatorial vertical plasma drifts, three-dimensional modelling of equatorial plasma bubble development, coupling processes that control the development of equatorial spread F (ESF) irregularities, equatorial bubble development conditions diagnosed from long term optical data set, midlatitude ionospheric irregularities and medium scale travelling ionospheric disturbances (MSTIDs) as studied by radars, optical imagers and GPS receivers, and midlatitude plasma instabilities that arise from E-and F region coupling.

Part IV concerns Thermosphere–Ionosphere coupling, dynamics and trends. The focus here is on ion-neutral coupling and its important role in the dynamics and large scale features, such as the longitudinal wave structure found in this height region. The first article presents an overview of new aspects of the coupling between thermosphere and ionosphere that have emerged from CHAMP satellite mission results. The subsequent articles address the following topics: Influence of thermospheric winds on the Equatorial Ionization Anomaly (EIA) and secular variation (long term trend) of the EIA in the Brazilian longitude, the whole atmosphere model (GCM) simulation of temperature and density structure of equatorial thermosphere, tide induced longitudinal wave structure in equatorial thermospheric zonal wind as observed by CHAMP, connection between such structure in plasma density and the vertical plasma drift of the equatorial ionosphere, longitudinal structure of mid- and low latitude ionosphere as observed by space borne GPS receivers, and ionosphere–thermosphere coupling in low latitudes. This section further presents a tutorial on midlatitude sporadic E layers and two articles on long term trend in the upper atmosphere and on the use of F layer parameters to determine long term trends in the thermosphere dynamics.

Part V on ionosphere–thermosphere disturbance and modelling starts with an overview article on storm time responses of the thermosphere–ionosphere system. Related topics presented in subsequent articles are concerned with outstanding questions on ionospheric data assimilation and limitations on the model due to missing physics, magnetospheric electric field penetration to low latitudes during storms, modelling of the storm time electrodynamics, and a discussion of possible physical mechanism for positive ionospheric storms over low and middle latitudes.

As editors, we wish to thank all the authors of the articles for their dedicated efforts that made possible the realization of this book. We also express our gratitude to all the reviewers listed below for their help in the evaluation of the science incorporated in the present book.

Sao Jose dos Campos, Brazil — Mangalathayil Ali Abdu
Sofia, Bulgaria — Dora Pancheva
Navi Mumbai, India — Archana Bhattacharyya

Contents

Contributors

Mangalathayil Ali Abdu National Institute for Space Research, São Jose dos Campos, SP, Brazil, maabdu@dae.inpe.br

David Anderson Cooperative Institute for Research in Environmental Sciences, University of Colorado, Boulder, CO, USA; Space Weather Prediction Center, National Oceanic and Atmospheric Administration, Boulder, CO, USA, David.Anderson@noaa.gov

Graham J. Bailey Department of Applied Mathematics, Sheffield University, Sheffield, UK, G.Bailey@sheffield.ac.uk

S.M. Bailey Electrical & Computer Engineering, Virginia Polytechnic Institute and State University, Blacksburg, VA, USA

Nanan Balan Control and Systems Engineering, University of Sheffield, Sheffield, UK, B.Nanan@sheffield.ac.uk

Archana Bhattacharyya Indian Institute of Geomagnetism, Navi Mumbai, India, archana@iigs.iigm.res.in; abh@iigs.iigm.res.in

Inez S. Batista National Institute for Space Research, São José dos Campos, SP, Brazil, inez@dae.inpe.br

Tom Blix Norwegian Defence Research Establishment (FFI), Kjeller, Norway, tab@ffi.no

Phillip Bracikowski Dartmouth College, Hanover, NH, USA, pbracikowski@gmail.com

R.A. Buriti Universidade Federal de Campina Grande (UFCG), Campina Grande, Brazil, rburiti@df.ufcg.edu.br

J. Carstens Electrical & Computer Engineering, Virginia Polytechnic Institute and State University, Blacksburg, VA, USA

D. Chakrabarty Physical Research Laboratory, Ahmedabad, India, dipu@prl.res.in

C.H. Chen Department of Geophysics, Graduate School of Science, Kyoto University, Kyoto, Japan, koichi@kugi.kyoto-u.ac.jp

Mihail V. Codrescu Space Weather Prediction Center, National Oceanic and Atmospheric Administration, Boulder, CO, USA

Alexey D. Danilov Institute of Applied Geophysics, Moscow, Russia, adanilov99@mail.ru

Eurico R. de Paula Instituto Nacional de Pesquisas Espaciais (INPE), São José dos Campos, SP, Brazil, eurico@dae.inpe.br

R. Dhanya Equatorial Geophysical Research Laboratory, Indian Institute of Geomagnetism, Tirunelveli, India, dhanya.ramani@gmail.com

Erica M. Diogo National Institute for Space Research, São José dos Campos, SP, Brazil; Universidade Estadual do Rio de Janeiro – Campus Regional de Resende, Resende, RJ, Brazil, ericadiogo@gmail.com

Martin Friedrich Graz University of Technology, Graz, Austria, martin.friedrich@tugraz.at

David C. Fritts Colorado Research Associates Division, NorthWest Research Associates, Boulder, CO, USA, dave@cora.nwra.com

Hitoshi Fujiwara Department of Geophysics, Tohoku University, Sendai, Japan, fujiwara@pat.gp.tohoku.ac.jp

Timothy J. Fuller-Rowell Cooperative Institute for Research in Environmental Sciences, University of Colorado, Boulder, CO, USA; Space Weather Prediction Center, National Oceanic and Atmospheric Administration, Boulder, CO, USA, tim.fuller-rowell@noaa.gov

Sixto A. González Arecibo Observatory, Arecibo, PR, USA, sixto@naic.edu

Jörg Gumbel Stockholm University, Stockholm, Swedon, gumbel@misu.su.se

S. Gurubaran Equatorial Geophysical Research Laboratory, Indian Institute of Geomagnetism, Tirunelveli, India, gurubara@iigs.iigm.res.in

Christos Haldoupis Physics Department, University of Crete, Heraklion, Crete, Greece, chald@physics.uoc.gr

Kumiko K. Hashimoto Kyushu University of Health and Welfare, Nobeoka Miyazaki, Japan, khashi@phoenix.ac.jp

Kathrin Häusler GFZ German Research Centre for Geosciences, Telegrafenberg, Potsdam, Germany, kathrin@gfz-potsdam.de

Ove Havnes University of Tromso, Tromso, Norway, ove.havnes@uit.no

Russell B. Hedden Department of Earth and Atmospheric Sciences, Cornell University, Ithaca, NY, USA, rbh92@cornell.edu

Ulf-Peter Hoppe Norwegian Defence Research Establishment (FFI), Kjeller, Norway, uph@ffi.no

J.D. Huba Naval Research Laboratory, Washington, DC, USA, huba@ppd.nrl.navy.mil

David L. Hysell Department of Earth and Atmospheric Sciences, Cornell University, Ithaca, NY, USA, dlh37@cornell.edu

Hidekatsu Jin National Institute of Information and Communications Technology, Tokyo, Japan, jin@nict.go.jp

G. Joyce Icarus Research, Inc., Bethesda, MD, USA, joyce@ppd.nrl.navy.mil

Y. Kakinami Department of Earth Science, National Cheng Kung University, Tainan, Taiwan; Institute of Space Science, National Central University, Chung-Li, Taiwan, kaki@jupiter.ss.ncu.edu.tw

E. Alam Kherani National Institute for Space Research, São Jose dos Campos, SP, Brasil, alam@dae.inpe.br

Takashi Kikuchi Solar-Terrestrial Environment Laboratory, Nagoya University, Nagoya, Aichi, Japan, kikuchi@stelab.nagoya-u.ac.jp

Hyosub Kil Applied Physics Laboratory, The Johns Hopkins University, Laurel, MD, USA, hyosub.kil@jhuapl.edu

Scott Knappmiller University of Colorado, Boulder, CO, USA, scott.knappmiller@colorado.edu

Tsutomu Kondo Department of Earth and Planetary Sciences, Hokkaido University, Hokkaido, Japan, kondou@ep.sci.hokudai.ac.jp

N. Kotake Solar-Terrestrial Environment Laboratory, Nagoya University, Nagoya, Japan; Mitsubishi Electric, Tokyo, Japan

J. Krall Naval Research Laboratory, Washington, DC, USA, krall@ppd.nrl.navy.mil

Miguel F. Larsen Department of Physics and Astronomy, Clemson University, Clemson, SC, USA, mlarsen@clemson.edu

Jan Laštovička Institute of Atmospheric Physics ASCR, Prague, Czech Republic, jla@ufa.cas.cz

C.H. Lin Department of Earth Science, National Cheng Kung University, Tainan, Taiwan; Earth Dynamic System Research Center, National Cheng Kung University, Tainan, Taiwan, charles@mail.ncku.edu.tw

C.H. Liu Acedemia Sinica, Taipei, Taiwan, chliu2@gate.sinica.edu.tw

Han-Li Liu National Center for Atmospheric Research (HAO/NCAR), Boulder, CO, USA, liuh@ucar.edu

Huixin Liu Research Institute for Sustainable Humanosphere, Kyoto University, Uji, Japan, huixin@rish.kyoto-u.ac.jp

J.Y. Liu Institute of Space Science, National Central University, Chung-Li, Taiwan, jyliu@jupiter.ss.ncu.edu.tw

Hermann Lühr Helmholtz Centre Potsdam, GFZ German Research Centre for Geosciences, Potsdam, Germany, hluehr@gfz-potsdam.de

J. Lumpe Computational Physics Inc., Boulder, CO, USA

Thomas S. Lund NorthWest Research Associates/Colorado Research Associates Division, Boulder, CO, USA, lund@cora.nwra.com

K.A. Lynch Dartmouth College, Hanover, NH, USA, lynch@birkeland.Dartmouth.EDU

Jonathan J. Makela Department of Electrical and Computer Engineering, University of Illinois at Urbana-Champaign, Urbana, IL, USA, jmakela@illinois.edu

Daniel R. Marsh Atmospheric Chemistry Division, National Center for Atmospheric Research, Boulder, CO, USA, marsh@ucar.edu

Naomi Maruyama Cooperative Institute for Research in Environmental Sciences, Space Weather Prediction Center, University of Colorado, Boulder, CO, USA; Space Weather Prediction Center, National Oceanic and Atmospheric Administration, Boulder, CO, USA, Naomi.maruyama@noaa.gov

Jeffrey C. Mast Science Systems and Application Inc., Hampton, VA, USA, jeffrey.c.mast@nasa.gov

Astrid Maute High Altitude Observatory, National Center for Atmospheric Research, Boulder, CO, USA, maute@ucar.edu

A.F. Medeiros Universidade Federal de Campina Grande (UFCG), Campina Grande, Brazil, afragoso@df.ufcg.edu.br

Ethan S. Miller Applied Physics Laboratory, The Johns Hopkins University, Laurel, MD, USA

George H. Millward Laboratory for Atmospheric and Space Physics, University of Colorado, Boulder, CO, USA, George.Millward@noaa.gov

Yasunobu Miyoshi Department of Earth and Planetary Sciences, Kyushu University, Fukuoka, Japan, miyoshi@geo.kyushu-u.ac.jp

Martin G. Mlynczak NASA Langley Research Center, Hampton, VA, USA, m.g.mlynczak@nasa.gov

Plamen Mukhtarov Geophysical Institute, Bulgarian Academy of Sciences, Sofia, Bulgaria, engpjm@abv.bg

Sevim Müller Norwegian Meteorological Institute, Tromsø, Norway, s.mueller@met.no

John Munro University of the Virgin Islands, St. Croix, USVI, jmunro@uvi.edu

Elnana Nossa Formerly at Cornell University, Department of Earth and Atmospheric Sciences, Cornell University, Ithaca, NY, USA, en45@cornell.edu

K. Nielsen Computational Physics Inc., Boulder, CO, USA

T. Ogawa National Institute of Information and Communications Technology, Tokyo, Japan

Y. Otsuka Solar-Terrestrial Environment Laboratory, Nagoya University, Nagoya, Japan, otsuka@stelab.nagoya-u.ac.jp

Dora Pancheva Geophysical Institute, Bulgarian Academy of Sciences, Sofia, Bulgaria, dpancheva@geophys.bas.bg

Jeahueng Park Helmholtz Centre Potsdam, GFZ German Research Centre for Geosciences, Berlin, Germany; Department of Physics, Korea Advanced Institute of Science and Technology (KAIST), Daejeon, Republic of Korea, park@gfz-potsdam.de

P.T. Patil MF Radar Observatory, Indian Institute of Geomagnetism, Shivaji University Campus, Kolhapur, India, ptpiigkop@yahoo.com

I. Paulino National Institute for Space Research, São José dos Campos, SP, Brazil, igo@dae.inpe.br

P.-D. Pautet Center for Atmospheric and Space Science, Utah State University, Logan, UT, USA, dominiquepautet@gmail.com

Larry J. Paxton Applied Physics Laboratory, The Johns Hopkins University, Laurel, MD, USA

Geetha Ramkumar Space Physics Laboratory, Vikram Sarabhai Space Centre, Indian Space Research Organization, Trivandrum, India, geetha_ramkumar@vssc.gov.in

C.E. Randall Laboratory for Atmospheric and Space Physics, University of Colorado, Boulder, CO, USA

Markus Rapp Department of Radars and Sounding Rockets, Leibniz-Institute of Atmospheric Physics, Rostock University, 18225 Kuehlungsborn, Germany, rapp@iap-kborn.de

Arthur D. Richmond High Altitude Observatory, National Center for Atmospheric Research, Boulder, CO, USA, Richmond@ucar.edu

James M. Russell III Center for Atmospheric Sciences, Hampton University, Hampton, VA, USA, james.russell@hamptonu.edu

A. Saito Graduate School of Science, Kyoto University, Kyoto, Japan

S. Sathiskumar Equatorial Geophysical Research Laboratory, Indian Institute of Geomagnetism, Tirunelveli, India, sathish@iigs.iigm.res.in

Stanislav Sazykin Physics and Astronomy Department, Rice University, Houston, TX, USA, sazykin@rice.edu

L. Scherliess Center for Atmospheric and Space Sciences, Utah State University, Logan, UT, USA, ludger.scherliess@usu.edu

R.W. Schunk Center for Atmospheric and Space Sciences, Utah State University, Logan, UT, USA, robert.schunk@usu.edu

R. Sekar Space and Atmospheric Science Division, Physical Research Laboratory, Ahmedabad, India, rsekar@prl.res.in

Hiroyuki Shinagawa National Institute of Information and Communications Technology, Tokyo, Japan, sinagawa@nict.go.jp

Atsuki Shinbori Solar-Terrestrial Environment Laboratory, Nagoya University, Nagoya, Aichi, Japan, shinbori@stelab.nagoya-u.ac.jp

K. Shiokawa Solar-Terrestrial Environment Laboratory, Nagoya University, Nagoya, Japan

Anne K. Smith Atmospheric Chemistry Division, National Center for Atmospheric Research, Boulder, CO, USA, aksmith@ucar.edu

Steven Smith Center for Space Physics, Boston University, Boston, MA, USA, smsm@bu.edu

Jonas R. Souza National Institute for Space Research, São José dos Campos, SP, Brazil, jonas@dae.inpe.br

Robert W. Spiro Physics and Astronomy Department, Rice University, Houston, TX, USA, spiro@rice.edu

P.A. Stamus Colorado Research Associates Division, NorthWest Research Associates, Boulder, CO, USA, pasconsulting@earthlink.net

J. Stegman Department of Meteorology, Stockholm University, Stockholm, Sweden

Boris Strelnikov Department of Radars and Sounding Rockets, Leibniz-Institute of Atmospheric Physics, Rostock University, Kuehlungsborn, Germany, strelnikov@iap-kborn.de

Irina Strelnikova Department of Radars and Sounding Rockets, Leibniz-Institute of Atmospheric Physics, Rostock University, Kuehlungsborn, Germany, strelnikova@iap-kborn.de

Michael P. Sulzer Arecibo Observatory, Arecibo, PR, USA, msulzer@naic.edu

Hisao Takahashi Instituto Nacional de Pesquisas Espaciais (INPE), São José dos Campos, SP, Brazil, hisaotak@laser.inpe.br

Michael J. Taylor Center for Atmospheric and Space Science, Utah State University, Logan, UT, USA, Mike.Taylor@usu.edu

Kaori Terada Department of Geophysics, Tohoku University, Sendai, Japan, kaneda@pat.gp.tohoku.ac.jp

D.C. Thompson Center for Atmospheric and Space Sciences, Utah State University, Logan, UT, USA, don.thompson@usu.edu

Frank R. Toffoletto Physics and Astronomy Department, Rice University, Houston, TX, USA, toffo@rice.edu

H.F. Tsai Central Weather Bureau, Taipei, Taiwan, hftsai@cwb.gov.tw

T. Tsugawa National Institute of Information and Communications Technology, Tokyo, Japan

Yuji Tsuji Solar-Terrestrial Environment Laboratory, Nagoya University, Nagoya, Aichi, Japan, ytsuji@stelab.nagoya-u.ac.jp

Sharon L. Vadas Colorado Research Associates Division, NorthWest Research Associates, Boulder, CO, USA, vasha@cora.nwra.com; vasha@co-ra.com

Shigeto Watanabe Department of Earth and Planetary Sciences, Hokkaido University, Hokkaido, Japan, shw@ep.sci.hokudai.ac.jp

Shin-Ichi Watari National Institute of Information and Communications Technology, Koganei, Tokyo, Japan, watari@nict.go.jp

Richard A. Wolf Physics and Astronomy Department, Rice University, Houston, TX, USA, rawolf@rice.edu

C.M. Wrasse Universidade de Vale do Paraíba (UNIVAP), São José dos Campos, Brazil, cristiano.wrasse@yahoo.com.br

Tatsuhiro Yokoyama NASA Goddard Space Flight Center, Greenbelt, MD, USA, tatsuhiro.yokoyama@gsfc.nasa.gov

Y. Zhao Center for Atmospheric and Space Science, Utah State University, Logan, UT, USA

List of Reviewers

Mangalathayil A. Abdu
Rachid Akmaev
David Anderson
Nanan Balan
Charles Barden
Santimay Basu
Inez S. Batista
Paulo P. Batista
Archana Bhattacharyya
Peter Dalin
Mathew DeLand
Douglas P. Drob
Bela Fejer
Jeffrey Forbes
Tim Fuller-Rowell
Subramanian Gurubaran
Chausong Huang
Joe Huba
Kil Hyosub
David Hysell
Thomas Immel
Martin Jarvis
Martin Kaufmann
Jan Lastovicka
Charles Lin
Guiping Liu
Manuel Lopez-Puertas
Jonathan Makela
Takashi Maruyama
Eugeny Merzlyakov
Andrei Mikhailov
Nicholas J. Mitchell
Banjamin Murray
Stefan Maus
Jens Oberheide
Y. Otsuka
Duggirala Pallamraju
Scott Palo
Dora Pancheva
Victor Pasko
Amit K. Patra
Nicholas M. Pedatella
Liying Qian
Susumu Saito
Ludger Scherliess
Gordon Shepherd
Venkataraman Sivakumar
R. Sekar
Anne Smith
Jose H.A. Sobral
Mike Stevens
Wes Swartz
Richard Walterscheid

Part I
Mesosphere-Lower Thermosphere Dynamics and Chemistry

Chapter 1

Chemical–Dynamical Coupling in the Mesosphere and Lower Thermosphere

Daniel R. Marsh

Abstract A chemistry climate model is used to illustrate several ways in which dynamics and chemistry are coupled in the mesosphere and lower thermosphere (50–120 km in altitude). First the mean chemical state of the modeled atmosphere is presented, including the distribution of key radiatively-active species that determine heating and cooling rates. This is followed by a description of the major dynamical modes of variability resolved by the model, which span timescales from hours to seasons, and includes the mean meridional circulation, the upper atmospheric response to a sudden stratospheric warming, and upward propagating tidal modes. It is demonstrated that all these have significant effects on composition, either through advection, or by the temperature and pressure dependence of chemical reaction rates. For those constituents modified by dynamics that play a role in the energy budget, the potential for a feedback on the dynamics exists.

1.1 Introduction

Apart from the major constituents, molecular nitrogen and oxygen, there are no "fully-mixed" constituents in the mesosphere and lower thermosphere (MLT). In other words, all minor constituents have mixing ratios μ, defined as the ratio of number density to the total number density, that vary with height or latitude to some degree. The flux of a species is proportional to the spatial gradient of its mixing ratio and the magnitude of the wind along that gradient. It follows, then, that atmospheric motions affect the concentration of all minor constituents. This is clear when examining the continuity equation for the mixing ratio of a single constituent:

$$\frac{\partial \mu}{\partial t} + u\frac{\partial \mu}{\partial x} + v\frac{\partial \mu}{\partial y} + w\frac{\partial \mu}{\partial z} = \left(\frac{\partial \mu}{\partial t}\right)_{\text{diff}} + S$$

where t, x, y, and z are time, latitude, longitude, and height; u, v, and w are zonal, meridional, and vertical velocities; and S represents chemical production and loss. The first term on the right hand side of the equation accounts for both molecular and eddy diffusive tendencies, which usually act to reduce the gradient in μ.

The effect of dynamics on a constituent is not limited to advection. Even species with very short chemical lifetimes are affected, since S will likely depend on temperature and density dependent kinetic rate constants and on the distribution of other species that are affected by advection. For example, although ozone (O_3) is rapidly destroyed by photolysis in the day-lit mesosphere, its production depends on the rate of recombination of atomic and molecular oxygen (O and O_2) in the presence of a "third-body" (usually molecular nitrogen, N_2):

$$O + O_2 + M \rightarrow O_3 + M \tag{1.1}$$

The rate of this reaction depends on temperature ($k_{O+O2+N2} = 6.0 \times 10^{-34}(300/T)^{2.4}$), and linearly on the number density of each reactant (n_i). Since

D.R. Marsh (✉)
Atmospheric Chemistry Division, National Center for Atmospheric Research, Boulder, CO, USA
e-mail: marsh@ucar.edu

M.A. Abdu, D. Pancheva (eds.), A. Bhattacharyya (Coed.), *Aeronomy of the Earth's Atmosphere and Ionosphere*,
IAGA Special Sopron Book Series 2, DOI 10.1007/978-94-007-0326-1_1,

a constituent's number density when viewed on a constant pressure surface is inversely proportional to temperature ($n_i = \mu_i P/kT$; k is the Boltzmann's constant, P is pressure, T is temperature), it is clear that ozone's production rate is very sensitive to temperature. In addition, while O_3 is short lived, O can have a lifetime of months, and be affected by advection by the mean wind. Smith and Marsh (2005) showed that the variability in ozone around 100 km depended equally on the temperature dependence of reaction rates and to the transport of atomic oxygen.

The coupling between chemistry and dynamics is not one way. Winds and temperatures depend on atmospheric heating rates, which in turn depend on the distribution of radiatively-active constituents. Species such as O_3, nitric oxide (NO), and carbon dioxide (CO_2) determine where and when solar energy is deposited into the atmosphere and also how the atmosphere cools. Radiation in the CO_2 15 μm infrared bands is the primary way the mesosphere cools (Fomichev et al., 2004). Higher in the atmosphere emission at 5.3 μm from NO becomes another important way in which the atmosphere cools, in particular, during enhanced geomagnetic activity (Mlynczak et al., 2003). Exothermic reactions also play an important role in heating the atmosphere and the redistribution of absorbed solar energy. For example, in the case of the energy released from the reaction of atomic and molecular oxygen described in Eq. (1.1) above, it could occur far removed and months after the initial photolysis of molecular oxygen that produces the atomic oxygen. Clearly, then, coupling is two-way, with the possibility that dynamically induced changes in composition could lead to changes in heating that damp or accelerate the original dynamical perturbation (i.e. positive or negative dynamical feedbacks).

Perhaps one of the most interesting aspects of the MLT is that the timescales for chemistry and dynamics are often comparable, which means both aspects must be considered simultaneously when attempting to understand the state of this part of the atmosphere. In the following sections, a chemistry climate model is used to illustrate various couplings between chemistry and dynamics. The timescales are arranged from long to short, beginning with the global and seasonal means, and ending with variations of less than a day. All examples shown have been observed to some degree in the MLT, either by satellite or ground-based observations.

1.2 Model Description

Version 3.5 of NCAR's Whole Atmosphere Chemistry Climate Model (WACCM) is a general circulation model coupled interactively with an atmospheric chemistry model that extends from the surface to approximately 135 km. The simulations presented here are based on an updated version of the model described in Marsh et al. (2007) and Garcia et al. (2007). It includes revisions to the chemical kinetic reaction rates recommended by Sander et al. (2006) and improvements in the specification of parameterized gravity waves (Richter et al., 2010). The specification of the solar spectral irradiance has also been updated using an empirical model of the wavelength-dependent sunspot and facular influences (Lean, 2000; Wang et al., 2005).

While WACCM is usually operated as a free-running climate model, results shown here use a version of the model that incorporates reanalysis observations in the troposphere and stratosphere. This is achieved by relaxing the horizontal winds and temperatures to version 5.2 reanalysis from the Goddard Earth Observing System Data Assimilation System (GEOS-5.2) (Rienecker et al., 2008). The relaxation to GEOS-5.2 occurs with a timescale of ~10 h from the surface to 40 km. Between 40 and 50 km the amount of relaxation is linearly reduced such that at 50 km the model is free-running. Therefore, in the domain of this study, the chemical–dynamical interactions are still consistent but, as is shown in a later section, it is possible to simulate the effects on the mesosphere of a particular dynamical event that occurred in the stratosphere, and also to improve the mean wind climatology. In this mode, the model has 88 levels in the vertical, with a resolution of approximately two grid-points per scale height in the mesosphere and lower thermosphere. In the horizontal, the resolution is 2.5° longitude by 1.9° latitude.

The number of chemical species (57) and reactions (211) has remained unchanged in the latest version of WACCM. The model includes a chemistry scheme suitable for simulating the chemistry of the middle and upper atmosphere, i.e., it includes the members of oxygen, nitrogen, and hydrogen families, an E-region ionosphere, and excited species such as $O(^1D)$ and $O_2(^1\Delta)$. Figure 1.1 shows the global mean distribution of several of the modeled constituents in WACCM averaged over the year 2004. In this case it

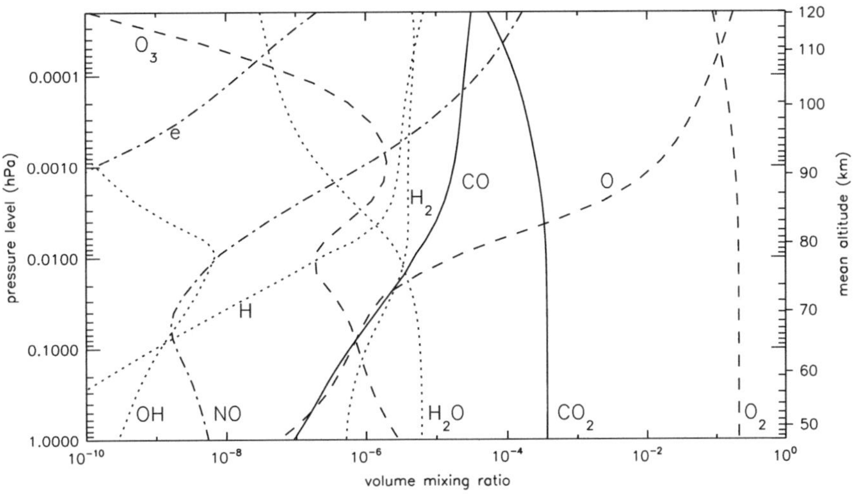

Fig. 1.1 WACCM simulated annually and globally averaged mean volume mixing ratios for constituents in the mesosphere and lower thermosphere

is reasonable to ignore the effects of advection by the resolved winds; the distributions are primarily determined by chemical production and loss and vertical diffusion. Molecular oxygen and CO_2 are essentially fully-mixed below 80–90 km, having no significant chemical sources or sinks. Above this level, however, their mixing ratios drop off through destruction from absorption of ultraviolet radiation and molecular diffusion. In WACCM transport by molecular diffusion is calculated using the formulation of Banks and Kockarts (1973), and depends on the mass of an atom or molecule relative to the mean mass of the atmosphere; heavier species have mixing ratios that tend to drop while lighter species rise.

The profile of atomic oxygen rapidly increases with height by a factor of 10^6, becoming a major constituent in the lower thermosphere. This is primarily due to the photodissociation of molecular oxygen by solar ultraviolet radiation. Water vapor, which has a mean mixing ratio of around 6 parts per million by volume (ppmv) below 0.1 hPa, is also photodissiociated by EUV radiation, specifically, Lyman-α radiation at 121.6 nm. Molecular and atomic hydrogen are produced from the products of the water vapor photolysis such that the sum 0.5 H + H_2 + H_2O is approximately constant. Hydroxyl (OH) is also produced from the destruction of H_2O, and it has a peak in mixing ratio around 0.01 hPa. Reactions with H and OH are important in determining the profile of O_3, and in particular cause the minimum that occurs at the peak of the OH layer. Ozone is lost rapidly via the following catalytic cycle:

$$H + O_3 \rightarrow OH + O_2$$
$$OH + O \rightarrow O_2 + H$$

These reactions are also important when considering the energy balance of the mesosphere, since they are exothermic and provide much of the heating between 0.01 and 0.001 hPa (Mlynczak and Solomon, 1993). In polar night, they are the only heat source and heating rates from these reactions provide in excess of 6 K/day heating (shown in Fig. 1.2b). Typically, in the diurnal mean, mesospheric heating from chemical reactions above 0.01 hPa, including the energy released in the quenching of excited species via collisions, exceeds that from the absorption of solar photons that does not go into breaking chemical bonds by factors of 2–3.

1.3 Seasonal Variability

The seasonal variation of constituents with long lifetimes in WACCM (and in the real atmosphere) depends to a large degree on the mean wind and temperature. It is therefore necessary to review the model dynamical climatology.

The modeled zonal-mean zonal winds and temperatures under solstice and equinox conditions are

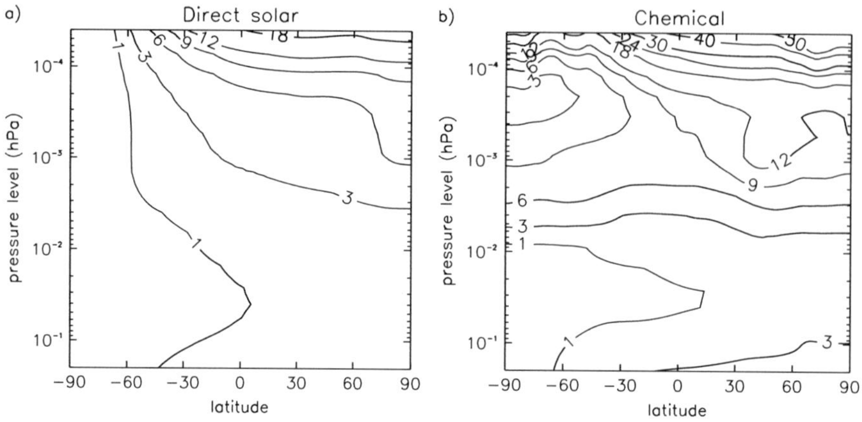

Fig. 1.2 WACCM heating rates (K/day) during July due to, (**a**) direct absorption of radiation, and (**b**) exothermic and quenching chemical reactions

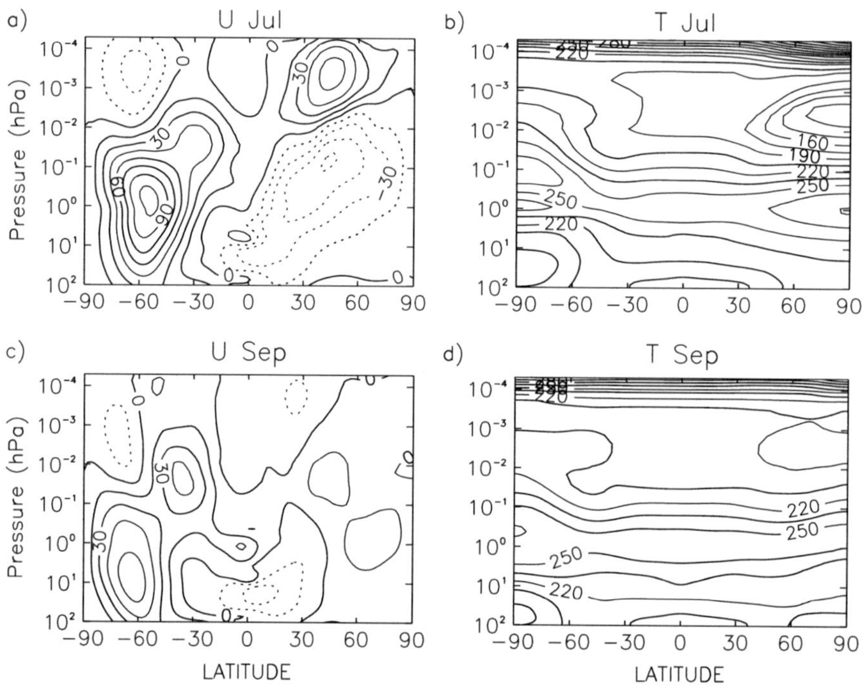

Fig. 1.3 WACCM simulated 2005 zonal-mean zonal winds and temperatures for July (**a**,**b**) and September (**c**,**d**). Units are m/s for winds and K for temperature

shown in Fig. 1.3. During July there is a strong westward jet centered at mid-latitudes around 0.1 hPa in the northern hemisphere. Constraining to reanalysis winds has led to an improvement in the location of the jet compared the free-running version of the model, which places the center of the jet in the sub-tropics (Richter et al., 2010). Directly above this jet, the winds reverse and an equally strong eastward wind maximum is predicted at 0.001 hPa. In the winter hemisphere the polar night jet peaks at the stratopause, and tilts towards the equator in the mesosphere. In September the mesospheric winds are more symmetric, with winds westward in the tropics, and eastwards at mid-latitudes. Peak wind amplitudes are weaker in September than during July. Overall, WACCM winds are in good agreement with the Upper

Atmosphere Research Satellite Reference Atmosphere Project (URAP) wind climatology (Swinbank and Ortland, 2003).

Mesopause temperatures in July (shown in Fig. 1.3b) show a two-level structure consistent with Sounding of the Atmosphere using Broadband Emission Radiometry (SABER) observations (Xu et al., 2007). The coldest temperatures (as low as 121 K) occur poleward of 40°N, around 0.003 hPa. Equatorward of this latitude and in the winter hemisphere mesopause temperatures are warmer and occur 2 scale heights higher in the atmosphere. As would be expected from the large heating rates in Fig. 1.2, temperatures above the mesopause rapidly increase into the thermosphere, driven by the absorption of UV radiation and also from diffusion of heat from higher in the thermosphere.

The cool mesopause temperatures and the closing off of the large mid-latitude jets in the lower mesosphere occur due to the dissipation of upwardly propagating gravity waves. Gravity waves are absorbed when their horizontal phase speed equals the local horizontal wind speed. Westward stratospheric winds in the summer hemisphere prevent waves with westward phase speed from propagating into the mesosphere, but allow eastward phase speed waves through where they will "break" and cause an eastward acceleration. The opposite happens in the winter hemisphere. The resulting zonal force on the mean wind is balanced by a meridional wind directed from the summer to winter hemisphere (Andrews et al., 1987). This, through mass continuity, leads to upward winds over the summer pole and descending winds over the winter pole. The associated adiabatic cooling leads in turn to the cool summer mesopause temperatures (Fig. 1.3b). These motions are best visualized in terms of residual mean meridional circulation, also called the transformed Eulerian-mean (TEM) circulation (Dunkerton et al., 1981), which are shown in Fig. 1.4 for July and September, 2005. TEM winds are consistently upward/downward over the whole mesospheric summer/winter pole; mean velocities can be as large as 5 cm/s. The summer to winter circulation peaks near 0.01 hPa, and a weaker winter to summer circulation occurs in the lower thermosphere (above 10^{-4} hPa). In September a weak, more symmetric, circulation pattern exists, with upwelling at the equator and downwelling over both polls. The absence of strong stratospheric jets allows gravity waves with eastward and westward phase speeds to propagate into the mesosphere. While their dissipation does not drive a strong residual mean meridional circulation, since the eastward and westward zonal wind forcing approximately cancels, the eddy turbulence created when they dissipate is still present, and exceeds that during July at mid-latitudes.

The effects of the residual mean meridional circulation on constituents are well documented. Satellite observations of NO (e.g. Siskind et al., 1997; Marsh and Russell, 2000), CO (e.g. López-Puertas et al., 2000; Clerbaux et al., 2005; Pumphrey et al., 2007), and H_2O (e.g. Chandra et al., 1997; Lossow et al., 2009) show large meridional gradients in the mixing ratio due to transport by the residual mean meridional circulation. WACCM distributions for these three constituents are shown in Fig. 1.5. NO and CO both have mixing ratios that increase with height. Their source region is the lower thermosphere, where they are produced through reactions that are initiated with the absorption of energetic photons (wavelengths $\leq$ 121.6 nm) or with ionization by energetic particles (primarily electrons with energies of ~1–6 KeV in the auroral regions). The horizontal gradients in NO and CO are more pronounced in July than September, as a result of the strong downwelling over the winter pole, which can bring air rich in NO and CO down to the lower mesosphere.

NO has a low mixing ratio in the mesosphere due to the following rapid loss mechanism:

$$NO + h\nu \rightarrow N + O$$
$$N + NO \rightarrow N_2 + O$$

The lifetime of NO in the sunlit mesosphere is a few days (Minschwaner and Siskind, 1993). In September, NO decreases with decreasing height uniformly with latitude, due to this loss process. However, this loss does not occur in the absence of sunlight in polar night, increasing the chemical lifetime of NO dramatically, and so allows a vertical wind of even a few cm/s to effectively transport NO into the mesosphere (Siskind and Russell, 1996; Marsh and Roble, 2002). It will be shown in the following section that the downward transport of NO in the polar night can be dramatically affected by short-term variability that originates in the stratosphere.

CO has a longer lifetime throughout the MLT and acts more like a passive tracer of transport. The winter

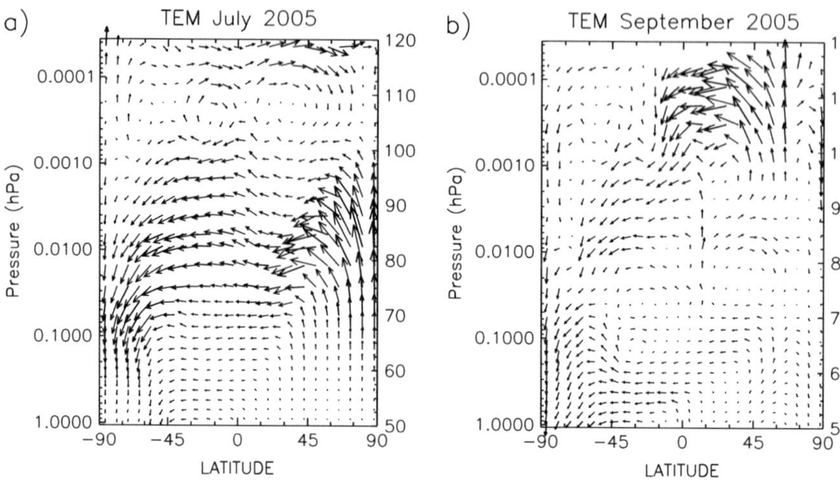

Fig. 1.4 The monthly-mean residual meridional circulation for (**a**) July and (**b**) September calculated from WACCM. Maximum wind amplitudes shown are 11.5 m/s in the horizontal direction, and 4.9 cm/s in the vertical direction, and occur in the Northern Hemisphere during July around 0.01 hPa

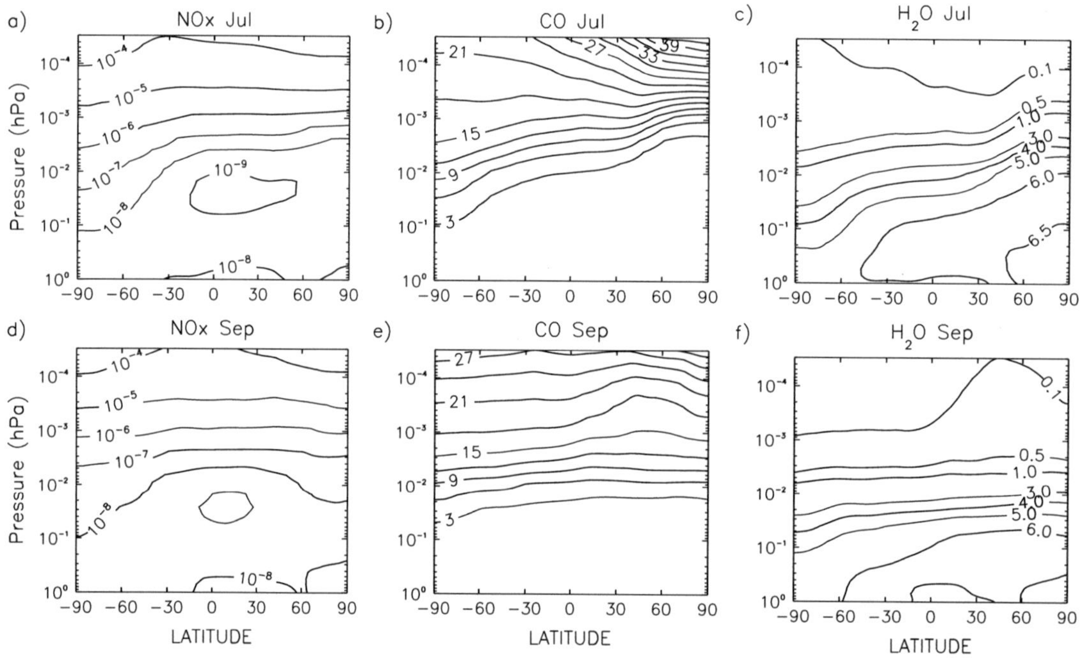

Fig. 1.5 Zonal-mean, July 2005 monthly-mean volume mixing ratios of (**a**) NO, (**b**) CO (ppm), and (**c**) H_2O (ppm). (**d–f**) as (**a–c**) for September 2005

pole downwelling effect is evident in its distribution, but also evident is the effect of the circulation in summer hemisphere. It appears as though air rich in CO is transported to the summer pole by meridional winds above 10^{-4} hPa, while CO-poor air is brought upward from the mesosphere below 10^{-3} hPa. The result is a steepening of the vertical gradient in CO at the summer mesopause. In September, the circulation centered at 30°N and 3×10^{-4} hPa clearly deforms the mixing ratio isolines.

The distribution of H_2O is almost the mirror image of CO, which is consistent with it having its source in the lower atmosphere and a loss mechanism that peaks near the mesopause. An interesting consequence of the meridional transport of water vapor is that the relatively high concentrations of water vapor and cold temperatures in the summer mesopause leads to the formation of ice clouds. A better understanding of these clouds is the goal of the current Aeronomy of Ice in the Mesosphere satellite mission (Russell et al., 2009).

Of course, transport by the residual mean meridional circulation is not limited to these constituents. A similar pattern to H_2O is seen in CO_2, and the meridional transport of atomic oxygen produced by EUV photolysis in the tropics and mid-latitudes to the polar night is key to creating the chemical heating shown in Fig. 1.2.

1.4 SSW Effects on the MLT

Stratospheric sudden warmings (SSW) are abrupt increases in the temperature of the polar stratosphere accompanied by a slowing or reversal of the eastward winds in the polar vortex. SSWs arise from the interaction of the stratospheric mean flow with upward propagating planetary waves originating in the troposphere. A SSW is considered a major warming if it leads to a reversal of the wind and an increasing temperature gradient towards the pole at or below 10 hPa (Andrews et al., 1987). The effects of an SSW in the MLT on temperatures and composition in a GCM were studied by Liu and Roble (2002). Here WACCM is used to examine the MLT response of one of the strongest major warmings observed, which began around the third week of 2006 (Siskind et al., 2007; Manney et al., 2008).

Figure 1.6 shows the variation in WACCM zonal-means of the zonal wind, temperature, CO, and NO_y ($NO + NO_2 + 2{*}N_2O_5 + HNO_3$) between the beginning of December 2005 and mid-March 2006. Recall, in the model used here the winds and temperature below 1 hPa are relaxed to reanalysis data, and so the stratospheric winds are very close to the observed winds for this period. In the lower-mesophere and upper-stratosphere there is a reversal of the zonal wind direction (shown between 55°N and 70°N) from greater than 60 m/s eastward in December to 40 m/s westward in January. Following this reversal, the winds revert to being eastward in February. In the upper mesosphere the opposite shift occurs with winds going from westward to eastward and returning to westward. Simultaneous with wind reversal is a warming of the polar upper stratosphere, and cooling of the mesosphere. This is followed by a dramatic shift in the height of the stratopause to around 0.01 hPa at the beginning of February. Over time the stratopause descends to around the height seen before the stratospheric warming.

The evolution of the zonal winds and temperatures in the mesosphere agrees very well with the Microwave Limb Sounder (MLS) and Sounding of the Atmosphere with Broadband Emission Radiometry (SABER) observations reported by Manney et al. (2008). WACCM is better at reproducing the elevated stratopause than either GEOS-5 or European Centre for Medium-Range Weather Forecasts analyses, presumably because the height of the elevated stratopause is close to the top of the models used to calculate the analyses, whereas the upper boundary of WACCM is much higher. Other factors could include the way these models handle non-orographic gravity waves and specification of ozone in radiative transfer calculations (Manney et al., 2008).

The temporal variations in CO and NOy mixing ratios (Fig. 1.6c, d) appear to be strongly correlated to the dynamical variations over this period. CO at 0.01 hPa drops from December values in excess of 10 ppmv to around 3.3 ppmv by mid-January. This is followed by a rapid increase of CO to values around 13 ppmv by early February. This variation is very similar to that seen by the Atmospheric Chemistry Experiment Fourier Transform Spectrometer and MLS instruments (Jin et al., 2009). The fact that WACCM reproduces the observed variability in temperature and CO gives confidence that the model is correctly reproducing the dynamical variability in the mesosphere.

As mentioned previously, CO has a long chemical lifetime and a mixing ratio that increases with height, and so it is sensitive to changes in the residual vertical wind ($\overline{w}^*$), which is shown averaged over the pole in Fig. 1.7c. Between 0.1 and 0.001 hPa there is a good correlation between upward tilt of the CO contours (decreasing CO over time) and positive (upward) vertical residual winds. This is most clearly seen in the second week of January when $\overline{w}^*$ exceeds 3 cm/s. This occurs shortly after the zonal winds reverse direction at

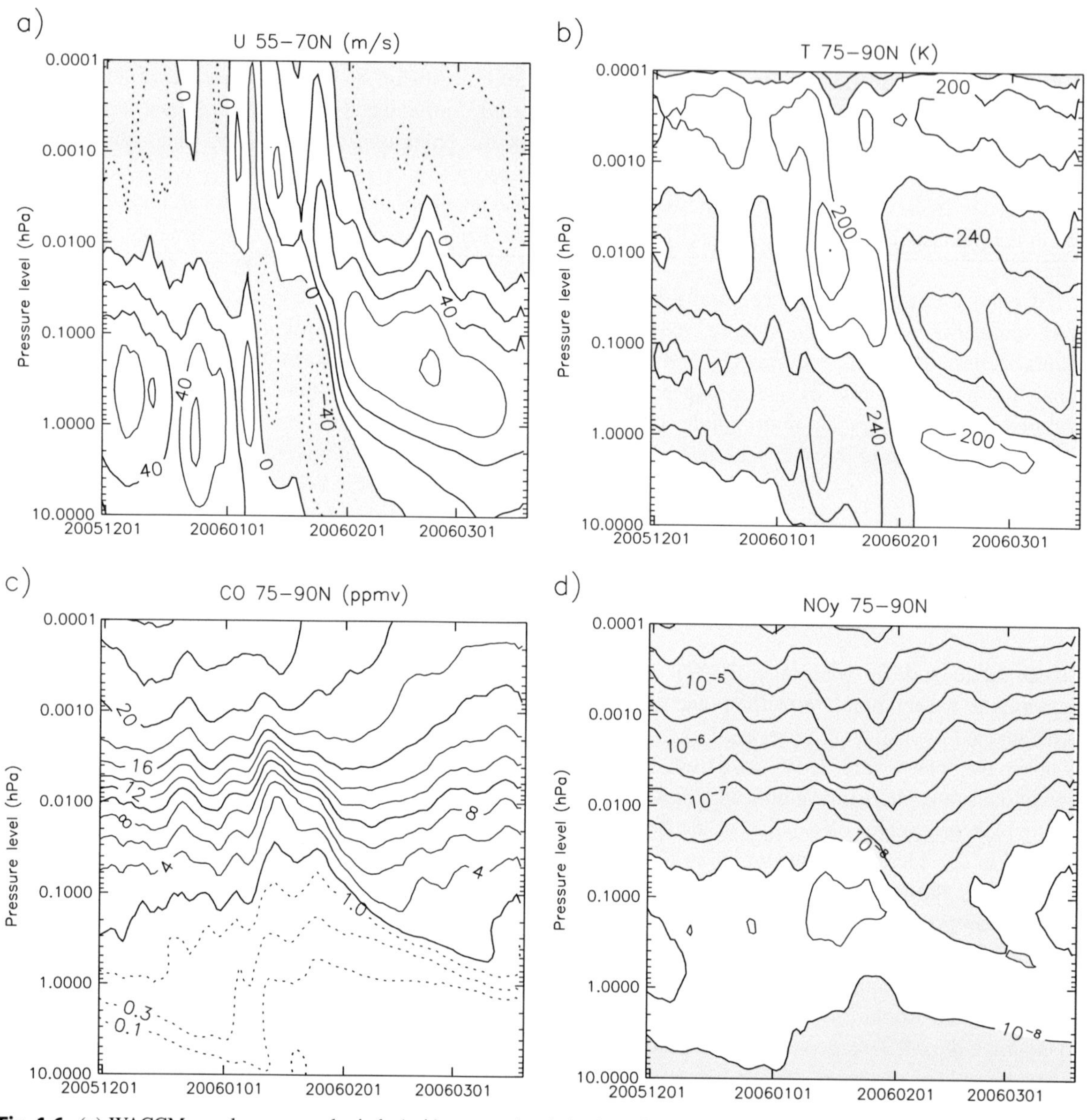

Fig. 1.6 (**a**) WACCM zonal-mean zonal winds (m/s) averaged over the latitude band 55°N–70°N over the period December 1, 2005 to March 15, 2006. (**b**), (**c**) and (**d**) are averages over latitudes poleward of 75°N of temperature (K), carbon monoxide vmr (ppm) and NOy vmr. All fields are daily means smoothed in time with a 5-day window

mid-latitudes, at which time the residual mean meridional winds switch from poleward to equatorward (Fig. 1.7b). The reversal of the zonal winds occurs because of a dramatic change in the acceleration due to the dissipation of breaking gravity waves (shown in Fig. 1.7a) which changes from more than 50 m/s/day westward in December to more than 75 m/s/day eastward in January. The westward stratospheric winds that occur during the SSW allow waves with eastward phase speed to propagate into the mesosphere where they dissipate and cause an eastward acceleration.

Following the SSW, and a return to eastward stratospheric winds in February, the GW forcing at 0.01 hPa is again westward, driving a polward and downward residual circulation and a warming of the mesosphere through adiabatic heating. Both NOy and

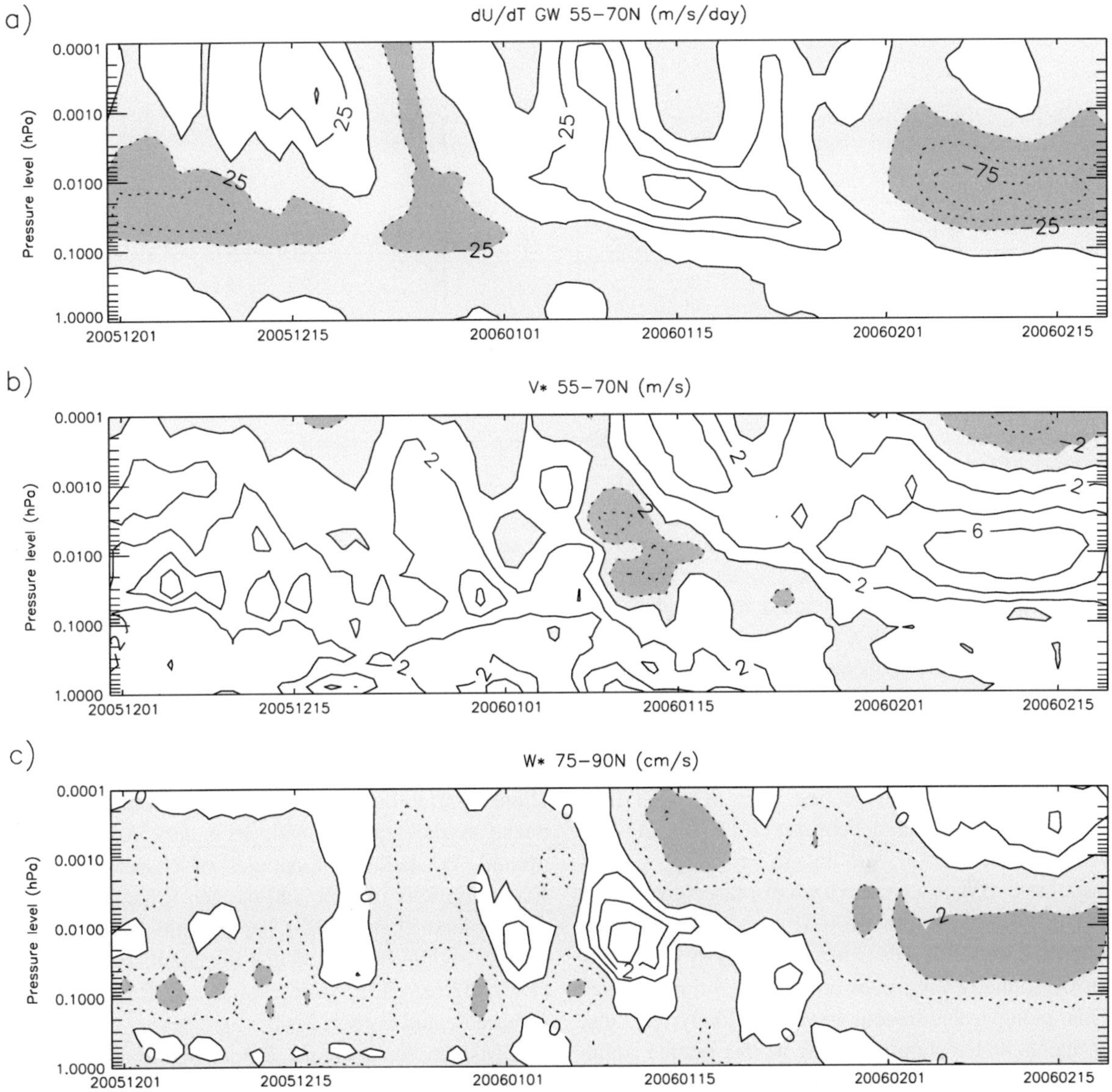

Fig. 1.7 (**a**) Acceleration of the zonal wind by parameterized dissipating gravity waves (m/s/day) and (**b**) the residual mean meridional wind (m/s) averaged over the latitude band 55°–70°N. (**c**) The residual mean vertical wind (cm/s) averaged over latitudes poleward of 75°N. All figures are for the same time period as Fig. 1.6

CO are now rapidly transported downward, leading to a significant increase in their mixing ratios in the lower mesosphere.

WACCM simulations seem better able to reproduce the elevated stratopause than the simulations from the NOGAPS-ALPHA (Navy Operational Global Atmospheric Prediction System – Advanced Level Physics High Altitude) GCM of Siskind et al. (2007). In their model, which only includes the effects of variations in the orographic gravity wave drag, they were not able to reproduce temperatures in excess of 250 K at 0.01 hPa. In WACCM, forcing from orographic gravity waves is typically much less than the non-orographic waves. Prior to the stratospheric warming orographic gravity wave forcing was ~4 m/s/day below 0.02 hPa, much smaller above that level and at all levels above 1 hPa after the stratospheric warming. Siskind et al. (2007) suggest the deficiencies in the

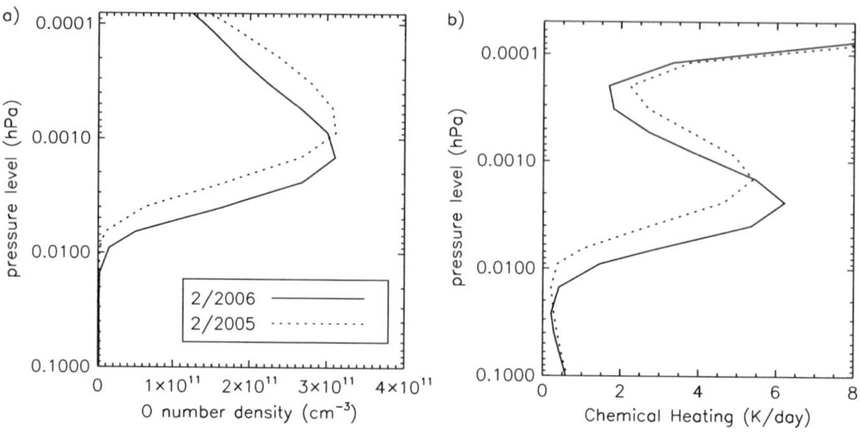

Fig. 1.8 (**a**) Atomic oxygen density (cm^{-3}), and (**b**) chemical heating rates (K/day) averaged over the polar cap (>70°N) for February 2005 and 2006

NOGAPS-ALPHA simulations could be due to not including non-orographic gravity waves or the chemical heating that would be caused by bringing thermospheric atomic oxygen into the mesosphere. This seems a reasonable conclusion, since WACCM has both of these processes, and can produce the observed warm temperatures around 0.01 hPa.

Figure 1.8 compares zonal mean atomic oxygen densities and chemical heating rates averaged from 70°N to 90°N between February 2006 and 2005 (a year without a major warming in February). Clearly the atomic oxygen layer has been displaced downward, and heating rates between 0.01 and 0.001 hPa have increased. At 0.007 hPa O has increased by over a factor of 5, and heating rates are almost 3 times higher. This, along with adiabatic heating is likely the cause of the high temperatures seen in the middle atmosphere following the stratospheric warming. Further corroborating WACCM predictions is a recent study that concluded that the descent and increase in atomic oxygen densities led to the lowering and brightening of the OH Meinel airglow emission in the winter of 2006 relative to 2005 observed by the SABER instrument (Winick et al., 2009). The downward transport of O was also the likely cause of the enhanced ozone seen during this period (Smith et al., 2009).

1.5 Tidal Variability

There is strong observational evidence that upward propagating atmospheric tides affect composition. This is seen both in direct observations (e.g., those made via occultation) and in the spontaneous emission from constituents in excited states (airglow). As with other atmospheric motions, tides advect constituents, but also affect reaction rates through changes in temperature and density. In the tropical and sub-tropical upper mesosphere the migrating (sun-synchronous) diurnal tide is one of the largest drivers or short-term constituent variability. The signature of this tide typically peaks at the equator, with secondary perturbations around 35° latitude. Examples of observed diurnal tidal variations include odd-oxygen (Shepherd et al., 1995; Marsh et al., 2002; Chapter 5, this volume), OH Meinel nightglow (Zhang and Shepherd, 1999; Marsh et al., 2006), nitric oxide (Marsh and Russell, 2000; Oberheide and Forbes, 2008).

Consider, for example, a tidal perturbation in the vertical wind. A temperature increase, due to adiabatic heating, will occur that lags the minimum in the vertical wind by a quarter of the wave period. If the chemical lifetime of the advected constituent is considerably longer than the period of the wave then it too will have its largest perturbation a quarter of the wave period after the vertical wind minimum (Marsh and Roble, 2002). The sign of the perturbation depends on the sign of the vertical gradient in the constituent mixing ratio. In the case of nitric oxide and atomic oxygen, which both rapidly increase with height in the MLT (see Fig. 1.1), the constituent variations are in phase with those of temperature. The amplitude of the perturbation will depend on the wind amplitude and the vertical gradient in the mixing ratio of the constituent. The temperature and density effects can work to enhance or diminish the tidal signal in the constituent. For example, Marsh et al. (2006) calculated that, at a specific local time, the tendency in OH Meinel 2 μm emission due to advection of atomic oxygen was an increase at 88 km of 50%, but that density and temperature effects reduced the net tidal response of the emission to around +22%. In the case of ozone at 95 km, it is thought the high ozone (>20 ppmv) observed in some circumstances at the equator is the result of very low temperatures caused by the diurnal tide (Smith et al., 2008).

Figure 1.9 presents WACCM longitude/latitude distributions at midnight Universal Time (UT00) averaged for all days in March 2006. These synoptic global views at 1.46×10^{-3} hPa (94 km in log-pressure altitude) show large variations around a latitude circle, particularly in the tropics. Temperatures at the equator vary by 35 K, and atomic oxygen mixing ratio varies between 0.005 and 0.019. The temperature and oxygen variations are positively correlated, while the water vapor variations are anti-correlated (Fig. 1.9c), which is consistent with tidal variability driven by tidal vertical motions as discussed before. At the equator the maximum in atomic oxygen occurs just before midnight local time (~330° longitude). This agrees with pre-midnight maximum observed in OH Meinel emissions, which can be shown to be proportional to atomic oxygen concentrations (Marsh et al., 2006). The total heating rate (chemical heating and solar absorption) is shown in Fig. 1.9d. The peak at noon (180° longitude) in excess of 20 K/day is to be expected, and occurs from the peak in the absorption of solar radiation and quenching of $O(^1D)$. However, there is peak in the heating of comparable magnitude shortly before midnight, which can only come from enhanced chemical heating due to atomic oxygen recombination and other exothermic reactions.

While the patterns seen in these synoptic maps are compatible with a tidally-driven perturbation mechanism, proof requires further analysis. After all, variations in longitude could be stationary, and not vary with local time at all. Using Fourier analysis, the temperature timeseries at 30-min resolution over the entire month can be approximated as the sum of a monthly mean and a series of migrating and non-migrating diurnal and semi-diurnal perturbations. Not all of these tidal components have significant amplitudes for the chosen altitude and month, so only those with large amplitudes need be retained in the sum. In Fig. 1.10, the reconstructed temperature field at UT00 (minus the mean) is presented. For a fixed universal time, longitudinal variations are equivalent to local time variations, and this is shown in the upper axis. It is clear that the majority of the variability is captured in the reconstruction, including the large equatorial temperature maximum just before midnight, proving that the variation in longitude is primarily tidal, and not a stationary feature (i.e. the spatial variability in temperature is a function of local time as well as longitude).

Figure 1.10 also shows the latitudinal structure of tidal amplitudes for all tides included in the reconstruction, separated into westward or stationary and eastward propagating modes. The largest amplitude calculated is for the westward zonal wavenumber 1 diurnal tide (designated DW1). This is the migrating diurnal tide, and has a peak amplitude at the equator of 10 K. Near the equator the semi-diurnal migrating tide (wavenumber 2) in temperature has its maximum amplitude of 4.5 K. There are secondary temperature maxima at ±30° latitude. The temperature structure and amplitudes are consistent with the SABER March mean measurements reported by Pancheva et al. (2009). Of the non-migrating tides, the diurnal westard-2 (DW2) and eastward-3 (DE3) have the largest amplitudes (3–5 K). At this height and month, the contribution of the non-migrating semi-diurnal tides (not shown) does not appear to be significant: temperature amplitudes for wave numbers between ±4 range between 0.3 and 0.9 K. Higher in the atmosphere the non-migrating semi-diurnal tides need to be included when considering the total variability.

The vertical wind tides associated with the temperature tides in Fig. 1.10a are shown in Fig. 1.10b. At the equator, there is a close correspondence between the times of zero vertical wind and maxima or minima in the tidal temperature. Following an extended period of downwelling beginning around 1600 h local time, there is a maximum in the tidal temperature at 2300 h, which is consistent with the phase lag described earlier. This perturbation comes mostly from the migrating diurnal tide, which has the largest wind amplitude at the equator, reaching almost 10 cm/s. The DW2 tide is also large at the equator, reaching around 5 cm/s. Combined, a 15 cm/s amplitude vertical wind diurnal tidal will produce approximately 20 K variation in temperature through adiabatic heating alone, so the

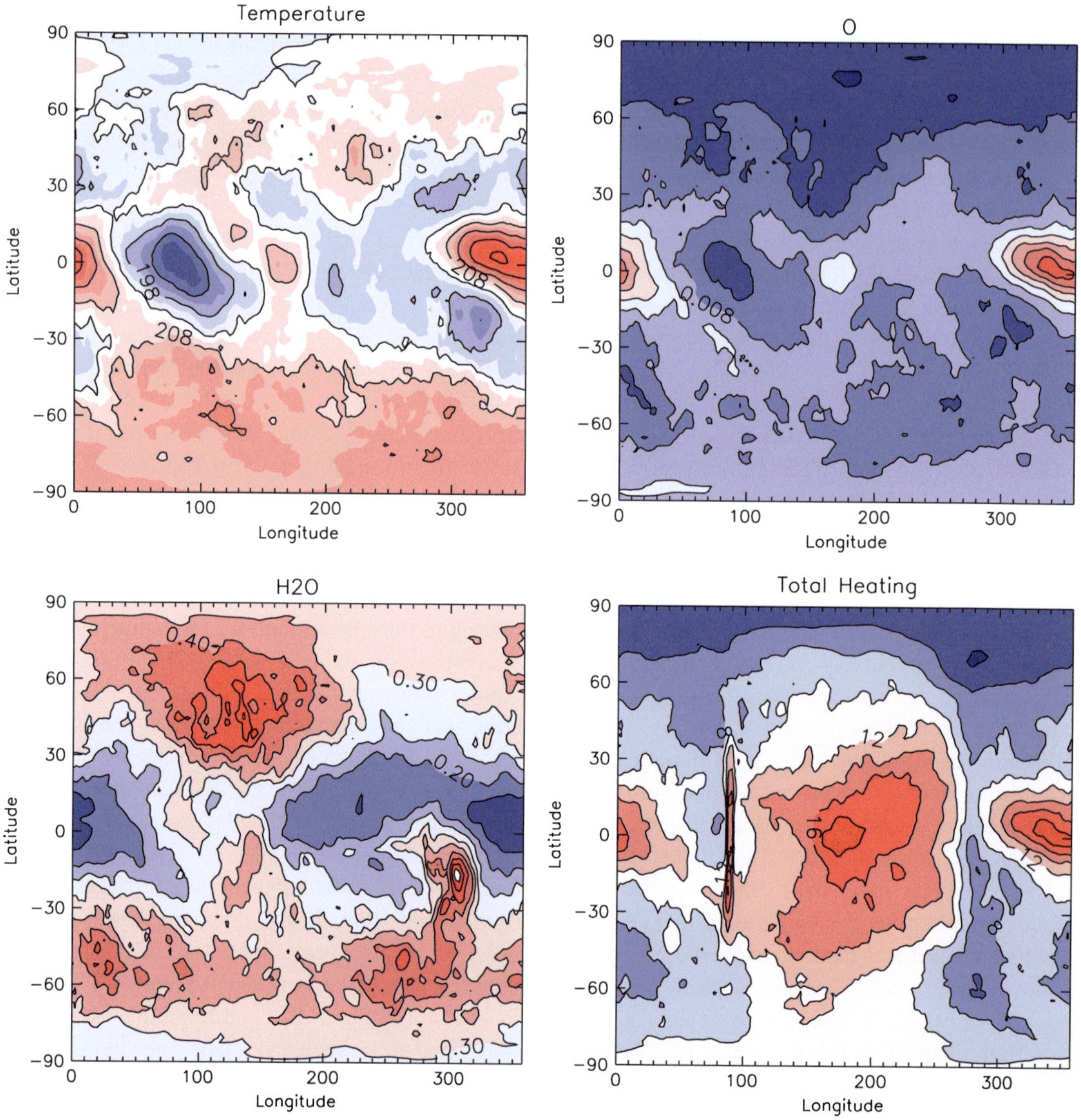

Fig. 1.9 The global distribution at UT00 of temperature, atomic oxygen, water vapor, and heating rates at 1.46×10^{-3} hPa, averaged over the month of March 2006

vertical tidal winds are of sufficient amplitude to create the temperature perturbations in Fig. 1.10a. Note that while the wind amplitude of the semi-diurnal tide (8 cm/s at the equator) is almost as large as the migrating diurnal tide, its temperature amplitude is half of that tide, simply because its frequency is double; vertical advection occurs over only 12 h and so the parcel displacement and total adiabatic heating or cooling is half. The tidal vertical wind structure is not only consistent with the temperature structure but also with the H_2O, O and chemical heating. The vertical gradients in H_2O and O are of opposite sign and so the perturbation induced by the vertical wind is also of opposite sign, and the chemical heating variation simply follows the O variation.

It is often the case when analyzing constituent and airglow data from satellites with slowly-precessing orbits that observations are averaged into local time bins with no account for the longitudinal variability. Day or night observations around a longitude circle

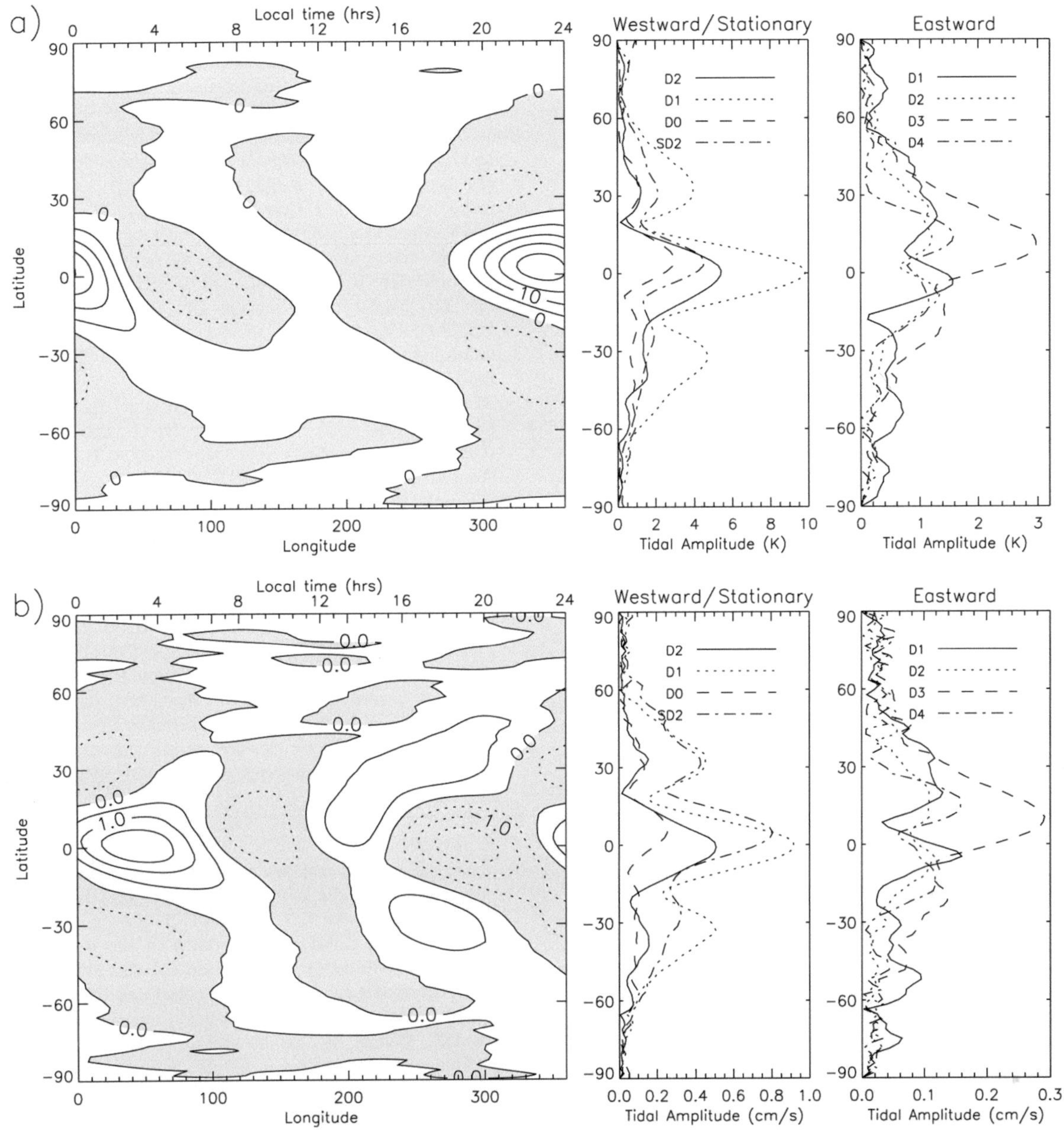

Fig. 1.10 (**a**) Tidal temperature and (**b**) vertical wind pertubations at UT00 averaged for March 2006, and amplitudes of largest diurnal and semi-diurnal tidal modes. All data at 1.46×10^{-3} hPa

will usually be at approximately one local time, so averaging into local time bins is essentially a zonal mean. This method isolates the migrating tide since it is only a function of local time, but non-migrating tides that are functions of local time and longitude are removed in the zonal mean. It is clear from the above analysis that non-migrating tides do play an important role in chemical variability in the mesopause region, as is seen in Figs. 1.9 and 1.10. This is particularly important in understanding the diurnal variation of a constituent or airglow when viewed from a single location. Certainly their effects must be considered when comparing satellite to ground-based observations.

1.6 Summary

This overview has attempted to illustrate some of the ways chemistry and dynamics are coupled in the MLT using a 3-dimensional chemistry climate model. The dynamical state of the atmosphere clearly responds to changes in heating and cooling related to the distribution of minor constituents. For example, chemical heating plays a critical role in the MLT energy budget in polar night and also during the nighttime hours at the equator, where it is observed indirectly in airglow observations. Conversely, the distribution of all minor constituents depends, to some degree, on dynamics, which can either transport the constituent directly, or affect them indirectly by modifying the rates of chemical reactions. The changes in chemistry wrought by dynamics in turn feed back on the winds and temperatures through modification of heating rates.

It is worth noting that the coupled system described above is additionally strongly coupled to the lower atmosphere. All the examples given here depend on wave fluxes from below: the residual circulation is driven by gravity waves originating in the troposphere, major SSWs dramatically affect high-latitude mesospheric descent rates and heating in the troposphere and stratosphere leads to large amplitude tides in the MLT. This review is far from all-inclusive, and it remains a significant challenge to understand the entire coupled system. Meeting that challenge will entail continued advances in both modeling and observations.

Acknowledgements The National Center for Atmospheric Research is sponsored by the National Science Foundation. Any opinions, findings and conclusions or recommendations expressed in the publication are those of the author and do not necessarily reflect the views of the National Science Foundation. Support for this research was also provided by the National Aeronautics and Space Administration, grants NNX08AL61G and NNX08AX28G. The author would like to thank Anne Smith for providing insightful comments on this manuscript.

References

Andrews DG, Holton JR, Leovy CB (1987) Middle atmosphere dynamics. Academic, New York, NY

Banks PM, Kockarts G (1973) Aeronomy, part B. Elsevier, New York, NY, 355 pp

Chandra S, Jackman CH, Fleming EL, Russell JM (1997) The seasonal and long term changes in mesospheric water vapor. Geophys Res Lett 24:639–642. doi:10.1029/97GL00546

Clerbaux C, Coheur P-F, Hurtmans D, Barret B, Carleer M, Colin R, Semeniuk K, McConnell JC, Boone C, Bernath P (2005) Carbon monoxide distribution from the ACE-FTS solar occultation measurements. Geophys Res Lett 32:L16S01. doi:10.1029/2005GL022394

Dunkerton T, Hsu C-PF, McIntyre ME (1981) Some Eulerian and Lagrangian diagnostics for a model stratospheric warming. J Atmos Sci 38:819–843

Fomichev VI, Fu C, de Grandpré J, Beagley SR, Ogibalov VP, McConnell JC (2004) Model thermal response to minor radiative energy sources and sinks in the middle atmosphere. J Geophys Res 109:D19107. doi:10.1029/2004JD004892

Garcia RR, Marsh DR, Kinnison DE, Boville BA, Sassi F (2007) Simulation of secular trends in the middle atmosphere, 1950–2003. J Geophys Res 112:D09301. doi:10.1029/2006JD007485

Jin JJ et al (2009) Comparison of CMAM simulations of carbon monoxide (CO), nitrous oxide (N_2O), and methane (CH_4) with observations from Odin/SMR, ACE-FTS, and Aura/MLS. Atmos Chem Phys 9:3233–3252

Lean J (2000) Evolution of the Sun's spectral irradiance since the maunder minimum. Geophys Res Lett 27(16):2425–2428. doi:10.1029/2000GL000043

Liu HL, Roble RG (2002) A study of a self-generated stratospheric sudden warming and its mesospheric-lower thermospheric impacts using the coupled TIME-GCM/CCM3. J Geophys Res 107(D23):4695. doi:10.1029/2001JD001533.

López-Puertas M, López-Valverde MA, Garcia RR, Roble RG (2000) A review of CO_2 and CO abundances in the middle atmosphere. In: Siskind DE, Eckermann SD, Summers ME (eds) Atmospheric science across the stratopause. Geophysical monograph series, vol 123. AGU, Washington, DC, pp 83–100

Lossow S, Urban J, Schmidt H, Marsh DR, Gumbel J, Eriksson P, Murtagh D (2009) Wintertime water vapor in the polar upper mesosphere and lower thermosphere: first satellite observations by Odin submillimeter radiometer. J Geophys Res 114:D10304. doi:10.1029/2008JD011462

Manney GL et al (2008) The evolution of the stratopause during the 2006 major warming: satellite data and assimilated meteorological analyses. J Geophys Res 113:D11115. doi:10.1029/2007JD009097

Marsh DR, Garcia RR, Kinnison DE, Boville BA, Sassi F, Solomon SC, Matthes K (2007) Modeling the whole atmosphere response to solar cycle changes in radiative and geomagnetic forcing. J Geophys Res 112:D23306. doi:10.1029/2006JD008306

Marsh DR, Roble RG (2002) TIME-GCM simulations of lower-thermospheric nitric oxide seen by the halogen occultation experiment. J Atmos Solar-Terr Phys 64(8–11):889–895. doi:10.1016/S1364-6826(02)000445

Marsh DR, Russell JM III (2000) A tidal explanation for the sunrise/sunset anomaly in HALOE low-latitude nitric oxide observations. Geophys Res Lett 27:3197–3200

Marsh DR, Skinner WR, Marshall AR, Hays PB, Ortland DA, Yee J-H (2002) High resolution Doppler imager observations of ozone in the mesosphere and lower thermosphere. J Geophys Res 107(D19):4390. doi:10.1029/2001JD 001505

Marsh DR, Smith AK, Mlynczak MG, Russell JM III (2006) SABER observations of the OH Meinel airglow variability near the mesopause. J Geophys Res 111:A10S05. doi:10.1029/2005JA011451

Minschwaner K, Siskind DE (1993) A new calculation of nitric oxide photolysis in the stratosphere, mesosphere, and lower thermosphere. J Geophys Res 98(D11):20, 401–420, 412

Mlynczak MG, Solomon S (1993) A detailed evaluation of the heating efficiency in the middle atmosphere. J Geophys Res 98:10517–10541

Mlynczak M et al (2003) The natural thermostat of nitric oxide emission at 5.3 mm in the thermosphere observed during the solar storms of April 2002. Geophys Res Lett 30(21):2100. doi:10.1029/2003GL017693

Oberheide J, Forbes JM (2008) Thermospheric nitric oxide variability induced by nonmigrating tides. Geophys Res Lett 35:L16814. doi:10.1029/2008GL034825

Pancheva D, Mukhtarov P, Andonov B (2009) Global structure, seasonal and interannual variability of the migrating semidiurnal tide seen in the SABER/TIMED temperatures (2002–2007). Ann Geophys 27:687–703

Pumphrey HC et al (2007) Validation of middle-atmosphere carbon monoxide retrievals from the microwave limb sounder on aura. J Geophys Res 112:D24S38. doi:10.1029/2007JD008723

Richter JH, Sassi F, Garcia RR (2010) Toward a physically based gravity wave source parameterization in a general circulation model. JAS. doi:10.1175/2009JAS3112.1

Rienecker MM et al (2008) The GEOS-5 data assimilation system—documentation of versions 5.0.1, 5.1.0, and 5.2.0. NASA/TM-2008-104606, vol 27. Technical Report Series on global modeling and data assimilation. Goddard Space Flight Center, Greenbelt, Maryland, 118 pp

Russell JM III et al (2009) Aeronomy of ice in the mesosphere (AIM): overview and early science results. J Atmos Solar-Terr Phys. doi:10.1016/j.jastp.2008.08.01

Sander SP et al (2006) Chemical kinetics and photochemical data for use in atmospheric studies. Evaluation 15, JPL Publ. 06-2

Shepherd GG, McLandress C, Solheim BH (1995) Tidal influence on $O(^1S)$ airglow emission rate distributions at the geo-graphic equator as observed by WINDII. Geophys Res Lett 22:275–278

Siskind DE, Bacmeister JT, Summers ME, Russell JM III (1997) Two-dimensional model calculations of nitric oxide transport in the middle atmosphere and comparison with halogen occultation experiment data. J Geophys Res 102(D3): 3527–3545

Siskind DE, Eckermann SD, Coy L, McCormack JP, Randall CE (2007) On recent interannual variability of the arctic winter mesosphere: implications for tracer descent. Geophys Res Lett 34:L09806. doi:10.1029/2007GL029293

Siskind DE, Russell JM III (1996) Coupling between middle and upper atmospheric NO: constraints from HALOE observations. Geophys Res Lett 23(2):137–140

Smith AK, Lopez-Puertas M, Garcia-Comas M, Tukiainen S (2009) SABER observations of mesospheric ozone during NH late winter 2002–2009. Geophys Res Lett 36:L23804. doi:10.1029/2009GL040942

Smith AK, Marsh DR (2005) Processes that account for the ozone maximum at the mesopause. J Geophys Res 110:D23305. doi:10.1029/2005JD006298

Smith AK, Marsh DR, Russell JM III, Mlynczak MG, Martin-Torres FJ, Kyrölä E (2008) Satellite observations of high nighttime ozone at the equatorial mesopause. J Geophys Res 113:D17312. doi:10.1029/2008JD010066

Swinbank R, Ortland DA (2003) Compilation of wind data for the upper atmosphere research satellite (UARS) reference atmosphere project. J Geophys Res 108(D19):4615. doi:10.1029/2002JD003135

Wang Y-M, Lean JL, Sheeley NR Jr(2005) Modeling the Sun's magnetic field and irradiance since 1713. Astrophys J 625: 522–538

Winick JR, Wintersteiner PP, Picard RH, Esplin D, Mlynczak MG, Russell JM III, Gordley LL (2009) OH layer characteristics during unusual boreal winters of 2004 and 2006. J Geophys Res 114:A02303. doi:10.1029/2008JA 013688

Xu J, Liu H-L, Yuan W, Smith AK, Roble RG, Mertens CJ, Russell JM III, Mlynczak MG (2007) Mesopause structure from thermosphere, ionosphere, mesosphere, energetics, and dynamics (TIMED)/sounding of the atmosphere using broadband emission radiometry (SABER) observations. J Geophys Res 112:D09102. doi:10.1029/2006JD007711

Zhang SP, Shepherd GG (1999) The influence of the diurnal tide on the $O(^1S)$ and OH emission rates observed by WINDII on UARS. Geophys Res Lett 26(4):529–532. doi:10.1029/1999GL900033

Chapter 2

Atmospheric Tides and Planetary Waves: Recent Progress Based on SABER/TIMED Temperature Measurements (2002–2007)

Dora Pancheva and Plamen Mukhtarov

Abstract The present paper is focused on the global spatial (altitude and latitude) structure, seasonal and interannual variability of the atmospheric tides (migrating and nonmigrating) and planetary waves (stationary and zonally traveling) derived from the SABER/TIMED temperature measurements for full 6 years (January 2002–December 2007). The mean wave amplitudes and phases are presented for the latitude range 50°N–50°S and from the lower stratosphere to the lower thermosphere (20–120 km). The main advantage of the results presented in this paper is that the migrating and nonmigrating tides as well as all significant planetary waves found in the SABER/TIMED temperatures are extracted simultaneously from the raw data (downloaded from the SABER web site temperatures). Therefore, using the same analysis techniques and the same data set makes it possible to get a consistent picture of the wave activity in the stratosphere-mesosphere-lower thermosphere system. Concerning the atmospheric tides, in addition to the migrating diurnal and semidiurnal tides the following nonmigrating tides also received significant attention: diurnal eastward propagating with zonal wavenumbers 2 and 3 and westward propagating with zonal wavenumber 2 and semidiurnal westward propagating with zonal wavenumber 3 and eastward propagating with zonal wavenumbers 2 and 3. A special attention is paid to the climatology and interannual variability of the temperature SPW1 and its origin in the lower thermosphere, as well as for following zonally propagating planetary waves: the ~5-day *Rossby* wave; ~6-day *Kelvin* wave, the ~10-day *W1* wave and ~16-day *W1* wave. The presented detailed picture of the spatial (altitude, latitude) structure and temporal variability of the considered atmospheric tides and planetary waves can serve as a benchmark and guide for future numerical modeling studies aimed at better understanding the stratosphere-mesosphere-lower thermosphere coupling by tidal and planetary wave patterns.

2.1 Introduction

The middle atmosphere dynamical regime, and particularly that of the mesosphere and lower thermosphere (MLT), is dominated by global scale waves with large amplitudes. Two important wave motions which are essential components of the general circulations and appear regularly are atmospheric (solar) tides and planetary waves. Much of the observed temporal and spatial variability in neutral winds, temperature, and trace species concentrations in the MLT is due directly or indirectly to wave motions. These waves play major roles also in maintaining the zonal mean momentum and temperature budgets.

Atmospheric (solar) tides are ubiquitous features of the Earth's atmosphere that are observed in all atmospheric dynamical fields, including wind, temperature, density and pressure. They are induced by the daily cyclic absorption of solar energy in the atmosphere and have periods that are an integral fraction of a day. The restoring force that acts on atmospheric tides is gravity, so tides are a special class of gravity waves,

D. Pancheva (✉)
Geophysical Institute, Bulgarian Academy of Sciences, Sofia, Bulgaria
e-mail: dpancheva@geophys.bas.bg

M.A. Abdu, D. Pancheva (eds.), A. Bhattacharyya (Coed.), *Aeronomy of the Earth's Atmosphere and Ionosphere*, IAGA Special Sopron Book Series 2, DOI 10.1007/978-94-007-0326-1_2,

which are affected by the Earth's rotation and sphericity because of their global scales. Like other waves the tidal components grow in amplitude with increasing altitude since atmospheric density decreases and energy must be conserved.

Atmospheric tides are normally divided into two types: migrating and nonmigrating tides. Migrating (or Sun-synchronous) solar tides are global-scale atmospheric waves which propagate westward with the apparent motion of the Sun at periods that are harmonics of a solar day, i.e. they are a function of local time alone. These tides are forced predominantly in the troposphere and stratosphere and propagate vertically to the lower thermosphere where they attain large amplitudes and dominate the large-scale wind and temperature fields. The absorption of solar radiation by a longitudinally invariant atmosphere is the primary source of migrating solar tides. The forcing of diurnal and semidiurnal tides due to solar infrared radiation absorption by tropospheric water vapor (H_2O) and the solar ultraviolet absorption by the zonal mean stratospheric and lower mesospheric ozone (O_3) are the most well-known migrating tidal excitation (Chapman and Lindzen, 1970; Forbes and Garret, 1979; Groves, 1982a, b). Other sources may contribute to tidal variations that depend explicitly on longitude and in this way resulting in nonmigrating tides (all tides propagating eastward, zonally symmetric tides and those propagating westward but with a different phase speed from that of the apparent motion of the Sun). It is now generally accepted that nonmigrating tides arise from at least two mechanisms: zonally asymmetric thermal forcing (surface topography, geographically varying heat sources, variation of solar heating with longitude) and nonlinear interactions between migrating tides and stationary planetary waves (SPWs) (Angelats i Coll and Forbes, 2002). For instance, the release of latent heat by deep convective clouds is dependent on universal time (UT), longitude, latitude and season hence it is a possible source that could generate a nonnegligible tidal migrating and nonmigrating response in the MLT (Hamilton, 1981; Williams and Avery, 1996; Hagan, 1996; Forbes et al., 1997; Hagan and Forbes, 2002, 2003; Oberheide et al., 2002). Mc Landress and Ward (1994) suggested that the interaction of a zonally asymmetric distribution of the gravity waves interacting with migrating tides could also generate nonmigrating tides.

Both ground-based and satellite observations have revealed many important features of the diurnal and semidiurnal tides. A major advantage of the ground-based methods is their ability to study in detail seasonal, intraseasonal and interannual variability of the tides at various geographic points (Bernard, 1981; Avery et al., 1989; Clark and Salah, 1991; Manson et al., 1990, 1999; Jacobi et al., 1999; Vincent et al., 1998; Pancheva et al., 2000, 2002), as well as their vertical structure including vertical wavelengths and direction of propagation (Mitchell et al., 2002; She et al., 2004; Yuan et al., 2008). A disadvantage of the ground-based methods is their inability to distinguish between global and local signatures and to provide only coarse information on spatial (latitudinal and longitudinal) structures of the tides.

The satellite observations from the HRDI (High Resolution Doppler Imager), WINDII (Wind Imaging Interferometer) and MLS (Microwave Limb Sounder) instruments on board the UARS (Upper Atmosphere Research Satellite) satellite have provided a global scale picture of the migrating diurnal and semidiurnal tides seen in the neutral wind and temperature fields of the mesosphere and lower thermosphere (Burrage et al., 1995; McLandress et al., 1994, 1996; Khattatov et al., 1997a, b; Forbes and Wu, 2006; Wu et al., 2008a). It is important to note that the satellite observations enable the researchers to separate the nonmigrating from migrating diurnal tides and to study their global structures. The presence of nonmigrating tides in the surface pressure observations was reported first by Chapman and Lindzen (1970) and further developed by Kato (1989). Their detection in the upper levels of the atmosphere however was possible only after the advent of global satellite measurements at altitudes where the tidal signal is significantly large in comparison with other variations (Lieberman, 1991; Talaat and Lieberman, 1999; Oberheide and Gusev, 2002; Forbes et al., 2003; Huang and Reber, 2004; Lieberman et al., 2004; Forbes and Wu, 2006; Wu et al., 2008b; Oberheide and Forbes, 2008).

Various models, as the Global Scale Wave Model (GSWM), the Canadian Middle Atmosphere Model (CMAM), Whole Atmosphere Model (WAM), Whole Atmosphere Community Climate Model (WACCM), and different General Circulation Models (GCM) have been able to reproduce the global distribution and seasonal changes in the tidal fields (Hagan et al., 1995, 1999a, 2001; McLandress, 1997, 2002a; Miyahara et al., 1999; Grieger et al., 2002; Hagan and Forbes, 2002, 2003; Akmaev et al., 2008; Chang et al., 2008).

Other waves that appear regularly, particularly during winter, are planetary waves. Planetary waves are global-scale oscillations that exist due to the Earth's rotation and the conservation of absolute vorticity. The restoring force for a planetary wave oscillation is ultimately the poleward gradient of planetary vorticity. As with tides, planetary waves are globally coherent oscillations in which the wave has an integer number of cycles "wrapped around" a circle of latitude. The planetary wave perturbations have wavelike form in the longitudinal and vertical directions and often also in the latitude direction. Air parcels in a planetary wave may be meridionally displaced by up to several thousand kilometres. Planetary waves, particularly in the stratosphere, are crucial in the transport of species, including ozone, from its region of production, in the tropics, around the rest of the world. In the mesosphere, planetary waves can reach very large amplitudes (up to ~50 ms^{-1} in wind and ~15 K in temperature) and interact strongly with gravity waves, tides and each other.

The most important are quasi-stationary Rossby waves and travelling normal modes, also known as free modes with periods around 2, 5, 10 and 16 days. There is also a class of planetary waves confined to the equatorial region and known as *Kelvin* waves. The interaction of the planetary waves and the zonal mean flow is known to be the major driver of winter stratospheric dynamics (Andrews et al., 1987). Classical studies showed that the zonal mean flow affects the planetary wave propagation by changing the refractive index (Charney and Drazin, 1961). Time-varying or dissipating planetary waves interact with the zonal mean flow and alter it dramatically, as happens in sudden stratospheric warmings (SSW). This mechanism behind the SSW was initially proposed by Matsuno (1971) but now is widely accepted. The interaction decelerates and/or reverses the eastward winter winds in the stratosphere-mesosphere system and also induces a downward circulation in the stratosphere causing adiabatic heating and an upward circulation in the mesosphere resulting in adiabatic cooling (Liu and Roble, 2002, 2005; Coy et al., 2005).

The time period preceding the onset of a SSW is usually characterized by high wave activity in the middle atmosphere during which more than one type of planetary waves (quasi-stationary Rossby waves, zonally symmetric or travelling normal modes) may be present simultaneously. The key role of wave forcing in preconditioning the atmosphere prior to the SSW is shown by Hartmann (1983), Hirota et al. (1990), Cevolani (1991), Shiotani et al. (1993), Sivjee et al. (1994) and Jacobi et al. (2003).

The problem of planetary wave coupling between the stratosphere-mesosphere-lower thermosphere has attracted significant attention only recently. It is supposed that much of the variability in the MLT is a result of upward propagation of disturbances from the stratosphere, particularly during the winter (Krüger et al., 2005; Espy et al., 2005; Azeem et al., 2005; Palo et al., 2005; Chshyolkova et al., 2006; Shepherd et al., 2007; Pancheva et al., 2007, 2008a, b). There are studies however which report in situ excitation of the planetary wave variability in the upper mesosphere and lower thermosphere by zonally asymmetric momentum deposition from gravity waves which have undergone filtering by the stratospheric planetary wave wind field (Smith, 1996, 1997, 2003; Forbes et al., 2002; Xiao et al., 2009).

Most of the above mention papers devoted to the tidal and planetary wave coupling between the stratosphere-mesosphere-lower thermosphere used different data sets. The utilization of different type of data for stratosphere and MLT region, however, does not allow a detailed study of the vertical and latitudinal structure of the waves present simultaneously in the both regions to be accomplished. The Sounding of the Atmosphere using Broadband Emission Radiometry (SABER) instrument (Russell et al., 1999) on the Thermosphere-Ionosphere-Mesosphere-Energetics and Dynamics (TIMED) satellite has provided continuous global temperature data for the latitude range 50°N–50°S from the lower stratosphere to the lower thermosphere since 2002 and in this way provides an unprecedented opportunity for studying in detail the atmospheric waves, as well as their role in coupling the lower and upper atmosphere. The present overview paper is going to report the recent progress in studying the climatological features of the diurnal and semidiurnal tides (migrating and nonmigrating) and the strongest planetary waves (stationary and zonally propagating) seen in 6 full years of the SABER/TIMED temperature measurements (2002–2007) for latitudes between 50°N and 50°S and altitudes between 20 km and 120 km. It is worth noting that several very nice papers have been already published on the tides and different planetary waves seen in the SABER/TIMED temperatures, as Zhang et al. (2006), Huang et al. (2006a, b), Riggin et al. (2006), Forbes et al. (2006, 2008), Palo et al. (2005, 2007), Oberheide and Forbes (2008), Xiao et al. (2009), Xu et al. (2009), Ern et al. (2009), and Forbes et al. (2009). The main advantage of the results presented here is that the migrating and nonmigrating tides, as well as all significant planetary waves, found in the SABER/TIMED temperatures, are extracted simultaneously from the raw data (downloaded from the SABER web site temperatures). Therefore, using the same analysis techniques and the same data set makes it possible to get a consistent picture of the wave activity in the stratosphere-mesosphere-lower thermosphere system.

2.2 Observations and Method for Data Analysis

2.2.1 SABER Temperature Data

The TIMED satellite was launched on 7 December 2001 and the SABER instrument began making observations in January 2002. It measures CO_2 infrared limb radiance from approximately 20 to 120 km altitude and kinetic temperature profiles are retrieved over these heights using LTE radiative transfer in the stratosphere and lowest part of the mesosphere (up to ~60 km), and a full non-LTE inversion in the MLT (Mertens et al., 2001, 2004). SABER views the atmosphere at an angle of 90° with respect to the satellite velocity vector in a 625 km height and 74° inclination orbit so that the latitude coverage on a given day extends from about 53° in one hemisphere to 83° in the other. About every 60 days, the latitude ranges flip as the spacecraft yaws to keep the instrument on the anti-sunward side of the spacecraft. In this way high latitude data are available only in 60-day segments, with no information for the 60 days preceding or following. This is the reason not to generate results poleward of 50°. The TIMED orbit precesses and covers 12 h of local time in each 60-day yaw period, so that ascending and descending data together give almost 24 h of local time sampling.

Our results are derived from the latest Version 1.07 of the SABER data, which were downloaded from the web site: http://saber.gats-inc.com. While there is a little difference below about 70 km between V 1.07 and V 1.06, there are substantial improvements for the upper MLT (García-Comas et al., 2008). The SABER V 1.07 kinetic temperature is validated by Remsberg et al. (2008) who provided the following systematic error due to CO_2 abundance uncertainty: 1.3 K at 80 km, 3.6 K at 90 km and 1.4 K at 100 km. The authors concluded that the SABER data set can be the basis for improved, diurnal-to-interannual-scale temperatures for the middle atmosphere and especially for the upper MLT region. García-Comas et al. (2008) have evaluated the systematic errors in the SABER operationally retrieved kinetic temperature due to the current uncertainties in the collisional rates of the non-LTE model for typical midlatitude and polar conditions. The authors found that the overall error in the kinetic temperature is around 1–2 K below 95 km and around 4 K at 100 km in midlatitudes.

We use 6 years of temperature data, from January 2002 to December 2007, to study the temporal variability and global spatial structure of the diurnal and semidiurnal tides and the planetary waves that are strongest during that period. The data were averaged into 5 km altitude and 10° latitude bins and each bin was independently fit. We work in UT and for each altitude (from 20 to 120 km in 5 km steps) and latitude (from 50°S to 50°N in 10° steps) the data were arranged into a matrix with 24 columns (this is the longitude with a step of 15°) and the number of rows is equal to the length of the considered period of time (January 2002–December 2007) in hours.

2.2.2 Method for Extracting Waves from the Saber Temperatures

Salby (1982) showed that asynoptic (observations made at a single location at a given time) space-time satellite data can be spectrally analyzed in the wavenumber-frequency domain via a rotation from synoptic (simultaneous observations at different locations) to asynoptic coordinates and performing a Fourier transform. A significant advantage of this approach is that it directly maps the asynoptic data set to the equivalent space-time spectrum. However, since this approach relies on the discrete Fourier transform operation, any irregularities in the sampling pattern, i.e. not evenly spaced sampling, will destroy the orthogonality of discrete projections of the observations onto the Fourier expansion functions (Azeem et al., 2000). To avoid the constraint of sampling uniformity we use a least squares fitting technique to spectrally analyze a two-dimensional space-time data set. This technique allows the determination of the space-time spectrum from the asynoptic sampling patterns within the allowed range of wavenumber-frequency space defined by the Asynoptic Sampling Theorem (Salby, 1982).

The data analysis method for calculating the monthly mean tidal characteristics is carried out in two steps: first, the daily characteristics of the waves for the entire 6-year period of time (2002–2007) are derived, and second, the monthly mean wave characteristics are calculated by vector averaging of the daily wave parameters for each calendar month. The seasonal behavior of the wave characteristics during the entire 6-year period of time (climatology of the waves) is estimated by the vector averaging of the monthly mean wave characteristics for each month of the year.

In order to extract the waves from the SABER data (i.e. to determine their amplitudes and phases) at a given latitude and altitude, we perform a linear two-dimensional (time-longitude) least-squares fitting including diurnal and semidiurnal tides with zonal wavenumbers up to 4, zonally travelling planetary waves with zonal wavenumbers up to 3, together with the first three modes of the stationary planetary waves (SPW), i.e. SPWs with zonal wavenumbers 1, 2 and 3. The well-known planetary wave normal modes with periods near 5, 10 and 16 days are included in the fitting procedure. The spectral analysis has been performed on each year of data in order to define the prevailing periods of the normal modes. The mean spectra indicated that the most well expressed periods are: 5.5 days for the 5-day wave, 11 days for the 10-day wave and 17 days for the 16-day wave. The ~24-day wave is also regularly observed in the stratosphere and mesosphere particularly in the winter (Pancheva and Mitchell, 2004; Pancheva et al., 2007, 2008a, 2009a, b), hence this planetary wave period is included in the fitting procedure as well. It is worth noting that: (i) the quasi-2-day wave is excluded from the fitting procedure because of its burst-like behavior with time scales near or less than a month and because of using a rather long window, a 60-day one (see the explanation below), for extracting the waves, and (ii) it has been checked that the variable period of the 10-(8–12 days) and 16-day (14–18 days) waves does not affect the tidal results. All oscillations are extracted simultaneously in order to avoid a possible distortion of the weaker waves by the stronger ones and to minimize some aliasing between the tides and the SPWs. Because it takes SABER 60 days to sample 24 h in local time by combining ascending and descending data together, the length of the sliding time window used for performing the least-squares fitting procedure should be 60 days. Then the 60-day window is moved through the time series with steps of 24 h in order to obtain the daily values of the wave characteristics. The temperature variations during a 60 day period can be due to seasonal or intraseasonal time scales. A possible effect of the semiannual oscillation (SAO) is expressed by including a linear term in the temperature decomposition. The waves at given latitude and height are extracted by using the following expression that is applied to 60-day windows:

$$\begin{aligned}\Theta(t,\lambda) = {} & \Theta_0 + \Theta_r t + \sum_{j=1}^{4}\sum_{s=-3}^{3} A_{js} \cos\left[\frac{2\pi}{24T_j}t - \frac{2\pi}{360}s\lambda - \frac{2\pi}{360}\varphi_{js}\right] + \\ & + \sum_{s=1}^{3} B_s \cos\left[\frac{2\pi}{360}s\lambda - \frac{2\pi}{360}\psi_s\right] + \\ & + \sum_{k=1}^{2}\sum_{s=-4}^{4} C_{ks} \cos\left[k\frac{2\pi}{24}t - \frac{2\pi}{360}s\lambda - \frac{2\pi}{360}\gamma_{ks}\right] \\ & + R(t,\lambda)\end{aligned} \tag{2.1}$$

where: t is time in hours counted from 0 UT of 1 January 2002, λ is longitude in degrees, s is wavenumber, k is number of the diurnal harmonic, T_j are the planetary wave periods, i.e. $T_1 = 24$ days, $T_2 = 17$ days, $T_3 = 11$ days, $T_4 = 5.5$ days and $R(t, \lambda)$ is the residual from the least squares best-fit procedure including planetary waves with periods less than 5 days, tides with periods less than 12 h and the noise of the data. Eastward (westward) propagation corresponds to $s > 0$ ($s < 0$). The five terms on the right hand side of the above expression are as follows: (i) the mean temperature Θ_0, (ii) a linear term with coefficient Θ_r responsible for zonally symmetric variations with period longer than 60 days; this term includes a possible effect of the SAO, (iii) travelling planetary waves (including zonally symmetric planetary waves) with amplitudes A_{js} and phases ϕ_{js}, (iv) stationary waves with amplitudes B_s and phases ψ_s, and (v) 24- and 12-h migrating and nonmigrating tides with amplitudes C_{ks} and phases γ_{ks}. In the above expression the phases of the travelling planetary waves (including zonally symmetric planetary waves as well) and the tides (including zonally symmetric tides as well) are defined as the time of the wave maxima at zero longitude and are presented in degrees, while the phase of the stationary wave is the longitude of the wave maximum. According to the formula (1) we have all together 100 coefficients that should be determined by solving the system with on the average about 1000 measurements within each 60-day window at each height and latitude.

Because of the satellite data sampling pattern and the variability of the planetary waves during the 60-day windows used for extracting the waves, a very important problem in the tidal assessment is how to determine the reference noise level. We thoroughly investigated the winter stratosphere levels where the strong and variable SPWs could affect the tidal results (Pancheva et al., 2009c). For each 60-day time window we generated 100 Gaussian noise datasets with length of 60 days and with data sampling pattern and variance as those of the true SABER measurements (Kong et al., 1998). Then the *2D Lomb-Scargle* periodogram has been performed on the resulting noise datasets. The reference noise spectrum for each zonal wavenumber is calculated as an average spectrum. The results from this analysis indicate that, due only to the data sampling pattern and the planetary wave variability in the winter, all of the noise spectra show a cluster of peaks near 24 h although their amplitudes never exceed ~0.95 K. As a result from the above analysis all tidal waves extracted from a given 60-day time window with amplitudes larger than 1 K are accepted as statistically significant and only these tides are considered in this analysis.

Therefore, by using a least squares fitting technique (to avoid the constraint of sampling uniformity) applied to a 60-day window (to sample 24 h in local time by combining ascending and descending data together) where the tides and all types of planetary waves are simultaneously extracted from the SABER temperatures the aliasing problems between the variable zonal mean, variable planetary waves and tides are significantly minimized.

2.3 Atmospheric Tides

The climatological features of the waves seen in the SABER/TIMED temperatures are investigated by the distribution of the monthly mean tidal amplitudes and phases in time and space (latitude and altitude). The results of data analysis cover the interval between February 2002 and December 2007. Throughout the remainder of this paper we utilize the notation *DWm* and *DEm* to denote a westward- or eastward-propagating diurnal tide, respectively, with zonal wavenumber $m = s$. For semidiurnal tide *S* replaces *D*. The zonally symmetric tides (with zonal wavenumber $s = 0$) are denoted *D0* and *S0*, and stationary planetary waves (*SPW*) with zonal wavenumber *m* are expressed as *SPWm*.

2.3.1 Migrating Diurnal (DW1) and Semidiurnal (SW2) Tides

The climatology of the *DW1* and *SW2* tides in this subsection will be presented as the multiyear (2002–2007) averaged amplitudes and phases which are a function of the altitude, latitude and season. Some of the main tidal features, particularly in the mesopause region, have been already known from both ground-based and satellite observations. Among them are the seasonal cycle and semiannual variability of the *DW1*, as well as its vertical wavelength of about 25–30 km, and the amplification of the *SW2* tide toward the middle latitudes (we remind that only the latitudes between 50°N and 50°S are considered in the present study) and the seasonal dependence of its vertical wavelength, changing from ~40–50 km in the winter to ~90–100 km, and even more, in the summer. Using ground-based measurements however we have to bear in mind that the registered 12- or 24-h oscillations are actually a superposition of all, migrating and nonmigrating, tides at a given geographical point and the assumption that the migrating component is a predominant one in most of the cases is not valid.

2.3.1.1 Climatology of DW1 Temperature Tide

Figure 2.1a shows latitude versus months contours of *DW1* temperature amplitudes in Kelvin at altitudes of 90 km (near mesopause, upper plot) and 50 km

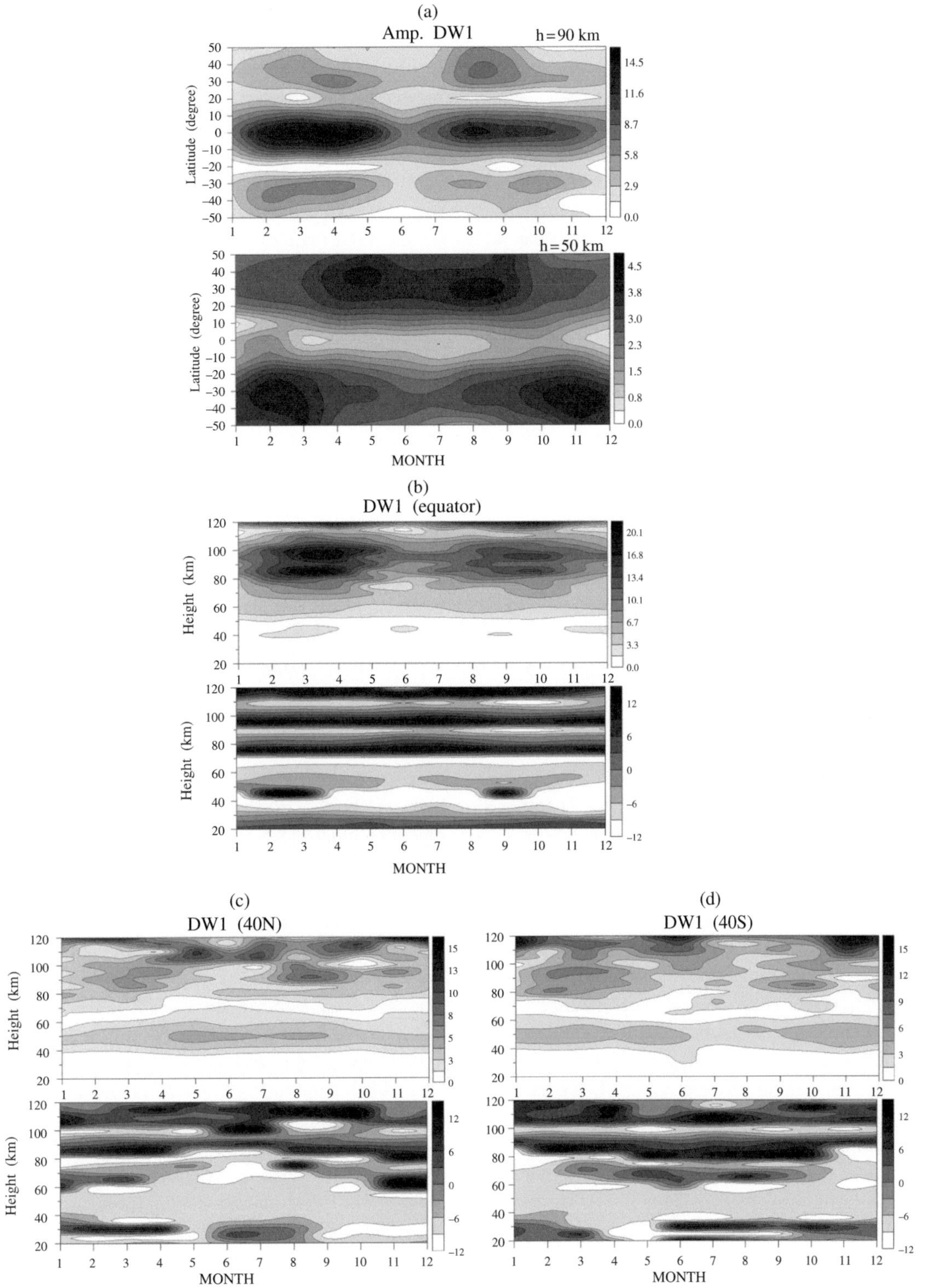

Fig. 2.1 (**a**) Latitude-time cross sections of *DW1* tidal amplitudes (in Kelvin) at altitudes of 50 km (*bottom*) and 90 km (*upper*), (**b**) Altitude-time cross sections of *DW1* amplitude (*upper*) and phase (in local time, LT) (*bottom*) at the equator, (**c**) The same as (b), but for 40°N, and (**d**) The same as (**b**), but for 40°S

(near stratopause, bottom plot). While the mesopause latitudinal structure is known from the previous findings with a main maximum at the equator and secondary maxima near ±35° latitude (Zhang et al., 2006; Huang et al., 2006a; Forbes et al., 2008), the stratopause structure was for the first time carefully examined by Mukhtarov et al. (2009). The *DW1* amplitude grows towards middle latitudes reaching maximum of ~4.5 K at the stratopause level near 40°. Their seasonal behavior is dominated by main annual variation with maximum during summer months and secondary semiannual one with equinoctial maxima. The results of Fig. 2.1a revealed some peculiarities of the *DW1* features over the equator and at middle latitudes. The altitude structure of the *DW1* tide will be considered in detail at both latitudinal regions.

Figure 2.1b shows altitude versus month contours of the *DW1* temperature amplitude and phase at the equator, while Fig. 2.1c and d show the same, but for 40°N and 40°S respectively. The tidal phase in LT is calculated by taking into consideration that $t_{UT} = t_{LT} - \lambda/15$. Clearly outlined double-peaked maxima centered near 85 and 100 km height can be distinguished at the equator as the average (for 6 years) amplitude reaches values of ~20 K. The ubiquitous double-peaked vertical structure of the *DW1* tidal amplitude over the equator was observed for the first time in the SABER temperatures and has been found also by Zhang et al. (2006) and Xu et al. (2009). The seasonal behavior is dominated by main semiannual oscillation with equinoctial maxima and secondary annual one with larger amplitude during vernal equinox. The altitude phase structure reveals a vertically upward propagating tide in the altitude range 55–110 km and a signature of trapped modes (that do not propagate vertically, i.e. with vertical phase gradient zero, and decay away from their levels of excitation) near 40–45 km height. Two different vertical phase gradients however could be distinguished in the altitude range 55–110 km: it is much smaller between 55 and 70 km, than that between 70 and 110 km altitude. In the latter height region two wave cycles are clearly visible indicating that the vertical wavelength is ~20 km.

The average vertical *DW1* amplitude and phase structures at middle latitudes (40°) are presented in Fig. 2.1c and d. The tidal characteristics are very similar in both hemispheres. There is an altitude region located between 40 and 60 km where the tidal amplitudes reach values of ~4–5 K and indicate a regular temporal variability with amplification in the summer months of both hemispheres. The region located near 50 km altitude coincides with the height of the maximum heating rate attributed to the stratospheric ozone absorption (Hagan, 1996; McLandress, 1997). The phase distribution indicates that these are vertically trapped diurnal tidal modes (Forbes and Garret, 1979). In the upper mesosphere (above 80 km) the seasonal behavior of the *DW1* tide, similarly to the equatorial region, is dominated by semiannual oscillation with equinoctial maxima. The phase structure indicates that the diurnal tide is upward propagating with vertical wavelength similar to that over the equator, ~20 km. During the summer however, the altitude range where the *DW1* tide is composed mainly of trapped modes is enlarged to altitudes between ~40 and ~80–85 km in both hemispheres.

The left column of plots in Fig. 2.2 shows the 6 year average for the altitude-latitude cross-sections of the tidal amplitude and phase for March, while the middle and right columns show the same but for September and December respectively. For March, both amplitude and phase plots indicate the predominance of the first symmetric propagating component of the *DW1* tide that has a main maximum at the equator and secondary maxima near ±35° latitude, with 180° phase shift between the two maxima. The mean equatorial maximum reaching ~18 K is almost a factor of 2 larger than the secondary maxima at ±35° latitude. The equatorial *DW1* tide undergoes some reduction near 90 km height and significant saturation at 115 km heights however it rapidly amplifies above this level. Zhang et al. (2006) suggested that a saturated tide might indicate the onset of convective instability. Between 40 and 60 km, where the solar radiation absorption by ozone has a dominant effect generating in-situ trapped diurnal modes, mean tidal amplitude of ~4 K is found that maximizes in middle latitudes. The amplitude and phase plots for September are similar to those for March, but some contribution of the antisymmetric modes on the derived *DW1* tide can be distinguished as well. The December amplitude and phase plots express the asymmetric nature of the SABER tidal amplitudes above 60 km. The mean equatorial maximum approaches ~11–12 K. The phase distribution indicates downward tilting of phase lines as the tide progresses from the summer to winter hemisphere. The summer maximum

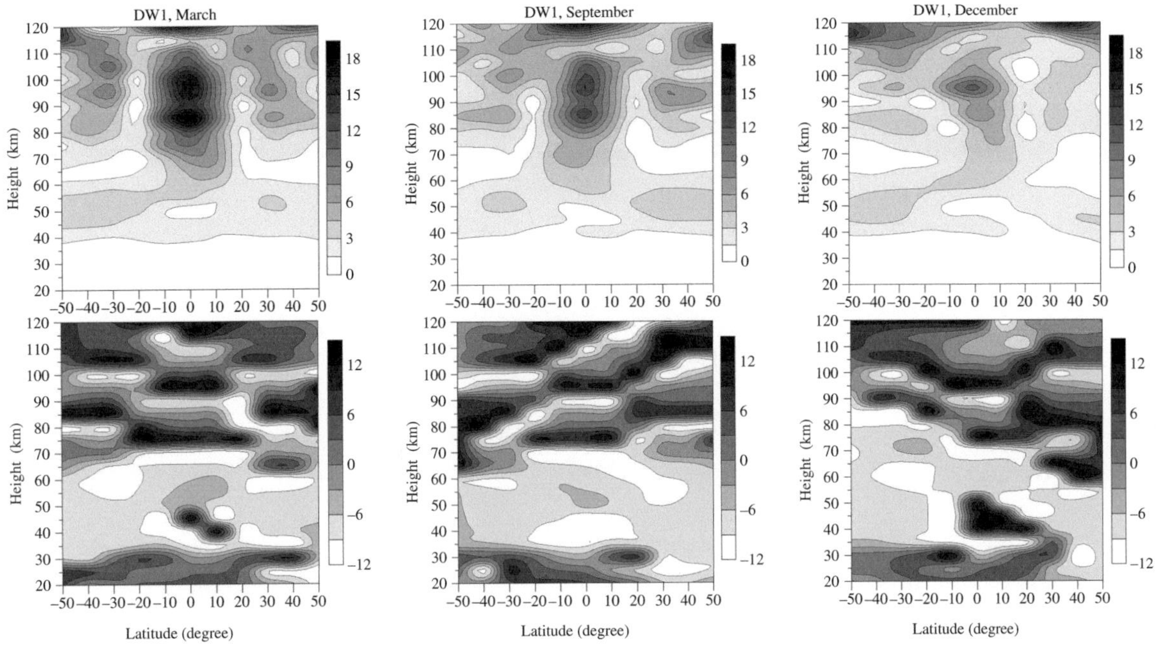

Fig. 2.2 Altitude-latitude cross sections of *DW1* tidal amplitudes (in Kelvin) (*upper row*) and phases (in LT) (*bottom row*) for March (*left column*), September (*middle column*) and December (*right column*)

of the trapped *DW1* tide near 50 km height is well defined with an average amplitude of ~4.5 K.

It is known from the previous studies and supported also by this one that above 70 km height the *DW1* tide reflects mainly the distinctive features of the first symmetric propagating (1,1) mode. A *Hough* mode analysis of the *DW1* temperature tide was performed in Mukhtarov et al. (2009) in order to clarify which *Hough* modes contribute to the observed SABER *DW1* tide below 70 km height. The analysis revealed that at altitudes near 50 km the *DW1* tide is trapped and is described mainly by the first symmetric (1,–2) mode that maximizes at this height. In the altitude range 35–70 km this trapped mode has a constant phase of ~16 LT and a mean amplitude of ~4 K. With increasing height the vertically propagating symmetric (1,1) mode rapidly grows while the trapped (1,–2) mode decays and above 70 km the propagating mode prevails over the trapped one.

The climatology of the *DW1* temperature tide and particularly: (i) the trapped diurnal component observed in the height range 40–60 km and (ii) the ubiquitous double-peak vertical structure observed in the MLT region over the equator could be of particular interest for a reexamination of the stratospheric tidal heating and the effects of eddy diffusivity on the numerical tidal simulations.

2.3.1.2 Climatology of SW2 Temperature Tide

Figure 2.3a shows latitude versus months contours of *SW2* temperature amplitudes at altitudes of 90 km (near mesopause) and 110 km (lower thermosphere). At the mesopause region the *SW2* maximizes in the middle latitudes (40°–50°) and is dominated by winter amplification. In the Northern Hemisphere (NH) there is an additional, secondary intensification observed in the tropical region (near 20°N) which is dominated by a summer, June-August, maximum. In the lower thermosphere (110 km) at both hemispheres the *SW2* tide indicates strong amplification at the tropics/subtropics (~20°–30°) and in June-August with an average (for 6 years) amplitude of ~28 K. In the middle latitudes (near 40°), particularly in the NH, the *SW2* temperature tide shows also a secondary winter amplification (November-February) which is at least 2 times weaker than the main tropical maximum. The results of Fig. 2.3a revealed some peculiarities of the *SW2* tidal features at tropical and middle latitudes and its altitude

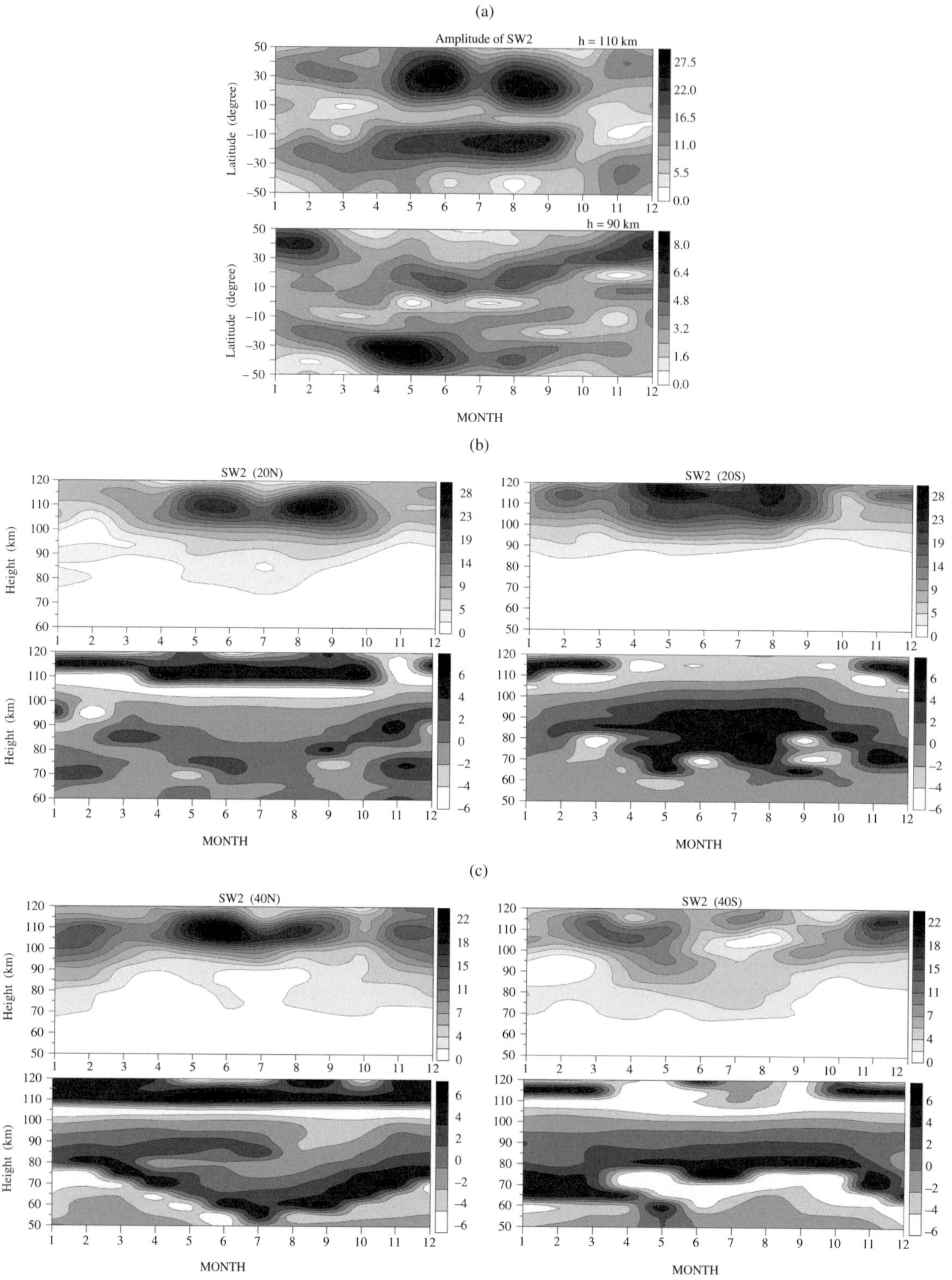

Fig. 2.3 (**a**) Latitude-time cross sections of *SW2* tidal amplitudes (in Kelvin) at altitudes of 90 km (*bottom*) and 110 km (*upper*), and (**b**) Altitude-time cross sections of *SW2* amplitudes (*upper row*) and phases (in LT) (*bottom row*) at 20°N (*left column*), 20°S (*right column*) and (**c**) The same as (b) but for 40°N (*left column*) and 40°S (*right column*)

structure will be considered in detail at both latitudinal regions.

Figure 2.3b shows altitude versus months contours of the *SW2* temperature amplitude and phase at 20°N and 20°S, while Fig. 2.3c show the same, but for 40°N and 40°S respectively. The figure shows the altitude range of 50–120 km even though the tidal amplitudes are insignificant below 70 km. The reason behind this is the seasonal variability of the tidal phase (particularly that in the NH) that can be distinguished clearly in this height range. The seasonal behavior of the *SW2* tide at the tropics and in both hemispheres reveals maxima exactly in May–June and August. The vertical wavelength, particularly in the NH, indicates seasonal variability; the wavelength is being larger in summer (~50–60 km) than in winter (~30–35 km). The vertical wavelength in the SH shows slight changes around a mean wavelength of ~35–40 km. The vertical *SW2* tidal structure in the middle latitudes (Fig. 2.3c) is similar to that in the tropics particularly for the NH. The seasonal behavior at both hemispheres besides the June–August maximum shows a secondary one in November–February. In the altitude range 70–80 km there is also a slight amplification of the tidal amplitude in June and August–October in the NH and March–September in the SH.

The left column of plots in Fig. 2.4 shows the 6 year average (2002–2007) altitude-latitude cross-sections of the tidal amplitudes and phases for June, while the middle and right columns show the same but for December and March respectively. Only the altitude range between 70 and 120 km is considered because the tidal amplitudes below 70 km are too small to be meaningful. In general a symmetric nature of the SABER tidal amplitudes is visible in June solstice with a maximum reaching ~27 K. While the summer maximum is located at latitude of ~30° and altitude near 110 km the winter one is near latitude of 20° and altitude of ~115 km. The phase distribution shows a downward tilting of phase lines as one progresses from the summer to winter hemisphere. The altitude versus latitude distribution of the tidal amplitude in December (middle column of plots) indicates more complicated structure than that in June. There are three amplitude maxima in the lower thermosphere: (i) summer maximum (~21 K) located near latitude of 30°S–40°S and altitude of 115 km, (ii) equatorial (0–10°N) maximum of ~16 K located at 115 km height, and (iii) winter maximum (~14 K) located at latitude of 40°N and centered between 100 and 110 km altitudes. Similarly to the June solstice the phase distribution in December shows a downward tilting of phase lines as the tide progresses from the summer to winter hemisphere. In general, a symmetric nature of the lower thermospheric tidal amplitudes is visible in March. The March amplitude maxima (~16–17 K) are located at latitudes near ±30° and altitudes of ~115 km in the SH and ~110 km in the NH. The phase distribution in March reveals a downward tilting of phase lines as one progresses from the NH to the SH.

The seasonal altitude-latitude structure of the tidal amplitudes and phases indicates that different Hough modes contribute to the tidal composition during different seasons and altitudes. The vertical wavelength of the SABER temperature tide revealed some seasonal dependence (larger in summer than in winter) but on the average it ranges between 30–35 km and 45–60 km. This means that the symmetric (2,2) and antisymmetric (2,3) modes, whose vertical wavelengths are larger than ~300 km and ~80 km respectively and which have been widely accepted to be the main contributors to *SW2*, must have small amplitudes or be present only over a limited altitude range. A *Hough* analysis was performed in response to the comparatively short vertical wavelengths of the SABER *SW2* temperature tide found by Pancheva et al. (2009d). The detailed results from the *Hough* analysis can be found in Pancheva et al. (2010b). Here only the most important outcome from it will be mentioned; it revealed that the modes with significantly shorter wavelengths, as (2,4), (2,5) modes and in winter (2,6) mode as well, contribute to the composition of the SABER *SW2* temperature tide. While the (2,3) mode contributes to summer and fall SABER *SW2* tide, the (2,2) mode becomes dominant only above ~110 km for solstice *SW2* tide.

2.3.1.3 Interannual Variability of DW1 and SW2 Temperature Tides

Interannual variability of the *DW1* and *SW2* temperature tides is considered on the basis of the monthly mean temporal distribution of the tidal amplitudes presented in detail by Mukhtarov et al. (2009) and Pancheva et al. (2009d). Part of these results will be summarized here.

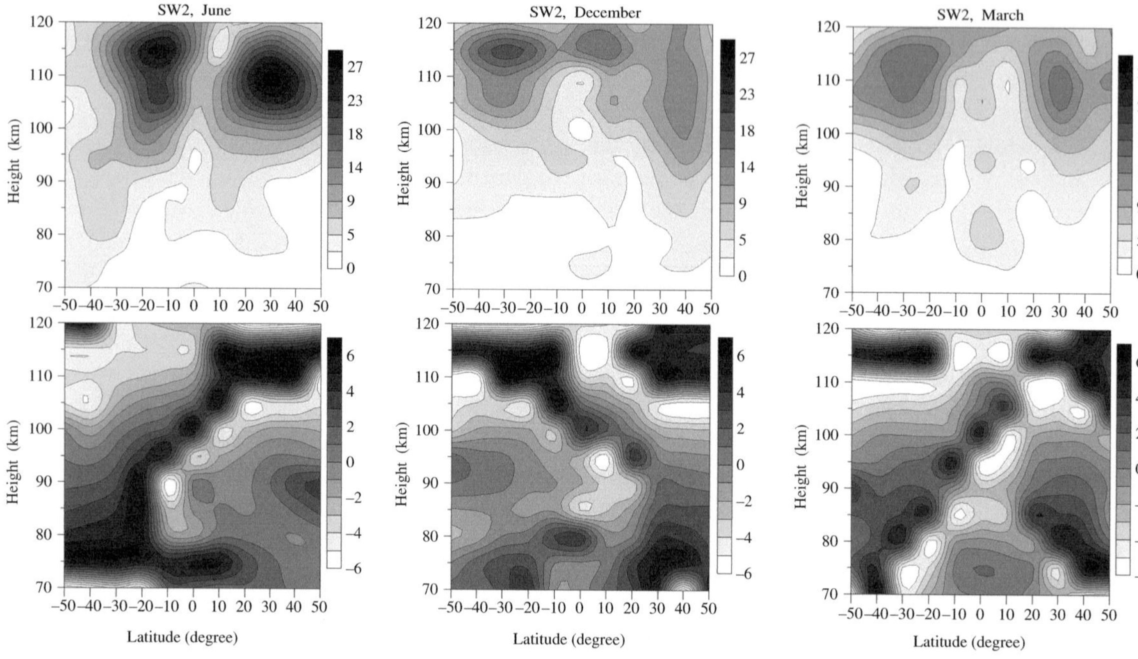

Fig. 2.4 Altitude-latitude cross sections of *SW2* tidal amplitudes (in Kelvin) (*upper row*) and phases (in LT) (*bottom row*) for June (*left column*), December (*middle column*) and March (*right column*)

It has been mentioned before that at altitude range of 40–60 km the *DW1* tide is dominated by the first symmetric trapped (1,–2) mode with a mean amplitude of ~4 K and phase of ~16 LT at 50 km height. It is known that the region located near 50 km altitude coincides with the height of the maximum heating rate attributed to the stratospheric ozone absorption (Hagan, 1996; McLandress, 1997). The mean ozone density however indicates some dependence on the solar activity; on the average the observed ozone changes through the solar cycle are ~3%. Something else, there is a possibility for solar cycle related changes in the thermal structure of the stratosphere, as well as thermal structure changes due to increasing greenhouse gases. Therefore, if the trapped *DW1* modes are generated in situ mainly by the stratospheric ozone, where the mean thermal structure depends on the solar cycle and increased greenhouse gases, then their amplitudes have to depend also on the solar activity. Figure 2.5a shows the monthly mean *DW1* tidal amplitudes at an altitude of 50 km and latitude of 40°N. The considered period of time 2002–2007 belongs to the declining phase of the solar activity. The amplitude of the *DW1* tide also indicates some reduction with the time; the negative trend is shown by the dashed line. We noted however that the above result has to be accepted only as a signature for possible solar cycle effect on the *DW1* tidal amplitude in the stratosphere because the observations cover only the declining phase of the solar cycle. The phase of the *DW1* tide is also shown in the figure and it is marked by empty crests; during the considered 6 years the tidal phase at 50 km height is almost stable and clusters near 15–16 LT.

Figure 2.5b shows the *DW1* tidal amplitude at an altitude of 80 km over the equator. This plot reveals a quasi-2-year variability which is particularly visible on the March equinoctial maximum, which in 2002, 2004 and 2006 is stronger than those in 2003, 2005 and 2007. The March amplifications are shown by tick arrows. The strongest 2-year effect is observed in the MLT region where the diurnal tide itself maximizes. It is worth noting that there is not only a general negative trend of the tidal amplitude denoted by a dashed line, but also the magnitude of the 2-year amplitude peaks diminishes over the considered 6-year period. Such 2-year variation of the SABER temperature diurnal tide have been reported by Huang et al. (2006b), Xu et al. (2009), as well by Burrage et al. (1995) based

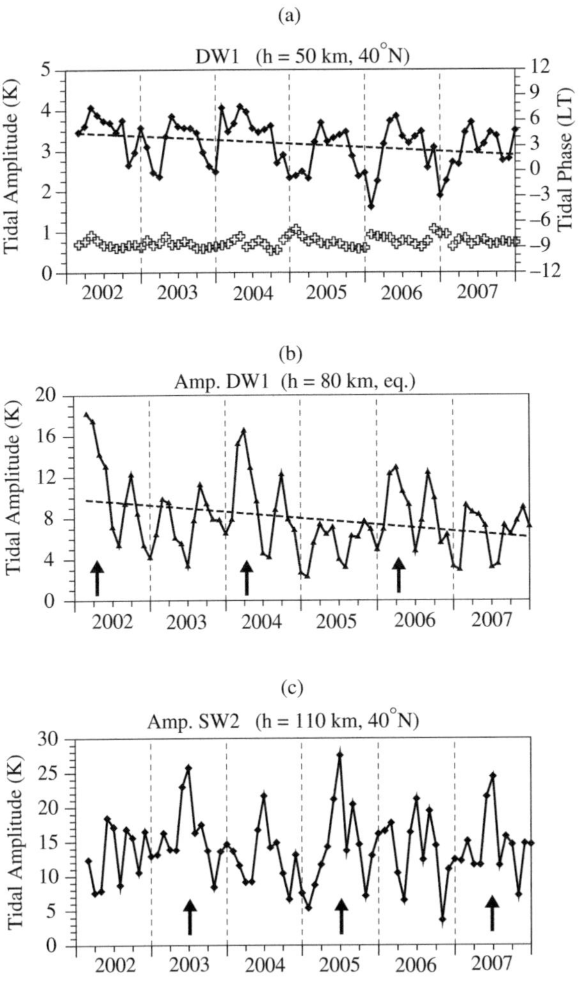

Fig. 2.5 (**a**) Temporal distribution of monthly mean *DW1* amplitudes (*solid line with diamonds*) and phases (*empty crests*) at altitude of 50 km and at 40°N latitude; (**b**) Monthly mean *DW1* amplitudes (*solid line with triangles*) at altitude of 80 km over the equator; *dashed lines* indicate the negative trend of tidal amplitude in plots (**a**) and (**b**), and (**c**) Temporal distribution of monthly mean *SW2* amplitudes (*solid line with diamonds*) at altitude of 110 km and at 40°N latitude; the thick arrows in plots (**b**) and (**c**) show the 2-year amplification of *DW1* and *SW2* amplitudes

on the wind measurements with the HRDI instrument on UARS and Wu et al. (2008a) based on the wind measurements with the TIDI instrument on TIMED.

Recently the interannual variations of atmospheric tides have received considerable attention, but have not been successfully explained, especially for the quasi biennial oscillation (QBO) in the diurnal tide. Two mechanisms have been proposed to explain the observations. The first mechanism is modulation of the tide by the QBO in the background wind in the stratosphere. Modeling studies (Hagan et al., 1999b; McLandress, 2002b) have found that the amplitude and phase of the diurnal tide are sensitive to variations in the zonal wind. Another mechanism that could force a QBO variation in the diurnal tide is momentum deposition from small-scale gravity waves (Mayr and Mengel, 2005). Their modeling studies found that the QBO modulation of the tide in the upper mesosphere is caused to a large extent by variations in gravity wave momentum deposition resulting from filtering by the background winds in the middle atmosphere.

The *SW2* tide also shows quasi-2-year variability that is better visible at the lower thermosphere of middle latitudes. Figure 2.5c shows the monthly mean *SW2* tidal amplitude at 110 km height and at 40°N. The 2-year amplification of the summer amplitudes is marked by tick arrows. While the *DW1* 2-year amplification is observed during the east phase of the stratospheric QBO that of the *SW2* tide is observed during the west QBO phase. This analysis indicates that the interannual variability of the semidiurnal tide observed predominantly in the NH lower thermosphere at mid-latitudes is at least partly generated by the QBO. It is worth noting that in contrast to the modeling study of Hagan et al. (1992) where the QBO signature has been evident in the January 12-h tide the SABER result revealed QBO oscillation well seen in the June semidiurnal tide.

2.3.2 Nonmigrating Temperature Tides

In this section we provide some perspective on the sufficiently stable (in amplitude and phase) nonmigrating diurnal and semidiurnal tides seen in full 6 years (2002–2007) of observations of the kinetic temperature measured by the SABER/TIMED. This topic deserves special attention because most of the nonmigrating tides amplify in the lower thermosphere (~110–115 km), i.e. in the dynamo region, where trough electrodynamical effects can have an impact on the variability of the ionosphere (Forbes et al., 2003, 2008; Forbes and Wu, 2006; Zhang et al., 2006; Oberheide et al., 2007; Oberheide and Forbes, 2008). As noted previously, the nonlinear interaction between the *SPWs* and the migrating tides (Angelats i Coll and Forbes, 2002) and the latent heating associated with deep tropical convection are zonally asymmetric sources known to be important for MLT tides. A qualitative visualization of how the asymmetry of the latent heating associated with deep tropical convection is manifested in the context of tidal theory has been described by Zhang et al. (2006) and Forbes et al. (2008) and will not be touched here.

Only the climatology of the following nonmigrating tides: *DE3, DE2, DW2, SW3, SE2* and *SE3* are presented in this section. These tides are not only sufficiently stable during the considered 6 years, but it has been found that have important contribution to the longitudinal structure and variability of the ionosphere (Immel et al., 2006; England et al., 2006a, b; Lin et al., 2007).

2.3.2.1 Nonmigrating Diurnal Tides

Our analysis of all nonmigrating tides with zonal wavenumber up to 4 observed in the SABER/TIMED temperatures for the considered 6 years revealed clear interannual variability only for *DE3* and *DE2* tides. In order to demonstrate this feature the monthly mean variability for the entire period of time January 2002–December 2007 will be presented below only for these two tides.

The *DE3* tide is established as a major contributor to the wave-4 structure seen in the ionosphere (Hagan et al., 2007) and it was found to be the largest of all the nonmigrating diurnal tidal components observed in the SABER temperatures. Figure 2.6a shows the latitude versus months cross sections of the DE3 amplitudes at altitudes of 80 km (bottom plot) and 105 km (upper plot), representing the tidal features in the mesosphere and lower thermosphere, respectively. The time-axes of Fig. 2.6a are in months started from January 2002 and each year is separated by a dashed line. The results for the latitudinal structure and temporal variability of the *DE3* tide can be summarized as follows: (i) the *DE3* in the mesosphere maximizes at ±10–20°, whereas in the lower thermosphere it maximizes over the equator, or more precisely it is centered between 0° and 10°S and can be detected in the latitude range of ±40°, (ii) while the seasonal course of the *DE3* tide is dominated by a strong main maximum in August–September and a secondary one in March–April in the mesosphere of the NH, in the SH mesosphere it amplifies only in December–January, i.e. in the austral summer, and (iii) the two main maxima in the NH mesosphere (80 km) and also at 105 km height show quasi-2-year variation with stronger *DE3* activity in 2002, 2004 and 2006 (marked by thick arrows at the bottom plot) than in 2003, 2005 and 2007.

The above results clearly indicate the dominance of the antisymmetric tidal modes in the mesosphere and the symmetric ones in the lower thermosphere. The appearance of the *DE3* maxima during local summer in the mesosphere (~80 km height) suggest an influence of the seasonal-latitudinal migration of the deep convective activity in the troposphere, mentioned by Forbes et al. (2008). In a modeling paper by Achatz et al. (2008) the influence of the background wind fields on the tidal propagation is shown as well.

Figure 2.6b shows the latitude-time cross sections of the *DE2* amplitudes observed in the upper mesosphere (90 km, bottom plot) and lower thermosphere (110 km, upper plot). The seasonal behavior is dominated by strong semiannual variation with maxima in the solstices. The main *DE2* maximum is observed during June solstice, however while in the upper mesosphere it is situated near 10°N, in the lower thermosphere it is located near 10°S, as it reaches amplitudes of ~11 K. During the considered (2002–2007) period of time it was found that the *DE3*, with maximum amplitude of ~18 K, is significantly stronger than the *DE2*, having maximum amplitude of ~11 K. The secondary December solstice *DE2* maximum is located at about 10°N with amplitudes reaching ~8.5 K in the lower thermosphere (upper plot). The latitudinal and seasonal structures of the *DE2* tidal amplitudes are presumably due to some combination of latent heat forcing and mean wind effects.

Both June solstice maxima near 10°S in the lower thermosphere (110 km, upper plot) and near 10°N in the upper mesosphere (90 km, bottom plot) indicate quasi-2-year variation, i.e. the tidal amplitudes in 2002, 2004 and 2006 (marked by thick arrows at the bottom plot) are stronger than those in 2003, 2005 and 2007. More details about the 2-year variability of the *DE3* and *DE2* tides can be found in Pancheva et al. (2010c). A possible reason for this interannual variability of both nonmigrating tides is the QBO variability of the zonal mean circulation in the low-latitude middle atmosphere that could affect the propagation conditions of the tides.

The multiyear climatology of the nonmigrating diurnal tides considered in this section is shown in the left column of Figs. 2.7 and 2.8a. While the left

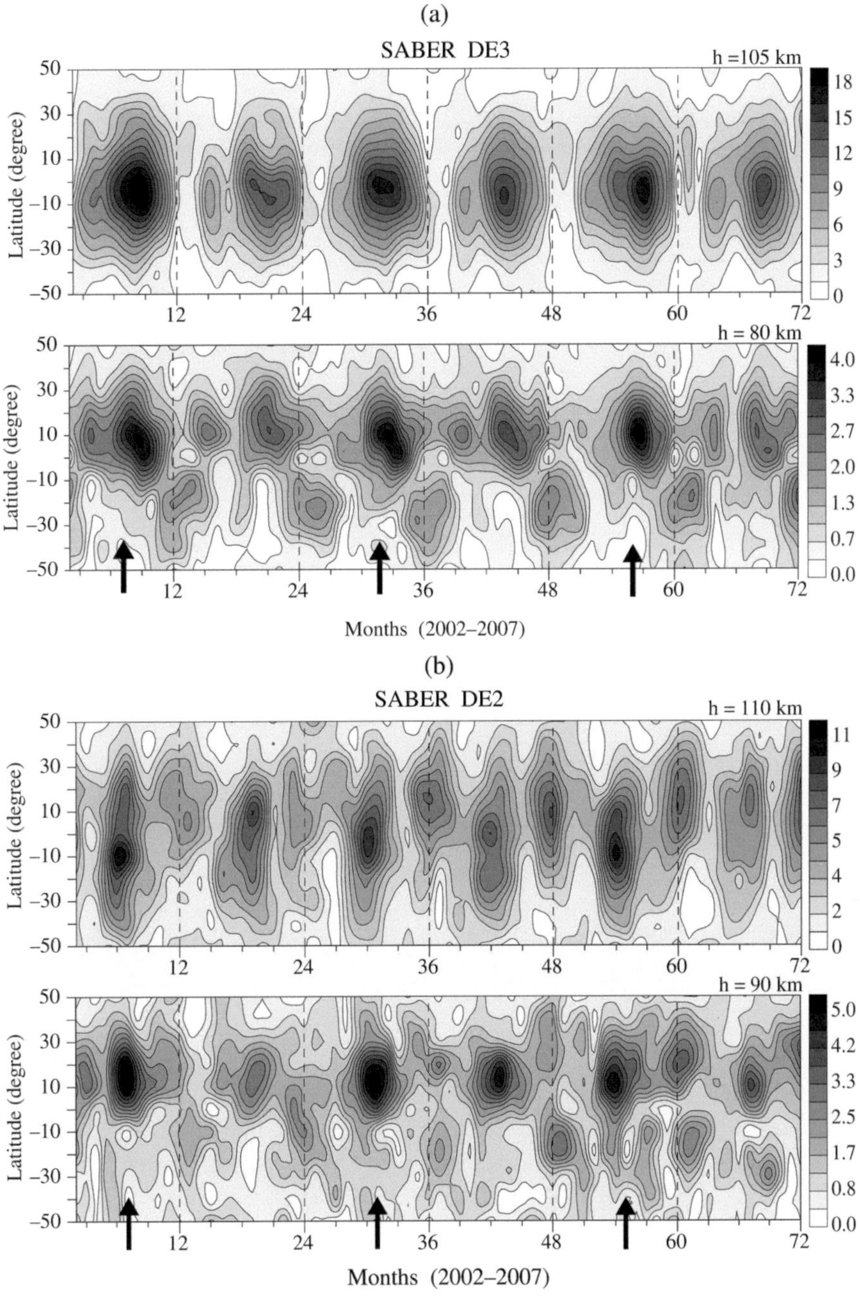

Fig. 2.6 (**a**) Latitude-time cross sections of the *DE3* amplitude (in Kelvin) at altitudes of 80 km (*bottom plot*) and 105 km (*upper plot*); (**b**) Latitude-time cross sections of the *DE2* amplitude at altitudes of 90 km (*bottom plot*) and 110 km (*upper plot*); the 2-year maxima are marked by thick arrows situated in the bottom side of the plots

column of Fig. 2.7 shows the latitude versus months contours of the *DE3* (upper) and *DE2* (middle) tides and *DW2* tide (bottom) at lower thermospheric altitudes where these tides maximize, Fig. 2.8a shows the altitude-latitude cross sections of the tidal amplitudes in Kelvin (left column) and phases in degrees (right column) for *DE3* (upper row) in August, *DE2* (middle row) in June and *DW2* (bottom row) in November, i.e. during the months when the considered tides have the largest amplitudes.

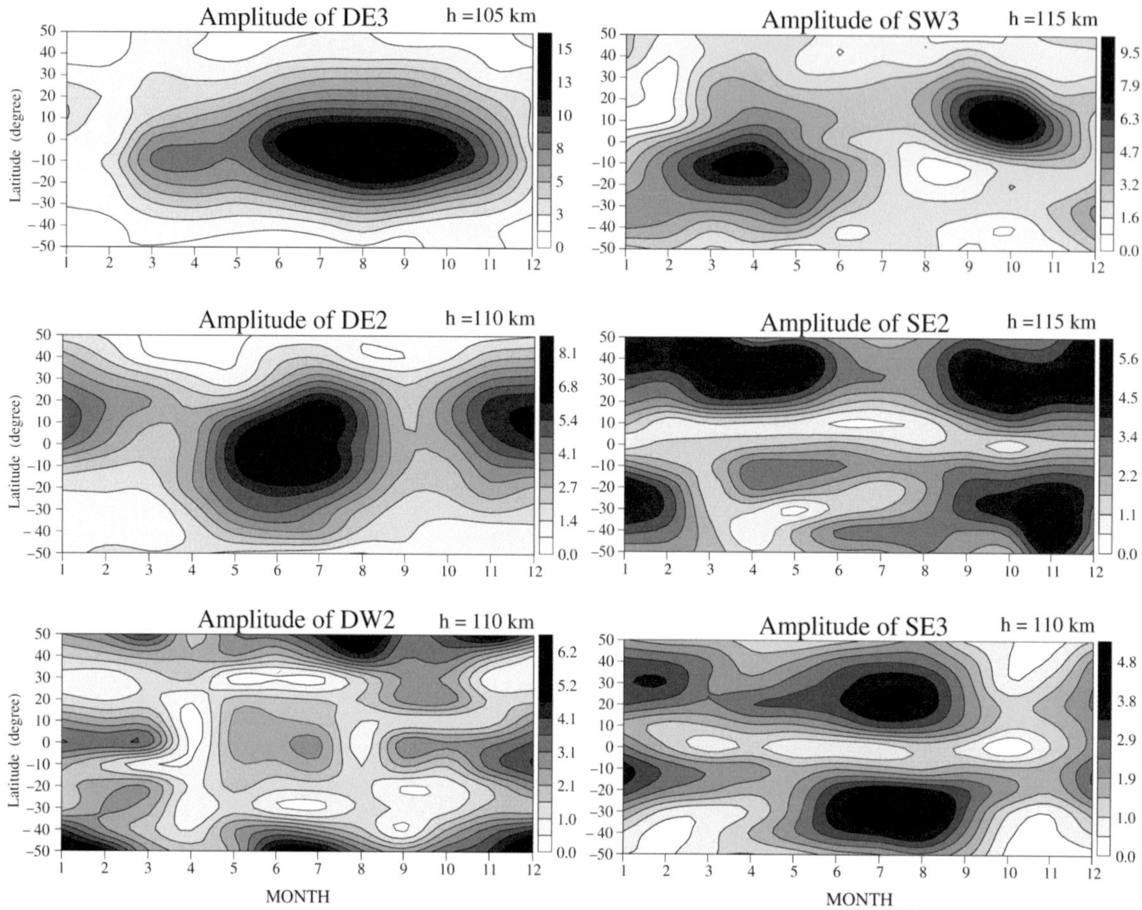

Fig. 2.7 Latitude-time cross sections of the following nonmigrating diurnal (*left column, from top to the bottom*): *DE3*, *DE2* and *DW2* and semidiurnal (*right column; from top to the bottom*): *SW3*, *SE2* and *SE3* tides

The very large average (for 6 years) *DE3* tidal amplitude of ~15 K in August–September (Fig. 2.7, upper left plot) indicates that this tide has very stable phase structure. The main tidal maximum is located in the lower thermosphere at 105 km height (Fig. 2.8a, upper left plot) over the equator (more precisely between 0°S and 10°S) and the tide can be detected in the latitude range of ±40°. The altitude tidal structure also indicates a secondary maximum situated near 85 km height and 10°N latitude which amplitude is ~5 K in August. This secondary maximum is better evident in April (Pancheva et al., 2010c). The latitudinal structure of the *DE3* tide above 95 km height is approximately symmetrical about 10°S while below 85–90 km some contribution of the antisymmetric *DE3* tidal modes has been found as well. The *DE3* is a vertically upward propagating tide with vertical wavelength of ~35–40 km. The latitude phase structure of the *DE3* indicates that this tide can be defined as an "equatorially trapped tide" (the latitudinal phase gradient is almost zero) only above ~95 km height and mainly during August–September.

The average (2002–2007) *DE2* tidal amplitude of ~8 K in June is also very large and similarly to the *DE3* tide this is an indication for a stable phase structure. As noted previously, this tide maximizes at ~10°S latitude during June solstice and at ~10°N during December (Fig. 2.7, middle left plot) indicating that the latitudinal and seasonal structures of the *DE2* are defined by interplay between the latent heat forcing and mean wind effects. The mean June solstice maximum is situated at ~110 km height (Fig. 2.8 middle left plot), while the December one is located at ~115 km (not shown result). The latitude structure of the tidal amplitudes

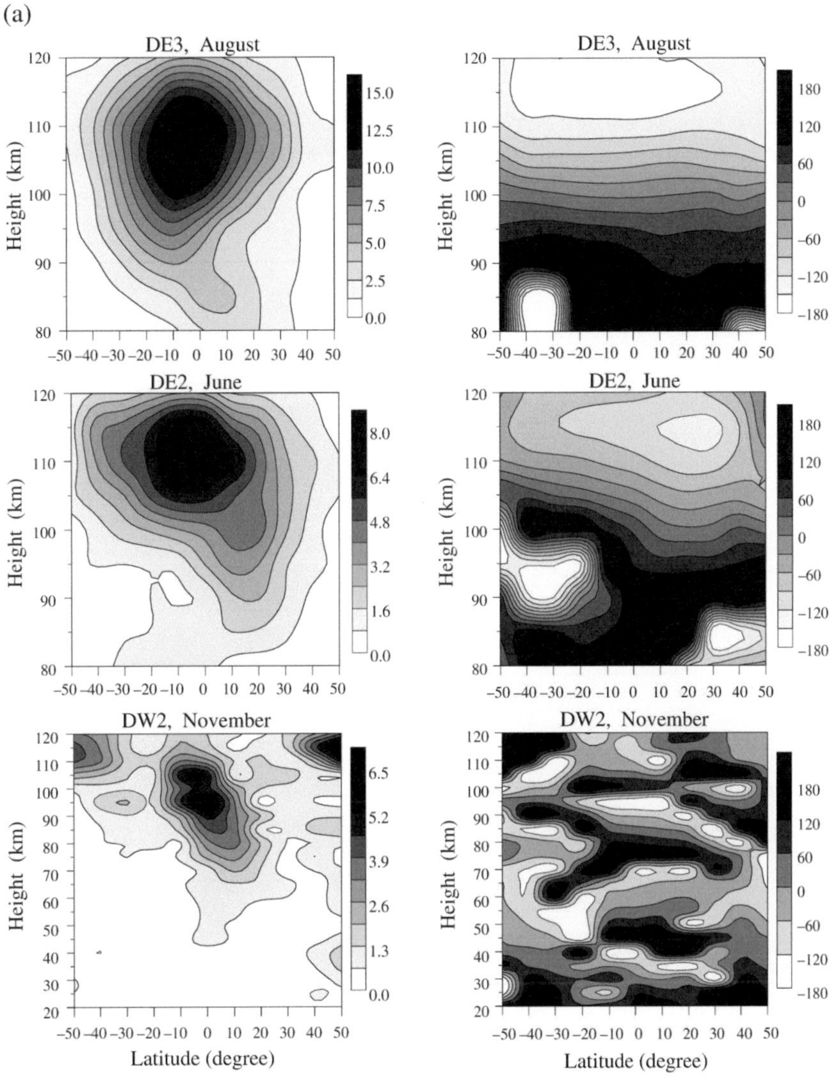

Fig. 2.8a Latitude-altitude cross sections of the amplitudes (in Kelvin) (*left column*) and phases (in degrees) (*right column*) of the nonmigrating diurnal tides shown in Fig. 2.7. *DW2* tide is shown for the altitude range 20–120 km, while *DE3* and *DE2* are shown for 80–120 km

above ~105 km height is symmetric about ~10°S, while below ~100 km some contribution of antisymmetric *DE2* tidal modes has been found. The phase distributions revealed downward tilting of phase lines (upward energy propagation) with average vertical wavelength of ~35–45 km, being larger in June than in December.

The latitudinal structure of the *DW2* tide (bottom left plot of Fig. 2.7) in the lower thermosphere (110 km height) reveals amplification over the equator and at middle latitudes as the tidal amplitude amplifies towards high latitudes. Having in mind that the largest monthly mean amplitude of this tide during 2002–2007 is ~8 K (Pancheva et al., 2010c), then its average amplitude (for 6 years) of ~6 K indicates that the *DW2* tide has also a stable phase structure. The seasonal behavior of the *DW2* tide in the lower thermosphere is dominated by a semiannual variability with solstice maxima. In the mesosphere (below ~90–95 km height) however it amplifies during winter. Numerical models showed that both interaction between *DW1* and *SPW1* (Hagan and Roble, 2001) and latent heating (Hagan and Forbes, 2003; Forbes at al., 2007) are the sources for the *DW2* tide. There are observations supporting particularly the nonlinear interaction mechanism (Lieberman et al.,

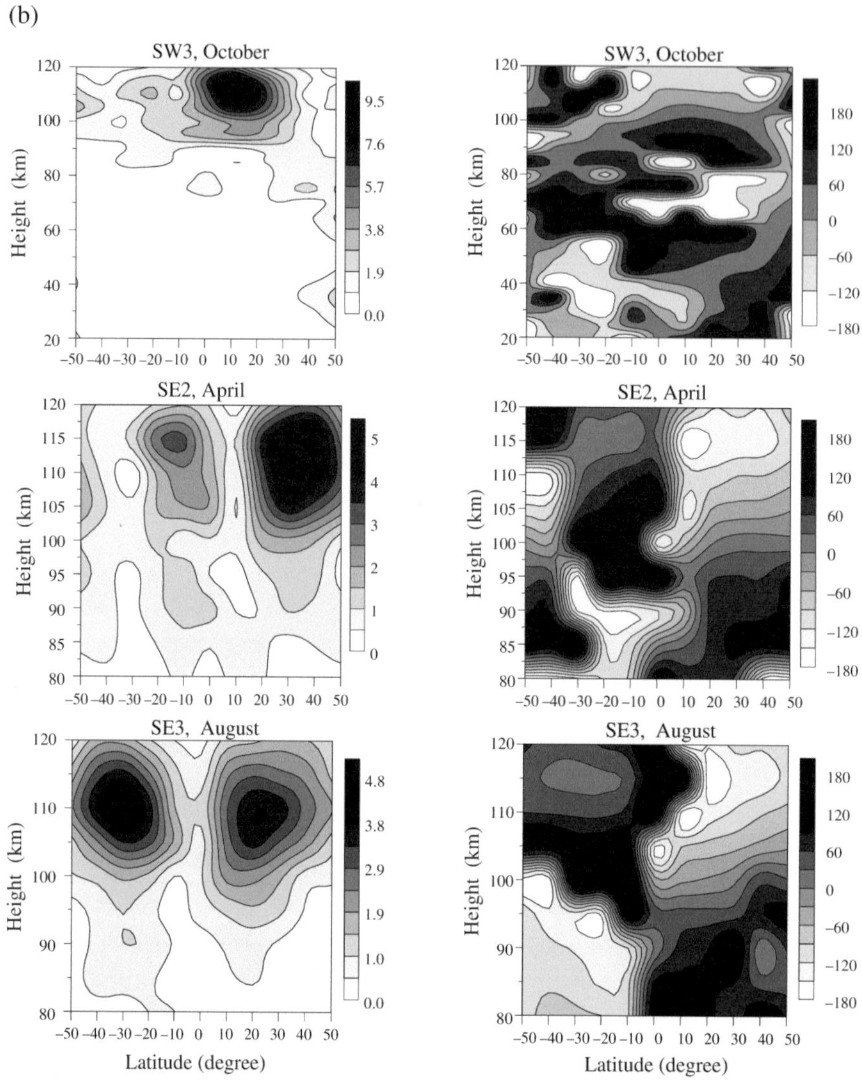

Fig. 2.8b Latitude-altitude cross sections of the amplitudes (in Kelvin) (*left column*) and phases (in degrees) (*right column*) of the nonmigrating semidiurnal tides shown in Fig. 2.7. *SW3* tide is shown for the altitude range 20–120 km, while *SE2* and *SE3* are shown for 80–120 km

2004; Pancheva et al., 2009c). Having in mind the *DW2* sources, the results from the bottom row of plots in Fig. 2.8a, showing the altitude-latitude cross section of the *DW2* amplitude (left) and phase (right) for November, can be interpreted in the following way: it seems that the *DW2* tidal activity is composed by three different types: (i) The amplification of the *DW2* tide in the stratosphere and mesosphere of the NH middle latitudes is forced by the coupling between *DW1* and *SPW1*. The average (for 6 years) tidal amplitude is small, ~2–3 K; we remind however that the DW1 in the stratosphere and lower mesosphere is composed mainly by trapped modes with an average amplitude of ~4–4.5 K. The phase structure (bottom right plot) indicates that this is vertically propagating tide with quite large vertical wavelength, ~60–65 km. There is no indication for inter-hemispheric propagation. (ii) The second amplification is confined mainly to low-latitudinal MLT region. There is a clear double-peak vertical structure located at ~95 and ~105 km with magnitude reaching ~6.5 K. The low-latitude *DW2* tide is also vertically propagating wave but with significantly shorter wavelength, ~25–30 km, than that of the middle latitude NH *DW2* tide. The low-latitude *DW2* tide is also forced by the coupling between *DW1* and *SPW1*, but most probably the process of coupling takes place in the low latitudes. There are observations showing that when the *SPW1* propagates vertically upward it propagates toward the equator as well (this will be shown in the section for *SPW*). Therefore, the coupling process amplifies significantly in the mesosphere where the two prime waves, *DW1* and *SPW1*, have significant amplitudes. Both double-peak vertical structure of the *DW2* tidal amplitude and its short-vertical wavelength confirm that the propagating component of the *DW1* tide participates in the coupling. (iii) The third amplification is observed in the mid-latitude lower thermosphere (∼115 km) of both hemispheres. The *DW2* amplitude amplifies towards high latitudes. The phase structure of this tide (bottom right plot of Fig. 2.8a) indicates that it is an in situ excited tide most probably by a zonally asymmetric momentum deposition from gravity waves which have undergone filtering by the stratospheric planetary wave field.

2.3.2.2 Nonmigrating Semidiurnal Tides

The right column of Fig. 2.7 shows the latitude versus month contours of the *SW3* (upper), *SE2* (middle) and *SE3* (bottom) tides at lower thermospheric altitudes where these tides maximize, while Fig. 2.8b shows the altitude-latitude cross sections of the tidal amplitudes in Kelvin (left column) and phases in degrees (right column) for *SW3* (upper row) in October, *SE2* (middle row) in April and *SE3* (bottom row) in August, i.e. during the months when the considered tides have large amplitudes.

The *SW3* tide is the strongest nonmigrating semidiurnal tide reaching an average (for 6 years) amplitude of ∼9.5 K in the lower thermosphere. It amplifies in the tropical latitudes (10–20°) during the autumn equinoxes in both hemispheres. The latitude structure of the *SW3* tide is dominated by antisymmetric modes. The altitude-latitude structure of the *SW3* tide in October indicates that its major source is interaction between *SW2* and *SPW1* that probably takes place in the winter (NH) lower atmosphere. The vertical phase structure, shown in the upper right plot of Fig. 2.8b, shows that this is a vertically upward propagating tide with long vertical wavelength (cannot be estimated precisely but is larger than 60 km) in the stratosphere and mesosphere. When the *SW3* propagates vertically upward in the stratosphere and mesosphere it propagates toward the equator as well, but this penetration is confined to ∼85–90 km height. The rapid amplification of the *SW3* tide in the tropical lower thermosphere (∼115 km height and near 10°N) can be caused by additional source of this tide originating somewhere in the tropical mesosphere.

The latitude structure of the *SE2* tide in the lower thermosphere (115 km height), shown in the right middle plot of Fig. 2.7, reveals a tidal amplification in both hemispheres between 30° and 40° latitudes. The average (for 6 years) tidal amplitude reaches 5.6 K; considering that the largest monthly mean amplitude is 7.5 K (Pancheva et al., 2010c) this means again that the *SE2* phase tidal structure is stable. The seasonal behavior in NH is dominated by semiannual variability with equinox maxima; there is however a secondary winter amplification as well. In the SH the seasonal behavior is dominated by main annual variability with summer (November–January) maximum; there is also a secondary one in winter that is located near 10°S and 40°S. The altitude-latitude structure of the *SE2* tide in April is shown in the middle row of plots in Fig. 2.8b, as the *SE2* amplitude is in the left plot while the phase is in the right plot. Only the altitude range between 80 and 120 km is considered because the tidal amplitudes below 80 km are too small to be meaningful. The latitudinal structure of the *SE2* amplitude indicates some prevalence of the antisymmetric modes upon the symmetric ones in April. The tidal amplitude grows with increasing height and reaches a maximum of about 5 K at 110–115 km height and latitudes 30°N–40°N. The phase distribution indicates downward tilting of phase lines in the considered altitude range. Some tidal progression in the NH toward the equator can be seen as well. The *SE2* tide is almost out of phase in both hemispheres at the altitude where it maximizes which is an indication for a contribution of antisymmetric *SE2* tidal modes. The average vertical wavelength calculated from the vertical phase gradient in the NH is ∼ 30–35 km.

The latitude structure of the *SE3* tide in the lower thermosphere (110 km height) is shown in the right bottom plot of Fig. 2.7. The SE3 tide grows towards tropical latitudes reaching average maximum of ∼4.8 K in the SH and ∼4 K in the NH at latitudes near 20°–30°. The seasonal behavior in both hemispheres has a main maximum during June solstice and a secondary one during December solstice. The average (2002–2007) altitude-latitude cross-section of the *SE3* tide for August is presented in the bottom row of plots in Fig. 2.8b. The tidal amplitude is shown in the left bottom plot while the tidal phase in the right bottom plot. The tidal amplitudes grow with increasing height and reach maxima which are situated at ~110 km height in both hemispheres. The phase distributions, particularly that for the NH, reveals downward tilting of phase lines in the considered altitude range (80–120 km) with vertical wavelength of ~35–40 km. Similarly to the *SE2* tide here also some tidal progression in the NH toward the equator can be seen.

Herein, we elucidate seasonal, spatial (altitude and latitude) and interannual temperature variabilities for some nonmigrating diurnal and semidiurnal tides covering January 2002–December 2007. All considered nonmigrating tides achieve maximum amplitudes in the lower thermosphere (~105–115 km) and in this way are potentially important for the modulation of plasma properties accompanying vertical drift of the equatorial ionosphere. It is worth remembering that the tidal amplitudes and phases are determined here by using a 60-day window, hence they represent 60-day means. The day-to-day variability of the tides however can be significantly stronger than that presented in this section. The same is valid also for their modulation effects on the ionospheric variability.

2.4 Planetary Waves

As noted before, the spectral analysis performed on the SABER/TIMED temperatures for the considered period of time 2002–2007 indicated that the regularly persistent planetary waves with large amplitudes have the following mean periods: 5.5, 11, 17 and ~24 days. In the decomposition procedure, besides the SPWs with zonal wavenumbers up to 3, all above mentioned zonally propagating and zonally symmetric planetary waves with zonal wavenumbers up to 3 were included. In this section the results for the following planetary waves are presented: (i) *SPW1*, (ii) ~5-day *Rossby* and ~6-day *Kelvin* waves (these are waves with a mean period of 5.5 days and zonal wavenumbers –1 and 1 respectively), and (iii) 10- and 16-day waves (these are waves with mean periods 11 and 17 days with zonal wavenumber –1).

2.4.1 Stationary Planetary Wave with Zonal Wavenumber 1 (SPW1)

The climatological features of the *SPW1* are investigated by the distribution of the monthly mean wave amplitudes and phases in the time and space (latitude and altitude). The *SPW1* phase is expressed in degrees and indicates the geographic longitude where the temperature maximum is located. Figure 2.9a shows the latitude-time cross sections of the *SPW1* amplitudes for three altitudes: 40, 70 and 115 km, indicating the latitude structure and time evolution of the wave amplitudes in the stratosphere, mesosphere and lower thermosphere. The scale of the plot (in Kelvin) for the lower thermosphere is different from that for stratosphere and mesosphere showing a rapid growth of the wave amplitudes above mesopause region. As it has been expected the *SPW* amplitude increases with increasing latitude and the wave maxima are apparently located at higher than 50° latitudes. Xiao et al. (2009) showed that the SPW1 is centered near 60°–70° in both hemispheres. There is only weak presence of the *SPW1* at tropical MLT as particularly in the mesosphere the wave most probably penetrates across the equator. While a winter *SPW* maximum dominates the seasonal variability in the strato-mesosphere there are several maxima in the lower thermosphere. An important feature of the *SPW1* in the lower thermosphere is its symmetrical latitudinal structure. However, the *SPW1* amplitudes in the SH are significantly stronger than those in the NH; while the largest amplitudes in the NH reach 20 K those in the SH are 30 K.

A careful inspection of the temporal variability of the wave amplitudes, particularly in the mesosphere ($h = 70$ km), shows some 2-year variability; it seems that in the NH the winters: 2003/2004 and 2005/2006 are stronger than 2002/2003, 2004/2005 and 2006/2007, while in the SH the austral winters of 2002, 2004 and 2006 are stronger than those in 2003, 2005 and 2007. Figure 2.9b shows the SPW1 amplitudes at $h = 70$ km and 50°N; in the SH (bottom plot) the result for 40°S is presented as this effect is better visible there. The 2-year maxima are marked by thick arrows situated in the bottom side of the plots. The average amplitude of the modulation is ~3 K, hence the interannual variability causes near 20–25% of the wave variability in the strato-mesosphere. This feature could be due to modulation by the QBO of the stratosphere zonal mean flow as the wave propagates from the stratosphere up to the mesosphere. The amplitudes of the *SPW1* are larger during the easterly phase of the QBO. This fact has some impact on the SSW phenomena. It has been found that major SSWs in the polar stratosphere are somewhat more likely in the easterly

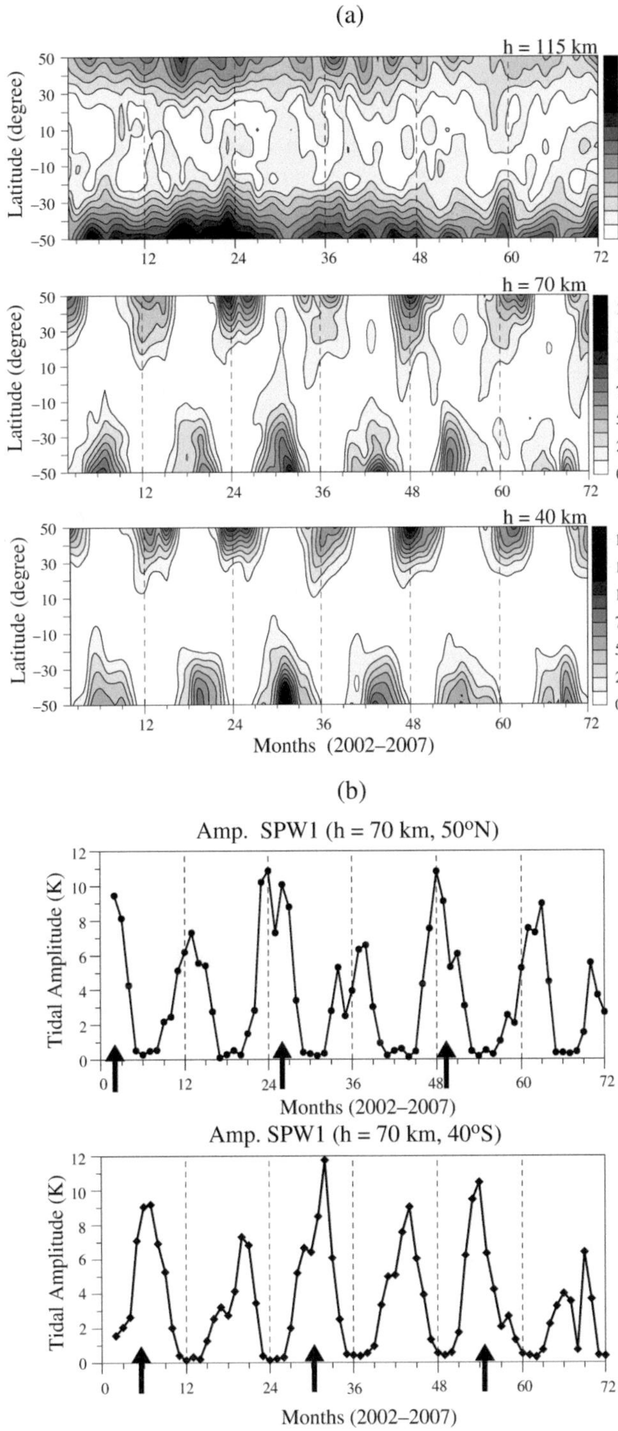

Fig. 2.9 (**a**) Latitude-time cross sections of the *SPW1* amplitudes (in Kelvin) at three altitudes: 40 km (*bottom plot*), 70 km (*middle plot*) and 115 km (*upper plot*); (**b**) Monthly mean *SPW1* amplitudes at altitude of 70 km and latitude of 50° N (*upper plot*) and 40° S (*bottom plot*); the 2-year maxima are marked by thick arrows situated in the bottom side of the plots

phase of the QBO (Labitzke and van Loon, 1999). In the lower thermosphere however the wave amplitudes in both hemispheres increase by the end of 2003 and then gradually decrease to the end of the considered period of time.

Figure 2.10a shows the average (2002–2007) seasonal behavior, i.e. the climatology of the *SPW1* in the NH (left plot) and SH (right plot) for altitude range 20–120 km. Clear climatological features can be distinguished in both hemispheres, i.e. winter amplification in the strato-mesosphere and equinoctial-winter amplification in the lower thermosphere. In the NH strato-mesosphere the *SPW1* is large between October and April while in the SH it is large between May and October. In the lower thermosphere the wave amplifies in April–May at both hemispheres and November–January in NH and near November in SH. The average *SPW1* phase plots display a propagating type wave in the strato-mesosphere with average vertical wavelength of ~70–80 km and a trapped type in the lower thermosphere. There is clear indication that the *SPW1* only in the NH propagates up to lower thermosphere in January–March, however its vertical wavelength significantly decreases above 80 km height.

The phase comparisons between both hemispheres reveal that while in the strato-mesosphere the phases during boreal and austral winters are very similar with a mean difference of ~30°, the waves in the lower thermosphere are almost out of phases at both hemispheres. Figure 2.10b shows the hemisphere comparison between the *SPW1* average (for 6 years) phases at altitude of 115 km. The mean result for the *SPW1* phases are on the average ~130°E in SH and ~110°W in NH. This is a new and important result found by Mukhtarov et al. (2010) that helps a lot in clarifying the main source of the *SPW1* in the lower thermosphere

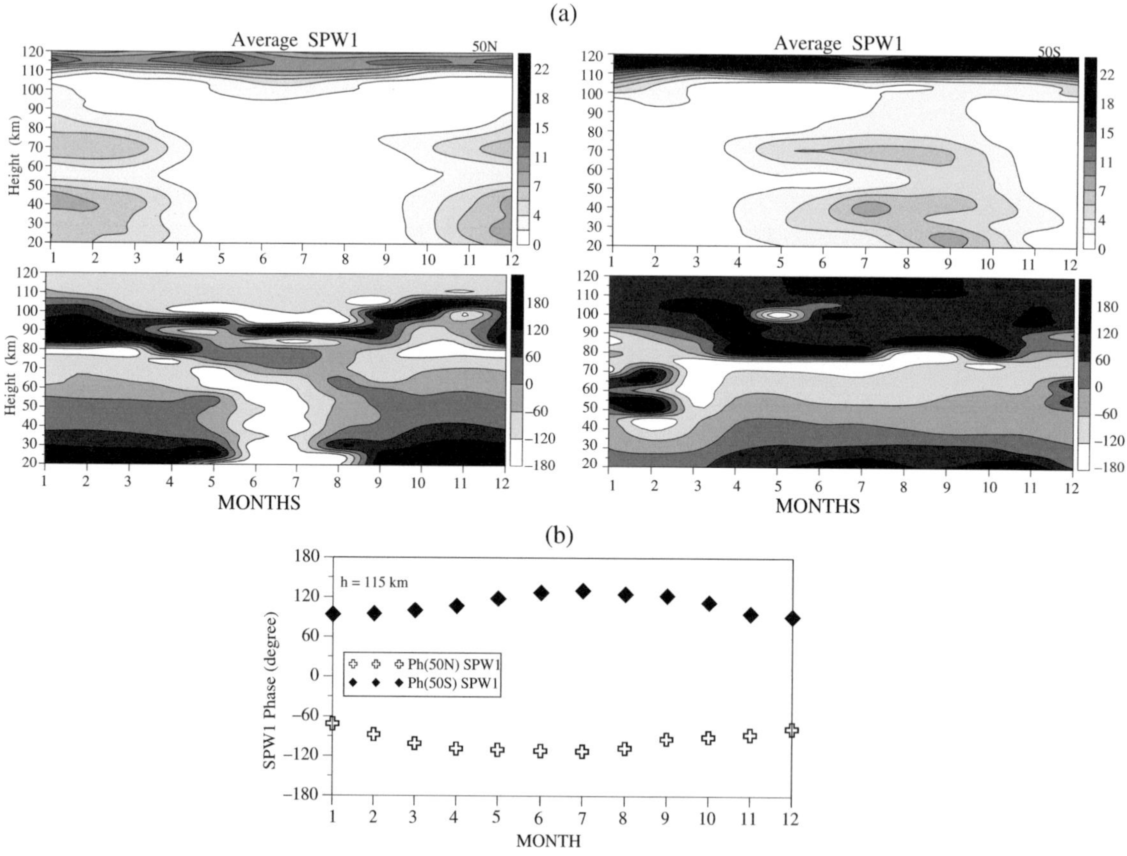

Fig. 2.10 (**a**) Average (2002–2007) altitude-time cross sections of *SPW1* at 50°N (*left column*) and 50°S (*right column*); (**b**) Hemispheric comparison of SPW1 average phases at altitude of 115 km

(above ~90–100 km), a question that has not yet received a definite answer.

The *SPW* field in the lower thermosphere is believed to be primarily associated with the generation in situ of the planetary waves by zonally asymmetric momentum deposition from gravity waves which have undergone filtering by the stratospheric planetary wave field (Smith, 1996, 1997, 2003; Forbes et al., 2002; Xiao et al., 2009). The above found phase relationship between the lower thermospheric *SPW1* phases in the NH and SH result does not support this mechanism. Another suggestion is related to ducting from the winter hemisphere (Pogoreltsev and Sukhanova, 1993; Forbes et al., 2002). The presented in Mukhtarov et al. (2010) detailed study of the temperature SPWs reveals that there is some propagation of the *SPW1* toward the equator but it is limited only up to ~90–100 km altitude; hence the SABER *SPW1* temperature results do not support this suggestion either.

Another in situ forcing can be related to the auroral heating. During geomagnetic activity large auroral energy is deposited into the polar atmosphere due to dissipation of electric currents, known as Joule heating, and due to the dissipation of kinetic energy from precipitating particles. This mechanism has been suggested by Xiao et al. (2009) however no evidence for its validity has been presented there. The result for the found steady (almost fixed) *SPW* phases in the lower thermosphere is evidence supporting the auroral heating as a main origin of the temperature *SPW* temperature field in the lower thermosphere because the auroral heating is centered in the magnetic pole.

Figure 2.11a presents the *SPW1* field in polar coordinate system with a pole coinciding with the geographic pole. The figure shows the wave field in April 2003 when a very strong *SPW1* has been observed in the lower thermosphere (see Fig. 2.9a, upper plot). The temperature *SPW1* field at $h = 115$ km is presented in the figure as the left plot is for the NH, while the right one for the SH. The *SPW1* temperature field resembles in some way the auroral oval in NH (left plot) and SH (right plot) particularly during strong geomagnetic activity when the auroral oval expands toward the equator. The auroral ovals are centered in Earth's magnetic poles. According to the global model International Geomagnetic Reference Field (IGRF) the North and South dipole poles have the following geographic coordinates for epoch 2005.0: (83.2°N, 118.3°W) and (64.5°S, 137.9°E) respectively; these coordinates are computed from the 10th Generation IGRF (http://www.geomag.bgs.ac.uk/poles.html). As it can be seen from Fig. 2.11a the longitudes of the dipole poles are very similar to the *SPW1* phases at $h = 115$ km. The coincidence of the dipole pole longitude with the lower thermospheric *SPW1* phase at both hemispheres is strong evidence indicating that the auroral heating is the main origin of the lower thermospheric *SPW* field. Knowing that the auroral heating excites the *SPW* field in the lower thermosphere then it is easy to explain why the *SPW1* in SH is significantly stronger than that in the NH. This is due to the fact that the geographic latitude of the dipole pole in SH (64.5°S) is significantly lower than that of the dipole pole in NH (83.2°N). Therefore in SH larger part of Joule heating penetrates toward the equator in comparison with that in NH.

If the lower thermospheric *SPW* field is excited mainly by auroral heating then the *SPW* amplitudes should indicate positive corellation particularly with geomagnetic activity at high latitudes which is strongly affected by charge precipitations. For this purpose we use the NOAA Hemispheric Power Data which present the estimated power in gigawatts deposited in the polar regions by energetic particles during transits over the poles by the NOAA Polar-orbiting Operational Environmental Satellite (POES) http://www.swpc.noaa.gov/. A cross-corellation analysis between the power data and *SPW1* amplitudes for each hemisphere has been performed. The comparison between both types of data is presented in Fig. 2.11b; the left plot is for NH, while the right one for SH. The *SPW1* amplitudes are shown with solid line and full diamonds, while the power data with dash line and empty diamonds. There is clear similarity between the variability of both types of data. The cross-corellation function with time lag up to 5 months has been calculated. The maximum cross-corellation coefficients have been found at time lag zero, as for NH and SH they are respectively 0.70 and 0.73 (shown in the upper right side of each plot). This high corellation is another strong evidence that supports the auroral heating as a main origin of the temperature *SPW* field in the lower thermosphere.

It is worth mentioning that first Riggin et al. (1999) and later Baumgaertner et al. (2006) have suggested

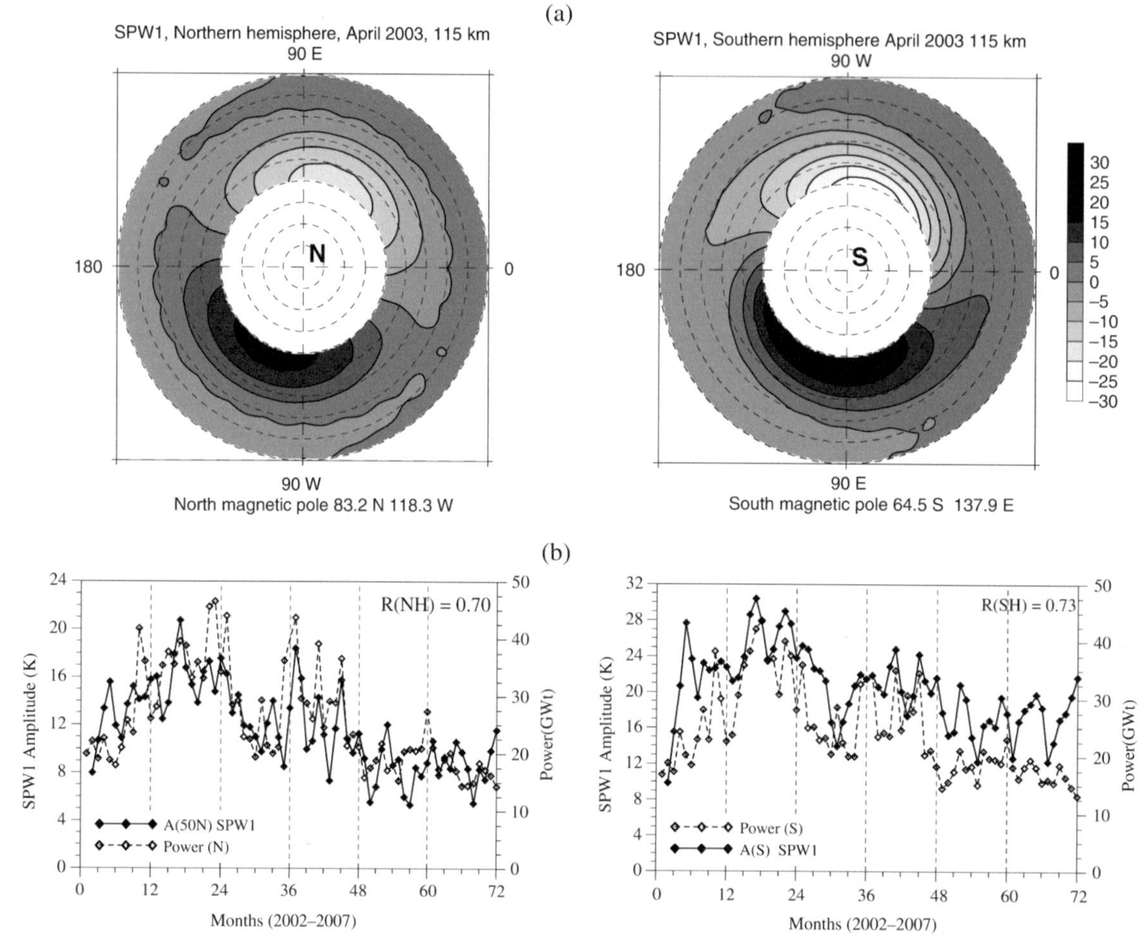

Fig. 2.11 (**a**) *SPW1* temperature field for April 2003 presented at polar coordinate system with a pole coinciding with the geographic poles; left plots presents SPW1 field in NH, while the right one in SH; (**b**) Comparison between the monthly mean hemispheric power data (in gigawatts) and *SPW1* amplitudes (in Kelvin) at $h = 115$ km for NH (*left plot*) and SH (*right plot*); *SPW1* amplitudes are shown with solid line and full diamonds, while the power data with dash line and empty diamonds; the cross corellation coefficients at time lag zero are shown in the upper right side of each plot

that zonally asymmetric heating via particle precipitation could create a wavenumber 1 temperature structure with respect to the geographic pole. The above shown results presented strong evidence in support of this idea.

In conclusion, we note that the obtained monthly mean *SPW* structures, which extend from the stratosphere into the lower thermosphere, can serve as a benchmark and guide for future numerical modeling studies aimed at better understanding the stratosphere-mesosphere-lower thermosphere coupling by stationary planetary wave patterns.

2.4.2 5.5-Day Westward and Eastward Propagating Waves with Zonal Wavenumber 1

The present subsection is focused on the global distribution and temporal variability of the ~5-day *Rossby* and ~6-day *Kelvin* waves seen in the SABER/TIMED temperatures. The ~5-day *Rossby* wave has been already explored before by Wu et al. (1994), Talaat et al. (2001), and Riggin et al. (2006). In this study the data analysis method ensures minimal distortion of the results due to aliasing or to the influence of the stronger waves on the weaker ones when they are present simultaneously in the data. To the authors' knowledge there is no other study representing the climatology of the ~6-day *Kelvin* wave from the stratosphere up to the lower thermosphere. We note in advance that the use of a long 60-day time segment for extracting the considered 5.5-day wave leads to some smoothing of the wave amplitudes. However, in order to be sure that the features of the considered planetary waves are not affected by the tides (through aliasing and the influence of the stronger wave on the weaker one), which are normally stronger than the planetary waves, particularly in the MLT, we have to be sure that the main tides are fully extracted from the data.

2.4.2.1 Climatology of the ~5-day Rossby (W1) Wave

Figure 2.12a shows the latitude-time cross sections of the 5.5-day *W1* amplitudes for altitudes 80 (bottom plot) and 105 km (upper plot) indicating the latitude structure and time evolution of the wave amplitudes in the mesosphere and lower thermosphere. The latitude structure of the temperature 5.5-day *W1* waves is similar at both height levels: the wave amplitudes maximize near 40°–50° at both hemispheres. This type of the symmetrical latitude structure resembles that of the 5-day *Rossby* waves. It is consistent with the gravest symmetric wave number-1 *Rossby* wave, i.e. the (1,1) mode. In addition, the temporal evolution of the 5-day wave in the MLT shows regular seasonal variability. The 5-day *Rossby* wave amplifies during the equinoxes at both hemispheres. In the NH the wave amplifies in March–April (reaching monthly mean amplitude of ~7.5 K) and September (~7 K), while in the SH it is strong in March (~5.8 K) and September–November (~6 K) (Pancheva et al., 2010a).

Figure 2.12b shows the average (2002–2007) seasonal behavior, i.e. the climatology of the ~5-day *Rossby* wave amplitudes in the NH (left plot) and SH (right plot) for altitude range 20–120 km. Clear climatological features can be distinguished in both hemispheres. In the NH the ~5-day *Rossby* wave amplifies in the equinoctial months, i.e. centered in March and September, while in the SH in March and September–November. The average altitude structure indicates double-peaked maxima. The main maximum is located in the lower thermosphere, 110 km, while the secondary one near 80 km; only the SH maximum in September is located at 90 km. On the average the main maximum is at least two times stronger than the secondary one.

Figure 2.12c shows the 6-year average altitude-latitude cross-sections of the ~5-day *Rossby* wave amplitudes (upper plots) and phases (bottom plots) for months March (left column) and September (right column). Only the altitudes above 70 km are considered since the wave amplitudes below this height are too small to be meaningful. Besides the double-peaked amplitude maxima the following wave features are seen: (i) the ~5-day *Rossby* wave propagates vertically up to 120 km altitude with a mean vertical wavelength calculated from the vertical phase gradient of ~50–60 km; only in March the ~5-day Rossby wave penetrates up to ~110 km. This means that the ~5-day Rossby wave seen in the SABER/TIMED temperatures is an internal mode; the deviation from external mode behavior for a resonant oscillation suggests that dissipation is acting on the wave, and (ii) some propagation toward the equator can be distinguished in both hemispheres. Similar result was obtained before by Riggin et al. (2006).

2.4.2.2 Climatology of the ~6-day Kelvin (E1) Wave

The spectral analysis performed on the SABER/TIMED temperatures for the considered 6 years clearly indicated the presence of the equatorially-trapped ~5.5-day *E1* wave in the stratosphere and mesosphere (Pancheva et al., 2010a). This wave exhibits a fair degree of equatorial symmetry and has been indentified as a "fast" ~6-day *Kelvin* wave. However, in order to prove this assertion the 5.5-day *E1* wave needs to be extracted from the temperature data and their spatial (latitude and altitude) structure to be carefully considered. It has been already mentioned that because of using a 60-day window the obtained amplitudes of the ~5-day *Rossby* wave are smoothed. The same effect is also applied to the ~6-day *Kelvin* wave.

Figure 2.13a shows the latitude-time cross sections of the 5.5-day *E1* wave at 40 and 80 km altitudes. Although the monthly mean amplitudes are small (do not exceed 1.8 K) the latitude wave structure in the

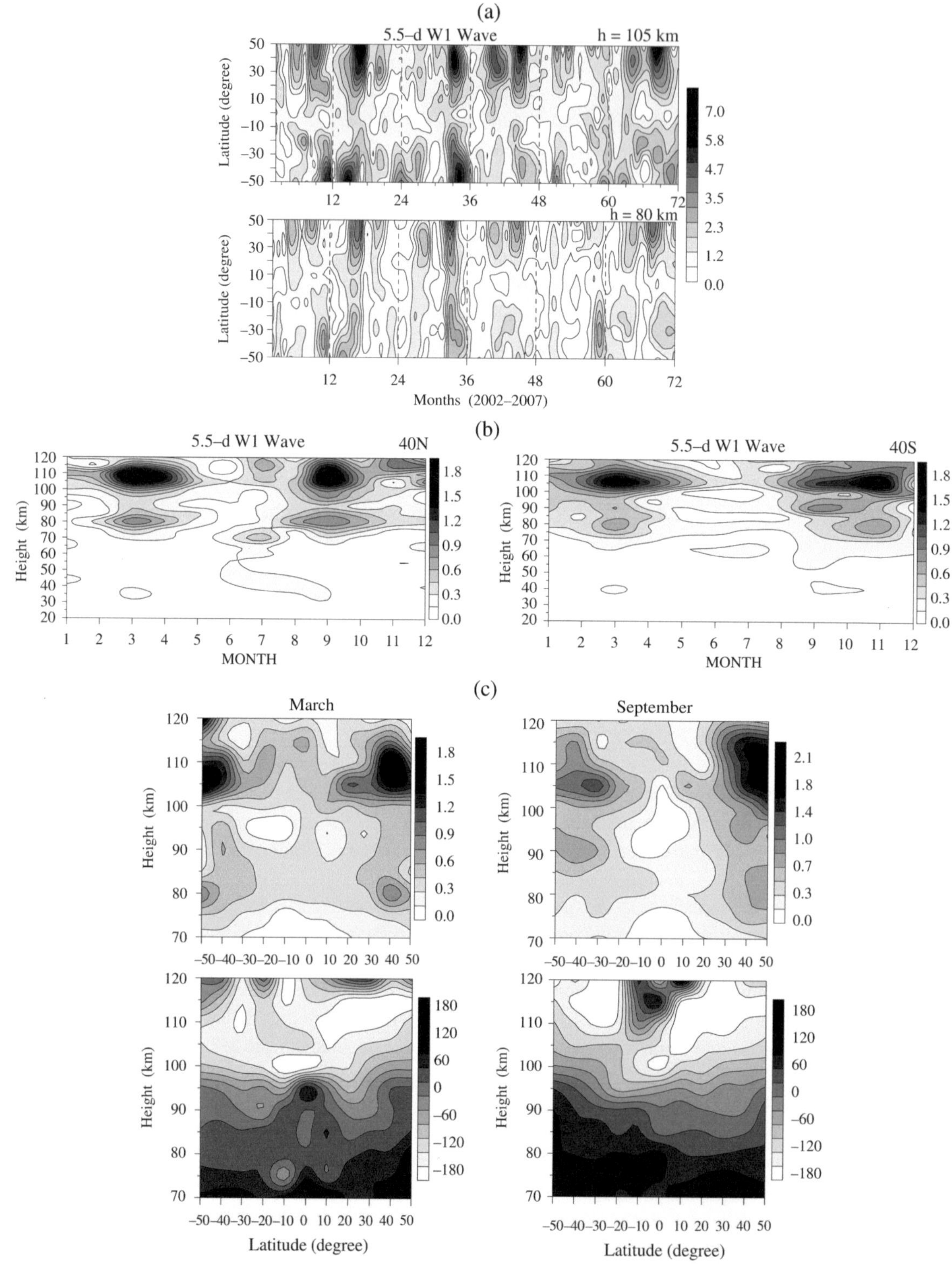

Fig. 2.12 (**a**) Latitude-time cross sections of 5-d *W1* amplitudes (in Kelvin) at altitudes of 80 km (*bottom*) and 105 km (*upper*); (**b**) Altitude-time cross sections of 5-d *W1* average (2002–2007) amplitudes at latitudes of 40°N (*left*) and 40°S (*right*), and (**c**) Altitude-latitude cross sections of 5-d *W1* average amplitudes (*upper row*) and phases (in degrees) (*bottom row*) for March (*left column*) and September (*right column*)

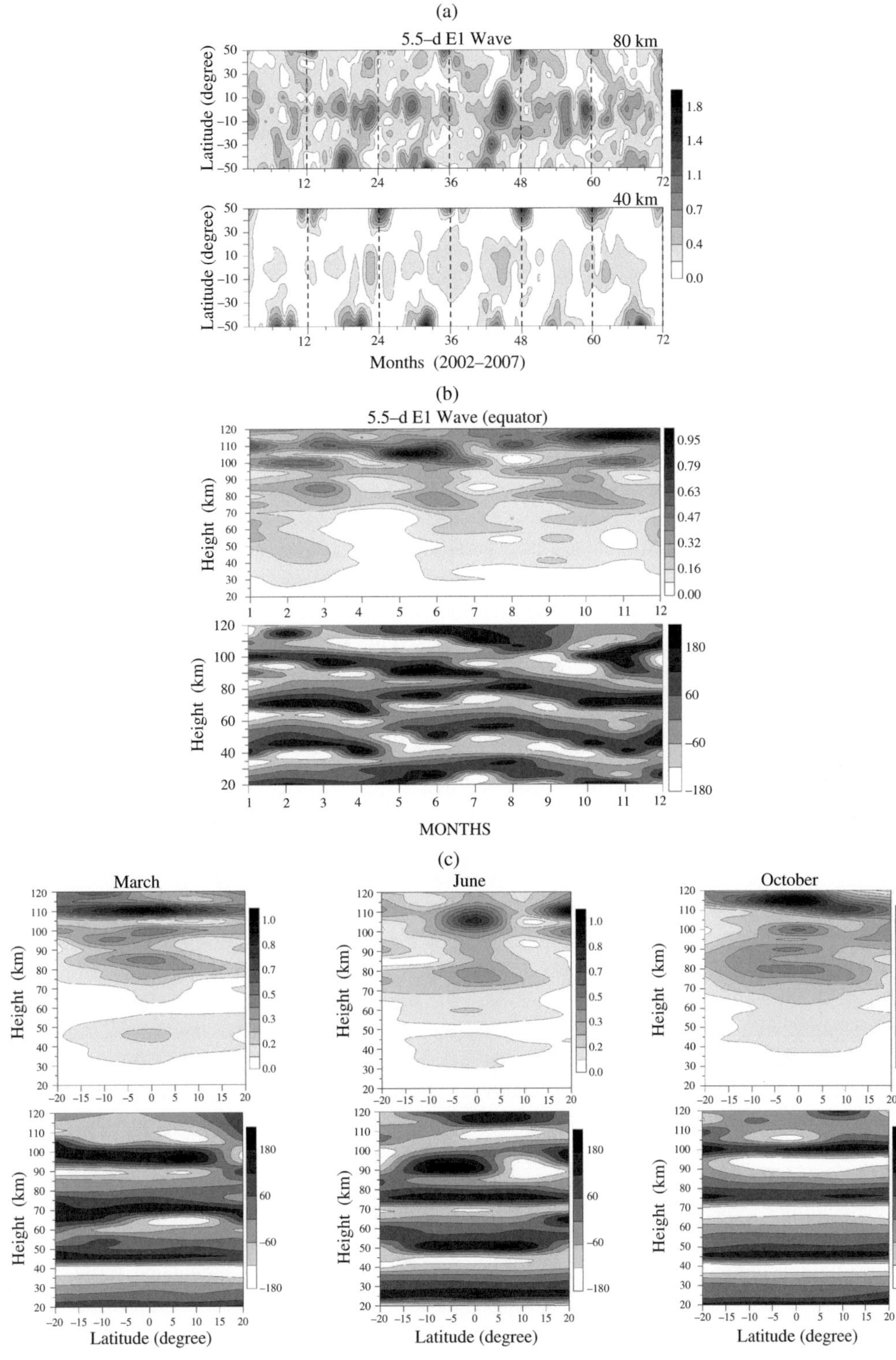

Fig. 2.13 (**a**) Latitude-time cross sections of 5-d *E1* amplitudes (in Kelvin) at altitudes of 40 km (*bottom*) and 80 km (*upper*); (**b**) Altitude-time cross sections of 5-d *E1* average (2002–2007) amplitudes at equator, and (**c**) Altitude-latitude cross sections of 5-d *E1* average amplitudes (*upper row*) and phases (in degrees) (*bottom row*) for March (*left column*), June (*middle column*) and October (*right column*); only the latitude range 20°N–20°S is shown

stratosphere and mesosphere is well pictured with two latitude ranges of wave amplifications. The first type of the 5.5-day E1 waves, centered at the equator and located between ±20°, is most probably the well known "fast" ~6-day *Kelvin* wave. This wave amplifies mainly in summer and equinoctial months. The second type of the 5.5-day *E1* waves maximizes toward the high latitudes. These waves have visible winter maxima, as the SH amplifications are stronger than those in the NH. This eastward-propagating wave observed in the winter hemisphere is most probably the well-known 4-day wave which is a ubiquitous feature in the polar winter upper stratosphere (Allen et al., 1997; Lait and Stanford, 1988; Azeem et al., 2005). The reports indicated that this wave consists of waves 1 through at least 4 all moving with the same phase speed, such that the period of wave 1 is near 4 days (Allen et al., 1997). These quasi-nondispersive 4-day wave "packets" with zonal wavenumbers 1–4 in the wintertime polar stratosphere are consistent with barotropic instability of the stratospheric polar night jet (Manney et al., 1988).

Figure 2.13b shows the average (2002–2007) altitude versus month contours of the ~6-day *Kelvin* wave amplitudes and phases. The average seasonal behavior of the wave amplitude indicates equinoctial (February–March and September–November) and June solstice (May–July) amplifications. A multi-peaked vertical structure can be distinguished in the average wave amplitudes. The maxima are located in the stratosphere, near 40–45 km altitude, in the mesosphere, near 75–85 km and in the lower thermosphere, near 100–110 km height. The average vertical phase structure (bottom plot) shows not only that the wave is vertically propagating with a mean vertical wavelength of ~25 km, but also that there is a clear seasonal variability of the wave phases. Usually the wave phases in equinoctial months occur earlier than those in June solstice.

Figure 2.13c shows the average, for the considered 6 years, altitude-latitude cross-sections of the ~6-day *Kelvin* amplitudes and phases for months March (left column), June (middle column) and October (right column). Only the latitudes between 20°N and 20°S are considered because Fig. 2.13a indicated that this wave is centered at the equator and located in the above mentioned latitude range. The average latitude structure of the ~6-day Kelvin wave amplitudes indicates that in the strato-mesosphere this wave is detectable in very narrow latitude band, ±10°, while in the lower thermosphere this band is larger, ±15–20°. The ~6-day *Kelvin* wave phases show no phase change with latitude, particularly well visible in the stratosphere and mesosphere. This result indicates that it is really an equatorially trapped mode. Three full cycles of the wave are seen up to 95 km height, confirming previously determined vertical wavelength of 25 km. The vertical phase structure for the considered in the Fig. 2.13c months reveals also that the ~6-day *Kelvin* wave propagates up to ~100–105 km altitude.

2.4.2.3 Interannual Variability of the ~5-day Rossby and ~6-day Kelvin Waves

A careful inspection of the ~5-day *Rossby* wave amplitude in MLT region shown in Fig. 2.12a can distinguish some amplitude modulation with a quasi-2-year period. The quasi-2-year effect is particularly visible on the October–November maximum in the SH and on the overall wave enhancement in the NH during 2003, 2005 and 2007. In order to emphasize the quasi-2-year variability of the wave amplitude, particularly for the NH), 5-month running mean amplitudes are calculated. Figure 2.14a shows the monthly mean amplitudes of the ~5-day *Rossby* wave at 105 km height and at 50°N (left plot) and 50°S (right plot) drawn by thin line. The 5-month running mean amplitudes are shown by thick lines on both plots. The 2-year maxima are marked by arrows situated at the bottom of the plots. This feature could be due to modulation by the QBO of the stratosphere zonal mean flow as the wave propagates from the stratosphere up to the lower thermosphere. It seems that the amplitudes of ~5-day *Rossby* wave are larger during the easterly phase of the QBO.

Similar result was found before by Miyoshi and Hirooka (2003) in investigating the 5-day *Rossby* wave seen in the NCEP reanalysis data. The authors showed that the 2-year variation of the wave amplitude is intimately related to the zonal wind oscillations associated with the equatorial QBO in the stratosphere, i.e. the 5-day wave is modulated through the propagation condition accompanied with the zonal wind variation of the QBO. The authors mentioned also that the quasi-2-year variability can be attributed also to the excitation mechanism of the 5-day *Rossby* wave. Miyoshi (1999) and Miyoshi and Hirooka (1999) indicated that the heating due to moist convection is very

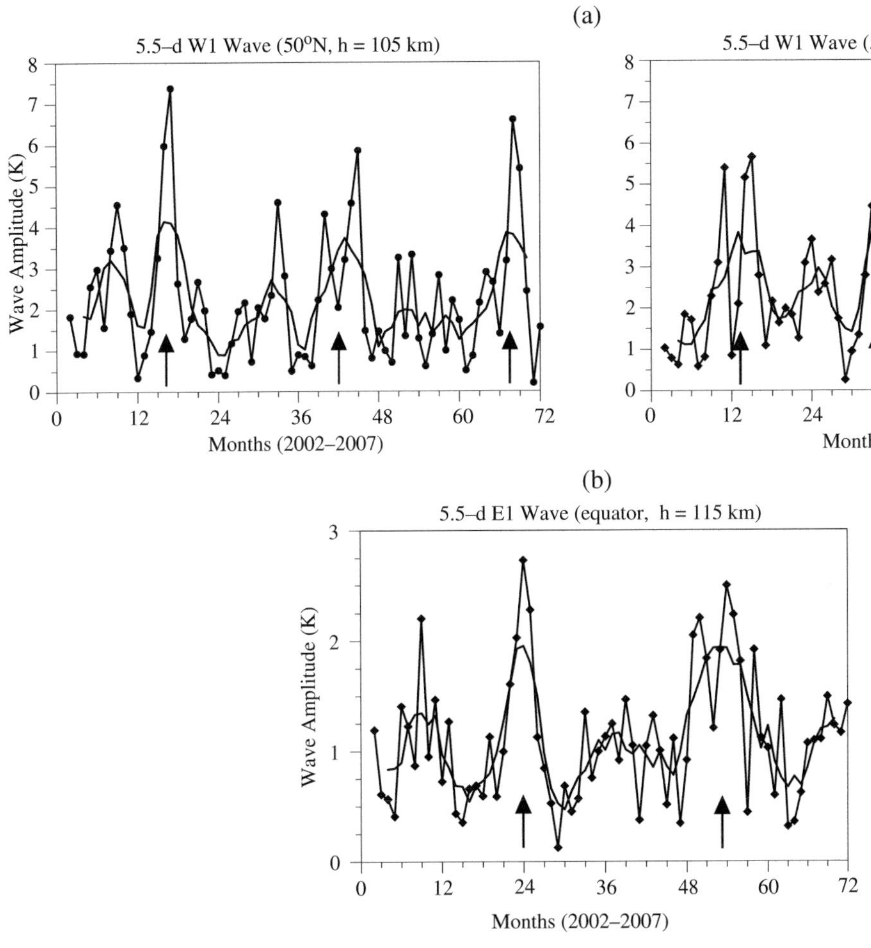

Fig. 2.14 (**a**) Monthly mean amplitudes of the ~5-day *Rossby* wave at altitude of 105 km and at 50°N (*left plot*) and 50°S (*right plot*) drawn by thin line, while the 5-month running mean amplitudes are shown by thick line; (**b**) The same as (**a**), but for the amplitudes of the ~6-day *Kelvin* wave at altitude of 115 km over the equator; the arrows situated at the bottom side of the plots indicated the 2-year maxima

important for excitation of the 5-day wave. Lau and Sheu (1988) however showed that the global precipitation fluctuates with timescale of 2–3 years. Therefore, the magnitude of heating due to the moist convection varies with the same period. As a result, the 5-day wave amplitude may be influenced by variations of the moist convective heating with a period of 2–3 years (Miyoshi and Hirooka, 2003).

Similarly to the ~5-day *Rossby* wave, the amplitudes of the ~6-day *Kelvin* wave also reveal interannual variability (see Fig. 2.13a). Figure 2.14b shows the monthly mean ~6-day *E1* wave amplitudes over the equator at 115 km altitude drawn by thin line, while the 5-month running mean amplitudes are shown by thick line. Again the amplitude maxima are indicated by arrows situated in the bottom side of the plot. In this case the interannual variability of the ~6-day *Kelvin* wave reveals a period longer than 2 years. As the considered period of time is too short, only 6 years, this topic needs to be investigated further.

In conclusion of this subsection, we note that the obtained climatological global structure and temporal variability of the ~5-day *Rossby* and ~6-day *Kelvin* waves seen in the SABER/TIMED temperatures (2002–2007) and particularly the result for their vertical propagation from the stratosphere up to the lower thermosphere sheds some light on the recently reported results about the coupling of the atmosphere-ionosphere system driven by the 5–7 day planetary waves (Tsunoda et al., 1998, Haldoupis and Pancheva, 2002; Takahashi et al., 2006; Pancheva et al., 2008c; Zuo and Wan, 2008). It is worth noting that the mechanism(s) behind the double-peaked or multi-peaked vertical structure of the ~5-day *Rossby* and ~6-day *Kelvin* temperature waves respectively is still unknown and most probably there is a necessity for re-examination of the wave excitation mechanisms, the effects of the eddy diffusivity and shears, mean wind distribution, etc. in the numerical wave simulations.

2.4.3 11- and 17-Day Westward Propagating Waves with Zonal Wavenumber 1(W1)

There are only a few studies about the global structure of the 10-day ((1,2) mode) and 16-day ((1,3) mode) waves particularly in the mesosphere. We remind that in the notation (*s, n–s*) mode of the normal mode *Rossby* waves, *s* is the zonal wavenumber, and *n* is the meridional index number such that *n–s–1* is the number of nodes between the poles. Forbes et al. (1995) investigated behaviors of the 16-day wave in the summer hemisphere of the mesopause region by using a GSWM. They showed that the 16-day appeared in the high-latitude summer hemisphere. However, seasonal variation and global structure of the 16-day wave in the mesopause region, and its relation to the stratospheric 16-day wave are not clear. Pancheva et al. (2008a) used UK Met Office (UKMO) assimilated data and radar measurements at 8 stations to determine planetary waves in MLT in the Arctic winter of 2003/2004. They showed that prior to the SSW, the stratosphere-mesosphere system was dominated by an upward and westward propagating 16-day wave that was detected simultaneously in the UKMO and MLT zonal and meridional wind data. The same winter, 2003/2004, has been investigated also by Shepherd et al. (2007) by using not only UKMO and radar data, but SABER/TIMED temperatures as well. They found not only 16-day wave but 10-day wave as well. Fedulina et al. (2004) analyzing UKMO data in order to study the variability of the traveling planetary waves in the upper stratosphere during 1992–2001 concluded that the longer-period planetary waves ($10 \leq T \leq 20$ days) have significant amplitudes mainly during winter in the both hemispheres.

The main focus of this subsection is to present briefly the global structure and seasonal variability of the ~11-*W1* and 17-day *W1* waves seen in the SABER/TIMED temperatures during 2002–2007.

Figure 2.15a shows the latitude versus time (number of months for the period January 2002–December 2007) contours of the 11-day *W1* wave amplitude (in Kelvin) at altitudes 40 km (bottom plot) and 80 km (upper plot). The wave amplifies toward high latitudes at both altitude levels. The seasonal behavior at both hemispheres is dominated by winter maximum; there is some enhancement in the equinoxes as well. While in the stratosphere the 11-day *W1* wave is confined to middle latitudes in the mesosphere the wave penetrates to lower latitudes as well. The largest monthly mean wave amplitude observed at mesosphere levels is ~4 K and the wave is slightly stronger in the NH than that in the SH.

Figure 2.15b presents the altitude-time cross sections of the 11-d *W1* amplitudes at latitudes of 50°N (upper plot) and 50°S (bottom plot). In the strato-mesosphere at both hemispheres there are two altitude levels where the wave amplifies; these are altitudes of ~40 km and ~80–90 km. In general, the wave amplitude amplifies with increasing the height, particularly in the mesosphere, reaching the largest amplitude in the lower thermosphere. The vertical phase distribution (not shown here) indicated that the 11-day *W1* wave is vertically propagating wave with a mean vertical wavelength of ~50 km. A careful inspection of the wave phase revealed that the 11-d *W1* wave propagates up to ~100–105 km altitude at both hemispheres.

The results for the 17-day *W1* wave are very similar to those for the 11-day *W1* wave. The latitude structure of the 17-day *W1* wave amplitude at strato-mesosphere is shown in Fig. 2.16a, while Fig. 2.16b presents the vertical structure of the wave amplitudes at latitudes of 50°N (upper plot) and 50°S (bottom plot). In this case the antisymmetric structure of the wave, particularly in the stratosphere (40 km), is well visible. The seasonal behavior is dominated by winter amplification, and again the wave is stronger in the NH (the largest amplitude is ~4.8 K) than that in the SH. The 17-day *W1*wave in winter is a vertically propagating wave (the wave phase is not shown here) with a vertical wavelength similar to that of the 11-day *W1* wave. The phase structures of both 11- and 17-day waves

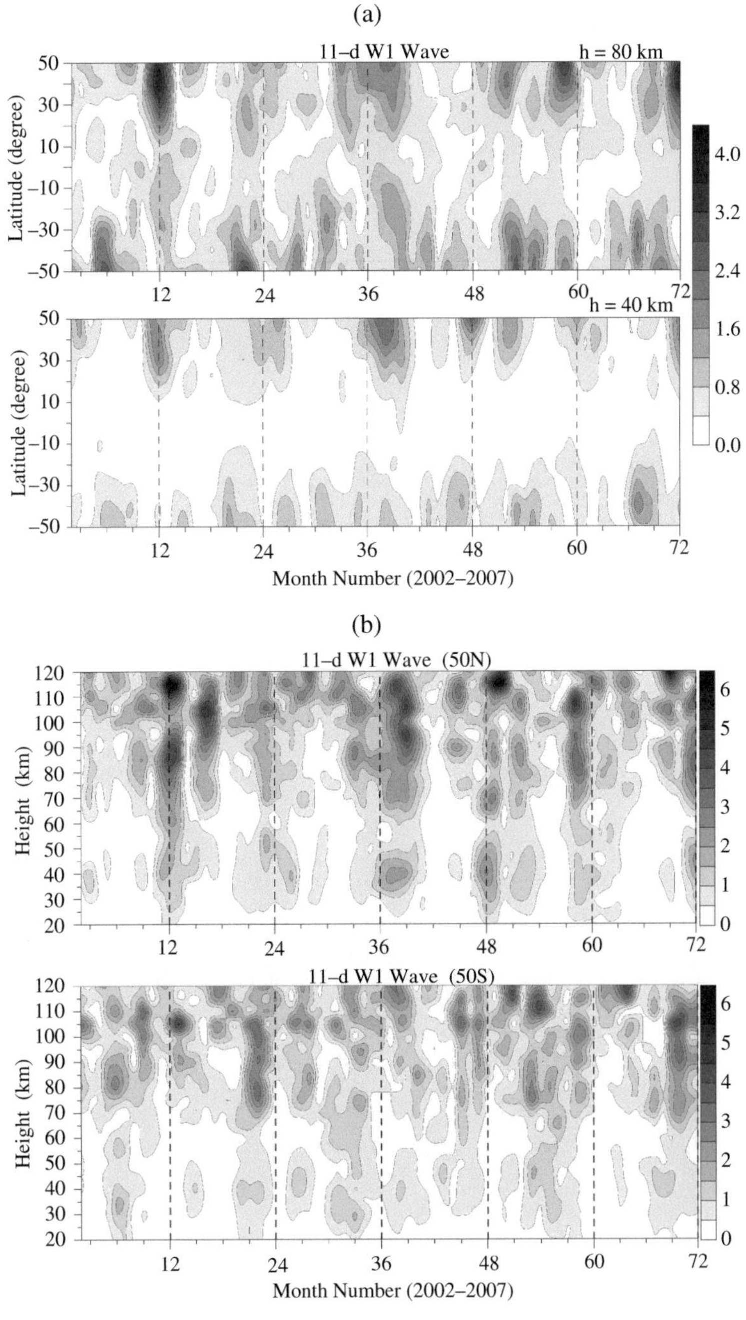

Fig. 2.15 (**a**) Latitude-time cross sections of 11-d *W1* amplitude (in Kelvin) at altitudes of 80 km (*upper plot*) and 40 km (*bottom plot*); (**b**) Altitude-time cross sections of 11-d *W1* amplitude at latitudes of 50°N (*upper plot*) and 50°S (*bottom plot*)

indicate that the source region is located somewhere below 20 km height.

The above described global structure and seasonal variability of the 17-day *W1* wave seen in the SABER/TIMED temperatures to a large extent support the numerical simulation of the 16-day wave in the mesopause region reported by Miyoshi (1999). The author found also that the 16-day wave is mainly excited by heating due to the moist convection in the troposphere.

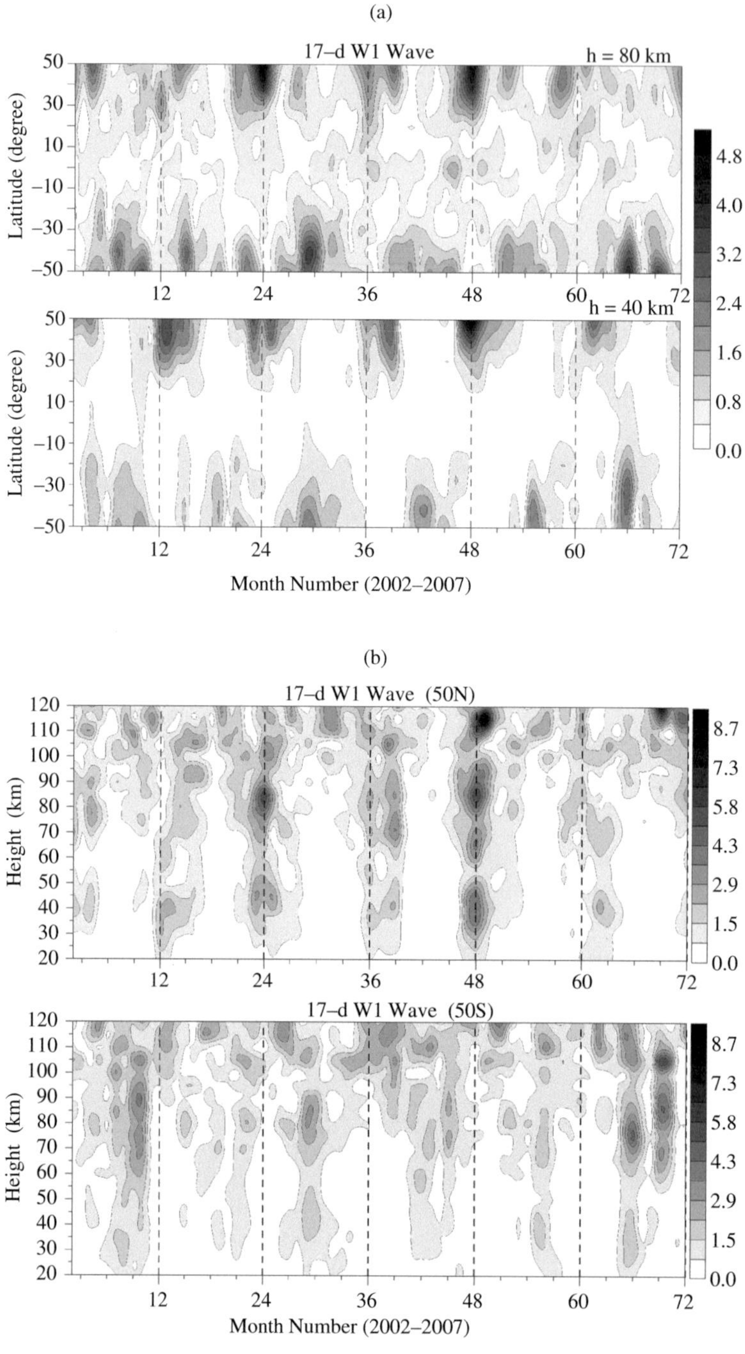

Fig. 2.16 (**a**) Latitude-time cross sections of 17-d *W1* amplitude (in Kelvin) at altitudes of 80 km (*upper plot*) and 40 km (*bottom plot*); (**b**) Altitude-time cross sections of 17-d *W1* amplitude at latitudes of 50°N (*upper plot*) and 50°S (*bottom plot*)

2.5 Summary

The present paper is focused on the global spatial (altitude and latitude) structure, seasonal and interannual variability of the atmospheric tides and planetary waves derived from the SABER/TIMED temperature measurements for full 6 years (January 2002–December 2007). The SABER instrument provides continuous global temperature data for the latitude range 50°N–50°S from the lower stratosphere to the lower thermosphere (20–120 km) and in this way provides an unprecedented opportunity for studying in detail the atmospheric waves, as well as their role in coupling the lower and upper atmosphere. We have presented a sampling of results that delineate the latitude, altitude, longitude, seasonal and interannual variations in atmospheric tidal and planetary wave temperature fields. The main advantage of the results presented in this paper is that the considered migrating and nonmigrating tides, as well as all significant planetary waves, found in the SABER/TIMED temperatures, are extracted simultaneously from the raw data (downloaded from the SABER web site temperatures). Therefore, using the same analysis techniques and the same data set makes it possible to get a consistent picture of the wave activity in the stratosphere-mesosphere-lower thermosphere system.

Concerning the atmospheric tides significant attention besides the migrating diurnal (*DW1*) and semidiurnal (*SW2*) tides received also the following nonmigrating diurnal: *DE3, DE2, DW2* and semidiurnal: *SW3, SE2, SE3* tides. About the *DW1* tide it is worth mentioning that the found trapped diurnal component observed in the height range of 40–60 km, as well as the ubiquitous double-peak vertical structure observed in the MLT region over the equator, could be of interest for reexamination of the stratospheric tidal heating, or the convective instabilities effects, in the numerical tidal simulations. The *SW2* temperature tide clearly indicated tropical amplifications in both hemispheres with a remarkable seasonal behavior at the altitude where it maximizes, ~110 km in the NH and ~115 km in the SH, indicating repeatable each year maxima exactly in May–June and August. The two migrating tides as well as the nonmigrating *DE3* and *DE2* tides revealed a quasi-2-year variability that partly could be caused by the QBO.

Special attention is paid to the climatology and interannual variability of the temperature SPW1 and particularly to its origin in the lower thermosphere. Strong evidence has been provided supporting the auroral heating as a main origin of the temperature SPW temperature field The paper presents the results also for: the ~5-day *Rossby* wave ((1,1) mode); ~6-day *Kelvin* wave, as well as for the ~10-day *W1* wave ((1,2) mode) and ~16-day *W1* wave ((1,3) mode).

The detailed picture of the spatial (altitude, latitude) structure and temporal variability of the atmospheric tides and planetary waves can serve as a benchmark and guide for future numerical modeling studies aimed at better understanding the stratosphere-mesosphere-lower thermosphere coupling by tidal and planetary wave patterns.

Acknowledgement We are grateful to the SABER team for the access to the data on http://saber.gats-inc.com.

References

Achatz U, Grieger N, Schmidt H (2008) Mechanisms controlling the diurnal solar tide: analysis using a GCM and a linear model. J Geophys Res 113:A8. doi:10.1029/2007JA9012967

Akmaev RA, Fuller-Rowell TJ, Wu F, Forbes JM, Zhang X, Anghel AF, Iredell MD, Moorthi S, Juang H-M (2008) Tidal variability in the lower thermosphere: comparison of Whole Atmosphere Model (WAM) simulations with observations from TIMED. Geophys Res Lett 35:L03810. doi:10.1029/2007GL032584

Allen DR, Stanford JL, Elson LS, Fishbein EF, Froidevaux L, Waters JW (1997) The 4-day wave as observed from the Upper Atmosphere Research Satellite Microwave Limb Sounder. J Atmos Sci 54:420–434

Andrews DG, Holton JR, Leovy CB (1987) Middle atmosphere dynamics. Academic, San Diego, CA, p 489

Angelats i Coll, M, Forbes JM (2002) Nonlinear interactions in the upper atmosphere: the s=1 and s=3 nonmigrating semidiurnal tides. J Geophys Res 107:A8. doi:10.1029/2001JA900179

Avery SK, Vincent RA, Phillips A, Manson AH, Fraser GJ (1989) High-latitude tidal behavior in the mesosphere and lower thermosphere. J Atmos Solar-Terr Phys 51:595–608

Azeem SMI, Killeen TL, Johnson RM, Wu Q, Gell DA (2000) Space-time analysis of TIMED Doppler Interferometer (TIDI) measurements. Geophys Res Lett 27(20):3297–3300

Azeem SMI, Talaat ER, Sivjee GG, Liu H-L, Roble RG (2005) Observational study of the 4-day wave in the mesosphere preceding the sudden stratospheric warming events during 1995 and 2002. Geophys Res Lett 32:L15804. doi:10.1029/2005GL023393

Baumgaertner AJG, Jarvis MJ, McDonald AJ, Fraser GJ (2006) Observations of the wavenumber 1 and 2 components of the semi-diurnal tide over Antarctica. J Atmos Solar-Terr Phys 68:1195–1214
Bernard R (1981) Variability of the semi-diurnal tide in the upper mesosphere. J Atmos Solar-Terr Phys 43: 663–674
Burrage MD, Hagan ME, Skinner WR, Wu DL, Hays PB (1995) Long-term variability in the solar diurnal tide observed by HRDI and simulated by the GSWM. Geophys Res Lett 22(19):2641–2644. doi:10.1029/95GL02635
Cevolani G (1991) Strato-meso-thermosphere coupling at mid-latitudes in the course of mid-winter stratwarmings during DYANA. Geophys Res Lett 18:1987–1990
Chang L, Palo S, Hagan M, Richter J, Garcia R, Riggin D, Fritts D (2008) Structure of the migrating diurnal tide in the Whole Atmosphere Community Climate Model (WACCM). Adv Space Res 41:1398–1407
Chapman S, Lindzen RS (1970) Atmospheric tides: thermal and gravitational. Gordon and Breach, New York, NY, 200 pp
Charney JG, Drazin PG (1961) Propagation of planetary-scale disturbances from the lower into the upper atmosphere. J Geophys Res 66:83–109
Chshyolkova T, Manson AH, Meek CE, Avery SK, Thorsen D, MacDougall JW, Hocking, W, Murayama Y, Igarashi K (2006) Planetary wave coupling processes in the middle atmosphere (30–90 km): A study involving MetO and MFR data. J Atmos Solar-Terr Phys 68:353–368
Clark RR, Salah JE (1991) Propagation of the solar semidiurnal tide in the mesosphere and lower thermosphere at midlatitudes. J Geophys Res 96:1129–1133
Coy L, Siskind DE, Eckermann SD, McCormack JP, Allen DR, Hogan TF (2005) Modeling the August 2002 minor warming event. Geophys Res Lett 32:L07808. doi:10.1029/2005GL022400
England SL, Maus S, Immel TJ, Mende SB (2006a) Longitudinal variation of the E-region electric fields caused by atmospheric tides. Geophys Res Lett 33:L21105. doi:10.1029/2006GL027465
England SL, Immel TJ, Sagawa E, Henderson SB, Hagan ME, Mende SB, Frey HU, Swenson CM, Paxton LJ (2006b) Effect of atmospheric tides on the morphology of the quiet time, postsunset equatorial ionospheric anomaly. J Geophys Res 111:A10S19. doi:10.1029/2006JA011795
Ern M, Lehmann C, Kaufmann M, Riese M (2009) Spectral wave analysis at the mesopause from SCIAMACHY airglow data compared to SABER temperature spectra. Ann Geophys 27:407–416
Espy PJ, Hibbins RE, Riggin DM, Fritts DC (2005) Mesospheric planetary waves over Antarctica during 2002. Geophys Res Lett 32:L21804. doi:10.1029/2005GL023886
Fedulina IN, Pogoreltsev AI, Vaughan G (2004) Seasonal, interannual and short-term variability of planetary waves in UKMO assimilated fields. Q J R Meteor Soc A 130(602):2445–2457
Forbes JM, Garret HB (1979) Theoretical studies of atmospheric tides. Rev Geophys 17:1951–1981
Forbes JM, Hagan ME, Miyahara S, Vial F, Manson AH, Meek CE, Portnyagin YI (1995) Quasi 16-day oscillation in the mesosphere and lower thermosphere. J Geophys Res 100:9149–9163
Forbes JM, Hagan M, Zhang X (2007) Seasonal cycle of nonmigrating diurnal tides in the MLT region due to tropospheric heating rates from the NCEP/NCAR Reanalysis Project. Adv Space Res 39:1347–1350. doi:10.1016/j.asr.2003.09.076
Forbes JM, Hagan ME, Zhang X, Hamilton K (1997) Upper atmosphere tidal oscillations due to latent heat release in the tropical troposphere. Ann Geophys 15:1165–1175
Forbes JM, Russell J, Miyahara S, Zhang X, Palo S, Mlynczak M, Mertens CJ, Hagan ME (2006) Troposphere-thermosphere tidal coupling as measure by the SABER instrument on TIMED during July–September 2002. J Geophys Res 111:A10S06. doi:10.1029/2005JA011492
Forbes JM, Wu D (2006) Solar tides as revealed by measurements of mesosphere temperature by the MLS experiment on UARS. J Atmos Sci 63:1776–1797
Forbes JM, Zhang X, Talaat ER, Ward W (2003) Nonmigrating diurnal tides in the thermosphere. J Geophys Res 108:A1, 1033. doi:10.1029/2002JA009262
Forbes JM, Zhang X, Ward W, Talaat ER (2002) Climatological features of mesosphere and lower thermosphere stationary planetary waves within ±40 latitude. J Geophys Res 107(D17):4322. doi:10.1029/2001JD001232
Forbes JM, Zhang X, Palo S, Russell J, Mertens CJ, Mlynczak M (2008) Tidal variability in the ionospheric dynamo region. J Geophys Res 113:A02310. doi:10.1029/2007JA012737
Forbes JM, Zhang X, Palo SE, Russell J, Mertens CJ, Mlynczak M (2009) Kelvin waves in stratosphere, mesosphere and lower thermosphere temperatures as observed by TIMED/SABER during 2002–2006. Earth Planets Space 61:447–453
García-Comas M, Lopez-Puertas M, Marshall BT, Wintersteiner PP, Funke B, Bermejo-Pantaleon D, Mertens CJ, Remsberg EE, Gordley LL, Mlynczak MG, Russell JM (2008) Errors in Sounding of the Atmosphere using Broadband Emission Radiometry (SABER) kinetic temperature caused by non-local-thermodynamic-equilibrium model parameters. J Geophys Res 113:D24106. doi:10.1029/2008JD010105
Grieger N, Volodin EM, Schmitz G, Hoffmann P, Manson AH, Fritts DC, Igarashi K, Singer W (2002) General Circulation Model results on migrating and nonmigrating tides in the mesosphere and lower thermosphere. Part 1: comparison with observations. J Atmos Solar Terr Phys 64:897–911
Groves GV (1982a) Hough components of water vapor heating. J Solar-Atmos Terr Phys 44:281–290
Groves GV (1982b) Hough components of ozone heating. J Solar-Atmos Terr Phys 44:111–121
Hagan ME, Burrage MD, Forbes JM, Hackney J, Randel WJ, Zhang X (1999a) GSWM-98: Results for migrating solar tides. J Geophys Res 104(A4):6813–6827. doi:10.1029/1998JA900125
Hagan ME, Burrage MD, Forbes JM, Hackney J, Randel WJ, Zhang X (1999b) QBO effects on the diurnal tide in the upper atmosphere. Earth Planets Space 51:571– 578
Hagan M, Forbes J, Vial F (1995) On modeling migrating solar tides. Geophys Res Lett 22:893–896
Hagan ME, Forbes JM (2002) Migrating and nonmigrating diurnal tides in the middle and upper atmosphere excited by tropospheric latent heat release. J Geophys Res 107:4754. doi:10.1029/2001JD001236
Hagan ME, Forbes JM (2003) Migrating and nonmigrating semidiurnal tides in the upper atmosphere excited by

tropospheric latent heat release. J Geophys Res 108:A2, 1062. doi:10.1029/2002JA009466
Hagan M, Vial F, Forbes J (1992) Variability in the upper propagating semidiurnal tide due to effects of QBO in the lower atmosphere. J Atmos Solar-Terr Phys 54:1465–1474
Hagan ME (1996) Comparative effects of migrating solar sources on tidal signatures in the middle and upper atmosphere. J Geophys Res 101:D16, 21213–21222
Hagan ME, Maute A, Roble RG, Richmond AD, Immel TJ, England SL (2007) Connections between deep tropical clouds and the Earth's ionosphere. Geophys Res Lett 34:L20109. doi:10.1029/2007GL030142
Hagan ME, Roble RG, Hackney J (2001) Modeling thermospheric tides. J Geophys Res 106:12739–12752
Hagan ME, Roble RG (2001) Modeling diurnal tidal variability with the NCAR TIME-GCM. J Geophys Res 106: 24869–24882
Haldoupis C, Pancheva D (2002) Planetary waves and midlatitude sporadic E layers; strong experimental evidence for a close relationship. J Geophys Res 107:A6. doi:10.1029/2001JA000212
Hamilton K (1981) Latent heat release as a possible forcing mechanism for atmospheric tides. Mon Weather Rev 109:3–17
Hartmann DL (1983) Baroclinic instability of the polar night jet stream. J Atmos Sci 36:1141–1154
Hirota I, Kuroi K, Shiotani M (1990) Midwinter warmings in the Southern hemisphere stratosphere in 1988. Q J R Meteor Soc 116:929–941
Huang FT, Mayr HG, Reber CA, Killeen T, Russell JM, Mlynczak M, Skinner W, Mengel JG (2006a) Diurnal variations of temperature and winds inferred from TIMED and UARS measurements. J Geophys Res 111:A10S04. doi:10.1029/2005JA011426
Huang FT, Mayr HG, Reber CA, Russell JM, Mlynczak M, Mengel JG (2006b) Stratospheric and mesospheric temperature variations for quasi-biennial and semiannual (QBO and SAO) oscillations based on measurements from SABER (TIMED) and MLS (UARS). Ann Geophys 24: 2131–2149
Huang FT, Reber CA (2004) Nonmigrating semidiurnal and diurnal tides at 95 km based on wind measurements from the High Resolution Doppler Imager on UARS. J Geophys Res 109:D10110. doi:10.1029/2003JD004442
Immel TJ, Sagawa E, England SL, Henderson SB, Hagan ME, Mende SB, Frey HU, Swenson CM, Paxton LJ (2006) Control of equatorial ionospheric morphology by atmospheric tides. Geophys Res Lett 33:L15108. doi:10.1029/2006GL026161
Jacobi C, Kürschner D, Muller HG, Pancheva D, Mitchell NJ, Naujokat B (2003) Response of the mesopause region dynamics to the February 2001 stratospheric warming. J Atmos Solar-Terr Phys 65:843–855
Jacobi Ch, Portnyagin Yu, Solovjova T, Hoffmann P, Singer W, Fahrutdinova A, Ishmuratov R, Beard G, Mitchell N, Muller G, Schminder R, Kürschner D, Manson A, Meek C (1999) Climatology of the semidiurnal tide at 52°N–56°N from ground-based radar wind measurements 1985–1995. J Atmos Solar-Terr Phys 61:975–991
Kato S (1989) Non-migrating tides. J Atmos Terr Phys 51: 673–682
Khattatov BV, Geller MA, Yudin VA, Hays PB (1997b) Diurnal migrating tide as seen by the high resolution Doppler imager/UARS: 2. Monthly mean global zonal and vertical velocities, pressure, temperature and inferred dissipation. J Geophys Res 102(D4):4423–4435. doi:10.1029/96JD03654
Khattatov BV, Yudin VA, Geller MA, Hays PB, Vincent RA (1997a) Diurnal migrating tide as seen by the high resolution Doppler imager/UARS: 1. Monthly mean global meridional winds. J Geophys Res 102(D4):4405–4422. doi:10.1029/96JD03655
Kong AKH, Charles PA, Kuulkers E (1998) Long-term X-ray variability in GX 354-0. New Astronomy 3(5):301–307
Krüger K, Naujokat B, Labitzke K (2005) The unusual midwinter warming in the Southern Hemisphere 2002: A comparison to Northern Hemisphere phenomena. J Atmos Sci 62:603–613
Labitzke K, van Loon H (1999) The stratosphere: phenomena, history and relevance. Springer, New York, NY
Lait LR, Stanford JL (1988) Fast, long-lived features in the polar stratosphere. J Atmos Sci 45(24):3800–3809
Lau K-M, Sheu PJ (1988) Annual cycle, quasibiennial oscillation and southern oscillation in global precipitation. J Geophys Res 93:10975–10988
Lieberman RS (1991) Nonmigrating diurnal tides in the equatorial middle atmosphere. J Atmos Sci 48:1112–1123
Lieberman R, Oberheide J, Hagan M, Remsberg E, Gordley L (2004) Variability of diurnal tides and planetary waves during November 1978–May 1979. J Atmos Solar-Terr Phys 66:517–528
Lin CH, Wang W, Hagan ME, Hsiao CC, Immel TJ, Hsu ML, Liu JY, Paxton LJ, Fang TW, Liu CH (2007) Plausible effect of atmospheric tides on the equatorial ionosphere observed by the FORMOSAT-3/COSMIC: three-dimensional electron density structures. Geophys Res Lett 34:L11112. doi:10.1029/2007GL029265
Liu H-L, Roble RG (2002) A study of a self-generated stratospheric sudden warming and its mesospheric-lower thermospheric impacts using the coupled TIME-GCM/CCM3. J Geophys Res 107(D23):4695. doi:10.1029/2001JD001533
Liu H-L, Roble RG (2005) Dynamical Coupling of the stratosphere and mesosphere in the 2002 Southern Hemisphere major stratospheric sudden warming. Geophys Res Lett 32:L13804. doi:10.1029/2005GL022939
Manney GL, Nathan TR, Stanford JL (1988) Barotropic stability of realistic stratospheric jets. J Atmos Sci 45:2545–2555
Manson AH, Meek CE, Hagan M, Hall C, Hocking W, MacDougall J, Franke S, Riggin D, Fritts D, Vincent R, Burrage M (1999) Seasonal variations of the semi-diurnal tides in the MLT: multi-year MF radar observations from 2 to 70°N, and the GSWM tidal model. J Atmos Sol-Terr Phys 61:809–828
Manson AH, Meek CE, Schminder R, Kürschner D, Clark RR, Müller HG, Vincent RA Phillips A, Fraser GJ, Singer W, Kazimirovsky ES (1990) Tidal winds from the MLT global radar network during the first LTCS campaign September 1987. J Atmos Solar-Terr Phys 52:175–183
Matsuno T (1971) A dynamical model of the stratospheric sudden warming. J Atmos Sci 28:1479–1494
Mayr HG, Mengel JG (2005) Interannual variations of the diurnal tide in the mesosphere generated by the

quasi-biennial oscillation. J Geophys Res 110:D10111. doi:10.1029/2004JD005055
McLandress C (1997) Seasonal variability of the diurnal tide: results from the Canadian middle atmosphere general circulation model. J Geophys Res 102:D25, 29747–29764
McLandress C (2002a) The seasonal variation of the propagating diurnal tide in the mesosphere and lower thermosphere, Part II: the role of tidal heating and zonal mean winds. J Atmos Sci 59:907–922
McLandress C (2002b) Interannual variations of the diurnal tide in the mesosphere induced by a zonal-mean wind oscillation in the tropics. Geophys Res Lett 29(9):1305. doi:10.1029/2001GL014551
McLandress C, Rochon CY, Shepherd GG, Solheim BH, Thuillier G, Vial F (1994) The meridional wind component of the thermospheric tides observed by WINDII on UARS. Geophys Res Lett 21:2417–2420
McLandress C, Shepherd GG, Solheim BH (1996) Satellite observations of thermospheric tides: results from the Wind Imaging Interferometer on UARS. J Geophys Res 101(D2):4093–4114. doi:10.1029/95JD03359
McLandress C, Ward WE (1994) Tidal/gravity wave interactions and their influence on the large scale dynamics of the middle atmosphere: model results. J Geophys Res 99:8139–8156
Mertens CJ et al (2001) Retrieval of mesospheric and lower thermospheric kinetic temperature from measurements of CO_2 15 μm earth limb emission under non-LTE conditions. Geophys Res Lett 28:1391–1394
Mertens CJ et al (2004) SABER observations of mesospheric temperature and comparisons with falling sphere measurements taken during the 2002 summer MaCWINE campaign. Geophys Res Lett 31:J03105. doi:10.1029/2003GL018605
Mitchell NJ, Pancheva D, Middleton H, Hagan M (2002) Mean winds and tides in the Arctic mesosphere/lower thermosphere region and comparison with the GSWM. J Geophys Res 106:A1. doi:10.1029/2001JA900127
Miyahara S, Miyoshi Y, Yamashita K (1999) Variations of migrating and nonmigrating tides simulated by the middle atmosphere circulation model at Kyushu University. Adv Space Res 24:1549–1558
Miyoshi Y (1999) Numerical simulation of the 5-day and 16-day waves in the mesopause region. Earth Planets Space 51: 763–772
Miyoshi Y, Hirooka T (1999) A numerical experiment of excitation of the 5-day wave by a GCM. J Atmos Sci 56:1698–1707
Miyoshi Y, Hirooka T (2003) Quasi-biennial variation of the 5-day wave in the stratosphere. J Geophys Res 108:D19, 4620. doi:10.1029/2002JD003145
Mukhtarov P, Pancheva D, Andonov B (2009) Global structure and seasonal and interannual variability of the migrating diurnal tide seen in the SABER/TIMED temperatures between 20 and 120 km. J Geophys Res 114:A02309. doi:10.1029/2008JA013759
Mukhtarov P, Pancheva D, Andonov B (2010) Climatology of the stationary planetary waves seen in the SABER/TIMED temperatures (2002–2007). J Geophys Res 115:A06315. doi:10.1029/2009JA015156
Oberheide J, Forbes JM (2008) Tidal propagation of deep tropical cloud signatures into the thermosphere from TIMED observations. Geophys Res Lett 35:L04816. doi:10.1029/2007GL032397
Oberheide J, Gusev OA (2002) Observations of migrating and nonmigrating diurnal tides in the equatorial lower thermosphere. Geophys Res Lett 29:2167. doi:10.1029/2002GL016213
Oberheide J, Hagan ME, Roble RG, Offermann D (2002) Sources of nonmigrating tides in the tropical middle atmosphere. J Geophys Res 107:4567. doi:10.1029/2002JD002220
Oberheide J, Wu Q, Killeen TL, Hagan ME, Roble RG (2007) A climatology of nonmigrating semidiurnal tides from TIMED Doppler Interferometer (TIDI) wind data. J Atmos Solar-Terr Phys 69:2203–2218
Palo SE, Forbes JM, Zhang X, Russell JM, Mertens CJ, Mlynczak MG, Burns GB, Espy PJ, Kawahara TD (2005) Planetary wave coupling from the stratosphere to the thermosphere during the 2002 Southern Hemisphere pre-stratwarm period. Geophys Res Lett 32:L23809. doi:10.1029/2005GL0242298
Palo SE, Forbes JM, Zhang X, Russell III JM, Mlynczak MG (2007) An eastward propagating two-day wave: evidence for nonlinear planetary wave and tidal coupling in the mesosphere and lower thermosphere. Geophys Res Lett 34:L07807. doi:10.1029/2006GL027728
Pancheva D, Mukhtarov P, Mitchell NJ, Beard AG, Muller HG (2000) Comparative study of neutral wind and tidal variability in the lower thermosphere above Bulgaria and UK. Ann Geophys 18:1304–1315
Pancheva D et al (2002) Global-scale tidal structure in the mesosphere & lower thermosphere during the PSMOS campaign summer-99 and comparison with the Global Scale Wave Model. J Atmos Solar-Terr Phys 64: 1011–1035
Pancheva DV, Mitchell NJ (2004) Planetary waves and variability of the semidiurnal tide in the mesosphere and lower thermosphere over Esrange (68°N, 21°E) during winter. J Geophys Res 109:A08307. doi:10.1029/2004JA010433
Pancheva DV, Mukhtarov PJ, Andonov BA (2007) Zonally symmetric oscillations in the Northern hemisphere stratosphere during the winter of 2003/2004. Geophys Res Lett 34:L04807. doi:10.1029/2006GL028666
Pancheva D, Mukhtarov P, Mitchell NJ, Merzlyakov E, Smith AK, Andonov B, Singer W, Hocking W, Meek C, Manson A, Murayama Y (2008a) Planetary waves in coupling the stratosphere and mesosphere during the major stratospheric warming in 2003/2004. J Geophys Res 113:D12105. doi:10.1029/2007JD009011
Pancheva D, Mukhtarov P, Mitchell NJ, Andonov B, Merzlyakov E, Singer W, Murayama Y, Kawamura S, Xiong J, Wan W, Hocking W, Fritts D, Riggin D, Meek C, Manson A (2008b) Latitudinal wave coupling of the stratosphere and mesosphere during the major stratospheric warming in 2003/2004. Ann Geophys 26:467–483
Pancheva D, Mukhtarov P, Mitchell N, Fritts D, Riggin D, Takahashi H, Batista P, Clemesha B, Gurubaran S, Ramkumar G (2008c) Planetary wave coupling (5–6-day waves) in the low latitude atmosphere-ionosphere system. J Atmos Solar-Terr Phys 70:101–122
Pancheva D, Mukhtarov P, Andonov B, Mitchell NJ, Forbes JM (2009a) Planetary waves observed by TIMED/SABER in coupling the stratosphere-mesosphere-lower thermosphere

during the winter of 2003/2004: Part 1, Comparison with the UKMO temperature results. J Atmos Solar-Terr Phys 71:61–74
Pancheva D, Mukhtarov P, Andonov B, Mitchell NJ, Forbes JM (2009b) Planetary waves observed by TIMED/SABER in coupling the stratosphere-mesosphere-lower thermosphere during the winter of 2003/2004: Part 2, Altitude and latitude planetary wave structure. J Atmos Solar-Terr Phys 71: 75–87,
Pancheva D, Mukhtarov P, Andonov B (2009c) Nonmigrating tidal activity related to the sudden stratospheric warming in the Arctic winter of 2003/2004. Ann Geophys 27: 975–987
Pancheva D, Mukhtarov P, Andonov B (2009d) Global structure, seasonal and interannual variability of the migrating semidiurnal tide seen in the SABER/TIMED temperatures (2002–2007). Ann Geophys 27:687–703
Pancheva D, Mukhtarov P, Andonov B, Forbes JM (2010a) Global distribution and climatological features of the 5–6-day planetary waves seen in the SABER/TIMED temperatures (2002–2007). J Atmos Solar-Terr Phys 72: 26–37
Pancheva D, Mukhtarov P, Andonov B (2010b) Reply to Manson et al's comment on "Global structure, seasonal and interannual variability of the migrating semidiurnal tide seen in the SABER/TIMED temperatures (2002–2007)". Ann Geophys 28:677–685
Pancheva D, Mukhtarov P, Andonov B (2010c) Global distribution, seasonal and interannual variability of the eastward propagating tides seen in the SABER/TIMED temperatures (2002–2007). Adv Space Res 46:257–274. doi:10.1016/j.asr.2010.03.026
Pogoreltsev AI, Sukhanova SA (1993) Simulation of the global structure of stationary planetary waves in the mesosphere and lower thermosphere. J Atmos Solar-Terr Phys 55(1): 33–40
Remsberg EE, Marshall BT, García-Comas M, Krueger D, Lingenfelser GS, Martin-Torres J, Mlynczak MG, Russell JM, Smith AK, Zhao Y, Brown C, Gordley LL, Lopez-Gonzales MJ, Lopez-Puertas M, She C-Y, Taylor MJ, Thompson RE (2008) Assessment of the quality of the Version 1.07 temperature-versus-pressure profiles of the middle atmosphere from TIMED/SABER. J Geophys Res 113:D17101. doi:10.1029/2008JD010011З
Riggin DM, Fritts DC, Jarvis MJ, Jones GOL (1999) Spatial structure of the 12-hour wave in the Antarctic as observed by radar. Earth, Planets Space 51:621–628
Riggin D, Liu H-L, Lieberman RS et al (2006) Observations of the 5-day wave in the mesosphere and lower thermosphere. J Atmos Solar-Terr Phys 68:323–339
Russell III JM, Mlynczak MG, Gordley LL, Tansock J, Esplin R (1999) An overview of the SABER experiment and preliminary calibration results. Proceedings of the SPIE, 44th annual meeting, Denver, CO, 18–23 July: v. 3756, pp 277–288
Salby ML (1982) Sampling theory for asynoptic satellite observations. Part 1: Space-time spectra, resolution, and aliasing. J Atmos Sci 39:2577–2600
She CY et al (2004) Tidal perturbations and variability in the mesopause region over Fort Collins, CO (41 N, 105 W): continuous multi-day temperature and wind lidar observations. Geophys Res Lett 31:L2411. doi:10.1029/2004GL021165
Shepherd MG, Wu DL, Fedulina IN, Gurubaran S, Russell JM, Mlynczak MG, Shepherd GG (2007) Stratospheric warming effects on the tropical mesospheric temperature field. J Atmos Sol-Terr Phys 69:2309–2337
Shiotani M, Shimoda N, Hirota I (1993) Inter-annual variability of the stratospheric circulation in the Southern Hemisphere. Q J R Meteor Soc 119:531–546
Sivjee GG, Walterscheid RL, McEwen DJ (1994) Planetary wave disturbances in the arctic winter mesopause over Eureka (80°N). Planet Space Sci 42:973
Smith AK (1996) Longitudinal variations in mesospheric winds: evidence for gravity wave filtering by planetary waves. J Atmos Sci 53:1156– 1173
Smith AK (1997) Stationary planetary waves in upper mesospheric winds. J Atmos Sci 54:2129– 2145
Smith AK (2003) The origin of stationary planetary waves in the upper mesosphere. J Atmos Sci 60(24):3033–3041
Takahashi H, Wrasse CM, Pancheva D, Abdu MA, Batista IS, Lima LM, Batista PP, Clemesha BR, Shiokawa K (2006) Signatures of 3–6-day planetary waves in the equatorial mesosphere and ionosphere. Ann Geophys 24: 3343–3350
Talaat ER, Lieberman RS (1999) Nonmigrating diurnal tides in mesospheric and lower thermospheric winds and temperatures. J Atmos Sci 56:4073–4087
Talaat ER, Yee J-H, Zhu X (2001) Observations of the 6.5-day wave in the mesosphere and lower thermosphere. J Geophys Res 106:20715–20723
Tsunoda R, Yamamoto M, Igarashi K, Hocke K, Fukao S (1998) Quasi-periodic radar echoes from midlatitude sporadic E and the role of the 5-day planetary wave. Geophys Res Lett 25:951
Vincent R, Kovalam AS, Fritts DD, Isler JR (1998) Long-term MF radar observations of solar tide in the low-latitude mesosphere: interannual variability and comparison with the GSWM. J Geophys Res 103(D8):8667–8684. doi:10.1029/98JD00482
Williams CR, Avery SK (1996) Diurnal nonmigrating tidal oscillations forced by deep convective clouds. J Geophys Res 101:4079–4091
Wu DL, Hays PB, Skinner WR (1994) Observations of the 5-day wave in the mesosphere and lower thermosphere. Geophys Res Lett 21:2733–2736
Wu Q et al (2008a) Global distribution and interannual variations of mesospheric and lower thermospheric neutral wind diurnal tide: 1. migrating tide. J Geophys Res 113:A05308. doi:10.1029/2007JA012542
Wu Q et al (2008b) Global distribution and interannual variations of mesospheric and lower thermospheric neutral wind diurnal tide: 2. nonmigrating tide. J Geophys Res 113:A05309. doi:10.1029/2007JA012543
Xiao C, Hu X, Tian J (2009) Global temperature stationary planetary waves extending from 20 to 120 km observed by TIMED/SABER. J Geophys Res 114:D17101. doi:10.1029/2008JD011349
Xu J, Smith AK, Liu H-L, Yuan W, Wu Q, Jiang G, Mlynczak MG, Russell III JM, Franke SJ (2009) Seasonal and quasi-biennial variations in the migrating diurnal tide observed by Thermosphere, Ionosphere, Mesosphere, Energetics and Dynamics (TIMED). J Geophys Res 114:D13107. doi:10.1029/2007JD011298

Yuan T, Schmidt H, She CY, Krueger DA, Reising S (2008) Seasonal variations of semidiurnal tidal perturbations in mesopause region temperature and zonal and meridional winds above Fort Collins, Colorado (41°N, 105°W). J Geophys Res 113:D20103. doi:10.1029/2007JD009687

Zhang X, Forbes JM, Hagan ME, Russell III JM, Palo SE, Mertens CJ, Mlynczak MG (2006) Monthly tidal temperatures 20–120 km from TIMED/SABER. J Geophys Res 111:A10S08. doi:10.1029/2005JA011504

Zuo X, Wan W (2008) Planetary wave oscillations in sporadic E layer occurrence at Wuhan. Earth Planets Space 60:647–652

Chapter 3

Dynamical Coupling Between Different Regions of Equatorial Atmosphere

Geetha Ramkumar

Abstract Based on the current understanding of the atmospheric dynamics, the prominent waves and oscillations present in the equatorial middle atmosphere and the wave–mean flow interaction are discussed in this chapter. The middle atmospheric dynamics studies conducted from Indian region during last few decades are summarized to some extent. The gravity waves and equatorial waves generated in the lower atmospheric region are found to couple different regions of atmosphere dynamically, as they propagate upward into mesospheric heights. The ground-based observational studies conducted at Indian stations to quantitatively estimate the forcing by gravity waves toward the generation of quasi-biennial oscillation, stratopause semi-annual oscillation, and mesopause semi-annual oscillation are also discussed in this chapter. The chapter concludes with the mention of importance of these observational studies from tropical region in climate-related issues.

3.1 Introduction

The uneven distribution of the incoming radiation on earth leads to differential heating of the atmosphere and this global temperature gradient in turn forces the atmosphere into motion. These motions or wave perturbations with spatial scales of tens to thousands kilometers and temporal scales of the order of few minutes to few years in variability manifest themselves in temperature and wind. The current understanding of atmospheric dynamics attributes the fluctuations in the middle atmosphere to the effect of coupling with the lower atmosphere. Middle atmosphere dynamics is indeed concerned with the elucidation of the nature of wave motions responsible for the vertical coupling. Middle atmosphere can be generally described as the layer between the level where convective–radiative equilibrium gives way to radiative equilibrium at the lower boundary and the level where molecular diffusion takes over from eddy diffusion and ionization begins to be a major source of heating at the upper boundary (Kelley, 1989). Vertically, this is the region between 10 and 100 km. Review work of Hamilton (1999) on dynamical coupling of the lower and middle atmosphere provides a historical background of the research for the last five decades.

Tropical middle atmosphere and its dynamics have distinctly different features like the small Coriolis parameter, strong latent heat release, and large-scale oscillations like quasi-biennial oscillation (QBO) and stratopause and mesopause semi-annual oscillations (SSAOs, MSAOs) compared to those elsewhere in the earth's atmosphere (Gray and Hamilton, 2003).

This chapter briefly explains the most important waves that influence the tropical middle atmospheric dynamics and mainly focuses on some of the recent studies from tropical stations that address the dynamical coupling between different regions of the middle atmosphere through wave activities.

G. Ramkumar (✉)
Space Physics Laboratory, Vikram Sarabhai Space Centre, Indian Space Research Organization, Trivandrum 695-022, India
e-mail: geetha_ramkumar@vssc.gov.in

M.A. Abdu, D. Pancheva (eds.), A. Bhattacharyya (Coed.), *Aeronomy of the Earth's Atmosphere and Ionosphere*, IAGA Special Sopron Book Series 2, DOI 10.1007/978-94-007-0326-1_3, © Springer Science+Business Media B.V. 2011

3.2 Present Scenario in Equatorial Middle Atmosphere Studies – Indian Perspective

Tropics hold the key to much of the global climate variability and hence tropical atmosphere has become the focus of recent scientific research activity. India carried out rocket and balloon-based measurements in campaign mode under the Indian counterpart of the Middle Atmosphere Program (IMAP) during 1980s and regular balloon and rocket flights during the period 1970–1991 from Thumba Equatorial rocket launching station (TERLS). Studies using the wind and temperature data collected from these observations could reveal the characteristics of equatorial waves, migrating and nonmigrating tides, annual oscillation, SAO, and QBO in the middle atmospheric region (Devarajan et al., 1985; Reddy et al., 1986; Sasi and KrishnaMurthy, 1991; Reddy and Lekshmi Vijayan, 1993; Raghava Reddi and Geetha Ramkumar, 1997). These long-term measurements could also lead to an empirical reference model atmosphere up to 80 km for Indian equatorial zone (Sasi, 1994). Though the campaigns conducted during 1984–1988 from Indian stations under the IMAP provided valuable information on equatorial wave characteristics, information on momentum fluxes of these waves could not be obtained from these campaigns. The vertical component of wind and/or temperature is essential for the estimation of the wave momentum fluxes besides the horizontal components of the wind. With the installation of MST radar and Rayleigh lidar in the early and late 1990s, respectively, at Gadanki (13.5°N, 79.2°E), vertical wind and temperature became accessible parameters for measurement. In the years 1999 and 2000, comprehensive experimental "Equatorial Wave Studies" (EWS 1999, 2000) campaigns were conducted by measuring temperatures and winds in the troposphere, stratosphere, and mesosphere up to about 80 km at two low-latitude stations, Gadanki (13.5°N, 79.2°E) and Sriharikota Altitude Range (SHAR (13.7°N, 80.2°E)), for 45 consecutive days to characterize equatorial waves and to estimate the momentum fluxes associated with these waves (Sasi et al., 2003). Sasi et al. (2005) review the studies on equatorial waves from Indian stations during three decades.

To divulge the characteristics and climatology of gravity waves and also to quantify the middle atmospheric variability at different time scales, an ambitious program called ISRO's Middle Atmospheric Dynamics Studies "MIDAS" was carried out during 2002–2007 (Ramkumar et al., 2006), making use of coordinated fortnightly observations of winds and temperatures by means of high-altitude balloon flights; Rohini Sounding RH-200 Rocket flights and meteor wind radar from Thumba/Trivandrum (8.5°N, 77°E); partial reflection radar from Equatorial Geophysical Research Laboratory, Tirunelveli (8.7°N, 77.8°E); and MST radar and lidar at National Atmospheric Research Laboratory, Gadanki (13.5°N, 79.2°E) (Kumar et al., 2007; Maria Antonita et al., 2007, 2008a, b). This well-debated program could bring out several first kind of results like quantitative estimation of contribution of gravity waves toward the generation of QBO and SAO through wave–mean flow interaction and dynamical coupling between different regions of atmosphere, which would be discussed in detail in Sections 3.4, 3.5, and 3.6.

3.3 Waves in the Equatorial Region

3.3.1 Equatorial Waves

Equatorially confined wave motions with periods in the range of $\sim$3 to 25 days have been observed to propagate vertically and horizontally through the equatorial middle atmosphere. These are of planetary scale in the zonal direction but are trapped latitudinally within $\pm 15°$ of the equator, and hence referred to as equatorial waves. Of these, Kelvin waves propagate in the eastward direction, while the Rossby gravity (RG) waves propagate in the westward direction. The Kelvin and RG waves are generally thought to be forced by geographically confined time variations in the large-scale cumulus convective heating in the equatorial troposphere. Randomly distributed sources excite these waves efficiently and the preferred vertical wavelength of the excited waves is about twice the vertical scale of the heat source. The relatively slow-moving Kelvin and RG wave modes observed in the upper troposphere and lower stratosphere are effectively damped out by thermal dissipation (radiative damping) in the upper stratosphere. However, fast moving Kelvin waves can penetrate into the mesosphere (Dhaka et al., 1995).

3.3.2 Gravity Waves

The major sources of gravity waves are topography, convection, wind shear, unbalanced flows in the vicinity of jet streams, frontal systems including wave–wave interaction, and local body forcing. These gravity waves propagate both horizontally (east–west; north–south) and vertically with horizontal scales of a few thousands of kilometers and vertical scales of a few tens of kilometers. Several observational studies have been conducted using radars, lidars, rockets, and satellites to explore the gravity wave characteristics right from troposphere to the mesospheric regions (Tsuda et al., 1994; Goldberg et al., 1997; Fritts et al., 1997). Compared to all other techniques, Rayleigh lidar, which provides high-resolution density or temperature fluctuations, proves to be a quite powerful means of routine monitoring of the middle atmospheric gravity waves (Chanin and Hauchecorne, 1981; Mitchell et al., 1991; Whiteway and Carswell, 1995).

Temperature observations in the 30–60 km altitude region from Rayleigh lidar at Gadanki could extensively be used to delineate the characteristic features of gravity waves at a tropical location (Rajeev et al., 2003; Siva Kumar et al., 2006). The relationship between the wave activity and the strength of typical sources, viz convection and wind shear that excite these gravity waves, down below in the troposphere could be established making use of simultaneous temperature and wind observations from co-located Lidar and MST radar at Gadanki (Ramkumar et al., 2006).

3.3.3 Quasi-biennial Oscillations (QBOs)

The equatorial middle atmospheric dynamics is dominated by QBO in the lower stratosphere, stratopause semi-annual oscillation (SSAO), and mesopause semi-annual oscillation (MSAO) of the zonal winds. QBO is a classical example of the wave–mean flow interaction in the equatorial middle atmosphere. QBO is manifested as alternating westerly and easterly winds with periods varying from 24 to 36 months. Baldwin et al. (2001) have extensively reviewed the characteristics of QBO and the related dynamics in the middle atmospheric region. Besides the stratospheric QBO, High-Resolution Doppler Imager (HRDI) measurements onboard Upper Atmosphere Research Satellite (UARS) along with MF radar observations at tropical stations revealed a QBO in the mesospheric region (MQBO) (Burrage et al., 1996).

3.3.4 Semi-annual Oscillations (SAOs)

Hirota (1978) using rocket observations over Ascension Island showed that the SAO consists of two oscillations with amplitude maxima near the stratopause and mesopause which are nearly out of phase. These oscillations are found to be confined to tropics. Maximum easterlies were found to occur during solstices while maximum westerlies occurred during equinoxes. Significant seasonal asymmetry in the amplitudes of the SAO with stronger SAO occurring during the first half of the calendar year is reported. The oscillation first appears in the mesosphere and propagates downward into the stratosphere. Transition from easterly to westerly occurs with downward descent rate of 5–10 km/month, while westerly to easterly transition occurs over a large depth nearly simultaneously.

3.4 Coupling Between Troposphere and Stratosphere

Equatorial waves and gravity waves propagate in all preferred directions both horizontally and vertically and while these waves propagate through a prevailing background atmosphere, the wave characteristics change with height and the wave may be dissipated at critical levels when the phase speed of the wave equals the background wind speed.

The basic mechanism responsible for the QBO and SAO is the wave–mean flow interaction wherein upward propagating equatorial waves and gravity waves interact with the stratospheric mean flow and deposit their momentum producing the alternating westerly and easterly regimes (Lindzen and Holton, 1968; Plumb, 1977). The eastward (westward) propagating waves force eastward (westward) mean flow acceleration at the critical levels where they are absorbed.

3.4.1 Forcing by Gravity Waves Toward the Generation of QBO

The extensive temperature observations in the 30–60 km altitude region from Gadanki (13.5°N, 79.8°E) over a decade could be used to delineate the momentum flux of gravity waves of period 30 min to 4 h. Using the gravity wave polarization relations, the vertical flux of the zonal momentum per unit density (F) of gravity wave could be estimated using the equation (Krishna Murthy et al., 2002).

$$F = \left\langle \overline{u'w'} \right\rangle = -\frac{k}{m}\left(\frac{g}{\overline{N}}\right)^2 \left\langle \frac{T'}{\overline{T}} \right\rangle^2 \approx \frac{\omega}{N}\left(\frac{g}{\overline{N}}\right)^2 \left\langle \frac{T'}{\overline{T}} \right\rangle^2 \tag{3.1}$$

where u' and w' are the wave perturbations in zonal and vertical winds, T' is the temperature perturbation, $\tilde{N}$ and $\check{T}$ are the mean Brunt–Vaisala frequency and temperature, and g is the acceleration due to gravity. The mean flow acceleration $\frac{\partial \bar{u}}{\partial t}$produced by the gravity waves observed was calculated from the divergence of momentum flux of gravity waves using the equation (Lindzen, 1984).

$$\frac{\partial \bar{u}}{\partial t} \approx -\frac{d}{dz}\left\langle \overline{u'w'} \right\rangle + \frac{\left\langle \overline{u'w'} \right\rangle}{H} \tag{3.2}$$

where H is the scale height calculated from the mean temperature.

Since the QBO exhibits latitudinal variation and due to the paucity of long-term wind measurements at Gadanki at the QBO height region, the background wind information could be collected from NCEP/NCAR reanalysis data. The monthly mean zonal winds over a decade from NCEP/NCAR reanalysis data corresponding to Gadanki latitude were used to study the QBO characteristics in the lower stratospheric region (20 and 10 hpa levels). Estimated forcing by gravity waves toward the generation of various cycles of QBO was compared with the observed mean flow acceleration from monthly mean winds. It can be seen from Fig. 3.1 that on an average 10–60% and 10–30% of the forcing to drive the westerly and easterly phases of QBO, respectively, is from gravity waves (Maria Antonita et al., 2008b). This is in well accordance with earlier modeling studies (Alexander and Holton, 1997; Piani et al., 2000). Though there are some discrepancies between the mean flow acceleration estimated from the divergence of gravity wave momentum fluxes (lidar) and that observed from monthly mean zonal wind values (NCEP) during few months, they are in good agreement during months wherever the continuous lidar data were available. Since daily observations in wind/temperature in the stratospheric region over a long period are not available in the equatorial region, the forcing of equatorial waves toward the generation of QBO could not be estimated quantitatively so far.

3.4.2 Forcing by Gravity Waves Toward the Generation of SSAO

The simultaneous temperature and wind measurements using lidar from Gadanki and rocket flights from Trivandrum under MIDAS program could be used to estimate the mean flow acceleration in the 30–60 km altitude region produced by the divergence of momentum flux of gravity waves and also from the measured winds. From the studies of Reddy et al. (1986) and Hopkins (1975), it is reasonable to assume that characteristics of SSAO over Gadanki are similar to that over Trivandrum. It is seen from Fig. 3.2a, b that the behavior of the mean flow acceleration calculated from two different means of measurements (temperature and wind) is in good agreement. The gravity wave forcing toward the mean flow acceleration varies significantly from cycle to cycle, which is attributed to the seasonal variations in the strength of gravity wave sources. From the above figure, it is clearly seen that on an average ~30–60% of the forcing to drive the westerly phase of SSAO is contributed by gravity waves. The discrepancy between the estimated and the observed acceleration could be due to the contribution from longer period gravity waves and planetary waves.

The simultaneous wind and temperature data collected during EWS 2000 campaign could be used to estimate the forcing of equatorial waves toward the generation of stratopause semi-annual oscillations (SSAOs), which revealed that Kelvin waves of periods ranging from 6 to 18 days contribute about 25% toward the generation of westerly phase of SSAO (Sasi et al., 2003). Due to paucity of data, contribution of equatorial waves toward easterly phase of SSAO could not be estimated from this study.

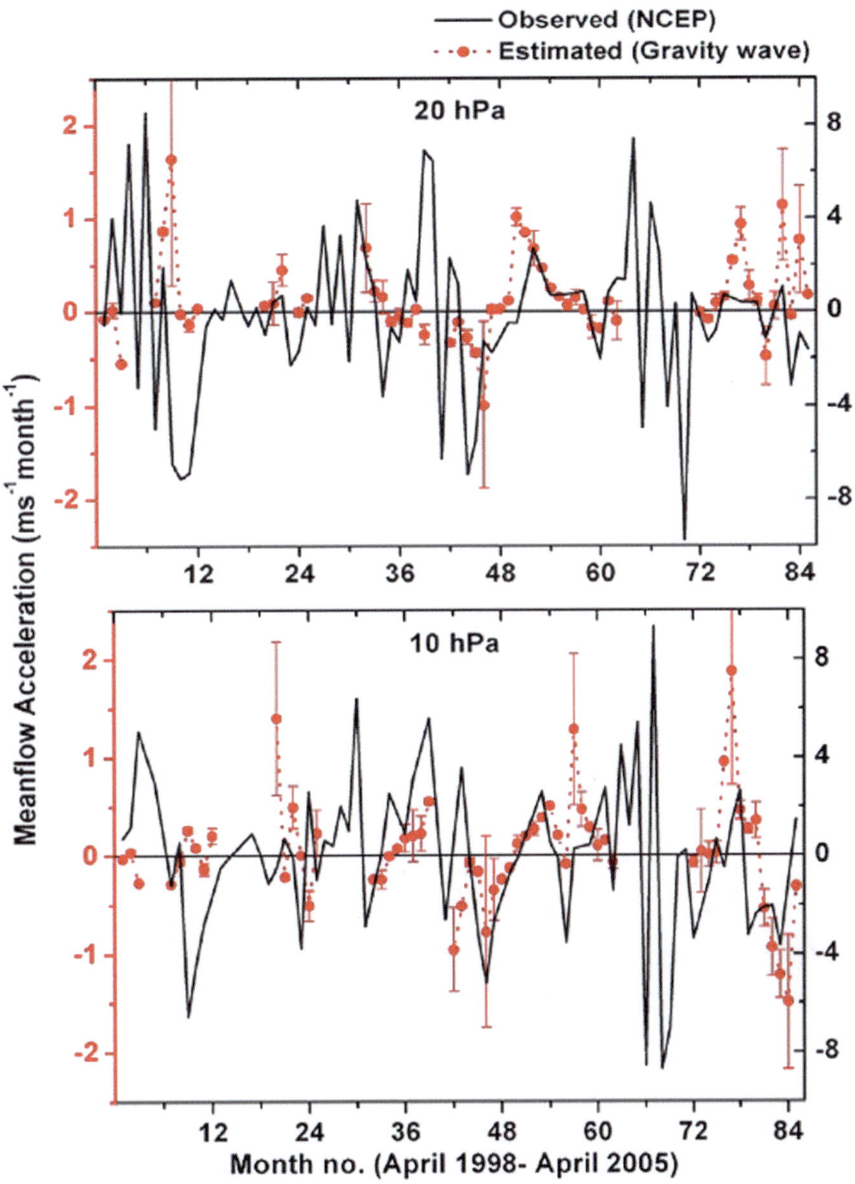

Fig. 3.1 Comparison of observed (*solid line*) and estimated (*dotted line with circle*) acceleration at 20 hPa and 10 hPa (Maria Antonita et al., 2008 Courtesy: American Geophysical Union)

3.5 Coupling Between Troposphere and Mesosphere

3.5.1 *Forcing by Gravity Waves Toward the Generation of MSAO*

Round the clock wind measurements in the MLT region from meteor wind radar over a low-latitude station – Trivandrum – could be very useful in studying the characteristics of gravity waves in the MLT region. The momentum flux of these mesoscale gravity waves of period 2–3 h was estimated employing a method proposed by Hocking (2005). The seasonal variation in the mesoscale short period gravity wave momentum fluxes showed a semi-annual variation with equinoctial maxima and solstitial minima in contrast to mid and high latitudes, which showed annual variation. The mean flow acceleration estimated from the divergence of gravity wave momentum fluxes was compared with observed mean flow acceleration computed using the monthly mean zonal winds in the 82–98 km height

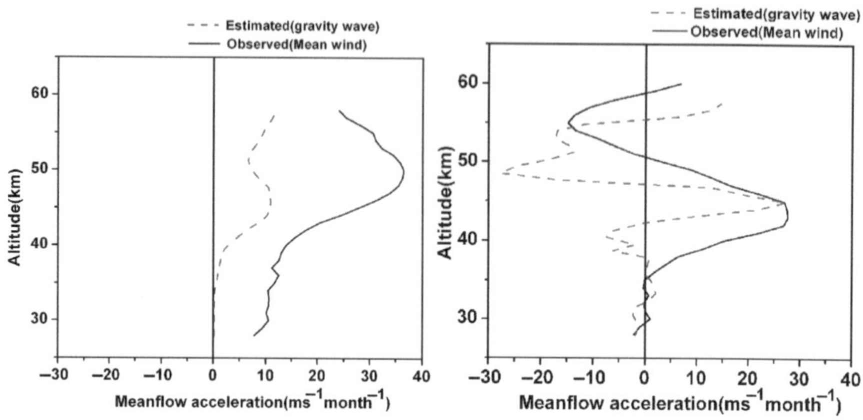

Fig. 3.2 Altitude profiles of observed (*solid line*) and estimated acceleration (*dotted line*), during typical cycles of SSAO (Maria Antonita et al., 2007 Courtesy: American Geophysical Union)

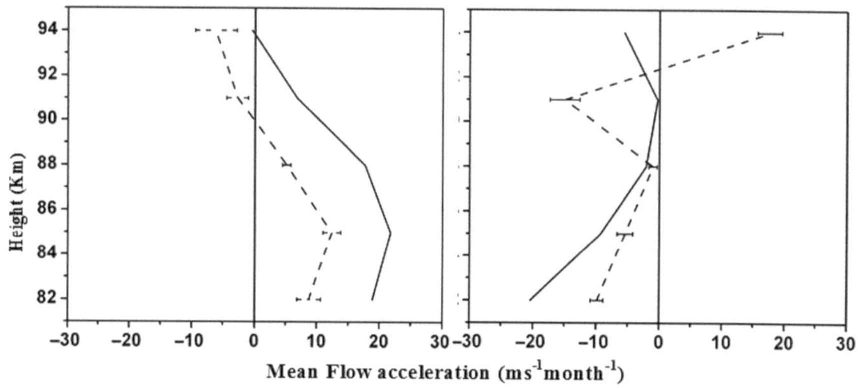

Fig. 3.3 Altitude profiles of observed (*solid line*) and estimated acceleration (*dotted line*) for westerly and easterly phase of MSAO (Maria Antonita et al., 2008 Courtesy: American Geophysical Union)

region. Figure 3.3a, b shows typical examples that depict the contribution of the gravity wave toward westerly and easterly phases of MASO. The contribution of gravity waves toward the westerly phase of MSAO ranges from ~30 to ~50% in the 82–88 km height region and that toward easterly phase of MSAO is almost ~50% in the 82–94 km region.

3.6 Coupling Between Troposphere/Stratosphere and Mesospheric/Lower Thermospheric (MLT) Region

Recent studies conducted using integrated observations from troposphere to MLT region in the tropics could establish the coupling between these regions through different types of wave activities. In an attempt to study the intra-seasonal oscillations (ISO) in the low-latitude MLT region using MF radar measured winds. Kumar et al. (2007) could link these oscillations with the lower atmospheric processes. They showed that the 50- to 70-day oscillations in the MLT region were caused by lower atmospheric convective activity, whereas 20- to 40-day oscillations caused by variability in the water vapor mixing ratio in the 0–4 km region. Figure 3.4a, b clearly depicts the 50–70 days ISO in the MLT region linked with convective activity (OLR) and 20–40 days ISO linked with water vapor mixing ratio in the lower atmospheric region. Pancheva et al. (2007), in a study using simultaneous wind observations from several equatorial/tropical stations, reported the presence of fast Kelvin waves of 6-day period in the equatorial stratospheric and MLT region with enhanced vertical wavelength than the theoretically expected value in the MLT region. This was attributed to the Doppler shifting of Kelvin waves to higher frequencies, favored by easterly phase of SSAO. The CAWSES-India Tidal Campaign conducted during March–April 2006, to determine the characteristics of tides in the troposphere–stratosphere–mesosphere and lower thermosphere region (0–100 km) making use of

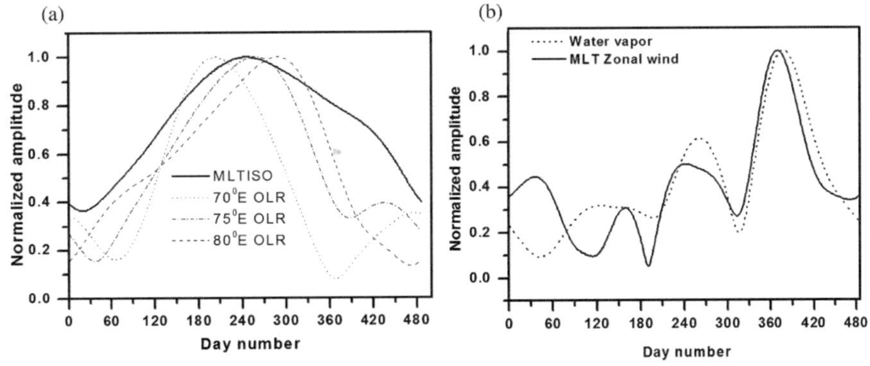

Fig. 3.4 Time evolution of (**a**) MLTISO (50–70 days) and tropospheric ISO and (**b**) MLTISO (20–40 days) and water vapor mixing ratio derived from wavelet analysis (Kishorekumar et al., 2007 Courtesy: American Geophysical Union)

MST radar, MF radar, Rocket and Meteor Radar observations, could bring out 15- to 20-day modulation of diurnal tide activity at MLT heights and a similar variation in the OLR fields in the western Pacific (120–160° longitude region) suggesting a possible link between the observed tidal variabilities and the variations in the deep tropical convection through the nonmigrating tides it generated (Gurubaran et al., 2008). In another study by Deepa et al. (2008) to look into the interannual variability of tidal oscillations in the MLT region at the equatorial station Trivandrum, the tidal amplitudes were found to be enhanced when the prevailing winds in the 30–35 km region and 50–65 km are in westerly phase. Under this condition, the tidal oscillations were found to accelerate the mean background wind above 90 km through momentum deposition, thus making it strong westward. Though there are studies of gravity waves in the MLT region using radar observations, coupling with lower atmospheric processes is not looked into due to the paucity of high-resolution data simultaneously available in the lower and upper atmospheric region.

3.7 Concluding Remarks

This chapter focused on the various aspects of middle atmospheric dynamics and coupling between different regions of equatorial middle atmosphere, achieved through limited ground-based observations over the equatorial region. It is true that all relevant studies could not be included here due to limitation of space.

The discussion on this topic will not be complete unless the downward coupling process is mentioned. Coupling at the tropopause between stratosphere and troposphere appears in the exchange of energy, momentum, and constituents like H_2O and O_3, and also in the influence of the stratospheric QBO on the troposphere. Studies on global warming showing enhancement of tropospheric ozone in tropics can be attributed to the enhanced atmospheric circulations both in stratosphere and in troposphere, leading to stratospheric intrusion of ozone to the troposphere. Emerging evidence indicates that stratosphere has got the capability of shaping not only its own evolution but that of tropospheric circulation and dynamics as well (Mohan Kumar, 2008). With the establishment of more ground-based and space-borne techniques for monitoring the middle atmosphere and attempts toward networking of the existing facilities, a better understanding of the link and the complex physical mechanism underlying the coupling processes, including dynamical, chemical, and electrical, is possible in the near future.

Acknowledgments I wish to acknowledge the encouragement given by Director, Space Physics Laboratory, VSSC in preparing this chapter. I express my sincere gratitude toward all investigators of the EWS 2000 and MIDAS program, scientists/engineers from Rohini Sounding Rocket project, Meteorological Facility, NARL, and EGRL.

References

Alexander MJ, Holton JR (1997) A model study of zonal forcing in the equatorial stratosphere by convectively induced gravity waves. J Atmos Sci 54:408–419

Antonita M, Ramkumar G, Kumar KK, Appu KS, Nambhoodiri KVS (2007) A quantitative study on the role of gravity waves

in driving the tropical stratospheric semiannual oscillation. J Geophys Res 112:D12115. doi:10.1029/2006jd008250
Antonita M, Ramkumar G, Kumar KK, Deepa V (2008a) Meteor radar observations of gravity wave momentum fluxes in the MLT region and their contribution towards mesospheric semiannual oscillation. J Geophys Res 113:D10115. doi: 10,1029/2007JD009089
Antonita M, Ramkumar G, Kumar KK, Sunilkumar SV (2008b) Quantification of gravity wave forcing in driving the stratospheric quasi-biennial oscillation. Geophys Res Lett 35:L09805. doi:10,1029/2008GL33960
Baldwin MP, Gray LJ, Dunkerton TJ, Hamilton K et al (2001) Quasi-biennial oscillation. Rev Geophys 39:79–229
Burrage MD, Vincent RA, Mayr HG, Skinner WR, Arnold NF, Hays PB (1996) Long- term variability of the equatorial middle atmosphere zonal wind. J Geophys Res 101: 12,847–12,854
Chanin ML, Hauchecorne A (1981) Lidar observation of gravity and tidal waves in the stratosphere and mesosphere. J Geophys Res 86:9715–9721
Deepa V, Ramkumar G, T Maria Antonita, Kishore Kumar K (2008) Meteor wind radar observations of tidal amplitudes over a low latitude station Trivandrum (8.5°N, 77°E): interannual variability and the effect of background wind on diurnal tidal amplitudes. J JASTP. doi:10, 1016/j, jastp, 2008,07,017
Devarajan M et al (1985) Rocket observations of Kelvin waves in the stratosphere over India. J Atmos Sci 42:1873–1879
Dhaka SK, Krishna Murthy BV, Nagpal OP, Raghava Rao R, Sasi MN, Sundaresan S (1995) A study of equatorial waves in the Indian zone. J Atmos Solar-Terr Phys 57: 1189–1202
Fritts DC, Carten JF, Riggin DM, Golberg RA, Lehmacher GA, Schmidlin FJ, McCarthy S, Khudeki E, Fawcett CD, Hitchman MH, Lieberman RS, Reid IM, Vincent RA (1997) Equatorial dynamics observed by rocket, radar, and satellite during the CADRE/MALTED campaign 2, mean and wave structures, coherence, and variability. J Geophys Res 102:26191–26216
Goldberg RA, Lehmacher G, Schmidlin FJ, Fritts DC, Mitchell JD, Croskey CL, Friedrich M, Swartz WE (1997) Equatorial dynamics observed by rocket, radar, and satellite during the CADRE/MALTED campaign, 1, programmatics and small scale fluctuations. J Geophys Res 102:26,179–26,190
Gray L, Hamilton K (2003) The physics of the equatorial atmosphere. Technical report RAL-TR-2003-008, CLRC
Gurubaran S, Narayana Rao D, Ramkumar G, Ramkumar TK, Dutta G, Krishna Murthy BV (2008) First results from the CAWSES-INDIA tidal campaign. Ann Geophys 26: 2323–2331
Hamilton K (1999) Dynamical coupling of the lower and middle atmosphere: historical background to current research. J Atmos Solar-Terr Phys 61:73–84
Hirota I (1978) Equatorial waves in the upper stratosphere and mesosphere in relation to the semi annual oscillation of the zonal wind. J Atmos Sci 35:714–722
Hocking WK (2005) A new approach to momentum flux determinations using SKiYMET meteor radars. Ann Geophys 23:1–7
Hopkins RH (1975) Evidence of polar-tropical coupling in upper stratospheric zonal wind anomalies. J Atmos Sci 32:712–719
Kelley MC (1989) The earth's ionosphere, plasma physics and electrodynamics. International geophysics series, vol 43. Academic, San Diego, CA
Krishna Murthy BV, Satheesan K, Parameswaran K, Sasi MN, Ramkumar G, Bhavanikumar Y, Raghunath K, Krishnaiah M (2002) A Study of equatorial waves in temperature from 4 to 70 km using MST radar and Lidar observations at Gadanki (13.5°N, 79.2°E). Quart J R Meteorol Soc 128: 819–838
Kumar KK, Antonita TM, Ramkumar G, Deepa V, Gurubaran S, Rajaram R (2007) On the tropospheric origin of mesosphere lower thermosphere region intra seasonal wind variability. J Geophys Re 112:D07109. doi:10,1029/2006JD007962
Lindzen RS (1984) Gravity waves in the mesosphere. In: Holton JR, Matsuno T (eds) Dynamics of the middle atmosphere.Terra, Tokyo, pp 3–18
Lindzen RS, Holton J (1968) A theory of the quasi-biennial oscillation. J Atmos Sci 25:1095–1107
Mitchell NJ, Thomas L, Marsh AKP (1991) Lidar observations of long-period gravity waves in the stratosphere Ann Geophys 9:588–596
Mohan Kumar K (2008) Stratosphere troposphere interactions-an introduction, 1st edn. Springer
Pancheva DV, Mukhtarov PJ, Mitchell NJ, Fritts DC, Riggin DM, Takahashi H, Batista PP, Clemesha BR, Gurubaran S, Ramkumar G (2007) Planetary wave coupling (5–6-day waves) in the low latitude atmosphere-ionosphere system. doi:10,1016/j.JASTP 2007,10,003
Piani C, Durran D, Alexander MJ, Holton JR (2000) A numerical study of three- dimensional gravity waves triggered by deep tropical convection and their role in the dynamics of the QBO. J Atmos Sci 57:3689–3702
Plumb RA (1977) The interaction of two internal waves with the mean flow: implications for the theory of the quasi-biennial oscillation. J Atmos Sci 34:1847–1858
Raghava Reddi C, Ramkumar G (1997) The annual and semi-annual wind fields in low latitudes. J Atmos Solar-Terr Phys 59:487–495
Rajeev K, Parameswaran K, Sasi MN, Ramkumar G, Krishna Murthy BV (2003) Gravity waves in the tropical middle atmosphere: characteristics and wave-mean flow interaction. Adv Space Res 32(5):807–812
Ramkumar G, Antonita TM, Bhavani Kumar Y, Venkata Kumar H, Narayana Rao D (2006) Seasonal variation of gravity waves in the equatorial middle atmosphere: results from ISRO's middle atmospheric dynamics (MIDAS) program. Ann Geophys 24:2471–2480
Reddy CA, Reddy CR, Mohan Kumar KG (1986) Zonal wind components in the stratosphere and troposphere over India. Quart J R Meteorol Soc 112:811–823
Reddy CA, Vijayan L (1993) Annual and semi annual oscillations in the equatorial middle atmospheric winds. Adv Space Res 13:373–376
Sasi MN (1994) A reference atmosphere for the Indian equatorial zone. I J Radio Space Phys 23:299–312
Sasi MN, Krishna Murthy BV (1991) A study of quasi biennial oscillation in the tropical stratosphere. J Atmos Solar-Terr Phys 53:1173–1180
Sasi MN, Krishna Murthy BV, Ramkumar G et al (2003) A study of equatorial wave characteristics using rockets, balloons, lidar and radar. Adv Space Res 32:813–818

Sasi MN, Ramkumar G, Krishna Murthy BV (2005) Studies on equatorial waves over the Indian zone. Curr Sci 89(3): 475–487

Shindell DT, Schmidt GA, Miller RL, Rind D (2001) Northern hemisphere winter climate response to green house gases, increase, ozone, solar and volcanic forcing. J Geophys Res 106:7193–7210

Siva Kumar V, Rao PB, Bencherif H (2006) Lidar observations of middle atmospheric gravity wave activity over a low-latitude site (Gadanki, 13.5°_ N, 79.2°_ E). AnnGeophys 24:823–834

Tsuda T, Murayama Y, Nakamura T, Vincent RA, Manson AH, Meek CE, Wilson RL (1994) Variations of the gravity wave characteristics with height, season, and latitude revealed by comparative observations. J Atmos Solar-Terr Phys 56: 555–568

Whiteway JA, Carswell AI (1995) Lidar observations of gravity wave activity in the upper stratosphere over Toronto. J Geophys Res 100:14,113–14,124

Chapter 4

Microphysical Properties of Mesospheric Aerosols: An Overview of In Situ-Results from the ECOMA Project

Markus Rapp, Irina Strelnikova, Boris Strelnikov, Martin Friedrich, Jörg Gumbel, Ulf-Peter Hoppe, Tom Blix, Ove Havnes, Phillip Bracikowski, K.A. Lynch, and Scott Knappmiller

Abstract Six sounding rockets were launched within the ECOMA (="Existence and Charge state Of Meteoric smoke particles in the middle Atmosphere") project to study the characteristics of meteoric smoke particles (MSPs) and mesospheric ice particles, as well as their possible microphysical relation. The launches were conducted during three campaigns from the Andøya Rocket Range (69°N, 16°E), one in September 2006, and the other two in the summers of 2007 and 2008. This chapter provides an overview of these observations and presents the corresponding geophysical results with special emphasis on our understanding of the micropyhsics of mesospheric ice particles. Most notably, we are able to confirm the existence of MSPs at all altitudes between 60 and 85 km in September, and a seasonal variation that is consistent with previous model studies in which MSP-variability is mainly driven by the global circulation. Together with these model studies as well as recent satellite observations of MSPs our results hence cast some doubt on a standard assumption of state-of-the-art microphysical models of mesospheric ice clouds, namely that ice nucleation mainly occurs heterogeneously on MSPs.

4.1 Introduction

The mesosphere is host to several aerosol species which are involved in a large variety of processes. Among these aerosol types, the most prominent ones are ice particles which nucleate and develop in the environment of the extremely low temperatures of the polar summer mesopause and are observed as noctilucent clouds/polar mesospheric clouds (NLC/PMC) or as polar mesosphere summer echoes (PMSE) (Thomas, 1991; Rapp and Lübken, 2004). Scientific interest in these ice clouds has intensified in recent years since it has been speculated that their properties should be severely modified by minute changes of background temperatures and water vapor such that they might be sensitive indicators for changes of the background mesospheric state (Thomas et al. 1989). While the most recent analyses of satellite PMC observations do indeed show trends of both PMC brightness as well as occurrence frequency (Deland et al., 2007, Shettle et al., 2009), the underlying causes for these changes are yet to be fully understood (Lübken et al., 2009, Marsh and Merkel, 2009).

Hence, it appears to be obvious that experimental efforts need to be targeted at unraveling the microphysical properties of these ice clouds, where one of the most important uncertainties is the issue of their actual nucleation mechanism (Rapp and Thomas, 2006). For the latter, the heterogeneous nucleation on another mesospheric aerosol species, namely meteoric smoke particles (MSP) (Hunten et al. 1980), has been the favored nucleation pathway in most previous studies of mesospheric ice microphysics even though solid experimental evidence for this or any other proposed nucleation mechanism has so far been elusive [(Rapp and Thomas, 2006) and references therein].

Motivated by the obvious need to gain a deeper understanding of mesospheric aerosol properties and related processes, the German-Norwegian-led

M. Rapp (✉)
Department of Radars and Sounding Rockets, Leibniz-Institute of Atmospheric Physics, Rostock University,
18225 Kuehlungsborn, Germany
e-mail: rapp@iap-kborn.de

M.A. Abdu, D. Pancheva (eds.), A. Bhattacharyya (Coed.), *Aeronomy of the Earth's Atmosphere and Ionosphere*,
IAGA Special Sopron Book Series 2, DOI 10.1007/978-94-007-0326-1_4,

ECOMA (="Existence and Charge state Of Meteoric smoke particles in the middle Atmosphere") project focussed on in situ observations of MSPs and mesospheric ice particles as well as their potential relation.

This chapter provides an initial overview of the results obtained during the first three out of a total of four major field campaigns involving the launching of a total of 9 sounding rockets and a multitude of ground based observations. In Section 4.2 we describe the instrumentation of the ECOMA-payloads and give an overview of the first three field campaigns. Turning to the scientific results, we then describe the results of our initial MSP-observations in September 2006, i.e., after the polar summer (Section 4.3.1), followed by several aspects of ice charge density measurements during the ECOMA summer flights (Sections 4.3.2 and 4.3.3), to finally focus on aspects of simultaneous observations of MSP and ice particle properties and corresponding implications for ice particle nucleation (Section 4.3.4). Finally, the most important results will be summarized in Section 4.4 including a short outlook regarding the final planned ECOMA campaign which is to be conducted in winter 2010.

4.2 The ECOMA Project: Payload Description and Campaigns

The concept of the ECOMA-payload is to gather a suite of instruments which allows simultaneous and common volume observations of mesospheric aerosol particles along with their most important ambient parameters such as electron and positive ion number densities, neutral densities, temperatures, turbulence parameters, and information about the charging state of the ECOMA vehicle with respect to the ambient plasma. A photo of the front deck of the ECOMA-payload as it was launched during flights ECOMA04 and ECOMA06 (see Table 4.1) is shown in Fig. 4.1. This picture shows the ECOMA particle detector (PD) to measure the charge number density of mesospheric aerosols (by means of the classical Faraday cup technique) along with the total particle volume density (by means of the newly developed active photoionization/photodetachment technique, see Rapp and Strelnikova (2009), Rapp et al. (2010) for details) in the center, surrounded by two fixed biased Langmuir probes on deployable booms (the electron probe, EP, and the positive ion probe, PIP) to measure relative altitude profiles of electrons and positive ions (Brattli et al., 2009). Further instruments on this deck are the MAGIC particle sampler to collect particles during the flight and return them to the ground for further laboratory analysis (Gumbel et al., 2005), the receiving antennas of the wave propagation experiment to obtain absolute electron number densities (Friedrich et al., 2009), and two more Faraday cups to collect particle charge number densities, i.e., the charged dust detector (CDD) from Tromsø University (Havnes et al., 1996), and the Dartmouth dust detector (DDD) from Dartmouth College (Lynch et al., 2005). Note that these two additional Faraday cups were replaced by simple Pirani gauges during all flights except ECOMA04 and ECOMA06 to obtain altitude profiles of neutral number densities and temperatures. In the rear, each ECOMA payload was equipped with a CONE (combined sensor for neutrals and electrons) ionization gauge for measuring neutral parameters and (relative) electron densities (Giebeler et al., 1993), as well as with two cold plasma probes (CPP) to measure the payload potential along with electron temperature (Strelnikova et al., 2009). Finally, a mid-section of the payload contained a photometer during all summer flights to detect and characterize mesospheric ice particles during NLC-passages of the payload (Megner et al., 2009).

Table 4.1 Dates, times, solar zenith angles, and launch conditions of the ECOMA sounding rocket flights. Note that during flight ECOMA02 there appeared a malfunction of the ECOMA particle detector

Label	Date	Time [UT]	Solar zenith angle°	Launch condition
ECOMA01	08 Sept. 2006	22:17	114.5	Moderate ionosphere (ios)
(ECOMA02)	17 Sept. 2006	21:07	112.0	Disturbed ios
ECOMA03	03 Aug. 2007	23:22	93.2	NLC and PMSE
ECOMA04	30 Jun. 2008	13:22	50.8	NLC and PMSE
ECOMA05	07 Jul. 2008	21:24	86.6	NLC, but no PMSE; quiet ios
ECOMA06	12 Jul. 2008	10:46	47.5	NLC and PMSE, disturbed ios

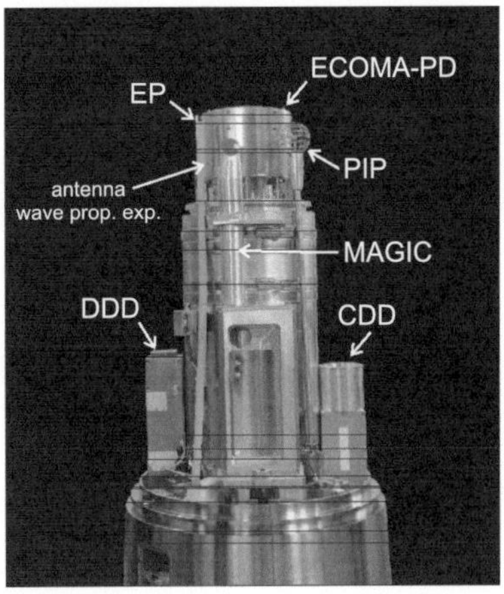

Fig. 4.1 Photo of the main instruments of the ECOMA payload located on the front deck of the 14 inch payload. Note that these instruments are located under the split-nosecone which is ejected at ~55 km on the rocket ascent. See text for more details regarding these instruments and corresponding acronyms

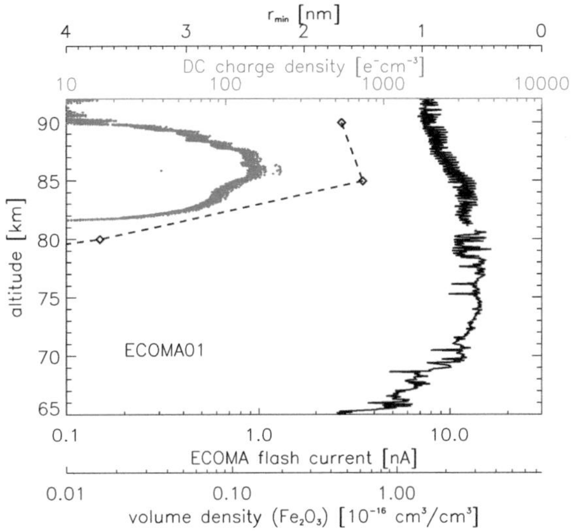

Fig. 4.2 Overview of the main results obtained with the ECOMA-PD from flight ECOMA01. The thick *black line* shows the altitude profile of peak photoelectron currents (denoted as ECOMA flash current) recorded by the instrument. The lowermost abscissa converts these currents to MSP volume densities assuming a Fe_2O_3-composition (see Rapp et al. 2010) for a detailed discussion on the composition of MSPs and corresponding implications). The *grey solid line* shows charge number densities of particles which penetrated into the ECOMA Faraday cup (denoted as DC charge density). Note that the limited altitude extent of this layer is determined by the aerodynamics-limited detection efficiency of Faraday-cup-type instruments as indicated by the altitude profile of minimum detectable MSP radius (*dashed line*, uppermost abscissa)

So far, six ECOMA payloads have been launched using a two stage vehicle consisting of a Nike-Improved Orion motor combination reaching typical apogees of about 130 km. All launches were made from the Andøya Rocket Range (69°N, 16°E) in the frame of three field campaigns in September 2006, and in the summers 2007 and 2008. Table 4.1 gives an overview of the basic dates of these rocket launches along with some short information about the situations in which the rockets were launched.

4.3 Scientific Results

4.3.1 *MSP Properties During September 2006*

Figure 4.2 summarizes the most important results of the first successful measurements with the ECOMA-PD during flight ECOMA01 in September 2006. This figure shows MSP volume densities derived from the flash current measurements (originating from photoelectrons emitted from the particles; see Rapp and Strelnikora (2009) and Rapp et al. (2010) for details including an in-depth discussion of cross-sensitivity to species other than MSPs) with the ECOMA-PD revealing evidence for the existence of MSPs in the entire altitude range from 60 to 85 km in September in satisfactory agreement with state-of-the-art MSP-models (Rapp et al. 2010).

In contrast, the direct collection of charged particles with the Faraday-cup part of the instrument reveals a narrow layer of negatively charged particles between 82 and 90 km. As shown in detail in Strelnikova et al. (2009) this apparent layering is caused by the aerodynamical properties of the instrument and does not reflect the true altitude distribution of MSPs. Note that similar sensitivity restrictions apply to all previous Faraday cup measurements of MSPs.

4.3.2 *Ice Particle Charge Densities in PMSE*

Figure 4.3 shows an overview plot of all summertime particle charge number densities measured with

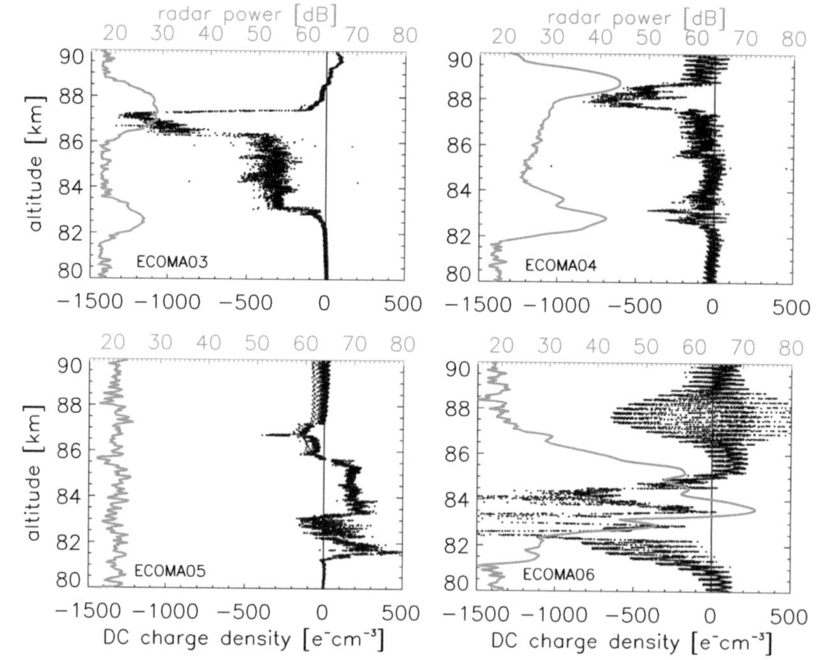

Fig. 4.3 Overview of all ECOMA-PD Faraday cup-measurements of particle charge number densities (*black dots*) during the summer flights of the ECOMA project. The *grey solid lines* show corresponding radar power profiles of PMSE averaged over 10 min around the rocket launches. Note that the strong modulation of the charge densities observed during flight ECOMA06 is likely caused by photoemission from the instrument walls, and that the feature above 86 km is probably caused by rocket coning and is not geophysical

the Faraday cup channel of the ECOMA-PD. This figure reveals that during all flights with PMSE detected by the ALWIN radar (Latteck et al., 1999), the ECOMA-PD observed net-negatively charged particles in the same altitude range where the radar received strongly enhanced backscatter. The exception to this is flight ECOMA05, where the rocket was launched into a situation with confirmed absence of PMSE.

In spite of this lack of radar backscatter, the onboard photometer clearly revealed the presence of an NLC layer when the rocket passed through an altitude range from ~80 to 86 km. Interestingly, the Faraday cup channel of the ECOMA-PD recorded apparently positively charged aerosol particles in this altitude range with charge number densities of the order of just a few hundreds of positive elementary charges/cm^3. Whether or not these positive charge signatures are truly caused by positive particles or rather hint at secondary charge production as suggested by Havnes and Næsheim (2007) is hard to judge from our data alone, but we note that current understanding of mesospheric particle charging cannot explain positive charges on pure ice particles (Rapp et al., 2010; Vondrak et al., 2006). While the definitive cause for this extraordinary particle charge signature is currently not understood, we still find it worthy to point out that the electron density during this rocket flight was exceptionally low as compared to all other ECOMA summer flights (see Fig. 4.6). This might on the one hand explain the absence of radar backscatter which has previously been shown to require a minimum electron number density of a few hundred electrons cm^{-3} (Rapp et al., 2002), and on the other hand this might indicate that in this low electron density plasma the charging and/or secondary charging properties of mesospheric aerosol particle are substantially different from "normal" situations with larger electron densities.

4.3.3 A Three Instrument Consistency Comparison

As noted before, the payloads ECOMA04 and ECOMA06 carried three instruments which are (at least partly) based on a Faraday cup design such that the corresponding measurements provide an ideal data set for intercomparison and investigation of effects like detection sensitivity and/or secondary charge production. Figure 4.4 compares altitude profiles of (negative) particle charge densities derived from observations with the ECOMA-PD, the CDD and the DDD from flight ECOMA04. This comparison reveals several interesting features: First of all note that the very large spin modulation of the CDD and DDD-data is likely

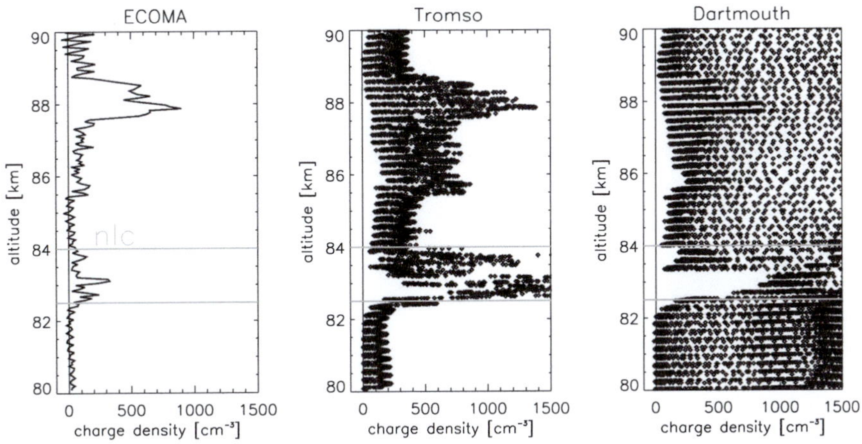

Fig. 4.4 Comparison of profiles of particle charge number densities measured during flight ECOMA04 with the ECOMA-particle detector (*left panel*), the charged dust probe of Tromsø University (*middle panel*), and the dust detector of Dartmouth College (*right panel*)

from photoelectrons from sunlight which are modulated as the instruments go in and out of the shadow of the ECOMA-PD (see Fig. 4.1). For comparison to the ECOMA-PD-results one should hence concentrate on the inner envelope of the data instead of individual data points. Secondly, it is evident that all three instruments observe similar features with a double layer of negatively charged particles located at around 83 and 88 km. However, when looking into greater detail, it is striking that the ECOMA-PD charge densities are larger than the others in the upper layer while they are smaller in the lower layer.

In an initial attempt to understand these differences we set up a full three-dimensional Direct Monte Carlo Simulation (DSMC) of the aerodynamical environment of these instruments using a freely available software package developed by G. A. Bird (www.gab.com.au) and used in many previous studies of the aerodynamics of sounding rocket measurements [Knappmiller et al., (2008) and references therein].

Figure 4.5 shows the relative density field (relative to its free-flow value) around the top deck of the ECOMA04 payload in a plane intersecting the three particle detectors for flight conditions at an altitude of 85 km. This simulation reveals that the density enhancement within the foremost ECOMA-PD is much larger than the enhancement in the CDD and DDD. This happens, because the CDD and DDD are behind the shock wave of the ECOMA instrument structure, i.e., behind this shock wave the

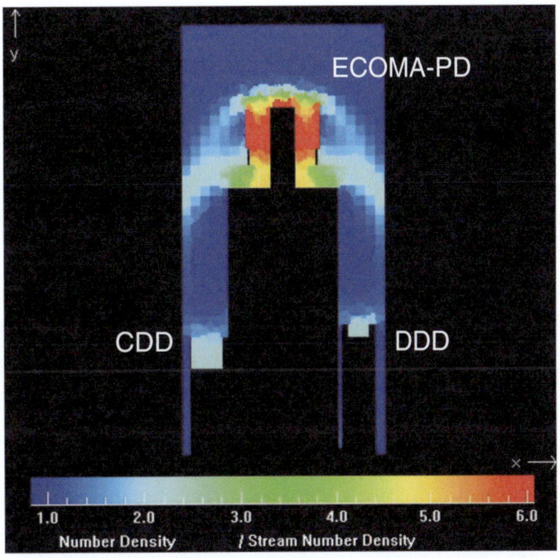

Fig. 4.5 Relative density field around the ECOMA04-payload for a background number density of $2\times10^{14}\text{cm}^{-3}$, a background temperature of 142 K, and a rocket velocity of 1000 m/s

flow is strongly decelerated from its free flow value of 1000 m/s to less than the speed of sound, i.e., ~300 m/s. While particle trajectory calculations within this density field are yet to be done, this implies that particles in the upper layer (which are smaller than particles in the lower layer as confirmed by onboard photometer measurements which detected NLC only in the lower layer) are so small that they are deflected by the shock wave extending from the edges of the ECOMA-PD such that only the largest of them can make it into the downstream CDD and DDD.

In the lower layer, however, particles are larger and presumably have such large inertia that they penetrate the above mentioned shock wave and reach the CDD and DDD. But how can this possibly explain that ECOMA-PD-densities are even smaller than the CDD and DDD-densities? Here we consider two possible scenarios: on the one hand it is at least conceivable that due to the strongly enhanced density and temperature at this altitude particles sublimate in the ECOMA-PD and are hence not detected.

On the other hand, it is also possible that the charge densities recorded by the CDD and DDD are actually dominated by secondary charges produced by particles which collided with the walls of the ECOMA-PD structure upstream of the CDD and DDD under grazing incidence. Such secondary impact effects have earlier been observed with a dust probe (Havnes and Næsheim, 2007) similar to the CDD but placed centrally in the front of the payload. An analysis based on the observed currents to a probe grid and the currents to a collector plate directly under the grid, show that electrons are rubbed off the grid during impacts and that the negatively charged collision fragments thereafter hit the collector plate and contribute to the currents to it. This analysis indicates that fragments of the large NLC particles can rub off some 50 elementary charges per impact thereby producing strong positive currents to the grid while the collector plate shows negative currents. Havnes and Næsheim (2007) suggested that a possible consequence of this is that the NLC particles may contain a substantial amount of meteoric smoke particles which survive the collision and are charged during the impact process. It could then have been these charge fragments which were recorded by the CDD and DDD and not the primary charge density of the mesospheric ice particles.

Whether one or the other scenario is correct cannot be judged based on the current state of analysis but will require more work in the future.

4.3.4 MSP and Ice Particle Properties Under Polar Summer Conditions

Finally, Fig. 4.6 summarizes all summertime flash current-profiles measured by the ECOMA-PD and compares them to corresponding electron densities derived from the wave propagation, PIP, and EP experiment. First of all, we note that the flash current peaks seen during all summer flights in the 80–85 km range have been shown to be direct evidence for NLC-layers by comparison to photometer observations and can actually be used to derive ice particle volume densities (Rapp et al., 2009, 2010). Secondly, we note the very different altitude extent of the flash current profiles. As shown in detail in Rapp et al. (2010), this is caused by the corresponding different background electron density-profiles. For example, detailed calculations of the charging of MSPs show that under sunlit conditions a minimum electron density of about 1000 cm^{-3} is required to turn a significant fraction of the MSPs from an average positive charge to an average neutral charge state. Outside the ice region, altitudes where the electron density reaches a value close to this limit, the flash current appears out of the background noise of the ECOMA-PD. This happens because the threshold energy for photoemission is increased by a positive particle charge owing to the additional Coulomb attraction of the photoelectron by this positive charge (Wood, 1981). See fig. 9 and corresponding discussion in Rapp et al. (2010) for more details. These results shed new light on the issue of

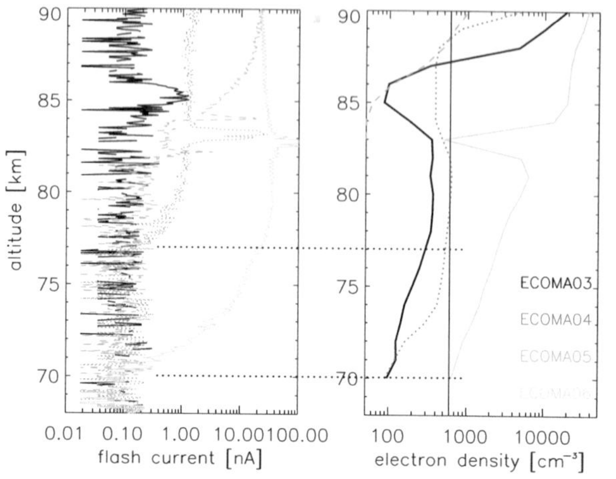

Fig. 4.6 Comparison of altitude profiles of flash currents measured during the ECOMA sounding rocket flights in summer (*left panel*) and corresponding electron densities from the wave propagation experiment. *Dotted horizontal lines* indicate altitudes with electron densities of 600 cm^{-3} (*vertical black line*) for flights ECOMA04 and ECOMA06. *Grey shading* indicates the altitude range with ice particles. Updated from Rapp et al. (2010), copyright by the AGU

MSP charging and imply that MSPs are either positively charged or neutral under sunlit conditions (i.e., in polar summer) and negatively charged in darkness. Finally, Rapp et al. (2010) have shown that the comparison of ECOMA-flights in September and during the summer months is consistent with model calculations of the seasonal variation of MSPs and hence reinforces earlier model results that MSP seasonal variation is mainly driven by the variation of the global circulation (Megner et al., 2008). Importantly, these model results are also largely supported by the first global observations of MSPs with the SOFIE instrument on AIM (Hervig et al., 2009).

Taking all these results together, this implies a minimum of MSP number densities during polar summer and hence casts some doubt on the feasibility of mesospheric ice nucleation on MSPs (Megner et al., 2008).

Future work on mesospheric ice nucleation should hence also consider alternative nucleation mechanisms such as heterogeneous nucleation on charged MSPs (Megner and Gumbel, 2009), homogeneous nucleation under conditions of extreme temperature variations owing to gravity waves (Lübken et al., 2009; Murray and Jensen, 2010), and other proposals as summarized and discussed in Rapp and Thomas (2006).

4.4 Summary

In summary, the main results of the ECOMA project are to date:

- Successful demonstration of the active photoemission/photodetachment technique for the detection of mesospheric particles.
- Proof of MSP existence in the entire altitude range from 60 to 85 km in September 2006 at 69°N, 16°E.
- Particle volume densities can be derived from the photoionisation measurements.
- Limitation of previous MSP observations by Faraday cup-type instruments are due to aerodynamics.
- Comparison of three Faraday cups on the same payload begins to validate aerodynamic models of how these detectors work and underlines the potential importance of secondary effects. Clearly though, more work is needed to understand Faraday cup detector function.
- Close correspondence between negatively charged ice particles and PMSE is confirmed.
- Observation of apparently positively charged particles under conditions of very low plasma densities and in the confirmed presence of NLC, likely influenced by secondary charging effects.
- MSP charging is influenced by photodetachment/photoionization processes. MSPs are likely positively charged or neutral under sunlit conditions and negatively charged during darkness.
- MSP observations in September and June/July/August imply MSP variability consistent with global scale model predictions. Taking all now available evidence from models, satellite observations and the here presented results together casts some doubt on the feasibility of predominant mesospheric ice nucleation on MSPs.

At this stage the final ECOMA campaign is scheduled for December 2010 to study the influence of the Geminid meteor shower on the properties of MSPs. It is planned to launch one ECOMA payload each before, during, and after the peak activity of the shower – hopefully allowing insight into the time scales of MSP formation and microphysics.

Acknowledgments This work was supported by the German Space Agency (DLR) under grants 50 OE 0301 and 50 OE 0801 (Project ECOMA). The Norwegian part of the project was supported by the Norwegian Space Centre and the Research Council of Norway (grant 177295).

References

Brattli A, Lie-Svendsen O, Svenes K, Hoppe U-P, Strelnikova I, Rapp M, Latteck R, Friedrich M (2009) The ECOMA 2007 campaign: rocket observations and numerical modellingof aerosol particle charging and plasma depletion in a PMSE/NLC layer. Ann Geophys 27(2):781–796

DeLand MT, Shettle EP, Thomas GE, Olivero JJ (2007) Latitude-dependent long-termvariations in polar mesospheric clouds from SBUV version 3 PMC data. J .Geophys Res 112:D10315. doi:10.1029/2006JD007857

Friedrich M, Torkar KM, Singer W, Strelnikova I, Rapp M, Robertson S (2009) Signaturesof mesospheric particles in ionospheric data. Ann Geophys 27(2):823–829

Giebeler J, Lübken F-J, M. Nägele (1993) CONE – a new sensor for in-situ observations ofneutral and plasma density fluctuations. Proceedings of the 11th ESA symposium on Europeanrocket and balloon programmes and related research, Montreux, Switzerland (ESA SP), pp 311–318

Gumbel J, Waldemarsson T, Giovane F, Khaplanov M, Hedin J, Karlsson B, Lossow S, Megner L, Stegman J, Fricke KH, Blum U, Voelger P, Kirkwood S, Dalin P, Sternovsky Z, Robertson S, Horányi M, Stroud R, Siskind DE, Meier RR, Blum J, Summers M, Plane JMC, Mitchell NJ, Rapp M (2005) The MAGIC rocket campaign: an overview. Proceedings of the 17th ESA symposium on European rocket and balloon programmes and related research, Sandefjord, Norway (ESA SP–590), pp 139–144

Havnes O, Næsheim LI (2007) On the secondary charging effects and structure of mesosphericdust particles impacting on rocket probes. Ann Geophys 25:623–637

Havnes O, Trøim J, Blix T, Mortensen W, Næsheim LI, Thrane E, Tønnesen T (1996) Firstdetection of charged dust particles in the Earth's mesosphere. J Geophys Res 101:10839–10847

Hervig ME, Gordley LL, Deaver LE, Siskind DE, Stevens MH, Russell JM III, Bailey SM, Megner L, Bardeen CG (2009) First satellite observations of meteoric smoke in theupper atmosphere. Geophys Res Lett 36:L18805. doi:10.1029/2009GL039737

Hunten DM, Turco RP, Toon OB (1980) Smoke and dust particles of meteoric origin in themesosphere and stratosphere. J Atmos Sci 37:1342–1357

Knappmiller S, Robertson S, Sternovsky Z, Friedrich M (2008) A rocket-borne mass analyzerfor charged aerosols in the mesosphere. Rev Sci Instrum 79:104502

Latteck R, Singer W, Bardey H (1999) The ALWIN MST radar: technical design and performance. Proceedings of the 14th ESA symposium on European rocket and balloon programmesand related research, Potsdam, Germany (ESA SP–437), pp 179–184

Lübken F-J, Berger U, Baumgarten G (2009) Stratospheric and solar cycle effects on longtermvariability of mesospheric ice clouds. J Geophys Res 114:D00I06. doi:10.1029/2009JD012377

Lübken F-J, Lautenbach J, Höffner J, Rapp M, Zecha M (2009) First continuous temperaturemeasurements within polar mesosphere summer echoes. J Atmos Solar-Terr Phys 71:453–463

Lynch KA, Gelinas LJ, Kelley MC, Collins RL, Widholm M, Rau D, MacDonald E, Liu Y, Ulwick J, Mace P (2005) Multiple sounding rocket observations of charged dust in thepolar winter mesosphere. J Geophys Res 110:A03302. doi:10.1029/2004JA010502

Marsh D, Merkel A (2009) 30-year PMC variability modeled by WACCM. Eos Trans AGU Spring Meet Suppl, pages Abstract SA33B–08

Megner L, Gumbel J (2009) Charged meteoritic particles as ice nuclei in the mesosphere: part 2 – a feasibility study. J Atmos Solar-Terr Phys 71:1236–1244

Megner L, Gumbel J, Rapp M, Siskind D (2008) Reduced meteoric smoke particle densityat the summer pole implications for mesospheric ice particle nucleation. Adv Space Res 41:41–49

Megner L, Khaplanov M, Baumgarten G, Gumbel J, Stegman J, Strelnikov B, Robertson S (2009) Large mesospheric ice particles at exceptionally high altitudes. Ann Geophys 27(3):943–951

Murray BJ, Jensen EJ (2010) Homogeneous nucleation of amorphous solid water particles inthe upper mesosphere. J Atmos Sol Terr Phys 72:51–61

Rapp M, Gumbel J, Lübken F-J, Latteck R (2002) D-region electron number density limitsfor the existence of polar mesosphere summer echoes. J Geophys Res 107(D14). doi:10.1029/2001JD001323

Rapp M, Lübken F-J (2004) Polar mesosphere summer echoes (PMSE): review of observationsand current understanding. Atmos Chem Phys 4:2601–2633

Rapp M, Strelnikova I (2009) Measurements of meteor smoke particles during the ECOMA-2006 campaign: 1. particle detection by active photoionization. J Atmos Sol Terr Phys 71:477–485

Rapp M, Strelnikova I, Strelnikov B, Hoffmann P, Friedrich M, Gumbel J, Megner L, Hoppe U-P, Robertson S, Knappmiller S, Wolff M, Marsh DR (2010) Rocket-borne in-situmeasurements of meteor smoke: charging properties and implications for seasonal variation. J Geophys Res 115:D00I16. doi:10.1029/2009JD012725

Rapp M, Strelnikova I, Strelnikov B, Latteck R, Baumgarten G, Li Q, Megner L, Gumbel J, Friedrich M, Hoppe U-P, Robertson S (2009) First in situ measurement of the verticaldistribution of ice volume in a mesospheric ice cloud during the ecoma/mass rocket-campaign. Ann Geophys 27(2):755–766

Rapp M, Thomas GE (2006) Modeling the microphysics of mesospheric ice particles: assessmentof current capabilities and basic sensitivities. J Atmos Solar-Terr Phys 68: 715–744

Shettle EP, DeLand MT, Thomas GE, Olivero JJ (2009) Long term variations in thefrequency of polar mesospheric clouds in the northern hemisphere from SBUV. Geophys Res Lett 36:L02803. doi:10.1029/2008GL036048

Strelnikova I, Rapp M, Strelnikov B, Baumgarten G, Brattli A, Svenes K, Hoppe U-P, Friedrich M, Gumbel J, B. Williams P (2009) Measurements of meteor smoke particles duringthe ECOMA-2006 campaign: 2. results. J Atmos Solar-Terr Phys 71:486–496

Thomas GE (1991) Mesopheric clouds and the physics of the mesopause region. Rev Geophys 29:553–575

Thomas GE, Olivero JJ, Jensen EJ, Schröder W, Toon OB (1989) Relation betweenincreasing methane and the presence of ice clouds at the mesopause. Nature 338:490–492

Vondrak T, Plane JMC, Meech SR (2006) Influence of submonolayer adsorption on thephotoemission of the Cu(111)/water ice surface. 125:224702

Wood DM (1981) Classical size dependence of the work function of small metallic spheres. PhysRev Lett 46:749

Chapter 5

SABER Observations of Daytime Atomic Oxygen and Ozone Variability in the Mesosphere

Anne K. Smith, Daniel R. Marsh, Martin G. Mlynczak, James M. Russell III, and Jeffrey C. Mast

Abstract Multiyear observations from the SABER (Sounding of the Atmosphere using Broadband Emission Radiometry) instrument on the TIMED (Thermosphere, Ionosphere, Mesosphere, Energetics and Dynamics) satellite are used to determine the daytime variations of ozone and atomic oxygen in the upper mesosphere. Both O_3 and O vary during the daylight hours in response to tidal variations in temperature and winds. Ozone around 85 km responds most strongly to the very large O variations and its variations are in phase with those of O and T. At 97 km, where the O variations are weaker, ozone responds more strongly to the temperature and its variations are out of phase with those of O and T.

5.1 Introduction

Ozone (O_3) and atomic oxygen (O) are closely linked throughout the middle atmosphere. They are much more reactive and less abundant than molecular oxygen (O_2). O and O_3 are called odd oxygen species; they make up the chemical family O_x. The family designation is a convenient way of treating a collection of species that interact rapidly among themselves but more slowly as a group. Over most of the middle atmosphere, the lifetime of the O_x family is much longer than the timescale for O and O_3 to come into photochemical equilibrium with each other (e.g., Brasseur and Solomon, 2005).

During daytime, the set of chemical reactions linking members of the oxygen family is fairly simple because there is a single reaction generating O_3 from O and there is a strongly dominant process converting the O_3 back to O (Chapman, 1930.). This system, given in Eqs. (5.1) and (5.2), represents the destruction of O_3 by photodissociation and the formation of O_3 by the reaction of O with O_2.

$$O_3 + h\nu \rightarrow O + O_2 \tag{5.1}$$

$$O + O_2 + M \rightarrow O_3 + M \tag{5.2}$$

These can be combined into an expression for the O_3 concentration during daytime

$$\frac{dO_3}{dt} = k_{O\text{-}O_2} \cdot O \cdot O_2 \cdot n^2 - J_{O_3} \cdot O_3. \tag{5.3}$$

where d/dt is the material derivative, O, O_2, and O_3 are the mixing ratios, $k_{O\text{-}O_2}$ is the reaction rate of the reaction in Eq. (5.2), n is the atmospheric number density, and J_{O_3} is the photolysis rate in Eq. (5.1). In the upper mesosphere, there is an abundance of O_x produced by the photodissociation of O_2 by ultraviolet radiation from the sun. The lifetimes of the O_x family and of its dominant component, O, are long (weeks to months) above about 85 km. The lifetime of O_3 above 80 km, however, is short during both day and night. The maximum lifetime of about 30 min occurs during night at about 100 km. The daytime lifetime ranges from a few seconds to a few minutes. With the short lifetime, direct transport of O_3 can be neglected and the

A.K. Smith (✉)
Atmospheric Chemistry Division, National Center for Atmospheric Research, Boulder, CO 80307, USA
e-mail: aksmith@ucar.edu

M.A. Abdu, D. Pancheva (eds.), A. Bhattacharyya (Coed.), *Aeronomy of the Earth's Atmosphere and Ionosphere*,
IAGA Special Sopron Book Series 2, DOI 10.1007/978-94-007-0326-1_5,

left side of (Eq. (5.3)) can be set equal to zero, giving the following expression for daytime ozone.

$$O_3 = \frac{k_{O\text{-}O_2} \times O \times O_2 \cdot n^2}{J_{O_3}} \quad (5.4)$$

Equation (5.4) depends on T in two ways. The reaction rate $k_{O\text{-}O_2}$ decreases with increasing temperature ($k_{O\text{-}O_2} = 6 \times 10^{-34}(300/T)^{2.4}$) (Sander et al., 2006). The number density on a pressure surface is inversely proportional to temperature ($n = p/(kT)$, where p is pressure and k is the Boltzmann constant). Equation (5.4) can just as easily be changed around to give an expression for O in terms of O_3

$$O = \frac{J_{O_3} \times O_3}{k_{O\text{-}O_2} \times O_2 \cdot n^2} \quad (5.5)$$

This paper presents the diurnal variability of atomic oxygen and ozone during daytime in the upper mesosphere and emphasizes the relationship between the two species. The nighttime O_x chemical system, which is much different from that during the day because of the lack of photolysis, is not discussed.

5.2 SABER Data and Analysis

The ozone and atomic oxygen presented here are based on version 1.07 data from the SABER (Sounding of the Atmosphere using Broadband Emission Radiometry) instrument on the TIMED (Thermosphere, Ionosphere, Mesosphere, Energetics and Dynamics) satellite. This data set, which includes temperature and ozone through the entire middle atmosphere for 8 years, is much more comprehensive than previously available. The data products used are Level 2 temperature and ozone. Remsberg et al. (2008) describe the non-LTE retrieval of temperature from CO_2 emission observations and give error analysis and comparisons with other observations. There are two independent retrievals of ozone, one using SABER measurements of an ozone emission at 9.6 μm and the other using a molecular oxygen emission at 1.27 μm (Mlynczak et al., 2007). Some results from the 1.27 μm O_3 are shown here but most of the results to be presented use O_3 from the 9.6 μm retrieval. The 9.6 μm O_3 is also used in the retrieval of the daytime O presented here, as described by Smith et al. (2010), although the 1.27 μm O_3 could also be used (Mlynczak et al., 2007).

The non-LTE retrieval of ozone from the 9.6 μm emission requires knowledge of the atomic oxygen profile. The O profiles used in the daytime non-LTE model are determined from the 1.27 μm emission (Mlynczak et al., 2007) below 95 km and are specified from the NRL-MSIS model (Picone et al., 2002) above that level.

Atomic oxygen is not available as a standard product in version 1.07 of the SABER data. We retrieved the O presented here with the assumption that ozone is in photochemical equilibrium with O (Eqs. (5.4) and (5.5)). The retrieval relies on temperature, pressure and ozone information from SABER files. To determine the O profiles used in the present analysis, the photolysis rate J_{O_3} was calculated using the TUV model (Madronich and Flocke, 1998) with the observed O_3 and an assumed (mean) profile of O_2. Ozone outliers that fall beyond the 99.9% range of occurrence are omitted, as described in Smith et al. (2010). O profiles are retrieved over the pressure range 0.01–0.0004 hPa. Above and below there, the equilibrium assumption (Eqs. (5.4) and (5.5)) begins to be unreliable.

Smith et al. (2010) discuss the reliability of the O determined from SABER. They show that the daytime and nighttime retrieved O values are in excellent agreement when the variations due to vertical transport by the diurnal tide are taken into account. The procedure for nighttime retrieval of O, which relies on the 2.0 μm emission from the OH Meinel bands, is independent of that during daytime. The day-night consistency gives confidence in the two O retrievals but does not prove that they are correct. Smith et al. (2010) also note that the SABER O is higher by factors of two to four than O found in previous measurements (e.g. Llewellyn and McDade, 1996; Russell et al., 2005) and in the NRL-MSIS empirical model (Picone et al., 2002). Most of the previous measurements available for comparison were made during night. The discrepancies point to ongoing uncertainties in the absolute amount of O in the upper mesosphere.

SABER profiles extend from the lower stratosphere into the lower thermosphere. Here we will concentrate on the altitude range 75–100 km. The TIMED satellite yaws around six times per year in an annually repeating pattern, every 60–65 days. The data span from about

53° in one hemisphere to about 83° in the other; the hemisphere with high latitude coverage changes with each yaw. The duty cycle is high so there is a large quantity of daytime profiles available for analysis. Data begin 25 January 2002; those processed in this study extend through 11 January 2010: 48 yaw periods.

5.3 Mean Concentrations

Figure 5.1 shows latitude-pressure cross-sections of daytime *T*, O, and the 9.6 μm and 1.27 μm O_3 for the periods near NH winter solstice (the Nov–Jan yaw period), averaged over 8 years. Data from all daytime local times are included. During the yaw periods that include solstices, SABER observations extend to high latitudes in the summer hemisphere. During this period, the O_3 distributions retrieved by the two methods (lower panels on Fig. 5.1) are similar in magnitude and structure. Strong upwelling by the wave-driven circulation causes very cold temperatures at the summer pole and also brings air that is low in O. The low O, in turn, leads to low O_3 by Eq. (5.4). The daytime secondary ozone maximum in mixing ratio spans all latitudes at 90–95 km and reaches mixing ratios of about 1.2 ppm. Smith and Marsh (2005) showed that two factors are primarily responsible for the altitude of this daytime maximum. The O_3 maximum occurs close to the altitude of the maximum in O density, in the altitude range 90–100 km, and also in the region of the very cold temperatures that defines the mean mesopause position (near 95–100 km, except in the high latitude summer).

The other (May–July) solstice period (Fig. 5.2) has some differences in the structure of the ozone and temperature. In particular, there is more daytime ozone at the secondary maximum and the NH summer temperature is colder than that in the SH summer. The hemispheric asymmetry and colder summer temperature in the NH have been reported previously (see a summary of observations in Xu et al. (2007a)) but there is not yet a consensus on its cause.

Near the vernal equinox (the March–May yaw period), *T* and O are more uniform in latitude (Fig. 5.3). There are also quite large differences in the ozone from the two retrieval methods. Zhu et al. (2007)

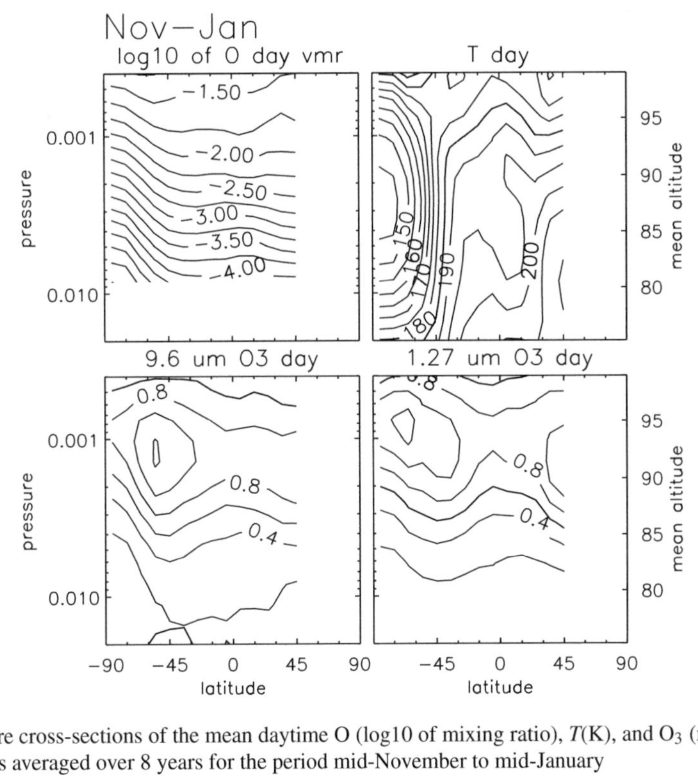

Fig. 5.1 Latitude-pressure cross-sections of the mean daytime O (log10 of mixing ratio), *T*(K), and O_3 (mixing ratio in ppm) from two independent retrievals averaged over 8 years for the period mid-November to mid-January

Fig. 5.2 As in Fig. 5.1 but for the period mid-May to mid-July

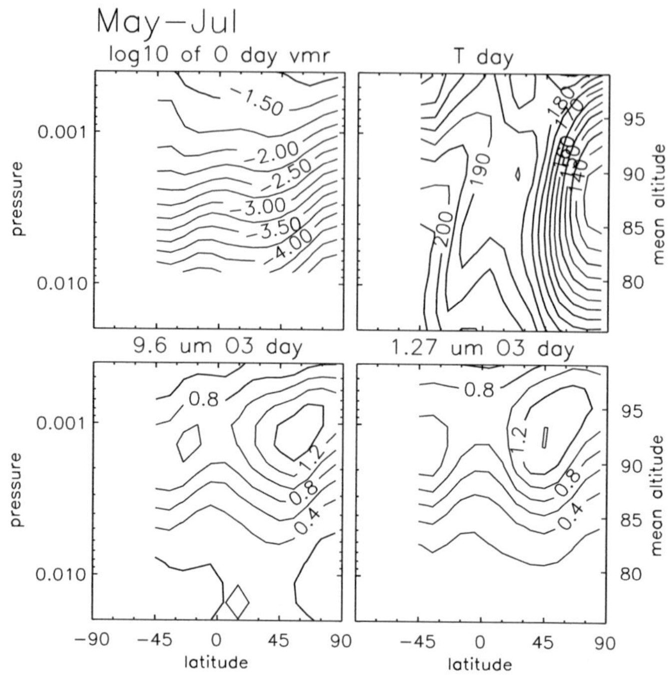

Fig. 5.3 As in Fig. 5.1 but for the period mid-March to mid-May

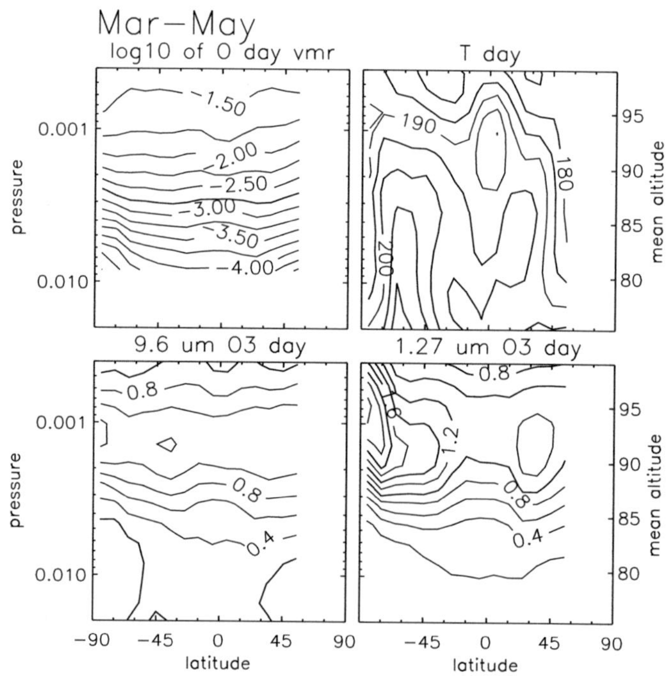

discuss the possible inaccuracies in the 1.27 μm O_3 retrieval near sunrise and during periods of strong vertical transport because $O_2(^1\Delta)$, which is the source of the 1.27 μm emission, has a lifetime against spontaneous emission of about 75 min. With the long lifetime, the emission could be related to the O_3 at a previous time rather than the instantaneous O_3 at the time of emission. We do not know if this problem is responsible for the high latitude differences in O_3 seen in Fig. 5.3. A similar, although smaller, difference is also seen in high latitudes during the September–November period (not shown). The 1.27 μm O_3 values are not used further in this study.

5.4 Impact of the Migrating Diurnal Tide on Daytime Concentrations

The diurnal cycle in O in the upper mesosphere is mainly driven by the diurnal tide and is described by Smith et al. (2010). Near the equator, where the tidal vertical winds are largest and the horizontal winds are small or zero, the transport in the vertical direction can be substantial. Equations (5.6) and (5.7) give a simplified description of the perturbations to O and T due to vertical transport.

$$\left(\frac{\partial}{\partial t}+\bar{u}\frac{\partial}{\partial x}\right)\mathrm{O}' + \frac{\partial \overline{\mathrm{O}}}{\partial z}w' = 0 \qquad (5.6)$$

$$\left(\frac{\partial}{\partial t}+\bar{u}\frac{\partial}{\partial x}\right)T' + \left(\frac{\partial \overline{T}}{\partial z}+\frac{g}{c_\mathrm{p}}\right)w' = 0 \qquad (5.7)$$

x and z are the zonal and vertical dimensions, w is vertical velocity and u is zonal velocity. Primes indicate tidal fields and overbars indicate zonal diurnal averages. Diabatic processes have been neglected in Eq. (5.7) because the radiative timescale in the upper mesosphere is more than a week while the tidal timescale is much shorter (24 h). Combining Eqs. (5.6) and (5.7) gives

$$\mathrm{O}' = \frac{\overline{\mathrm{O}}_z}{S}T' \qquad (5.8)$$

where S is the static stability $S = \overline{T}_z + g/c_\mathrm{p}$ and the subscript z indicates a derivative. Equation (5.8) indicates that, on timescales short enough that O is conserved and diabatic processes are not important, perturbations to O and T will be correlated. Since $\overline{\mathrm{O}}_z$ and S are almost always positive in this part of the atmosphere, the correlation is positive.

The behavior of O_3 is quite different. At all altitudes in the mesosphere, the diurnal cycle of O_3 is strongly dominated by the day/night difference; mixing ratios are much higher during night, by amounts that can be a factor of 10–15. The reason for this is well known: the loss during day by Reaction (5.1) is much faster than the kinetic reactions that destroy O_3 during night. Because of the difference in the leading chemical reactions and the large difference in concentrations, day and night O_3 are normally analyzed separately. Also note that the change in the ozone photolysis rate during the daylight hours is small (about 5% higher at midday than at sunrise and sunset) since the part of the solar spectrum that is responsible for the ozone photolysis is optically thin at these altitudes. The mean rate is about $0.009\ \mathrm{s}^{-1}$.

Tidal signals in 24-h O_3 data are masked by the large photochemically driven day-night difference but they become apparent when looking only at the daytime data (e.g. Marsh et al., 2002; Dikty et al., 2010). Figure 5.4 shows the daytime variations of O, O_3 and T averaged over the latitudes 10°S–10°N for the March-May yaw period. The amplitude of the migrating diurnal tide in temperature is largest in this latitude region and at this time of year (e.g. Xu et al., 2009). The tidal temperature variations are evident as the downward progression of the altitude of the maximum (from 103 km at 6 h to 97 km at 18 h) and the minimum (from 97 to 87 km over the 12 h period). The mixing ratio of O increases with altitude. The variations in O during the day follow those of T but have proportionally higher magnitude where the O vertical gradient is large (80–90 km) than where it is weaker (above 90 km). At 85–90 km, where the daily O variation is large, O_3 increases during the day as the O increases. This gives a positive relationship between the diurnal variations of T and O_3. At 95 km, where the O variation is smaller but the T variation is larger, O_3 also increases during the day although, here, O and T are both decreasing.

The tidal variation of O_3 with T is seen more clearly in Fig. 5.5, which shows scatter plots of daytime O and O_3 against T at two levels. Each point is the average of the data for a single local time bin (1 h) over a full yaw period (60–62 days), and either the 0°N –10°N or 10°S–0°S latitude bin. In the tropics, the diurnal

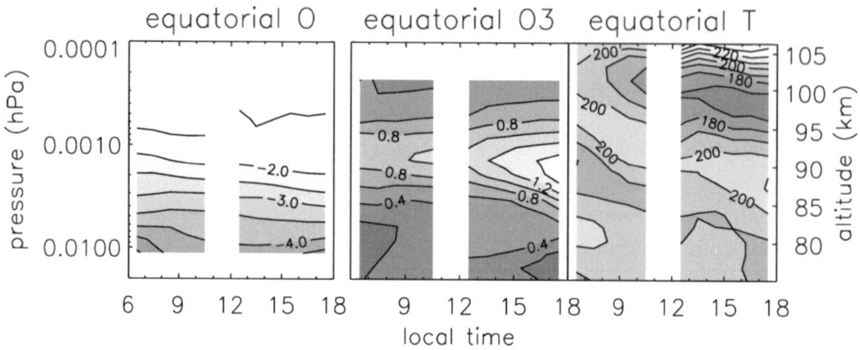

Fig. 5.4 Daytime variation of O (log10 of mixing ratio), O_3 (ppm), and $T(K)$ averaged over the latitude band 10°S–10°N averaged for the Mar-May yaw period over 8 years. O is retrieved only over the pressure range 0.01–0.0004 hPa

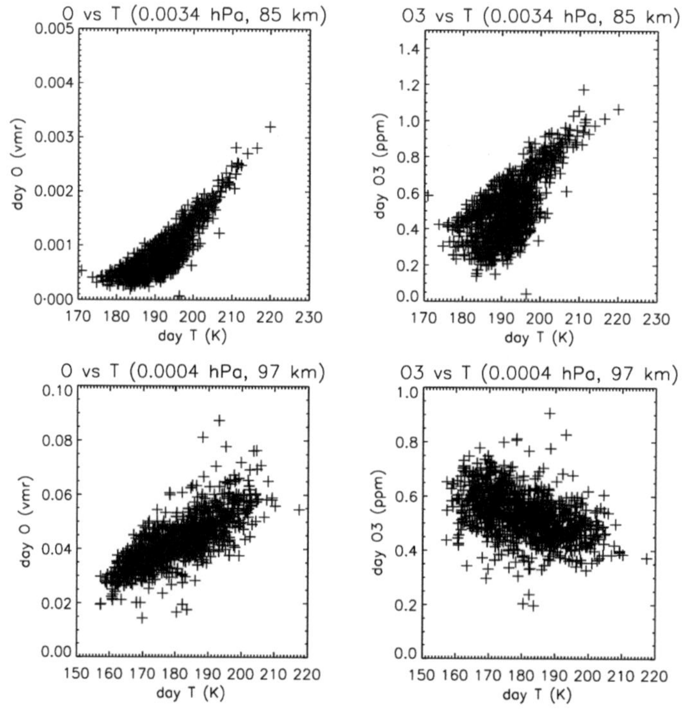

Fig. 5.5 Scatter plot of O (*left*) and O3 (*right*) against T at 0.0034 hPa (*top*) and 0.0004 hPa (*bottom*) for data near the equator that have been averaged in local time

variation of T is large compared to the annual variation and so it dominates the temperature variations. In the pressure range shown, O variations are always in phase with those of T, consistent with Eq. (5.8). Tidal variations in T are due to adiabatic temperature changes associated with vertical motion such that the warmest temperatures occur following downward motion and the coolest temperatures follow upward motion. The same vertical motion transports O. As shown in Smith et al. (2010), the mean diurnal variations of O derived from SABER data in the tropics at altitudes above about 83 km are very consistent with those predicted from tidal transport.

OH Meinel bands emissions, which are closely related to the O concentration, also vary strongly with the diurnal tide (e.g. Marsh et al., 2006). At the altitude of the emission peak (about 87 km), the Meinel emission is in phase with temperature. Ward (1999) used a simple model to show the impacts of the diurnal tide on several airglow features. The basic components of his model confirm the mechanism described above. That is, the relationship with temperature is controlled by the vertical transport of atomic oxygen and the temperature dependence of kinetic rates. The relationship between the Meinel emission and temperature holds up even for polar latitudes and longer time scales (e.g. Cho and Shepherd, 2006).

The relationship of daytime tropical O_3 with T varies with altitude. O_3 variations are in phase with those of T at 85 km (Figs. 5.4 and 5.5). The positive correlation occurs because O_3 is proportional to O and the O variations are large. Over the temperature range from 180 to 210 K, O increases by about a factor of 6. At 97 km, daytime O_3 is out of phase with T (Fig. 5.5). A negative correlation occurs because of the temperature dependence of the rate coefficients and number density ($k_{O\text{-}O_2}$ and n in Eq. (5.4)). It is evident from Figs. 5.4 and 5.5 that the temperature range at 97 km is much greater over the course of the daylight hours than that at 85 km. The change in the O for a given change in T is much less at 95 km (for example, less than a factor of 2 for a T increase from 180 to 210 K).

The strong relationships between O, O_3 and T are in part due to the strong, rapidly varying winds and temperature associated with the diurnal tide. Away from the equatorial region, the tide's role in transport is much reduced. Figure 5.6 shows the correlations of O_3 and O with T for an equatorial latitude band, for a subtropical band, and for a midlatitude band (both hemispheres). This profile-by-profile correlation includes all times of the year so the correlations reflect patterns on all scales from short period waves to the seasonal mean circulation. These results are for 2004 but other years are very similar. O and T are positively correlated in all regions. This correlation is expected because the chemical lifetime of O is long. Even on the timescale of the seasonal mean circulation, perturbations in O are associated with the vertical velocity. The temperature perturbations due to the migrating diurnal tide are large in the latitude band 20°–40° but are substantially reduced over those near the equator. Likewise, the correlations of O with T and O_3 with T have the same signs as the equatorial case but smaller magnitudes. However, in midlatitudes, where the migrating diurnal tide is small, the O_3-T correlation is weak throughout the upper mesosphere.

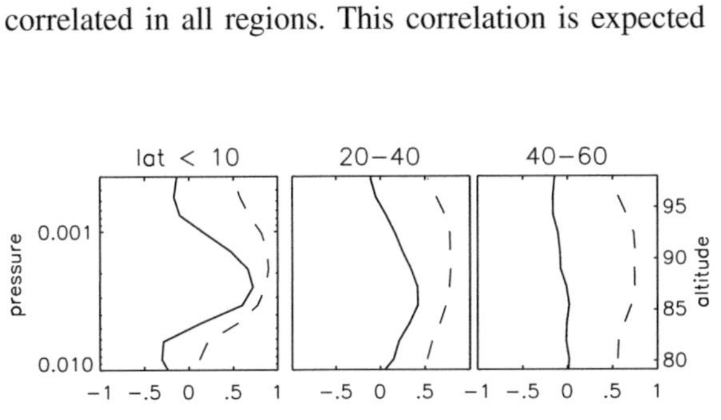

Fig. 5.6 Correlation coefficients of O_3 with T (*solid lines*) and O with T (*dashed lines*) for 2004. Latitude bands include NH and SH

5.5 Conclusions

Atomic oxygen and ozone in the upper mesosphere are quite variable during the daytime. SABER observations presented here show the importance of tidally driven changes over the course of the daylit hours. Tidal winds transport O, which has a long photochemical lifetime in the upper mesosphere. The resulting variations are in phase with those of temperature. O_3 has a short photochemical lifetime and therefore direct transport is not an important factor. Instead, O_3 variations are controlled by the transport of O and by the T, which affects photochemical reaction rates. These two processes imply opposite signs. Where O transport is dominant, over altitudes 83–95 km, O_3 is in phase with O and T. Where the T variations are the dominant effect, O_3 is out of phase with both O and T. Away from the tropical region, the correlation of O with T weakens somewhat but remains positive while the correlation of O_3 with T disappears in the annual mean.

Acknowledgments The National Center for Atmospheric Research is sponsored by the National Science Foundation. Any opinions, findings and conclusions or recommendations expressed in the publication are those of the authors and do not necessarily reflect the views of the National Science Foundation. Support for this research was also provided by the National Aeronautics and Space Administration, grants NNX09AG65G and NNX08AL87G.

References

Brasseur GP, Solomon, S (2005) Aeronomy of the middle atmosphere, 3rd edn. Springer, Dordrecht

Chapman S (1930) On ozone and atomic oxygen in the upper atmosphere. Phil Mag 10:369

Cho Y-M, Shepherd GG (2006) Correlation of airglow temperature and emission rate at Resolute Bay (74.68°N), over four winters (2001–2005). Geophys Res Lett 33:L06815. doi:10.1029/2005GL025298

Dikty S, Schmidt H, Weber M, von Savigny C, Mlynczak MG (2010) Daytime ozone and temperature variations in the mesosphere: a comparison between SABER observations and HAMMONIA model. Atmos Chem Phys Discuss 10:2005–2029

Llewellyn EJ, McDade IC (1996) A reference model for atomic oxygen in the terrestrial atmosphere. Adv Space Res 18: 209–226

Madronich S, Flocke S (1998) The role of solar radiation in atmospheric chemistry. In: Boule P (ed) Handbook of environmental chemistry. Springer, Heidelberg, pp 1–26

Marsh DR, Skinner WR, Marshall AR, Hays PB, Ortland DA, Yee J-H (2002) High resolution Doppler imager observations of ozone in the mesosphere and lower thermosphere. J Geophys Res 107:4390. doi:10.1029/2001JD001505

Marsh D, Smith A, Brasseur, G, Kaufmann M, Grossmann K (2001) The existence of a tertiary ozone maximum in the high-latitude middle mesosphere. Geophys Res Lett 28:4531–4534

Marsh DR, Smith AK, Mlynczak MG, Russell JM III (2006) SABER observations of the OH Meinel airglow variability near the mesopause. J Geophys Res 111:A10S05. doi:10.1029/2005JA011451

Mlynczak MG, Marshall BT, Martin-Torres FJ, Russell JM III, Thompson RE, Remsberg EE, Gordley LL (2007) Sounding of the atmosphere using broadband emission radiometry observations of daytime mesospheric $O_2(^1\Delta)$ 1.27 μm emission and derivation of ozone, atomic oxygen, and solar and chemical energy deposition rates. J Geophys Res 112:D15306. doi:10.1029/2006JD008355

Picone JM, Hedin AE, Drob DP, Aiken AC (2002) NRLMSIS-00 empirical model of the atmosphere: statistical comparisons and scientific issues. J Geophys Res 107:1468. doi:10.1029/2002JA009430

Remsberg EE et al (2008) Assessment of the quality of the retrieved temperature versus pressure profiles in the middle atmosphere from TIMED/SABER. J Geophys Res 113:D17101. doi:10.1029/2008JD010013

Russell JP, Ward WE, Lowe RP, Roble RG, Shepherd GG, Solheim B (2005) Atomic oxygen profiles (80–115 km) derived from Wind Imaging Interferometer/Upper Atmospheric Research Satellite measurements of the hydroxyl and greenline airglow: local time latitude dependence. J Geophys Res 110:D15305. doi:10.1029/2004JD005570

Sander SP et al (2006) Chemical kinetics and photochemical data for use in atmospheric studies. Jet Propulsion Laboratory, Pasadena, CA. http://jpldataeval.jpl.nasa.gov/

Smith AK, Marsh DR (2005) Processes that account for the ozone maximum at the mesopause. J Geophys Res 110:D23305. doi:10.1029/2005JD006298

Smith AK, Marsh DR, Mlynczak MG, Mast JC (2010) Temporal variations of atomic oxygen in the upper mesosphere from SABER. J Geophys Res 115:D18309. doi:10.1029/2005JD013434

Ward WE (1999) A simple model of diurnal variations in the mesospheric oxygen nightglow. Geophys Res Lett 26: 565–3568

Xu J, Liu H-L, Smith AK, Roble RG, Mertens CJ, Russell JM III, Mlynczak MG (2007a) Mesopause structure from thermosphere, ionosphere, mesosphere energetics and dynamics (TIMED)/sounding of the atmosphere using broadband emission radiometry (SABER) observations. J Geophys Res 112:D09102. doi:10.1029/2006JD007711

Xu J, Smith AK, Liu H-L, Yuan W, Wu Q, Jiang G, Mlynczak MG, Russell JM III, Franke SJ (2009) Seasonal and quasi-biennial variations in the migrating diurnal tide observed by thermosphere, ionosphere, mesosphere, energetics and dynamics (TIMED). J Geophys Res 114:D13107. doi:10.1029/2008JD011298

Xu J, Smith AK, Yuan W, Liu H-L, Wu Q, Mlynczak MG, Russell JM III (2007b) Global structure and long-term variations of zonal mean temperature observed by TIMED/SABER. J Geophys Res 112:D24106. doi:10.1029/2007JD008546

Zhu X, Yee J-H, Talaat ER (2007) Effect of dynamical-photochemical coupling on oxygen airglow emission and implications for daytime ozone retrieved from 127 μm emission. J Geophys Res 112:D20304. doi:10.1029/2007JD008447

Chapter 6

In Situ Measurements of Small-Scale Structures in Neutrals and Charged Aerosols

Boris Strelnikov and Markus Rapp

Abstract The ECOMA sounding rocket program began in 2006 and involved the launching of six instrumented rockets for studying properties of mesospheric aerosols and related phenomena from the north-Norwegian Andøya Rocket Range (69°N; 16°E). Among other things, the ECOMA payloads carried the CONE instrument and the ECOMA particle detector to measure densities of neutral air and charged aerosols, respectively. These measurements were done with very high spatial resolution and precision which allows us to study small-scale structures in those species at spatial scales down to one meter. While small-scale fluctuations in the plasma may originate from either electrodynamics or neutral dynamical processes, neutral density fluctuations are a unique tracer for turbulent velocity fluctuations. Such measurements can be used to derive the spectral content of the turbulence field from which, in turn, the turbulent energy dissipation rate can be reliably derived. Accompanying ground based measurements by VHF radar that, in particular, continuously monitor polar mesosphere summer echoes (PMSE), allow us to investigate connection between those phenomena and small-scale structures in charged aerosols. Simultaneous measurements of the densities of neutral air and charged aerosols make it possible to derive Schmidt numbers with a high spatial resolution. Our measurements show that the charged aerosols inside and between the PMSE layers are highly structured down to spatial scales of a few meters. The Schmidt numbers derived for the charged aerosols fall within the range from ~6 to ~4500 which implies particle radii from ~1 to ~26 nm.

6.1 Introduction

It is now commonly accepted that the polar summer mesopause, the coldest region of Earth's atmosphere, is populated by different types of aerosols. In this extremely cold environment where temperature can be as low as ~100 K (Lübken et al., 2009) ice aerosol layers occur. This type of aerosols is nowadays established to be involved in different atmospheric phenomena like polar mesosphere summer echoes (PMSE) and noctilucent clouds/polar mesospheric clouds (NLC/PMC) (Hervig et al., 2009b). It is known that ice particles in this region can grow up to sizes of ~80 nm (Baumgarten et al., 1998). Another sort of aerosols in the mesosphere/lower thermosphere (MLT) region, so-called meteoric smoke particles (MSP) have recently been in the focus of several experimental studies (Gelinas et al., 1998; Lynch et al., 2005; Rapp et al., 2005; Strelnikova et al., 2009c; Hervig et al., 2009a) such that their existence in the MLT region is also turning into accepted knowledge.

All these mesospheric aerosol particles are surrounded by the ionospheric plasma and can gain a charge because of e.g., electron and ion capture processes or photoionization by solar UV radiation. Charged aerosols change properties of the ionospheric plasma and thereby play a crucial role in layering phenomena like PMSE (Rapp and Lübken, 2004).

Despite significant advances in recent years that improved our understanding of the MLT layering,

B. Strelnikov (✉)
Department of Radars and Sounding Rockets, Leibniz Institute of Atmospheric Physics at the Rostock University, 18225 Kuehlungsborn, Germany
e-mail: strelnikov@iap-kborn.de

M.A. Abdu, D. Pancheva (eds.), A. Bhattacharyya (Coed.), *Aeronomy of the Earth's Atmosphere and Ionosphere*, IAGA Special Sopron Book Series 2, DOI 10.1007/978-94-007-0326-1_6, © Springer Science+Business Media B.V. 2011

dynamics, and structure little is known about the microphysical details of these aerosol related phenomena. In particular, there is an obvious need to gain a better knowledge of the aerosol properties like composition, charge state, size distribution, etc. (Rapp et al., 2006).

To fill this lack of our knowledge, the ECOMA sounding rocket program was started in 2006 by the Leibniz-Institute of Atmospheric Physics (IAP) in Germany and the Norwegian Defense Research Establishment (FFI) with contributions from Sweden, Austria and the US. ECOMA stands for "Existence and Charge state Of meteoric smoke particles in the Middle Atmosphere".

Initial results of this program from rocket flights in 2006, 2007, and 2008 are presented in Rapp and Strelnikova (2009); Strelnikova et al. (2009c), Rapp et al. (2009), Strelnikov et al. (2009a), Friedrich et al. (2009), Brattli et al. (2009) and Megner et al. (2009), and Rapp et al. (2010b) respectively. For a deeper review on the entire ECOMA project the reader is referred to Rapp et al. (2010a).

In this chapter we summarize results of in situ measurements of *small-scale* structures in neutral air and charged aerosols during the ECOMA project. These measurements allow us to derive Schmidt numbers for the charged aerosols.

The chapter is organized as follows. A brief description of the measurement technique is present in Section 6.2. In Section 6.3 we present the measurements results and in Section 6.4 summarize the main findings.

6.2 Measurement Technique

Our studies of small-scale structures are based on high resolution in situ measurements of the densities of the different species followed by a spectral analysis of their relative fluctuations. The relative fluctuations are derived by subtracting a running mean of a length of about 1–2 km from the measured densities. As shown by (Lübken, 1987), relative fluctuations of neutral air density are an excellent tracer for mesospheric turbulence measurements, which were successfully used to derive energy dissipation (or heating) rates for 40 rocket soundings.

The ionospheric plasma in the MLT region, which consists of electrons, positive ions, and charged aerosols, can be disturbed by the neutral air turbulence through the frequent plasma-neutral collisions. Following from the theory, initially proposed by Cho et al. (1992) and later extended by Rapp et al. (2003) and Lie-Svendsen et al. (2003), the presence of charged aerosols should ultimately reduce the diffusivity of all charged plasma species, which are electrostatically coupled with each other, thereby changing their spectral characteristics. In turbulence theory, this reduced diffusivity can be expressed in terms of a high Schmidt numbers $Sc = \nu/D$, a dimensionless number defined as the ratio of momentum diffusivity (viscosity) and mass (molecular) diffusivity. Therefore, the power spectrum of the measured plasma density fluctuations and its dependence on turbulence characteristics is also a function of the Schmidt number.

In this chapter, we focus on in situ density measurements of neutral air and charged aerosols. For this purpose, the ECOMA payloads used in all the rocket soundings during the entire ECOMA project may be considered as identical. The neutral density measurements were done with the CONE instrument (Giebeler et al., 1993) and the aerosol density measurements were conducted with the ECOMA particle detector (ECOMA-PD) (Rapp and Strelnikova, 2009).

6.2.1 Instrumentation

6.2.1.1 CONE

Briefly, the CONE (COmbined sensor for Neutrals and Electrons) is an ionization gauge designed for density measurements of neutral air surrounded by an electrostatic probe for electron density measurements. CONE can resolve structures in neutral air density down to scales of some centimeters (Giebeler et al., 1993), which are caused by neutral air turbulence. During the ECOMA project the CONE was always mounted on the rear deck of the payload and, therefore, was in a favorable (for most precise density measurements) aerodynamical condition during the descent of the sounding rocket (Rapp et al., 2001).

6.2.1.2 ECOMA-PD

The ECOMA particle detector is a rather complicated instrument (Rapp and Strelnikova, 2009), which is designed to simultaneously measure both charged as well as neutral aerosols with very high spatial resolution. The ECOMA-PD has two data channels, one with a sampling of 16 Hz only and other with high enough time resolution to study spectral behavior. The first (which relies on artificial photoionization of the particles) can potentially also detect neutral particles but does not have sufficient resolution for spectral analysis. In this work we only consider the second data channel, the measurements of the naturally charged fraction of the mesospheric aerosols which are referred to in other ECOMA-papers as DC-channel measurements, e.g. Rapp and Strelnikova, (2009) and Strelnikova et al. (2009c). This part of the ECOMA-PD is similar to the probe first used by Havnes et al. (1996) and is known as a Faraday cup. The ECOMA-PD was always mounted on the front deck of the payload, such that high quality charged aerosol data were gathered on the ascent of the flight trajectory.

6.2.2 Data Analysis

The first in situ measurements of Sc at PMSE heights were done by Lübken et al. (1994) and Lübken et al. (1998) and utilized measurements of small-scale *electron* and neutral density fluctuations. They used the same CONE instrument with resulting values between ~6 and 500. For the measurements of the Sc of charged aerosols we use the same technique, the so called spectral model method introduced by Lübken (1992) for turbulence measurements, which is briefly summarized below.

First, we derive the turbulence energy dissipation rate, ε, from the high resolution neutral density measurements obtained with the CONE (see Lübken (1997) and Strelnikov et al. (2003), for more details). This is done by fitting a theoretical model of Heisenberg (1948) (Fig. 6.1, dashed line) or Tatarskij (1961), to the power spectrum of measured relative density fluctuations (Fig. 6.1, black line) of neutral air.

Making use of the high resolution density measurements yielded by the ECOMA-PD, we derive power spectrum of the charged aerosols, which is resolved down to meter scales (Fig. 6.1, grey line). This spectrum reflects the reduced diffusivity (high Sc-value) of the tracer (charged aerosols) in that it reveals the viscous-diffusive subrange, i.e. the k^{-1} power law marked in Fig. 6.1. By fitting a model of Driscoll and Kennedy (1985), shown by dotted line in Fig. 6.1, which includes Sc as a free fitting parameter, to this measured spectrum we derive the Schmidt number of the charged aerosols. Importantly, as pointed out by Lübken et al. (1994, 1998) for the robust derivation of the Schmidt number this method requires as an

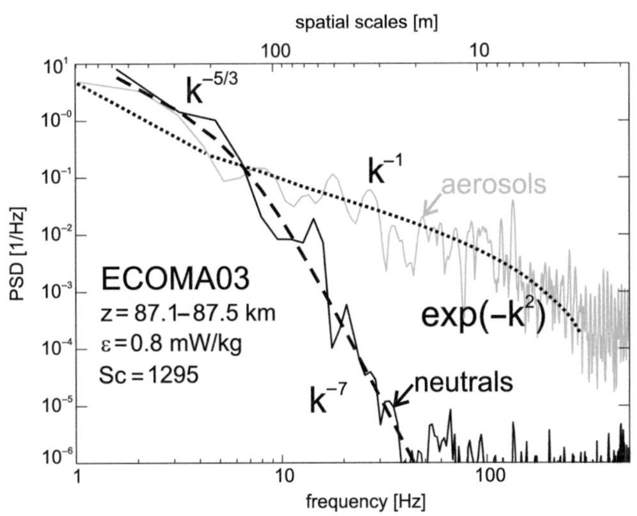

Fig. 6.1 *Solid lines*: Power spectral densities (PSD) of a relative density fluctuations of neutrals (*black*) and charged aerosols (*grey*). *Dashed line*: Fitted model of Heisenberg (1948). It indicates the Kolmogorov $k^{-5/3}$ power law in the inertial subrange, a k^{-7} decay in the viscous subrange, and energy dissipation rate $\varepsilon = 0.8$ mW/kg. *Dotted line*: Fitted model of Driscoll and Kennedy (1985). In addition to the inertial and viscous subranges, it includes the viscous-diffusive subrange with the k^{-1} power law and results in Sc = 1295

input an independently measured turbulence energy dissipation rate, ε (which we obtain from the independent CONE-measurements).

Thus, in the example shown in Fig. 6.1, the turbulence energy dissipation rate derived from the neutral density fluctuations is equal to 0.8 mW/kg, and the Schmidt number derived for the charged aerosol density fluctuations is equals to 1295.

6.3 Scientific Results

6.3.1 Small-Scale Structures

Figures 6.2 and 6.3 show small-scale structures observed in charged aerosols during two sounding rocket campaigns which took place in summer seasons 2007 and 2008, respectively. Upper panels show relative density fluctuations of the charged aerosols and the lower panels represent their power spectra derived using wavelet analysis (Strelnikov et al., 2003).

All these measurements were done under PMSE and NLC conditions. PMSE were observed by ALWIN VHF radar with high temporal and spatial resolution and were used as rocket launch criteria throughout the ECOMA campaigns. The ALWIN radar (Latteck et al., 1999) operating at 53.5 MHz is located close to the rocket launch site at the Andøya Rocket Range. The NLC were detected by both onboard photometers (Megner et al., 2009) and by ground based lidars (Baumgarten et al., 2009) during all three flights discussed in this Chapter (Rapp et al. 2010b). For more details on flight ECOMA03 (Fig. 6.2) the reader is referred to Rapp et al. (2009) and on flights ECOMA04 and ECOMA06 to Rapp et al., (2010a, b). The PMSE observed during ECOMA03 and ECOMA04 flights had a double layer structure with lower layer around 83 km and upper layers around 87–88 km (see also PMSE profiles in Fig. 6.4 shown by dashed grey lines). In case of the ECOMA06 flight, the ALWIN radar detected a single layer PMSE around 83 km. The most pronounced small-scale structuring in charged aerosols occurred inside these PMSE layers as it is clearly visualized by the wavelet spectra. In these regions the spectra extend down to 1 m spatial scales, which indicates a reduced diffusivity of charged aerosols (high Sc-value) at those heights. Otherwise, as shown by Lübken et al. (1998), an extension of the spectra down to a scale of several meters assuming that $Sc = 1$ would imply very large turbulence energy dissipation rate of ~20 kW/kg and hence, unrealistic heating rate of approximately ~1.6 million Kelvin per day.

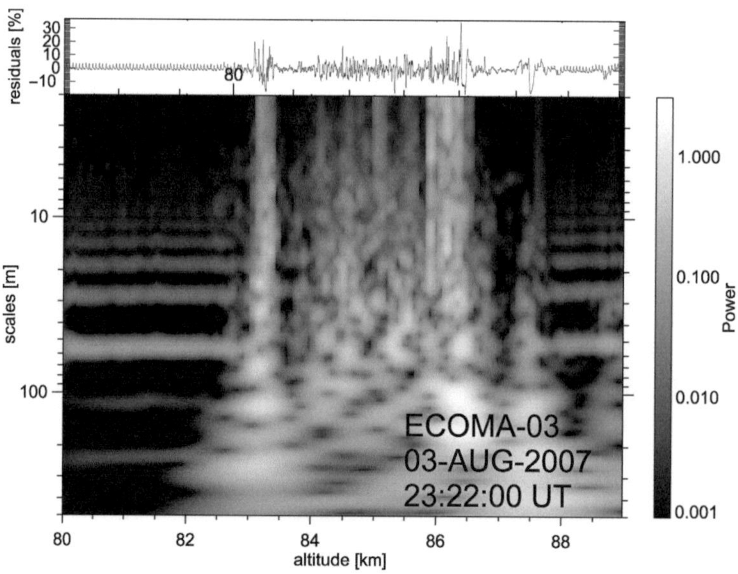

Fig. 6.2 Measurements of charged aerosols by ECOMA-PD during flight ECOMA03 done on 03-AUG-2007 at 23:22:00 UT from Andenes, 69°N, 16°E. *Upper panel*: Relative density fluctuations (residuals). *Lower panel*: Wavelet spectrogram of the residuals derived using Morlet-12 wavelet function

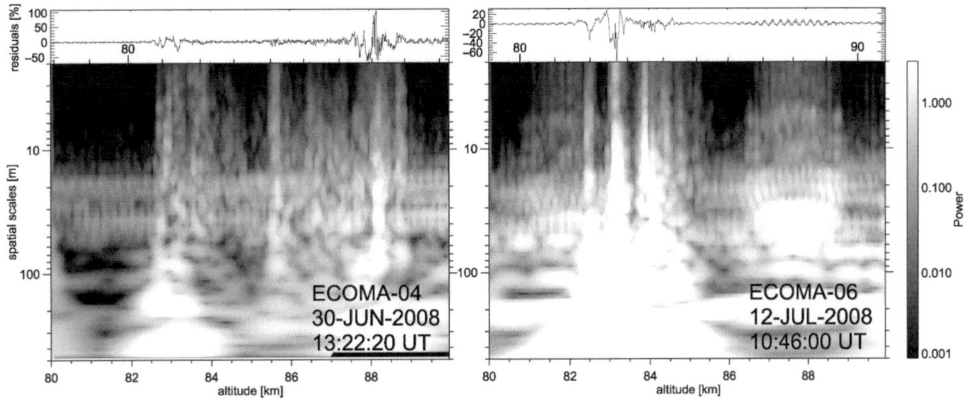

Fig. 6.3 The same as Fig. 6.2 but for the flights ECOMA04 and ECOMA06

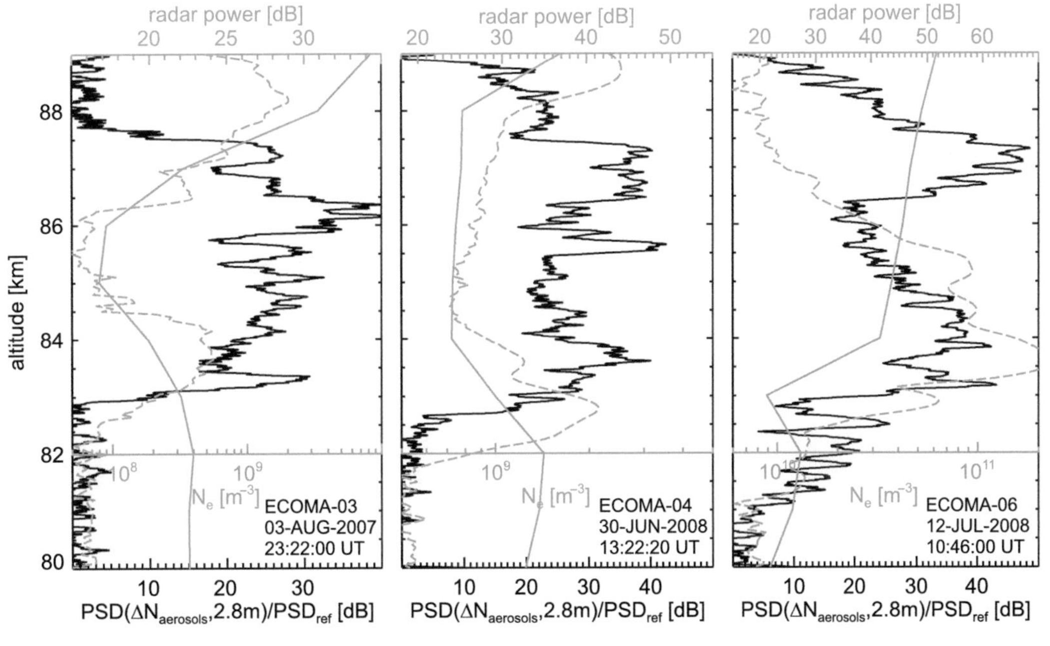

Fig. 6.4 Measurement results for the flights: ECOMA03, ECOMA04, and ECOMA06, respectively. *Black*: Profiles of relative density fluctuations taken at the ALWIN radar Bragg scale (2.8 m). *Dashed grey*: Power profiles measured by ALWIN VHF radar. *Solid grey*: Absolute electron density measured by radio wave propagation technique

Our turbulence measurements with the CONE instrument onboard the same payload result in ε-values which do not exceed 1 W/kg in that height region (see Fig. 6.5, left panel for corresponding ε-profiles). For the flights ECOMA03 and ECOMA04, between the PMSE layers, there is also a lot of small-scale structuring detected in situ in charged aerosols. The spectra also extend down to a spatial scale of several meters and, thereby also indicate enhanced Schmidt numbers for the aerosols in that region. In the case of the ECOMA06 flight, there are also some artificial fluctuations between 86 and 89 km, caused by rocket coning which is clearly seen in both residuals and spectrum as periodical harmonic oscillations (Fig. 6.3, right panel). This coning and other signatures are well separated in frequency space so that such a well defined artificial perturbation does not affect the overall quality of the spectral data.

Further exploring small-scale features of the charged aerosols we show in Fig. 6.4 comparison of

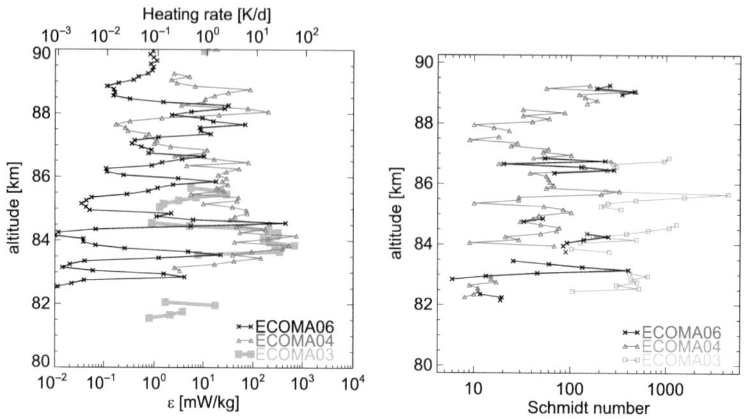

Fig. 6.5 *Left panel*: turbulence energy dissipation rates derived from in situ measurements by CONE. *Right panel*: Schmidt numbers derived for charged aerosols from relative density fluctuations measured in situ by ECOMA-PD

the VHF radar PMSE observations with the small-scale aerosol density fluctuations at 2.8 m scale, which is ALWIN radar's Bragg scale (i.e. half wavelength). Figure. 6.4 shows PMSE signatures (dashed grey lines) measured close to the rocket launch times by radar, the profiles of the power spectral densities of the charged aerosols, $PSD(\Delta N_{\text{aerosol}}, 2.8\,\text{m})$ in dB (black lines) derived as a slice of the spectrogram from Figs. 6.2 and 6.3 at 2.8 m, and measured absolute electron density profiles (solid grey) by radio wave propagation technique (Friedrich et al., 2009) onboard the same payload. As shown in Rapp et al. (2003) and Lie-Svendsen et al. (2003), the quantity $PSD(\Delta N_{\text{aerosol}}, 2.8\,\text{m})$ is ultimately proportional to the radar power and, therefore should reproduce the PMSE signature as it is observed by the VHF radar. The in situ measured profiles of the small-scale fluctuations, however, reproduce the general height range of the observed PMSE, but do not reveal the pronounced double-layer structure seen by the radar during flights ECOMA03 and ECOMA04. Note also, that we do not expect to see an exact overlapping of that measurements because the radar volume (beam diameter of about 10 km at 85 km height) is much larger than the centimeter scale horizontal volume probed by the rocket.

Actually, this data simply supports our current understanding of PMSE which, as shown by Rapp et al. (2002), can only be observed if background electron density exceeds a threshold of a few hundreds of electrons/cm^3. This is demonstrated by means of concurrent electron density measurements (Fig. 6.2, solid grey lines), which show an electron density depletion between the two PMSE layers preventing detectable radar backscatter. Note that radar waves are scattered from electron irregularities and not aerosol density structures. That is, regardless of the fact that both presence of aerosols was confirmed by photometer and lidar and irregularities in aerosol density were observed by the ECOMA-PD, no radar echo could be detected from the region with the low electron density. For a review of such previous observations see Friedrich and Rapp (2009). For the case of the ECOMA06 flight, however, the PMSE signature in the in situ measured aerosol density fluctuations, closely reproduces the radar power profile if we exclude the artifact between 86 and 89 km mentioned above.

A more detailed discussion of different measurements during the ECOMA flights can be found in Rapp et al. (2009, 2010b), Strelnikova et al. (2009a), and Rapp and Strelnikova (2009). We only note here, that based on the onboard photometer and ground based lidar measurements and taking into account that simultaneous onboard temperature measurements revealed supersaturation with respect to ice (Rapp et al., 2010b), the aerosols measured by the DC-channel of the ECOMA-PD during these summer flights were clearly identified as ice particles.

6.3.2 Schmidt Numbers

Since, both the charged aerosol density measurements by the ECOMA-PD and the turbulence measurements were conducted during all the ECOMA flights, it

Table 6.1 Range of values for Sc/r_{ice},nm for ECOMA flights

	Min	Mean	Median	Max
ECOMA03	103/4	650/10	415/8	4463/26
ECOMA04	8/1	72/4	49/3	478/9
ECOMA06	6/1	131/5	90/4	465/9

allows us to derive the Schmidt number, Sc, for the charged aerosols as described in Section 6.2.2. However, because ECOMA-PD measurements were done on the upleg and CONE measured turbulence on the downleg, there is an uncertainty in the Sc-values owing to the horizontal distance between those measurements of about 50 km. In most cases during the ECOMA project turbulence measurements were done both in situ and using ground-based radar observations (Strelnikov et al., 2009a, b; Engler et al., 2009). This allows us to estimate the uncertainty of the ε-values and its influence on our Sc-derivation. We varied the strength of turbulence in the range of values that we estimated from both rocket-borne and radar measurements (see Strelnikov et al. (2009a) where the technique is described in detail for the flight ECOMA03). The resultant change in the derived Sc-values lies within ~60% error bar. Moreover, the Sc-profiles derived using the radar and the in situ turbulence measurements qualitatively agree (see Strelnikov et al. (2009a) and Strelnikov et al. (2009b) for more details on flights ECOMA03 and ECOMA04, respectively).

The derived Sc-profiles for the flights ECOMA03, ECOMA04, and ECOMA06 are shown in Fig. 6.5 and are summarized in Table 6.1. They mostly reveal relatively high values, i.e. Sc >> 1. According to Lübken et al. (1998), the Schmidt number is related to the particle size by $Sc \approx 6.5r^2$, where r is the particle radius in nm. This implies that the detected particle radii fall within the size range from ~1 to ~9 nm for ECOMA04 and ECOMA06 flights and, for the ECOMA03 flight reach values of ~26 nm. The latest were also confirmed by another independent measurements by Rapp et al. (2009). We summarize the derived Sc-values in Table 6.1 where for each flight are shown the minimum, mean, median, and maximum values. The radius derived from a Schmidt number can be viewed as a weighted mean of the size distribution, where the weighting is by the radius to the second power. This is because the underlying physics of the Schmidt number effect is particle-neutral collisions which are modeled as hard sphere collisions that scale as r^2 (Cho et al., 1992). Also note that the large uncertainty of ~60% in the Sc estimate only leads to about 3% error in particle radii assessment.

It is interesting to note that the largest Sc-value of ~4500 indicating the presence of ice particles with radii of ~26 nm was detected during ECOMA03 flight at around 85.8 km height, that is between the PMSE layers. However, the onboard photometer observed a broad NLC layer from 82 to 88 km height range during upleg. During the descent of that flight, an ice layer centered around 85.5 km was detected by the onboard photometer (Megner et al., 2009).

6.4 Conclusions

The ECOMA particle measurements showed that the charged aerosols inside and between the PMSE layers, i.e. ice particles, are highly structured down to a few meters spatial scales. The Schmidt numbers derived for the charged ice fall within the range from 6 to 4463 (with an uncertainty of ~60%) which implies particle radii from ~1 to ~26 nm. The biggest uncertainty in the Schmidt number estimate from the small-scale density fluctuations is because of the horizontal separation between in situ turbulence and aerosol density measurements. For a more precise Schmidt number derivation true common volume neutral and plasma measurements are needed.

Acknowledgements The ECOMA project is sponsored by the German Space Center under DLR-grants 50OE0301 and 50OE0801. We would like to thank Martin Friedrich for supplying the electron density data and Ralph Latteck for provision of the radar data employed in this work.

References

Baumgarten G, Fiedler J, Fricke KH, Gerding M, Hervig M, Hoffmann P, Müller N, Pautet P, Rapp M, Robert C, Rusch D, von Savigny C, Singer W (2009) The noctilucent cloud (NLC) display during the ECOMA/MASS sounding rocket flights on 3 August 2007: morphology on global to local scales. Ann Geophys 27:953

Baumgarten G, Fiedler J, Lübken FJ, von Cossart G (2008) Particle properties and water content of noctilucent clouds and their interannual variation. J Geophys Res. doi:10.1029/2007JD008884

Brattli A, Lie-Svendsen Ø, Svenes K, Hoppe U, Strelnikova I, Rapp M, Latteck R, Friedrich M (2009) The ECOMA 2007 campaign: rocket observations and numerical modelling of aerosol particle charging and plasma depletion in a PMSE/NLC layer. Ann Geophys 27:781

Cho JYN, Hall TM, Kelley MC (1992) On the role of charged aerosols in polar mesosphere summer echoes. J Geophys Res 97:875

Driscoll RJ, Kennedy LA (1985) A model for the spectrum of passive scalars in an isotropic turbulence field. Phys Fluids 28:72

Engler N, Singer W, Latteck R, Rapp M, Strelnikov B (2009) A case study of extrem aspect sensitive vhf radar backscatter in the vicinity of pmse during the ECOMA 2008 rocket campaign. Proceedings of the 19th ESA symposium on European rocket and balloon programmes and related research, Bad Reichenhall, Germany (ESA SP-671), 7–11 June 2009

Friedrich M, Rapp M (2009) News from the Lower Ionosphere: A Review of Recent Developments. Surv Geophys 30:525. doi:10.1007/s10712-009-9074-2

Friedrich M, Torkar KM, Singer W, Strelnikova I, Rapp M, Robertson S (2009) Signatures of mesospheric particles in ionospheric data. Ann Geophys 27:823

Gelinas LJ, Lynch KA, Kelley MC, Collins S, Baker S, Zhou Q, Friedman JS (1998) First observation of meteoric charged dust in the tropical mesosphere. Geophys Res Lett 25(21):4047. doi:10.1029/1998GL900089

Giebeler J, Lübken FJ, Nägele M (1993) CONE - a new sensor for in-situ observations of neutral and plasma density fluctuations. Proceedings of the 11th ESA symposium on European rocket and balloon programmes and related research, Montreux, Switzerland (ESASP), pp 311–318

Havnes O, Trøim J, Blix T, Mortensen W, Næsheim LI, Thrane E, Tønnesen T (1996) First detection of charged dust particles in the Earth's mesosphere. J Geo-phys Res 101: 10839

Heisenberg W (1948) Zur statistischen Theorie der Turbulenz. Z Physik 124:628

Hervig ME, Gordley LL, Deaver LE, Siskind DE, Stevens MH, Russell JM, Bailey SM, Megner L, Bardeen CG (2009a) First Satellite Observations of Meteoric Smoke in the Middle Atmosphere. Geophys Res Lett 36:18805. doi:10.1029/2009GL039737

Hervig ME, Gordley LL, Russell JM, Bailey SM (2009b) SOFIE PMC observations during the northern summer of 2007. J Atmos Solar-Terr Phys 71:331. doi:10.1016/j.jastp.2008.08.010

Jübken FJ, Rapp M, Blix T, Thrane E (1998) The ALWIN MST radar - technical design and performances. Geophys Res Lett 25:893. doi:10.1029/98GL50479

Latteck R, Singer W, Bardey H (1999) Modeling the plasma response to small-scale aerosol particle perturbations in the mesopause region. Proceedings of the 14th ESA symposium on European rocket and balloon programmes and related research, Potsdam, Germany (ESA SP–437), pp 179–184

Lie-Svendsen Ø, Blix TA, Hoppe U, Thrane EV (2003) First continuous temperature measurements within polar mesosphere summer echoes. J Geophys Res (Atmo-spheres) 108:8442. doi:10.1029/2002JD002753

Lübken F, Lautenbach J, Höffner J, Rapp M, Zecha M (2009) TOTAL: A new instrument to study turbulent parameters in the mesosphere and lower thermosphere. J Atmos Solar-Terr Phys 71:453. doi:10.1016/j.jastp.2008.06.001

Lübken FJ (1987) On the extraction of turbulent parameters from atmospheric density fluctuations. ESA SP-270: European rocket & balloon programmes and related research, pp 215–218

Lübken FJ (1992) Seasonal variation of turbulent energy dissipation rates at high latitudes as determined by insitu measurements of neutral density fluctuations. J Geophys Res 97:20,385

Lübken FJ (1997) In situ measurement of the Schmidt number within a PMSE layer. J Geophys Res 102:13,441

Lübken FJ, Giebeler J, Blix TJ, Thrane E, Singer W, Bremer J (1994) Microphysical and turbulent measurements of the Schmidt number in the vicinity of polar mesosphere summer echoes. Geophys Res Lett 21:1651

Lynch KA, Gelinas LJ, Kelley MC, Collins RL, Widholm M, Rau D, Mac-Donald E, Liu Y, Ulwick J, Mace P (2005) Multiple sounding rocket observations of charged dust in the polar winter mesosphere. J Geophys Res 110(A3):A03302. doi:10.1029/2004JA010502

Megner L, Khaplanov M, Baumgarten G, Gumbel J, Stegman J, Strelnikov B, Robert-son S (2009) Large mesospheric ice particles at exceptionally high altitudes. Ann Geophys 27:943

Rapp M, Hedin J, Strelnikova I, Friedrich M, Gumbel J, Lübken FJ (2005) Absolute density measurements in the middle atmosphere. Geophys Res Lett 32:L23821

Rapp M, Gumbel J, Lübken FJ (2001) D region electron number density limits for the existence of polar mesosphere summer echoes. Ann Geophys 19:571

Rapp M, Gumbel J, Lübken FJ, Latteck R (2002) Observations of positively charged nanoparticles in the nighttime polar mesosphere. J Geophys Res 107(D19):4187. doi:10.1029/2001JD001323

Rapp M, Lübken F (2003) On the nature of PMSE: Electron diffusion in the vicinity of charged particles revisited. J Geophys Res 108:8437. doi:10.1029/2002JD002857

Rapp M, Lübken FJ (2004) Polar mesosphere summer echoes (PMSE): Review of observations and current understanding. Atmos Chem Phys 4:2601

Rapp M, Strelnikova I (2009) Measurements of meteor smoke particles during the ECOMA-2006 campaign. J Atmos Solar-Terr Phys 71:477. doi:10.1016/j.jastp.2008.06.002

Rapp M, Strelnikova I, Strelnikov B, Friedrich M, Gumbel J, Hoppe UP, Blix T, Havnes O, Bracikowski P, Lynch K, Knappmiller S (2010a) Microphysical properties of mesospheric aerosols: an overview of in situ-results from the ECOMA-project.

Rapp M, Strelnikova I, Strelnikov B, Hoffmann P, Friedrich M, Gumbel J, Megner L, Hoppe UP, Robertson S, Knappmiller S, Wolff M, Marsh DR (2010b) First in situ measurement of the vertical distribution of ice volume in a mesospheric ice cloud during the ECOMA/MASS rocket-campaign. J Geophys Res p. accepted

Rapp M, Strelnikova I, Strelnikov B, Latteck R, Baumgarten G, Li Q, Megner L, Gum-bel J, Friedrich M, Hoppe U,

Robertson S (2009) Rocket-borne in-situ measurements of meteor smoke: charging properties and implications for seasonal variation. Ann Geophys 27:755

Rapp M, Thomas GE (2006) Modeling the microphysics of mesospheric ice particles - Assessment of current capabilities and basic sensitivities. J Atmos Solar-Terr Phys 68:715. doi:10.1016/j.jastp.2005.10.015

Strelnikov B, Rapp M, Lübken FJ (2003) A new technique for the analysis of neutral air density fluctuations measured in situ in the middle atmosphere. Geophys Res Lett 30:2052. doi:10.1029/2003GL018271

Strelnikov B, Rapp M, Strelnikova I, Engler N, Latteck R (2009a) Small-scale structures in neutrals and charged aerosol particles as observed during the ECOMA/MASS rocket campaign. Ann Geophys 27:1449

Strelnikov B, Rapp M, Strelnikova I, Latteck R, Engler N, Blix TA, Hoppe UP, Friedrich M (2009b) Small-scale structures in neutral and plasma species in the middle atmosphere as observed during the ECOMA rocket campaigns. Proceedings of the 19th ESA symposium on European rocket and balloon programmes and related research, Bad Reichenhall, Germany (ESA SP-671), 7–11 June 2009

Strelnikova I, Rapp M, Strelnikov B, Baumgarten G, Brattli A, Svenes K, Hoppe UP, Friedrich M, Gumbel J, Williams BP (2009c) Measurements of meteor smoke particles during the ECOMA-2006 campaign. J Atmos Solar-Terr Phys 71:486. doi:10.1016/j.jastp.2008.07.011

Tatarskii VI (1961) Wave propagation in a turbulent medium. McGraw-Hill Book, New York, NY

Chapter 7

High-Latitude Gravity Wave Measurements in Noctilucent Clouds and Polar Mesospheric Clouds

Michael J. Taylor, P.-D. Pautet, Y. Zhao, C.E. Randall, J. Lumpe, S.M. Bailey, J. Carstens, K. Nielsen, James M. Russell III, and J. Stegman

Abstract Distinct wave forms traditionally observed in Noctilucent Cloud (NLC) photographs and most recently captured in high-resolution panoramic images of Polar Mesospheric Clouds (PMC) provide an exceptionally rich resource for quantifying gravity wave activity and properties in the high-latitude summer mesopause region. Using extensive image data on PMC structures obtained by the Cloud Imaging and Particle Size (CIPS) ultraviolet instrument onboard the NASA Aeronomy of Ice in the Mesosphere (AIM) satellite, we have investigated the properties of prominent quasi-monochromatic waves imaged over the northern hemisphere polar region during summer 2007. Our two-dimensional spectral analysis has focused on the peak season, July period and over 450 events have been measured. The PMC field was found to contain a broad spectrum of gravity waves with horizontal wavelengths ranging from at least 20–400 km. The smallest scale wave events (<50 km), exhibit the highest occurrence frequency, but substantial evidence for larger-scale (>100 km) wave activity was also found. The direction of motion of the waves (both large and small-scale) deduced from their orientations (with 180° ambiguity) was predominantly zonal (with a small meridional component), and differed significantly from recent NLC Type II band measurements which were dominated by strong near poleward wave motions. Evidence was also found for a reduction in gravity wave activity over the Europe/North Atlantic sector during the July 2007 period, as compared to other longitudes. These results build significantly on an initial seasonal investigation of quasi-monochromatic events by Chandran et al. (2009), and further demonstrate the high potential of CIPS data for detailed gravity wave studies. A more comprehensive investigation of the strong spatial alignments of the wave events and their longitudinal variability is currently in progress.

7.1 Introduction

For well over a century, scientists and amateur observers alike have marveled at the beauty and structure of noctilucent clouds (NLC). First reported by Leslie (1885) and triangulated on by Jesse (1896), it is now well established that NLC are very tenuous ice clouds that appear in a thin layer in the upper mesosphere at ~82 km, making them the highest clouds on earth (Gadsden and Schröder, 1989). They are the result of intense dynamically driven cooling that occurs in the high-latitude summertime mesopause region when the ambient temperature falls well below 150 K, allowing microscopic ice crystals to nucleate and grow to observable sizes (Gadsden, 1981; Jensen and Thomas, 1988; Thomas, 1991). NLC are almost exclusively observed during the months May-August in the Northern Hemisphere and November-February in the Southern Hemisphere (with a typical season of ~90 days). As they are optically thin, NLC can only be seen from the ground during the hours of twilight, when the observer and atmosphere below them are in darkness, while the clouds themselves remain sunlit. This condition occurs for solar depression angles of ~6°–15°. Sightings of NLC over the past 100 years

M.J. Taylor (✉)
Center for Atmospheric and Space Science, Utah State University, Logan, UT, USA
e-mail: Mike.Taylor@usu.edu

M.A. Abdu, D. Pancheva (eds.), A. Bhattacharyya (Coed.), *Aeronomy of the Earth's Atmosphere and Ionosphere*,
IAGA Special Sopron Book Series 2, DOI 10.1007/978-94-007-0326-1_7,

have usually been confined to the latitude range ~50°–65° (Gadsden and Schröder, 1989). The lower latitude limit representing a statistical bound to the clouds southerly extension, while perpetual daylight during the summer season creates a natural upper latitude limit.

NLC are truly striking in their appearance, usually containing a variety of luminous silvery blue or opal white cloud structures that glow in the twilight sky. The very distinctive and regular wave patterns that often occur in NLC were commented on right from their first observations and resemble glowing cirrostratus clouds. Figure 7.1 shows an example of extensive NLC waves, termed "Type II Bands" (WMO, 1970; Gadsden and Parviainen, 1995), imaged from Stockholm, Sweden (59.3°N, 18.1°E) on 29/30 July, 2007. The picture was taken using a digital SLR camera with a medium field of view (53° horizontal × 41° vertical), aimed at low elevation (~20°) toward the northern twilit sky. Two geographically extensive sets of NLC bands are seen intersecting each other almost at right angles. The most prominent pattern contained over 10 wave crests orientated near E-W and was observed to propagate towards the north. The second wave event was fainter (primarily due to perspective) and appeared as several diverging wave crests radiating out of the northern horizon. Similar type wave patterns have been imaged in the Mesospheric and Lower Thermospheric (MLT) nightglow emissions (e.g. Taylor et al., 1997; Nakamura et al., 2003; Pautet et al., 2005; Nielsen et al., 2006; Suzuki et al., 2007) and have been shown to be associated with freely propagating or ducted short-period (<1 h) gravity waves, most likely of tropospheric origin (e.g. Witt, 1962; Taylor and Hapgood, 1988; Yue et al., 2009). Figure 7.2 shows a "satellite view" computer projection of the wave data of Fig. 7.1 imaged over central Sweden assuming an NLC height of 83 km. This analysis technique has been developed for measuring wave events imaged in the MLT nightglow emissions and has recently been adapted for detailed studies of waves in NLC (Pautet et al., 2010). The data are plotted over the elevation range 3.0°–40.5° (corresponding to a ground range of ~700 km) and clearly show the two intersecting wave patterns which exhibited similar horizontal wavelengths of 24 and 30 km. The velocities of the waves are measured from time sequences of such maps and are shown by the two arrows indicating their observed directions of motion relative to the ground which were 22 m/s at 16°N, and 16 m/s at 75°N, respectively.

Fig. 7.1 Example photograph of Type II band patterns imaged in NLC from Stockholm, Sweden (59.3°N, 18.1°E) on 29/30 July 2007. Two intersecting band events are evident in this low elevation image which has a field of view of 53° horizontal ×41° vertical

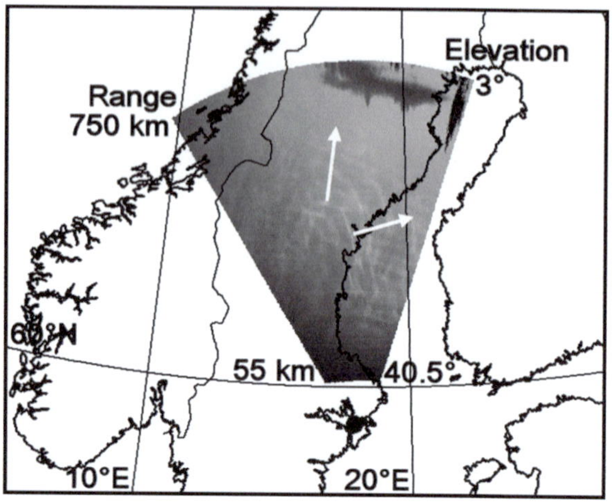

Fig. 7.2 A "topside" computer projection view of the NLC (Fig. 7.1) imaged over central Sweden, assuming a cloud height of 83 km. This technique clearly shows the two intersecting band patterns while the arrows depict their observed directions of motion. (Note the dark structure at the top of the picture is due to meteorological cloud)

Although there has been a remarkable increase in NLC research as daytime lidar measurements and radar soundings (termed Polar Mesospheric Summer Echoes, PMSE) have been developed (e.g. von Zahn and Bremer, 1999; Gerding et al., 2007; Rapp and Lubken, 2004; Nicolls et al., 2007; Taylor et al., 2009), there have been surprisingly few studies focused on NLC wave characteristics. Fogle and Haurwitz (1969) used available data in the late 1960's literature to summarize the properties of Type II Bands (and Type III Billows which are much smaller-scale, transient wave patterns associated with instabilities in the upper mesosphere, e.g., Fritts et al., 1993). Although the number of events reported on was relatively few, they found that bands exhibited lifetimes of typically a few hours, horizontal wavelengths of 10–70 km and phase speeds of 10–60 m/s. Furthermore, they all exhibited predominantly northward motion, often with some small zonal (E-W) component of motion. Table 7.1 summarizes these results and includes new measurements from Dalin et al. (2004) of a well-defined gravity wave event that was simultaneously measured by radar, and over 30 band events imaged from Scandinavia and reported recently by Pautet et al. (2010).

Since the early 1970's, opportune observations from a number of satellites have significantly extended the realm of ground-based NLC studies (which has been mainly limited to the Northern Hemisphere). These photometric observations have revealed the frequent presence of mesospheric ice clouds, termed Polar Mesospheric Clouds (PMC) in the perpetually sunlit summer polar regions, and often encompassing the whole polar cap region (e. g., Carbary et al., 2000; Deland et al., 2006; Evans et al., 1995). For example, the SBUV instrument on the NOAA fleet of satellites has continuously monitored PMC over the past 30 years and has shown that the occurrence and brightness of the clouds has been steadily growing. This is a new result of key interest for investigating long-term changes in the mesosphere (Deland et al., 2006). Apart from sketches and occasional photographs of NLC observed from space by Astronauts and Cosmonauts (Gadsden and Schröder, 1989), the first image measurements of NLC were obtained by the WINDII limb viewing interferometer aboard the UARS (Upper Atmosphere Research Satellite) which recorded patchy structures at visible wavelengths in the cloud layer as observed in the earth's limb at ~83 km on 17 July, 1993 (Evans et al., 1995). However, the instrument resolution was insufficient to identify any wave properties.

The UVISI (Ultraviolet and Visible Imaging and Spectrographic Imaging) instrument on the MSX (Mid-Course Space Experiment) satellite provided the first glimpse of large-scale wave structure in PMC over the summer polar regions. Video images were obtained on a limited number of transpolar orbits (26), mainly over Antarctica, providing clear evidence for "transverse wave structures" in the PMCs imaged in the earth's sub-limb. Their result suggested a predominance for large-scale waves of ~500–1000 km (Carbary et al., 2000) but later analyses of the image data also revealed structures with scales less that 100 km, more akin to those evident in ground-based NLC data (Carbary et al., 2003). However, it was not until the launch of the NASA Aeronomy of Ice in the Mesosphere (AIM) satellite in April, 2007 that detailed image measurements of the PMC structures within the summer polar region were regularly obtained. AIM was developed to investigate the formation and variability of PMC, and a key instrument was included: the Cloud Imaging and Particle Size (CIPS) Experiment, which is a high-resolution UV imager able to take large-field snapshots of the nadir PMC field as the satellite traversed over the summer polar region. Investigations of a variety of structures evident

Table 7.1 Summary of NLC wave measurements reported by Haurwitz and Fogle (1969) and new results by Dalin et al. (2004) and Pautet et al. (2010)

References	Date	No. of events	λ Range (km)	λ mean (km)	Orientation of crest normal ϕ (deg.)	Average observed speed (m/s)	Direction of motion
Witt (1962)	10 Aug. 1958	–	30–75	–	–	10–15	NE
Grishin (1967)	20 July 1954	–	20–30	–	–	30	NW
Haurwitz and Fogle (1969)	29 June 1965	2	51–55	53	320	–	–
	15 July 1965	6	11–57	27	41	31	NE
	26 July 1965	5	10–23	15	345	57	WNW
	27 July 1965	3	10–26	18	0	–	–
Dalin et al. (2004)	11 Aug. 2000	1	33–47	40	348	16	NNW
Pautet et al. (2010)	2004–2008	30	10–50	25	24	27	NNE

in PMCs have recently been reported by Rusch et al. (2009) and Russell et al. (2009) using the high quality CIPS/AIM image data. In particular, Chandran et al. (2009) have utilized the first season 2007 CIPS data to make an initial investigation of the seasonal characteristics of gravity waves seen in the PMC layer. In this paper we build and extend on their measurement capabilities, using a two-dimensional spectral analysis to investigate the statistical properties of quasi-monochromatic wave events evident in the CIPS data set. For this study we have focused on mid-season data from July 2007 yielding over 450 events and compared their spatial properties with Type II Band events commonly observed at lower latitudes in the NLC zone.

7.2 Instrumentation

The Cloud Imaging and Particle Size (CIPS) experiment on the AIM satellite is a panoramic UV imager designed to measure radiance and morphology of PMC near 83 km altitude over a wide range of scattering angles (Russell et al., 2009). The CIPS instrument consists of an array of four near identical intensified CCD cameras each fitted with a narrow band (~15 nm FWHM) filter centered at 265 nm. The cameras are arranged in a "cross pattern" with overlapping fields resulting in a wide angle view of 120° (along orbit track) by 80° (across orbit track). Images of the PMC field are obtained by the fore and aft looking cameras using integration times of 0.73 s, while the two side by side nadir cameras operate in a time-delayed integration mode with an effective integration time of 0.75 s (Rusch et al., 2009; McClintock et al., 2009).

CIPS provides "snapshot" images of PMCs with a pixel spatial resolution of ~2 km in the nadir and about 5 km at the edges of the forward and aft cameras. The combination of each set of four images is termed a "scene". CIPS records scenes of cloud radiance in the summer hemisphere from the terminator to ~40° latitude along the sunlit portion of the orbit. The instrument takes a new set of snap shots every 43 s resulting in ~26 scenes per orbit. These data are then combined to create a ~1000 km wide by ~9000 km long swath with a pixel resolution 5 × 5 km, summarizing the PMC field for each orbital pass. AIM's near-polar orbit and the wide cross-track field of view of CIPS creates 15 PMC swaths per day that overlap at latitudes higher than ~70°, and the entire polar cap region (up to ~85° latitude) is mapped daily. The exceptionally high spatial resolution of CIPS provides a major improvement in the horizontal resolution of previous PMC measurements obtained with nadir-viewing space experiments such as by the Solar Backscatter Ultra Violet (SBUV) instruments on the NOAA satellite series, which has a pixel footprint of 150 × 150 km (DeLand et al., 2006; Rusch et al., 2009). Further details of the CIPS instrument and its operation are given in McClintock et al. (2009).

7.3 Observations and Data Analysis

Figure 7.3 is a typical example of the CIPS PMC data set and shows part of an orbital swath obtained on 24 July (orbit 1336) which started at ~11:04 UT. This is a CIPS/AIM level 4 data product and plots the cloud albedo in 5 by 5 km bins. The swath shown

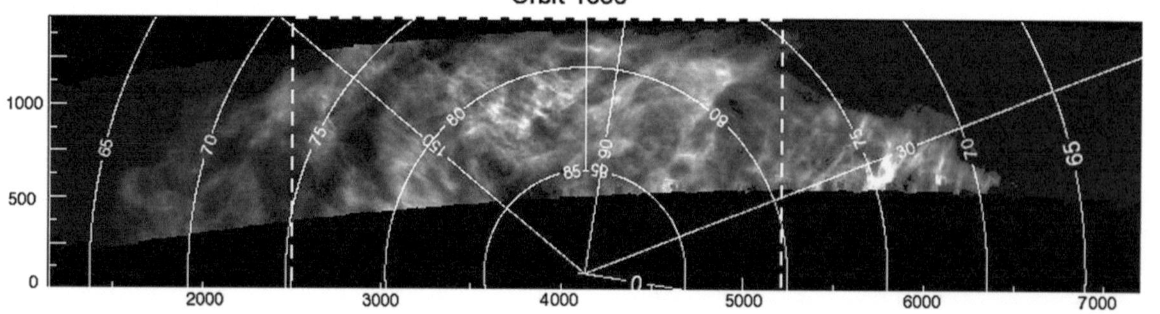

Fig. 7.3 A typical example of the CIPS PMC image data showing part of an orbital swath obtained on 24 July (orbit 1336) at ~11 UT. This is a CIPS/AIM level 4 data product and plots the cloud albedo in 5 km by 5 km bins. The swath shown is ~7000 km long and the satellite motion is from *left* to *right*. A geographic grid is superimposed onto the swath and shows a continuous PMC field over northern Alaska, the Arctic Ocean and northern Russia as the satellite traversed over the northern hemisphere summer polar region. The *dashed vertical lines* mark the 2400 km selected PMC region used in this wave study

is ~7000 km long and the satellite motion is from left to right. A geographic grid is superimposed onto the swath which shows a continuous PMC layer as the satellite traversed over northern Alaska, the Arctic Ocean and northern Russia. The large spatial extent of the PMC throughout this region is typical for the July period and a variety of PMC structures are evident within the swath extending to the lowest latitudes that CIPS can effectively detect ~68° (Bailey et al., 2009; Merkel et al., 2009). The dashed vertical lines on the figure bound a selected ~2400 km long region of the PMC field which contains a concentration of complex structures including some quasi-monochromatic waves. This region spans the latitude range ~75°N–85°N and for this investigation, we have focused our analysis on wave data that occur only within this area. Using this data set, the limiting resolution of the CIPS imagery for investigating periodic structures in the PMC radiance is ~10 km, but in practice we have restricted our analyses to waves with horizontal scales >20 km.

To investigate prominent wave events present in the PMC, we have used a robust method developed to quantify ground-based measurements of similar-type wave patterns imaged in the MLT nightglow emissions which yield accurate information on their horizontal wavelengths, observed phase velocities and apparent wave periods (e. g., Taylor and Garcia, 1995; Coble et al., 1998; Garcia et al., 1997; Pautet et al., 2005; Taylor et al., 1997). Due to the rapid motion of the AIM spacecraft (7 km/s), we are not able to measure the wave motion and our PMC analyses are limited to determining the horizontal wavelength, wave pattern orientation and geographical location of each event. The Level 2 data were first processed to isolate the region of interest (i.e., spanning the latitude range ~75°N–85°N) for each orbital swath. The data were then visually inspected to identify areas within each region that contained prominent quasi-monochromatic waves.

This is illustrated in Fig. 7.4 which shows selected regions from three orbital swaths (orbits 1141, 1172 and 1336), observed on 11 July, 13 July and 24 July, 2007 respectively, including that of Fig. 7.3. In each example, the white boxes (640 × 640 km) identify the locations of several well-defined wave patterns analyzed in this study.

In many cases, the patterns were complex and more than one wave event was measured within this large sample region as illustrated in Fig. 7.4c. In other cases the wave events were masked by additional dynamical process and accurate measurement of the wave parameters were not possible.

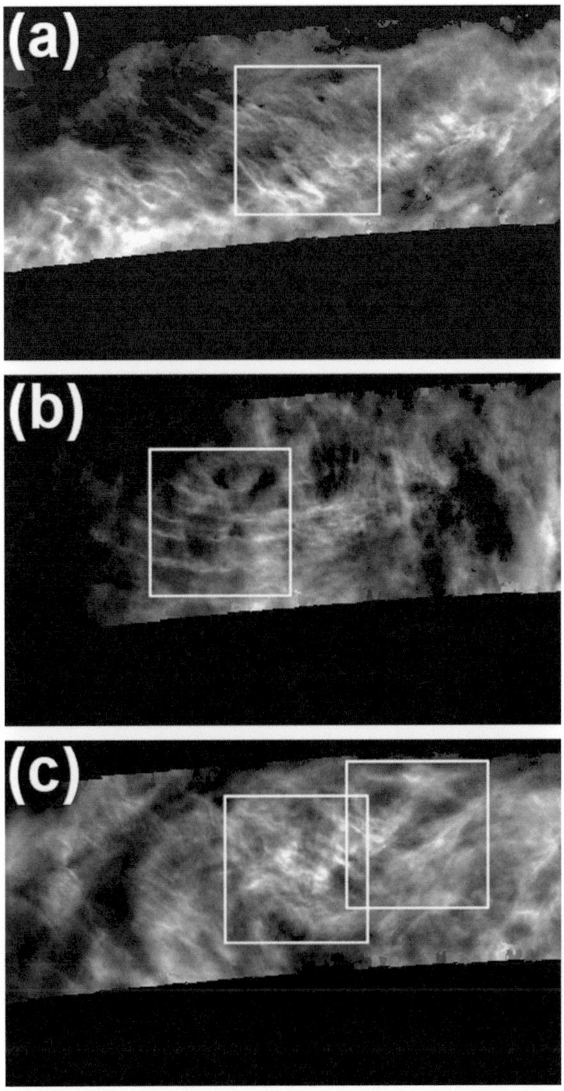

Fig. 7.4 Selected regions from three PMC swaths (orbit 1141, 1172, 1336) including that of Fig. 7.3. The white boxes (640 × 640 km) identify several well-defined wave patterns analyzed in this study. In many cases, the patterns were complex and more than one wave event was identified within this large sample region. This is illustrated in (c) where the boxes identify adjacent small-scale (*left*) and large-scale (*right*) wave events

Figure 7.5 illustrates our method of wave analysis applied to the large-scale wave structures identified in Fig. 7.4c. A standard 2-dimensional (ambiguous) Fast Fourier Transform (FFT) process was applied to the data within this selected region to determine the 2-D power spectrum. The results are shown in Fig. 7.5a,

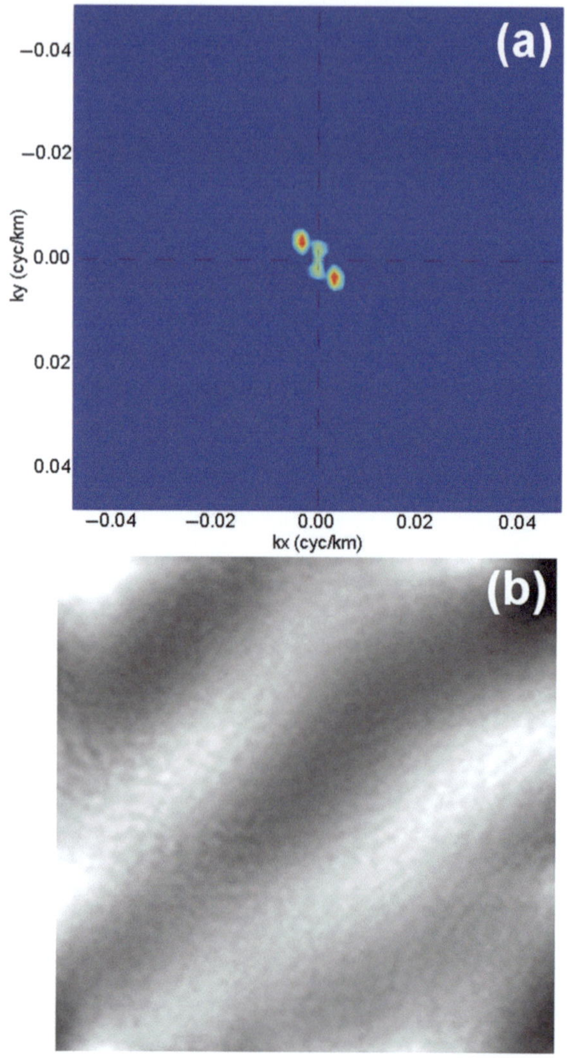

Fig. 7.5 Illustrates our method of wave analysis and shows (**a**) an example 2-D FFT power spectrum derived from the large-scale PMC wave identified in Fig. 7.4c, and characterized by two symmetric peaks in the power spectrum, and (**b**) the spatial properties of the reconstructed wave for direct comparison with the original event

where the prominent symmetric peaks identify the dominant wave frequency. The horizontal wavelength of this event and its orientation are given by the spatial location of the peaks in the power spectrum. In this case the horizontal wavelength was determined to be 207 ± 5 km and its direction of motion 126°±5° (with 180° ambiguity). To ensure we have identified the correct wave event, the peaks in the spectrum were then isolated and an inverse FFT was applied to the reduce spectrum. The results are shown in Fig. 7.5b, which shows the spatial characteristics of the wave extracted from the complex background. Comparison of this reconstructed wave and the original wave identified visually in the data of Fig. 7.4c confirms the correct measurements of this event and its spatial characteristics. The results of five events identified in Fig. 7.4 are listed in Table 7.2. The uncertainties in their horizontal wavelengths were ± 5 km and wave orientation ± 5°. In general, waves evident in the CIPS data varied considerably in their horizontal scales and geographical extent, and in most cases were near linear in nature. However, as shown in Fig. 7.4b, sometimes the wave crests were strongly curved which can help considerably in the determination of their sources (e.g., Dewan et al., 1998; Suzuki et al., 2007; Taylor and Hapgood, 1988; Sentman et al., 2003; Yue et al., 2009).

7.4 Results

Using this analysis method, we have identified and measured over 450 prominent quasi-monochromatic wave events in the CIPS UV imagery of PMC over the northern polar region during July 2007. As described earlier, only the central section of the CIPS PMC swath was used in this study. Nevertheless, this region covers an exceptional large geographic area of ~1.5 × 10^6 km^2, a latitude range of ~75°–85°, and spans all longitudes.

Table 7.2 Wave parameters derived from the CIPS data shown in Fig.7.5

Event	Date, 2007	Latitude	Longitude	λ (±5 km)	ϕ (±5°)
1	July 11 (orbit 1141)	81	168	146	124
2				41	116
3	July 13 (orbit 1172)	79	151	128	80
4	July 24 (orbit 1336)	82	123	37	46
5		81	98	207	126

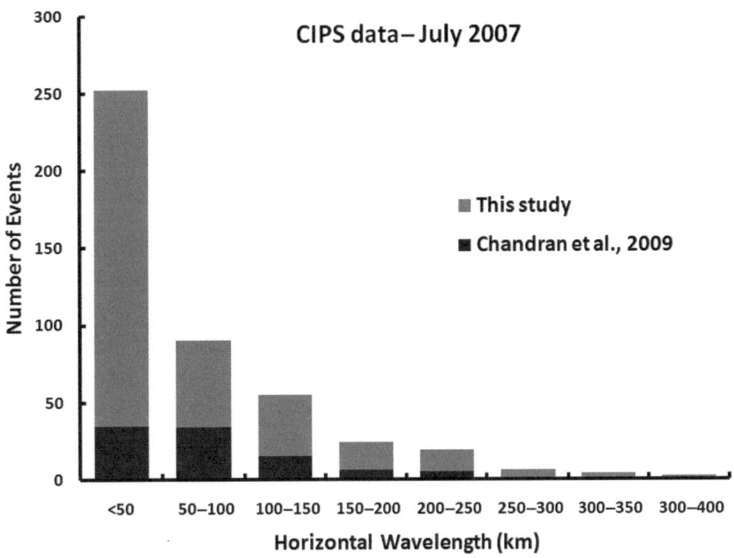

Fig. 7.6 Histogram plot of 452 quasi-monochromatic wave events as function of their measured horizontal wavelength. The data have been summed into 50 km wide bins and span the wavelength range <50–400 km. The distribution peaks at the shortest wavelength bin and over 75% of the events exhibited wavelengths <100 km. Adjacent larger-scale wave bins contained a successively smaller number of events. The July results from Chandran et al. (2009) are superimposed onto this plot for comparison

Figure 7.6 (light grey) is a histogram plot of the number of wave events versus their measured horizontal wavelengths. The data have been summed into 50 km wide bins and spans the wavelength range <50–400 km. The distribution is characterized by a strong peak at the shortest wavelength bin, and over 75% of the events exhibited wavelengths <100 km. Adjacent larger wavelength bins contain a successively smaller number of events that decrease almost exponentially with increasing wavelength. Thus, while larger-scale waves up to ~400 km were readily detectable in the PMC data sets, they were much less common than the smaller scale wave events (as is also apparent visually in the example CIPS imagery in Fig. 7.4). Also plotted in Fig. 7.6 (dark grey) are the 2007 results of Chandran et al. (2009) who used a wavelet technique applied to one dimensional intensity scans across selected wave events to perform an initial investigation of seasonal variability of wave evident in the CIPS data set. Although the number of events detected in July was comparatively few (a total of 102), their distribution also revealed a strong preference for wavelengths <100 km, with a peak occurrence around 45 km.

To investigate day-to-day variability in wave occurrence, Fig. 7.7 plots the number of events as a function of day number during July. The daily bins comprise wave measurements from 15 orbits spanning all longitudes, and each bin has been divided to show the relative contribution of small-scale waves (λ<100 km, black) to the larger-scale waves ($\lambda \geq 100$ km, white). Note, no CIPS data are currently available for days 193 and 209. A range of both large and small-scale waves were detected on almost every day, with a dominant contribution of smaller-scale waves, as indicated in Fig. 7.6. During the first ~third of the month (days 182–191) the total number of events detected was typically ~11 per day, with a variable, but much smaller contribution (~3 events per day) from the larger scale ($\geq$100 km) waves. Thereafter, the total number of events increased to typically ~20 per day (up to day

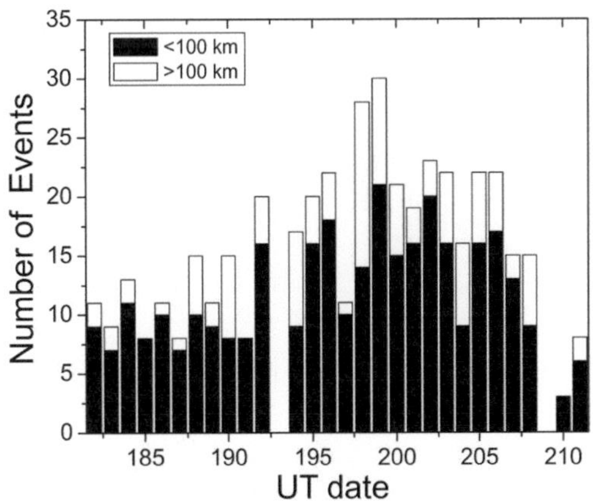

Fig. 7.7 The number of wave events as a function of day number during July. The data have been divided to show the dominant contribution of the smaller scale waves <100 km (*black*) relative to the number of larger wave events $\geq$100 km events (*white*) Considerably more events were measured during the later 2/3 of the month

~208) and exhibited more day-to-day variability. A slight increase in the number of larger-scale waves (~5 events/day) was also detected but the ratio of large to small-scale waves remained approximately the same. This suggests significant development of the gravity wave activity (over all scales) during the course of July. However, these data are for only 1 month and we do not yet know if such variability is normal.

The other key wave parameter that we are able to extract from the CIPS PMC imagery is the orientation of the wave crests for each of the identified events.

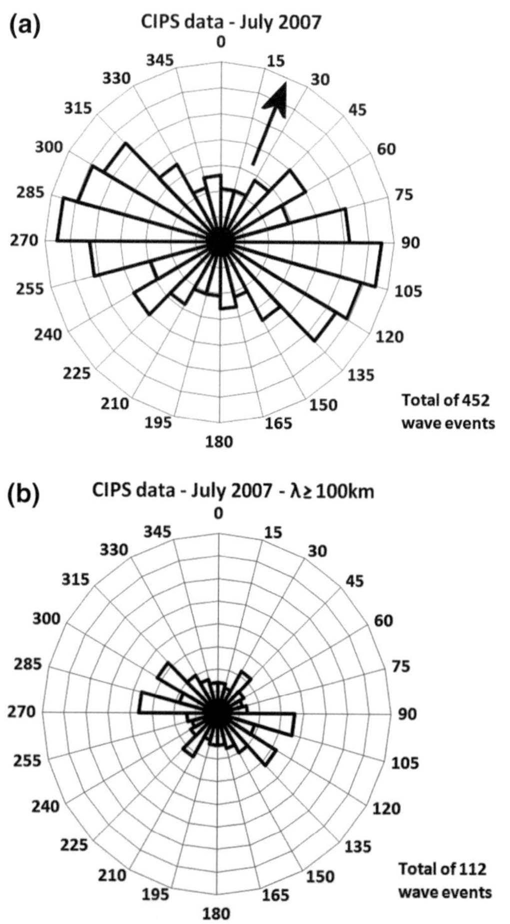

Fig. 7.8 Azimuthal distributions showing the normal direction to each wave event summed into 15° wide bins. There is a 180° ambiguity in the direction of motion of the wave events (as evident in the symmetry of these plots). All 452 wave events are shown in (**a**) while (**b**) shows directionality of the 112 wave events ($\lambda \geq 100$ km). Both plots indicate marked anisotropy suggesting a strong preference for zonal wave propagation. The arrow indicates the dominant direction of motion for NLC type II bands (Pautet et al., 2010)

Figure 7.8a, b plots the azimuthal distributions showing the normal (i.e. perpendicular) direction to each wave event summed into 15° wide bins. This is well within the estimated uncertainty of the orientation measurement (± 5°) and provides useful information on the propagation of the waves. As mentioned earlier, due to the rapid motion of the spacecraft, we are not able to measure the actual motion of the waves, so there is a 180° ambiguity in their inferred direction of motion (as evident in the symmetry of these plots). Figure 7.8a shows the azimuthal distribution for all 452 wave events. Despite the ambiguity in the wave headings, these data reveal a strong and persistent anisotropy in the wave propagation during July, with a clear preference for zonal E-W (or W-E) wave motion. This plot is dominated by the azimuthal contributions from the smaller-scale (λ<100 km) waves, however, a similar shaped distribution is seen in Fig. 7.8b which plots only the larger-scale ($\lambda \geq 100$ km) waves.

7.5 Discussion

The CIPS PMC imagery is a remarkably valuable data set for investigating structure and dynamics in the high latitude summer mesopause region. While MLT airglow measurements continue to provide a good method for quantifying gravity wave activity and dynamics mainly at lower latitudes, they are confined to night sky measurements and observations at high latitudes have been hindered by auroral emissions. In particular, the large panoramic field of view of CIPS coupled together with its high spatial resolution provides a much needed capability for investigating the occurrence and properties of a broad spectrum of wave scales.

Our results show a strong preference for smaller-scale structures with 75% <100 km. Chandran et al. (2009) also reported a higher frequency of smaller-scale waves shown in Fig. 7.6. To further investigate the distribution of waves <100 km, Fig. 7.9a re-plots our data of Fig. 7.6 for horizontal wavelengths from 20 to 70 km, for direct comparison with recent airglow and NLC data. A total of 306 events were summed into 5 km wide bins, revealing a strong peak around 25–40 km. About 75% of these smaller-scale events have wavelengths <50 km. Note, wavelengths

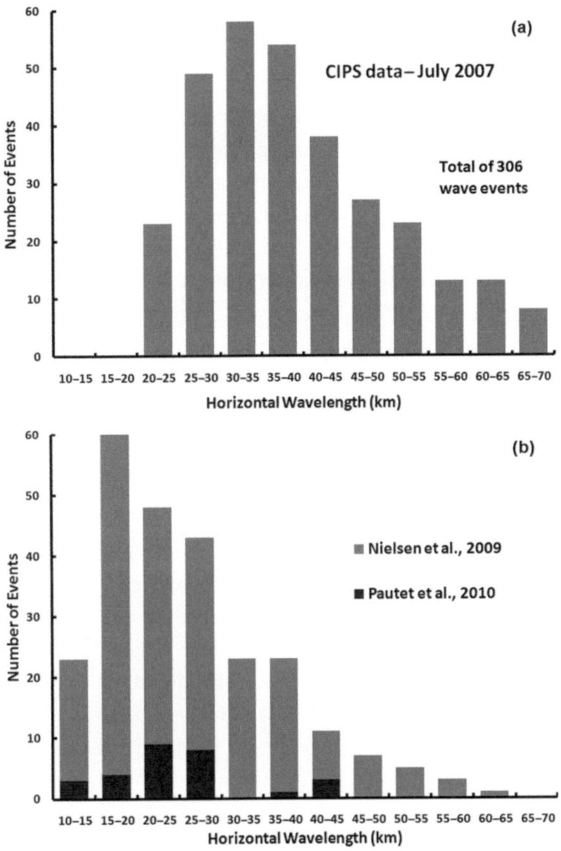

Fig. 7.9 Histogram plots comparing (**a**) the distribution of wave events (λ < 70 km) measured in the CIPS data set compared with (**b**) wavelength distribution of small-scale gravity waves measured at high latitudes by Nielsen et al. (2009) in OH airglow during the winter months over Antarctica, and Pautet et al. (2010) in NLC over Northern Scandinavia

of <20 km represent a practical limit for the CIPS wave measurements (using the Level 2 data product with 5 km/pixel resolution).

Figure 7.9b plots results of small-scale gravity wave measurements at high latitudes recently reported by Nielsen et al. (2009) and Pautet et al. (2010). The measurements by Nielsen et al. comprise over 200 events and are one of the largest data sets on quasi-monochromatic gravity wave measurements at polar latitudes. The measurements were made from Halley Station, Antarctica (76°S) at a similar latitude to our PMC wave measurements, but during the southern hemisphere winter months (May-Sept, 2000–2001). The distribution peaks around 15–30 km, slightly shorter than the CIPS data, and exhibits a decreasing number of events at successively larger horizontal wavelengths. The dark grey histogram plot superimposed on the Nielsen et al. data shows the distribution of 30 wave events measured in NLC by Pautet et al. (2010). The observations were made from Stockholm, Sweden (59°N) during summer time 2004–2008 which overlap with the CIPS measurements. This distribution, although limited, also shows a preponderance for waves in the 20–30 km range. (Note, the dip in the distribution around 30–40 km wavelength was most probably due to the relatively low number of events measured.) These results suggest that the spatial resolution of the CIPS data has most probably caused us to underestimate the shortest gravity wave contributions that are captured in ground-based measurements of NLC and mesospheric nightglow data. In contrast, ground-based airglow and NLC measurements (particularly of NLC which are confined to the twilit portion of the sky) are more limited in their field of view than CIPS and larger-scale gravity waves although present are not readily measured. Thus, within the observational constraints of the NLC, PMC and airglow measurements, they all exhibit the same general distribution for the small-scale waves, with the majority of events <50 km wavelength.

The CIPS measurements of waves ≥100 km provide new information on much larger sale waves in the summer polar MLT region. These waves were readily detectable in the PMC data set and exhibited horizontal wavelengths up to ~400 km (limited by the width of the CIPS swath), and constituted ~25% of the total number of events. This said, they were often masked by the prolific abundance of the smaller-scale structures (as seen in Fig. 7.4). Carbary et al. (2000) reported somewhat larger scale waves (~500–1000 km) detected during several transpolar passes by the MSX satellite. Such wave scales may also be present in the CIPS orbital swaths but have yet to be investigated. Thus, the AIM mission has provided the first clear observations of a broad range of gravity waves in the summer polar mesosphere.

The 2-D spectral analysis provides important additional information on the orientation of the wave crests for each of the identified events from which the direction of motion can be inferred (with 180° uncertainty). The plots shown in Fig. 7.8a, b indicate the observed distributions of the normal to the wave crests. In (a), the azimuthal distribution of all July wave events (large and small) reveal a clear preference for near E-W (or W-E) wave propagation. As discussed earlier, this plot is dominated by the smaller-scale (λ<100 km) waves (~75% of events). Recent investigations of gravity wave propagation heading as observed in NLC (Pautet et al., 2010), have revealed a dominant northward motion (towards NNE) for the waves of similar horizontal wavelengths (which agrees well with previous NLC wave measurements in the literature as listed in Table 7.1). As pointed out by Pautet et al. (2010), this preferred direction of motion is consistent with other reports of short-period gravity waves observed in the MLT airglow emissions at lower latitudes which show poleward wave motion during the summer months. This preferred direction for the NLC waves is marked on Fig. 7.8a by the arrow which highlights the difference between the NLC and our PMC wave results. The origin of this difference is not yet clear, but it may be related to the difference in latitudes sampled in these studies. The NLC were imaged from Stockholm (59.3°N), while the PMC waves shown here were measured at a significantly higher latitude range from ~75°N–85°N.

By using the background wind field derived from the HWM-07 wind model (Drobs et al., 2008), potential filtering effects on freely propagating waves at NLC heights have been investigated by Pautet et al. (2010). They showed the limited meridional wind component (<10 m/s), at all altitudes below the NLC layer, was insufficient to block the observed strong northward propagating waves which exhibited phase speeds of typically 10–60 m/s. A similar investigation of the background wind field at 80°N using the HWM-07 model indicates an overall weaker wind field up to PMC height (zonal <20 m/s, meridional <10 m/s) suggesting that the winds are too small to explain any changes in wave orientation for gravity waves propagating at these high latitudes. This assumes a similar range of wave phase speeds to those seen in the NLC. The origin of the strong preference for zonally progressing waves at PMC heights therefore remains uncertain. However, it has been noted by many observers, even from the earliest NLC observations (e.g., Jesse, 1896; Witt, 1962), that prominent band type waves tend to propagate against the overall motion of NLC cloud field. Such waves propagating upstream would be Doppler shifted to shorter intrinsic periods while their vertical wavelengths and perturbation amplitudes would increase (potentially leading to instability and wave breaking). The net effect would be to enhance their visibility over other waves propagating in different directions. The zonally averaged background wind at PMC altitudes (~82 km) was ~22 m/s at azimuth of ~300°N (i.e., towards WNW). This aligns well with the observed, near zonal wave motion, as indicated in Fig. 7.8. Assuming these waves also tended to propagate into the background wind flow, their motions were most probably towards ESE (azimuth ~115°). This hypothesis is consistent with our PMC wave observations (both large and small-scale) but remains speculative at this time. We are currently investigating the potential to estimate wave velocities for some of the larger-scale, long-lived wave events that are detectable in adjacent orbits separated by ~1.5 h (e.g. Chandran et al., 2009). This will not be possible for the short events which evolve significantly from orbit-to-orbit.

To further investigate reported longitudinal variability in the occurrence of the waves, Fig. 7.10 plots the distribution of July wave events (of all scale sizes) summed into 15° wide bins versus their longitude of detection over the latitude range 70°N to 85°N. On average, the wave data were evenly distributed in longitude with ~20 events in the sectors corresponding to North America (~120°W) and Asia (~120°E). However in the European/North Atlantic

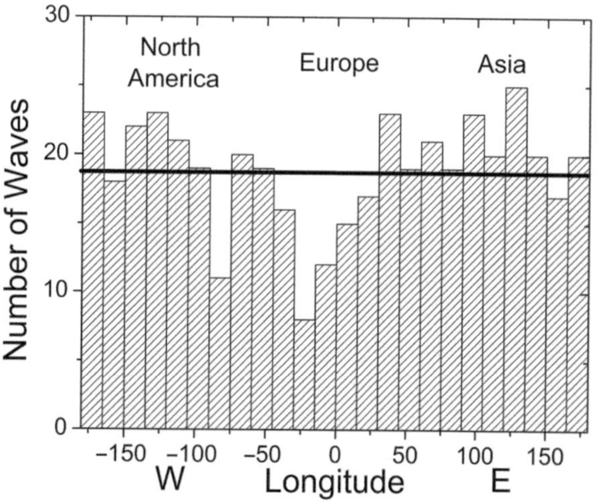

Fig. 7.10 Distribution of July wave events (of all scale sizes) summed into 15° wide bins versus their longitude of detection over the latitude range 70–85°N. On average, the wave data were evenly distributed in longitude with ~20 events in the sectors corresponding to North America (~120°W) and Asia (~120°E). A reduction in the number of wave events is evident in the European/North Atlantic sector (~0°). The horizontal line shows the average number of waves

sector (~0° longitude) there appears to be a reduction in the number of events in 3–4 adjacent bins down to about ~8–15 events/bin (i.e. by as much a ~50%). This result complements the measurements of Chandran et al. (2009) who have also examined the seasonal variability of gravity waves for 2007. Over the 3 month period (June-August) they identified 240 wave events of which 102 events were from July. They reported marked concentrations of wave events over Greenland (45°W), Eastern Canada (135°W) and over the Arctic Ocean (120°E) just north of central Siberia. These regions correspond well to our results of sustained wave activity as shown in Fig. 7.10. Importantly, they reported only one event over the European/North Atlantic sector in June and none in July or August. In comparison, our data of Fig. 7.10 only suggest evidence of reduced quasi-monochromatic wave activity at this longitude range. They also detected very little wave activity at 180° longitude where our analyses indicated equally high wave occurrences as observed over North America and Asia. Thus, both of these studies, which used different analysis techniques to independently measure the wave events, have identified reduced polar (75°–85° latitude) PMC wave activity in the ~0° longitude sector (at least during the July period). However, it remains to be seen if this feature is present throughout the summer 2007 season (or re-occurs in other seasons). The lack of agreement between our analyses at 180° longitude may be due to the lower number of wave events reported in the initial study presented by Chandran et al. (2009).

Finally, this study only identified prominent quasi-monochromatic waves, but still resulted in over 450 events in July. While other wave events were also evident, they were often masked by other dynamical process and accurate measurements of their wave parameters were not made. These results therefore underestimate the actual wave content in the CIPS data but are nevertheless a good measure of the wealth, diversity and characteristics of the wave forms present in the summer polar mesosphere.

7.6 Summary

The CIPS data set is an exceptionally high quality, growing archive of considerable importance for quantifying gravity wave properties and variability in the northern and southern polar summer mesopause regions. Key results of this limited gravity wave study (July 2007 period) build on an initial seasonal investigation of quasi-monochromatic events by Chandran et al. (2009). The PMC field contained a broad spectrum of gravity waves spanning horizontal wavelengths from at least 20 to 400 km, with the highest occurrence frequency for the smallest scale waves (<50 km), but substantial evidence for larger-scale (>100 km) wave activity. The direction of motion of the waves (both large and small-scale) deduced from their orientations (with 180° ambiguity) was predominantly zonal (with a small meridional component), and differed significantly from recent NLC Type II band measurements which were dominated by strong near poleward wave motions. Further evidence was found for a possible reduction in gravity wave activity over the Europe/North Atlantic sector during the July 2007 period, as compared to other longitudes.

Comparison of our results with the HWM wind model suggests that the most prominent waves reported in this study may have propagated mainly into the background winds at the PMC level, which would have enhanced their vertical scales and hence their detection by CIPS. If so, we then suggest that the dominant wave motions over the latitude range ~75°N–85°N were mainly towards the ~ESE. This hypothesis, remains to be tested and further analysis of the extended CIPS data set (2007-to date) is in progress using a more comprehensive spectral analysis method.

Acknowledgements We are most grateful to the NASA Small Explorer program for supporting the development of the AIM satellite mission under contract NAS5-03132 and for providing the support for this research. We would also like to acknowledge the efforts of the entire development, engineering, science and operation teams. The USU analysis of the CIPS data was supported via a sub-contract from Hampton University.

References

Bailey, SM, Thomas GE, Rusch DW, Merkel AW, Jeppesen C, Carstens JN, Randall CE, McClintock WE, Russell JM III (2009) Phase functions of polar mesospheric cloud ice as observed by the CIPS instrument on the AIM satellite. J Atmos Solar-Terr Phys. doi:10.1016/j.jastp.2008.09.039

Carbary JF, Morrison D, Romick GJ (2000) Transpolar structure of polar mesospheric clouds. J Geophys Res 105:24,763–24,769

Carbary JF, Morrison D, Romick GJ (2003) Maps of polar mesospheric clouds. J Geophys Res 108(D8):8446. doi:10.1029/2002JD002255
Chandran A, Rusch D, Palo SE, Thomas GE, Taylor M (2009) Gravity wave observation from the cloud imaging and particle size (CIPS) experiment on the AIM spacecraft. J Atmos Solar-Terr Phys. doi:10.1016/j.jastp.2008.09.041
Coble MR, Papen GC, Gardner CS (1998) Computing two-dimensional unambiguous horizontal wavenumber spectra from OH airglow images. IEEE Trans Geosci Remote Sens 36(2):368–382
Dalin P, Kirkwood S, Moström A, Stebel K, Hoffmann P, Singer W (2004) A case study of gravity waves in noctilucent clouds. Ann Geophys 22:1875–1884
Deland MT, Shettle EP, Thomas GE, Olivero JJ (2006) A quarter-century of satellite polar mesospheric cloud observations. J Atmos Solar-Terr Phys 68, 9–29
Dewan EM, Picard RH, O'Neil RR, Gardiner HA, Gibson J, Mill JD, Richards E, Kendra M, Gallery WO (1998), MSX satellite observations of thunderstorm-generated gravity waves in mid-wave infrared images of the upper stratosphere. Geophys Res Lett 25:939–942. doi:10.1029/98GL00640
Drobs DP, Emmert JT, Crowley G, Picone JM, Shepherd GG, Skinner W, Hays P, Niciejewski RJ, Larsen M, She CY, Meriwether JW, Hernandez G, Jarvis MJ, Sipler DP, Tepley CA, O'Brien MS, Bowman JR, Wu Q, Murayama Y, Kawamura S, Reid IM, Vincent RA (2008) An empirical model of the Earth's horizontal wind fields: HWM07. J Geophys Res 113:A12304. doi:10.1029/2008JA013668
Evans WFJ, Laframboise LR, Sine KR, Wiens RH, Shepherd GG (1995) Observation of polar mesospheric clouds in summer, 1993 by the WINDII instrument on UARS. Geophys Res Lett 22:793–2796
Fogle B, Haurwitz B (1969) Wave forms in noctilucent clouds. Deep-Sea Res 16:85–95
Fritts DC, Isler JR, Thomas GE, Andreassen Ø (1993) Wave breaking signatures in noctilucent clouds. Geophys Res Lett 20:2039–2042
Gadsden, M. (1981) The silver-blue cloudlets again: nucleation and growth of ice in the mesosphere. Planet Space Sci 29:1079–1087
Gadsden M, Parviainen P (1995) Observing noctilucent clouds. International Association of Geomagnetism & Aeronomy, pp 5
Gadsden M, Schröder W (1989) Noctilucent clouds. Springer, Berlin
Garcia FJ, Taylor MJ, Kelley MC (1997) Two-dimensional spectral analysis of mesospheric airglow image data. Appl Opt 36:7374–7385
Gerding M, Hoffner J, Rauthe M, Singer W, Zecha M, Lubken F-J (2007) Simultaneous observations of noctilucent clouds, mesospheric summer echoes, and temperature at a midlatitude station (54°N). J Geophys Res 112: D12111, doi:10.1029/2006JD008135
Grishin NI (1967) Morphologicheskie Issledovaniya i Primda Serebristikh Oblakov. In "Nabludeniya Serebristikh Oblakov". Izdatelstvo Nauka, Moscow, pp 5–34
Jesse O (1896) Die Ho¨he der leuchtenden nachtwolken. Astronomische Nachrichten 140:161–168
Jensen EJ, Thomas GE (1988) A growth-sedimentation model of polar mesospheric clouds: comparison with SME measurements. J Geophys Res 93:2461–2473
Leslie R (1885) Sky glows. Nature 32:245
Merkel, WA, Rusch DW, Palo SE, Russell JM III, Bailey SM (2009) Mesospheric planetary wave activity inferred from AIM-CIPS and TIMED-SABER for the northern summer 2007 PMC season. J Atmos Solar-Terr Phys. doi:10.1016/j.jasp.2006.05.01
McClintock W, Rusch DW, Thomas GE, Merkel AW, Lankton MR, Drake VA, Bailey SM, Russell JM III (2009) The cloud imaging and particle size experiment on the aeronomy of ice in the mesosphere mission: instrument concept, design, calibration, and on-orbit performance. J Atmos Solar-Terr Phys. doi:10.1016/j.jastp.2008.10.011
Nakamura TA, Tsuda T, Admiranto T, Achmad AG, Suranto E (2003) Mesospheric gravity waves over a tropical convective region observed by OH airglow imaging in Indonesia. Geophys Res Lett 30(17):1882
Nicolls MJ, Heinselman CJ, Hope EA, Ranjan S, Kelley MC, Kelly JD (2007) Imaging of polar mesosphere summer echoes with the 450 MHz Poker Flat advanced modular incoherent scatter radar. Geophys Res Lett 34:L20102
Nielsen K, Taylor MJ, Hibbins RE, Jarvis MJ (2009) Climatology of short-period mesospheric gravity waves over Halley, Antarctica (76°S, 27°W). J Atmos Solar-Terr Phys 71:991–1000
Nielsen K, Taylor MJ, Pautet P-D, Fritts DC, Mitchell N, Beldon C, Williams BP, Singer W, Schmidlin FJ, Goldberg RA (2006) Propagation of short-period gravity waves at high-latitudes during the MaCWAVE winter campaign. Ann Geophys 24:1227–1243 SRef_ID:1432-0576/ag/2006- 24-1227
Pautet P-D, Stegman J, Wrasse CM, Takahashi H, Taylor MJ (2010) Analysis of gravity waves structures visible in noctilucent cloud images, global perspectives on the aeronomy of the summer mesopause region, 8th international workshop on layered phenomena in the mesopause region. J Atmos Solar-Terr Phys. doi:10.1016/j.jastp.2010.06.001
Pautet P-D, Taylor MJ, Liu AZ, Swenson GR (2005) Climatology of short-period gravity waves observed over northern Australia during the Darwin area wave experiment (DAWEX) and their dominant source regions. J Geophys Res 110:D03S90
Rapp M, Lubken F-J (2004) Polar mesosphere summer echoes (PMSE): review of observations and current understanding. Atmos Chem Phys 4:2601–2633
Rusch DW, Thomas GE, McClintock W, Merkel AW, Bailey SM, Russell JM III, Randall CE, Jeppesen C, Callan M (2009) The cloud imaging and particle size experiment on the aeronomy of ice in the mesosphere mission: cloud morphology for the northern 2007 season. J Atmos Solar-Terr Phys. doi:10.1016/j.jastp.2008.11.005
Russell JM III, Bailey SM, Horanyi M, Gordley LL, Rusch DW, Hervig ME, Thomas GE, Randall CE, Siskind DE, Stevens MH, Summers ME, Taylor MI, Englert CR, Espy PJ, McClintock WE, Merkel AW (2009) Aeronomy of ice in the mesosphere (AIM): overview and early science results. J Atmos Solar-Terr Phys. doi:10.1016/j.jastp.2008.08.011
Sentman DD, Wescott EM, Picard RH, Winick JR, Stenbaek Nielsen HC, Dewan EM, Moudry DR, Sao Sabbas FT, Heavner MJ, Morrill J (2003) Simultaneous observations of mesospheric gravity waves and sprites generated by a midwesterm thunderstorm. J Atmos Terr Phys 65:537–550. doi:10.1016/S1364-6826(02)00328-0

Suzuki S, Shiokawa K, Otsuka Y, Ogawa T, Nakamura K, Nakamura T (2007) A concentric gravity wave structure in the mesospheric airglow images. J Geophys Res 112:D02102. doi:10.1029/2005JD006558

Taylor MJ, Garcia FJ (1995) A two-dimensional spectral analysis of short period gravity waves imaged in the OI (557.7 nm) and near-infrared OH nightglow emissions over Arecibo, Puerto Rico. Geophys Res Lett 22:2473–2476

Taylor MJ, Hapgood MA (1988) Identification of a thunderstorm as a source of short period gravity waves in the upper atmospheric nightglow emissions. Planet Space Sci 36:975–985. doi:10.1016/0032-0633(88)90035-9

Taylor MJ, Pendleton WR Jr, Clark S, Takahashi H, Gobbi D, Goldberg RA (1997) Image measurements of short-period gravity waves at equatorial latitudes. J Geophys Res 102(D22):26283–26299

Taylor MJ, Zhao Y, Pautet P-D, Nicolls MJ, Collins RL, Baker-Tvedtness J, Burton CD, Thurairajah B, Reimuller J, Varney RH, Heinselman CJ, Mizutani K (2009) Coordinated optical and radar image measurements of noctilucent clouds and polar mesospheric summer echoes. J f Atmos Solar-Terr Phys. doi:10.1016/j.jastp.2008.12.005

Thomas GE (1991) Mesospheric clouds and the physics of the mesopause region. Rev Geophys 29:553–576

von Zahn U, Bremer J (1999) Simultaneous and common-volume observations of noctilucent clouds and polar mesosphere summer echoes. Geophys Res Lett 26: 1521–1524

Witt G (1962) Height, structure and displacements of noctilucent clouds. Tellus 14:1–18

WMO (1970) International Noctilucent Cloud Manual, World Meteorological Organization, 250.TP.138, Geneva

Yue J, Vadas SL, She C-Y, Nakamura T, Reising SC, Liu H-L, Stamus P, Krueger DA, Lyons W, Li T (2009) Concentric gravity waves in the mesosphere generated by deep convective plumes in the lower atmosphere near Fort Collins, Colorado. J Geophys Res 114:D06104. doi:10.1029/2008JD011244

Part II
Vertical Coupling by Upward Propagating Waves

Chapter 8

Gravity Wave Influences in the Thermosphere and Ionosphere: Observations and Recent Modeling

David C. Fritts and Thomas S. Lund

Abstract Observational and theoretical studies have suggested gravity wave propagation and influences in the thermosphere and ionosphere for half a century. Gravity waves contribute, or are believed to contribute, to a variety of neutral and electrodynamic phenomena ranging from vertical coupling, energy and momentum transport and deposition, neutral perturbations and accelerations, traveling ionospheric disturbances, ionospheric irregularities, and plasma instabilities under quiet conditions to strong coupling from high to low latitudes and accompanying electrodynamics under storm-time conditions. Our goals here are to briefly review what has been learned to date, to illustrate some of the more recent results indicative of gravity wave effects, and to identify some aspects of neutral dynamics not previously considered that we expect may also have significant influences on neutral dynamics and electrodynamics in the thermosphere and ionosphere.

8.1 Introduction

Gravity waves (GWs) were first proposed to account for "irregular motions" in the thermosphere and ionosphere (hereafter the TI) in the pioneering work by Hines (1960) a half century ago. Since then, our understanding of the sources, propagation, and effects of GWs in the TI has expanded significantly, often accompanying research focused on lower altitudes (see Fritts and Alexander, 2003, for a recent review). Despite these advances, major unknowns remain and current research spans a wide range of measurement, modeling, and theoretical efforts. Indeed, new impetus for studies of GWs in the TI has come from needs as diverse as predicting climate change, responses of atmospheric circulation and structure to variable solar forcing, and ionospheric influences on communication and navigation systems. Our purpose here is to review recent evidence for GW influences in the TI and current and new modeling capabilities that are enabling a more quantitative understanding of the various GW effects.

Observations over the last few decades have revealed the presence and persistence of GWs extending to very high altitudes (~500 km and above), defined representative GW scales and frequencies as functions of altitude, and suggested sources of TI GWs ranging from deep convection at tropical latitudes to aurora at high latitudes (e.g., Mayr et al., 1990; Hocke and Schlegel, 1996; Mendillo et al., 1997; Oliver et al., 1997; Djuth et al., 1997, 2004; Innis et al., 2001; Innis and Conde, 2002; Livneh et al., 2007; Fritts et al., 2008; Vadas and Nicolls, 2008, 2009; Abdu et al., 2009). Other observations have revealed coupled neutral and plasma structures, especially the GW ionospheric manifestations referred to as "traveling ionospheric disturbances" (TIDs) and suggested potential or significant influences of neutral motions in plasma instability processes (e.g., Klostermeyer, 1969, 1978; McClure et al., 1977, 1998; Hysell et al., 1990; Kirchengast et al., 1995; Ma et al., 1998; Nicolls et al., 2004; Earle et al., 2008; Fritts et al., 2009a). Recent modeling and theoretical studies have defined the environmental conditions enabling GW

D.C. Fritts (✉)
Colorado Research Associates Division, NorthWest Research Associates, Boulder, CO, USA
e-mail: dave@cora.nwra.com

M.A. Abdu, D. Pancheva (eds.), A. Bhattacharyya (Coed.), *Aeronomy of the Earth's Atmosphere and Ionosphere*,
IAGA Special Sopron Book Series 2, DOI 10.1007/978-94-007-0326-1_8,

propagation to high altitudes, anticipated secondary GWs arising from nonlinear interactions or primary wave dissipation at lower altitudes, and quantified viscous influences on GW structures and penetration altitudes (Klostermeyer, 1972, 1991; Sidi and Teitelbaum, 1978; Hickey and Cole, 1987, 1988; Walterscheid and Schubert, 1990; Chimonas et al., 1996; Vadas and Fritts, 2002, 2004, 2005, 2006, 2009; Vadas, 2007; Fritts and Vadas, 2008; Snively and Pasko, 2008; Huang et al., 2009; Vadas et al., 2009). Modeling and theoretical studies addressing neutral-plasma coupling have examined how GWs contribute to TID structures and GW contributions to plasma bubble seeding and identified conditions under which they appear to enable or enhance plasma bubble growth (Huang et al., 1993; Huang and Kelley, 1996; Kudeki et al., 2007; Abdu et al., 2009; Kherani et al., 2009). Thus, our understanding of GW sources, characteristics, and effects in the TI is considerably advanced over what it was a decade or two earlier. There nevertheless remain major unknowns, especially where observations and modeling fail to capture all of the relevant physical processes and the spatial and temporal scales on which these occur.

Ray tracing studies by Vadas and colleagues, in particular, identified the spatial and temporal scales arising from GW sources in the lower atmosphere and in the mesosphere and lower thermosphere (MLT) most likely to impact the TI, largely explained the observed variations of GW scales and observed periods with altitude, and highlighted expectations that GW amplitudes and spatial scales should increase dramatically with altitude prior to dissipation. These authors also recognized that increasing amplitudes and scales pushed the ray tracing assumptions (particularly WKB theory and linearity) near, or beyond, their limits of applicability, and that other methods would be needed to address the non-WKB and nonlinear GW dynamics that they imply. To address these needs, a new modeling capability addressing deep, nonlinear, 3D GW dynamics was recently developed, and has been applied to several idealized problems that demonstrate the need to account for these dynamics in furthering our understanding of the influences of large-scale and large-amplitude GWs in the TI (Lund and Fritts, 2011).

We will begin by reviewing the observational and theoretical evidence for the importance of deep, viscous, large-amplitude GWs in the TI arising from lower atmosphere sources. We will then review the benefits, as well as the assumptions and potential pitfalls, of ray tracing methods in order to understand where they serve us well, where they likely fail, and how they might be refined to account for finite-amplitude or nonlinear effects that they cannot address directly. We will conclude with a brief summary of some of the new dynamics accompanying GWs that attain large spatial scales and large amplitudes, and exhibit strong nonlinear responses, in the TI.

8.2 Evidence of Deep, Large-Amplitude GWs in the TI

Observations of electron density fluctuations by the Arecibo Observatory (AO) incoherent scatter radar (ISR), the MU radar in Japan, the Poker Flat ISR (PFISR) in Alaska, and the EISCAT ISR in Norway provide persuasive evidence that GWs extending from $\sim$100 to 500 km or higher occur essentially continuously (Kirchengast et al., 1995; Oliver et al., 1997; Djuth et al., 1997, 2004; Ma et al., 1998; Livneh et al., 2007; Vadas and Nicolls, 2008, 2009). Additional evidence of large-scale and large-amplitude neutral GW motions, or TIDs, comes from ionosondes, all-sky imagers, GPS receivers, in situ measurements, and satellite remote sensing (e.g., Klostermeyer, 1969; Mendillo et al., 1997; Innis et al., 2001; Ogawa et al., 2002; Nicolls et al., 2004; Earle et al., 2008; Fritts et al., 2008; Kamalabadi et al., 2009). While the sources of this GW "continuum" are still under debate, observed GW scales and periods appear to be relatively well defined, and there is general agreement that the primary GW sources occur in the lower neutral atmosphere under "quiet" TI conditions. At these times, GW vertical wavelengths increase from $\sim$10 km or greater at $\sim$100 to $\sim$200 km or greater at $\sim$300 km, while periods vary from $\sim$10 min to several hours at $\sim$100 to $\sim$20 min to $\sim$2 h at $\sim$300 km. The distribution of vertical wavelengths with altitude obtained by Oliver et al. (1997) with the MU radar under quiet conditions is shown in Fig. 8.1. Observed variations of wavelength and period with altitude suggest that GWs, likely arising from different sources in the lower atmosphere, contribute preferentially at different altitudes.

Observed wavelength and period variations with altitude under quiet conditions appear to be in good

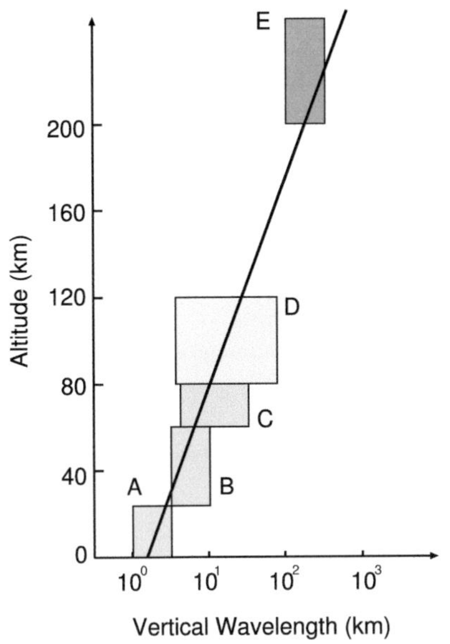

Fig. 8.1 Distributions of GW vertical wavelengths with altitude observed by the MU radar. (After Oliver et al., 1997)

agreement with predictions of GW scales and frequencies based on the viscous dissipation theory of Vadas and Fritts (2005) and Vadas (2007) applied to GWs arising in the lower atmosphere. The approximate viscous dispersion relation of Vadas and Fritts (2005), assuming a Prandtl number Pr = 1 (which differs from the real value of Pr ∼0.7) is given by

$$(\omega_i + \nu m/H)^2 = k_h^2 N^2 (|\mathbf{k}|^2 + 1/4H^2), \qquad (8.1)$$

where $\omega_i = k_h(c - U_h)$ is the real GW intrinsic frequency, $\omega = k_h U_h$ is the real GW ground-based frequency, U_h is the component of mean wind in the plane of GW propagation, k_h and m are the GW horizontal and vertical wavenumbers, $k_h = 2\pi/\lambda_h$ and $m = 2\pi/\lambda_z$, λ_h and λ_z are the GW horizontal and vertical wavelengths, the total GW wavenumber is $\mathbf{k} = (k_h, m)$, H is the density scale height, and N is the buoyancy frequency. Where viscosity is negligible, Eq. (8.1) yields the usual inviscid anelastic dispersion relation of Marks and Eckermann (1995), which may be written as

$$m^2 = k_h^2(N^2/\omega_i^2 - 1) - 1/4H^2. \qquad (8.2)$$

Equation (8.2) illustrates more conveniently the influences of increasing or decreasing N, ω_i, and k_h on GW vertical structure. As examples, decreasing k_h for constant N and ω_i reduces m^2, resulting in GWs having large horizontal wavelengths being evanescent throughout the lower atmosphere; likewise, decreasing N or increasing $|c - U_h|$ may cause GWs having large k_h and small horizontal wavelengths to become evanescent where these conditions occur; finally, GWs approaching critical levels, where $\omega_i = k_h(c - U_h)$ approaches zero, causes these GWs to cease vertical propagation. The consequences of these processes (and viscosity) for GWs arising from general sources in the lower atmosphere for various solar conditions and thermospheric mean winds were evaluated by Fritts and Vadas (2008). These results are illustrated in Fig. 8.2 for three GW propagation directions with a thermospheric temperature of 1000 K, an eastward thermospheric wind of 100 ms^{-1}, and Pr = 0.7.

The results displayed at top in Fig. 8.2 show that long horizontal wavelengths are expected to be evanescent in the lower atmosphere, with more severe limits occurring for higher initial GW frequencies. Decreasing $|c - U_h|$ results in critical levels for all but the longest horizontal wavelengths for GWs propagating eastward (right panels), while decreasing N and increasing $|c - U_h|$ result in decreasing m^2 and an approach to turning levels (and reflection) for GWs propagating westward at the smaller horizontal wavelengths (left panels). Decreasing N also causes reflection for the higher initial frequencies (N/2 and N/3) and smaller horizontal wavelengths even without changes in the intrinsic phase speeds for those GWs propagating meridionally (middle panels). The consequences of these various effects are a decreasing range of horizontal and vertical wavelengths achieving higher altitudes for each initial GW frequency considered (bottom panels of Fig. 8.2). GWs having horizontal and vertical wavelengths of ∼100–1000 km and propagating westward (against the thermospheric wind) will retain ∼3% or more of their initial momentum flux at ∼250 km altitude, with a smaller range of wavelengths surviving to this altitude without Doppler shifting. For the conditions displayed here, only wavelengths of ∼150–500 km survive to 300 km altitudes, and only for up-shifted GWs propagating against the mean wind. As shown by Fritts and Vadas (2008), similar results are obtained for smaller and larger thermospheric temperatures, with smaller temperatures yielding higher viscosities, lower penetration, and more restricted wavelengths and higher temperatures yielding lower viscosities, higher penetration, and a

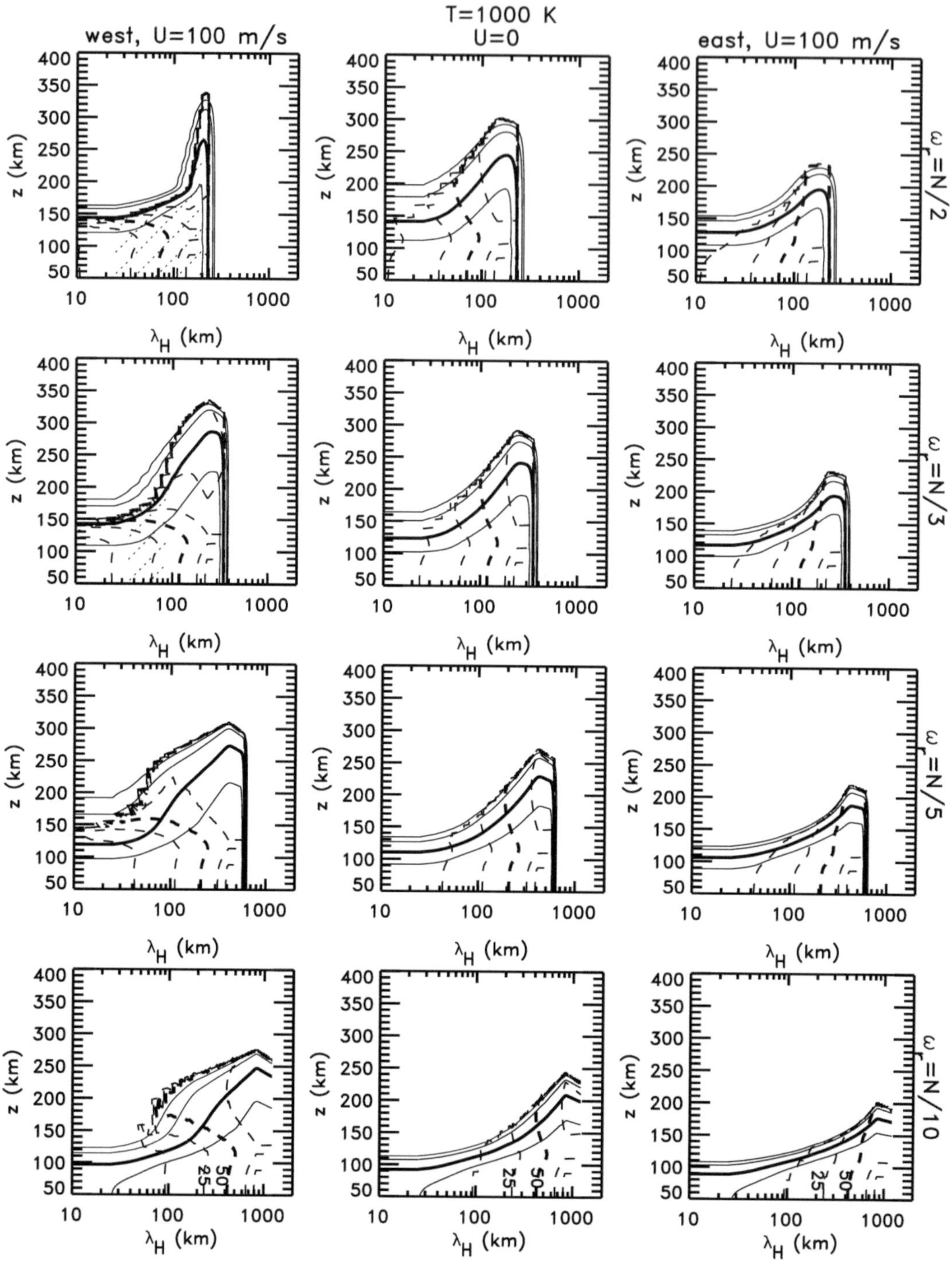

Fig. 8.2 *Top panels* show gravity wave penetration altitudes for horizontal wavelengths varying from 10 to ~1000 km for initial GW frequencies of *N*/2, *N*/3, *N*/5, and *N*/10 (*top* to *bottom*, with $N = 0.02\ s^{-1}$) and for GW propagation against, across, and along (*left* to *right*) an eastward thermospheric wind of 100 ms^{-1}. Bottom panels show horizontal and vertical wavelengths achieving altitudes of 150, 200, 250, and 300 km (*top* to *bottom*) for GW propagation directions as at *left*. A Prandtl number Pr = 0.7 and a thermospheric temperature of 1000 K are assumed. *Solid lines* denote surviving GW momentum flux, $\rho_0(z) < u'_h w' >$, magnitudes with altitude for each wavenumber and frequency at the source altitude, with contour intervals of 0.9, 0.5 (*bold*), 0.1, and 0.03. *Dashed lines* in top panels are contours of vertical wavelength (in km), with 50 km in *bold*. *Dashed lines* in bottom panel indicate

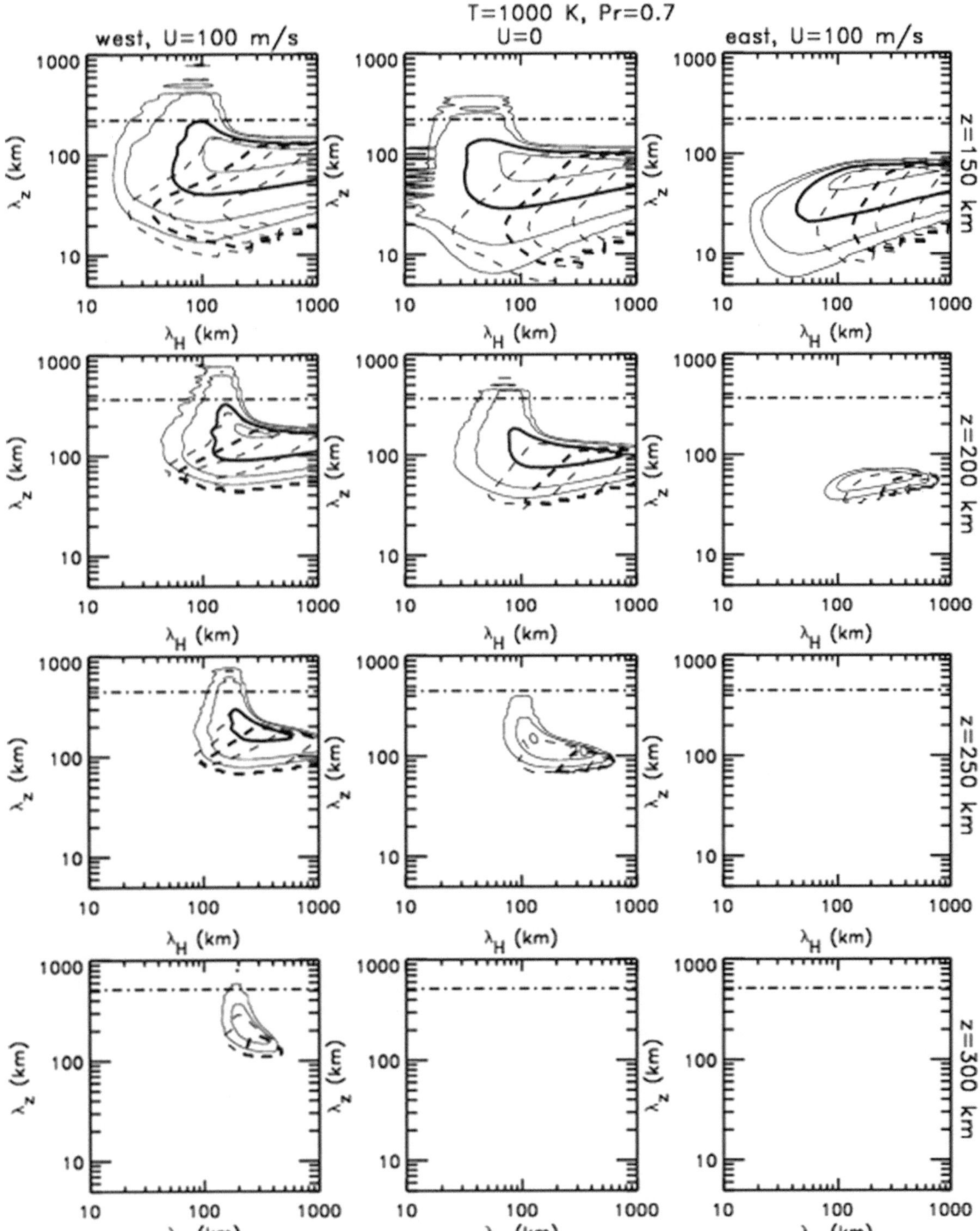

Fig. 8.2 (continued) observed GW periods of 10, 20 (*bold*), 30, and 60 min. Vertical wavelengths in the lower panels are not reliable near the value of $4\pi H$ (shown with the horizontal dashed line in each panel) because the WKB assumption is violated at these scales. (After Fritts and Vadas, 2008)

somewhat expanded range of wavelengths. The spatial and temporal scales observed by various radars and predicted by viscous ray theory are largely consistent with inferences from satellite, digisonde, and GPS TEC data, where these data are available (Innis and Conde, 2002; Earle et al., 2008; Fritts et al., 2008).

Corresponding GW amplitude estimates in the TI come from several sources. In situ measurements of neutral and ion velocities (and perturbation densities, temperatures, and pressures) obtained by Dynamics Explorer 2 (DE-2) at middle and high latitudes have demonstrated consistency with the expected GW polarization relations at wavelengths consistent with ISR observations at DE-2 altitudes (Earle et al., 2008; Innis and Conde, 2002). Vertical velocities of $\sim$20 ms^{-1} at mid-latitudes and altitudes of $\sim$250 to 300 km were observed by Earle et al. (2008), while significantly larger amplitudes were reported by Innis and Conde (2002) at high latitudes. Similar vertical velocities to those seen by Earle et al. (2008) were also inferred from digisonde electron densities (see Fig. 8.3, left) during the 2005 Spread F Experiment (SFX), $\sim$10 to 30 ms^{-1} (Fig. 8.4), where electron density perturbations as high as $\sim$40% also implied large neutral GW amplitudes, downward phase motions, and periods of $\sim$20 min to 2 h (Fritts et al., 2008; Abdu et al., 2009). Large electron density perturbations were also inferred from GUVI 1356 nm measurements during SFX extending from $\sim$100 km to much higher altitudes (Kamalabadi et al., 2009, Fig. 8.3, right), with the perturbations below $\sim$250 km not attributable to plasma bubbles and horizontal wavelengths of $\sim$200 to 500 km comparable to those expected for GWs at these altitudes (Fig. 8.2). Finally, we note that GW horizontal velocities are almost always larger than vertical velocities, with typical factors of $\sim$2 to 5 for those GWs expected to reach the highest altitudes (Fritts and Vadas, 2008). Thus, GW momentum fluxes (per unit mass), $< u'_h w' > = u'_{h0} w'_0/2$ (where angle brackets denote an average over GW phase and u'_{h0} and w'_0 are the GW horizontal and vertical velocity amplitudes), and the body forces and secondary wave radiation accompanying wave dissipation, are also expected to be substantial.

Additional evidence for large-amplitude GWs entering the TI from below comes from extensive ray tracing studies of GWs assumed to arise from specific sources in the lower atmosphere and the MLT by Vadas and colleagues using the same methodology as the results of Fritts and Vadas (2008) discussed above. Most attention has focused on GWs arising from deep convection because this is the most obvious lower atmosphere source that readily yields large-scale GWs having large ground-relative phase speeds (Vadas and Fritts, 2004, 2009; Vadas, 2007; Vadas et al., 2009). Large phase speeds are required to penetrate the $\sim$100 ms^{-1} mean and tidal winds in the MLT that remove essentially all lower phase speed waves from the spectrum through critical-level interactions and related instability dynamics. Similar indications of GWs arising from deep convection and penetrating into the MLT have come from direct simulations of convection and the propagation of GWs that result (e.g., Lane et al., 2001; Horinouchi et al., 2007; Horinouchi, 2004), but these simulations typically have not extended to significantly higher altitudes.

Further evidence for the importance of GWs arising from tropical convection in the TI comes from observed correlations of regions of enhanced convection with (1) enhanced ionospheric irregularities in the upper mesosphere and lower thermosphere seen in GPS radio occultation by Hocke and Tsuda (2001) and (2) a higher statistical occurrence of equatorial plasma bubbles (EPBs) at the topside F layer by McClure et al. (1998). Large-scale GWs arising from deep convection were invoked as the coupling mechanism in each case. McClure et al. (1998) also noted a secondary EPB occurrence maximum accompanying convection in the eastern Pacific during an El Nino period. The correlations identified by Hocke and Tsuda (2001) are shown in Fig. 8.5. Both studies provide compelling evidence of a dynamical link between deep convection and ionospheric perturbations or instabilities suggesting GWs as the coupling mechanism. Indeed, recent analyses and modeling in support of measurements made during the Spread F Experiment (Fritts et al., 2009a) suggest that observed GWs can play a role in EPB seeding when background ionospheric conditions are conducive (Abdu et al., 2009; Kherani et al., 2009). In contrast, other lower atmosphere sources, especially topography and jet stream winds and shears, yield GW phase speeds that typically do not enable penetration into the TI.

Ray tracing of GWs arising from deep convection suggests that smaller spatial scale GWs with larger amplitudes are often unstable and dissipate in the MLT, while GWs with larger spatial scales (and larger phase speeds), but smaller MLT amplitudes, can more readily

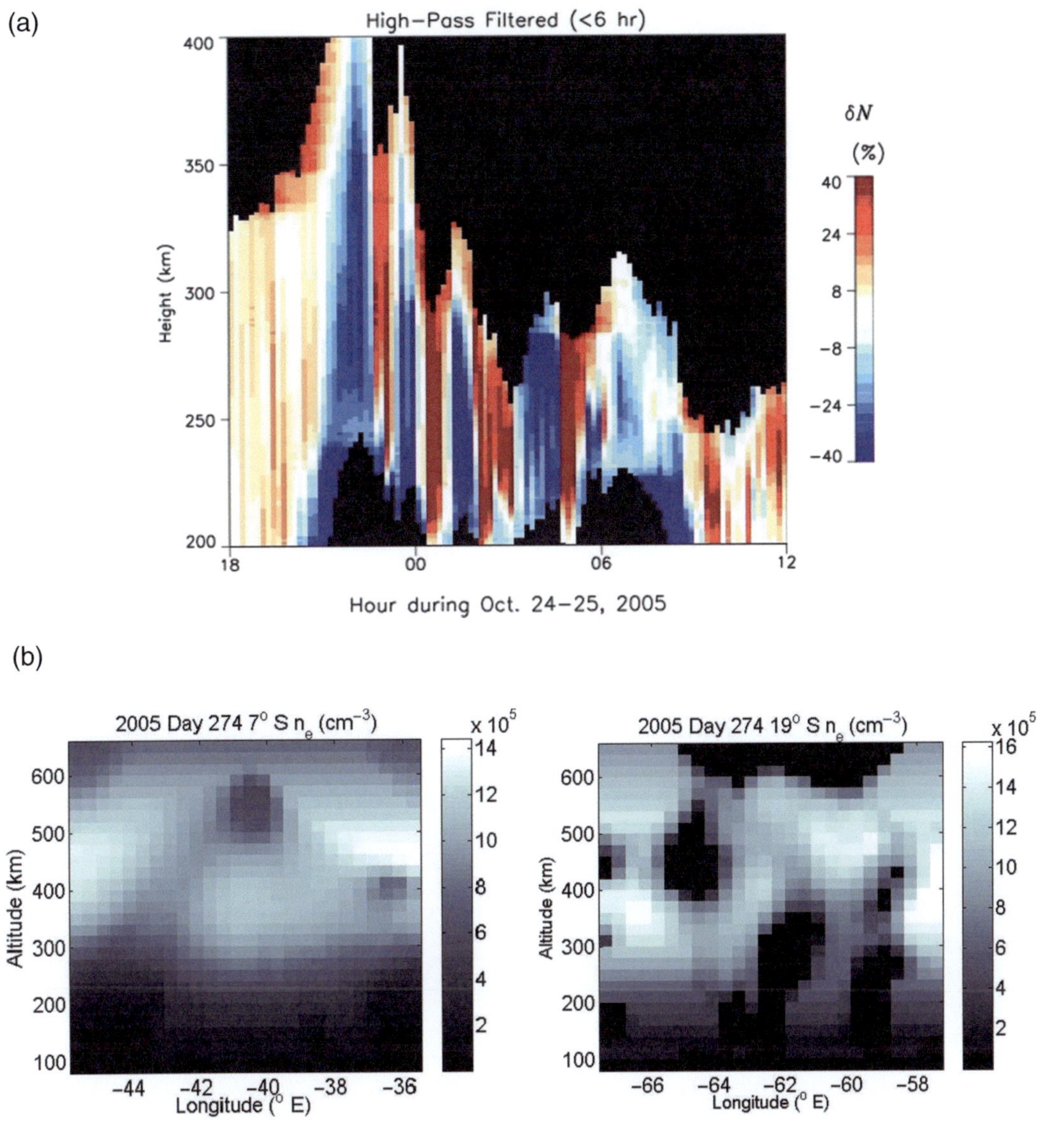

Fig. 8.3 Electron density perturbations inferred from digisonde measurements at Fortaleza, Brazil (*top*) and GUVI 1356 nm emissions (*bottom*) during SFX. (After Fritts et al., 2008; Kamalabadi et al., 2009)

penetrate to significantly higher altitudes. Thus it is likely the large-scale, small-amplitude "tail" of the convective GW spectrum that has the dominant direct effects in the TI (Vadas and Fritts, 2004, 2009; Vadas, 2007; Fritts and Vadas, 2008; Vadas et al., 2009). Indeed, Vadas et al. (2009) predict maximum GW horizontal velocities of ~10 to 50 ms^{-1} in response to individual, but significant, convective plumes at spatial scales and phase speeds that cannot penetrate typical MLT winds, while Fritts and Vadas (2008) and Fritts et al. (2008) show that GWs having nominal horizontal winds (~1 ms^{-1} at 80 km) can experience amplitude increases of ~100 or more prior to dissipation at altitudes above ~250 to 350 km (under suitable propagation conditions). The implication is that even GWs having large spatial scales and horizontal phase speeds, but relatively small amplitudes at MLT altitudes, may yield very much larger amplitudes and TI influences if they avoid instability and dissipation at intermediate altitudes. Ray tracing methods, however, cannot account for either linear behavior when assumptions on which it is based (i.e. the WKB approximation) are not

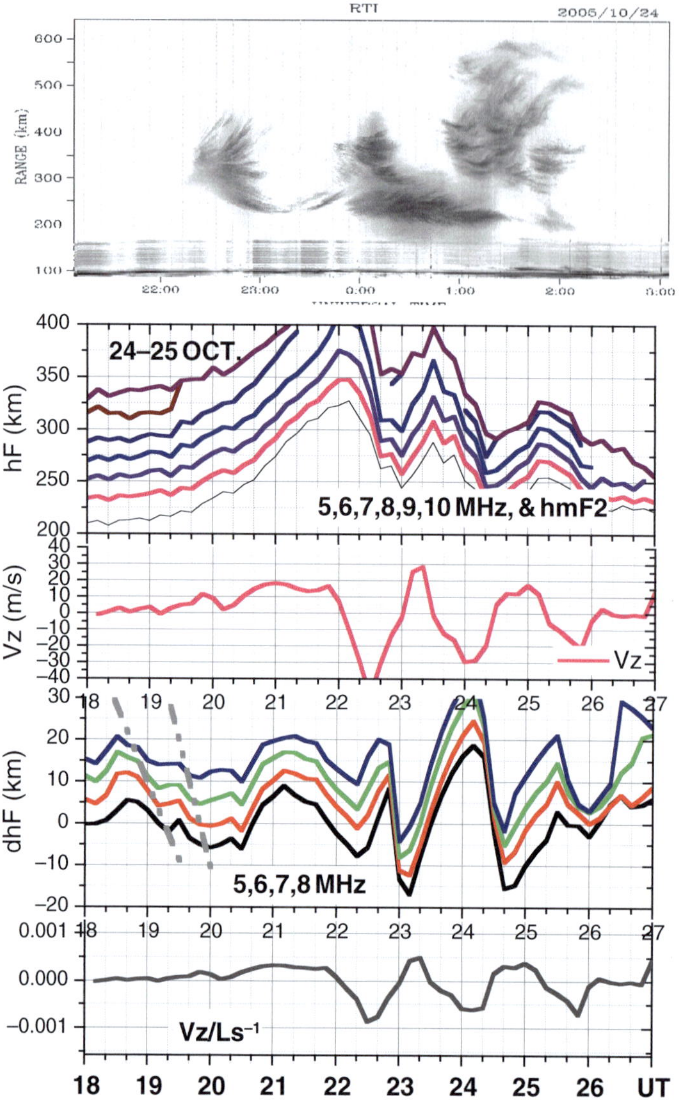

Fig. 8.4 RTI map of 5-m irregularities observed with the 30 MHz radar at Sao Luis for 24–25 October 2005 (*top*), F layer true heights at plasma frequencies of 5, 6, 7, and 8 MHz (*second panel*), mean drift velocity computed from d(*hF*)/d*t* (*third panel*), band-pass filtered (20 min to 3 h) height oscillations at the four frequencies (*fourth panel*), and instability growth rate factor due to the V_Z/L term (*bottom panel*). (After Abdu et al., 2009)

valid or when GW amplitudes are sufficiently large that nonlinear behavior is expected. Both of these concerns are appropriate for GWs having larger vertical scales at higher altitudes in the TI and will be addressed in greater detail below.

Another apparent source of deep, fast GWs that readily penetrate into the TI from below is body forcing accompanying dissipation (and spatially and temporally localized momentum flux divergence) of GWs having amplitudes, horizontal scales, and phase speeds that cause them to become unstable in the MLT (Vadas and Fritts, 2001, 2002; Vadas, 2007; Vadas and Liu, 2009). These "primary" GWs typically (1) have rapid vertical propagation, (2) achieve large amplitudes and large momentum fluxes, and (3) induce significant forcing on the spatial scales of these GW packets at the altitudes at which they become unstable. These characteristics yield body forces that are deep and horizontally limited, but at larger spatial scales that the primary GW wavelengths, which act

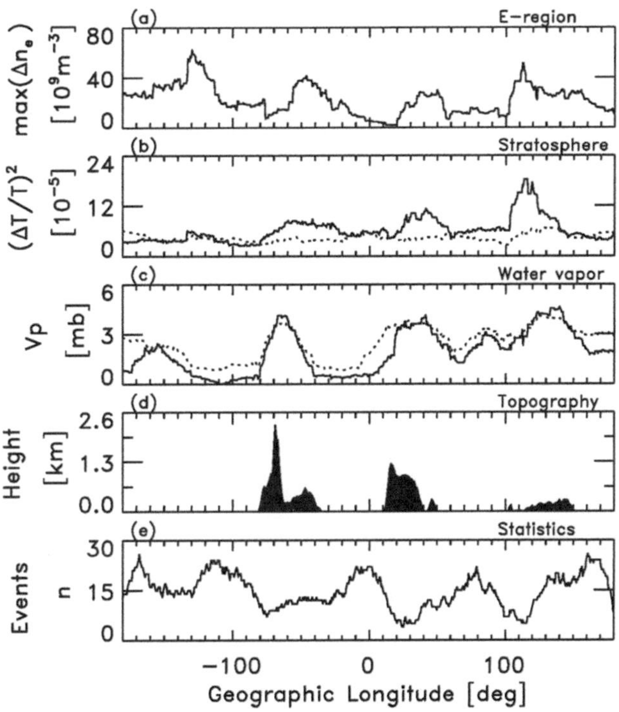

Fig. 8.5 Measurements of small-scale (< 7 km) electron density fluctuations (**a**), stratospheric fractional temperature variance, with *solid* and *dashed* lines denoting altitudes of 22–28 and 32–38 km, respectively (**b**), water vapor pressure averaged from 4–6 km, with *solid* and *dashed* lines denoting GPS/MET and ECMWF data, respectively (**c**), average terrain height from 5–25°S (**d**), and the number of occultation events. Data in (**a**), (**b**), (**c**), and (**e**) were smoothed over 10° of longitude. (After Hocke and Tsuda, 2001)

as preferential sources for "secondary" GWs having larger vertical scales and higher phase speeds than the primary GWs and which themselves readily propagate to much higher altitudes.

An example of the response to local body forcing assumed to arise from rapid GW dissipation at ~180 km simulated with the TIME GCM by Vadas and Liu (2009) reveals a GW with horizontal wavelength ~2000 km and maximum phase speed ~500 ms^{-1} quickly achieving much higher altitudes, propagating globally, and yielding large TEC perturbations (see Fig. 8.6). This thermospheric response is reminiscent of mesospheric responses to deep convection observed when intervening winds do not disrupt the nearly circular GW patterns seen in mesopause airglow emissions (i.e., Sentman et al., 2003; Yue et al., 2009). Such TI motions are typically characterized as TIDs, which are plasma manifestations of neutral GWs occurring on spatial and temporal scales ranging from 100's to 1000's of km and 10's of minutes to hours and having phase speeds of 100's of ms^{-1}. The example shown in Fig. 8.6 would be considered a large-scale TID, and such specific responses to body forcing in the MLT have yet to be identified in observational studies. We nevertheless expect them to be an important component of the thermospheric motion field, and its variability, and a key momentum input from lower altitudes. Large-scale TIDs also arise accompanying large energy and momentum inputs in the auroral zones during geomagnetic storms. These inputs result in strong heating, flow accelerations, and potentially very large amplitudes in the neutral and plasma density, temperature, and wind fields extending to high altitudes that readily propagate to equatorial latitudes and beyond (Fuller-Rowell et al., 1994, 2002; Hocke and Schlegel, 1996; Shiokawa et al., 2002; Nicolls et al., 2004). Small- and medium-scale TIDs having smaller phase speeds and amplitudes are more common, and are more likely to be manifestations of primary GWs entering the TI from sources in the lower atmosphere.

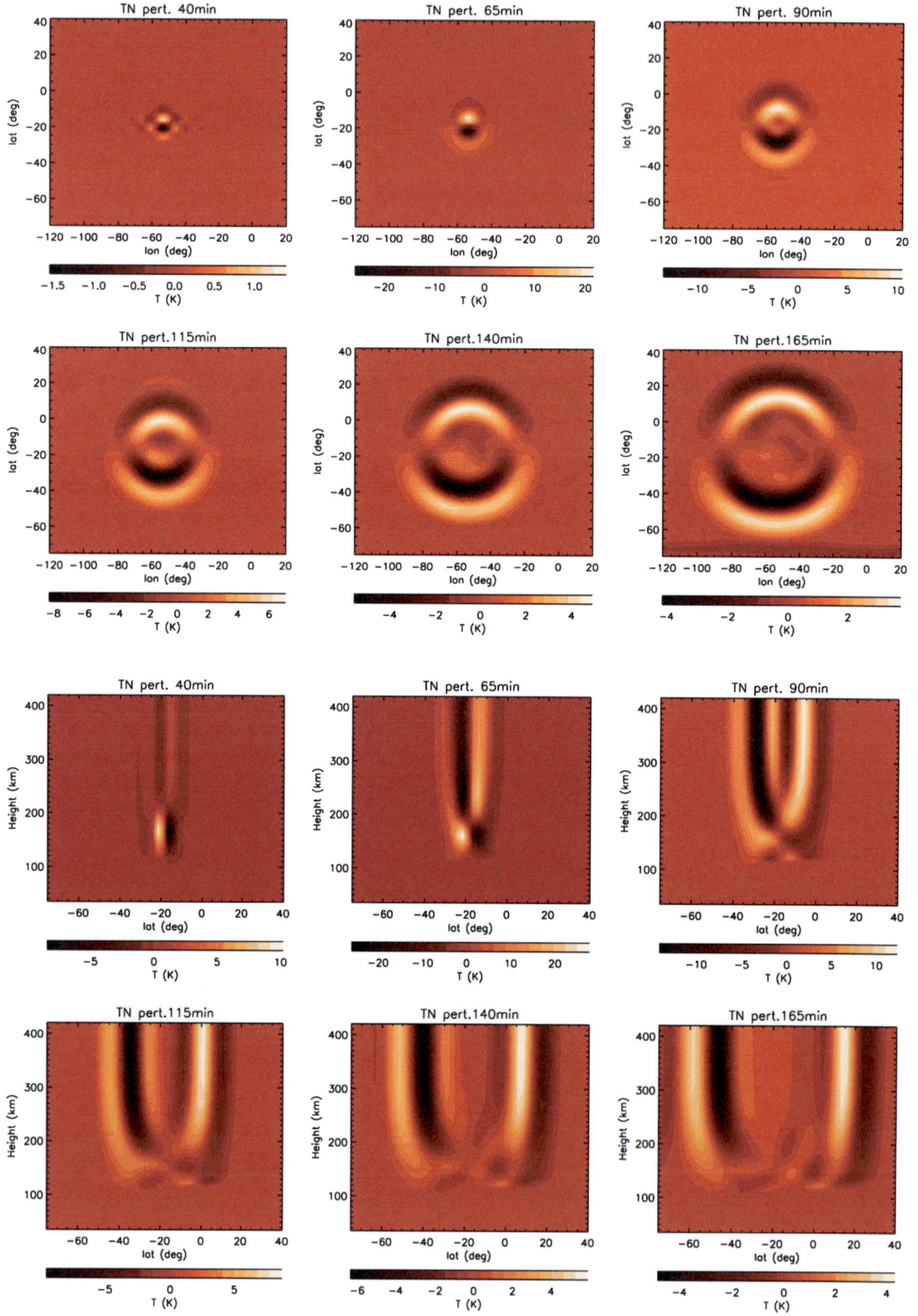
TN pert. 40min
TN pert. 65min
TN pert. 90min
TN pert.115min
TN pert.140min
TN pert.165min
lat (deg)
lon (deg)
Height (km)
T (K)

Fig. 8.6

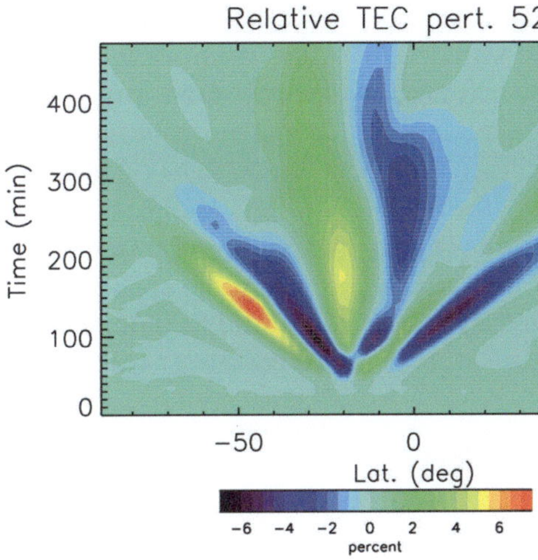

Fig. 8.6 TIME GCM responses in neutral temperature at 250 km (*top*) and in a latitude-height cross section (*second panels*) to a local body force at ~180 km assumed to arise from GW breaking and momentum deposition. Times (*left* to *right*) are 115, 140, and 165 min following the body force centered at 180 km. Corresponding latitude-time TEC perturbations induced by this deep GW response (*lower panel*) reveal outward and upward propagation with a horizontal wavelength of ~2000 km and a maximum phase speed of ~500 ms^{-1}. (After Vadas and Liu, 2009)

8.3 Validity of Ray Tracing Assumptions in the TI

8.3.1 *The WKB Approximation*

The viscous dispersion relation and ray equations developed by Vadas and Fritts (2005) and employed by Vadas (2007) and colleagues for predictions of GW propagation, refraction, and dissipation are based on the WKB approximation, which assumes that $m^2 >> 1/4H^2$, or $\lambda_z < 4\pi H$. This assumption is readily satisfied for the smaller vertical wavelengths that appear to dominate the GW spectrum in the MLT, say ~100 km and below (see Fig. 8.1). At higher altitudes, however, the larger vertical wavelengths typically approach or exceed $\lambda_z \sim 4\pi H$, thus apparently violating the WKB assumption and the conclusions based on its validity. Because of this, Vadas and Fritts (2005) employed the analysis of WKB validity by Einaudi and Hines (1970) to examine its validity in our applications more carefully. Einaudi and Hines (1970) showed that WKB validity requires a small "residue", defined as

$$R_2 = (1/2m^3)\, d^2m/dz^2 - (3/4m^4)\,(dm/dz)^2 \quad (8.3)$$

Vadas and Fritts (2005) displayed values of $R_2 = 1$ for predictions of GW propagation and dissipation for various GW scales and frequencies to indicate where the WKB approximation is clearly violated. This typically occurs only for larger vertical wavelengths and at altitudes near the predicted "dissipation altitude", defined by Vadas and Fritts (2005) as

$$z_{diss} \approx z_1 + H \ln(\omega_{Ir}/2H|m|^3\, \nu_1), \quad (8.4)$$

where z_{diss} is the altitude at which the momentum flux $<u'_h w'>$ peaks, z_1 is a reference altitude ~2 to 3 H below z_{diss}, and ν_1 is the kinematic viscosity at z_1.

Unfortunately, neither Vadas and Fritts (2005) nor Vadas (2007) displayed values of R_2 ~0.1 to 0.3 (for which the WKB approximation is more likely approximately valid), so the larger ranges of altitude over which the WKB assumption may be questionable have not been clearly defined for a range of GW wavelengths relevant to the thermosphere. While the altitudes at which $R_2 = 1$ move downward towards z_{diss} as vertical wavelengths increase (see Vadas and Fritts, 2005, Fig. 8.3), further analysis (Vadas, personal communication, 2010) indicates that in a majority of cases the altitudes at which R_2 exceeds 0.1, 0.3, and 1 are nearly the same, suggesting that the $R_2 = 1$ results of Vadas and Fritts (2005) may be a reasonable guide. There are also cases in which R_2 exceeds 0.1–0.3 at lower and intermediate altitudes (well below dissipation altitudes) for larger initial vertical wavelengths, but these are more likely to result in partial reflection rather than GW dissipation. Nevertheless, significant GW dissipation and momentum deposition may occur at altitudes above where R_2 ~0.1 to 0.3 for the larger vertical wavelengths at higher altitudes. This likely does not invalidate the prediction of most of the dissipation occurring near z_{diss}, likely within a scale height or so. But it very likely *does* influence the specific spatial dependence of these processes, and the vertical distributions of the implied body forces.

8.3.2 Quasi-Linear and Nonlinear Effects

It is well known that GWs exhibit interactions with other motions at small and large amplitudes, including mean flows and other structures having either 2D or 3D character, and that such interactions cause significant departures from "linear" behavior (e.g., Whitham, 1965, 1974; Hasselmann, 1967; Bretherton, 1969; Meid, 1976; Drazin, 1977; Grimshaw, 1977; Lighthill, 1978; Yeh and Liu, 1981; Klostermeyer, 1991; Lombard and Riley, 1996; Sonmor and Klaassen, 1997; Sutherland, 1999, 2000, 2001, 2006). Interactions with mean flows are typically referred to as "quasi-linear", while interactions among various components of the motion spectrum having spatial and temporal variations are referred to as nonlinear. Quasi-linear dynamics are often assumed to be "one-way", with the GW influencing the mean flow, but not the reverse. However, numerous studies indicate that such interactions may influence the GW strongly, altering its spatial and temporal character, its vertical propagation and momentum deposition, and enabling various instabilities, even in a Boussinesq fluid (McIntyre, 1973; Grimshaw, 1977; Fritts and Dunkerton, 1984; Walterscheid and Schubert, 1990; Sutherland, 2000, 2001; Huang et al., 2002).

GW packets that are localized only in the vertical transport momentum that resides in the mean flow and accompanies wave packet propagation via mean flow variations that are transient and reversible if the dynamics are linear and conservative, but that may become irreversible and enable various instabilities if quasi-linear or nonlinear effects become important (Grimshaw, 1977; Fritts and Dunkerton, 1984; Fritts et al., 1994, 1996, 1998; Sutherland, 2000). Fritts and Dunkerton (1984) argued that "self acceleration" can significantly increase the GW phase speed at the leading edge of a wave packet, thus enabling higher wave packet propagation, and penetration to altitudes above an initial critical level. Fritts et al. (1994, 1996, 1998) found transient induced mean flows in a mean shear (1) to become permanent due to instability and dissipation, (2) to enhance mean shears at the trailing edge of the wave packet, and (3) to contribute to the character of wave breaking dynamics. Sutherland (2000) proposed that a vertically-localized GW packet (1) will become unstable to overturning if the induced mean flow exceeds the GW horizontal group velocity (a "self-acceleration condition") and (2) will exhibit self-acceleration effects at infinitesimally small amplitudes if its frequency is very close to the local buoyancy frequency. Sutherland (2006) further demonstrated that GW packets that are suitably localized vertically and have sufficiently high frequencies will become unstable to self-acceleration effects rather than parametric instabilities.

GW packets that are localized both horizontally and vertically exhibit more complex mean responses and both quasi-linear and nonlinear influences on wave propagation and instability (Grimshaw, 1977; Sutherland, 1999, 2001). Mean flows induced by 2D (or 3D) wave packets necessarily have 2D (3D) structure to preserve continuity, even for conservative motions, with an expectation of secondary GW radiation accompanying induced vertical motions (Luo and Fritts, 1993; Vadas and Fritts, 2001, 2002). Wave packets that are dissipating, as a result of viscous dissipation, turbulence, or instabilities that erode wave amplitudes, exhibit momentum flux divergence with altitude that is not fully reversible and leave some momentum behind as the packet propagates to higher (or lower) altitudes. If the GW exhibits strong instability and large amplitude reductions, resulting, for example, from wave "breaking" rather than weaker and slower wave-wave interactions (Fritts et al., 2009b, c), the majority of the momentum is "deposited" (or more correctly, fails to be transported elsewhere) and spread over the spatial extent of the breaking region, again implying 2D or 3D mean responses and a potential for radiation of secondary GWs as discussed above.

While these various quasi-linear and nonlinear dynamics are anticipated in Boussinesq fluids, their potential importance is likely much greater in the atmosphere where GWs can propagate over many scale heights, with amplitudes and momentum fluxes (per unit mass) potentially increasing by orders of magnitude. GWs arising from sources in the lower atmosphere may increase in amplitude by ~100 times or more in reaching the MLT, while those waves having very large spatial scales and relatively high frequencies may increase in amplitude by another ~2 decades prior to dissipation far into the thermosphere (Vadas and Fritts, 2005; Vadas, 2007; Fritts and Vadas, 2008). Quasi-linear effects are amplified in such cases because the increment of momentum per unit area accompanying a wave packet may be written as

$$\Delta M = \int \rho \Delta U \, dz \quad (8.5)$$

with an induced mean flow change at altitude z and time t given by

$$\Delta U(z,t) = -\int \left[(1/\rho)\, d\left(\rho < u'_h w' >\right)/dz\right] dt \quad (8.6)$$

where the integrals are over the depth of the wave packet and from prior to wave packet arrival to an intermediate time, t, before it has passed, respectively, $< u'_h w' >$ is a function of z and t, and the GW momentum flux increases as $1/\rho$ following the wave packet for conservative motions. Assuming that the wave packet propagates over a number of scale heights, the increment of momentum remains constant, but $\Delta U(t)$ also increases as $\sim 1/\rho$ for constant vertical group velocity (see Fig. 8.7). As seen previously for Boussinesq fluids (Fritts and Dunkerton, 1984; Sutherland, 1999, 2001, 2006), accelerations (decelerations) at the leading (trailing) edge of the wave packet yield increases (decreases) in the phase speed of the GW at these locations that increase in magnitude with decreasing density. This results in a temporally-varying distribution of phase speed across the wave packet and a potential for "self-acceleration" instabilities that increases with increasing wave amplitudes and induced mean flows (Sutherland, 2000, 2006).

The combination of increasing wave amplitudes, momentum fluxes, induced mean flows, and phase speed variations as GWs propagate from sources in the lower atmosphere and MLT to higher altitudes essentially guarantees that the linear and WKB assumptions implicit in the ray tracing methodology must fail to provide a quantitative description of GWs to varying degrees in the MLT and at higher altitudes. Given the very large scales and amplitudes anticipated at the highest altitudes to which GWs penetrate prior to strong viscous dissipation, we expect that methods that specifically account for quasi-linear and nonlinear effects in an atmosphere with realistic kinematic viscosity and thermal diffusivity profiles will be required to define such dynamics as (1) GW refraction due to variable mean wind and stability profiles, especially behavior near turning levels (where the real part of the intrinsic frequency $\omega_i = N$), (2) the effects of "self acceleration" on wave packet propagation, (3) the various instabilities to which GW packets succumb at large amplitudes and spatial scales and high frequencies, and (4) the altitudes over which GW instability dynamics leads to significant turbulence and mixing.

8.4 Studies of Large-Scale GW Dynamics in the Thermosphere

Studies specifically addressing deep GW dynamics in the MLT and higher in the thermosphere that are unencumbered by WKB or linear assumptions must enable solutions of either the fully compressible or the (approximate) anelastic Navier-Stokes equations. Numerical simulations of compressible dynamics, however, demand enormous computational resources for high-resolution studies because of the large acoustic velocities in the thermosphere, even below those altitudes at which viscosity dominates the Courant-Friedrichs-Lewy (CFL) constraints in explicit numerical schemes. Numerical solutions of the optimal anelastic equations achieve up to an ~10 times increase in the CFL condition, but are constrained in the same manner by viscous diffusion. Thus, to date at least, the greatest insights into quasi-linear and nonlinear neutral dynamics for large-scale GW motions in the MLT have come from recent anelastic simulations. Here we describe anelastic DNS results intended to illustrate the influences of both self acceleration and nonlinearity accompanying strong GW amplitude growth with increasing altitude.

8.4.1 GW Self-Acceleration Effects

GWs that are spatially and temporally localized will exhibit self acceleration, but the magnitude and effects of self acceleration will depend on the GW parameters,

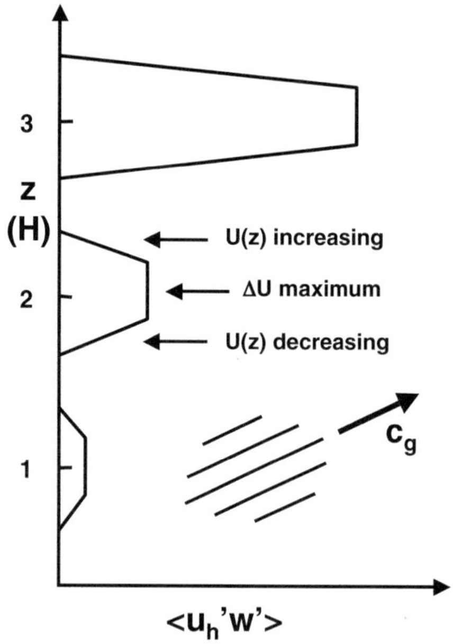

Fig. 8.7 Schematic of the variations of the induced mean flow following a gravity wave packet with $\rho\langle u'_{\mathrm{h}}w'\rangle$ increasing linearly at the leading edge of the wave packet, remaining constant at the center, and decreasing linearly at the trailing edge for a gravity wave with propagation upward and to the right with u'_{h} and w' in phase

the rate at which (or altitude interval over which) the GW momentum flux varies, the degree of spatial localization of the GW packet, and whether it is localized in 1, 2, or 3 dimensions. Maximum self acceleration occurs for GW packets localized in only the vertical, as this precludes induced horizontal and/or vertical circulations and distribution of the packet momentum increment over larger spatial scales than the scale of the packet itself.

To illustrate self-acceleration effects, we assume a GW packet that is localized only in the vertical, such that its momentum flux and self acceleration approximate those at the center of a horizontally-extended packet. The environment is assumed to be initially at rest, isothermal, with constant scale height, $H = 7$ km, and buoyancy frequency, $N = 0.02\ \mathrm{s}^{-1}$ (buoyancy period $T_{\mathrm{b}} = 314$ s) in order to illustrate self-acceleration effects without refraction due to changes of vertical wavelength with varying N), and to have a (true) kinematic viscosity of $\nu \sim 1.5 \times 10^{-5}\ \mathrm{m^2s^{-1}}$ at the Earth's surface. These imply kinematic viscosities of $\sim$1.4 and $\sim$400 $\mathrm{m^2s^{-1}}$ at altitudes of 80 and 120 km. We also assume the GW packet is excited at the Earth's surface ($z = 0$) with an intrinsic frequency of $\omega_{\mathrm{i}} = N/2^{1/2}$ (initial intrinsic period of $T_{\mathrm{i}} = 2\pi/\omega_{\mathrm{i}} =$ 444 s), horizontal and vertical wavelengths of $\lambda_{\mathrm{h}} = \lambda_{\mathrm{z}} = 20$ km, a Gaussian amplitude variation in time of half width $\sigma_{\mathrm{t}} = T_{\mathrm{i}}$, and maximum horizontal and vertical velocity perturbation amplitudes of $u'_{\mathrm{h}} = w' = 0.0225\ \mathrm{ms^{-1}}$ occurring at $t = 4T_{\mathrm{i}}$ (but in anti-phase with negative momentum flux). These GW parameters imply initial horizontal and vertical phase velocities of $c_{\mathrm{px}} = c_{\mathrm{px}} \sim 45\ \mathrm{ms^{-1}}$.

Our description of the evolution of this GW packet employs the finite-volume anelastic (FVA) DNS model recently introduced by Lund and Fritts (2011) for a 2D simulation employing a doubly periodic domain of width 40 km and depth 160 km, with isotropic spatial resolution of 200 m. This evolution is displayed with streamwise-vertical cross sections of vertical velocity at seven times in Fig. 8.8 at times $\sim$7, 9, 10.5, 11, 11.7, 11.9, 12.1 T_{i} after maximum forcing.

The wave packet evolution up to the first image in Fig. 8.8 is essentially linear, as the GW amplitude, momentum flux, and induced mean motions are still very small at this time and altitude (maximum $u'_{\mathrm{h}} = w' = 2.5\ \mathrm{ms^{-1}}$are $<< c_{px}$). This is confirmed by the nearly uniform phase structure of the GW packet at this time. As the GW propagates to higher altitudes, however, its amplitude increases as $e^{z/2H}$ (w'maxima

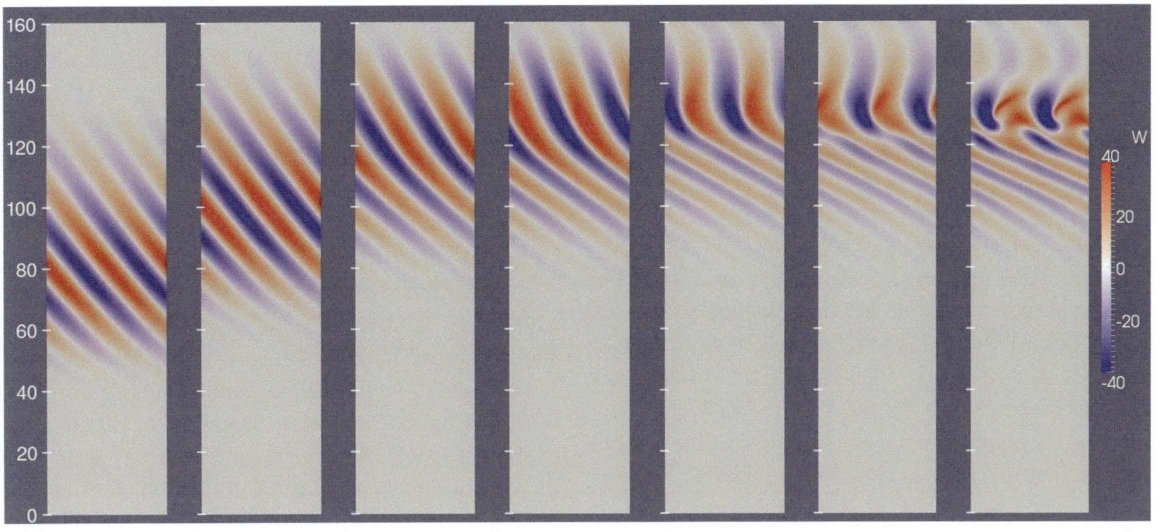

Fig. 8.8 Streamwise-vertical cross sections of vertical velocity at times of ~7, 9, 10.5, 11, 11.7, 11.9, 12.1 T_i after maximum forcing at $z = 0$ in the FVA DNS of GW self acceleration. Maximum vertical velocities at these times are w' are ~2.5, 8, 20, 40, 60, 60, and 40 ms^{-1}, respectively

are ~8, 20, 40, and 60 ms^{-1} in the 3rd through 6th images), its momentum flux and induced mean motions increase as $e^{z/H}$, and significant distortions of the phase structure begin to emerge. These are due to the accelerations of c_{px} spanning the leading edge of the GW packet that resides in the mean flow undergoing accelerations in the direction of GW propagation because of (negative) GW momentum flux that is increasing in time and decreasing with increasing altitude; opposite tendencies for c_{px} and mean motions occur where GW amplitudes are decreasing with decreasing altitude. Succeeding images reveal increasing phase distortions, including development of a "kink" where the induced mean motions and altered c_{px} are maximum, and continuing, but slowing, vertical propagation of the maximum response. By $t = 11.7\ T_i$ (5th image), the GW above the kink exhibits approximately vertical phase structure and the largest vertical velocities, both because GW velocities increase with decreasing density and because vertical velocities increase relative to horizontal velocities as $w' \sim k_h u'_h/m \sim (\lambda_z/\lambda_h)\, u'_h$

The last three images also exhibit differential c_{px} accelerations with altitude that result in the phase structure being advected through the vertical, with evidence of smaller-scale structures and apparent instability arising quickly thereafter. For reference, the evolution up to the time exhibiting vertical phase structures has required 11.7 T_i, but the subsequent evolution of small-scale structures within the velocity field spans only an additional 0.4 T_i. We take this as direct evidence of the "self-acceleration" instability anticipated by Sutherland (2000, 2006), occurring where the induced mean flow exceeds the GW horizontal group velocity of the GW. Given the large momentum fluxes and accelerations implied by the limited observations suggesting very strong and localized body forces that are available to date (e.g., Yamada et al., 2001; Fritts et al., 2002), we anticipate that such dynamics may be most common in the lower thermosphere, but that similar "self-acceleration" dynamics may also readily extend to lower altitudes whenever and wherever large-amplitude, localized GW packets arise. We also note that these self-acceleration effects arise largely due to temporal (and spatial) variations in mean flow forcing accompanying GW transience, rather than primarily spatial (altitude) variations of GW momentum flux and mean flow variations accompanying GW dissipation due to viscous effects or instability. This statement is based on the occurrence of very similar evolutions of GW packet phase structures seen to occur for stronger GW forcing at lower altitudes, where viscous effects must play a much smaller role (not shown). These various self-acceleration dynamics will be discussed at greater length elsewhere.

8.4.2 GW Breaking and Turbulence Generation in the Lower Thermosphere

8.4.2.1 Problem Specification

We now consider the consequences of nonlinearity for a GW expected to penetrate well into the thermosphere, based on the ray tracing results of Vadas (2007) and Fritts and Vadas (2008), and that achieves an amplitude expected to lead to instability and turbulence based on the DNS of GW breaking in the MLT employing the Boussinesq incompressible Navier-Stokes equations (Fritts et al., 2009a). For this purpose, we assume a GW horizontal wavelength of $\lambda_h = 60$ km, a horizontal phase speed of $c = 55$ ms^{-1}, and a maximum horizontal velocity amplitude of $u'_h = 5.5$ ms^{-1} at 100 km. These GW parameters are well within the ranges anticipated to arise due to deep convection by Vadas and Fritts (2009) and Yue et al. (2009). We also assume a mean temperature profile described by Vadas (2007) for mean solar conditions, with the temperature increasing from ~280 K at 100 km to ~900 K at 200 km, the corresponding buoyancy frequency decreasing from $N \sim 0.022$ s^{-1} at 100 km to ~0.011 s^{-1} at 200 km, and mean density decreasing by nearly 10^3 from 100 to 200 km. These assumptions imply a GW intrinsic frequency of $\omega_i = k_h(c - U_h) = N/10^{1/2}$, where U_h in this case is the mean flow induced by the GW discussed in Section 8.4.1, a vertical wavelength of $\lambda_z \sim 20$ km at 100 km, and a kinematic viscosity of $\nu \sim 14$ m^{2}s^{-1} at 100 km.

In order to examine nonlinear dynamics in the absence of complicating self acceleration, we also assume that the GW packet is sufficiently localized in the horizontal that the induced mean flow is $U_h \sim 0$ and the self-acceleration effects described in Section 8.4.1 may be neglected. Due to the strong density decrease with altitude, the nondimensional GW amplitude of $a = u'_h/|c - U_h| \sim u'_h/c = 0.1$ at 100 km increases to $a \sim 1$ near 160 km, suggesting a potential for GW breaking and instability that increases with altitude if the Reynolds number $Re = c\lambda_z/\nu$ is sufficiently large and the Boussinesq stability analysis of Fritts et al. (2009b) is approximately valid. Note that $Re \sim \lambda_z^2/\nu T_b$ for approximately hydrostatic GWs (with $T_b = 2\pi/N$), making its scale dependence more explicit. Thus, Re decreases with altitude as $\sim\rho$, but increases with altitude as λ_z^2, which largely offsets the variation with ρ, based on the variations with altitude of λ_z observed by Oliver et al. (1997, see Fig. 8.1). Clearly, then, the altitudes over which a particular GW might exhibit nonlinear dynamics and instability will depend on its spatial scales, intrinsic frequency, and amplitude, with larger scales and amplitudes favoring higher altitudes.

The nonlinear behavior of this GW is described with the FVA model of Lund and Fritts (2011) with a computational domain having a horizontal dimension $\lambda_h = 60$ km, a vertical dimension extending from 100 to 200 km, and a spanwise dimension of 30 km so as to allow possible 3D structures to arise. We also specify an isotropic resolution of 200 m so as to resolve well any structures having spatial scales larger than ~1 km. A linear increase with time (over 2.5 T_b) of the horizontal velocity perturbation to a final amplitude of $u'_h = 5.5$ ms^{-1} at 100 km generates a GW having a structure conforming to the viscous anelastic dispersion relation that propagates to higher altitudes with increasing amplitude with time.

8.4.2.2 Nonlinear GW Evolution

The temporal evolution of the GW is displayed in Fig. 8.9 with cross sections of vorticity magnitude in the streamwise-vertical plane of GW propagation (with propagation upward and to the left) at times of ~23, 25, 36, and 41 T_b following the start of the linear ramp of the GW amplitude at 100 km. At the earliest time shown, the underlying GW structure is clearly seen, with alternating phases exhibiting larger (smaller) vorticity magnitudes and lesser (greater) modulation of the initial plane wave structure at a common spanwise position. It requires only another 2 T_b, however, for significant initial overturning to occur, beginning as high at ~160 km where the GW amplitude is a maximum at this time. Once overturning begins, the instabilities exhibit a cascade to smaller scales of motion, display clear 3D structure (see below), and gradually extend down along the unstable GW phases to lower altitudes. As these instability dynamics penetrate to lower altitudes and higher densities, Re increases and enables an expanded range of turbulence scales, with the smallest scales at ~1 to 2 km in the last image. Also accompanying the descent of the instabilities is a cessation of instability at higher altitudes. This occurs because instability at lower altitudes limits the

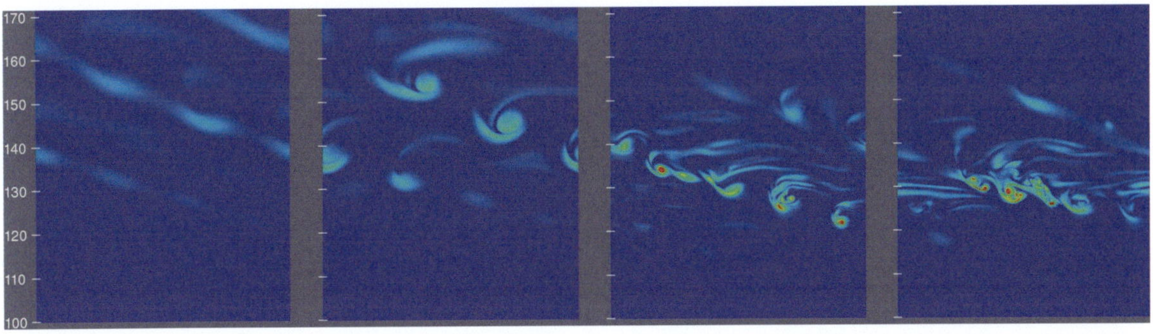

Fig. 8.9 Streamwise-vertical cross section of vorticity magnitude from an anelastic 3D simulation of GW breaking between 100 and 170 km altitude spanning a time of ~90 min (following Lund and Fritts, 2011). Overturning and turbulence generation extend to altitudes of ~160 km, approximately 3 scale heights above the "turbopause". The GW had a horizontal wavelength of $\lambda_x = 60$ km, an initial vertical wavelength of $\lambda_z \sim 20$ km, a horizontal perturbation velocity of $u' \sim 5.5$ ms^{-1} at 100 km, a horizontal phase speed of 55 ms^{-1}, and an intrinsic frequency of $\omega = N/10^{1/2}$ at 100 km. Background temperature increased from ~280 to ~900 K from 100 to 200 km

GW amplitudes above. Indeed, we expect typical GW breaking events in the thermosphere (and more generally where density decreases contribute to rapid amplitude growth and the tendency for instability) to exhibit such downward penetration. Instability continues to penetrate to lower altitudes until the GW amplitude is no longer sufficient to initiate overturning.

These dynamics bear a striking resemblance to the initial stages of the GW breaking events described by Fritts et al. (2009b, c) employing a Boussinesq incompressible code at much higher *Re*. Common features include

1. initial instability structures composed of streamwise-aligned counter-rotating vortices closed at their upstream (leftward) ends,
2. strong initial vortex dynamics in the high-shear zones at the lower edge of the least stable phase of the GW,
3. 3D vortex interactions that drive an enstrophy cascade to smaller scales of motion,
4. significant GW amplitude reductions accompanying GW breaking and instability, and
5. continued vertical propagation of the surviving GW having a smaller amplitude.

Additional evidence for the location and character of these instability dynamics within the GW field is provided with 3D images spanning the last 12 T_b of the evolution shown in Fig. 8.10. These provide views of isosurfaces of the second (negative) eigenvalue, λ_2, of the tensor defined as $\mathbf{L}=\mathbf{\Omega}^2+\mathbf{S}^2$, where $\mathbf{\Omega}$ and $\mathbf{S}$ are the rotation and strain tensors, with anti-symmetric and symmetric components $\mathbf{\Omega}_{ij}=(\partial_i u_j-\partial_j u_i)/2$ and $\mathbf{S}_{ij}=(\partial_i u_j+\partial_j u_i)/2$ (see Fig. 8.9). These define the most rapidly rotating vortex structures within a turbulent flow (Jeong and Hussain, 1995). The isosurfaces are also colored by vorticity magnitude, confirming that the lower edges of the initial, larger-scale structures are within the highly-sheared region beneath the upward- and leftward-moving least stable GW phase. They also reveal the "tube-like" character of the turbulence throughout the enstrophy cascade. But perhaps the most surprising aspect of this GW breaking DNS is the suggestion that turbulence generation should be expected to accompany large-scale and large-amplitude GWs penetrating well into the thermosphere.

8.5 Discussion and Conclusions

This paper began with a brief review of some of the considerable evidence for GW influences in the thermosphere and ionosphere. Observations have come from many measurement systems, among them ground-based radars, GPS receivers, ionosondes, airglow imagers, and FPIs, in situ measurements, and remote sensing from satellites. Theoretical and modeling studies have likewise indicated a potential for deep GW penetration and diverse effects. Collectively,

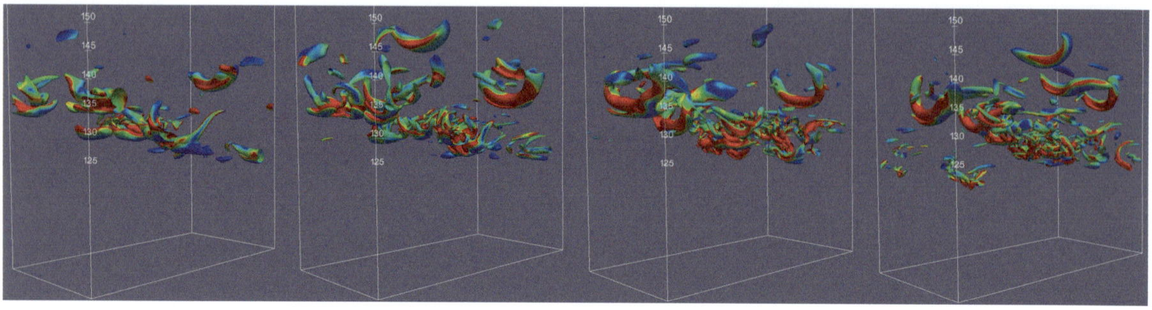

Fig. 8.10 3D views of λ_2 isosurfaces colored by vorticity magnitude showing the vortex scales and orientations extending to altitudes of ~150 km spanning the last ~60 min of the evolution displayed in Fig. 8.8

measurements, modeling, and theory have suggested GW influences including

1. penetration from sources in the lower atmosphere to ~500 km or above, depending on GW scales and frequencies,
2. large neutral and plasma density and velocity perturbations,
3. contributions to seeding of plasma instabilities,
4. local momentum deposition, strong mean forcing, and heating, and
5. radiation of large-scale GWs from body forcing events and auroral sources that propagate over very large scales.

Until recently, however, available tools lacked the ability to define the quasi-linear and nonlinear influences of wave-mean flow interactions, wave-wave interactions, and local instability dynamics accompanying GW attaining large amplitudes, momentum fluxes, and mean-flow accelerations at various altitudes in the TI. In particular, ray tracing theory makes predictions of GW scales and amplitudes that ultimately render the assumptions underlying the theory, and its predictions, suspect. This is particularly true for the largest GW scales predicted to achieve large amplitudes and have strong influences extending to the highest altitudes in the TI.

The recent development of a high-resolution 3D finite-volume anelastic model allow studies of large-scale GW propagation, wave-wave and wave-mean flow interactions, and instability dynamics in vertical domains spanning many scale heights (Lund and Fritts, 2011). Initial applications of this model to (1) GW self acceleration accompanying induced mean flows for transient, but horizontally-extended, GW packets and (2) instability and turbulence accompanying attainment of large GW amplitudes have revealed dynamics having surprising and significant implications for the MLT and at higher altitudes. Those discussed here include

1. significant self acceleration of GW packets that are vertically localized and horizontally extended,
2. corresponding changes in the GW structure, with steepening phase structure at the leading edge and reduced phase slopes at the trailing edge of the GW packet, with "self-acceleration" instability occurring for sufficiently strong events,
3. attainment of large amplitudes for large-scale GWs entering the TI that lead to GW breaking, amplitude reductions, instability, and 3D turbulence that may extend to significant altitudes above the "turbopause", and
4. instability characteristics that closely resemble the same dynamics anticipated for higher-Re flows at lower altitudes.

The implications of these anelastic simulation results are significant for the MLT because they reveal both (1) that GW breaking and turbulence generation should be anticipated well into the TI for sufficiently large GW amplitudes, but with minimum spatial scales dependent on the local Re, and typically at much larger scales than associated with turbulence at lower altitudes, and (2) that quasi-linear GW-mean flow interactions accompany large-amplitude GWs and will influence GW behavior, MLT and TI penetration altitudes, and the potential for GW instability at high altitudes.

The "turbopause", above which neutral turbulence is believed to largely cease due to increasing kinematic viscosity and thermal diffusivity, is generally regarded to occur at altitudes between ~105 and 110 km, with variability that reflects latitudinal and seasonal variations in GW sources and filtering conditions and lower thermospheric temperatures (Danilov, 1984; Hall et al., 1998). It has also been recognized that the turbopause at any location and time likely represents an altitude range over which turbulence varies from frequent to less frequent because of the intermittent nature of the underlying GW field and increasing viscous constraints on GW instabilities with increasing altitude (Hines, 1991). We are unaware of any studies suggesting a turbopause altitude at ~150 km or higher, however, so the strong GW instability and large-scale turbulence structures seen in Figs. 8.9 and 8.10 for quite reasonable GW scales and amplitudes at lower altitudes will come as a surprise to many. Indeed, the expected lower altitude of the turbopause based on measurements of atmospheric composition suggests that GW breaking and turbulence at much higher altitudes is an intermittent process. While we have not yet examined the potential instability of larger-scale GWs achieving even higher altitudes, it is quite possible that we should anticipate such events occurring even higher when propagation conditions allow large GW amplitudes at large spatial scales.

Future anelastic simulations will also enable evaluation of the GW dissipation and momentum flux profiles anticipated by viscous ray tracing, the efficiency of continued GW propagation to higher altitude following instability and turbulence below, and the scales and effects of secondary GW generation accompanying instability and momentum deposition, among other studies. A greater appreciation of these various dynamics will also enable a parameterization of these influences in the TI that is unavailable to modelers at present.

Acknowledgments Preparation of this paper was supported by AFOSR contract FA9550-09-C-0197, NASA contract NNH09CF40C, and NSF grant ATM-0836407. We also acknowledge the DoD High Performance Computing Modernization Office for computational resources employed for the simulations described.

References

Abdu MA, Kherani EA, Batista IS, de Paula ER, Fritts DC, Sobral JHA (2009) Gravity wave initiation of equatorial spread F/plasma bubble irregularities based on observational data from the SpreadFEx campaign. Ann Geophys (SpreadFEx special issue) 27:2607–2622

Bretherton FP (1969) Momentum transport by gravity waves. Quart J R Meteorol Soc 404:213–243

Chimonas G, Hauser HM, Bennett RD (1996) The excitation of ducted modes by passing internal waves. Phys Fluids (8):1486. doi:10.1063/1.868925

Danilov AD (1984) Direct and indirect estimates of turbulence around the turbopause. Adv Space Res 4(4):67–78

Djuth FT, Sulzer MP, Elder JH, Wickwar VB (1997) High-resolution studies of atmosphere-ionosphere coupling at Arecibo Observatory, Puerto Rico. Radio Sci 32: 2321–2344

Djuth FT, Sulzer MP, Gonzales SA, Mathews JD, Elder JH, Walterscheid RL (2004) A continuum of gravity waves in the Arecibo thermosphere? Geophys Res Lett 31. doi:10.1029/2003GL019376

Drazin, PG (1977) On the instability of an internal gravity wave. Proc R Soc Lond 356:411–432

Earle GD, Musumba AM, Vadas SL (2008) Satellite-based measurements of gravity wave-induced midlatitude plasma density perturbations. J Geophys Res 113:A03303. doi:10.1029/2007JA012766

Einaudi F, Hines CO (1970) WKB approximation in application to acoustic-gravity waves. Can J Phys 48(12):1458–1471

Fritts DC, Abdu MA, Batista BR, Batista IS, Batista PP, Buriti R, Clemesha BR, Dautermann T, de Paula E, Fechine BJ, Fejer B, Gobbi D, Haase J, Kamalabadi F, Kherani EA, Laughman B, Lima PP, Liu H-L, Medeiros A, Pautet D, Sao Sabbas F, Sobral JHA, Stamus P, Takahashi H, Taylor MJ, Vadas SL, Vargas F, Wrasse C (2009a) Overview and summary of the Spread F Experiment (SpreadFEx). Ann Geophys 27:2141–2155

Fritts DC, Alexander MJ (2003) Gravity dynamics and effects in the middle atmosphere. Rev Geophys 41. doi:10.1029/2001RG000106

Fritts DC, Arendt S, Andreassen O (1998) Vorticity dynamics in a breaking internal gravity wave, 2. Vortex interactions and transition to turbulence. J Fluid Mech 367:47–65

Fritts DC, Dunkerton TJ (1984) A quasi-linear study of gravity wave saturation and self acceleration. J Atmos Sci 41: 3272–3289

Fritts DC, Garten JF, Andreassen O (1996) Wave breaking and transition to turbulence in stratified shear flows. J Atmos Sci 53:1057–1085

Fritts DC, Isler JR, Andreassen O (1994) Gravity wave breaking in two and three dimensions, 2. Three-dimensional evolution and instability structure. J Geophys Res 99: 8109–8123

Fritts DC, Vadas SL (2008) Gravity wave penetration into the thermosphere: sensitivity to solar cycle variations and mean winds. Ann Geophys (SpreadFEx special issue) 26: 3841–3861

Fritts DC, Vadas SL, Riggin DM, Abdu MA, Batista IS, Takahashi H, Medeiros A, Kamalabadi F, Liu HL, Fejer BJ, Taylor MJ (2008) Gravity wave and tidal influences on equatorial spread F based on observations during the Spread F Experiment (SpreadFEx). Ann Geophys (SpreadFEx special issue) 26:3235–3252

Fritts DC, Vadas SA, Yamada Y (2002) An estimate of strong local gravity wave body forcing based on OH airglow and meteor radar observations. Geophys Res Lett 29(10). doi:10.1029/2001GL013753

Fritts DC, Wang L, Werne J, Lund T, Wan K (2009b) Gravity wave instability dynamics at high Reynolds numbers, 1: wave field evolution at large amplitudes and high frequencies. J Atmos Sci 66:1126–1148. doi:10.1175/2008JAS2726.1

Fritts DC, Wang L, Werne J, Lund T, Wan K (2009c) Gravity wave instability dynamics at high Reynolds numbers, 2: turbulence evolution, structure, and anisotropy. J Atmos Sci 66:1149–1171. doi:10.1175/2008JAS2727.1

Fuller-Rowell TJ, Codrescu MV, Moffett RJ, Quegan S (1994) Response of the thermosphere and ionosphere to geomagnetic storms. J Geophys Res 99(A3):3893–3914

Fuller-Rowell TJ, Millward GH, Richmond AD, Codrescu MV (2002) Storm-timechanges in the upper atmosphere at low latitudes. J Atmos Solar-Terr Phys 65:1383–1391

Grimshaw RHJ (1977) The modulation of an internal gravity wave packet and the resonance with the mean motion. Stud App Math 56:241–266

Hall CM, Manson AH, Meek CE (1998), Seasonal variation of the turbopause: One year of turbulence investigation at 69°N by the joint University of Tromsø/University of Saskatchewan MF radar. J Geophys Res 103(D22): 28,769–28,773

Hasselmann K (1967) A criterion for non-linear wave stability. J Fluid Mech 30:737–739

Hickey MP, Cole KD (1987) A quartic dispersion relation for internal gravity waves in the thermosphere. J Atmos Solar-Terr Phys 49:889–899

Hickey MP, Cole KD (1988) A numerical model for gravity wave dissipation in the atmosphere. J Atmos Solar-Terr Phys 50(8):689–697

Hines CO (1960) Internal gravity waves at ionospheric heights. Can J Phys 38(11):1441–1481

Hines CO (1991) The saturation of gravity waves in the middle atmosphere. Part III: formation of the turbopause and of turbulent layers beneath it. J Atmos Sci 48(11):1380–1385

Hocke K, Schlegel K (1996) A review of atmospheric gravity waves and traveling ionospheric disturbances: 1982–1995. Ann Geophys 14:917–940

Hocke K, Tsuda T (2001) Gravity waves and ionospheric irregularities over tropical convection zones observed by GPS/MET radio occultation. Geophys Res Lett 28(14):2815–2818

Horinouchi T (2004) Simulated breaking of convectively generated mesoscale gravity waves and airglow modulation. J Atmos Solar-Terr Phys 66(6–9):755–767

Horinouchi T, Nakamura T, Kosaka J (2007) Convectively generated mesoscale gravity waves simulated throughout the middle atmosphere. Geophys Res Lett 29(21). doi:10.1029/2002GL016069

Huang CS, Kelley MC (1996) Nonlinear evolution of equatorial spread-F. 2. Gravity wave seeding of Rayleigh-Taylor instability J Geophys Res 101:293

Huang CS, Kelley MC, Hysell DL (1993) Nonlinear Rayleigh-Taylor instabilities, atmospheric gravity waves, and equatorial spread-F. J Geophys Res 98:15,631

Huang C, Zhang S, Yi F (2002) A numerical study of the nonlinear propagation of gravity-wave packets excited by temperature disturbance. Chi J Space Sci 22:330–338

Huang C, Zhang S, Yi F (2009) Gravity wave excitation through resonant interaction in a compressible atmosphere. Geophys Res Lett 36(1):L01803. doi:10.1029/2008GL035575

Hysell DL, Kelley MC, Swartz WE, Woodman RF (1990) Seeding and layering of equatorial spread-F. J Geophys Res 95:17,253

Innis JL, Conde M (2002) Characterization of acoustic–gravity waves in the upper thermosphere using dynamics explorer 2 wind and temperature spectrometer (WATS) and neutral atmosphere composition spectrometer (NACS) data. J Geophys Res 107(A12):1418. doi:10.1029/2002JA009370

Innis JL, Greet PA, Dyson PL (2001) Evidence for thermospheric gravity waves in the southern polar cap from ground-based vertical velocity and photometric observations. Ann Geophys 19:533–543

Jeong J, Hussain F (1995) On the identification of a vortex. J Fluid Mech 285:69–94

Kamalabadi F et al (2009) Electron densities in the lower thermosphere from GUVI 1356 tomographic inversions in support of SpreadFEx. Ann Geophys (SpreadFEx special issue) 27:2439–2448

Kherani EA, Abdu MA, de Paula ER, Fritts DC, Sobral JHA, de Meneses FC Jr (2009) The impact of gravity waves rising from convection in the lower atmosphere on the generation and nonlinear evolution of equatorial plasma bubbles. Ann Geophys (SpreadFEx special issue) 27:511–523

Kirchengast G, Hocke K, Schlegel K (1995), Gravity waves determined by modeling of traveling ionospheric disturbances in incoherent scatter radar measurements. Radio Sci 30(5):1551–1567

Klostermeyer J (1969) Gravity waves in the F-region. J Atmos Solar-Terr Phys 31(1):25–45. doi:10.1016/0021-9169(69)90079-8

Klostermeyer J (1972) Numerical calculation of gravity wave propagation in a realistic thermosphere. J Atmos Solar-Terr Phys 34(5):765–774. doi:10.1016/0021-9169(72)90109-2

Klostermeyer J (1978) Nonlinear investigation of the spatial resonance effect in the nighttime equatorial F region. J Geophys Res 83:3753–3760

Klostermeyer J (1991) Two-and three-dimensional parametric instabilities in finite amplitude internal gravity waves. Geophys Astrophys Fluid Dyn 61:1–25

Kudeki E, Akgiray A, Milla M, Chau JL, Hysell DL (2007) Equatorial spread-F initiation: post-sunset vortex, thermospheric winds, gravity waves. J Atmos Solar-Terr Phys 69:2416–2427

Lane TP, Reeder MJ, Clark TL (2001) Numerical modeling of gravity wave generation by deep tropical convection. J Atmos Sci 58(10):1249–1274

Lighthill MJ (1978) Waves in fluids. Cambridge University Press, Cambridge

Livneh DJ, Seker I, Djuth FT, Mathews JD (2007) Continuous quasiperiodic thermospheric waves over Arecibo. J Geophys Res 112:A07313. doi:10.1029/2006JA012225

Lombard, PN, Riley JJ (1996) On the breakdown into turbulence of propagating internal waves. Dyn Atmos Ocean 23: 345–355

Lund T, Fritts DC (2011) Gravity wave breaking and turbulence generation in the thermosphere. Geophys Res Lett (to be submitted)

Luo Z, Fritts DC (1993) Gravity wave excitation by geostrophic adjustment of the jet stream, part ii: three-dimensional forcing. J Atmos Sci 50:104–115

Ma SY, Schlegel K, Xu JS (1998) Case studies of the propagation characteristics of auroral TIDS with EISCAT CP2 data using maximum entropy cross-spectral analysis. Ann Geophys 16(2):161–167. doi:10.1007/s00585-998-0161-3

Marks CJ, Eckermann SD (1995) A three-dimensional nonhydrostatic ray-tracing model for gravity waves: formulation and preliminary results for the middle atmosphere. J Atmos Sci 52(11):1959–1984

Mayr HG, Harris I, Herrero FA, Spencer NW, Varosil F, Pesnell WD (1990) Thermospheric gravity waves: observations and interpretation using the transfer function model (TFM). Space Sci Revs 54:297–375. doi:10.1007/BF00177800

McClure JP, Hanson WB, Hoffman JF (1977) Plasma bubble and irregularities in the equatorial ionosphere. J Geophys Res 82:2650

McClure JP, Singh S, Bamgboye DK, Johnson FS, Kil H (1998) Occurrence of equatorial F region irregularities: evidence for tropospheric seeding. J Geophys Res 103:29,119–29,135

McIntyre ME (1973) Mean motions and impulse of a guided internal gravity wave packet. J Fluid Mech 60:801–811

Mendillo M, Baumgardner J, Nottingham D, Aarons J, Reinisch B, Scali J, Kelley M (1997) Investigation of thermospheric-ionospheric dynamics with 6300-Å images from the Arecibo observatory. J Geophys Res 102(A4):7331–7343

Mied RR (1976) The occurrence of parametric instabilities in finite-amplitude internal gravity waves. J Fluid Mech 78:763–784

Nicolls MJ, Kelley MC, Coster AJ, Gonzalez SA, Makela JJ (2004) Imaging the structure of a large-scale TID using ISR and TEC data. Geophys Res Lett 31:L09812. doi:10.1029/2004GL019797

Oliver WL, Otsuka Y, Sato M, Takami T, Fukao S (1997) A climatology of F region gravity wave propagation over the middle and upper atmosphere radar. J Geophys Res 102:14,499–14,512

Ogawa T, Balan N, Otsuka Y, Shiokawa K, Ihara C, Shimomai T, Saito A (2002) Obervations and modeling of 630 nm airglow and total electron content associated with traveling ionospheric disturbances over Shigaraki, Japan. Earth Planets Space 54:45–56

Sentman DD, Wescott EM, Picard RH, Winick JR, Stenbaek-Nielson HC, Dewan EM, Moudry DR, São Sabbas FT, Heavner MJ (2003) Simultaneous observation of mesospheric gravity waves and sprites generated by a Midwestern thunderstorm. J Atmos Solar-Terr Phys 65:537–550

Shiokawa K et al (2002) A large-scale traveling ionospheric disturbance during the magnetic storm of 15 September 1999. J Geophys Res 107(A6):1088. doi:10.1029/2001JA000245

Sidi C, Teitelbaum H (1978) Thin shear turbulent layers within the lower thermosphere induced by non-linear interaction between tides and gravity waves. J Atmos Solar-Terr Phys 40(5):529–540. doi:10.1016/0021-9169(78)90090-9

Snively JB, Pasko VP (2008) Excitation of ducted gravity waves in the lower thermosphere by tropospheric sources. J Geophys Res 113:A06303. doi:10.1029/2007JA012693

Sonmor LJ, Klaassen GP (1997) Toward a unified theory of gravity wave stability. J Atmos Sci 54(22):2655–2680

Sutherland BR (1999) Propagation and reflection of large amplitude internal gravity waves. Phys Fluids 11:1081–1090

Sutherland BR (2000) Internal wave reflection in uniform shear. Quart J R Meteorol Soc 126(570):3255–3286

Sutherland BR (2001) Finite-amplitude internal wavepacket dispersion and breaking. J Fluid Mech 429:343–380

Sutherland BR (2006) Internal wave instability: wave-wave versus wave-induced mean flow interactions. Phys Fluids 18:074107. doi:10.1063/1.2219102

Vadas SL (2007) Horizontal and vertical propagation, and dissipation of gravity waves in the thermosphere from lower atmospheric and thermospheric sources. J Geophys Res 112:A06305. doi:10.1029/2006JA011845

Vadas SL, Fritts DC (2001) Gravity wave radiation and mean responses to local body forces in the atmosphere. J Atmos Sci 58:2249–2279

Vadas SL, Fritts DC (2002) The importance of spatial variability in the generation of secondary gravity waves from local body forces. Geophys Res Lett 29(20). doi: 10.1029/2002GL015574

Vadas SL, Fritts DC (2004) Thermospheric responses to gravity waves arising from mesoscale convective complexes. J Atmos Solar-Terr Phys 66:781–804

Vadas SL, Fritts DC (2005) Thermospheric responses to gravity waves: influences of increasing viscosity and thermal diffusivity. J Geophys Res 110:D15103. doi:10.1029/2004JD005574

Vadas SL, Fritts DC (2006) The influence of increasing temperature and solar variability on gravity wave structure and dissipation in the thermosphere. J Geophys Res (TIMED special issue) 111:A10812. doi:10.1029/2005JA011510

Vadas SL, Fritts DC (2009) Reconstruction of the gravity wave field from convective plumes via ray tracing. Ann Geophys (SpreadFEx special issue) 27(1):147–177

Vadas SL, Liu H-L (2009) Generation of large-scale gravity waves and neutral winds in the thermosphere from the dissipation of convectively generated gravity waves. J Geophys Res 114:A10310. doi:10.1029/2009JA014108

Vadas SL, Nicolls M (2008) Using PFISR measurements and gravity wave dissipative theory to determine the neutral, background thermospheric winds. Geophys Res Lett 35:L02105. doi:10.1029/2007GL031522

Vadas SL, Nicolls M (2009) Temporal evolution of neutral, thermospheric winds and plasma response using PFISR measurements of gravity waves. J Atmos Solar-Terr Phys 71:740–770

Vadas SL, Taylor MJ, Pautet P-D, Stamus P, Fritts DC, Liu H-L, Sao Sabbas F, Rampinelli VT, Batista P, Takahashi H (2009) Convection: the likely source of the medium-scale gravity waves observed in the OH airglow layer near Brasilia, Brazil, during the SpreadFEx campaign. Ann Geophys (SpreadFEx special issue) 27:231–259

Walterscheid RL, Schubert G (1990) Nonlinear evolution of an upward propagating gravity wave: overturning, convection, transience and turbulence. J Atmos Sci 47:101–125

Whitham GB (1965) A general approach to linear and nonlinear dispersive waves using a Lagrangian. J Fluid Mech 22: 273–283

Whitham GB (1974) Linear and nonlinear waves. Wiley, New York, NY

Yamada Y, Fukunishi H, Nakamura T, Tsuda T (2001) Breakdown of small-scale quasi-stationary gravity wave and transition to turbulence observed in OH airglow. Geophys Res Lett 28:2153–2156

Yeh KC, Liu CH (1981) The instability of atmospheric gravity waves through wave-wave interactions. J Geophys Res 86(C10):9722–9728

Yue J, Vadas SL, She C-Y, Nakamura T, Reising S, Krueger D, Liu H-Li, Stamus P, Thorsen D, Lyons W, Li T (2009) A study of OH imager observed concentric gravity waves near Fort Collins on 11 May 2004. J Geophys Res 114:D06104. doi:10.1029/2008JD011244

Chapter 9

Neutral Winds and Densities at the Bottomside of the F Layer from Primary and Secondary Gravity Waves from Deep Convection

Sharon L. Vadas and Han-Li Liu

Abstract We model the primary gravity waves (GWs) excited by overshooting convective plumes in a 20° × 20° region in central Brazil for 6 h during the evening on 01 October, 2005. We ray trace these GWs into the thermosphere. At $z = 250$ km (near the bottomside of the F layer), the surviving GWs have horizontal wavelengths of $\lambda_H \sim 200-300$ km, horizontal velocity amplitudes of $u', v' \leq 15$ m/s, and neutral density perturbations of $\rho'/\overline{\rho} \leq 2\%$. We compute the thermospheric body forces, and input them into the Thermosphere-Ionosphere-Mesosphere-Electrodynamics General Circulation Model (TIME-GCM). These forces create a "mean" eastward wind perturbation at $z \sim 150$ km of 200–340 m/s which lasts for ~3 h. These forces also excite large-scale secondary GWs in all directions for at least 3 h with horizontal wavelengths of $\lambda_H \sim 4000-5000$ km, horizontal phase speeds of $c_H \sim 500-600$ m/s, and density perturbations of 5–15% at $z = 250$ km. Prior to and after sunset, the wind perturbations are eastward with $u' \sim 75-150$ m/s at $z = 250$ km. This occurs over the convective region, where the winds from 2 large vortices (created by the forces) converge. In other areas, the winds are west, south, or northward. It is likely that these forces (and the GWs they excite) significantly affect the F region dynamo and the seeding of equatorial plasma bubbles (EPBs).

S.L. Vadas (✉)
Colorado Research Associates Division, NorthWest Research Associates, Boulder, CO, USA
e-mail: vasha@cora.nwra.com; vasha@co-ra.com

9.1 Introduction

GWs have been observed in the thermosphere for decades as TIDs (traveling ionospheric disturbances) (Hocke and Schlegel, 1996; Oliver et al., 1997; Djuth et al., 1997, 2004). From observations and modeling, it has been thought that large-scale, fast GWs are created from the aurora (Hocke and Schlegel, 1996; Richmond, 1978; Hickey and Cole, 1988), while medium scale GWs with $c_H < 250\,\mathrm{ms}^{-1}$ are created in the lower atmosphere (Georges, 1968; Waldock and Jones, 1986; Crowley et al., 1987; Ogawa et al., 1987). GWs excited by convection are frequently observed in the stratosphere (e.g., (Pfister et al., 1993; Dewan et al., 1998)). If the winds are small, these GWs appear as concentric rings in airglow layers near the mesopause, a stunning confirmation of their source (Taylor and Hapgood, 1988; Sentman et al., 2003; Suzuki et al., 2007; Yue et al., 2009). These patterns can last for hours, with $\lambda_H \sim 20-150$ km. Near the equator, tropical convection is the most important source of GWs which can propagate into the thermosphere.

Although models show that convective "primary" GWs can propagate to the upper mesosphere and break there (Holton and Alexander, 1999; Horinouchi et al., 2002), some convective GWs with large vertical wavelengths λ_z and initially small amplitudes can propagate into the thermosphere (Vadas and Fritts, 2006; Vadas, 2007). Primary GWs with $\lambda_H \geq 100$ km and $c_H \geq 200$ m/s can propagate to $z = 200-300$ km before dissipating (Vadas, 2007). Less energetic GWs dissipate at $z \sim 110-220$ km, creating horizontal thermospheric body forces (Vadas and Fritts, 2006; Vadas and Liu 2009). A single convective plume creates an acceleration of $\sim 1\,\mathrm{m/s}^2$ and "mean" winds of ~400 m/s at

M.A. Abdu, D. Pancheva (eds.), A. Bhattacharyya (Coed.), *Aeronomy of the Earth's Atmosphere and Ionosphere*, IAGA Special Sopron Book Series 2, DOI 10.1007/978-94-007-0326-1_9, © Springer Science+Business Media B.V. 2011

$z \sim 180$ km, and excites large-scale "secondary" GWs with $\lambda_H \sim 2200$ km and $c_H \sim 500$ m/s which propagate up to at least $z \sim 420$ km (Vadas and Liu, 2009). A recent ray trace study showed that the TIDs observed at the bottomside of the F layer near Wallops Island, USA (75.47° W and 37.95° N) were secondary GWs excited by the thermospheric body forces created by deep convection in Tropical Storm Noel (Vadas and Crowley, 2010).

Models show that GWs at the bottomside of the F layer can seed equatorial spread F (ESF) and EPBs through the Rayleigh-Taylor (RT) instability (Kelley et al., 1981; Hysell et al., 1990; Huang and Kelley, 1996a). Radar results have correlated wave structure in the bottomside ionosphere with the separations of plumes that develop thereafter (Tsunoda and White, 1981; Hysell et al., 2006). One study found wavelengths of 200–400 km (Miller et al., 2010). Another study showed a linear relationship between the wavelengths of the GWs observed near the mesopause with EPB spacings (Takahashi et al., 2009). Rottger (1973) found a median spacing of ~380 km for periodic equatorial irregularities. A recent study found that the spacings between multiple, periodic EPBs (100–800 km) agreed well with the spectrum of secondary GWs from the Wallops Island study (Makela et al., 2010).

At the dip (magnetic) equator, GW seeding altitudes are $z = 200-300$ km. These altitudes can be lower farther from the dip equator where the field lines are lower, if electric field perturbations are created by the GWs which then map to the magnetic equator and create the RT instability (Fritts et al., 2008). GW perturbations must be aligned with the magnetic field in order for the field-integrated perturbations to be sufficiently large (Huang and Kelley, 1996b). When ESF occurs, it is nearly always observed after sunset, when the ionosphere descends. During this descent, spatial resonance can occur with approximately eastward propagating GWs, since plasma drifts in the eastward direction (Kelley, 1989). An eastward neutral wind is required to drive the F layer dynamo which creates the pre-sunset prereversal enhancement electric field (PRE)/vertical drift (Rishbeth, 1971; Abdu et al., 1995), although a recent study questions the need for a strong PRE in generating ESF (Tsunoda et al., 2010).

Although (Vadas and Liu, 2009) calculated the primary and secondary GWs that result from a single convective plume, convection generally lasts for hours and involves tens to hundreds of plumes. The purpose of this chapter is to calculate the primary and secondary GWs near the bottomside of the F layer that result from 6 h of deep convection.

9.2 Primary GWs from Convective Overshoot

Convective sources of gravity waves can be described equivalently as heating or momentum sources, as these sources are coupled through the vertical momentum equation. Many linear models of GW excitation from convection have been developed (Stull, 1976; Salby and Garcia, 1987; Alexander et al., 1995; Walterscheid et al., 2001; Beres, 2004). Here, we use an analytic, linear, Boussinesq model which describes the excitation of GWs from a convective plume envelope with diameter $\mathcal{D}_H$ via an upward acceleration of air (Vadas and Fritts, 2009). This model neglects moisture processes, and assumes that the air above the tropopause is stationary within the frame of the mean horizontal wind at the tropopause until a convective plume overshoots the tropopause and pushes the stratospheric air upwards. It solves the linear solutions in a locally unsheared environment with a constant buoyancy frequency.

GOES-12 satellite images covering 45–65° W and 0–20° S in central Brazil were available every 30 min during the 6 h period from 18:22 to 23:53 UT on 01 October 2005 (e.g. (Vadas et al., 2009)). Since the local time (LT) is LT=UT−3 h, this includes sunset at ~2230 UT. Deep convection peaked at 19–22 UT, and weakened rapidly after 23 UT. Using balloon soundings, the tropopause temperature was determined to be ~209 K. All plumes cooler than this temperature likely underwent convective overshoot (i.e., punched into the stratosphere where the temperature increases), and therefore were identified as generating GWs. One hundred and thirty six convective objects are identified; each is classified as a single convective plume, a convective cluster containing 2 or more tightly-clumped convective plumes, or a convective complex with 2 or more less-tightly-clumped convective plumes. The average plume diameter in each object ranges from $\mathcal{D}_H = 5$–20 km. The plume updraft velocities, w_{pl}, are estimated from Convective Available Potential Energy (CAPE) maps, and range from $w_{pl} \simeq 0$ to 63 m/s. Because the primary and secondary GW amplitudes are proportional to w_{pl} and w_{pl}^2, respectively, we only simulate those objects with $w_{pl} \geq 10$ m/s here. Examples of primary GW spectra excited by plumes and clusters are shown in Vadas et al. (2009). Complexes are defined as 4 plumes at the corners of a square, with adjacent plume centers separated by $4.5\mathcal{D}_H$.

The primary GWs are ray traced into the upper atmosphere using an anelastic GW formulation (Vadas and Fritts, 2005). The wind and temperatures are determined from balloon soundings, meteor winds, and the TIME-GCM (Vadas et al., 2009). The GW parameters are saved in a 4D box with grid spacings of 50 km × 50 km × 4 km × 10 min in x, y, z, and t. The GW fields (such as the velocity perturbations (u', v', w') and density perturbations ρ') are then reconstructed (Vadas and Fritts, 2009). We now also include the effects of parameterized wave breaking from self and wave-wave interactions via the use of Lindzen's saturation condition (Lindzen, 1981), as these effects are now known to be important in the thermosphere for large-amplitude waves (Yiğit et al., 2008, 2009). Without this effect, the non-dimensional wave amplitudes in this study would have been unrealistically large, ~5–20, because of the multiple clusters with large updraft velocities that occurred on this day. Including this effect decreases the altitude where the thermospheric body forces are maximum. We include this effect by ray tracing the GWs from each convective object, reconstructing the GW solution, and calculating the sum of the non-dimensional amplitude squared, A^2, in each (x, y, z, t) bin from all convective objects. We then run the ray trace model for all convective objects again, and reduce the amplitude of each GW which enters a bin with $A > 1$ to ensure that $A = 1$ (Smith et al., 1987). This also has the effect of altering the background flow via the deposition of momentum.

Figure 9.1 shows the primary GW density perturbations at $z = 250$ km, ρ', divided by the neutral background density $\overline{\rho}$. Lines of propagating GWs are seen. This is in contrast to $z = 150$ km, which shows mostly partial concentric rings (not shown). This difference occurs because of wind and dissipative filtering. The GWs propagate in a direction perpendicular to the phase lines, which in general is east and southeastward. A GW's horizontal wavelength, λ_H, is the perpendicular distance between the white (or black) contour lines. From 1945 to 2155 UT, λ_H decreases from ~300 to 200 km.

The horizontal velocity perturbations of the primary GWs are shown at 1945 UT and $z = 250$ km in Fig. 9.2. The horizontal and vertical velocity perturbations, (u', v', w'), are as large as 10–15 m/s. These values are typical of other times. Note that w' is

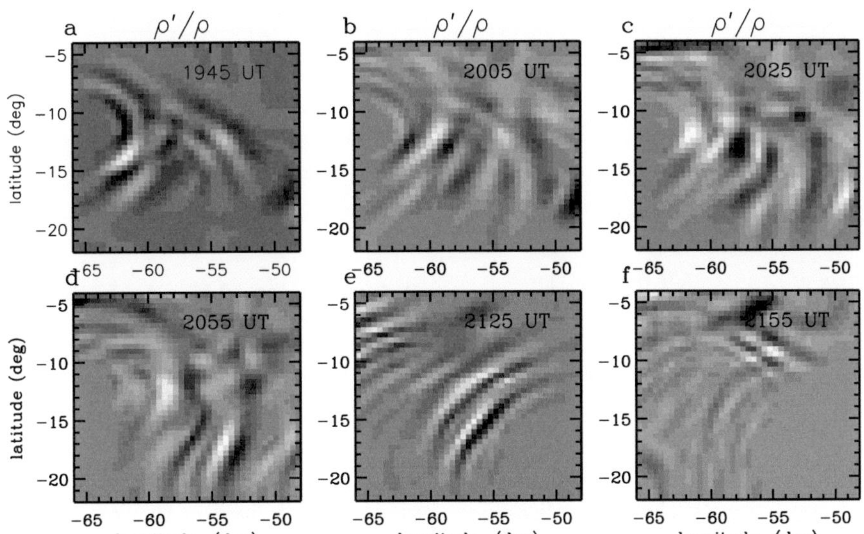

Fig. 9.1 Primary GW density perturbations, $\rho'/\overline{\rho}$, at $z =$ 250 km from ray tracing. (**a–f**): 1945, 2005, 2025, 2055, 2125, and 2155 UT, respectively. Maximum positive (negative) values are *white* (*black*). The maximum values of $|\rho'/\overline{\rho}|$ are (**a–c**): 2, 2, and 1%, respectively. (**d–f**): 2, 2, and 0.3%, respectively

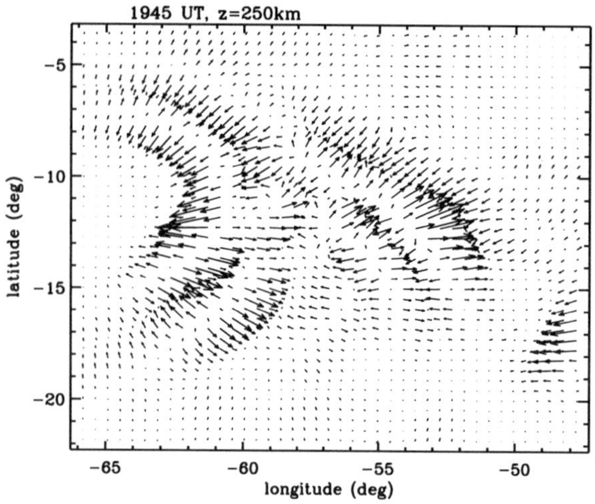

Fig. 9.2 Vector plot of the primary GW horizontal velocity perturbations at 1945 UT and $z = 250$ km from ray tracing. The maximum values of $|u'|$, $|v'|$, and $|w'|$ on this slice are 14, 7, and 12 m/s, respectively, where u', v', and w' denote the GW zonal, meridional and vertical velocity perturbations in geographic coordinates

comparable to u', v' because the primary GW intrinsic frequency, ω_{Ir}, is nearly equal to the buoyancy frequency.

9.3 Thermospheric Body Forces from Primary GW Dissipation

We determine the body forces which accompany primary GW dissipation and/or saturation in the thermosphere by computing the vertical divergences of the zonal and meridional momentum fluxes (per unit mass). The zonal and meridional components of the body force in geographic coordinates are

$$F_x = -\frac{1}{\overline{\rho}}\frac{\partial(\overline{\rho}\overline{u'w'^*})}{\partial z}, F_y = -\frac{1}{\overline{\rho}}\frac{\partial(\overline{\rho}\overline{v'w'^*})}{\partial z}, \quad (9.1)$$

respectively (Andrews et al., 1987). Here, overlines denote averages over 1–2 wave periods and wavelengths, and * denotes the complex conjugate. If the GWs are not dissipating or saturating, $(\overline{\rho}\overline{u'w'^*})$ and $(\overline{\rho}\overline{v'w'^*})$ are constant with altitude, so that $F_x = F_y = 0$.

Because the TIME-GCM includes a GW parameterization scheme up to $z = 120$ km, we smoothly zero our body forces below 120 km in order to avoid double counting the GW effects. We choose a smoothing range of $\delta = 20$ km. For altitudes below $z_i = 100$ km, we set the body forces equal to zero. For altitudes between z_{i} and $z_i + \delta$, we multiply the body forces by

$$1 - \cos^2\left[\pi(z - z_i)/(2\delta)\right]. \quad (9.2)$$

We find that the largest value of F_x is eastward, $F_x = 0.83$ m/s^2, and occurs at 60°W, 12°S, $z = 152$ km, and 1915 UT. The largest value of F_y is southward, $F_y = -0.89$ m/s^2, and occurs at 53°W, 20°S, $z = 164$ km, and 1925 UT. These accelerations are consistent with previous results from a single convective plume (Vadas and Liu, 2009). However, the altitude of the thermospheric body force is lower here by ~30–40 km because of wave saturation.

Figure 9.3 shows horizontal slices of (F_x, F_y) at $z = 150$ km. The strongest accelerations last for only 2 h, from 19 to 21 UT, and are east, north, and southward. The accelerations are patchy and variable because of constructive and destructive interference between GWs from different convective objects. They have partial ring shapes because the GW phase lines are partial rings at this altitude.

GWs moving against the background wind propagate to higher altitudes in the thermosphere (Fritts and Vadas, 2008). Those primary GWs which dissipate at $z = 150$ km are propagating in a direction which is opposite to the background wind direction 1–2 neutral density scale heights below $z = 150$ km (i.e., at $z \sim 120–130$ km) (Vadas and Liu, 2009). The dominant winds which filter the GWs in the lower thermosphere are semidiurnal and diurnal tides, which change on time scales of 6–12 h. This is why the body forces in Fig. 9.3 are directed in a similar direction from 19 to 21 UT.

9.4 Generated Neutral Mean Winds and Secondary Gravity Waves

Thermospheric body forces excite secondary GWs and create "mean" neutral winds (Vadas and Liu, 2009; Vadas, 2010a, b). Because the spatial variability of the forces in Fig. 9.3 are 1–5°, we expect the secondary GWs to have horizontal wavelengths of $\lambda_H \sim 100–5000$ km, with a peak at $\lambda_H \sim 200–500$ km. (Note that the secondary GWs at the bottomside of

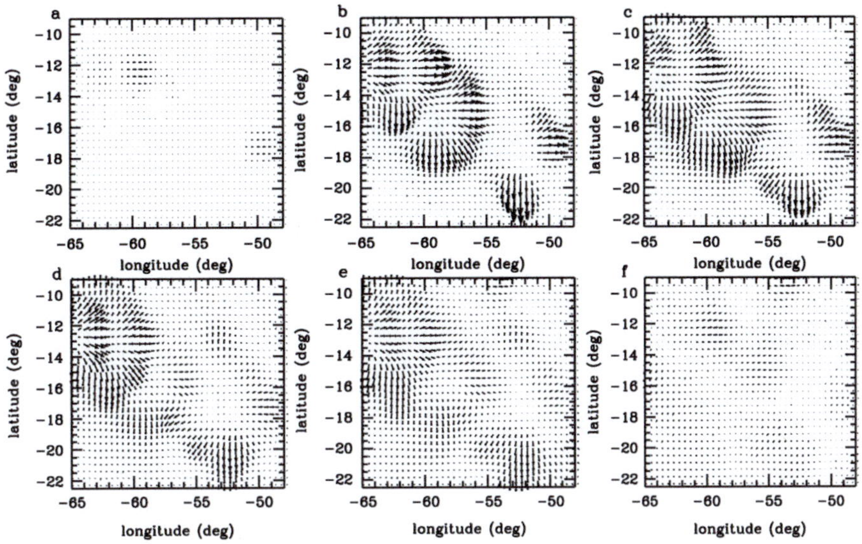

Fig. 9.3 (**a–f**): Vector plot of F_x and F_y from Eq. (9.1) (calculated from the ray trace solutions) every 30 min from 1855 to 2125 UT at $z = 152$ km. All panels are normalized by the same factor. The largest forces are $F_x = 0.68$ m/s^2 and $F_y = 0.83$ m/s^2 in (**b**)

the F layer in the Wallops Island study had $\lambda_H \sim 100–2000$ km, and peaked at $\lambda_H \sim 100–300$ km (Vadas and Crowley, 2010).) We insert F_x and F_y from Eq. (9.1) (as a function of x, y, z, t) into the high resolution TIME-GCM. Because of the 2.5° horizontal grid spacing of this model, however, GWs with $\lambda_H < 2000$ km are not resolvable (Vadas and Liu, 2009). We calculate the difference between this "perturbed" solution and the "unperturbed" high resolution TIME-GCM solution (i.e., with no thermospheric body forces). We show the perturbed minus the unperturbed TIME-GCM solutions in Figs. 9.4, 9.5, and 9.6.

Figure 9.4 shows the neutral density perturbations of the excited secondary GWs that are resolved by the TIME-GCM (i.e., having $\lambda_H \geq 2000$ km) at $z = 250$ km. There is a density decrease of 8–12% over central and western Brazil which moves southward in time. There is also a density enhancement of 7–9% over eastern Brazil which moves southeastward in time. These occur because (1) a density decrease is created near the back-half of a body force at $z \sim 150$ km as fluid is pushed away from this region, and (2) a density enhancement is created near the front-half of a body force as fluid is pushed into a region containing existing fluid. The density perturbation

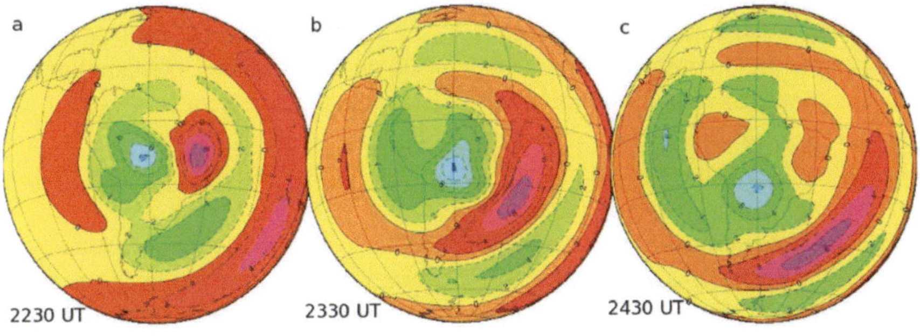

Fig. 9.4 Neutral density perturbations of the secondary GWs at $z = 250$ km from the TIME-GCM. (**a–c**) show 2230, 2330, and 2430 UT, respectively. *Green* (*purple*) colors show negative (positive) perturbations. Minimum and maximum perturbation amplitudes are (**a**): −10 and 7%. (**b**): −12 and 9%. (**c**): −8 and 7%. The latitude and longitude lines on Earth are separated by 20°

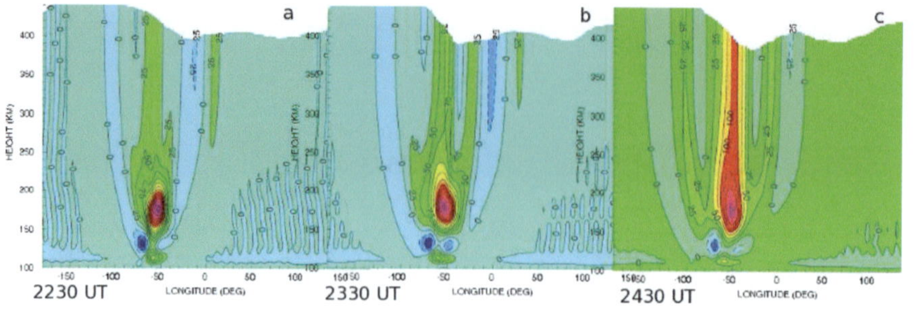

Fig. 9.5 Neutral zonal velocity perturbations at 14° S from the TIME-GCM. (**a–c**) show 2230, 2330, and 2430 UT, respectively. *Green* (*purple*) colors show negative (positive) perturbations. Minimum and maximum amplitudes are (**a**): −85 and 340 m/s. (**b**): −70 and 280 m/s. (**c**): −60 and 185 m/s

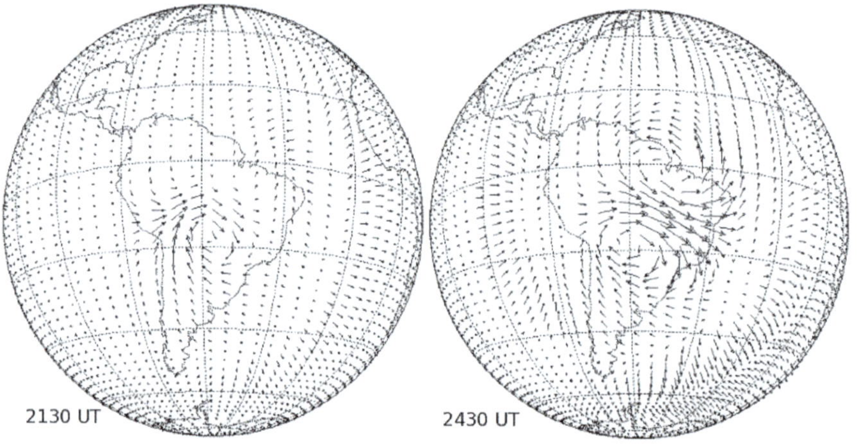

Fig. 9.6 Horizontal velocity perturbations of the secondary GWs at $z = 250$ km from the TIME-GCM. (**a**): 2130 UT. (**b**): 2430 UT. Maximum perturbation amplitudes are (**a**): 75 m/s. (**b**): 150 m/s. The latitude and longitude lines on Earth are separated by 20°

amplitudes are smaller after 24 UT because the body forces are weaker then (see Fig. 9.3). Co-aligned with the density enhancement is a density decrease (nearer to the south pole) which also moves southeastward in time. These constitute the ± phases of an excited large-scale secondary GW with $c_{\rm H} \sim 500-600$ m/s and $\lambda_{\rm H} \sim 4000-5000$ km. There are also large-scale secondary GWs which move north and westward in time, although their amplitudes are significantly smaller than the southeastward GWs before 24 UT.

The result that the northwestward and southeastward GWs have significantly smaller amplitudes before 24 UT differs from previous results; Vadas and Liu (2009) found that the secondary GWs have similar amplitudes in and opposite to the force direction. If the background winds are small, secondary GWs excited by horizontal body forces have equal amplitudes in and against the force direction (for the same intrinsic frequency $\omega_{\rm Ir}$) in the intrinsic reference frame (Vadas, 2010b; Vadas et al., 2003). In (Vadas and Liu, 2009), the "mean" winds and secondary GWs were created at the same time. Therefore, the background winds were relatively small at the excitation time. This is not the case prior to 24 UT here. Figure 9.5 shows the zonal wind perturbations (induced "mean" plus secondary GWs) at 14° S. The induced zonal mean wind is eastward from $z \sim 130$ to 230 km over a zonal extent of ∼2000 km, with a maximum value of ∼340 m/s at 2230 UT. There is also a large southward "mean" component (not shown). Therefore, before 24 UT, the secondary GWs are excited in a large southeastward wind created by previous thermospheric body forces. Because deep convection weakens rapidly after 2300 UT, the induced mean winds from the forces are weak after 24 UT (see Fig. 9.5).

We now discuss the influence of the background wind on the secondary GW amplitudes at $z = 250$ km. The ground based frequency ($\omega_{\rm r}$) of a GW is

$$\omega_{\rm r} = \omega_{\rm Ir} + kU + lV, \tag{9.3}$$

where U and V are the zonal and meridional background winds (which include the tides and the mean winds induced by the body forces), and (k, l, m) is the wavevector. We consider an eastward body force. If the background wind is eastward at the excitation altitude, $\omega_{\rm r}$ is larger (smaller) in the forcing region for secondary GWs (with the same $\omega_{\rm Ir}$) propagating eastward (westward). These east and westward GWs have the same initial amplitudes. Above the forcing region (at $z \geq 200$ km), U and V are much smaller, because the induced mean winds are small. Since ω_r is approximately constant along a GW's ray path, ω_{Ir} is therefore larger (smaller) at $z \geq 200$ km for the secondary GW propagating east (west), using Eq. (9.3). From the dispersion relation, λ_z is larger (smaller) for a GW with a larger (smaller) ω_{Ir}. Since a GW with smaller (larger) λ_z and ω_{Ir} dissipate at a lower (higher) altitude (Vadas, 2007), the eastward GW amplitude is larger than the westward GW amplitude at $z = 250$ km prior to 24 UT. After 24 UT, the background wind at $z = 150$ km is much smaller because the body forces are weaker (see Fig. 9.3). This causes ω_r to be approximately equal for these same secondary GWs, causing their amplitudes to be nearly the same at $z = 250$ km since they are dissipating similarly with altitude.

Figure 9.5 also shows that some secondary GWs propagate up to at least $z \sim 410-450$ km. Additionally, the amplitude difference between the eastward and westward secondary GWs prior to 24 UT does not occur at 2430 UT, in agreement with Fig. 9.4.

Figure 9.6 shows the horizontal velocity perturbations at $z = 250$ km prior to and after sunset. At this altitude, the "mean" wind created by the body forces is small (see Fig. 9.5). Therefore, these are primarily secondary GWs perturbations. Two large vortices swirl the neutral fluid in opposite directions above the convective region. Persistent east and southeastward motions of $u' \sim 75-150$ m/s occur where these vortices converge. The perturbations last for at least 3 h because convection occurs for ~4 h this night; as secondary GWs propagate upward from the forces, new secondary GWs are excited. These horizontal velocity perturbations are much larger than that of the primary GWs at this altitude (see Fig. 9.2).

9.5 Conclusions

In this chapter, we calculated the primary GWs excited by 6 h of deep convection on 01 October, 2005, in central Brazil. We then calculated the large-scale secondary GWs excited by the thermospheric body forces created by the dissipation of these primary GWs. We found that both primary and secondary GWs are present at the bottomside of the F layer (assumed to be at $z = 250$ km): (1) the primary GWs have $\lambda_H \sim 200-300$ km $u', v' \sim 10-15$ m/s, and neutral density perturbations of 1–2%; (2) the secondary GWs resolvable by the TIME-GCM (i.e., having $\lambda_H \geq 2000$ km) have $\lambda_H \sim 4000-5000$ km, zonal wind perturbations of $u' \sim 75-150$ m/s, and neutral density perturbations of 10–15%. The velocity perturbations are eastward in central Brazil for 20° in longitude where the winds from 2 large vortices (created by the forcings) converge, with $u' \sim 75-150$ m/s. This occurs prior to and after sunset above the region of deep convection. Other regions had north, south, or westward velocities. Note that the eastward GW velocities were larger than the background tidal winds.

The pre-sunset eastward neutral winds are necessary for driving the F layer dynamo which creates the PRE. The eastward wind component of the tides were weak at and before 2100 UT at F-layer altitudes, and became eastward thereafter, with amplitudes of ~100 m/s by 23–24 UT (Vadas and Liu, 2009; Fritts et al., 2008). Thus, the tidal wind amplitude is smaller than the eastward velocity perturbation created by the body force prior to sunset ($u' \sim 75$ m/s from Fig. 9.6). Additionally, the tidal wind varies spatially over tens of degrees, whereas the perturbation velocities vary over horizontal scales of 500–1000 km. Since only localized regions have eastward wind perturbations, the "mean" winds and large-scale secondary GWs induced by thermospheric body forces likely significantly influence the F layer dynamo and EPB seeding.

Acknowledgements SLV would like to thank Pete Stamus for the convective plume/cluster parameters, Fernando Sao Sabbas for the GOES-12 satellite images, and Ben Foster for implementing the body forces into the TIME-GCM. SLV was supported by NASA contract NNH07CC81C. HLL was supported in part by NASA contract NNH07CC81C. The National Center for Atmospheric Research is sponsored by the National Science Foundation.

References

Abdu MA, Batista IS, Walker GO, Sobral JHA, Trivedi NB, de Paula ER (1995) Equatorial ionospheric electric fields during magnetospheric disturbances: local time/longitude dependences from recent EITS campaigns. J Atmos Solar-Terr Phys 57:1065–1083

Alexander MJ, Holton JR, Durran DR (1995) The gravity wave response above deep convection in a squall line simulation. J Atmos Sci 52:2212–2226

Andrews DG, Holton JR, Leovy CB (1987) Middle atmosphere dynamics. Academic, Orlando, FL

Beres JH (2004) Gravity wave generation by a three-dimensional thermal forcing. J Atmos Sci 61:1805–1815

Crowley, G, Jones TB, Dudeney JR (1987) Comparison of short period TID morphologies in Antarctica during geomagnetically quiet and active intervals. J Atmos Terr Phys 49:1155–1162

Dewan EM, Picard RH, ONeil RR, Gardiner HA, Gibson J, Mill JD, Richards E, Kendra M, Gallery WO (1998) MSX satellite observations of thunderstorm-generated gravity waves in mid-wave infrared images of the upper stratosphere. Geophys Res Lett 25 939–942

Djuth FT, Sulzer MP, Elder JH, Wickwar VB (1997) High-resolution studies of atmosphere-ionosphere coupling at Arecibo observatory, Puerto Rico. Radio Sci 32: 2321–2344

Djuth FT, Sulzer MP, Gonzales SA, Mathews JD, Elder JH, Walterscheid RL (2004) A continuum of gravity waves in the Arecibo thermosphere? Geophys Res Lett 31. doi: 10.1029/2003GL019376

Fritts DC, Vadas SL (2008) Gravity wave penetration into the thermosphere: Sensitivity to solar cycle variations and mean winds. Ann Geophys 26:3841–3861

Fritts DC, Vadas SL, Riggin DM, Abdu MA, Batista IS, Takahashi H, Medeiros A, Kamalabadi F, Liu H-L, Fejer BJ, Taylor MJ (2008) Gravity wave and tidal influences on equatorial spread F based on observations during the Spread F Experiment (SpreadFEx). Ann Geophys 26: 3235–3252

Georges TM (1968) HF Doppler studies of traveling ionospheric disturbances. J Atmos Terr Phys 30:735–746

Hickey MP, Cole KD (1988) A numerical model for gravity wave dissipation in the thermosphere: J Atmos Terr Phys 50:689–697

Hocke K, Schlegel K (1996) A review of atmospheric gravity waves and traveling ionospheric disturbances: 1982–1995. Ann Geophys 14:917–940

Holton JR, Alexander MJ (1999) Gravity waves in the mesosphere generated by tropospheric convection. Tellus 51(A–B):45–58

Horinouchi T, Nakamura T, Kosaka J (2002) Convectively generated mesoscale gravity waves simulated throughout the middle atmosphere. Geophys Res Lett 29. doi:10.1029/2002GL016069

Huang C-S, Kelley MC (1996a) Nonlinear evolution of equatorial spread F 2. Gravity wave seeding of the Rayleigh-Taylor instability. J Geophys Res 101(A1):293–302

Huang C-S, Kelley MC (1996b) Nonlinear evolution of equatorial spread-F. 4. Gravity waves, velocity shear, and day-to-day variability. J Geophys Res 101:24, 521–524, 532

Hysell DL, Kelley MC, Swartz WE, Woodman RF (1990) Seeding and layering of equatorial spread F. J Geophys Res 95:17253–17260

Hysell D, Larsen M, Swenson C, Barjatya A, Wheeler T, Bullett T, Sarango M, Woodman R (2006) Rocket and radar investigation of background electrodynamics and bottom type scattering layers at the onset of equatorial spread F. Ann Geophys 24(5):1387–140

Kelley MC (1989) The Earth's ionosphere. Academic, San Diego, CA, p 487, 130

Kelley MC, Larsen MF, LaHoz C, McClure JP (1981) Gravity wave initiation of equatorial spread F: a case study. J Geophys Res 86:9087–9100

Lindzen RS (1981) Turbulence and stress owing to gravity wave and tidal breakdown. J Geophys Res 86:9707–9714

Makela JJ, Vadas SL, Muryanto R, Duly T, Crowley G (2010) Periodic spacing between consecutive equatorial plasma bubbles. Geophys Res Lett 37:L14103. doi:10.1029/2010GL043968

Miller ES, Makela JJ, Groves KM, Kelley MC, Tsunoda RT (2010) Coordinated study of coherent radar backscatter and optical airglow depletions in the central Pacific. J Geophys Res 115:A06307. doi:10.1029/2009JA014946

Ogawa, T, Igarashi K, Aikyo K, Maeno H (1987) NNSS Satellite observations of medium scale traveling ionospheric disturbances at southern high-latitudes. J Geomag Geoelectr 39:709–721

Oliver WL, Otsuka Y, Sato M, Takami T, Fukao S (1997) A climatology of F region gravity wave propagation over the middle and upper atmosphere radar. J Geophys Res 102:14,499–14,512

Pfister L, Chan KR, Bui TP, Bowen S, Legg M, Gary B, Kelly K, Proffitt M, Starr W (1993) Gravity waves generated by a tropical cyclone during the STEP tropical field program: a case study. J Geophys Res 98:8611–8638

Richmond AD (1978) Gravity wave generation, propagation, and dissipation in the thermosphere. J Geophys Res 83:4131–4145

Rishbeth H (1971) Polarization fields produced by winds in the equatorial F region. Planet Space Sci 19:357–369

Rottger J (1973) Wave-like structures of large-scale equatorial spread-F irregularities. J Atmos Terr Phys 35(6):1195–1206

Salby ML, Garcia RR (1987) Transient response to localized episodic heating in the tropics. Part 1. Excitation and short-time near-field behavior. J Atmos Sci 44:458–497

Sentman DD, Wescott EM, Picard RH, Winick JR, Stenbaek-Nielsen HC, Dewan EM, Moudry DR, Sao Sabbas FT, Heavner MJ, Morrill J (2003) Simultaneous observations of mesospheric gravity waves and sprites generated by a mid western thunderstorm. J Atmos Solar-Terr Phys 65:537–550

Smith SA, Fritts DC, VanZandt TE (1987) Evidence for a saturated spectrum of atmospheric gravity waves. J Atmos Sci 44:1404–1410

Stull RB (1976) Internal gravity waves generated by penetrative convection. J Atmos Sci 33:1279–1286

Suzuki S, Shiokawa K, Otsuka Y, Ogawa T, Nakamura K, Nakamura T (2007) A concentric gravity wave structure in the mesospheric airglow images. J Geophys Res 112:D02102. doi:10.1029/2005JD006558

Takahashi H, Taylor MJ, Pautet P-D, Medeiros AF, Gobbi D, Wrasse CM, Fechine J, Abdu MA, Batista IS, Paula E, Sobral JHA, Arruda D, Vadas SL, Sabbas FS, Fritts DC (2009) Simultaneous observation of ionospheric plasma bubbles and mesospheric gravity waves during the SpreadFEx Campaign. Ann Geophys 27:1477–1487

Taylor MJ, Hapgood MA (1988) Identification of a thunderstorm as a source of short period gravity waves in the upper atmospheric nightglow emissions. Planet Space Sci 36:975–985

Tsunoda RT, Bubenik DM, Thampi SV, Yamamoto M (2010) On large-scale wave structure and equatorial spread F without a post-sunset rise of the F layer. Geophys Res Lett 37:L07105. doi:10.1029/2009GL042357

Tsunoda RT, White BR (1981) On the generation and growth of equatorial backscatter plumes: 1. Wave structure in the bottom side F layer. J Geophys Res 86:3610–3616

Vadas SL (2007) Horizontal and vertical propagation and dissipation of gravity waves in the thermosphere from lower atmospheric and thermospheric sources. J Geophys Res 112: A06305. doi:10.1029/2006JA011845

Vadas SL (2010a) Compressible f-plane solutions to local body forces and heatings; Part I: Initial value and forced/heated solutions, and polarization relations. J Atmos Sci (submitted)

Vadas SL (2010b) Compressible f-plane solutions to local body forces and heatings; Part II: gravity wave, acoustic wave, and mean responses. J Atmos Sci (submitted)

Vadas SL, Crowley G (2010) Sources of the traveling ionospheric disturbances observed by the ionospheric TIDDBIT sounder near Wallops Island on October 30, 2007. J Geophys Res 115:A07324. doi:10.1029/2009JA015053

Vadas SL, Fritts DC (2005) Thermospheric responses to gravity waves: influences of increasing viscosity and thermal diffusivity. J Geophys Res 110:D15103. doi:10.1029/2004JD005574

Vadas SL, Fritts DC (2006) Influence of solar variability on gravity wave structure and dissipation in the thermosphere from tropospheric convection. J Geophys Res 111:A10S12. doi:10.1029/2005JA011510

Vadas SL, Fritts DC (2009) Reconstruction of the gravity wave field from convective plumes via ray tracing. Ann Geophys 27:147–177

Vadas SL, Fritts DC, Alexander MJ (2003) Mechanism for the generation of secondary waves in wave breaking regions. J Atmos Sci 60:194–214

Vadas SL, Liu H-L (2009) The generation of large-scale gravity waves and neutral winds in the thermosphere from the dissipation of convectively-generated gravity waves. J Geophys Res 114:A10310. doi:10.1029/2009JA014108

Vadas SL, Taylor MJ, Pautet P-D, Stamus P, Fritts DC, Liu H-L, Sao Sabbas FT, Rampinelli VT, Batista P, Takahashi H (2009) Convection: the likely source of the mediumscale gravity waves observed in the OH airglow layer near Brasilia, Brazil, during the Spread-FEx campaign. Ann Geophys 27:231–259

Waldock JA, Jones TB (1986) HF Doppler observations of medium-scale travelling ionospheric disturbances at mid-latitudes. J Atmos Terr Phys 48:245–260

Walterscheid RL, Schubert G, Brinkman DG (2001) Small-scale gravity waves in the upper mesosphere and lower thermosphere generated by deep tropical convection. J Geophys Res 106(D23):31,825–31,832

Yiğit E, Aylward AD, Medvedev AS (2008) Parameterization of the effects of vertically propagating gravity waves for thermosphere general circulation models: Sensitivity study. J Geophys Res 113:D19106. doi:10.1029/2008JD010135

Yiğit E, Medvedev AS, Aylward AD, Hartogh P, Harris MJ (2009) Modeling the effects of gravity wave momentum deposition on the general circulation above the turbopause. J Geoph Res 114:D07101. doi:10.1029/2008JD011132

Yue J, Vadas SL, She C-Y, Nakamura T, Reising S, Liu H-Li, Stamus P, Krueger D, Lyons W, Li T (2009) Concentric gravity waves in the mesosphere generated by deep convective plumes in the lower atmosphere near Fort Collins, Colorado. J Geophys Res 114:D06104. doi:10.1029/2008JD011244

Chapter 10

The Acoustic Gravity Wave Induced Disturbances in the Equatorial Ionosphere

E. Alam Kherani, Mangalathayil Ali Abdu, Dave C. Fritts, and Eurico R. de Paula

Abstract The role of acoustic gravity waves (AGWs) to excite atmospheric and ionospheric disturbances is examined in this work. These waves are launched in the atmosphere by tropospheric thermal sources and convective activity. An alternative fully time-spatial dependent nonlinear wave equation of acoustic gravity wave is derived and solved numerically using implicit finite-difference scheme. Their propagation in the atmosphere through mesopause thermal duct and lower thermosphere density duct, the role of nonlinear viscous effect to limit the amplitude of these waves in the density duct and to allow them to escape to higher altitude where they attain large amplitude in the bottomside F region Ionosphere, and the role of the mean zonal wind to reduce their amplitude are investigated in present study. To study AGW induced disturbances in the equatorial Ionosphere, the AGW model is coupled with hydromagnetic equations in Ionosphere. This coupling is explored in the context of the collisional interchange instability (CII) in the F region leading to the formation of equatorial F region plasma bubbles. To do so, AGW model is coupled with the CII model and simultaneously solved numerically. The possible role of the AGW to act as a seeding perturbation for equatorial plasma bubbles under varying nature of mean zonal wind and tropospheric thermal source are also investigated.

E.A. Kherani (✉)
National Institute for Space Research, São Jose dos Campos, SP, Brasil
e-mail: alam@dae.inpe.br

10.1 Introduction

The Earth's atmosphere hosts a variety of dynamical process triggered by neutral waves such as the Planetary waves, the tidal waves and the Acoustic Gravity Waves (AGWs). Among them, the AGWs find relatively higher priority to influence the atmosphere owing to the wide range of natural sources that may excite them abundantly and the wide spectral range they occupy in the atmosphere (Yeh and Liu, 1974; Fritts and Alexander, 2003). They also propagate in a wide height range from the troposphere to the thermosphere and from high-latitude to equatorial and low latitudes (Hines, 1960; Richmond, 1978).

Hines (1960) was the first to formulate a theory for AGW propagation in the atmosphere and to recognize their role in vertical atmospheric coupling and Atmosphere-Ionosphere coupling. His work demonstrated that the AGWs excited in the lower atmosphere may escape to the thermosphere. The theory further suggested that, for energy conservation, the amplitude of these waves should be inversely proportional to the atmospheric density so that an upgoing gravity wave may attain large amplitude at thermospheric and ionospheric heights and may produce identifiable signatures in the ionosphere. Based on this thermosphere-ionosphere coupling mechanism, the regularly observed traveling ionospheric disturbances were interpreted (Hines, 1967).

More rigorous theoretical frameworks were formulated to study the vertical atmospheric coupling due to the AGWs of tropospheric origin which addressed the issues such as the ducting of AGWs by the thermal/Doppler ducts in stratosphere/mesopause region, nonlinear nature of the AGWs, their saturation

M.A. Abdu, D. Pancheva (eds.), A. Bhattacharyya (Coed.), *Aeronomy of the Earth's Atmosphere and Ionosphere*, IAGA Special Sopron Book Series 2, DOI 10.1007/978-94-007-0326-1_10, © Springer Science+Business Media B.V. 2011

mechanism and the excitation of non-thermal atmospheric fluctuations (Pitteway and Hines, 1965; Fritts, 1984; Weinstock, 1982, 1990; Fritts and Rastogi; 1985). Various numerical studies were pursued to examine the propagation of AGWs through thermal and density ducts in the stratosphere, mesosphere and lower thermosphere (Hickey et al., 1998; Piani et al., 2000; Zhang and Yi, 2002; Yu and Hickey, 2007). These studies confirm the observations of strong velocity shear in the mesospheric thermal ducts where short wavelengths are trapped (Fritts et al., 1997). Theoretical studies also identify the possibility of longer wavelength to escape to the upper thermosphere (Yu and Hickey, 2007; Vadas, 2007, 2009; Snively and Pasko, 2008).

The penetration of large amplitude AGWs to ionospheric heights may have serious implications since the ionosphere hosts a variety of currents and electric fields driven by neutral winds (Haerendel et al., 1992; Abdu, 1999). The winds associated with the AGWs may drive ionospheric currents, in addition to the currents driven by much stronger tidal and global thermospheric winds (Harendel et al., 1992; Kherani et al., 2009a). The response of the equatorial ionosphere to AGWs driven currents and electric fields remains a subject of intensive research since a variety of nonthermal disturbances are triggered from tiny fluctuations in the currents and electric fields. These nonthermal disturbances include the phenomenon such as the electrojet plasma irregularities producing the Type-I and Type-II radar echoes, the quasi-periodic 150-km echoes and the equatorial plasma bubble (Fejer et al., 1976; Woodman and LaHoz, 1976; Kudeki and Fewcett, 1993).

Radar observations of equatorial night-time F region often reveal the existence of rising plumes or bubbles that may penetrate to the topside F layer attaining very high altitudes (Kelley et al., 1981). The collisional interchange instability (CII) is found to play a crucial role in the onset and development of the plasma bubble (Haerendel, 1973; Sultan, 1996). Numerous theoretical and modeling studies have been performed to assess the linear and nonlinear aspects of the plasma bubble development (Huang et al., 1993; Keskinen et al., 2003; Kherani et al., 2005; Huba et al., 2007; Keskinen and Vadas, 2009). These studies have revealed the generation of rising bubbles initiated by a seed perturbation at bottomside of a rising F layer.

The precise nature of the seed perturbations is one of the issues that has remained elusive. Many researchers have experimentally suggested that AGWs propagating in the equatorial thermosphere play an important role in seeding the bubbles (Rottger, 1981; Kelley et al., 1981; Abdu et al., 1981; Sobral et al., 1981; Hysell et al., 1990). Huang et al. (1993) and Huang and Kelley (1996) numerically studied the role of AGWs to seed the CII and investigated the evolution of plasma bubbles under different seeding and ambient Ionospheric conditions. Their studied showed that although plasma bubbles may be produced in all the situations they considered, the production initiated by gravity waves takes much shorter time than that resulting from twodimensional initial density perturbations. The study also showed that the timing of the seed gravity wave is critical to ESF generation. A gravity wave present in an ascending F region can initiate the CII leading to the formation of topside bubbles. In contrast, a gravity wave induced perturbations in a descending F layer evolve into largescale, wavelike structures but not bubbles.

The recent Spread F experimental campaign (SpreadFEx) conducted in the Brazilian equatorial-low-latitude region had the objective to study the role of AGWs to seed the CII/bubble generation (Fritts et al., 2009). The campaign results identified several cases of AGWs of tropospheric origin that gave rise to ionospheric structures in the lower E region and bottomside F region and that may have triggered non-thermal spread F phenomena (Takahashi et al., 2010; Abdu et al., 2009). Based on these results, Abdu et al. (2009) and Kherani et al. (2009b) have quantified the role of AGWs to seed the CII during a few selected days. Recently, Keskinen and Vadas (2009) have presented a three dimensional simulation of plasma bubble with AGWs seeding based on parameters obtained from the SpreadFEx campaign. They found that the AGWs seeding could give rise to plasma bubble growth provided the variation in the meridional wind of AGWs, which is parallel to the Earth's magnetic field lines, should have a longer wavelength. Similar effect was previously reported from a three-dimensional simulation of plasma bubble for density variation along the field line (Kherani et al., 2005).

The aforementioned theoretical and numerical studies and the SpreadFEx campaign results have identified the important role of AGWs of convective origin to excite the non-thermal fluctuations in the night-time equatorial ionosphere. The present work focuses on the issues related to the AGWs-ionosphere interaction addressing those issues as explained below:

Numerous simulation attempts to study the AGWs seeding mechanism for the CII appear to successfully explain the observational results on F region plasma bubble development. However, most studies treat AGWs without considering the possible effects of a time-varying atmosphere and time-varying tropospheric source during the evolution of the CII and plasma bubbles. The numerical studies of Huang and Kelley (1996) and Kherani et al. (2009b) had suggested the importance of temporal variations of the AGWs on the evolution of plasma bubble during the evening-time when atmospheric and Ionospheric conditions vary rapidly. Fritts and Vadas (2009) have shown the damping nature of the atmospheric mean wind on the AGW amplitude. The atmospheric mean wind is highly variable in nature during sunset time (Kudeki et al., 1981; Fejer et al., 1991) and therefore may have severe implications on the seeding mechanism. In addition, the tropospheric thermal source, which launches AGWs in thermosphere, is found to be highly dynamic in time and space (Sao sabbas et al., 2009). Thus to study the seeding mechanism under varying atmospheric/thermal-source conditions, one needs to consider the fully time-spatial dependent governing equations for the AGWs which should be solved simultaneously with the governing equations of the CII. This aspect is addressed in the present work where the governing wave equation for AGW is derived and solved numerically together with the nonlinear CII model developed by Kherani et al. (2004, 2005).

The paper is organized as follows: In Section 10.2 and appendix 1, the wave equation of AGWs is derived and the corresponding numerical results are presented which include study of excitation of AGWs from prescribed tropospheric thermal excitation, nonlinear effect of viscosity to limit the amplitude of the AGW in ducted region and permit them to propagate to upper thermosphere and role of the mean zonal wind to oppose the propagation of the AGW. In Section 10.3, the AGW-Ionospheric coupling is studied by solving hydromagnetic equations (Appendix 2) in the Ionosphere. Simultaneous simulations of the AGWs and the CII are carried out and various aspects of AGW seeding of CII namely effects of mean zonal wind and effects of life-time of thermal excitation source are investigated.

10.2 Numerical Simulation of Acoustic-Gravity Waves (AGWs)

10.2.1 *Wave Equation of AGWs*

In Appendix 1, the wave equation of AGWs is derived using Navier-Stokes equations namely the continuity, momentum and energy equations ((10.3), (10.4), and (10.5)) in the atmosphere. The wave equation for AGWs is derived in following from Eq. (10.7):

$$\begin{aligned}\frac{\partial^2 \mathbf{W}}{\partial t^2} - W_o^2 \nabla_m^2 \mathbf{W} &= \frac{\gamma p}{\rho}\nabla(\nabla.\mathbf{W}) + (\gamma-1)\frac{\nabla p}{\rho}\nabla.\mathbf{W} \\ &- \frac{\nabla p}{\rho}(\mathbf{W}.\nabla)\log\rho + \frac{1}{\rho}\nabla(\mathbf{W}.\nabla)p \\ &+ \frac{\mu}{\rho}\nabla^2\mathbf{W}\left[(\mathbf{W}.\nabla)\log\rho + \nabla.\mathbf{W} + \mu\nabla^2\frac{1}{\rho}\right] \\ &- R\Theta_w\nabla\log\rho\end{aligned}$$

or

$$\begin{aligned}\frac{\partial^2 \mathbf{W}}{\partial t^2} - W_o^2 \nabla_m^2 \mathbf{W} &= \mathrm{AWI} + \mathrm{AWII} + \mathrm{GWI} + \mathrm{GWII} \\ &+ (\nu_\rho + \nu_g)\nabla^2\mathbf{W} - R\Theta_w\nabla\log\rho\end{aligned} \quad (10.1)$$

where

$$\nu_\rho = \mu\frac{\mu}{\rho}\nabla^2\frac{1}{\rho},\ \nu_g = \frac{\mu}{\rho}\left[(\mathbf{W}.\nabla)\log\rho + \nabla.\mathbf{W}\right]$$

This equation describes fully compressible, non-hydrostatic wave motion where (ρ, $\mathbf{W}$, p) are the atmospheric density, the wind field associated with the AGWs and the atmospheric pressure respectively, $\nu = \mu/\rho, \mu$ are the kinematic and dynamic viscosity, Θ_w is the thermal wave excitation associated with tropospheric convection, $R = \frac{k}{m}$ and (k, m) are Boltzmann's constant and the atmospheric mass. Equations ((10.1) (10.3), and (10.5)) form a closed set of nonlinear governing equations for (ρ, $\mathbf{W}$, p) in the atmosphere. In the present study, Eq. (10.1) is solved for prescribed ρ, p and Eqs. (10.3) and (10.5) are not solved though they are used to derive wave equation for wind field $\mathbf{W}$. In the right hand side of Eq. (10.1), the first two terms, (AWI, AWII) represent the contributions from compressible flow ($\nabla.\mathbf{W} \neq 0$) i.e from acoustic wave, the third/fourth terms, (GWI, GWII) represent the contributions from advections of density and pressure force and correspond to gravity wave, the fifth/sixth terms represent the contributions from viscosity. As may be noted in Appendix 1, the ν_ρ arises owing to $\frac{\partial W}{\partial t}$ i.e, ν_ρ is associated with inertial dynamics and acts against the formation of strong velocity shear. The ν_g term is a nonlinear term and arises owing to $\frac{\partial \rho}{\partial t}$ i.e it is associated with convection dynamics and acts against the formation of density gradient. On left-hand-side of Eq. (10.1), the second term represents the contribution from the dynamics of the mean wind, W_o. ∇_m stands for the space derivative along the direction of the mean wind W_o when considered.

The description of AGW propagation in the form of wave Equation (10.1) brings out various features such as the identification of separate terms for acoustic and gravity wave components, identification of the different role of viscosity associated with $\nu_{g,\rho}$ (as studied in Section 10.2.2.1), and the identification of the wave-like effects associated with the mean wind (as studies in Section 10.2.2.3).

The wave Equation (10.1) is solved numerically using finite difference Cranck-Nicholson implicit scheme used earlier by Kherani at al (2005). The model domain is in two spatial dimension in Cartesian y-x plane corresponding to the altitude-longitude plane. The lower boundary is located at tropospheric height (10 km) and the model extends up to 510 km altitude, and it horizontally spans one horizontal wavelength ($\lambda_w = 150km$) allowing periodic boundary conditions to be imposed. A rigid boundary condition is imposed at lower boundary. It is conventional to use Rayleigh friction sponge layers to damp the waves as they approach the upper boundaries, hence reducing the magnitudes of any reflected waves (Snively and Pasko, 2007). However, the present model includes viscosity effects that significantly damp the waves in the thermosphere, thereby reducing artificial reflection and the influence of reflected wave energy (Snively and Pasko, 2007). The radial and horizontal grid spacing are 5 km. The time step is decided by vertical grid and maximum acoustic wave speed using Fredrick-Courant stability condition

$$\Delta t < \Delta t_o = \frac{\Delta r}{c_{sm}}$$

and is taken as $\Delta t = 0.25\Delta t_o$. Here c_{sm} represents the maximum sound speed.

The ambient atmospheric density and temperature are obtained using the NRLMSISE model (Huba et al., 2001). The atmospheric pressure is obtained using the gas state equation. The kinematic viscosity ν is estimated using a free-path equation described by Kherani et al. (2009a). While the dynamic viscosity μ does not change significantly with altitude, the kinematic viscosity $\nu = \mu/\rho$ is inversely proportional to density and therefore increases exponentially with altitude. In Fig. 10.1, the atmospheric mass density ρ (10.1a), the temperature, T (10.1b) and the kinematic viscosity ν (10.1c) used in the calculations are shown. The density and temperature profiles in Fig. 10.1a, b will be referred as multi-scale-exponential and non-iso-thermal profiles in the discussion to follow. The mean wind profiles which are used for a few cases, are plotted in Fig. 10.1d. The profiles have strong shear in the lower thermosphere with magnitude in 30–50 m/s range below100 km and with magnitude in 150–300 m/s range above 100 km. These profiles are qualitatively similar to the sheared mean wind profile used in previous studies (Keskinen and Vadas, 2009).

The thermal wave excitation Θ_w associated with the tropospheric convection represents the AGW source near the lower boundary (10 km) (Yu and Hickey, 2007; Snively and Pasko, 2008). Θ_w near 10 km is such that it varies sinusoidally and periodically over horizontal wavelength in longitude (zonal) direction and the periodic boundary condition at the zonal boundaries implies an infinite wave train in the horizontal direction. The prescribed source is a Gaussian envelope over an altitude of half-width $\sigma_y = 10$ km, centered at an altitude $y_o = 10$ km, and a Gaussian envelop over a time of half-width $\sigma_t = 10$ min, centered at time $\tau = 18$ min, and with an amplitude of $10^{-5}Ks^{-1}$. It is described analytically as:

$$\begin{aligned}\Theta_w(x,y) &= 10^{-5}e^{-(t-\tau)^2/2\sigma_t^2}e^{-(y-y_o)^2/2\sigma_y^2} \\ &\times cos(k_o x - \omega_o t); \qquad (10.2)\\ k_o &= 2\pi/\lambda_o,\ \lambda_o = 150\text{ km};\end{aligned}$$

In Fig. 10.2, the time variation of $\Theta_w(x = 0, y = 10)$ at lower boundary (10 km) is shown for $\omega_o = 0.16$ min^{-1} which will be used in present study under different cases described in Table 10.1.

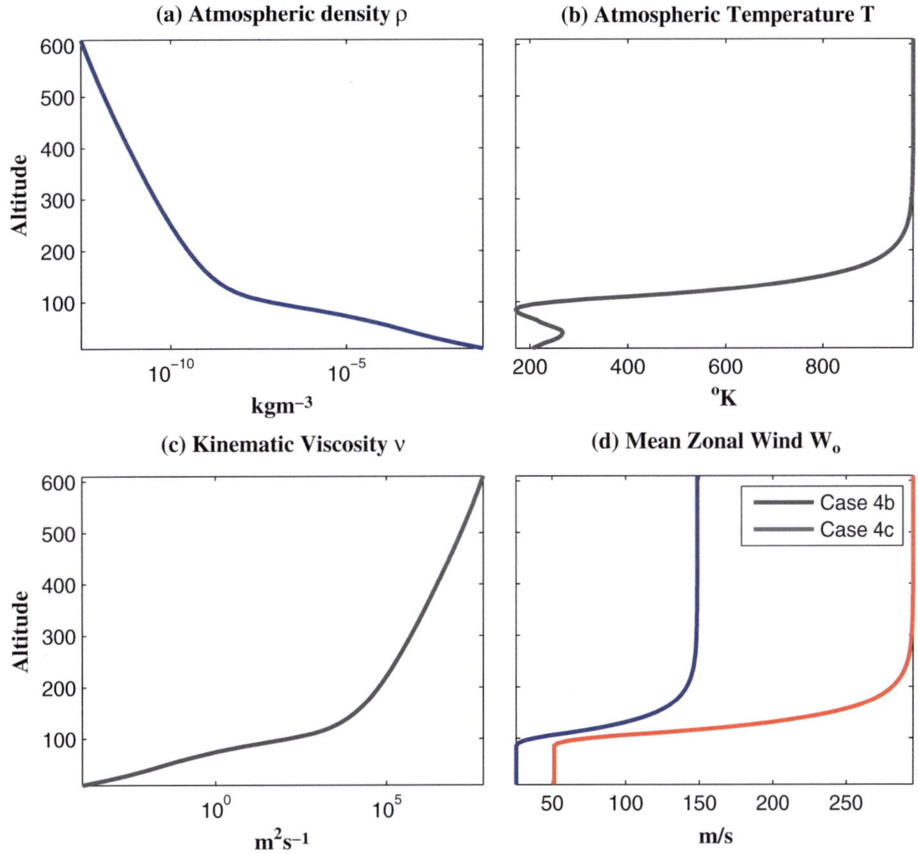

Fig. 10.1 Atmospheric Density (**a**), temperature (**b**), kinematic viscosity (**c**) and mean wind (**d**) profiles

10.2.2 Results and Discussion

10.2.2.1 One-Dimensional (Altitude) Simulation of AGWs

In order to study the propagation characteristics of the AGWs in the atmosphere, Eq. (10.1) may be first examined in vertical direction in which the atmosphere has dominant variations. With $k = 0$ in Eq. (10.2) implying uniform thermal excitation in zonal direction, the following cases are chosen:

Case 1a: Iso-thermal atmosphere with exponential density profile without viscous terms in Eq. (10.1),

Case 1b: Same as 1a except that the viscous term ν_ρ is included,

Case 2a: Non-isothermal/exponential atmosphere with temperature profile as shown in Fig. 10.1b,

Case 2b: Realistic Atmosphere with thermal and density profile as shown in Fig. 10.1a–b,

Case 3: Same as Case 2b and with the inclusion of another viscous term ν_g in Eq. (10.1).

The Eq. (10.1) is solved for vertical W_y and horizontal or zonal W_x wind components associated with the AGW. For $k = 0$ in Θ_w in Eq. (10.2), only vertical wind component W_y is excited and horizontal or zonal component W_x remains zero. In the top three panels of Fig. 10.3, the vertical wind W_y (normalized to its maximum value) associated with the AGW are plotted for Case 1a and 1b at three selected times denoted at the top of figure. The maximum value of W_y at each selected time is also denoted in each figure. It may be noted that thermal excitation near 10 km launches the wave-like fluctuation in the atmosphere with the following characteristics: (1) In Case 1a–1b, the AGW is excited which propagate vertically upward traveling upto 400 km within 20 min, i.e. it propagates with velocity equals to 330 m/s; (2) In Case 1a, the AGW is excited whose amplitude increases exponentially with

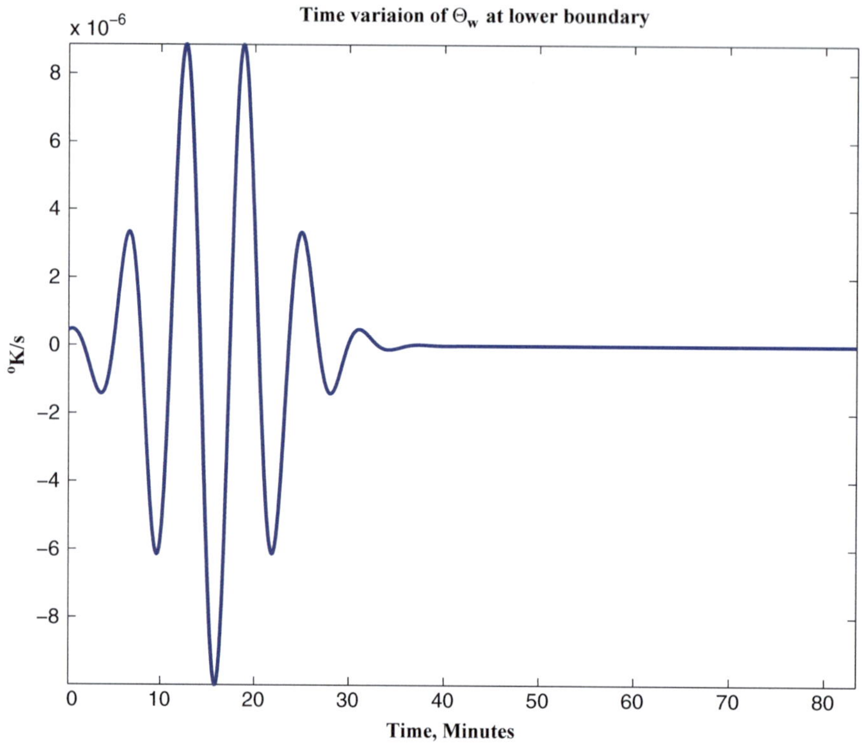

Fig. 10.2 The time variation of tropospheric thermal source, Θ_w, at lower boundary (=10 km)

Table 10.1 Description of different cases

Cases	Tropospheric source (Gaussian)	Atmospheric conditions	Viscosity ($\nu\rho$, ν_g)	Mean Wind (W_o)
Case 1a	$\sigma_t = 10$ min, $k_o = 0$	Isothermal/exponential	$\nu_\rho = 0,\ \nu_g = 0$	$W_o = 0$
Case 1b	$\sigma_t = 10$ min, $k_o = 0$	Isothermal/exponential	$\nu_\rho \neq 0,\ \nu_g = 0$	$W_o = 0$
Case 2a	$\sigma_t = 10$ min, $k_o = 0$	Non-Isothermal/exponential	$\nu_\rho \neq 0,\ \nu_g = 0$	$W_o = 0$
Case 2b	$\sigma_t = 10$ min, $k_o = 0$	Non-Isothermal/Multi-scale-exponential	$\nu_\rho \neq 0,\ \nu_g = 0$	$W_o = 0$
Case 3	$\sigma_t = 10$ min, $k_o = 0$	Non-Isothermal/Multi-scale-exponential	$\nu_\rho \neq 0,\ \nu_g \neq 0$	$W_o = 0$
Case 4a	$\sigma_t = 10$ min, $k_o \neq 0$	Non-Isothermal/Multi-scale-exponential	$\nu_\rho \neq 0,\ \nu_g \neq 0$	$W_o = 0$
Case 4b	$\sigma_t = 10$ min, $k_o \neq 0$	Non-Isothermal/Multi-scale-exponential	$\nu_\rho \neq 0,\ \nu_g \neq 0$	$W_o \neq 0$
Case 4c	$\sigma_t = 10$ min, $k_o \neq 0$	Non-Isothermal/Multi-scale-exponential	$\nu_\rho \neq 0,\ \nu_g \neq 0$	$W_o \neq 0$
Case 5a	$\sigma_t = 2.5$ min, $k_o \neq 0$	Non-Isothermal/Multi-scale-exponential	$\nu_\rho \neq 0,\ \nu_g \neq 0$	$W_o = 0$
Case 5b	$\sigma_t = 2.5$ min, $k_o \neq 0$	Non-Isothermal/Multi-scale-exponential	$\nu_\rho \neq 0,\ \nu_g \neq 0$	$W_o \neq 0$
Case 5c	$\sigma_t = 2.5$ min, $k_o \neq 0$	Non-Isothermal/Multi-scale-exponential	$\nu_\rho \neq 0,\ \nu_g \neq 0$	$W_o \neq 0$

altitude and attains the maximum (~6.7 m/s) at the upper boundary within 30 min; (3) The altitude of maximum amplitude of the AGW at 30 min lowers down to 300–400 km altitude region in Case 1b when viscosity term ν_ρ is included.

It is evident that ν_ρ term in Eq. (10.1) leads to the decrease in the AGW amplitude above 400 km altitude in Case 1b which otherwise would increase exponentially in the absence of ν_ρ term as in Case 1a. This aspect is known from previous studies of acoustic wave and gravity waves at thermospheric heights (Kherani et al., 2009; Vadas and Liu, 2009) and forms the first step to validate the simulation code formulated in the present study. In the middle and bottom panels of Fig. 10.3, the vertical wind W_y associated with the AGW are plotted for the Case 2a and Case 2b respectively. It may be noted that (4) in the non-isothermal/exponential atmosphere of the Case 2a,

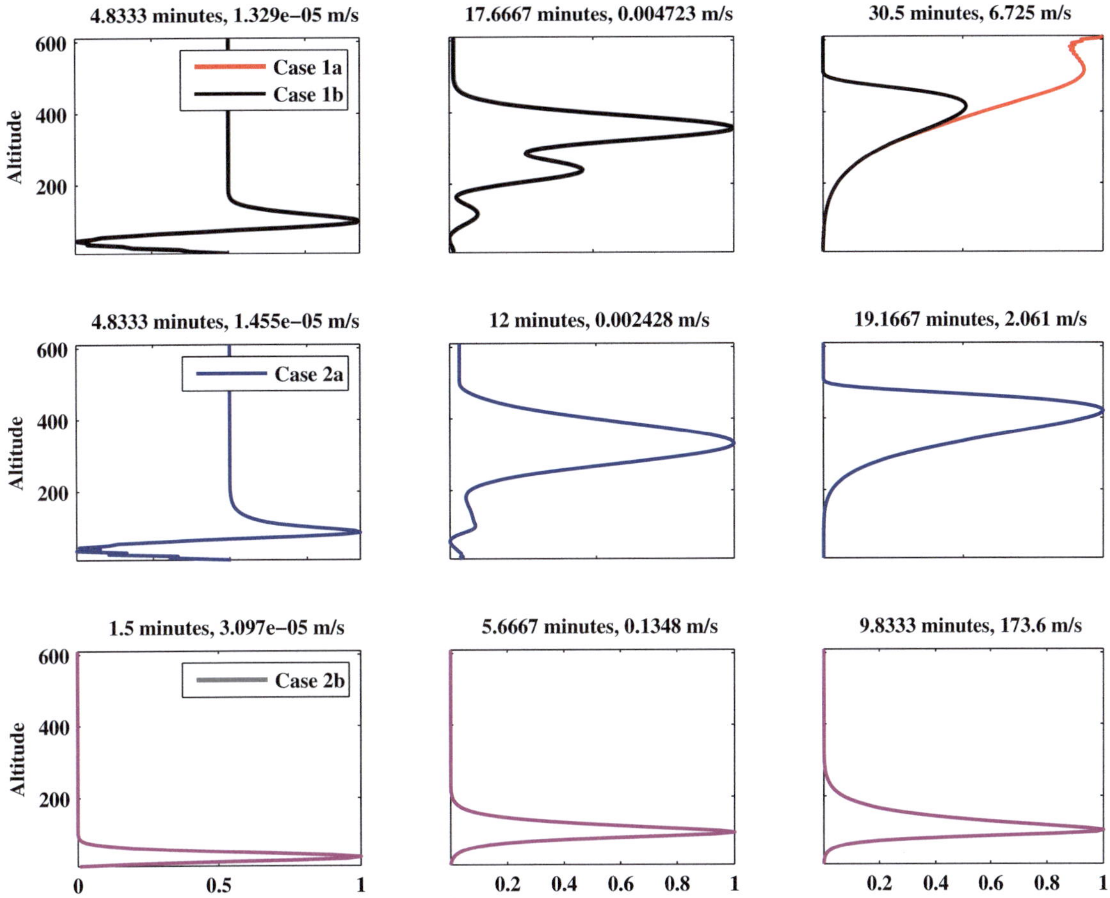

Fig. 10.3 Vertical wind component W_y of the AGWs for Case 1a–1b (*Upper-panel*), Case 2a (*Middle-panel*), Case 2b (*Lower-Panel*). They are normalized to corresponding maximum value of the W_y denoted at the top of the respective panels

the AGW acquires primary maxima in 300–400 km altitude region similar to the Case 1b and acquires secondary maxima (seen clearly at $t = 12$ min) centered at mesopause altitude near 95 km where large temperature gradient exists as noted in Fig. 10.1b; (5) In the non-isothermal/mutli-scale-exponential atmosphere in Case 2b, the amplitude of the AGW builds up rapidly (within 10 minutes) around 110 km where density scale height changes i.e., where gradient in the density scale height has a maximum as noted in Fig. 10.1a; (6) In Case 2b, the AGW remains confined around 120 km altitude and does not propagate to higher altitude.

The maxima near 95 km and 110 km in Case 2a and Case 2b respectively correspond to density/thermal ducts respectively as reported from earlier numerical studies (Yu and Hickey, 2007; Snively and Pasko, 2008). In Case 2a, the AGW escapes thermal duct and reach thermospheric heights attaining maximum amplitude in 300–400 km altitude region similar to Case 1b. In contrast to the thermal ducting in Case 2a, ducting of AGW in density duct near 120 km in Case 2b is severe where they attain large amplitude and AGW does not escape to higher altitude. This suggests that the density duct in Cases 2b is much more efficient to duct the AGW, limiting their propagation to higher altitudes, than thermal duct at mesopause in Case 2a.

In Case 2b, the realistic atmosphere shown in Fig. 10.1 with viscous term ν_ρ is considered. However, another viscous term ν_g in Eq. (10.1), which is proportional to amplitude of AGW and is a nonlinear term (proportional to W^2), is not taken into account. As can be noted in the bottom panel of Fig. 10.3, inside the density duct in Case 2b, the amplitude of the AGW tends to increase rapidly and thus ν_g may

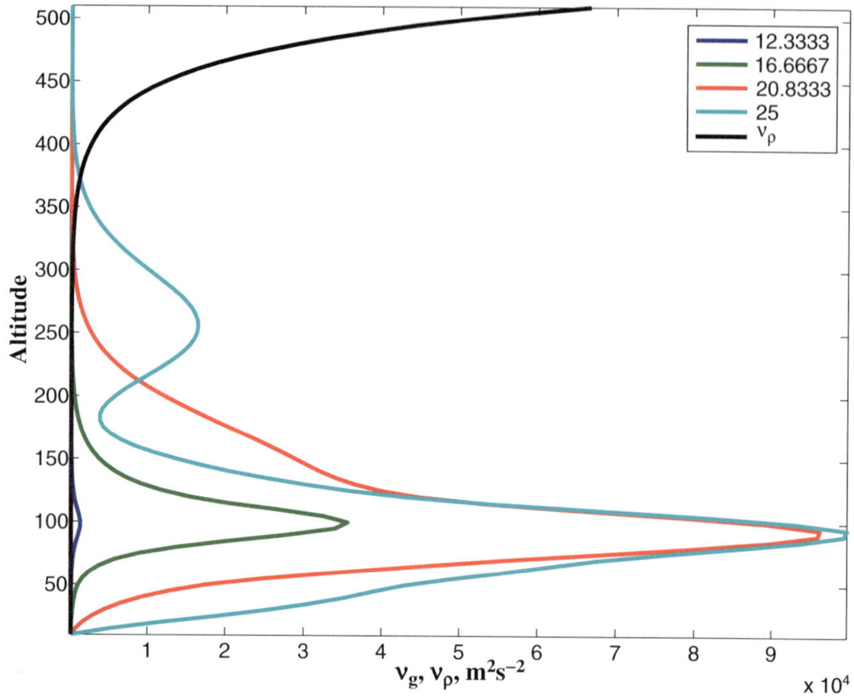

Fig. 10.4 Altitude profile of ν_g at different time (as shown in box in minutes) and ν_ρ

be of considerable importance within short time. The large amplitude inside the ducts may also bring other non-linear effects where continuity and energy equations may have to be solved simultaneously (Snively and Pasko, 2008). In the present study, however, the nonlinearity imposed by ν_g term in Eq. (10.1) is given the priority as this term is proportional to W^2 which increases very rapidly inside the density duct. In Fig. 10.4, the altitude variation of ν_g at different time and ν_ρ are plotted. It may be noted ν_g increases rapidly with time in the regions of density and thermal ducts . Moreover, ν_g dominates over ν_ρ below 300–400 km while ν_ρ remains the dominant frictional term above 400 km.

In Fig. 10.5, the wind amplitude W_y is plotted for the Case 3 when the ν_g term in Eq. (10.1) is also included. It may be note that W_y acquires a wave structure in altitude, oscillating in time with the following characteristics: (7) The AGW acquires maximum amplitude ($\sim$10 m/s) in the thermal and density ducts within 10 min and W_y remains insignificantly small above 200 km at this time; (8) After 10 min, the AGW acquires another maximum ($\sim$30 m/s) in 300–400 km altitude region:

These characteristics indicate the drastic change in the nature of AGWs propagation in the Case 3 with nonlinear viscous term ν_g included as compared to the Case 2b that does not include the ν_g term. Inside the density/thermal ducts where W_y grows rapidly, the nonlinear ν_g which is proportional to W_y, becomes large (Fig. 10.4) and introduces large friction over time which resists the amplitude W_y of AGW to grow in these ducts. As a result, W_y does not grow indefinitely in these ducts in Case 3 as it tended to grow in Case 2b. The frictional damping offered by ν_g and the growth in W_y compete and a state of equilibrium reaches when they attain similar order of magnitudes. Thereafter W_y oscillates with constant mean magnitude inside these ducts which may be noted in Fig. 10.5 when W_y acquires constant mean value ($\sim$20 m/s) near 100 km after 10 min. This equilibrium allows AGW to escape to higher altitude as is evident from Fig. 10.5 when after 10 min, oscillations with height and time are seen above 200 km.

The oscillating nature of the density duct is more evident in Fig. 10.6 wherein W_y is plotted in altitude-time plane which represents the propagation characteristics of the AGW. It may be noted that W_y near 110 km oscillates in time with mean value equals to 20 m/s. It is also evident that waves are launched to higher altitude from the density ducted region. The oscillation of density-ducted AGW is owing to following reason:

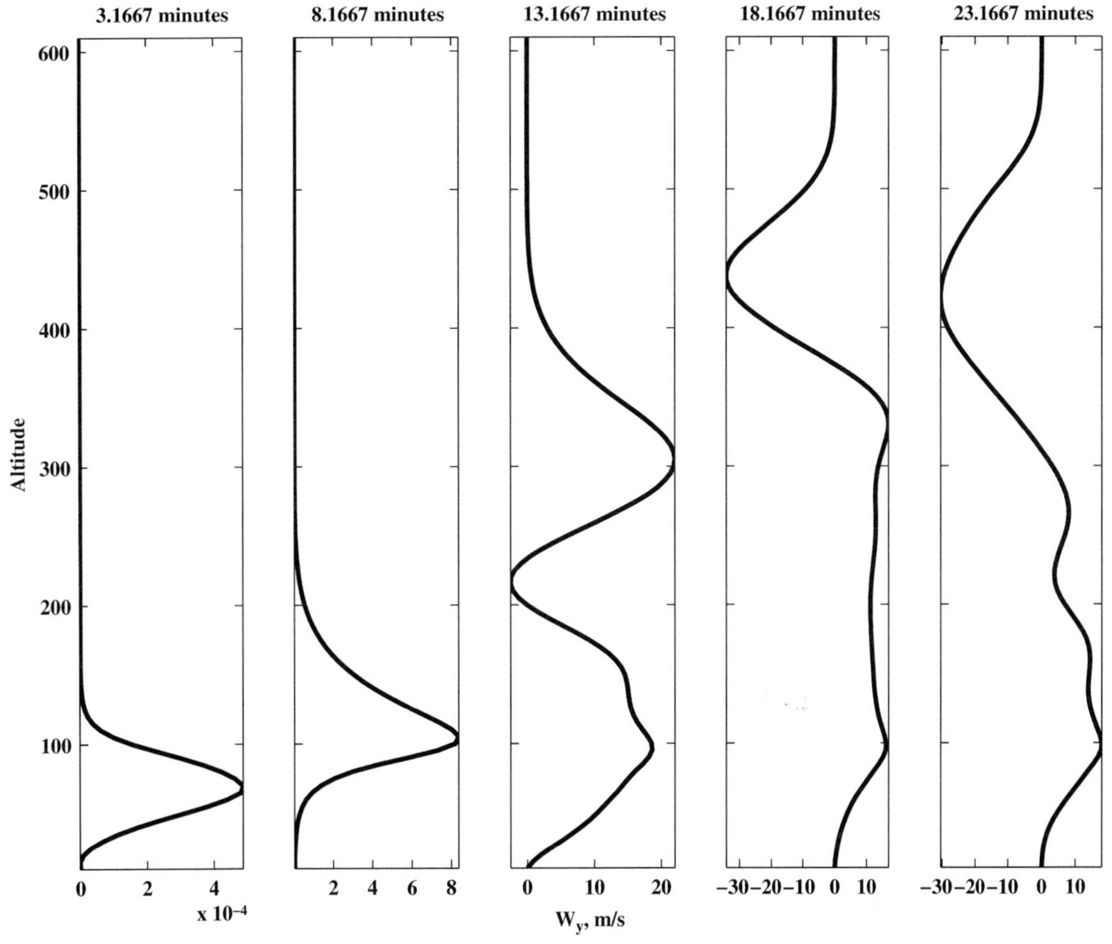

Fig. 10.5 Vertical wind component W_y of AGWs for Case 3 at a few selected time

The momentary balance between ν_g and the growth in W_y leads to an equilibrium which allows the AGW to escape the density duct, resulting in a reduction in both W_y and ν_g. Since forcing from Θ_w is always available, W_y again builds up, resulting in growth of ν_g also. This leads again to momentarily equilibrium inside the density duct and the same process repeats as long as forcing from Θ_w is available. In other words, the equilibrium inside the density duct acquires an oscillating owing to competition between tropospheric forcing and ν_g damping.

As also noted from Fig. 10.6, a maximum in the amplitude of AGWs always remains in 350–450 km altitude region after AGW escapes from the density duct. As noted from Fig. 10.4, this altitude region is decided by the relative magnitudes of decreasing (with altitude) atmospheric density ρ and ν_g on one side and increasing ν_ρ on other side and they have cross-over in 350–450 km altitude. It may be said from Fig. 10.6 that cross-over leads to the formation of neutral velocity layer in 350–450 km associated with the AGWs. The presence of such a layer may have important implications for the bottomside of F region where ions may gain considerable momentum/energy from this wind layer. This aspect is examined in detail in Section 10.3.

It may be pointed out that the role of viscosity to limit the amplitude of AGW inside ducts and the subsequent escape of AGW to higher altitude are the crucial aspects which the present study brings out. The identification of the different kinds of viscosity effects associated with ν_g, ν_ρ is addressed for the first time though they may have been present in previous numerical studies.

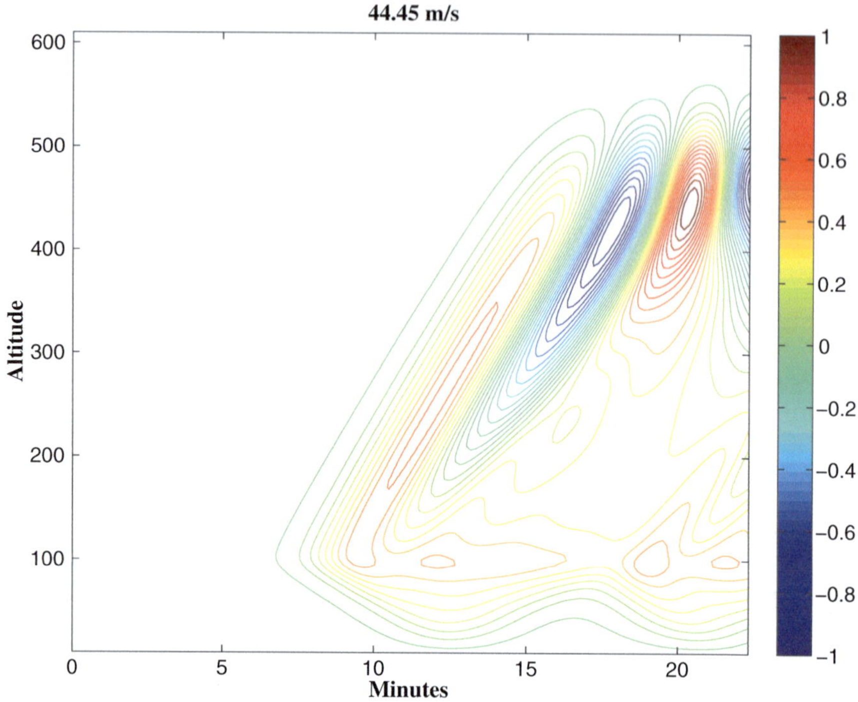

Fig. 10.6 Vertical propagation of the vertical wind component W_y of AGWs for the Case 3. At the *top* of the plot, the maximum value of the wind is depicted. The same color coding is used for Fig. 10.7, 10.8, 10.9, and 10.11

10.2.2.2 Two-Dimensional (Altitude-Longitude) Simulation of AGWs

In order to examine the propagation of the AGW in altitude-longitude plane, Case 4a is investigated wherein the wave number $k_o = 2\pi/\lambda$; $\lambda = 150$ km is introduced in Θ_w in Eq. (10.2) while atmospheric conditions are kept identical to the Case 3. In Figs. 10.7 and 10.8 respectively, the vertical W_y and the horizontal W_x components of the wind amplitude of the AGWs (normalized to corresponding maximum value) are plotted at a few selected time in altitude-longitude plane for the Case 4a. It may be noted that (9) the AGW propagates vertically upward and has both vertical W_y and horizontal W_x components; (10) In the first 10 min, when the wave is trapped below 200 km (Figs. 10.7 and 10.8), W_y and W_x have similar magnitude; (11) After 10 min, when the wave escape to higher altitudes, W_x becomes dominant over W_y above 200 km while remains of similar magnitude below 200 km; (12) After 20 min, W_x increases rapidly attains maximum value ($\sim$200 m/s) at 50 min and remains almost constant afterwards; (13) W_x always has dominant maximum above 250 km altitude unlike W_y which has another dominant maxima in density duct near 120 km altitude.

These characteristics indicate that above 200 km, wind vector associated with the AGWs is dominantly horizontal perpendicular to the vertical direction of wave propagation and thus wave is dominantly transverse in nature. The dominant transverse (longitudinal) nature of AGW above (below) 200 km may be explained as follows: In Eq. (10.1), the gravity wave contribution comes from GWI and GWII terms. In the presence of the horizontal varying tropospheric source, the horizontal and vertical components of these terms are inversely-proportional to the horizontal wavelength associated with tropospheric source and vertical scale height of the atmosphere respectively. Since the scale height of the atmosphere increases with the altitude, the horizontal components of GWI and GWII terms become dominate at higher heights while the vertical components dominate at lower heights. This leads to the altitude variation in the relative magnitudes of W_x and W_y as noted in Figs. 10.7 and 10.8.

In Figs. 10.7 and 10.8, we may note a compression of large vertical wavelength to smaller scales in the 300–500 km altitude region after 25 min. This may be owing to the following reason: There is always a close competition between ν_g and ν_ρ in the 300–500 km region after 25 min as shown in Fig. 10.4. Also noted in Fig. 10.4 is that ν_g has an oscillating pattern in this

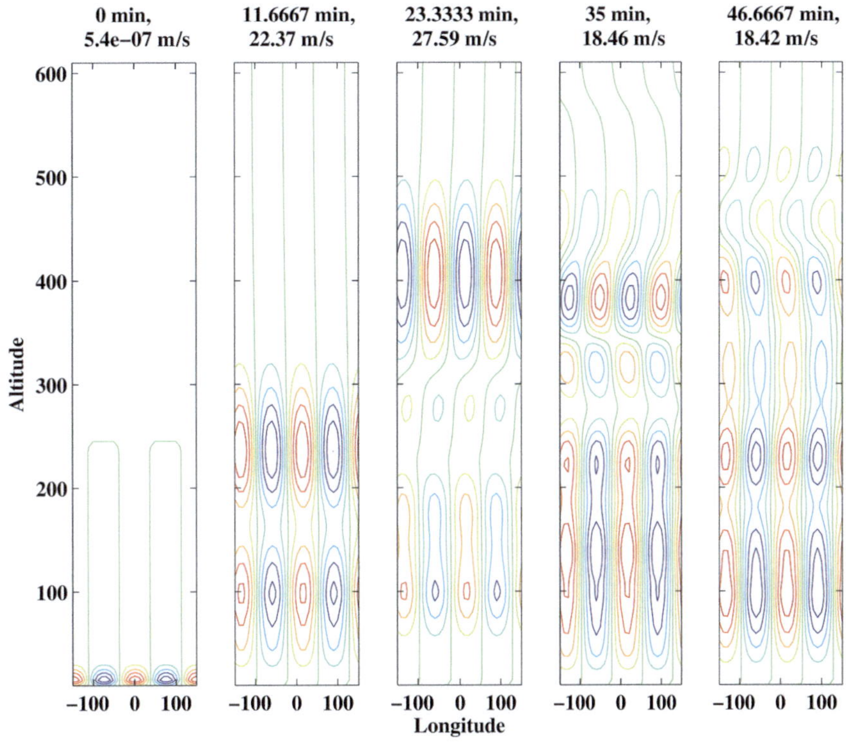

Fig. 10.7 The vertical wind component W_y, (normalized to its maximum) associated with the AGW for the Case 4a. At the *top* of each plot, the time and the maximum value of the wind are shown

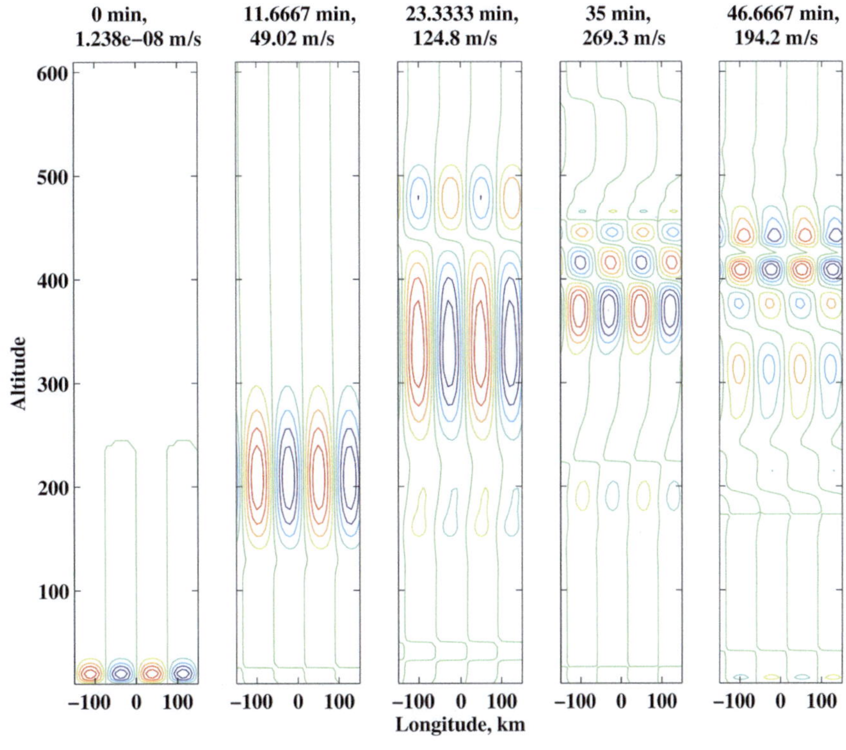

Fig. 10.8 The horizontal wind component W_x (normalized to its maximum) associated the with AGW for the Case 4a. At the *top* of each plot, the time and the maximum value of wind are shown

altitude region which of course is due to oscillating pattern in AGW amplitude. Thus there may be more than one altitude where a cross over takes place. In other words, a competition between ν_g and ν_ρ introduces spatial interference which leads to the formation of smaller wavelengths than principle wavelength. The shorter vertical structures are more prominent in W_x than W_y since W_x is much larger than W_y above 200 km leading to strong interference as noted in Fig. 10.8. It should be pointed out that the transverse component was not excited in the Case 3 of horizontally homogeneous Θ_w. Thus, the horizontal inhomogeneity in the tropospheric thermal source leads to excitation of horizontal component and stronger AGWs in the Case 4a. Similar response is expected from horizontal variation in density/temperature though this aspect is not pursued in this work. The horizontal variations in density/pressure arises from nonlinearity and it may also alter the propagation direction of the AGWs which may then propagate in oblique direction (Yu and Hickey, 2007; Snively and Pasko, 2008).

10.2.2.3 Effects of Finite Mean Horizontal Wind $W_o \neq 0$

The results presented above for the Case 4a correspond to the ambient atmosphere without a mean horizontal wind flow $W_o = 0$. In order to study the mean wind effects on the propagation of AGW, the following cases are chosen:

Case 4b: $\mathbf{W}_o = W_o\hat{x}$ as shown in Fig. 10.1d.
Case 4c: $\mathbf{W}_o = 2W_o\hat{x}$ as shown in Fig. 10.1d.

The dominant zonal wind component W_x of the AGW at t = 50 min are plotted for the Case 4b and 4c in left and right panels of Fig. 10.9 respectively. It may be noted that (14) The amplitude of the AGW reduces significantly (by one order of magnitude) in Case 4b in the presence of zonal mean wind as compared to zero-wind Case 4a; (15) In the strong-wind Case 4c, the amplitude of AGW reduces severely at ionospheric heights so that only maxima remains confined to the lower thermosphere near the density duct.

The reduction in the AGW amplitude in the presence of a mean zonal wind $W_o \neq 0$ in the Cases 4b and 4c suggests that the zonal wind has a damping effect on the AGW propagation. In the wave Eq. (10.1), we may note that the W_o term appears with $\nabla_m^2 = \frac{d^2}{dx^2}$ which means that this term alone introduces a zonal propagating wave with velocity equal to W_o. Introduction of this horizontally propagating wave should oppose the dominantly vertically propagating AGW thus leading to the reduction in the amplitude of AGWs. This aspect is more evident for the strong-wind Case 4c when the maxima near 350–450 km altitude disappears and the primary maxima remains confined to lower thermosphere near the density duct. The damping nature of the mean zonal wind is known from previous studies (Fritts and Vadas, 2009) and in present study, this aspect will be explored more in detail in the context of seeding of plasma bubble.

10.3 AGW-Collisional Interchange Instability (CII) Coupling

10.3.1 Atmosphere-Ionosphere Coupling and Motion in the Equatorial Ionosphere

The ionosphere is a weakly ionized plasma immersed in the neutral atmosphere and the Earth magnetic field. The dynamics of the ionosphere is governed by the hydromagnetic Eqs. ((10.8), (10.9), and (10.10)) described in Appendix 2. In the steady-state momentum Eq. (10.10), **W** is the neutral wind which may be associated with AGWs. The form of B3 suggests that momentum transfer from the AGWs to ions/electrons is determined by ion-neutral and electron-neutral collision frequencies $\nu_{i,\,en}$, the ion and electron gyro-frequencies $\Omega_{i,\,e}$ and magnitude of the wind field **W**. For the wind components $W_{x,\,y}$ associated with the AGWs, the velocities of ions may be obtained from Eq. (10.10). In the present study, the ionospheric parameters are derived using the SAMI2 model (Huba, 2001) and plasma number density and κ_i are shown in Fig. 10.10. In Fig. 10.11, the horizontal and vertical components of the ion velocity u_i, derived from Eq. (10.10), are plotted in the 110–510 km altitude region at t = 50 min for AGW corresponding to Case 4a. We may note that (16) Both the $u_{ix,\,iy}$ components maximize in the E region having similar magnitudes ~10 m/s and have secondary maxima in the F region; (17) The vertical component is much larger than the

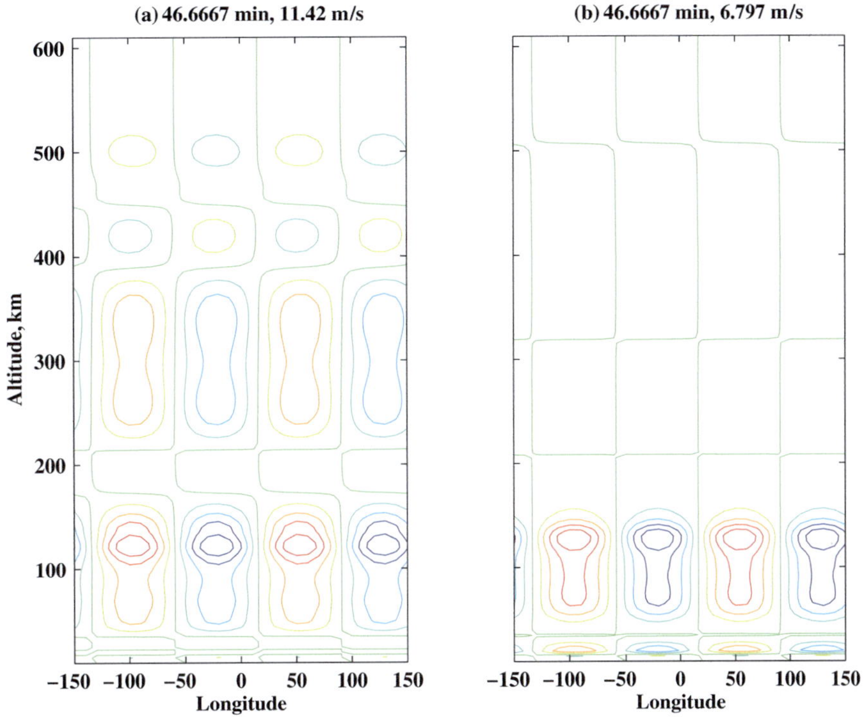

Fig. 10.9 The horizontal wind amplitude W_x (normalized to its maximum value mentioned in the top of each plots) associated with the AGW, at t = 47 min, for the Cases 4b and 4c in panels (**a**) and (**b**) respectively

horizontal component in the F region i.e. the ion velocity is dominantly vertical in the F region.

The ion velocity in the F region is considerably smaller than in the E region despite the AGW amplitude being larger in F region than in E region (Figs. 10.7 and 10.8). This is due to the smaller ν_{in} in the F region which is 2–3 order smaller than its value in the E region leading to weaker coupling of the AGWs and the ions in the F region than in the E region.

10.3.2 Simultaneous Simulation of AGWs and Plasma Bubble

10.3.2.1 Effects of Mean Zonal Wind and of Thermal Source Life-Time

From Fig. 10.11, we have noted that in the F region, ions dominantly move vertical owing to the AGWs-ionosphere coupling and has magnitude of order of 1–2 m/s. During the evening prereversal enhancement in zonal electric field, the ambient F region moves upward attaining largest velocity of order of 30 m/s. It is thus expected that the vertical velocity of the ionosphere is perturbed by 3–5% during the passage of an AGWs of tropospheric origin and may acts as a seeding perturbation for the excitation of plasma bubble.

In order to understand the seeding aspect leading to plasma bubble development, the AGW propagation model and the CII model described in Appendixs 1 and 2 respectively should be coupled together. The coupling between the two models is established through the Atmosphere-ionosphere coupling model described in the previous subsection. In general, the evening-time ionospheric upward drift increases from a low value (5 m/s) to 30 m/s within 1 h and then decreases turning downward within 40 min after prereversal peak (Abdu et al., 2003). The effects of a time-varying ionospheric vertical drift on the instability seeding mechanism was investigated in our previous study (Kherani et al., 2009b). In the present study, an average (over 2 h centered at prereversal time) constant vertical drift of 20 m/s is used. The initial density profile as shown in Fig. 10.10a has bottomside scale height=10 km and F_{peak}=325 km at t=0 second. With

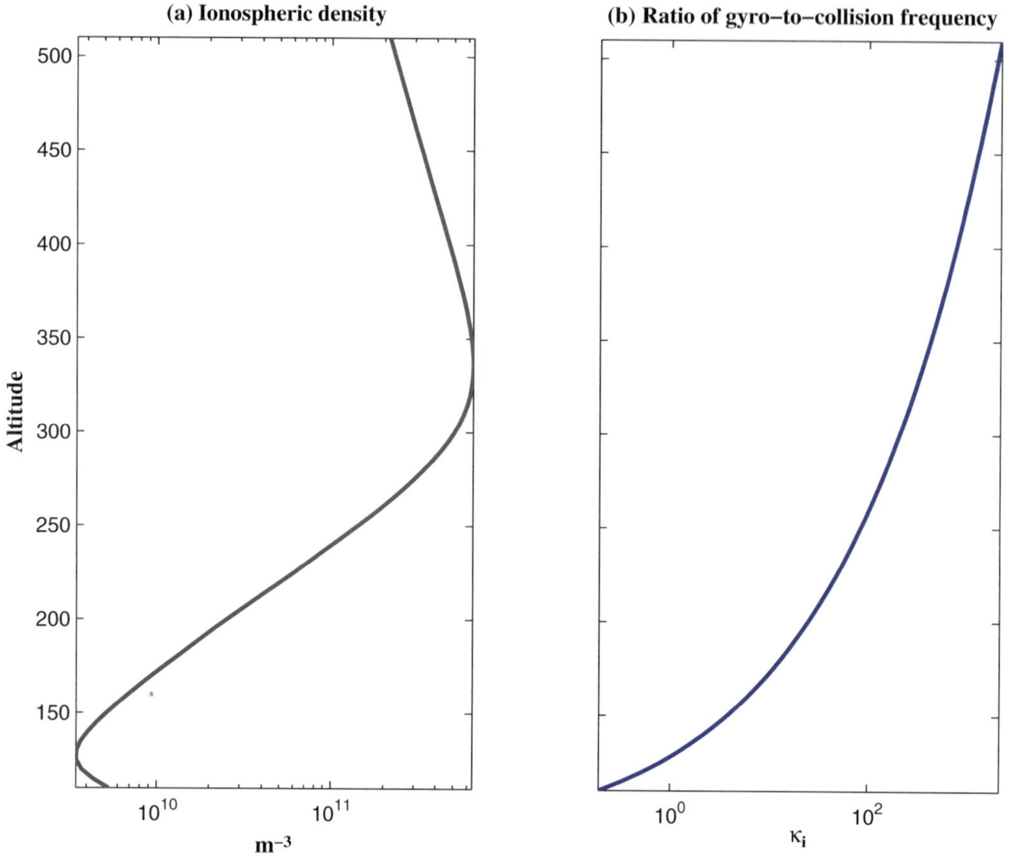

Fig. 10.10 The electron density profile (**a**) and κ_i (**b**)

these ionospheric conditions, the AGW-CII coupled model is studied for atmospheric conditions corresponding to Case 4a–c. In Fig. 10.12, the ionospheric iso-density contours in the simulation plane are plotted at different time during the passage of an AGW for the Case 4a. It may be noted that the density contours lying in the bottomside and in certain zonal position at t=0 second are stretched to higher altitudes with time leading to the formation of plasma bubble within 100 min. The bubble evolution may also be represented in terms of time variation of maximum upward velocity inside the bubble (Kherani et al., 2009b) as plotted in Fig. 10.13 for the Case 4a. It may be noted that velocity starts increasing exponentially and then grows multi-exponentially with time indicating the transition from linear to nonlinear evolution of the bubble (Kherani et al., 2009b). In Fig 10.13, the time variations of the bubble velocity is also plotted for the Cases 4b–4c. It may be noted that the evolution of bubble for Case 4b is much delayed as compared to that for the Case 4a. The delay bubble evolution in Case 4b is owing to the reduced amplitude of the AGW in the presence of a mean zonal wind $W_o \neq 0$ as compared to the wind-less cases 4a. It may be pointed out that the amplitude of the AGW for the Case 4b is one order less than its value in the Case 4a (Fig. 10.9). From Fig. 10.11, it may be expected that the maximum vertical ion velocity in the F region should be of order of 0.1 m/s leading to 0.5% perturbation in the vertical ionospheric velocity which is equal to 20 m/s. This degree of perturbation may still be sufficient to seed the bottomside and to gives rise to bubble growth as can be noted in Fig. 10.13. In contrast to the Cases 4a and 4b, the Case 4c does not lead to formation of plasma bubble till 160 min. From Fig. 10.9, we may note that the AGW amplitude in F region bottomside is considerably reduced for the Case 4c as compared to the Case 4b and this may lead to considerably weak or inefficient seeding

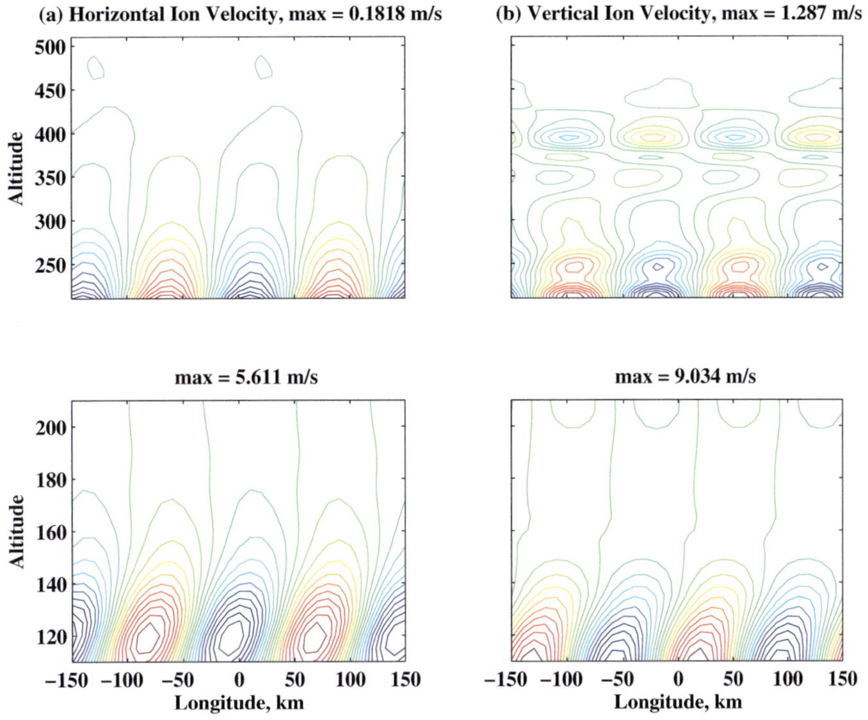

Fig. 10.11 (**a**) *Horizontal* or *Zonal* (u_{ix}) and (**b**) *Vertical* (u_{iy}) components of ion velocity for AGW corresponding to Case 4a at t = 50 min. In the lower (*upper*) panels, 110–210 km (210–510 km) altitude region are plotted

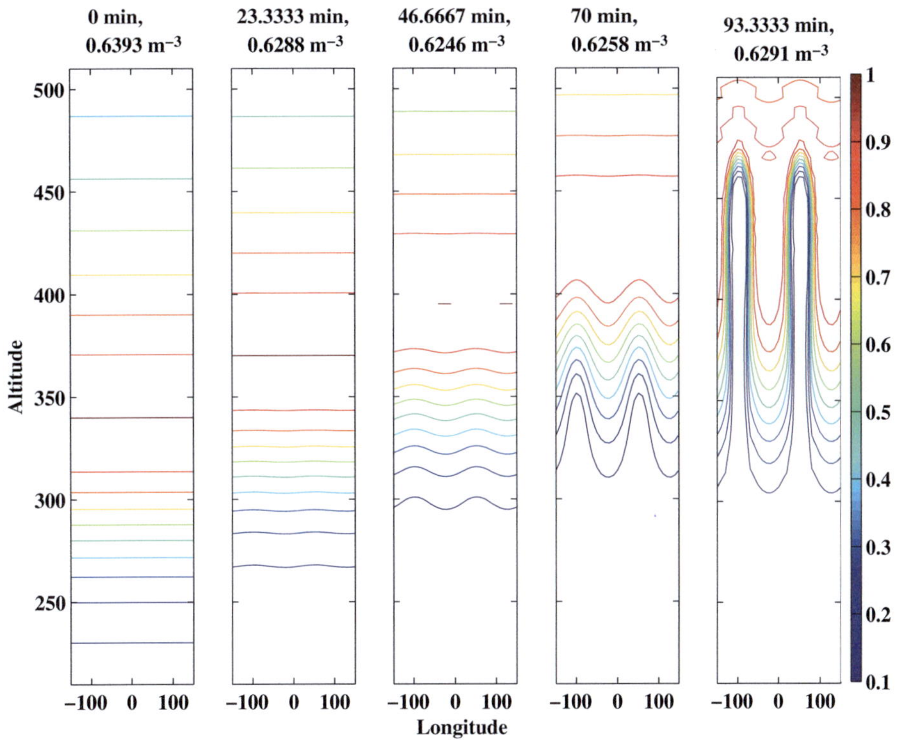

Fig. 10.12 Ionospheric Iso-density contours at different time corresponding to Case 4a. At *top* of each plot, time and maximum density divided by 10^{12} are depicted

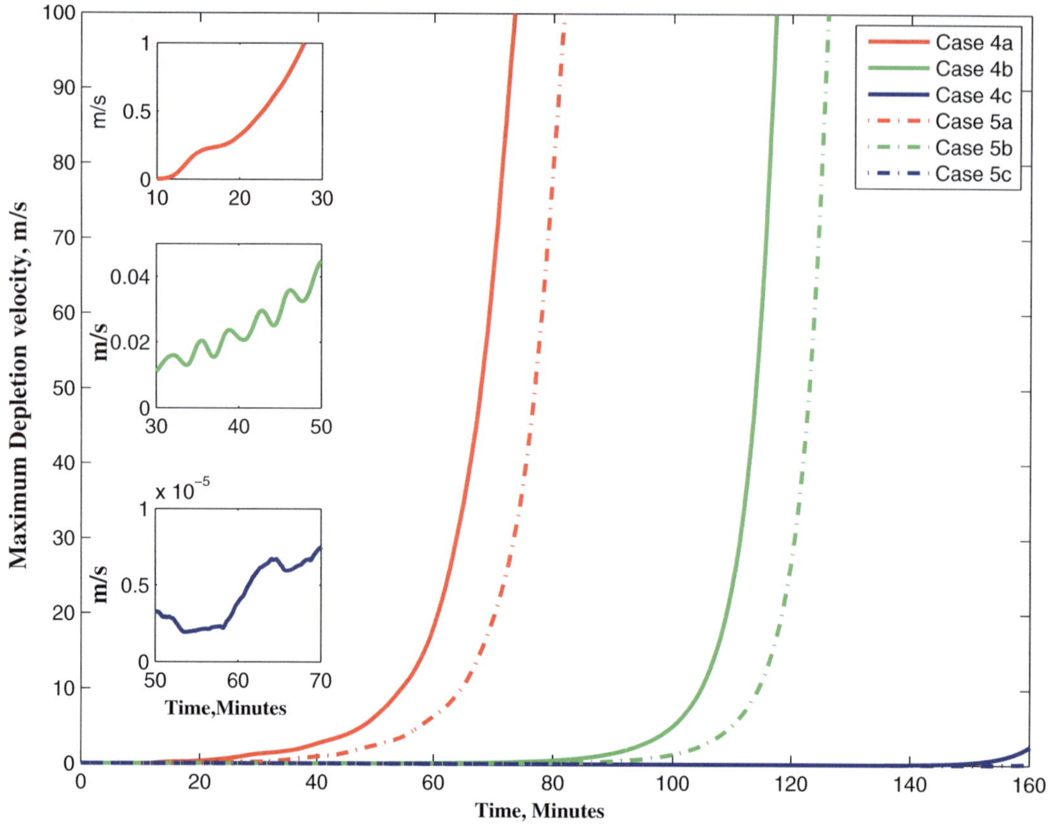

Fig. 10.13 Time variation of maximum velocity of plasma bubble (*depletion*) for Cases 4a-c and Case 5a,5c

of the CII in Case 4c. It should be pointed out that the mean wind profile for Case 4b (Fig. 10.1d) represents realistic thermospheric wind during the evening-time and the composite Case 4a and Case 4b reflect the time-variation of the thermospheric wind from zero to peak value that occur during 16–20 LT over the equator, consistent with the results by Fejer et al., (1985). On the other hand, the mean wind profile chosen for Case 4c (Fig. 10.1d) has large mean wind value (=300 m/s) in the Ionosphere. Large thermospheric winds of order of 300–400 m/s have been observed in the ionosphere during storm-time ($k_p > 6$–7) (Forbes et al., 1995). Thus, the Case 4c can be considered as representative of a storm-time mean wind behavior and the corresponding AGW and plasma bubble dynamics demonstrate the effects of storm-time mean wind. It may be noted that apart from altering the AGW amplitude in the ionosphere, the mean wind can significantly alters the prereversal enhancement in zonal electric field and may also excite the CII on horizontal gradients (Abdu et al., 1995; Kudeki et al., 2007; Huba et al., 2009). Of these effects, only that of the mean wind on the amplitude of the AGW is considered in present study.

In three insets in Fig. 10.13, the time variation of the upward velocity is zoomed during 10–30 min for the Case 4a, 30–50 min for the Case 4b and 50–70 min for the Case 4c. For the Cases 4a–4b, it reveals the presence of oscillating pattern in the upward velocity 30–50 min prior to the bubble evolution, in contrast to the later evolution when the velocity increases exponentially and then multi-exponentially without exhibiting oscillating behavior. Moreover, the mean value of the oscillating velocity also increases with time. The oscillating behavior is owing to the AGW forcing while exponential and multi-exponential increase are owing to the linear and nonlinear phases of the CII. It may thus be said that some 30–50 min prior to bubble development, ionospheric density or velocity perturbation builds-up linearly owing to the AGW forcing and thereafter, the CII dynamics dominate leading to the plasma bubble growth. The time variation of F layer heights, deduced from digisonde observations, often reveals height oscillations 1–2 hour prior to bubble development (Abdu et al., 2009). These oscillations have the characteristics of upward propagating gravity waves and therefore present evidence of AGW seeding of bottomside F layer and argued as precursor for plasma bubble development (Abdu et al., 2009). The velocity oscillations, 30–50 min prior to plasma bubble development for Case 4a–4b seem to corroborate these measurements and suggests that AGW induced modulations may indeed be a suitable precursor condition for plasma bubble development. In the lower-most inset in Fig. 10.13, time variation of the upward velocity is zoomed for Case 4c when bubble development was suppressed till 160 min. For this case too, the velocity oscillations are noted with the mean value increasing with time. However, the increase rate is very small compared to the increase rates in upper and middle insets. This situation leads to the suppression of plasma bubble development in Case 4c till 160 min. These characteristics suggest that the velocity oscillations with large increase rates in mean value lead to the plasma development within 1–2 hour as in Case 4a–4b. On the contrary, velocity modulation with low increase rate in mean value does not lead to the plasma bubble development. In other words, the velocity oscillations with rapid increase in the mean value is an indicator of an efficient AGW forcing to gives rise to plasma bubble and may be suggested as a suitable precursor for plasma bubble development.

The results corresponding to the Cases 4a–4c as shown in Fig. 10.13 correspond to the identical tropospheric source oscillating with the period of 6 min. Even with the same tropospheric source, the oscillations in the ion upward velocity in Cases 4a–4c, as noted in insets in Fig. 10.13, show different periodicity. This difference arises owing to the influence of the mean wind W_o and its varying effect on frequency and amplitude of the AGWs. In presence of the mean wind, the frequency of the oscillation of AGWs acquires Doppler shift $\omega = \omega_o \pm \kappa_y W_o$ where ω_o and k_y are frequency and wavenumber associated with tropospheric source. This leads to the appearance of high frequency (permitted by dispersion relation) in Case 4b as compared to Case 4a. On this ground, one may expect much higher frequency to be present in the Case 4c as compared to Case 4b since W_o is twice larger in Case 4c than Case 4a. However, large W_o leads to the severe reduction of the AGW amplitude in Case 4c than Case 4b as noted in Fig. 10.9 and high frequencies remain confined below 150 km.

In the present study, the plasma bubble simulation is carried out at the equator in a plane perpendicular to the Earth's magnetic field and the effects of parallel conductivity is ignored. Previous three-dimensional studies (Keskinen et al., 2004, Kherani et al., 2005) suggested that the bubble growth can be significantly delayed once the parallel conductivity is taken into account. Kherani et al. (2005) have shown that the nonlinear three-dimensional growth of bubble is delayed almost 2 times from the two-dimensional nonlinear growth if no density perturbation is considered along field line. In the presence of density perturbation, which may be caused by meridional wind associated with AGW, the three-dimensional nonlinear growth is further delayed. Thus the nonlinear growth time obtained from Fig. 10.13 for different Cases should be multiplied by factor of 2 to accommodate effects of parallel conductivity.

The threshold amplitude of a AGW required to produce a plasma bubble is mainly decided by the ionospheric conditions such as the bottomside density gradient and the nature of the vertical drift of ionosphere (Abdu et al., 2009; Kherani et al., 2009b). In the present study, the effect of atmospheric condition, namely, the nature of the mean wind, is studied. Since presence of a mean wind reduces seeding velocity perturbation from 5% in Case 4a to 0.5% i.e by one order of magnitude, in the Case 4b, and reduces much more severely in Case 4c for strong-wind case, its role may be considered to be critical in the evolution of bubble. Fejer et al. (1985) have shown that the reversal of the thermospheric wind from westward to eastward occurs during 16-17 LT over Jicamarca and then the wind builds-up from 20 to 150 m/s during 18 to 20 LT. Thus during the evening-time, the thermospheric wind may have nature of a composite of the Cases 4a and 4c leading to a situation where the bubble formation, which also takes 1–2 h, may depend critically on the mean zonal wind so that its time variation should be properly taken into account. This suggests the importance of performing simultaneous simulation of the AGW-CII under time-varying atmospheric condition.

The results presented so far correspond to Θ_w which has a Gaussian envelop over time with the life-time

$\sigma'_t = \sqrt{2}\sigma_t = 14$ min (Fig. 10.2). The growth time of the CII in the bottomside F region is also of order of 15 min. Thus, during the linear growth of the CII, thermal wave excitation with significant amplitude always remained available in the atmosphere. To examine the effect of the nature of the thermal source with different life-time scale, a case (case 5) is studied where $\sigma'_t = \sqrt{2}\sigma_t = 3.5$ min i.e the thermal source decays 16 times faster than in the previous case, after attaining maximum at t=18 min. In Fig. 10.13, the time variation of the maximum bubble velocity is plotted for Case 5a, 5b and 5c which correspond to wind-less, normal-wind and strong-wind cases respectively. It may be noted that as compared to the Case 4a, the bubble growth is delayed in Case 5a. Both Cases 4a and 5a correspond to wind-less case and the delay in the bubble growth in the Case 5a demonstrates the importance of the life-time of the thermal source to seed the CII.

The effect becomes much more pronounced in the presence of a mean zonal wind as in Cases 5b and 5c where the CII does not reach to nonlinear phase in 120 min i.e the bubble formation is delayed or suppressed significantly. It may be pointed out that highly favorable ionospheric conditions for bubble formation were chosen for Case 4 and 5. In spite of these conditions, the bubble formation is delayed or suppressed depending on the atmospheric condition and the nature of the tropospheric thermal source namely, the life-time of thermal source that preceded a potential bubble development process. These results suggest that a short lived thermal source does not favor the formation of bubble in spite of highly favorable ionospheric conditions. A thermal source with life-time similar to the growth time of CII is a suitable candidate for the formation of bubble. Huang and Kelley (1996) had found the suppression of bubble under the condition when AGW reaches the Ionosphere during the descent phase of the ionosphere. This was later found also in simulation of few case studies under SPREAFEx campaign (Kherani et al., 2009b). In the present study, inspite of the ionosphere ascending steadily at 20 m/s, the formation of plasma bubble is delayed or suppressed for a thermal source with life-time smaller than the growth time of CII. This suggests the critical and competing role of the life-time of thermal source and the growth-time of CII to decide the formation of plasma bubble. It should be mentioned that a typical life-time of tropospheric convection cells is of order of 30 min or more (Sao Sabbas et al., 2009). Thus the life-time taken under Case 5 seems not realistic. However, these cells are highly variable in space-time, having large horizontal growth and moving horizontally not being static in space (Sao Sabbas et al., 2009). Their motion may effectively reduce the life-time at a given longitude. These effects are not considered in the present study but will be the focus of a future investigation. These characteristics as also the results under the Cases 4 and 5 demonstrate the importance of simultaneous simulation of the AGW and the CII processes to properly accommodate the possible effects of a time-varying thermal source. The results under the Case 4 suggest that the wind amplitude of an AGW may become 100 m/s in the 350–450 km altitude. The wind amplitude also becomes large in the density/thermal ducts. For such large amplitudes, nonlinear effects become important and Eqs. (10.3) and (10.5) should be solved together with the nonlinear wave equation (10.7) i.e a time-varying atmosphere should be considered. Similar to time-varying thermal source, the time-varying atmospheric density and pressure may introduce additional features to the AGW-CII coupling. This will also be a topic of our future study.

It should be pointed out that other than the AGW seeding, shear instability (SI) (Hysell and Kudeki, 2004) and wind-driven instability (WI) (Kudeki et al., 2007) have been proposed as seeding mechanism for the CII. Both the SI and the WI rely on the post sunset vortex structures driven by thermospheric zonal wind (Haerendel et al., 1992; Kudeki and Bhattacharya, 1999). The growth rate of the SI is found to be proportional to vertical shear in zonal ion velocity (through vertical shear length) while the growth rate of the WI is found to be proportional to thermospheric wind and horizontal ionospheric density gradient scale length. During the evening-time when the thermospheric wind builds up from zero to large value, the amplitude of the AGW associated wind in bottomside F region may attain value in 10–200 m/s range as noted for the Case 4a–4d. Thus, the AGW itself may offer large zonal winds, associated wind shear and ion velocity shear (Fig. 10.8 and 10.10) and horizontal density gradient (Fig. 10.11 at t = 60 min) to excite the SI and the WI. In particular, both vertical shear length and horizontal density gradient scale offered by AGW could be significantly small compared to the corresponding scales offered by the vortex dynamo associated with the thermospheric wind. The growth rate of SI and WI are found to be proportional to vertical shear length and horizontal density gradient scale length respectively (Hysell and Kudeki, 2004; Kudeki et al., 2007). Thus the growth of SI/WI may be accelerated in the presence of small velocity shear scale and horizontal density gradient scale offered by AGW. This proposal should be examined, however, in the framework of linear perturbation analyses formulated by Hysell and Kudeki (2004) and Kudeki et al., (2007).

10.4 Summary

In this work, nonlinear wave equation of AGW in atmosphere is derived. The equation distinctly identifies acoustic wave and gravity wave contributions, a linear and a non-linear viscous terms and wave-like contribution from mean zonal wind. This equation is solved numerically to study the propagation characteristics of AGW in altitude and altitude-longitude plane. The linear viscous term resist the exponential growth of AGW above 300–400 km altitude region. The nonlinear viscous term resist the growth of AGW inside the density and thermal ducts and allows AGW to escape to higher altitudes. The mean zonal wind introduces a zonal wave propagating with velocity equals to mean zonal wind and opposes the vertical propagation of AGW.

The role of AGW to seed CII in F region and possible formation of plasma bubble is studied by coupling AGW and CII model. This coupling is established through atmosphere-ionosphere coupling model where wind associated with AGW set ionosphere into motion. The growth of plasma bubble under different atmospheric and thermal source conditions are studies. It is found that in the presence of mean zonal wind, seeding perturbation is reduced by one order of magnitude as compared to wind-less atmosphere and formation of bubble is considerably delayed. The strong-wind case which usually prevails during magnetic stormtime is found to suppress the CII. The nature of thermal source, namely life-time is found to have severe impact on the formation of plasma bubble. It is found that to have efficient seeding from AGW, thermal source should have life-time comparable to linear growth time of CII. The CII growth is found to be delayed for thermal source with life-time much shorter than growth time of CII even for highly favorable ionospheric conditions. These results demonstrate importance of simultaneous simulation of AGW/CII and may account for day-to-day variability in the occurrence of plasma bubble, among other atmospheric/ionospheric effects.

Acknowledgements EAK wish to acknowledge the supports from FAPESP through the process 07/00104-0.

Appendix 1: Governing Nonlinear Wave Equation of AGW

$$\frac{\partial \rho}{\partial t} + \nabla.(\rho \mathbf{W}) = 0 \tag{10.3}$$

$$\frac{\partial \mathbf{W}}{\partial t} + (\mathbf{W}.\nabla)\ \mathbf{W} = -\frac{\nabla p}{\rho} + F + \frac{\mu}{\rho}\nabla^2\mathbf{W};\ \mathbf{F} = \mathbf{g} \tag{10.4}$$

$$\frac{\partial p}{\partial t} + (\mathbf{W}.\nabla)p + \gamma p \nabla.\mathbf{W} = R\rho\Theta_w \tag{10.5}$$

Here (ρ, $\mathbf{W}$, p) are the atmospheric density, wind field associated with AGWs and atmospheric pressure respectively, $\mathbf{g}$ is the gravitational acceleration, $\nu = \mu/\rho, \mu$ are the kinematic and dynamic viscosity, Θ_w is the thermal wave excitation associated with tropospheric convection, $R = \frac{k}{m}$ and (k, m) are Boltzmann's constant and atmospheric mass. In (10.3), Θ_w term is similar to the term taken by Yu and Hickey (2007). In (10.4), Coriolis force and ion drag forces are neglected for wave period lower than 1 h. The dissipation effects enter through viscosity and other dissipation effect namely thermal conduction is neglect for linear case. The nonlinear advective second term in L.H.S. in (10.4) is linearized to accommodate effects of mean wind $\mathbf{W}_o$. It means that the mean wind is not allowed to change with wave momentum transport and instead of nonlinear $(\mathbf{W}_o + \mathbf{W}).\nabla$ term, linear $(\mathbf{W}_o.\nabla)$ term is considered. The time derivative of (10.4) then leads to following equation:

$$\frac{\partial^2 \mathbf{W}}{\partial t^2} + (\mathbf{W}_o.\nabla)\frac{\partial \mathbf{W}}{\partial t} = \frac{\nabla p}{\rho^2}\frac{\partial \rho}{\partial t} - \frac{1}{\rho}\nabla\frac{\partial p}{\partial t} - \frac{\mu}{\rho^2}\nabla^2 \mathbf{W}\frac{\partial \rho}{\partial t} + \frac{\mu}{\rho}\nabla^2\frac{\partial \mathbf{W}}{\partial t} \tag{10.6}$$

With the substitutions of Eqs. (10.3) and (10.5), various terms in right-hand-side of (10.6) may be simplified as follows:

$$\frac{\nabla p}{\rho^2}\frac{\partial \rho}{\partial t} = -\frac{\nabla p}{\rho}\left[(\mathbf{W}.\nabla)\log\rho + \nabla.\mathbf{W}\right]$$

$$-\frac{1}{\rho}\nabla\frac{\partial p}{\partial t} = \frac{1}{\rho}\left[\nabla(\mathbf{W}.\nabla)p + \gamma p\nabla(\nabla.\mathbf{W}) + \gamma\nabla p\nabla.\mathbf{W}\right]$$
$$-R\Theta_w\nabla\log\rho$$

$$-\frac{\mu}{\rho^2}\nabla^2\mathbf{W}\frac{\partial \rho}{\partial t} = \frac{\mu}{\rho}\nabla^2\mathbf{W}\left[(\mathbf{W}.\nabla)\log\rho + \nabla.\mathbf{W}\right]$$

$$\frac{\mu}{\rho}\nabla^2\frac{\partial \mathbf{W}}{\partial t} = -\frac{\mu}{\rho}\nabla^2\left[\frac{-\nabla p}{\rho} + \mathbf{g} + \frac{\mu}{\rho}\nabla^2\mathbf{W} - \mathbf{W}_o.\nabla\mathbf{W}\right]$$
$$\approx \frac{\mu}{\rho}\mu\nabla^2\mathbf{W}\nabla^2\frac{1}{\rho}$$

$$\mathbf{W}_o.\nabla\frac{\partial \mathbf{W}}{\partial t} = \mathbf{W}_o.\nabla\left[\frac{-\nabla p}{\rho} + \mathbf{g} + \frac{\mu}{\rho}\nabla^2\mathbf{W} - \mathbf{W}_o.\nabla\mathbf{W}\right]$$
$$\approx -W_o^2\nabla^2\mathbf{W}$$

In deriving above terms, third and fourth order space derivatives are ignored. With the substitution of these terms in (10.6), following wave equation for the wind field **W** is obtained:

$$\frac{\partial^2\mathbf{W}}{\partial t^2} - W_o^2\nabla^2\mathbf{W} =$$
$$\frac{\gamma p}{\rho}\nabla(\nabla.\mathbf{W}) + (\gamma-1)\frac{\nabla p}{\rho}\nabla.\mathbf{W} - \frac{\nabla p}{\rho}(\mathbf{W}.\nabla)\log\rho$$
$$+\frac{1}{\rho}\nabla(\mathbf{W}.\nabla)p$$
$$+\frac{\mu}{\rho}\nabla^2\mathbf{W}\left[(\mathbf{W}.\nabla)\log\rho + \nabla.\mathbf{W} + \mu\nabla^2\frac{1}{\rho}\right]$$
$$-R\Theta_w\nabla\log\rho \tag{10.7}$$

Appendix 2: Hydromagnetic Equations in Ionosphere

Ionospheric dynamics and CII are governed by hydromagnetic equations, namely Ion continuity equation, divergence-free current density equation and steady-state momentum equation:

$$\frac{\partial n}{\partial t} + \nabla.(n\mathbf{u}_e) = -\beta n - \alpha n^2 \tag{10.8}$$

$$\nabla.\mathbf{J} = e\nabla.[n(\mathbf{u}_i - \mathbf{u}_e)] = 0. \tag{10.9}$$

$$\mathbf{u}_{i,e} = \frac{\kappa_{i,e}}{1+\kappa_{i,e}^2}\mathbf{v}_{i,e}\times\hat{b} + \frac{1}{1+\kappa_{i,e}^2}\mathbf{v}_{i,e} \tag{10.10}$$

where

$$v_{i,e} = \frac{-c_{si,e}^2}{\nu_{i,en}}\nabla\log n + \frac{\mathbf{g}}{\nu_{i,en}} + b_{i,e}\mathbf{E} + \mathbf{W}$$

The subscripts 'e' and 'i' refer to the electrons and ions, respectively. Eqs. (10.8) and (10.9) are the ion continuity and divergence free current (**J**) equations respectively, while $\mathbf{u}_{i,e}$ in Eq. (10.10) are the ion and electron's steady state velocities. The plasma is assumed to be charge-neutral ($n_e = n_i = n$), and it is ensured by (10.9). The terms in the right-hand side of (10.8) correspond to the chemical loss of electrons by charge exchange process (β) and dissociative recombination process (α) respectively. The notations $\kappa_{i,e}$ represent the ratios of the gyro frequencies, $\Omega_{i,e}$, to collision frequencies, $\nu_{i,en}$ of the corresponding species, $b_{i,e} = e/m_{i,e}\nu_{i,en}$ represent the corresponding mobilities, **E** is the electric field in the neutral wind (**W**) frame, $\hat{b}$ is the unit vector in the magnetic field, B, direction and $c_{si,e}$ are the ions and electrons thermal velocities respectively. For further analysis, the thermal effects i.e., diffusion effects are neglected restricting our investigation to be applicable only for large scales. The diffusion effects are important to study the saturated turbulent state of the instability. However, in the present investigation, the nonlinear evolution of the CII rather than the saturated turbulent state is investigated. The chemical loss-production effects are retained in the present study via effective recombination rate R_e defined by Kherani et al. (2004). Eqs. (10.8), (10.9), and (10.10) are solved at magnetic equator in the Cartesian coordinate system where $\hat{x}$, $\hat{y}$ and $\hat{z}$ correspond to the westward, upward and Earth's magnetic field directions.

References

Abdu MA (1999) Coupling and energetics of the equatorial ionosphere-thermosphere system: advances during the STEP period. J Atmos Solar-Terr Phys 61(1–2):153–165

Abdu MA, Alam Kherani E, Batista IS, de Paula ER, Fritts DC, Sobral JHA (2009) Gravity wave initiation of equatorial spread F/plasma bubble irregularities based on

observational data from the SpreadFEx campaign. Ann Geophys 27:2607–2622
Abdu MA, Batista I, Bittencourt J (1981) Some characteristics of spread F at the magnetic equatorial station Fortaleza. J Geophys Res 86:A8. doi:10.1029/JA086iA08p06836
Abdu MA, Batista IS, Sobral JHA, Jayachandran PT (2003) Equatorial evening prereversal electric field enhancement and sporadic E-layer disruption: a manifestation of E and F-region coupling. J Geophys Res 108(A6):1254
Abdu MA, Batista IS, Walker GO, Sobral JHA, Trivedi NB, Paula ER (1995) Equatorial ionospheric electric fields during magnetospheric disturbances: local time longitude dependence from recent EITS campaigns. J Atmos Solar-Terr Phys 57(10):1065–1083
Fejer B, Farley D, Balsley B, Woodman R (1976) Radar observations of twodimensional turbulence in the equatorial electrojet, 2. J Geophys Res 81(1):130–134
Fejer BG (1991) Low latitude electrodynamic plasma drifts: a review. J Atmos Solar-Terr Phys 53:677
Fejer BG, Kudeki E, Farley DT (1985) Equatorial F-region zonal plasma drifts. J Geophys Res 90:12249
Fritts DC, Abdu MA, Batista BR et al (2009) Overview and summary of the spread F experiment (SpreadFEx). Ann Geophys 27:2141–2155
Fritts DC, Alexander MJ (2003) Gravity dynamics and effects in the middle atmosphere. Rev Geophys 41(1):1003. doi:10.1029/2001RG000106
Fritts DC, Rastogi P (1985) Convective and dynamical instabilities due to gravity wave motions in the lower and middle atmosphere: theory and observations. Radio Sci 20:6. doi:10.1029/RS020i006p01247
Fritts DC, Vadas SL (2008) Gravity wave penetration into the thermosphere: sensitivity to solar cycle variations and mean winds. Ann Geophys 26:3841–3861
Fritts DC et al (1997) Equatorial dynamics observed by rocket, radar, and satellite during the CADRE/MALTED campaign 2. Mean and wave structures, coherence, and variability. J Geophys Res 102(26):191–26216
Forbes JM, Roble RG, Marcos FA (1995) Equatorial penetration of magnetic disturbance effects in the thermosphere and ionosphere. J Atmos Solar-Terr Phys 57:1085–1093
Haerendel G (1973) Theory of equatorial spread F, report. Maxplanck-Institut fur extraterre. Physics Garching, Germany
Haerendel G, Eccles J, Cakir S (1992) Theory for modeling the equatorial evening ionosphere and the origin of the shear in the Horizontal Plasma Flow. J Geophys Res 97(A2): 1209–1223
Hickey MP, Taylor MJ, Gardner CS, Gibbons CR (1998) Full-wave modeling of small-scale gravity waves using Airborne Lidar and observations of the Hawaiian Airglow (ALOHA-93) O(1S) images and coincident Na wind/temperature lidar Measurements. J Geophys Res 103:6439–6453
Hines CO (1960) Internal atmospheric gravity waves at ionospheric heights. Can J Phys 38:1441–1481
Hines CO (1967) On the nature of traveling ionospheric disturbances launched by low-altitude nuclear explosions. J Geophys Res 72:1877–1882
Huang CS, Kelley MC (1996) Nonlinear evolution of equatorial spread F 4: gravity waves, velocity shear and day-to-day variability. J Geophys Res 101:24521
Huang CS, Kelley MC, Hysell DL (1993) Nonlinear Rayleigh-Taylor instabilities, atmospheric gravity waves, and equatorial spread F. J Geophys Res 98:15631
Huba JD, Joyce G (2007) Equatorial spread modeling: multiple bifurcated structures, secondary instabilities, large density biteouts, and supersonic flows. Geophys Res Lett 34:L07105. doi:10.1029/2006GL028519
Huba J, Joyce G, Fedder J (2000) Sami2 is another model of the ionosphere (SAMI2): a new lowlatitude ionosphere model. J Geophys Res 105(A10):23035–23053
Huba JD, Ossakow SL, Joyce G, Krall J, England SL (2009) Threedimensional equatorial spread modeling: zonal neutral wind effects. Geophys Res Lett 36:L19106. doi:10.1029/2009GL040284
Hysell DL, Kelley MC, Swartz WE, Woodman RF (1990) Seeding and layering of equatorial spread F by gravity waves. J Geophys Res 95:17253
Hysell DL, Kudeki E (2004) Collisional shear instability in the equatorial F region ionosphere. J Geophys Res 109:A11301. doi:10.1029/2004JA010636
Kelley MC, Larsen MF, LaHoz C, Mc Clure JP (1981) Gravity wave initiation of equatorial spread F: a case study. J Geophys Res 86:9087
Keskinen MJ, Ossakow SL, Fejer BG (2003) Three-dimensional nonlinear evolution of equatorial ionospheric spread-F bubbles. Geophys Res Lett 30(10). doi:1029/2003 GL017418
Keskinen MJ, Vadas SL (2009) Threedimensional nonlinear evolution of equatorial ionospheric bubbles with gravity wave seeding and tidal wind effects. Geophys Res Lett 36:L12102. doi:10.1029/2009GL037892
Kherani EA, Lognonne P, Kamath N, Crespon F, Garcia R (2009a) Response of the ionosphere to the seismic triggered acoustic wave: electron density and electromagnetic fluctuations. Geophys J Int 176:1–13
Kherani EA, Abdu MA, de Paula ER, Fritts DC, Sobral JHA, de Meneses FC Jr (2009b) The impact of gravity waves rising from convection in the lower atmosphere on the generation and nonlinear evolution of equatorial bubble. Ann Geophys 27:1657–1668
Kherani EA, Mascarenhas M, Sobral JHA, de Paula ER, Bertoni FC (2005) A three dimension simulation model of collisional interchange instability. Space Sci Rev 121: 253–269
Kherani EA, de Paula ER, Bertoni FCP (2004) Effects of the fringe field of Rayleigh-Taylor instability in the equatorial E and valley regions. J Geophys Res 109:A12310. doi:10.1029/2003JA010364
Kudeki E, Akgiray A, Milla M, Chau JL, Hysell DL (2007) Equatorial spread-F initiation: post-sunset vortex, thermospheric winds, gravity waves. J Atmos Solar-Terr Phys 69:2416–2427
Kudeki E, Bhattacharyya S (1999) Post sunset vortex in equatorial F-region plasma drifts and implications for bottom-side spread-F. J Geophys Res 104:28163–28170
Kudeki E, Fawcett CD (1993) High resolution observations of 150 km echoes at Jicamarca. Geophys Res Lett 20(18): 1987–1990
Kudeki E, Fejer BG, Farley DT, Ierkic HM (1981) Interferometer studies of equatorial F-region irregularities and drifts. Geophys Res Lett 8:377

Piani C, Durran D, Alexander MJ, Holton JR (2000) A numerical study of three-dimensional gravity waves triggered by deep tropical convection. J Atmos Sci 57:3689–3702

Pitteway MLV, Hines CO (1965) The reflection and ducting of atmospheric gravity waves. Can J Phys 43:2222–2245

Richmond AD (1978) Gravity wave generation, propagation, and dissipation in the thermosphere. J Geophys Res 83:4131–4145

Rottger J (1981) Equatorial spread F by electric fields and atmospheric gravity waves generated by thunderstorms. J Atmos Solar-Terr Phys 43:453

Sao Sabbas FT, Rampinelli VT, Santiago J, Stamus P, Vadas SL, Fritts DC, Taylor MJ, Pautet PD, Dolif Neto G, Pinto O (2009) Characteristics of sprite and gravity wave convective sources present in satellite IR images during the SpreadFEx 2005 in Brazil. Ann Geophys 27:1279–1293

Snively JB, Pasko VP (2008) Excitation of ducted gravity waves in the lower thermosphere by tropospheric sources. J Geophys Res 113:A06303. doi:10.1029/2007JA012693

Sobral JHA, Abdu M, Batista I, Zamlutti C (1981) Wave disturbances in the low latitude ionosphere and equatorial ionospheric plasma depletions 1981. J Geophys Res 86:A3. doi:10.1029/JA086iA03p01374

Sultan P (1996) Linear theory and modelling of the Rayleigh-Taylor instability leading to the occurrence of equatorial spread F. J Geophys Res 101:26875

Takahashi H et al (2010) Equatorial ionosphere bottomtype spread F observed by OI 630.0 nm airglow imaging. Geophys Res Lett 37:L03102. doi:10.1029/2009GL041802

Tsunoda RT (1994) Enhanced velocities and a shear in daytime Esq over Kwajalein and their relationship to 150 km echoes over the dip equator. Geophys Res Lett 21:2741

Vadas SL (2007) Horizontal and vertical propagation, and dissipation of gravity waves in the thermosphere from lower atmospheric and thermospheric sources. J Geophys Res 112:A06305. doi:10.1029/2006JA011845

Vadas SL, Liu H (2009) Generation of largescale gravity waves and neutral winds in the thermosphere from the dissipation of convectively generated gravity waves. J Geophys Res 114:A10310. doi:10.1029/2009JA014108

Weinstock J (1982) Nonlinear theory of gravity waves: momentum deposition, generalized Rayleigh friction, and diffusion. J Atmos Sci 39:1698–1710

Weinstock J (1990) Saturated and unsaturated spectra of gravity waves and scale-dependent diffusion. J Atmos Sci 47:2211–2225

Woodman R, La Hoz C (1976) Radar observations of F region equatorial irregularities. J Geophys Res 81(31):5447–5466

Yeh KC, Liu CH (1974) Acoustic-gravity waves in the upper atmosphere. Rev Geophys 1 (2):193–216

Yu Y, Hickey MP (2007) Numerical modeling of a gravity wave packet ducted by the thermal structure of the atmosphere. J Geophys Res 112:A06308. doi:10.1029/2006JA012092

Zhang SD, Yi F (2002) A numerical study of propagation characteristics of gravity wave packets propagating in a dissipative atmosphere. J Geophys Res 107. doi:10.1029/2001JD000864

Chapter 11

Mesosphere–Ionosphere Coupling Processes Observed in the F Layer Bottom-Side Oscillation

Hisao Takahashi, Sharon L. Vadas, C.M. Wrasse, Michael J. Taylor, P.-D. Pautet, A.F. Medeiros, R.A. Buriti, Eurico R. de Paula, Mangalathayil Ali Abdu, Inez S. Batista, I. Paulino, P.A. Stamus, and David C. Fritts

Abstract During the Spread FEx campaign, under the NASA Living with a star (ILWS) program which was carried out in the South American Magnetic Equatorial region from September to November 2005, we observed formation of the bottom-type spread F and simultaneous occurrence of mesospheric gravity wave events. The events were monitored by the ionosonde, coherent radar and airglow OI 630.0 nm and OH imager. It is found that the bottom-type scattering layer has a wave form generated most probably by local gravity waves. Reverse ray-tracing of the observed gravity waves indicate their possible sources in the troposphere or thermosphere. Forward ray-tracing indicates their penetration into the ionosphere. The present work summarizes the observational evidence and results of the data analysis and discusses the mesosphere–ionosphere coupling processes.

11.1 Introduction

The equatorial region of the earth's ionosphere undergoes rapid changes after sunset. The electron density in the lower part of the F-layer decreases due to a lack of ionization in the absence of sunlight and quick electron-ion recombination. In addition, the eastward component of the neutral wind blowing from the dayside to the night side (tides) produces upward plasma drift (F-region dynamo effect). These processes make the F-layer bottom side much steeper and unstable thus providing favorable conditions for the generation of plasma instabilities (Rayleigh-Taylor instability). During this condition, if any kind of perturbations (pressure, zonal wind or external electric field) penetrates in the bottom heights, the unstable condition grows and generates plasma irregularities which appear as a Spread F trace in the ionogram and a bottom-type scattering layer in the VHF coherent radar.

The ionospheric Spread F has been studied extensively since IGY (International Geophysical Years, 1957–1958). Especially studies on the equatorial Spread F (ESF) made significant progress after Jicamarca coherent and incoherent scatter radar started to provide novel radar images (Woodman and La Hoz, 1976). Abdu (2001, 2005) has reviewed previous works on this subject. If one observes the bottom-type spread F by back scatter radar image (Range-Time Intensity), it has a diffused form, confined to a narrow range of altitude (10–50 km) at an altitude around 250–350 km. It normally appears soon after sunset and lasts from a few tens of minutes to ~ hours depending on the amplitude and development of the instability. Hysell and Burcham (1998) have reported that bottom-type layers are a common feature in the equatorial F region and frequently extend to topside forming plasma bubbles. Hysell et al. (2004) presented two cases of the bottom-type scattering layers observed at Jicamarca, and discussed the seeding mechanism. They concluded that formation of a bottom-type layer is closely related to horizontal wind components during the sunset to evening time zone. From the horizontal structures (~30 km) observed they pointed out that the bottom-type layers are composed of primary waves excited

H. Takahashi (✉)
Instituto Nacional de Pesquisas Espaciais (INPE), São José dos Campos, SP, Brazil
e-mail: hisaotak@laser.inpe.br

M.A. Abdu, D. Pancheva (eds.), A. Bhattacharyya (Coed.), *Aeronomy of the Earth's Atmosphere and Ionosphere*,
IAGA Special Sopron Book Series 2, DOI 10.1007/978-94-007-0326-1_11,

by horizontal wind-driven gradient drift instabilities rather than the Rayleigh-Taylor or E×B instabilities.

Importance of the wind structure during the sunset for seeding of the F layer bottom-side instability has been discussed by Tsunoda (1981) and Kudeki and Bhattacharyya (1999). Tsunoda (1981) showed that the bottom-side flow is dominated by strong vertical wind shear. Kudeki et al. (2007) also mentioned that F-region zonal wind during the post sunset period has a crucial role in controlling the F-region plasma instability growth. They argued that if the zonal wind is of the order of 200 m/s, some external factors such as gravity waves should not be critical to generate the instability. Spatial (zonal) structures of the F-layer bottom side have been reported by several authors. From Kwajalein Atoll, using a VHF/UHF coherent/incoherent scatter radar, Hysell et al. (2006) observed a bottom-type coherent scattering layer at 200–250 km altitude. The layer had a patchy form, separated by 150–200 km. Takahashi et al. (2010) showed that the longitudinal extension of the bottom-type scattering layer could be more than 1500 km along the magnetic equator.

It seems that the occurrence of bottom-type scattering layer could act as a precursor for Spread F and consequent plasma bubbles formation. The evolution of the bottom-type Spread F depends on the local condition of instability generated by the zonal wind variability and the amplitude, as mentioned by Hysell et al. (2004). Whether the zonal wind variability is generated by gravity waves or by any other sources is not known and no conclusive assessment has come yet (Hysell et al., 2006). From an empirical model calculation, Vadas and Fritts (2004) demonstrated a possible penetration of the gravity waves generated in the lower atmosphere into the thermosphere (150–200 km). Fritts et al. (2009) and Abdu et al. (2009) presented observational evidence of gravity waves in the thermosphere that are propagating upwards (phase downward). Takahashi et al. (2009) presented a close relation between the observed plasma bubble spacing and horizontal wavelength of the mesospheric gravity waves. The Spread FEx campaign revealed that gravity waves have an important role in seeding plasma bubbles (Fritts et al., 2009; Vadas et al., 2009).

Only a few works have reported on the bottom-type scattering layer (Hysell et al., 2004, 2006; Takahashi et al., 2010). Its formation mechanism is still an open question. In order to understand it we need further observational evidences. In the present study we report on three cases of bottom-type Spread F, similar to what Hysell et al. (2004) observed. In addition to the VHF coherent radar, two ionosondes and two all-sky airglow imagers were used to diagnose the phenomenon. The coherent radar detects vertical structure of the scattering layer and the ionosonde measures vertical profile of the bottom-side F layer electron density, and the imager observes OI 630.0 nm intensity depletion in the two dimensional form. Using the three instruments, we observe the bottom-type scattering layers in the 3-D form. Then we investigate the seeding mechanism of the bottom-type scattering layer. For this purpose we use forward and reverse ray-tracing techniques.

11.2 Observation

The Spread FEx campaign was carried out from 22 September to 08 November 2005 in the Brazilian low latitude and equatorial regions. Figure 11.1 shows location of the observation sites. The instruments were deployed at São Luís (SL) (2.6°S, 44.2°W, 1.5°S geomag.): VHF coherent radar and Ionosonde (DPS), at Fortaleza (FZ) (3.9°S, 38.4°W, 6.2°S geomag.): Ionosonde (DPS), at São João do Cariri (CA) (7.4°S, 36.5°W, 11.5°S geomag.): Airglow all-sky imager and Meteor radar, and at near Brasilia (BR) (14.8°S, 47.6°W, 10°S geomag.): Airglow imager and GPS receivers. The aim of the campaign was to investigate the possible gravity wave seeding process of equatorial ionospheric bubbles. For this purpose it was essential to monitor the F-layer uplifting by the ionosonde, onset and evolution of the F-layer bottomside irregularity by the coherent radar, and spatial variation of the F-layer bottomside by OI 630.0 nm all-sky imager. Further, mesospheric gravity wave activities were monitored by OH imaging and wind profiles from 80 to 99 km by the meteor radar.

Digital ionospheric sounder (DPS-4) is a wide-band coded pulsed radar system with 500 W peak power and a fast-switching frequency synthesizer, covering a frequency range from 0.5 to 30 MHz. An ionogram is taken with a 10-minute observational interval. The ionospheric parameters used in this work are the virtual height at fixed frequency, the critical frequency of the F2 layer, foF2, and the minimum virtual height, h'F. F-layer bottom-side irregularities can be seen as a diffused and scattered form in the ionogram trace.

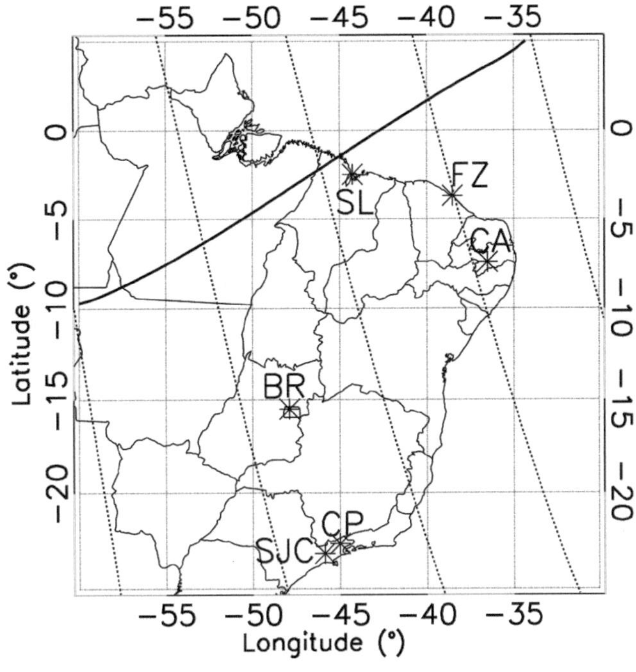

Fig. 11.1 Spread FEx campaign observation site map. SL: São Luís, FZ: Fortaleza, CA: Cariri, BR: near Brasilia, CP: Cachoeira Paulista, SJC: São José dos Campos

The 30 MHz coherent scatter radar provides a scintillation back scatter image as a function of height and time (RTI). Plasma irregularities with a scale size of 5 m can be detected. The radar echo images can show a localized structure with a field of view of 16°, corresponding to ~76 km horizontal extension at 270 km of altitude. Vertical evolution of plasma bubbles can be followed by this technique. Characteristics of the radar have been presented elsewhere (de Paula and Hysell, 2004).

An all sky imager with a 180° wide-angle fish-eye lens followed by a telecentric lens system and narrow band optical filters was used to get a monochromatic image on a CCD camera. A wide area (2.5 × 2.5 cm) CCD camera with 1024 × 1024 pixels and thermoelectrically cooled (–36°C) was mounted. The 630.0 nm images were obtained with a time integration of 90 s, while the OH near infra-red (NIR) (715–930 nm) images were obtained by 15 s. Each filter was introduced into the operation with a time interval of about 5 min. The 630.0 nm image presents spatial irregularities of the ionospheric plasma in a two dimensional form with a horizontal extension of ~2000 km (at a zenith angle of 80°) at an altitude of 250 km. On the other hand, the field of view of the OHNIR image is around 800 km (at ~80° zenith angle) at an altitude of 90 km. For image analysis, FFT power spectrum was used to find out characteristics of the wave motions. For the waves with long horizontal wavelength, longer than 200 km for example, the FFT analysis did not work well. In this case, Keogram technique was used. The wind structure in the mesopause region from 80 to 99 km was monitored by a meteor radar (SkiYmet) at Cariri.

As can be seen in Fig. 11.1, the geomagnetic equator is largely declined against the geographic equator in the eastern side of South America. The declination over Fortaleza is around –20°. Hence, Cariri and Fortaleza are located under approximately a same magnetic meridian. On the other hand, São Luís is located ~5° West of Fortaleza.

11.3 Results

During the first observation period from 22 September to 3 October, passing through the equinox, OI 630.0 nm and mesospheric OHNIR all-sky image data were obtained almost every night except some nights with the cloudy condition in the evening. The overview of the image data has been reported by Taylor et al. (2009).

11.3.1 Ionospheric F-Layer Bottomside Scattering During the Period of 22–25 September

Figure 11.2 shows dynamic day to day variability, from 22 to 25 September, of the F-layer condition, i.e., time variation of the fixed frequency virtual height, h'F, hmF2 and foF2. It is interesting to note that the evening uplifting of the F-layer was varying day by day. In the evening of 22 September, the upward drift was very rapid, the hmF2 reached 430 km and h'F at 350 km at 21:40 UT. Immediately, the Spread F started and developed into high altitude plasma bubbles. The coherent radar RTI image at São Luís showed a sudden scattered layer in the height range of 350–450 km at 22:06 UT. In the evening of September 23, however, the evening uplift was moderate and less than the previous evening, h'F reached only 280 km. On the night of September 24, there was no clear uplifting.

In Fig. 11.3 the coherent radar RTI image, ionogram and OI 630.0 nm image in the evening of 23 September are shown. At 23:00 UT (20:00 LT) a thin scattering layer could be seen at around 280 km of altitude, which then moved upward gradually and reached 350 km at around 24:00 UT. This is a typical form of F-layer bottom-type scattering layer (Hysell et al., 2004). Prior to it, at 22:20 UT the ionogram at Fortaleza registered a satellite trace form. According to Tsunoda (2008) this is formed when the F layer bottom-height has a spatially different height structure over the ionosonde viewing area, indicating a signature of large scale wave structure. Our present result is similar to it and suggests that the F-layer bottom-side had a spatial gradient. The height difference of the satellite trace to the main trace reaches about 50 km. Note that Fortaleza is located geomagnetically ~5° East of São Luís. Simultaneously, the OI 630.0 nm image map depicts spatial intensity depletions that are aligned to the magnetic field line. The depletions presented a periodic spacing with a distance of ~120 km. It seems that the three different observation techniques detected the same phenomenon in different forms. The Radar observed vertical extension of the irregularity. The

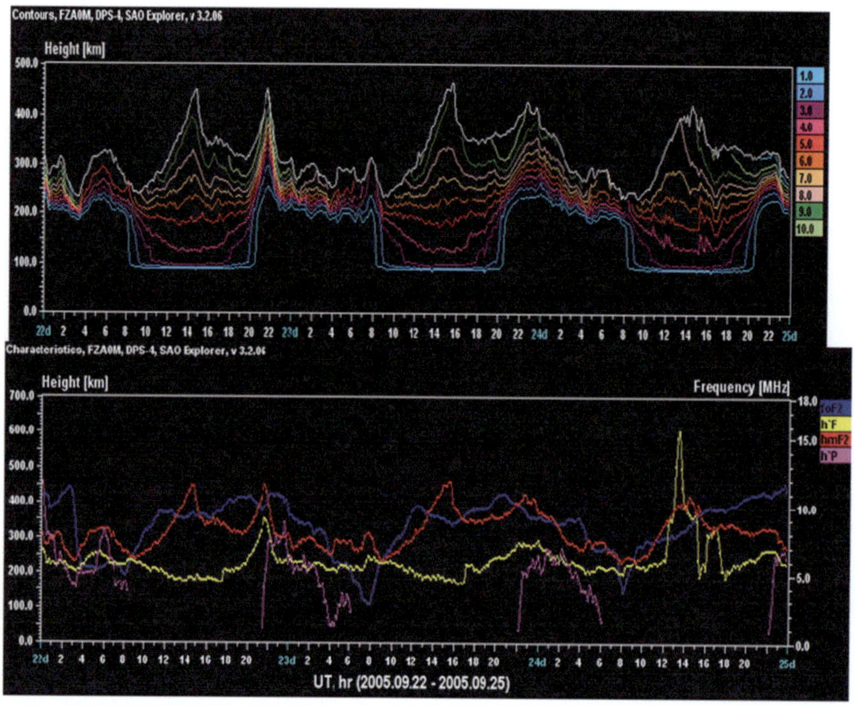

Fig. 11.2 Fortaleza ionogram fixed frequency virtual height variations (*top*) and ionospheric parameters, foF2, h'F, hmF2, h'P (*bottom*) from 22 to 25 September 2005

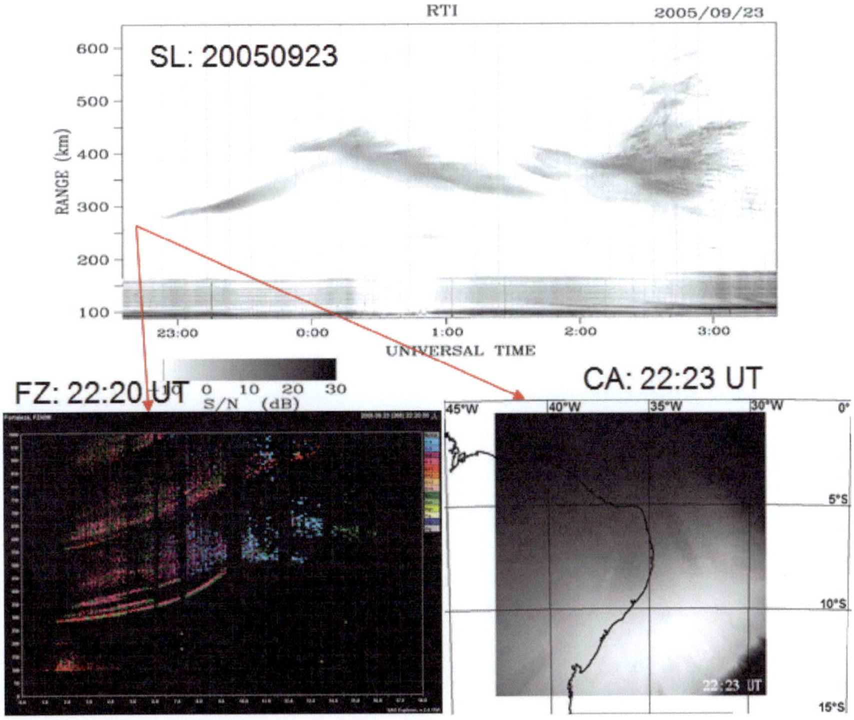

Fig. 11.3 VHF radar RTI image at São Luís (*top*), ionogram at Fortaleza (*bottom left*) and OI 630.0 nm image at Cariri (*bottom right*) on the evening of 23 September 2005

ionosonde observed horizontal gradient of the layer. On the other hand, OI 630.0 nm imager observed horizontal extension of the F-layer bottom-type scattering layer in a 2-D form.

A similar type of bottom type spread F was observed in the next evening of 24 September. Figure 11.4 shows the RTI, ionogram and OI 630.0 nm image. The bottom side scattering started at ~22:00 UT and lasted up to 22:50 UT, then, another scattering layer started above and formed a thick scattering layer (>100 km of thickness). The ionogram showed satellite traces in a diffused form and the OI 630.0 nm image presented depletions aligned along the magnetic field line, similar to the previous night. The spacing of depletion was around 200 km, wider than the previous evening. In addition to the wide spacing, a narrow distance of depletion by around 90 km was also observed from a sequence of the images. Both the nights, however, vertical development of Spread F were relatively weak. As seen in Fig. 11.2 the uplifting of the F-layer of the two evenings was moderate and not favorable to develop the Rayleigh Taylor (R-T) instability. Abdu et al. (2009) estimated that the threshold level of the F layer bottom-side vertical drift velocity of 20–30 m/s would be necessary in order to develop the R-T instability. In the present data the mean vertical drift velocity during the period of 21–22 UT were 25 (±5) m/s, 9.7 (±2) m/s, and 5.6 (±2) m/s for 22, 23 and 24 September, respectively. The vertical drift velocity of 22 September was much higher than the two following nights and was favorable to generate instability. The day to day variability of the evening uplifting is an interesting matter to further investigate.

The bottom-type Spread F was well developed and prolonged in the evening of 30 September. In Fig. 11.5, the RTI image shows that the irregularity started at 240 km at 22:40 UT and lasted to 24:00 UT, and the bottom height was gradually upward drifting to ~300 km. A well developed plume (plasma bubble) then appeared. The ionogram indicated horizontal gradient of the layer and the OI 630.0 nm image presented an oscillation form. Takahashi et al. (2010) reported that this is a signature of sinusoidal oscillation of the F-layer bottom height. The OI 630.0 nm image depicts a zonal spacing of the depletion of ~120 km.

Fig. 11.4 Same as Fig. 11.3 but for 24 September 2005

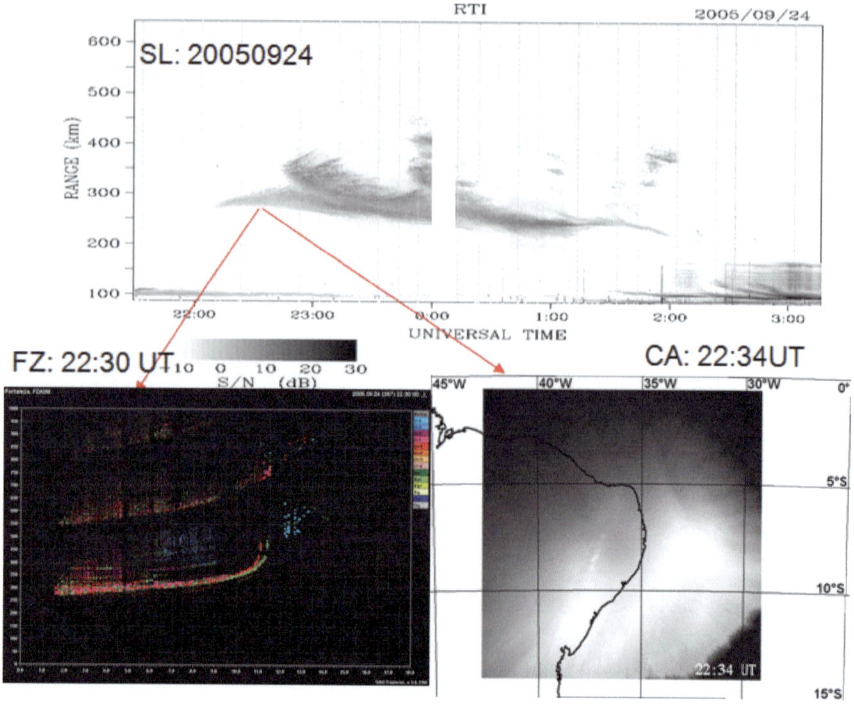

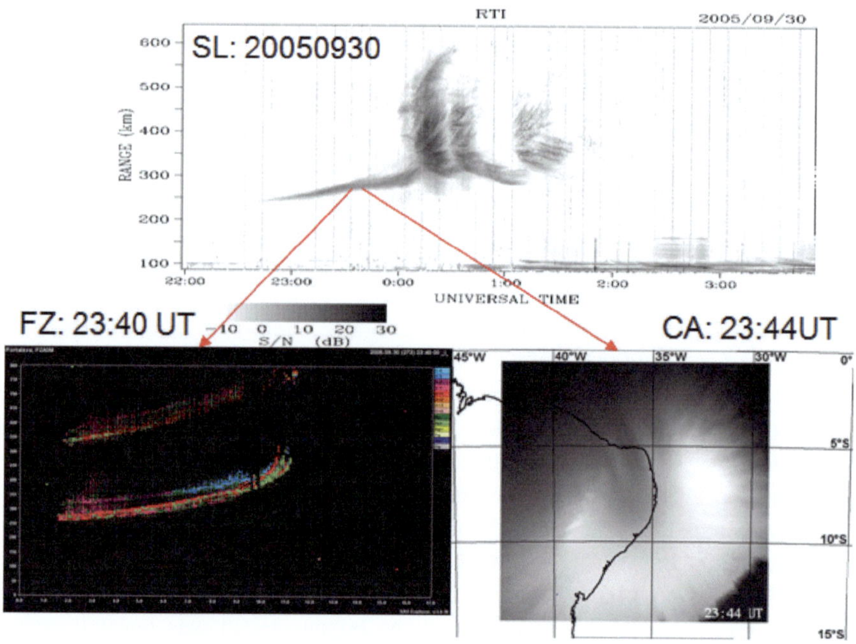

Fig. 11.5 Same as Fig. 11.3 but for 30 September 2005

A common feature of the ionospheric irregularity in the three evenings is periodic oscillation of the F layer bottom-side during the evening uplifting phase. During this period, the ionosonde observed bottom side gradient by 20–50 km. The OI 630.0 nm images revealed that the irregularity had a sinusoidal form with a zonal spacing of 100–200 km. The OI 630.0 nm emission rate is sensitive to the emission height because of its dissociative recombination process, $O_2^+ + e \rightarrow O(^1D) + O^*$, where * indicates any excited state. Therefore, the sinusoidal oscillation of the airglow intensity variation should be due to vertical displacement of the OI 630.0 nm emission layer as Takahashi et al. (2010) pointed out. Although a possibility of spatial variation of the electron density caused by gravity wave perturbation can not be ruled out as another responsible process for the OI 630.0 nm depletion, the vertical oscillation is much plausible if one looks into the F layer bottom height structure presented above. These observational facts lead us to believe that the sinusoidal depletions should be signature of vertical oscillation of the F layer bottom-side.

11.3.2 OI 630.0 nm Depletions and OH Gravity Waves

In order to determine possible sources of these sinusoidal variations, we investigate the contribution of atmospheric gravity waves in the lower thermosphere. A zonal wind oscillation produced by gravity waves could modulate the ionospheric layer upward and downward through the F-region dynamo effect. Unfortunately there are no means to measure neutral wind structure in the lower thermosphere (120–200 km) from ground. In this work, therefore, we used OH airglow image data in order to determine the gravity wave activity in the upper mesosphere and lower thermosphere (MLT) region.

In Fig. 11.6, a keogram of the OI 630.0 nm and the MLT OH emission images on the night of 24 September 2005 is shown as an example. A keogram is formed by N-S and E-W direction sliced images as a function of time. This technique has been used in the airglow image analysis when the scale of spatial variation is larger than the image view distance (1500 km in the case of OI 630.0 nm and 500 km for OH). The northern sky periodic depletions of OI 630.0 nm can be seen in the figure. It is interesting to note that the OH keograms depict a variety of wavelike oscillations during the occurrence of the OI 630.0 nm depletion. We further investigate the similarity of the keograms of the two emissions.

From the N-S and E-W propagation pattern of the intensity variation, the period, phase velocity and horizontal wavelength were calculated for the image data from 22 to 30 September 2005. We also used the Fast Fourier Transform (FFT) to calculate the wave parameters from the sequence of flat-field images for the horizontal wavelengths less than 150 km. In Table 11.1, the gravity wave characteristics obtained by the two methods are listed. The table shows only the gravity wave horizontal wavelength longer than 50 km; these are of interest in this work. The horizontal spacing of

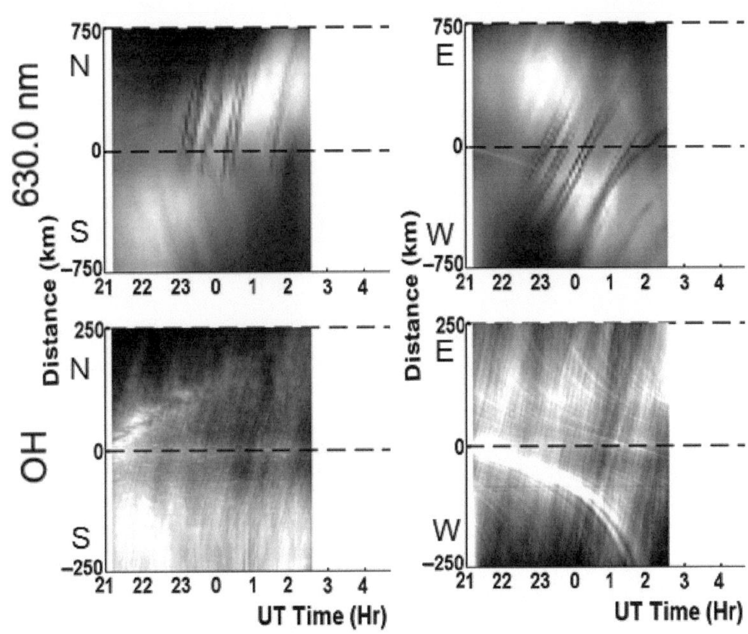

Fig. 11.6 OI 630.0 nm (*top*) and OHNIR (*bottom*) flatfield keogram of 24 September 2005

Table 11.1 Mesospheric gravity wave characteristics observed by OHNIR and OI 630.0 nm depletion spacing observed by imagers during the period from 23 to 30 September 2005

Day	Event #	λ (km)	τ (min.)	vp (m/s)	θ (° N)	Start (UT)	Duration (h)	6300 DS(1) (km)	6300 DS(2) (km)
23 Sep	(1)	95.3	33.6	47.3	29.7	2150	1.2	85	*
23 Sep	(2)	170.4	37	76.8	36.1	2200	3	*	121
24 Sep	(1)	98.3	34.2	47.9	39.8	2200	1.2	90	*
24 Sep	(2)	240.2	49	81.7	39.4	2200	2	*	207
25 Sep	(1)	189.1	70	45	40.9	2200	3	194	242
28 Sep	(1)	207.8	60	57.7	51.4	2130	2.5	125	175
29 Sep	(1)	78.1	32	40.7	135.3	2100	5.5	90	120
30 Sep	(2)	135	30.7	73.7	135	2245	1.2	130	*
30 Sep	(3)	220.4	53	69.3	138.6	2230	2.5	*	220

λ: horizontal wavelength (km), τ: period (min.), vp: phase velocity (m/s), θ: propagation direction from North (°), DS: OI 630.0 nm depletion spacing (km), *: no data. The estimated error range is ±10% for GW measurements and OI 630.0 nm depletion spacing.

the OI 630.0 nm depletions were also obtained from the flat-field images and are also listed in the table.

In Fig. 11.7 the horizontal wavelength of the mesospheric gravity waves and OI 630.0 nm depletion distances observed during the period from 23 to 30 September are plotted. Each plot corresponds to simultaneous observation of the two events. Linear correlation suggests a close relation between the two parameters. Takahashi et al. (2009) presented the same results for plasma bubbles. The range of the observed horizontal wavelength, from 100 to 200 km, is in agreement with a gravity wave propagation model calculation by Vadas (2007). According to them, gravity waves with the horizontal (vertical) wavelength longer than 50 km (20–30 km) can penetrate in the thermosphere to 150–200 km.

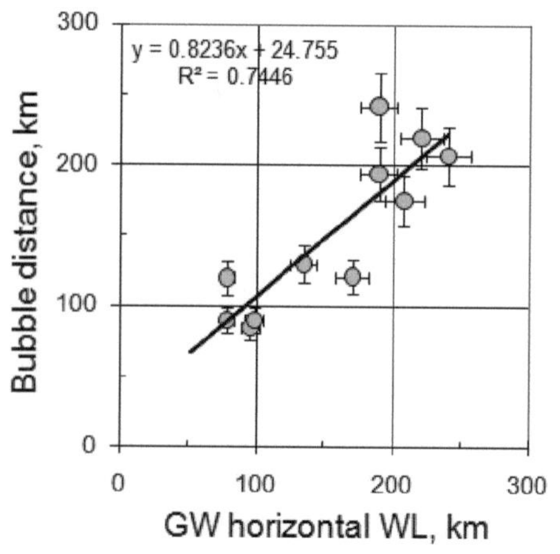

Fig. 11.7 Mesospheric gravity wave horizontal wavelengths and OI 630.0 nm depletion spacing

11.4 Discussion

11.4.1 Evidence on the F layer Bottomside Vertical Oscillation

From the three different data sets, VHF coherent radar back scattering images (RTI), Ionograms and all-sky OI 630.0 nm and OH images, we determine several new aspects. The bottom-type scattering layer is a signature of the vertical oscillation of the F-layer bottomside. From the ionogram satellite traces, we see a vertical gradient of the bottom height. Additionally, the sinusoidal intensity variation of the OI 630.0 nm indicates that the F-layer bottom-side (around 250 km) was affected by the vertical oscillation of some external force. The oscillation is sinusoidal and longitudinally extended by more than 800 km. This bottom-side oscillation might be one of the seeding processes of plasma bubbles. Whether the bottom-side oscillation develops vertically or not depends on the other factors, most plausibly, the F layer upward drift velocity (Abdu et al., 2009). The difference between the evening of 22, 23 and 24 September is a good example if one compares the ionograms between them. On the evening of 22 September the F layer uplifting at Fortaleza was very rapid and the irregularity started immediately in a wide height range as mentioned in the previous section (Fig. 11.2). In this case no clear signature of the bottom-side oscillation was observed. The rapid uplifting made a favorable condition of the bubble formation without forming the bottom-type layer in this evening.

The one to one relationship between the OI 630.0 nm sinusoidal depletion spacing and the upper mesosphere gravity wave horizontal wavelength suggests the upward propagation of waves into the lower thermosphere around 150–200 km altitude. Gravity waves generate horizontal wind structure superposed on back-ground winds in the thermosphere. The wind perturbation, eastward and westward, respectively, could cause upward and downward motion of the F-layer by the wind driven dynamo effect (V×B). Therefore it is essential to know the wind structure in the height region between 100 and 200 km. However, there is no tool to diagnose the wind structure to find out the effect of gravity waves in this region. In order to understand the gravity wave upward propagation condition, we instead used forward and reverse ray tracing techniques for gravity waves.

11.4.2 Forward and Reverse Ray Tracing of Gravity Waves

The temperature and wind models used in the present work for forward and reverse ray tracing of the gravity waves are nearly the same as that discussed in Vadas et al. (2009). However, the wind information between 80 and 100 km was obtained from the meteor radar at Cariri. The temperature and wind profiles below 30 km were obtained from WXP weather analysis package based on the balloon sounding data set. The TIME-GCM model was at all other altitudes. Because these model winds are somewhat uncertain, we also ray trace the waves through zero winds in order to assess the dependence of our results on the winds.

We consider several sources for the gravity waves. The first is that the gravity waves are "primary" waves that are excited directly by deep convective overshoot. The second are that the gravity waves are "secondary" waves excited by mesospheric or thermospheric body forces (Fritts et al., 2002; Vadas and Fritts, 2004, 2006). These body forces are horizontal accelerations of the fluid, created when gravity waves dissipate, break, or saturate. The body force is then the vertical divergence of the horizontal flux of vertical momentum. Although the primary waves are always upward propagating, the secondary waves can be upward or downward-propagating through the OH airglow layer (Vadas, 2010). Indeed, both upward and downward-propagating secondary waves from thermospheric body forces were identified at the bottom-side of the F layer (Vadas and Crowley, 2010). In the present work, however, it is difficult to find out whether the gravity waves observed in the OH airglow layer were upward or downward propagating. Therefore, we assume in the ray tracing model that the observed gravity waves are all upward-propagating primary or secondary waves. We initialize each gravity wave at 87 km altitude, and reverse ray trace it back to the tropopause. Unlike Vadas et al. (2009), where they stopped reverse ray tracing at the tropopause, we let the gravity wave reflect near the ground, and continue reverse ray-tracing the wave up to 180 km altitude. This is the maximum altitude that a thermospheric body force can be located (Vadas and Fritts, 2004, 2006). We then forward ray trace each wave upwards from 87 km until it reflects, dissipates, or meets a critical level.

In Figs. 11.8, 11.9, and 11.10, we show the ray trace results for the night of 23, 24 and 30 September, respectively. The paths of the gravity waves as a function of time and altitude are shown in the left column, and the ray-paths as a function of longitude and latitude are shown in the middle column. The altitude where the momentum flux is maximum is shown as squares (diagonals) for the model (zero) winds. The altitude where each gravity wave's momentum flux is half of its maximum value is shown as asterisks. Thus, the asterisks denote the approximate highest altitude where each gravity wave amplitude is significant. The right column shows the horizontal model winds at the location of the all-sky imager at the gravity wave's start time. The black dots denote where the reverse ray tracing hits ground/tropopause and reflects upward. They also denote where the ray tracing reaches the maximum height in the thermosphere. In the evening of 23 and 24 September, the reverse ray tracing reached an altitude of 140–170 km and 100–130 km region, respectively. The thermospheric body forces could have been at or below these altitudes. In the evening of 30 September, there were two gravity waves with the horizontal wavelength of 135 and 220 km. The former one showed the possible presence of a thermospheric body force at around 150 km. On the other hand the latter one did not show reverse ray tracing above the stratosphere.

On the nights of September 23 and 24, the gravity waves only forward ray traced as high as 120–165 km

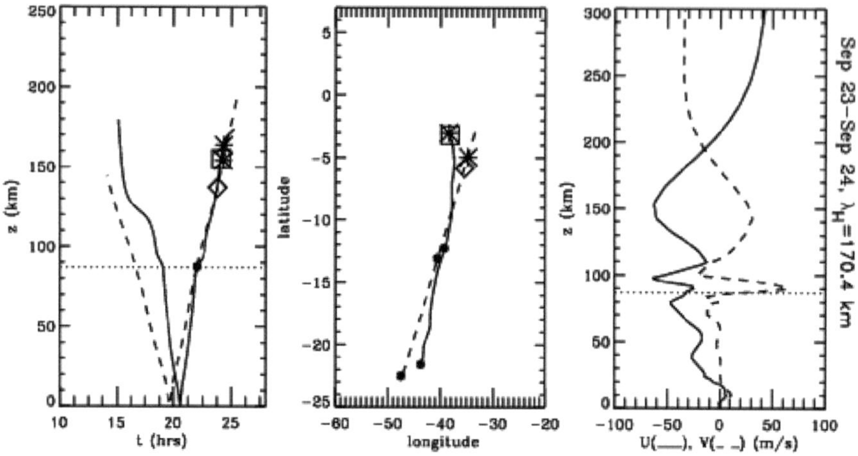

Fig. 11.8 Reverse and forward ray trace results for the medium-scale gravity waves observed on 23–24 September 2005. Gravity wave with $\lambda_h = 170.4$ km. Column 1 (*left side*): Altitude as a function of time. *Dash* and *solid lines* show the ray paths for zero winds and the model winds, respectively. The *dotted line* shows OH airglow altitude. Column 2 (*middle*): Flight path in latitude and longitude. *Solid* and *dash lines* are the same as in column 1. *Black dots* show the locations of the sources: the tropopause and a thermospheric body force. *Diamonds* and *squares* denote for zero winds and the model winds, respectively. The *asterisk* denotes the location of the all-sky imager. Column 3 (*right side*): Zonal (*solid*) and meridional (*dash*) model winds at the location of the all-sky imager at the start time for each gravity wave. Positive zonal (meridional) winds are eastward (northward)

Fig. 11.9 Same as Fig. 11.8 but for the evening of 24 September 2005. Gravity wave with $\lambda_h = 240.2$ km

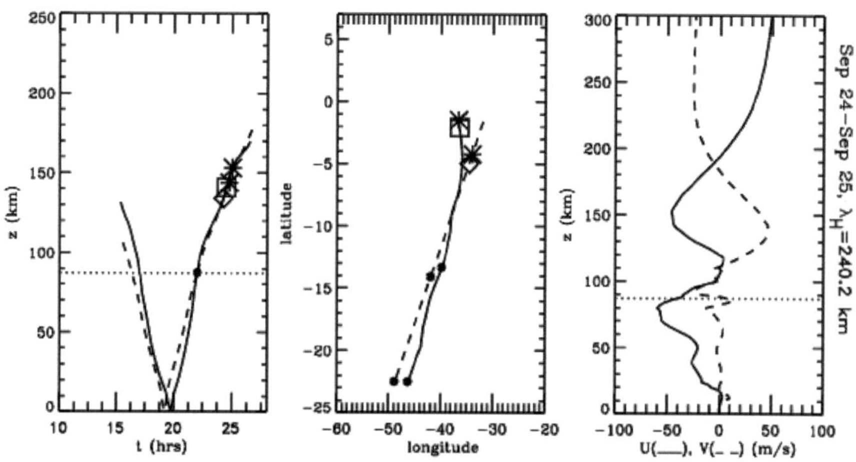

and 110–155 km, respectively. On these nights, the waves were propagating NE-ward into a NW-ward wind in the lower thermosphere. However, on the night of September 30, the gravity waves only forward ray trace as high as 130–190 km. On this night, the gravity waves were propagating SE-ward into a wind directed NW-ward in the lower thermosphere. Propagating against the wind enables the waves to penetrate into much higher altitudes (Fritts and Vadas, 2008).

These gravity waves may have been primary or secondary waves. Because thermospheric body forces excite both upward and downward-propagating secondary waves with similar horizontal and vertical scales (Vadas, 2010), we also consider the possibility that the F layer bottom-side oscillation may be seeded by similar upward-propagating secondary waves from thermospheric body forces. However, these waves would have similar dissipation altitudes as the gravity waves shown here. Therefore, the altitudes shown in

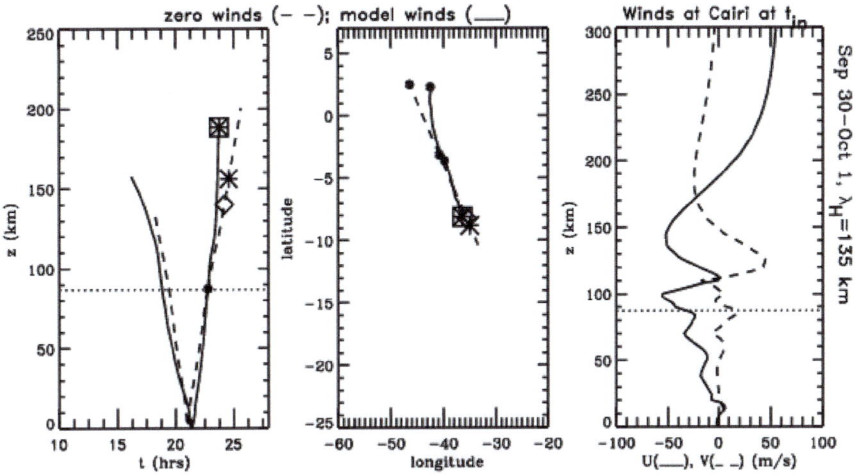

Fig. 11.10 Same as Fig. 11.8, but for the evening of 30 September 2005. Gravity wave with $\lambda_h = 135$ km

Figs. 11.8, 11.9, and 11.10 should be representative of the altitudes achievable by both primary and secondary gravity waves with these scales.

In order to further investigate gravity wave seeding process arising from convection in the troposphere, we overplot the ray-tracing result on the cloud map in Figs. 11.11, 11.12 and 11.13, for 23, 24 and 30 September 2005, respectively. In the evening of 23 September the trace passes in the troposphere, 700–800 km SW-ward away from Cariri (around 12 S, 39 W) where there was a convection system extending from NW to SE. The gravity wave (λ_h:170 km) might be excited in this area. Reverse ray-tracing from troposphere to the lower thermosphere reached ~140 km altitude crossing a large scale convection area (20°S–25°S, 40°W–45°W). This also suggests a possibility of thermospheric body forces. In the case of 24 September (Fig. 11.12) no strong convective activity was found in the area where the trace hits the ground, 800 km SW-ward of Cariri (dotted mark in the figure at 13°S, 40°W). Reverse ray-tracing from this point to the lower thermosphere reached ~130 km altitude. There was a deep convection system over the area ~500 km west of the ray-tracing terminating point (23°S, 47°W), suggesting a possible presence of mesospheric or thermospheric body force. The ray tracing of 30 September, again, hits the ground (sea surface) where no cloud activity occurred. It again suggests a presence of a thermospheric body force. There was a deep convection system over the Atlantic Ocean (around 8°N, 36°W). Thus, in terms of direct seeding of the F layer bottom-side oscillation by primary or secondary gravity waves, the night of 30 September is much more likely because the altitudes attained were higher than the case of 24 September 2005.

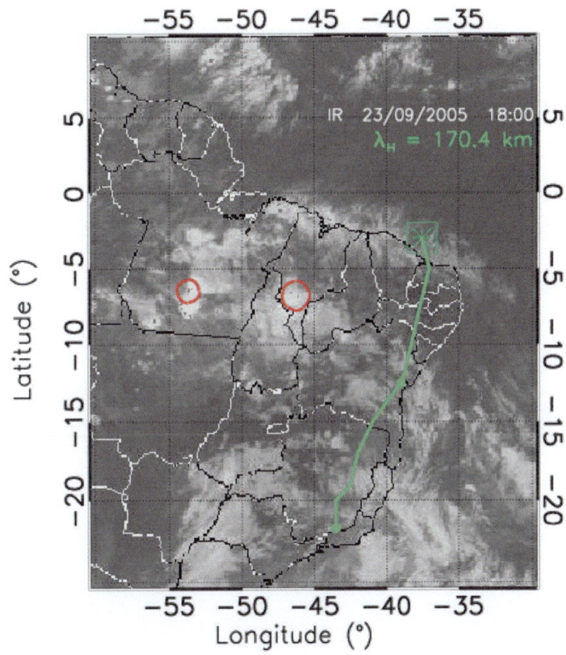

Fig. 11.11 Tropospheric cloud map and mesospheric gravity wave ray-tracing (*green line*) in the evening of 23 September 2005 at 20:00 UT(17:00 LT). *Red circles* indicate location of convections

These forward and backward ray tracing results suggest a direct or indirect contribution of the mesospheric

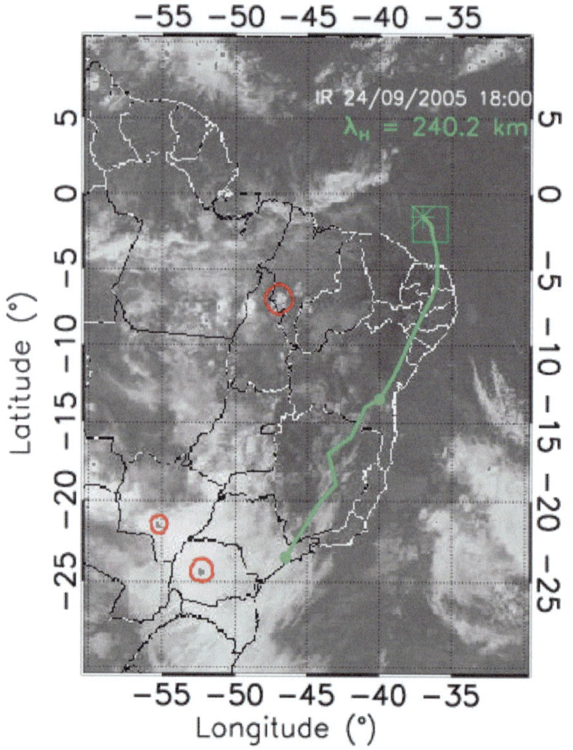

Fig. 11.12 Same as Fig. 11.11, but for the evening of 24 September 2005 at 20:00 UT(17:00 LT)

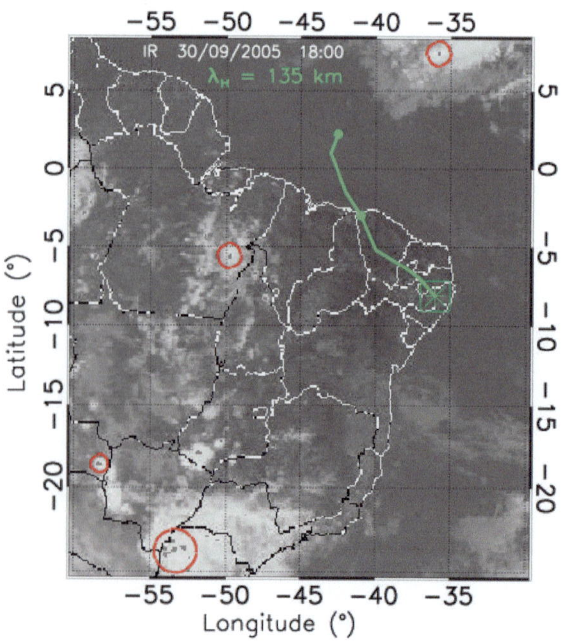

Fig. 11.13 Same as Fig. 11.11, but for the evening of 30 September 2005 at 20:00 UT(17:00 LT)

gravity waves for the formation of F-layer bottom-side oscillation. The source of the mesospheric gravity waves could be tropospheric convection nearby (typically less than 500 km of distance) or thermospheric body force sources (typically more than a few hundred km from the observation site) as suggested by Vadas and Fritts (2006). As another possible seeding process of the F layer bottom-side oscillation, one should consider meso-scale ionospheric traveling waves (MSTID) from the high latitudes. During the period from 22 to 30 September, however, the Kp index was less than 4, indicating less possibility of a disturbance from high latitudes. Additionally, MSTIDs with λ_h<300 km cannot generally propagate more than 3000 km (Vadas, 2007): only large-scale TIDs can propagate to the equator from geomagnetic disturbances at high latitudes.

11.5 Conclusion

We analyzed the ionospheric F-layer bottom-type scattering layer observed at near equatorial region over South America. From the measurements by ionosonde, VHF coherent scatter radar and OI 630.0 nm all-sky imager, we conclude that the bottom-type scattering layer has a form of sinusoidal variation of the bottom height, some cases extending more than 1500 km in the zonal direction. The depletion has a horizontal spacing of 80–250 km. The spacing presents a close similarity to the horizontal wavelength of the mesospheric gravity wave. The gravity wave forward ray-tracing demonstrates that the observed mesospheric gravity waves could propagate upwards at least up to 150–200 km of altitude, depending mainly on the thermospheric wind conditions. The reverse ray-tracing suggests possible tropospheric convection as a gravity wave source. The reverse ray-tracing on one of the cases on 30 September (and possibly 24 September) suggests a thermospheric body force as a gravity wave source. Close relation between the F-layer bottom-side dynamics and mesospheric gravity waves suggests further investigation of the mesosphere–ionosphere coupling processes.

Acknowledgements The authors thank for several institutions and observatory staffs by whom the ground based observations were carried out. Thanks are also due to Maria Goreti S. Aquino who reduced the ionogram data. São Luis VHF radar was developed and installed by the support of FAPECP under the process 2004/01065-0. The RTI images were prepared by Eng. Lazaro A. P. Camargo. We thank for him. The SpreadFEx field program and data analysis were supported by NASA under contracts NNH04CC67C and NAS5-02036. The present project was also partially supported by CNPq (Conselho Nacional de Desenvolvimento Cietífico e Tecnológico) under contract 301876/2007-0.

References

Abdu MA (2001) Outstanding problems in the equatorial ionosphere-thermosphere system relevant to spread F. J Atmos Solar-Terr Phys 63(9):869–884

Abdu MA (2005) Equatorial ionosphere-thermosphere system: electrodynamics and irregularities. Adv Space Res 35: 771–787

Abdu MA, Kherani EA, Batista S et al (2009) Gravity wave initiation of equatorial spread F/plasma bubble irregularities based on observational data from the SpreadFEx campaign. Ann Goephys 27:2607–2622

de Paula ER, Hysell DL (2004) The São Luís 30 MHz coherent scatter ionospheric radar: system description and initial results. Radio Sci 39:RS1014. doi:10.1029/2003RS002914

Fritts DC, Abdu MA, Batista IS et al (2009) Overview and summary of the spread F experiment (SpreadFEx). Ann Geophys 27:2141–2155

Fritts DC, Vadas SL (2008) Gravity wave penetration into the thermosphere: sensitivity to solar cycle variations and mean winds. Ann Geophys 26:3841–3861

Fritts DC, Vadas SL, Yamada Y (2002) An estimate of strong local body forcing and gravity wave radiation based on OH airglow and meteor radar observations. Geophys Res Lett 29(10). doi:10.1029/2001GL013753

Hysell DL, Burcham JD (1998) JULIA radar studies of equatorial spread F. J Geophys Res 103(A12):29155–29167

Hysell DL, Chun J, Chau JL (2004) Bottom-type scattering layers and equatorial spread F. Ann Geophys 22:4061–4069

Hysell DL, Larsen MF, Swenson CM et al (2006) Rocket and radar investigation of background electrodynamics and bottom-type scattering layers at the onset of equatorial spread F. Ann Geophys 24:1387–1400

Kudeki E, Akgiray A, Milla M et al (2007) Equatorial spread-F initiation: post-sunset vortex, thermospheric winds, gravity waves. J Atmos Solar-Terr Phys 69:2416–2427

Kudeki E, Bhattacharyya S (1999) Post-sunset vortex in equatorial F-region plasma drifts and implications for bottomside spread-F. J Geophys Res 104:28163–28170

Takahashi H, Abdu MA, Taylor MJ et al (2010) Equatorial ionosphere bottom-type spread F observed by OI 630.0 nm airglow imaging. Geophys Res Lett 37. doi:10.1029/2009GL041802

Takahashi H, Taylor MJ, Pautet P-D et al (2009) Simultaneous observation of ionospheric plasma bubbles and mesospheric gravity waves during the SpreadFEx campaign. Ann Geophys 27:1477–1487

Taylor MJ, Pautet PD, Medeiros AF et al (2009) Characterizing mesospheric gravity waves near the magnetic equator, Brazil during the SpreadFEx campaign. Ann Geophys 27: 461–472

Tsunoda RT (1981) Time evolution and dynamics of equatorial backscatter plumes, 1. Growth phase. J Gephys Res 86:139

Tsunoda RT (2008) Satellite traces: an ionogram signature for large-scale wave structure and a precursor for equatorial spread F. Geophys Res Lett 35. doi:101029/2008GL035706

Vadas SL (2007) Horizontal and vertical propagation and dissipation of gravity waves in the thermosphere from lower atmospheric and thermospheric sources. J Geophys Res 112. doi:10.1029/2006JA011845

Vadas SL (2010) Downward-propagating secondary gravity waves in the OH airglow layer from thermospheric body forces resulting from deep convection. J Geophys Res. submitted

Vadas SL, Crowley G (2010) Sources of the traveling ionospheric disturbances observed by the ionospheric TIDDBIT sounder near Wallops Island on October 30, 2007. J Geophys Res 115. doi:10.1029/2009JA015053

Vadas SL, Fritts DC (2004) Thermospheric responses to gravity waves arising from mesoscale convective complexes. J Atmos Solar-Terr Phys 66:781–804

Vadas SL, Fritts DC (2006) Influence of solar variability on gravity wave structure and dissipation in the thermosphere from tropospheric convection. J Geophys Res 111. doi:10.1029/2005JA011510

Vadas SL, Taylor MJ, Pautet P-D et al (2009) Convection: the likely source of the medium-scale gravity waves observed in the OH airglow layer near Brasilia, Brazil, during the SpreadFEx campaign. Ann Geophys 27:231–259

Woodman RF, LaHoz C (1976) Radio observations of F-region equatorial irregularities. J Geophys Res 81: 5447–5466

Chapter 12

A Case Study of Tidal and Planetary Wave Coupling in the Equatorial Atmosphere-Ionosphere System Over India: Preliminary Results

S. Gurubaran, R. Dhanya, S. Sathiskumar, and P.T. Patil

Abstract The present study examines the role of planetary waves and tides observed at upper mesospheric heights (~90 km) in the day-to-day variabilities of the quiet time ionospheric current system as manifested in ground magnetic field records at the dip equatorial and low latitude stations during the three month period, February–April 2008, when an experimental campaign under CAWSES-India program was conducted. The MF radars at Tirunelveli (8.7°N, 77.8°E) and Kolhapur (16.8°N, 74.2°E) provided the wind observations. The study reveals simultaneous presence of 5- to 6-day and 12- to 16-day oscillations in the radar wind fields and the strength of the equatorial electrojet (EEJ) derived from the geomagnetic field records from Tirunelveli and Alibag. Tidal winds and the amplitude of diurnal variation in the EEJ strength show a reasonably good correlation when 5-day averaged data sets are used. Though limited data were used, the present work will motivate further studies on the atmosphere – ionosphere coupling at low latitudes with extended data sets.

12.1 Introduction

Planetary-scale waves like tides, global-scale normal modes and equatorial Kelvin waves play an important role in the dynamics of the mesosphere-lower thermosphere (MLT) region at low latitudes (Vincent, 1993). It has been hypothesized in the past that if a global-scale wave with large amplitude and fairly long vertical wavelength propagates into the ionosphere from below, it should drive an electric current system at E-region heights through the dynamo action with a period of the global-scale wave (Richmond, 1995; Forbes, 1996, for reviews on this subject). This wave-like feature causes perturbations in geomagnetic field that could be recorded on ground (Forbes and Levoroni, 1992; Kohsiek et al., 1995; Gurubaran et al., 2001; Pancheva et al., 2006, 2008; Ramkumar et al., 2006, 2009, to state a few). The F-region of the ionosphere is also known to respond to dynamical forcing occurring at planetary scale wave periods (Forbes and Leveroni, 1992; Chen, 1992; Parish et al., 1994; Forbes et al., 1997; Altadill et al., 1998; Takahashi et al., 2006; Abdu et al., 2006; Pancheva et al., 2006, to state a few). Recent work suggests a role for the E-region dynamo electric fields influenced by nonmigrating tides to cause longitudinal variations in brightness and latitude of F-region airglow bands as well as F-region electron density distribution at low latitudes (England et al., 2008, for example).

Understanding the variabilities of ionospheric parameters at E-region dynamo and F-region heights in terms of variabilities in tides and planetary waves during magnetically quiet times has been a major unresolved issue during the past. It is not known what conditions exist at various altitudes in the MLT region and below, that permit these large-scale waves to reach the dynamo heights and have an influence on the ionospheric variablilities. Whether eddy dissipation due to waves propagating from below undergoing long-period modulation and whether composition changes taking place at turbopause altitudes can drive such variabilities are issues that need to be addressed. It is also not

S. Gurubaran (✉)
Equatorial Geophysical Research Laboratory, Indian Institute of Geomagnetism, Tirunelveli, India
e-mail: gurubara@iigs.iigm.res.in

M.A. Abdu, D. Pancheva (eds.), A. Bhattacharyya (Coed.), *Aeronomy of the Earth's Atmosphere and Ionosphere*, IAGA Special Sopron Book Series 2, DOI 10.1007/978-94-007-0326-1_12, © Springer Science+Business Media B.V. 2011

known whether any part of the wave spectrum that contributes to the ionospheric variabilities is composed of waves that are generated in-situ.

Aiming to address the vertical coupling of atmosphere-ionosphere system by means of tides and planetary waves, we initiated an analysis of simultaneous observations of MLT winds from the two low latitude MF radar sites, Tirunelveli (8.7°N, 77.8°E) and Kolhapur (16.8°N, 74.2°E), available for about 2 years (2008–2009). Confining the analysis to the Indian sector for the time period February–April 2008 when an experimental campaign under the national CAWSES program was conducted, this work examines whether long-period oscillations at planetary wave time scales in the radar winds are accompanied by variations at similar periodicities in the strength of the equatorial electrojet current system as manifested in the ground geomagnetic field data sets obtained from the two magnetometer sites in India, Tirunelveli, the dip equatorial station and Alibag (18.6°N, 72.9°E, geographic; 9.75°N magnetic) located off the dip equator. We also examine whether any variations in the diurnal and semi-diurnal tide amplitudes are correlated with the daily range in the EEJ strength during this period.

The objective of this exercise was not to perform an extensive analysis of radar winds and geomagnetic field variations but to focus on one small time segment during which simultaneous wind observations from two low latitude sites in the same longitude sector are available. Further analysis to be carried out by the authors will extend this exercise to much larger global data sets.

12.2 Selection of Data

The medium frequency (MF) radar at Tirunelveli has been providing continuous data on winds in the height region 80–98 km since 1992 (Rajaram and Gurubaran, 1998). A similar radar has been operating continuously at the low latitude station, Kolhapur, since December 2007. A CAWSES-India experimental campaign was conducted during February–April 2008. Results to be presented in this work were obtained from the radar and magnetometer data sets collected during this campaign.

The wind measurements, sampled for every 2 min, are averaged for each hour and used for further analysis. Though winds are measured every 2 km, the data acceptance rate, which is based on several rejection criteria (Briggs, 1984b), is sufficiently large only above 88 km. Radar measurements at Tirunelveli made at higher heights (above 94 km) are constrained by the influence of electric fields and the radar tends to measure the electron drift motion driven by intense electric field rather than the neutral wind (Gurubaran and Rajaram, 2000). Traditional harmonic and Morlet wavelet analysis methods are applied to the radar data to retrieve information on tidal and planetary wave characteristics. Data sets from the two sites are also subjected to cross-wavelet analysis to extract common periodicities representing distinct wave events occurring at the same time over the two sites.

The variations in the three components of the geomagnetic field are being recorded by a network of magnetic observatories of Indian Institute of Geomagnetism located from the magnetic equator to the northernmost Indian latitudes. Tirunelveli is a station located close to magnetic dip equator and is under the influence of the equatorial electrojet (EEJ), which is an intense band of eastward electric current flowing at an altitude of $\sim$105 km during daytime in a narrow latitudinal belt ($\pm 3°$) around the dip equator and superposed over the normal Sq current system, whereas Alibag is a station outside the electrojet belt and is solely under the influence of the Sq current system. It is a usual practice to take the difference in the horizontal magnetic field variations over the two sites as representing the strength of the EEJ (Kane, 1973; Anderson et al., 2002). By considering the difference in the horizontal magnetic field variations, any contribution from the distant magnetospheric currents that vary over a much larger spatial scale than the distance between the two low latitude stations is minimized.

Pancheva et al. (2006) described a novel method designed for studying the variability of the geomagnetic field with periods of planetary waves. With the supposition that the variability of the geomagnetic components with the period of the planetary waves appears as a modulation of the quiet-day diurnal cycle, this method invokes a decomposition of the data using a concrete mean diurnal variation instead of a standard sine function. The perturbation amplitude then is evaluated as a deviation from the mean diurnal cycle thus defining how the diurnal cycle is modulated at planetary wave periods. In a later study Pancheva et al. (2008) adopted this method to extract 5- to 6-day periodicities in the geomagnetic field variations. In this work we use this method to extract the signatures of planetary waves in the geomagnetic data sets.

12.3 Results

We begin by showing results from the Morlet wavelet transforms that were performed on the hourly mean data on winds at 90 km over Tirunelveli and Kolhapur and examine the dominant long-period planetary-scale waves that might be present over the two radar sites. For the sake of convenience, we have divided the scale of the long-period wind spectra into two period ranges: (i) 2- to 10-day period range and (ii) 10- to 20-day period range. Because the comparative analysis with geomagnetic data is restricted to February–April 2008 period, we examine the results from this exercise for this period in Figs. 12.1 and 12.2.

In the period range of 2–10 days (Fig. 12.1), two bursts of wave occurrences in both zonal and meridional wind components are noticed at Kolhapur (bottom panels) at periods close to 7 and 6 days, respectively, around day numbers 60 and 90. The meridional wind over Tirunelveli (top right panel) exhibits peaks at similar periods noticed in the corresponding wind field over Kolhapur. In the zonal component over Tirunelveli (top left panel), intense bursts of wave activity are seen at periods close to 6 days around day numbers 35 and 60. The 5-day wave is known to be a global scale Rossby mode (gravest symmetric Hough mode of westward propagating zonal wavenumber 1) detected in radar and satellite observations (Riggin et al., 2006, for example), whereas the 6-day wave is a Kelvin wave signature often observed at low latitudes (Pancheva et al., 2008, for example). With the limited data sets presented here, however, we are not able to identify what planetary wave modes drive the observed wind perturbations.

In the 10- to 20-day period range (Fig. 12.2), a large-amplitude wave at ~16-day period in the zonal wind is clearly evident over both Tirunelveli and Kolhapur (top left and bottom left panels) between day numbers 20 and 70 days. In the meridional component (top right and bottom right panels), a somewhat weaker wave event can be detected at periods closer to 12 days near day number 90 over both sites. Between day numbers 30 and 50, enhancement in wave intensity can be noticed at periods close to 12 days over Tirunelveli and at ~14 days over Kolhapur. The quasi-16-day wave is known to be one of the classical normal modes with an eigen period of 12.5 days that are expected to be Doppler shifted to greater periods in the realistic non-zero background flow (Forbes, 1995). The quasi-16-day wave, its characteristics and its long-term variability over Tirunelveli were studied earlier (Sridharan et al., 2003).

We next examine the geomagnetic field variations during the period February–April 2008 with the objective of detecting long-period oscillations in them and examining whether they are associated with planetary waves observed in radar winds at those times. The chosen period falls during the quiet phase of the 11-year sunspot cycle, though there was 27-day recurrent geomagnetic activity with moderate geomagnetic storms accompanied by sudden storm commencement occurring around 26 March 2008 (day number 85) and 22 April 2008 (day number 112). The geomagnetic activity index, Ap, registered largest values (in the range 20–40) around 2, 10 and 29 February (day numbers 33, 41 and 60, respectively), 9 and 27 March (day numbers 68 and 85, respectively) and 6 and 23 April (day numbers 96 and 113, respectively). The geomagnetic field at low latitudes would then show enhanced variations at these times different from their quiet-time behavior.

In Fig. 12.3a, b we depict the 1-min values of the variations in the horizontal component (ΔH) of the geomagnetic field recorded at Tirunelveli and Alibag, the difference field representing the strength of the equatorial electrojet and the Dst representing the strength of the magnetospheric ring current that would contribute to the low latitude geomagnetic field variations. Apart from the significant day-to-day variations, we also notice the disturbed time response of the geomagnetic field at these latitudes whenever Dst decreased below –50 nT (the days between vertical lines in Fig. 12.3a, b indicate those disturbed periods). However, these disturbance signatures are greatly suppressed in the difference fields representing the EEJ strength (shown in the bottom panels). We utilize the latter parameter in our further analysis.

Past results on the short-term periodic variation in the solar wind have revealed a broad spectral peak around 27-days and distributed over a range of 22–32

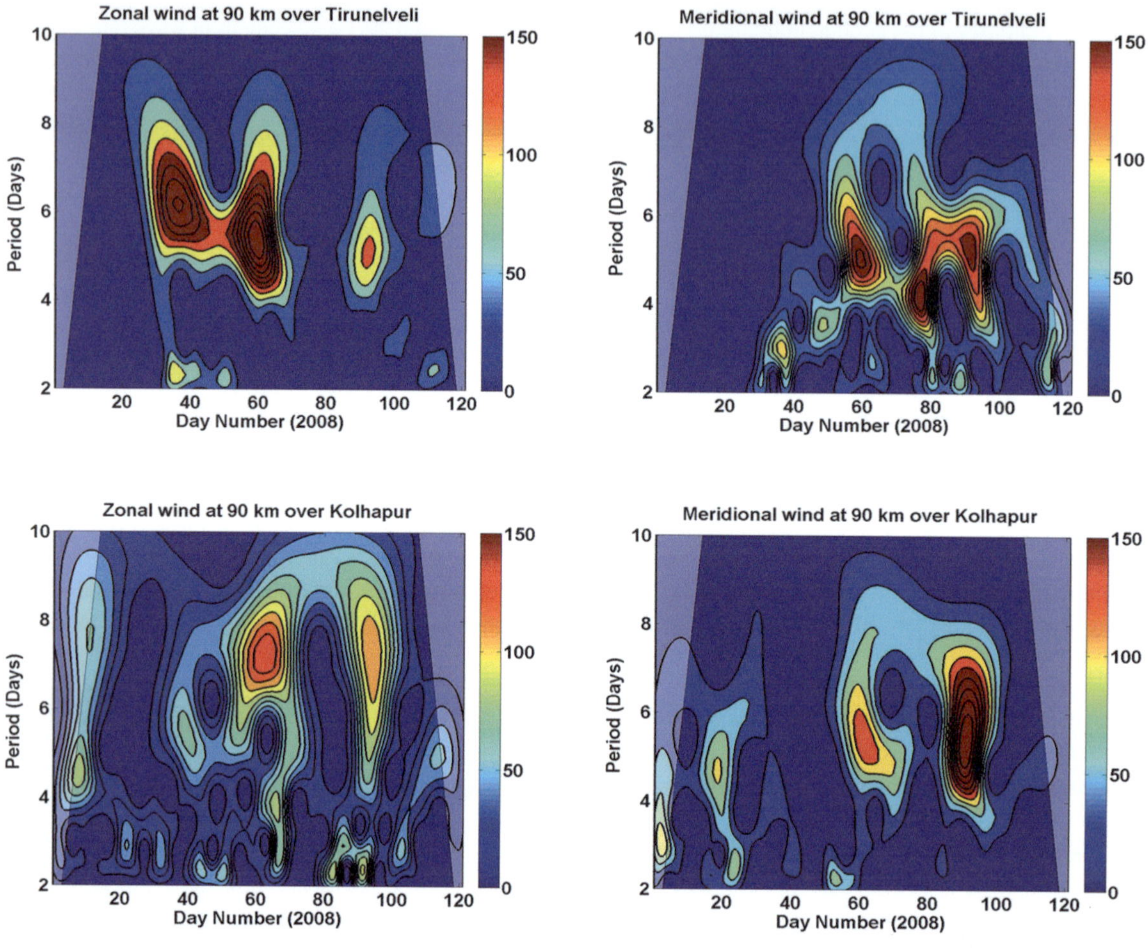

Fig. 12.1 Wavelet power spectra for zonal (shown on the *left panels*) and meridional winds (panels on the *right*) at 90 km over Tirunelveli (*top panels*) and Kolhapur (*bottom panels*) for the period range 2–10 days

days and a sharp and distinct peak at 13.5 and 9 days (Nayar, 2006; Lei et al., 2008, for example). Specifically, Lei et al. (2008) report a 9-day recurrence of fast streams in the solar wind having origin in solar coronal holes distributed roughly 120° apart in longitude and that modulates geomagnetic activity and thermospheric densities at 9-day period. It is expected that the geomagnetic activity would exhibit these spectral features in its response to solar wind variations. Any analysis attempting to determine planetary wave influence on ionospheric current systems and geomagnetic field variations will have to take into account this aspect of the terrestrial response to changes in solar output. In particular, the geomagnetic field variations at lower latitudes are dominated by magnetospheric current systems that get amplified during the main phase of the geomagnetic storms and are influenced by electric fields and resulting current systems due to direct penetration of polar cap electric fields to the equator (Kikuchi et al., 1996) and disturbance wind dynamo triggered by auroral Joule heating and ion-drag acceleration (Blanc and Richmond, 1980). It is anticipated that such disturbances of solar origin would produce a zonally symmetric response in the geomagnetic field whereas travelling planetary waves would induce zonally propagating features in the ground geomagnetic field variations (Pancheva et al., 2008).

A periodogram analysis of geomagnetic field variations recorded at Tirunelveli and Alibag reveal a 13.5-day oscillation accompanied by an equally strong 9-day oscillation occurring during the campaign period (not shown here). These peaks have been identified to

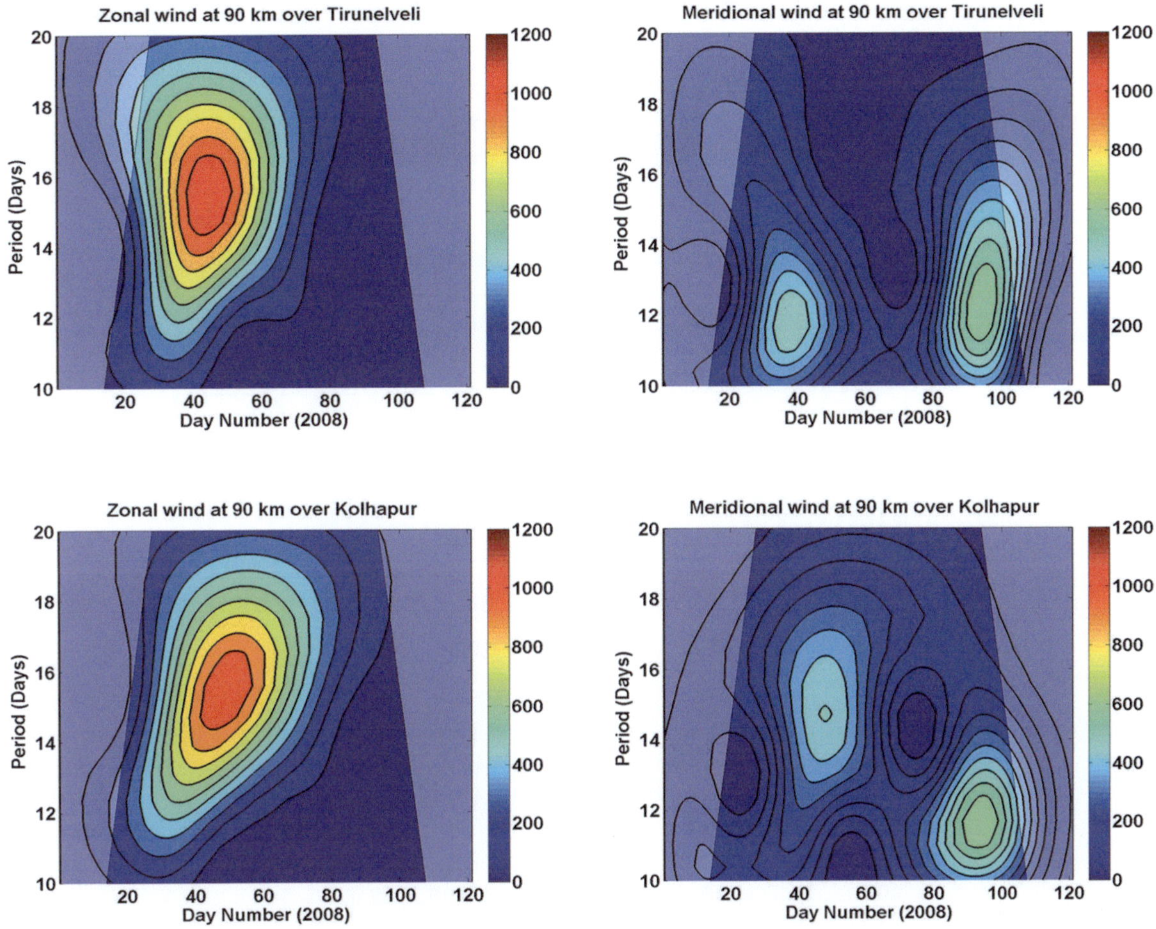

Fig. 12.2 Same as Fig. 12.1 but for the period range 10–20 days

arise due to the impact of intermittent streams of solar wind of coronal hole origin. It may be noted that amplitude spectra in the difference field representing the strength of the EEJ do not show these peaks (Fig. 12.4, bottom panel).

In the analysis of geomagnetic field variations, the method of Pancheva et al. (2006) is first applied to the difference in the horizontal component of the geomagnetic field recorded at Tirunelveli and Alibag, yielding a time series in which the diurnal variations are removed in advance and only the departures from the daily variations are represented in the data sets. The hourly differences between the resulting time series for Tirunelveli and Alibag (representing the strength of the EEJ) are then subjected to traditional least squares and wavelet analysis.

Figure 12.4 depicts the amplitude spectra for the zonal and meridional winds at 90 km over Tirunelveli and Kolhapur compared with the amplitude spectrum for the EEJ strength for the period February–April 2008. The 95% confidence levels using a standard method (Nawroozi, 1967) are shown as dashed/dash-dot lines in each of the panels in Fig. 12.4. Examining the zonal wind spectra (middle panels), we notice the presence of pronounced quasi-16-day wave over both sites (estimated amplitudes of 9 m/s over Tirunelveli and 5 m/s over Kolhapur). The amplitudes in the meridional component in this period are somewhat weaker (4 m/s over Kolhapur and below 95% confidence level over Tirunelveli). Examining the EEJ spectrum, we notice a peak with the largest amplitude at 5 days.

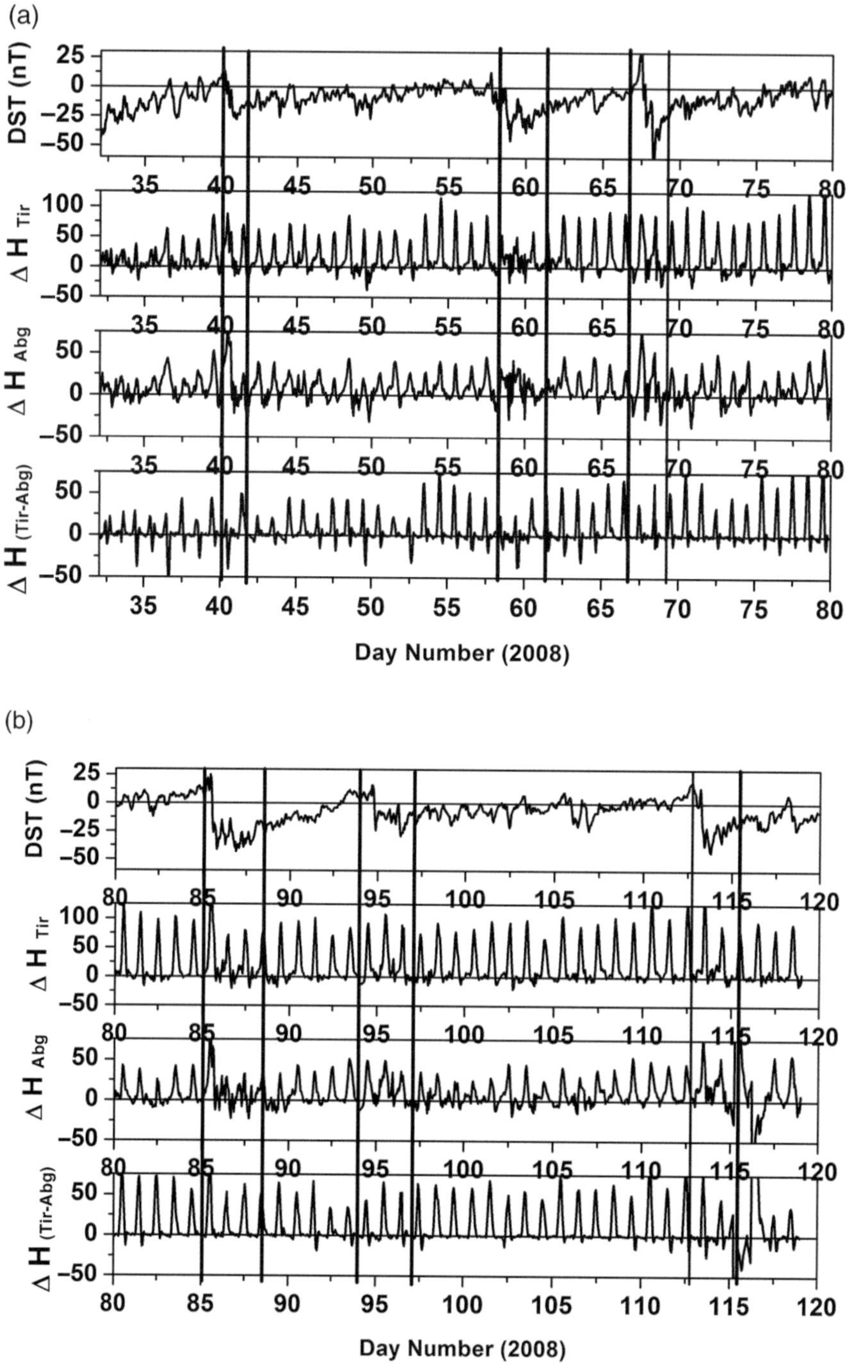

Fig. 12.3 (**a**) The variations in the horizontal component of the geomagnetic field measured at Tirunelveli and Alibag, the difference between the two components and the Dst index (*top panel*) for the period between the day numbers 30 and 80. (**b**) Same as Fig. 12.3a but for the period between the day numbers 80 and 120

In Figs. 12.5 and 12.6 we depict the wavelet spectra for the EEJ strength along with the cross-wavelet spectra for wind fields over Tirunelveli and Kolhapur. A noticeable feature in the wavelet power spectrum for the ground geomagnetic field records shown as colored contours in Fig. 12.5 is the presence of short-period oscillations in the period range of 3–8 days occurring near day numbers 50, 70 and 95. A closer examination

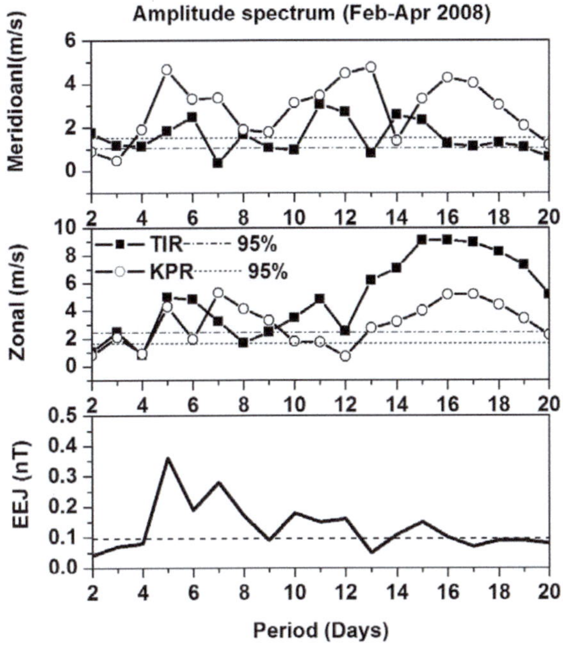

Fig. 12.4 Amplitude spectra for meridional (*top panel*) and zonal (*middle panel*) winds at 90 km over Tirunelveli and Kolhapur and EEJ strength (*bottom panel*) derived from data sets for the period February–April 2008

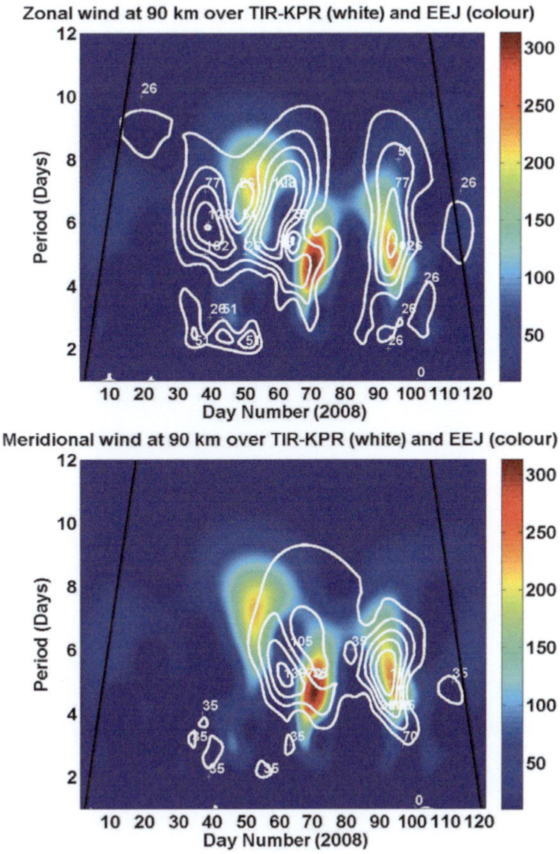

Fig. 12.5 Cross-wavelet spectra for the zonal and meridional winds over Tirunelveli and Kolhapur (shown as *white* contours) and wavelet spectra for the EEJ strength (shown as *colored* contours) for the period range 2–10 days

of results obtained from the wind observations (shown as white contours in Fig. 12.5, zonal wind in the top panel and meridional wind in the bottom panel) reveals the presence of a 5-day wave signature in the MLT meridional wind component around day numbers 60 and 95. In the zonal component, there are peaks close to 6 days near day numbers 40 and 60 that are not seen in the EEJ strength.

In the 10–20 day period range results for which are shown in Fig. 12.6, a 10- to 12-day oscillation is predominantly present in the EEJ strength between day numbers 60 and 100. The cross-wavelet spectrum for the meridional wind field does reveal a peak close to 12 days near day number 90. A quasi-16-day wave was detected earlier in the zonal wind component centered at day number 45. Though the EEJ strength initially reveals a peak close to 15 days near day number 45, the peak occurrence shifted to smaller periods as days progressed.

As the last exercise in this work, we examined the tidal winds at 90 km over Tirunelveli and compared the amplitude variations with the corresponding variation in the diurnal range in the EEJ strength. The range or the maximum amplitude of the diurnal variation in EEJ is expected to represent the strength of the electrojet on any given day. Figure 12.7 shows the results from this exercise. In the top panels, we compare the diurnal range with the amplitude of the diurnal tide in meridional wind at 90 km on the left and on the right we compare the same with the amplitude of the semi-diurnal tide. The scatter plots for the two parameters are shown in the bottom panels. Lines joining circled symbols in the top panels represent the diurnal range and lines joining filled squares represent the tidal amplitudes. The tidal amplitudes were obtained from composite hourly averages from a 5-day window sliding at one-day intervals. The values representing daily range in EEJ are 5-point running averages to the daily values obtained from ground magnetometer data.

The comparison plot shown in Fig. 12.7 reveals a reasonably good correlation (correlation coefficients

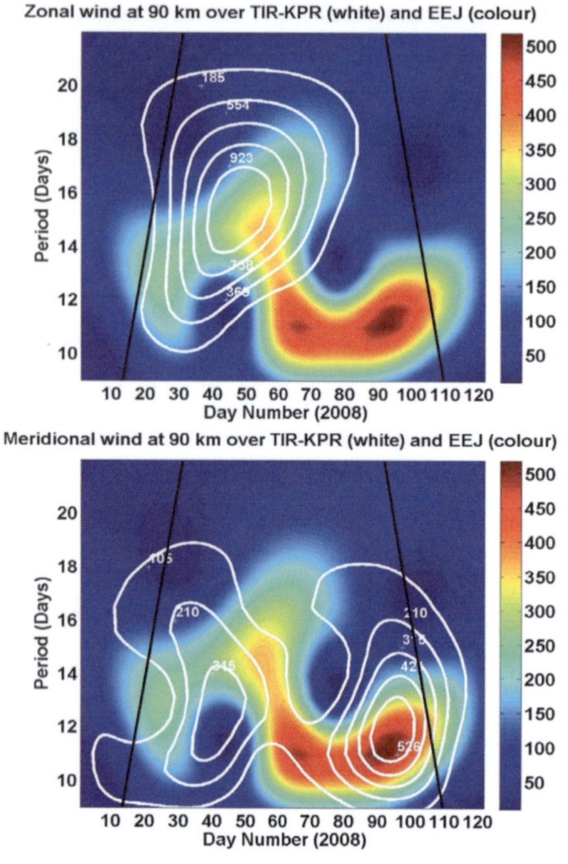

Fig. 12.6 Same as Fig. 12.5 but for the period range 10–20 days

indicated in the bottom panels are statistically significant at 95% confidence level) between the tidal wind fields at 90 km over Tirunelveli and the daily range in the horizontal component of the geomagnetic field representing the EEJ strength. Part of the uncorrelated field variations could be ascribed to a few moderate geomagnetic storms that occurred during this 3-month period. Though in this exercise we have not removed those days from the correlation analysis performed, we do not believe that exclusion of those days will drastically modify the correlation coefficients noticed in Fig. 12.6.

12.4 Discussion and Conclusion

Several reports in the past have emphasized the role of tides and planetary waves in contributing to the day-to-day variations in ionospheric parameters (Briggs, 1984a, for a review on this aspect). Apart from the solar flux variations that involve a periodicity at 27 days and its harmonics, the planetary waves at periods between 2 and 20 days are either expected to participate directly in the dynamo action or modulate the tidal winds at lower altitudes and the modulated tides then induce variations in electric fields and current systems at planetary wave time scales (Forbes, 1996). It is also likely that changes in the neutral composition at turbopause altitudes induced by planetary waves might influence the conductivity variations at dynamo altitudes (Mikhailov, 1983) but this aspect needs to be explored in detail.

In this work for the first time we have exploited the availability of simultaneous MLT radar wind observations from two low latitude sites in the Indian sector to address the vertical coupling of atmosphere-ionosphere system at equatorial and low latitudes during an experimental campaign conducted during February–April 2008 under CAWSES-India program. The goal of this initial study was to identify common periodicities between the radar observations of winds and geomagnetic field variations representing the strength of the equatorial electrojet with the supposition that some of the variabilities of the EEJ current system could be linked to the variabilities of tides and planetary waves observed at MLT heights. Results from this exercise are anticipated to provide motivation to carry out further elaborate studies on the dominant wave events detected during this campaign using global data sets as were carried out by Pancheva et al. (2006, 2008).

For the sake of convenience, the time scales considered in this work were divided into (i) 2- to 10-day and (ii) 10- to 20-day period ranges. The diurnal variations in the north-south component of the geomagnetic field were removed in advance and the residual time series were subjected to Morlet wavelet transform. The analysis clearly identifies the signature of a planetary wave of period 5 days occurring during the first fortnights of March and April (centered at day numbers 70 and 95 as revealed by the colored contours in Fig. 12.3). The cross-wavelet spectra in winds reveal enhanced spectral energy at a period of 6 days during the first fortnight of February but this period is not well expressed in the EEJ strength.

Pancheva et al. (2008) examined the 6 day wave extracted from global data sets on MLT winds and ground geomagnetic field variations. That analysis revealed the simultaneous presence of a 6-day

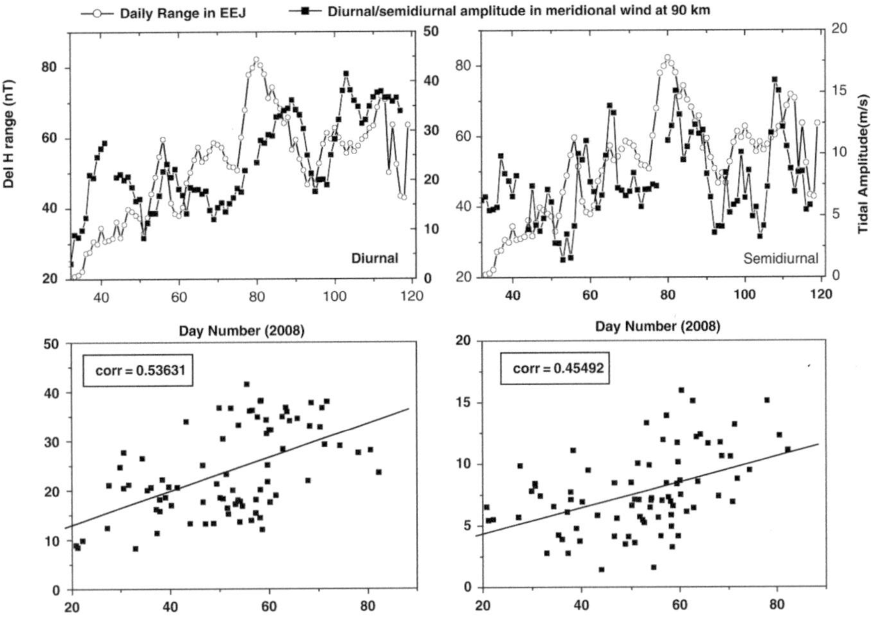

Fig. 12.7 Comparison for the diurnal range in the EEJ strength and diurnal and semi-diurnal amplitudes of meridional wind at 90 km over Tirunelveli for the period February–April 2008 (*top panels*). The *bottom panels* depict the corresponding scatter plots

wavenumber 1 westward propagating Rossby wave and a wavenumber 1 eastward propagating Kelvin wave during the chosen observation interval. The presence of these waves at stratospheric heights was confirmed in the National Center for Environmental Prediction (NCEP) reanalysis data sets on geopotential height and zonal wind at 30 and 10 hPa levels. The limited data sets used in the present analysis cannot establish whether the 6-day and 5-day perturbations in the geomagnetic field were caused by either of the two travelling waves that might have been present at those times.

A quasi-16-day wave was the dominant dynamical feature in the MLT zonal wind during the February–March 2008 period as expressed in the cross-wavelet results depicted in Fig. 12.6. The period of this wave in the meridional wind occurring around the same time was close to 12 days. Further, there was one more burst of this wave activity in the meridional wind centered around day number 95. It may be noted in this context that the wave period of the quasi-16-day wave was earlier reported to vary with height and time in a range of 12–24 days (Luo et al., 2002). There were also differences in amplitudes and periods between zonal and meridional wind components and between various sites noticed in the previous report (Luo et al., 2002). A variety of atmospheric conditions like temperature and background winds and the location of the node positions as prescribed by the Hough functions seem to govern the wave characteristics observed over any site. Whether the 16-day type of oscillation noticed in EEJ strength in the present work is related to the quasi-16-day wave observed in MLT wind observations will be a matter of investigation in a future study to be carried out by the authors.

Much of the day-to-day changes in the ionospheric current systems are expected to be driven by the day-to-day variations in the upward propagating tidal modes (Briggs, 1984a) but past studies were not able to bring out any clear relationship between the tidal winds observed at upper mesospheric heights and the strength of the ionospheric current system in day-to-day time scales because of the random nature of the variations (Phillips and Briggs, 1991, for example). The results presented in Fig. 12.7 demonstrating the correlation between the tidal amplitudes and the amplitude of the diurnal variation in the EEJ strength (expressed as "range") are therefore important in this context. It was possible to correlate the two parameters by considering 5-day composite 24-h means for estimating the tidal amplitudes and 5-day running averages of the range in EEJ strength. This apparent correlation, though observed in a qualitative manner, between the tidal winds and EEJ strength will encourage further studies on short-term tidal variabilities and their influence on EEJ variabilities in similar time scales. It is also worthwhile to examine whether these tidal variabilities extend to other latitudes and longitudes as it is believed that the Sq and the EEJ current systems are driven by large-scale tidal winds though local winds can modify the height and latitude distribution of the current system at certain latitudes (Anandarao and Raghavarao, 1979, for example).

From the Indian sector, Sridharan et al. (2002), for the first time, observed that the normal electrojet reversed (referred to as counterelectrojet (CEJ)) when the semi-diurnal tide amplitudes were enhanced during the solstice months of June and July 1995. A recent analysis from the same group has shown enhancement of semi-diurnal tidal amplitudes and occurrence of CEJ during high latitude northern hemispheric major sudden stratospheric warming events (Sridharan et al., 2009). Such an attempt to correlate the tidal activity with the strength of the afternoon electrojet was successful considering the fact that CEJ's usually occur in groups of 3–5 consecutive days and therefore it is possible to somewhat smooth out the day-to-day variations.

In conclusion, the present work examines the role of tides and planetary waves in driving variabilities of ionospheric dynamo region, though with limited data sets obtained during an experimental campaign. A 5-day oscillation was observed simultaneously in the meridional wind field at 88 km over Tirunelveli/Kolhapur and the EEJ strength during the campaign period, though the zonal winds over the two sites exhibit peaks at 6 days. The EEJ strength also reveals enhanced spectral densities at a period of $\sim$12 days that was observed in the meridional wind component over the two radar sites at around the same time. It is not clear whether the wind oscillations in the meridional wind represent the global planetary waves that might have contributed to the observed variabilities in geomagnetic field records at those planetary wave periods. Further analysis to be carried out by the authors with global data sets will provide insights of this proposed linkage. The tidal variabilities are correlated at least over time scales of 5 days and greater with corresponding variations in the amplitude of the diurnal variation or range in the EEJ strength. Indeed, a recent report that made use of MLT wind observations from Tirunelveli found evidence for the enhancement of semi-diurnal tide amplitude on days marked by afternoon counterelectrojet occurring during polar sudden stratospheric warmings (Sridharan et al., 2009).

References

Abdu MA, Ramkumar TK, Batista IS, Brum CGM, Takahashi H, Reinisch BW, Sobral JHA (2006) Planetary wave signatures in the equatorial atmosphere-ionosphere system, and mesosphere-E-F-region coupling. J Atmos Solar-Terr Phys 68:509–522

Altadill D, Apostolov EM (1998) Vertical displacement of the 2-day wave in the midlatitude ionospheric F region. J Geophys Res 103:29199–29206

Anandarao BG, Raghavarao R (1979) Effects of vertical shears in the zonal winds on the electrojet. Space Res 19:283–286

Anderson DN, Anghel A, Yumoto K, Ishituka M, Kudeki E (2002) Estimating daytime, vertical E×B drift velocities in the equatorial F-region using ground-based magnetometer observations. Geophys Res 29. doi:10.1029/2001GL014562

Blanc M, Richmond AD (1980) The ionospheric disturbance dynamo. J Geophys Res 85:1669–1686

Briggs BH (1984a) The variability of ionospheric dynamo currents. J Atmos Solar-Terr Phys 46:419–424

Briggs BH (1984b) The analysis of spaced sensor records by correlation techniques. Handb MAP 13:166–186

Chen PR (1992) Two-day oscillations of the equatorial ionization anomaly. J Geophys Res 97:6343–6357

England SL, Immel TJ, Huba JD (2008) Modeling the longitudinal variation in the post-sunset far-ultraviolet OI airglow using the SAMI2 model. J Geophys Res 113:A01309. doi:10.1029/2007JA012536

Forbes JM (1995) Tidal and planetary waves. In: The upper mesosphere and lower thermosphere: a review of experiment and theory. Geophysical monograph series, vol 87. American Geophysical Union, Washington DC, USA, vol. 97:pp 67–88

Forbes JM (1996) Planetary waves in the thermosphere-ionosphere system. J Geomag Geoelectr 48:91–98

Forbes JM, Guffe R, Zhang X, Fritts D, Riggin D, Manson A, Meek C, Vincent RA (1997) Quasi-2-day oscillation of the ionosphere during summer 1992. J Geophys Res 102: 7301–7305

Forbes JM, Levoroni S (1992) Quasi-16-day oscillation in the ionosphere. Geophys Res Lett 19:981–984

Gurubaran S, Rajaram R (2000) Signatures of equatorial electrojet in the mesospheric partial reflection drifts over magnetic equator. Geophys Res Lett 27:943–946

Gurubaran S, Ramkumar TK, Sridharan S, Rajaram R (2001) Signatures of quasi-2-day planetary waves in the equatorial electrojet: results from simultaneous observations of mesospheric winds and geomagnetic field variations at low latitudes. J Atmos Solar-Terr Phys 63:813–821

Kane RP (1973) A critical appraisal of the method of estimating equatorial electrojet strength. Proc Indian Acad Sci 78(A):149–158

Kikuchi T, Lühr H, Kitamura T, Saka O, Schlegel K (1996) Direct penetration of the polar electric field to the equator during a DP 2 event as detected by the auroral and equatorial magnetometer chains and the EISCAT radar. J Geophys Res 101:17161–17173

Kohsiek A, Glassmeier KH, Hirooka T (1995) Periods of planetary waves in geomagnetic variations. Ann Geophys 13:168–176

Lei J, Thayer JP, Forbes JM, Sutton EK, Nerem RS (2008) Rotating solar coronal holes and periodic modulation of the upper atmosphere. Geophys Res Lett 35:L10109. doi:10.1029/2008GL033875

Luo Y, Manson AH, Meek CE, Meyer CK, Burrage MD, Fritts DC, Hall CM, Hocking WK, MacDougall J, Riggin DM, Vincent RA (2002) The 16-day planetary waves: multi-MF radar observations from the arctic to equator and comparisons with HRDI measurements and the GSWM modeling results. Ann Geophys 20:691–709

Mikhailov AV 1983, Mechanism of in phase variations of electron concentration between E and F2 ionospheric layers (in Russian). Geomagn Aeron 23:557–561

Nawroozi AA (1967) Table for fisher's test of significance in harmonic analysis. Geophys J R Astr Soc 12:517–520

Nayar SRP (2006) Periodicities in solar activity and their signature in the terrestrial environment. In: Gopalswamy N, Bhattacharyya A (eds) Solar influence on the heliosphere and earth's environment: recent progress and prospects. Quest Publications. pp 170–177

Pancheva DV, Mukhtarov PJ, Mitchell NJ, Fritts DC, Riggin DM, Takahashi H, Batista PP, Clemesha BR, Gurubaran S, Ramkumar G (2008). Planetary wave coupling (5-6-day waves) in the low latitude atmosphere-ionosphere system. J Atmos Solar-Terr Phys 70:101–122

Pancheva DV, Mukhtarov PJ, Shepherd MG, Mitchell NJ, Fritts DC, Riggin DM, Franke SJ, Batista PP, Abdu MA, Batista IS, Clemesha BR, Kikuchi T (2006) Two-day wave coupling of the low-latitude atmosphere-ionosphere system. J Geophys Res 111:A07313. doi:10.1029/2005JA011562

Parish HF, Forbes JM, Kamalabadi F (1994) Planetary wave and solar emission signatures in the equatorial electrojet. J Geophys Res 99:355–368

Phillips A, Briggs BH (1991) The day to day variability of upper atmosphere tidal winds and dynamo currents. J Atmos Terr Phys 53:39–47

Rajaram R, Gurubaran S (1998) Seasonal variabilities of low latitude mesospheric winds. Ann Geophys 16:197–204

Ramkumar TK, Bhavanikumar Y, Narayana Rao D, Gurubaran S, Narendra Bab A, Ghosh AK, Rajaram R (2006) Observational evidences on the influence of tropical lower atmospheric $\sim$20 day oscillation on the ionospheric equatorial electrojet. J Atmos Solar-Terr Phys 68:523–538

Ramkumar TK, Gurubaran S, Rajaram R (2009) Mesospheric planetary wave signatures in the equatorial electrojet. J Geophys Res 114:A03309. doi:10.1029/2007JA012935

Richmond AD (1995) The ionospheric wind dynamo: effects of its coupling with different atmospheric regions. In: The upper mesosphere and lower thermosphere: a review of experiment and theory. Geophysical monograph series, Washington DC, USA, vol 87. pp 49–65

Riggin DM, Liu H, Lieberman RS, Roble RG, Russell JM III, Mertens CJ, Mlynczak MG, Pancheva D, Franke SJ, Murayama Y, Manson AH, Meek CE, Vincent RA (2006) J Atmos Solar-Terr Phys 68:323–339

Sridharan S, Gurubaran S, Rajaram R (2002) Structural changes in the tidal components in mesospheric winds as observed by the MF radar during afternoon counter electrojet events. J Atmos Solar-Terr Phys 64:1455–1463

Sridharan S, Gurubaran S, Rajaram R (2003) QBO influences on the variability of planetary waves in the equatorial mesopause region. Earth Planets Space 55:687–696

Sridharan S, Sathishkumar S, Gurubaran S (2009) Variabilities of mesospheric tides and equatorial electrojet strength during major stratospheric warming events. Ann Geophys 27: 4125–4130

Takahashi H, Wrasse CM, Pancheva D, Abdu MA, Batista IS, Lima LM, Batista PP, Clemesha BR, Shiokawa K (2006) Signatures of 3-6-day planetary waves in the equatorial mesosphere and ionosphere. Ann Geophys 24: 3343–3350

Vincent RA (1993) Long period motions in the equatorial mesosphere. J Atmos Solar-Terr Phys 55:1067–1080

Part III
Ionospheric Electrodynamics and Structuring

Chapter 13

Electrodynamics of Ionosphere–Thermosphere Coupling

Arthur D. Richmond

Abstract An overview of ionosphere-thermosphere electrodynamic coupling is presented. Collisions between the charged and neutral constituents of the upper atmosphere couple their respective dynamics and energetics. Magnetic stresses readily transfer momentum and energy over long distances along geomagnetic-field lines, accompanied by electric fields and currents. Consequently, the E and F regions of the ionosphere are strongly coupled, and momentum is transferred between the lower and upper thermosphere through the currents and their associated ion drag. Electrical conductivity mediates the degree of ion-neutral coupling. Conductivity is highly variable, and is itself affected by the electric field in various ways. Thermospheric winds drive the ionospheric wind dynamo. The winds are created by daily absorption of solar radiation in the thermosphere, by upward-propagating solar and lunar tides, by ion-drag acceleration at high latitudes, and by Joule heating at high latitudes. Electric current flows globally in the ionosphere and along geomagnetic-field lines through the magnetosphere. Interactions between the ion and neutral motions produce feedbacks that affect the dynamics of both components. Simulation models of thermosphere-ionosphere-electrodynamic interactions provide powerful tools for investigating the nature of these interactions, and for testing how well the uncertain model inputs and the physics incorporated in the models are able to predict observed features of the ionosphere and thermosphere.

A.D. Richmond (✉)
High Altitude Observatory, National Center for Atmospheric Research, Boulder, CO 80307-3000, USA
e-mail: richmond@ucar.edu

13.1 Introduction

Interactions between the ionospheric plasma and the much denser neutral air in the upper atmosphere have a strong influence on both constituents, and give rise to a wide variety of electrodynamic phenomena. While the dynamics of the neutrals is dominated by hydrodynamic forces, the dynamics of the plasma is dominated by electromagnetic forces. Collisional momentum and energy exchange between the plasma and neutral gas affects both components. Even though the mass density of the plasma is much less than that of the neutral gas, the plasma can transfer significant momentum to the neutrals through collisions, because the electric and magnetic forces experienced by the ions and electrons are large. The momentum to maintain the plasma motion often comes from afar, because energy and momentum are readily transferred over large distances, primarily along the direction of the magnetic field, via magnetic stresses associated with electric currents. The ionospheric plasma, especially the electron component, tends to convect in the presence of an electric field $\mathbf{E}$ as though all particles were forced to remain on a common geomagnetic-field line as they move, according to the "frozen-in" magnetic flux concept. In this case the motion of all charged particles perpendicular to the magnetic field $\mathbf{B}$ is essentially the velocity $\mathbf{E} \times \mathbf{B}/B^2$. Over the thickness of the ionosphere this velocity is nearly constant along geomagnetic-field lines, as is $\mathbf{E}$, and so the ionospheric E and F regions are strongly coupled electrodynamically. Only at lower altitudes do frequent collisions between ions and neutrals decouple the ion and electron motions in the direction perpendicular to $\mathbf{B}$, and allow significant current to flow in this direction.

M.A. Abdu, D. Pancheva (eds.), A. Bhattacharyya (Coed.), *Aeronomy of the Earth's Atmosphere and Ionosphere*, IAGA Special Sopron Book Series 2, DOI 10.1007/978-94-007-0326-1_13, © Springer Science+Business Media B.V. 2011

The electric field **E** and current density **J** are created in response to the forces acting on the plasma. The energy to generate them comes primarily from the kinetic energy of thermospheric winds and from the kinetic and thermal energy of magnetospheric plasma. Gravity usually plays only a minor role, although it can produce non-negligible electrodynamic effects in the night-time ionosphere. The geomagnetic field, which is dominated by the field created in the Earth's molten core, strongly organizes **E** and **J**. This organization greatly simplifies the analysis of electrodynamics in the ionosphere, unlike the situation in most space plasmas, where the magnetic field is not determined a priori, but co-evolves with the dynamics of the medium. The fact that **E** is nearly perpendicular to **B**, and is nearly constant along geomagnetic-field lines, means that a number of electrodynamic effects can be analyzed by considering field-line-integrated conductivities and currents transverse to **B**. The generation of **E** and **J** by thermospheric winds is referred to as the ionospheric wind dynamo, or ionospheric dynamo, for short. Observations of **E** and **J** give important diagnostic information about the processes driving them. Although ionospheric current is difficult to measure directly, considerable information about it can be derived from observations of the geomagnetic perturbations associated with the current, both on the ground and in space.

The frictional force that neutrals exert on the plasma through collisions is essentially balanced by the Ampère force $\mathbf{J} \times \mathbf{B}$ on the plasma. In effect, the collisions transfer the $\mathbf{J} \times \mathbf{B}$ force almost entirely to the neutrals. Electric current perpendicular to **B** is therefore a measure of the momentum transfer between the neutral upper atmosphere and the plasma in space. Since the perpendicular current is coupled with current flowing along **B** (field-aligned current or FAC) through the condition of current continuity, the forces are readily transferred long distances along geomagnetic-field lines. The interaction region in the upper atmosphere where the perpendicular currents flow and where the transfer of momentum and energy occurs is closely linked with the electrical conductivity of the ionosphere. In the sunlit ionosphere and in regions of significant auroral ionization the conductivity is largest below 150 km. Under these conditions it is below 150 km that the ionospheric wind dynamo is most effective, and that most of the energy and momentum transfer from the magnetosphere occurs; the F-region ionosphere above primarily responds to the electrodynamics, without being a major source of energy. On the other hand, when solar and auroral ionization is weak, the F region can often dominate the field-line-integrated conductivity, and upper-thermospheric winds can be a dominant generator of **E** and **J**.

13.2 Conductivity

The conductivity of the ionosphere has not been directly measured, but can be calculated from a combination of other observations and theoretical considerations. It is a tensor quantity that relates **J** to the electric field $\mathbf{E}'$ in a reference frame moving at the velocity **u** of the local medium. $\mathbf{E}'$ is related to the electric field in the Earth-fixed frame, **E**, by

$$\mathbf{E}' = \mathbf{E} + \mathbf{u} \times \mathbf{B}. \qquad (13.1)$$

The current density **J** is related to the velocities $\mathbf{v}_j$, number densities n_j, and charges q_j of the charged species:

$$\mathbf{J} = \sum_j n_j q_j \mathbf{v}_j. \qquad (13.2)$$

A relation between the velocities $\mathbf{v}_j$ and **E** is found by assuming the charged particles to be essentially in force balance, considering only the Lorentz force $q_j(\mathbf{E} + \mathbf{v}_j \times \mathbf{B})$ and the collisional force $m_j \nu_{jn}(\mathbf{u} - \mathbf{v}_j)$, where m_j is the charged-particle mass and ν_{jn} is its mean momentum-transfer collision frequency with neutrals. (Collisions between different charged-particle species can usually be neglected, as done here, except for electron motions parallel to **B**). Under these conditions Ohm's Law relating **J** to $\mathbf{E}'$ can be written as

$$\mathbf{J} = \sigma_{\parallel} \mathbf{E}'_{\parallel} + \sigma_P \mathbf{E}'_{\perp} + \sigma_H \mathbf{b} \times \mathbf{E}'_{\perp}, \qquad (13.3)$$

where **b** is a unit vector in the direction of **B**, subscripts $\parallel$ and $\perp$ on vectors denote components parallel and perpendicular to **B**, respectively, and $\sigma_{\parallel}$, σ_P and σ_H are the parallel, Pedersen, and Hall conductivities, respectively. The conductivities can be calculated from observed or modeled electron densities, together with models of the ion composition and of the collision frequencies. The collision frequencies are primarily dependent on the neutral density, which decreases exponentially with increasing height, and varies with location and time.Empirical models like the NRLMSISE00 model (Picone et al., 2002) are typically used to estimate the densities. The ion-neutral collision cross sections are obtained from a combination of laboratory measurements, theoretical considerations, and, sometimes, inferences derived from ionospheric dynamics. A summary is provided by Richmond (1995b).

Figure 13.1 shows typical midlatitude day-time and night-time Pedersen and Hall conductivities as a function of height for low (Min) and high (Max) levels of solar activity. The day-time conductivities maximize in the E region, where they are fairly predictable functions of solar zenith angle and level of solar activity. The F-region Pedersen conductivity, which is much larger than the F-region Hall conductivity, is not a simple function of solar zenith angle, but varies as the F-region electron density. It varies strongly with the solar cycle, as it depends on the product of the plasma and neutral densities, both of which have large solar-cycle variations (Takeda and Araki, 1985). At night, E-region conductivities decrease by about two orders of magnitude, while F-region conductivities decrease considerably less. The night-time F-region conductivities often contribute much more to the total geomagnetic-field-line-integrated conductivities than do the E-region conductivities, especially at solar maximum (e.g. Burnside et al., 1983; Takeda and Araki, 1985). The night-time E-region electron densities are difficult to measure, but are known to have considerable variability (e.g., Rowe and Mathews, 1973; Harper and Walker, 1977; Bauske et al., 1997). The F-region conductivities also have considerable night-to-night variability (e.g., Harper and Walker, 1977; Burnside et al., 1983). As a consequence, night-time ionospheric electrodynamics is more variable and complicated than day-time electrodynamics.

Auroral ionization at high latitudes often dominates the high-latitude conductivities, especially at night. The aurora varies strongly with geomagnetic activity, and has very complex spatial structure. Statistical models of the height-integrated auroral conductivities have been developed (e.g., Fuller-Rowell and Evans, 1987), which are valuable tools for studying high-latitude electrodynamics and magnetosphere-ionosphere coupling. However, in order to characterize properly the effects of small-scale auroral structure on the large-scale electrodynamics we also need to understand quantitatively the positive and negative correlations between aurorally produced conductivity and electric fields and currents (e.g., Evans et al., 1977; Foster et al., 1983).

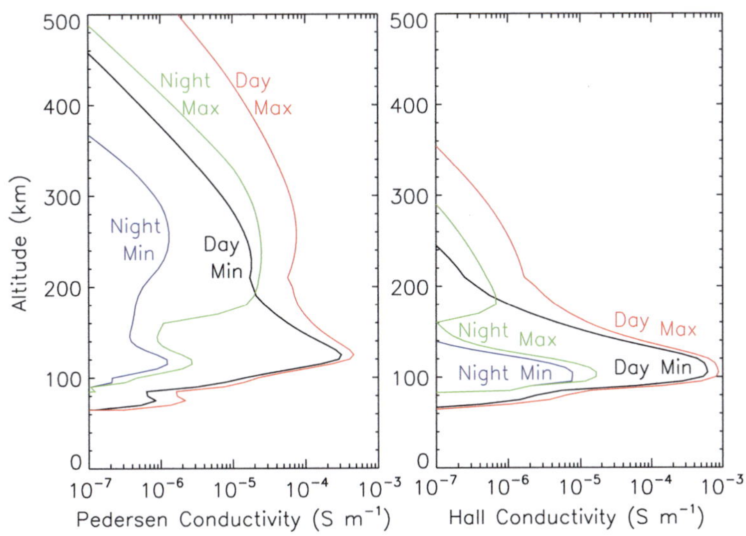

Fig. 13.1 Pedersen and Hall conductivities calculated for the location of the Arecibo radar (18.3° N, 66.75° W) for magnetically quiet equinox conditions ($A_p = 4$), local midnight (Night) and noon (Day), for low solar activity (Min, $F_{10.7} =$ 75 sfu) and high solar activity (Max, $F_{10.7} = 188$ sfu), using the International Reference Ionosphere (Bilitza, 1990), with the O^+–O collision frequency from Pesnell et al. (1993), and other collision frequencies as compiled by Richmond (1995b)

Ohm's Law assumes a linear relationship between **J** and **E**′. However, strong electric fields can alter this relationship, so that it is effectively no longer linear. Various types of effects occur. Strong electric fields and their associated rapid plasma drifts alter plasma densities through changes in loss rates (e.g., Banks et al., 1974; Merkin et al., 2005) and through plasma redistribution (e.g., Banks and Yasuhara, 1978; Footitt et al., 1983; Raghavarao et al., 1984; Stening, 1986; Singhal, 1991; Maruyama et al., 2005; Maruyama and Nakamura, 2007). Furthermore, conductivities that are effective on large scales (kilometres and larger) can be strongly influenced by small-scale irregularities (e.g., Rogister, 1971; Ronchi et al., 1990). This is believed to have a strong influence on the equatorial electrojet (e.g., Gagnepain et al., 1977; Alken and Maus, 2010b), and may significantly affect auroral electrodynamics (e.g., Buchert et al., 2006, 2008). All these effects need to be taken into account in order to achieve a complete understanding of ionosphere/thermosphere electrodynamics.

13.3 Thermospheric Winds

The thermospheric winds that help create the ionospheric dynamo are driven primarily by four sources. First, the daily absorption of solar ultraviolet radiation in the thermosphere creates a diurnally varying temperature, creating pressure gradients that accelerate and decelerate winds as the Earth rotates. The acceleration due to the horizontal pressure gradients generally increases with altitude, because the increased scale height in the afternoon and the decreased scale height in the early morning produce exponentially increasing differences with altitude of the constant-pressure-surface heights between these times. Consequently, the associated wind amplitudes generally increase with height throughout the thermosphere, and so wind dynamo influences tend to be weighted toward the higher regions of significant conductivity. At low latitudes the resulting winds tend to be predominantly westward during the day and eastward at night.

The second major source of thermospheric winds is the upward propagation of atmospheric tides generated by diurnally varying heating in the troposphere and stratosphere (Volland, 1988; Forbes, 1995). The tides can propagate vertically in the form of global-scale inertia-gravity waves. Ozone absorption of ultraviolet radiation in the stratosphere and water-vapor absorption of infrared solar radiation in the troposphere produce tides that migrate westward around the Earth with the apparent motion of the Sun. Latent heat release in clouds is another important source of tides. Because the development of clouds is strongly influenced by land-sea differences, this source generates tidal components that do not necessarily follow the Sun, but rather can propagate both westward and eastward at a range of phase velocities. The ability of tides to propagate into the thermosphere is influenced by their direction and speed. Tidal components with long effective vertical wavelengths, like the westward-propagating Sun-synchronous semidiurnal tide with wavenumber 2 (e.g., Forbes and Vial, 1991) and the eastward-propagating diurnal tides with longitudinal wavenumbers -2 and -3 (e.g., Hagan et al., 2009), can propagate up to 120 km and above before viscosity and thermal conduction dissipate them strongly. These components are particularly effective for the ionospheric dynamo, because they have substantial amplitudes through much of the region where dayside Pedersen and Hall conductivities are large. The amplitude and phase of the tides in the dynamo region are influenced not only by the strength and phase of their heating sources, but also by propagation conditions through the middle atmosphere, especially by the winds there. These conditions change with season, and are also modulated by global-scale planetary waves with periods of multiple days (e.g., Pancheva et al., 2002), which can produce modulations of similar periodicity in the dynamo (e.g., Pedatella and Forbes, 2009). Tides with lunar periods are also present in the upper atmosphere. When sampled once per day, the lunar semidiurnal tide produces a variation with a 14.3 day period, that is clearly seen in the ionosphere at low latitudes (e.g., Bartels and Johnston, 1940; Matsushita, 1967).

The third major source of thermospheric winds is acceleration by the rapidly drifting ions at high latitudes (e.g., Axford and Hines, 1961; Roble, 1992). This acceleration is greatest where the ion density is large, and where the ions are sufficiently decoupled from the neutrals that they are able to drift at speeds near the $\mathbf{E} \times \mathbf{B}/B^2$ velocity. The acceleration depends strongly on magnetospheric conditions, being much stronger during magnetic storms. Generally, the acceleration increases with altitude above 120 km up to the peak of the ionospheric F layer, and can generate winds of hundreds of metres per second (e.g., Killeen and Roble, 1984; Kwak and Richmond, 2007). Observations (e.g., Killeen and Roble, 1988; Richmond et al., 2003a) show strong westward winds in the afternoon and evening at magnetic latitudes of $60° - 75°$, and day-to-night winds over the magnetic polar region, that are dependent on the strength and direction of the interplanetary magnetic field.

The fourth major source of thermospheric winds is pressure gradients produced by high-latitude Joule heating (e.g., Volland and Mayr, 1971; Dickinson et al., 1975). This is particularly important during magnetic storms, when the heating can increase by one or two orders of magnitude over quiet-time values. The increased high-latitude pressure drives equatorward winds above about 120 km that increase with altitude and tend to be turned westward at midlatitudes by the Coriolis acceleration. Simulations indicate that the strongest winds are found on the night side of the Earth (e.g., Richmond et al., 2003b), but also extend into the day-side dynamo region. Observations of disturbance winds on the day side (e.g., Emmert et al., 2002) show a strong equatorward component above 120 km at upper-midlatitudes, together with a westward component at upper-midlatitudes and an eastward component at low latitudes. Through inertia the winds persist hours or even days after the high-latitude heating subsides, so that residual post-storm disturbance dynamo effects are observed globally (e.g., Huang et al., 2005).

13.4 Ionospheric Electric Fields and Currents

For a given distribution of thermospheric wind and conductivity, and for a specified manner in which the magnetosphere provides field-aligned current or high-latitude plasma convection to the ionosphere, it is possible to calculate the electric field and current at middle and low latitudes that are generated by the ionospheric dynamo. In addition to Ohm's Law (13.3), dynamo models assume that the current is divergence-free and that the electric field is electrostatic:

$$\nabla \cdot \mathbf{J} = 0 \tag{13.4}$$

$$\mathbf{E} = -\nabla \Phi \tag{13.5}$$

where Φ is the electric potential.

Figure 13.2 shows results of such a simulation using the National Center for Atmospheric Research Thermosphere-Ionosphere-Electrodynamics General-Circulation Model (TIE-GCM [Roble et al., 1988; Richmond et al., 1992]). It is only at middle and low latitudes that the electrodynamics are calculated internally in the model; at high latitudes an empirical model of electric fields is imposed to represent magnetospheric effects. Because of the very large field-aligned conductivity, field-aligned current (FAC) can readily flow from one hemisphere to the other at middle and low latitudes. In Fig. 13.2 this FAC appears as radial

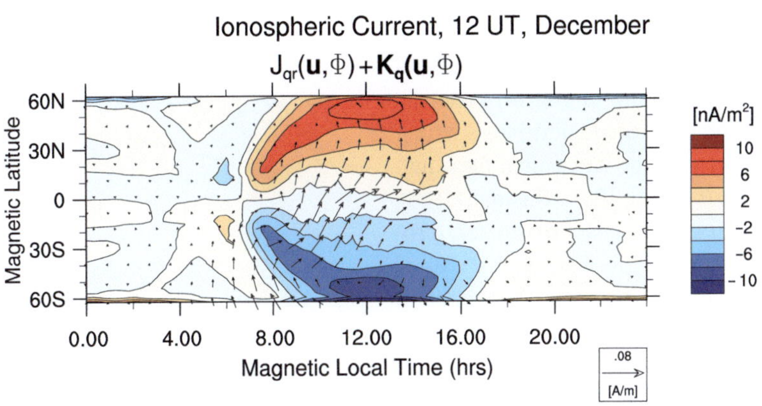

Fig. 13.2 Height-integrated horizontal ionospheric current density $\mathbf{K}_q$ driven by the combined neutral wind $\mathbf{u}$ and electric potential Φ (vectors), and upward component of field-aligned current density at 300 km altitude, J_{qr}, (contours and colors) in magnetic-latitude/magnetic-local-time coordinates between $\pm 63°$ latitude, for the December solstice, at 12 UT. Horizontal current flows from the southern to the northern hemisphere, while vertical current at 300 km flows out of the northern hemisphere and into the southern hemisphere

current J_{qr} at the top of the dynamo region. For this December example the FAC flows predominantly out of the northern (winter) hemisphere and into the southern (summer) hemisphere during the day time. There is a net return current flow within the ionosphere across the magnetic equator from summer to winter. Associated with the FAC is an east-west component of magnetic perturbation above the ionosphere, but below the FAC, which has been observed (e.g., Yamashita and Iyemori, 2002).

The larger variability of night-side than day-side conductivities, and the greater relative importance of F-region effects at night, results in greater variability of ionospheric electric fields at night than at day (e.g., Fejer, 1993; Fejer et al., 1995, 2005). It also makes accurate modeling of electrodynamics difficult on the night side of the Earth. In addition to the dynamo effects of winds, at night there is a non-negligible current associated with the gravitational force on the F-region plasma, which is balanced by a $\mathbf{J} \times \mathbf{B}$ force. Although small, this gravity-driven current can be large enough to affect the night-side electric field (e.g., Eccles, 2004).

At low latitudes, much of the electrodynamics can be considered in a simplified two-dimensional altitude-latitude picture. Current continuity means that relatively little net current flows upward and poleward across any geomagnetic-field line, if we neglect the relatively small convergence or divergence of east-west current below the field line. Near the magnetic equator in the E region, a strong vertical polarization electric field can develop to prevent any significant vertical current from flowing, since current cannot flow into or out of the neutral atmosphere below. The polarization field is associated with the large ratio of Hall to Pedersen conductivity around 100 km altitude. It drives a strong eastward Hall current during the day, called the equatorial electrojet (EEJ) (Forbes, 1981; Rastogi, 1989; Reddy, 1989). The EEJ current is approximately proportional to the eastward electric field, and it causes a large enhancement of the magnetic perturbation below. In fact, the ease of measuring the magnetic perturbation under the EEJ, and its close relation to the eastward ionospheric electric field, have prompted researchers to estimate the eastward electric field from observations of the EEJ magnetic perturbation (e.g., Anderson et al., 2002, 2006; Alken and Maus, 2010b). Somewhat complicating the relation, however, is the fact that variable low-latitude winds tend to produce apparent EEJ variations that are not directly related to the eastward electric field (e.g., Fambitakoye et al., 1976; Reddy and Devasia, 1981; Fang et al., 2008; Alken and Maus, 2010a).

At night, for low-latitude geomagnetic field lines that reach into the ionospheric F layer, the F-region conductivity usually dominates over the E-region conductivity in contributing to the total field-line-integrated conductivity, and ion-neutral momentum transfer is largely confined to the F-region portion of the field line. If the eastward wind varies little along the conducting portion of a field line, then it will generate an electric field of such a magnitude to cause the plasma to drift eastward at an $\mathbf{E} \times \mathbf{B}/B^2$ velocity that equals the mean wind velocity (e.g., Rishbeth, 1971). This can lead to an interesting effect, illustrated in Fig. 13.3, if an eastward wind velocity increases in strength with longitude. At each longitude the plasma tends to drift with the wind, and so the eastward plasma velocity increases with longitude. This eastward plasma velocity corresponds to a downward/equatorward electric field that increases in strength toward the east. The electric field, however, must be curl-free, and so there must be an altitude gradient of the eastward electric-field component such that the field becomes more strongly eastward with decreasing altitude (Rishbeth, 1971; Eccles, 1998). If other constraints on the electric field prevent it from being large at some apex height, such as an apex height

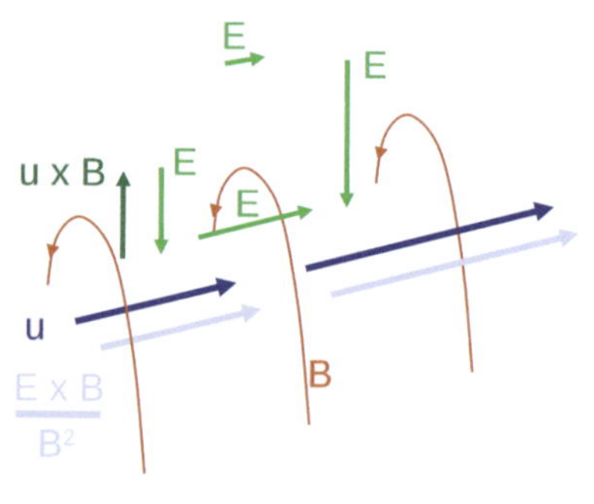

Fig. 13.3 A schematic of dynamo conditions in the evening F region. *Curved lines* represent geomagnetic-field lines $\mathbf{B}$. The eastward wind $\mathbf{u}$ increases in strength toward the east. A vertical component of polarization electric field $\mathbf{E}$ develops such that the horizontal component of electrodynamic drift velocity $\mathbf{E} \times \mathbf{B}/B^2$ is similar to $\mathbf{u}$, meaning that the vertical component of $\mathbf{E}$ tends to offset $\mathbf{u} \times \mathbf{B}$. This vertical component of $\mathbf{E}$ also increases toward the east. For the full vector electrostatic electric $\mathbf{E}$ to be curl-free, there must be a vertical gradient of the eastward field component of $\mathbf{E}$, such that this component becomes less eastward with increasing altitude

for which the field line descends through a region of relatively large Pedersen conductivity around the latitudes of the equatorial ionization anomaly, then the eastward field at lower heights can become increasingly strong with descending height. This eastward electric field corresponds to upward plasma drifts. Note that although we have estimated the meridional electric field using a two-dimensional geometry, neglecting longitude gradients, it is necessary to consider the longitude gradient of the meridional field to understand the origin of the eastward field in this case.

In the evening ionosphere, where such conditions occur, a strong upward drift can occur in the equatorial ionosphere after sunset, called the pre-reversal enhancement (PRE) of the vertical drift. This upward drift further modifies the electrodynamics, because by lifting the plasma it reduces the conductivity (which in the F region is proportional to the product of the ion and neutral densities), which further modifies the current path and the polarization field. In particular, by reducing the conductivity the uplift reduces the drag on flux-tube motions, thereby facilitating the uplift through a positive feedback effect.

13.5 Electric Current Effects on Winds

The ion-drag acceleration on neutrals is

$$\mathbf{J} \times \mathbf{B}/\rho = (\sigma_P B^2/\rho)(\mathbf{v}_\mathrm{E} - \mathbf{u}_\perp) + (\sigma_H B^2/\rho)\mathbf{b} \times (\mathbf{v}_\mathrm{E} - \mathbf{u}), \tag{13.6}$$

where ρ is the mass density, and $\mathbf{v}_\mathrm{E}$ is the electrodynamic velocity $\mathbf{E} \times \mathbf{B}/B^2$. This expression neglects the relatively small ion-drag acceleration due to gravitational and pressure-gradient forces on the ions. Typical values of the Pedersen and Hall ion-drag coefficients, $\sigma_P B^2/\rho$ and $\sigma_H B^2/\rho$, respectively, are shown in Fig. 13.4, for the same conditions as in Fig. 13.1. The term $(\sigma_P B^2/\rho)(\mathbf{v}_E - \mathbf{u}_\perp)$ in Eq. (13.6) represents an acceleration that tends to push the neutral velocity $\mathbf{u}$ towards the velocity $\mathbf{v}_E$ in the direction perpendicular to $\mathbf{B}$. As the difference between $\mathbf{u}$ and $\mathbf{v}_E$ is reduced, the current density and the ion-drag acceleration are also reduced, analogous to a reduction in the effective conductivity (e.g., Baker and Martyn, 1953; Akasofu and DeWitt, 1965; Denisenko et al., 2008). Volland (1976a, b) showed theoretically how, as the conductivity and ion-neutral coupling become stronger, there is a limit to the strength of the current. The term $(\sigma_H B^2/\rho)\mathbf{b} \times (\mathbf{v}_\mathrm{E} - \mathbf{u})$ represents an acceleration that tends to turn the wind direction. If $\mathbf{v}_\mathrm{E}$ is negligible the turning is analogous to the Coriolis acceleration of the wind, except that the horizontal component of acceleration is generally opposite to the direction of Coriolis acceleration. Other forces on the neutrals, including the gravitational, pressure-gradient, Coriolis, and viscous forces, also contribute in a major way to the net acceleration of the air. A characteristic value of both the daily acceleration and of the Coriolis acceleration is the angular rotation rate of the Earth, Ω, which is shown in Fig. 13.4 for reference. Whereas the Pedersen drag coefficient usually exceeds Ω in the F region, and is comparable to Ω down to 120 km at day, the Hall drag coefficient is generally less than Ω everywhere except in the auroral E-region ionosphere under strong particle precipitation.

As discussed earlier, the transfer of forces associated with the electric currents strongly couples plasma motions all along geomagnetic-field lines. Through the ion drag effect, it also couples neutral motions at different altitudes. For example, when $\mathbf{u}_\perp$ is larger than $\mathbf{v}_\mathrm{E}$, ion drag retards $\mathbf{u}_\perp$. The retarding force on upper-thermospheric winds at F-region heights, where $\sigma_P B^2/\rho$ is large, is transferred from the dense air at lower altitudes. Because of the large difference of neutral densities between E- and F-region heights, the ion-drag acceleration is much less at low than at high altitudes. At night, when the F region often dominates the field-line-integrated conductivity, the F region can become largely decoupled from the E region in terms of momentum transfer. In the low-latitude nighttime ionosphere, the fact that the ionosphere tends to drift eastward at a velocity similar to the neutral velocity means that ion drag on the neutral wind is relatively small, leading to faster winds. This effect contributes to the strong night-time eastward winds at low latitudes (Rishbeth, 1971).

Ion-neutral feedback effects are also important at high latitudes, where high-speed winds are generated by the rapid ion convection driven by magnetospheric processes. Average wind velocities at altitudes where Pedersen conductivity is large have magnitudes on the order of 10–50% the typical $\mathbf{E} \times \mathbf{B}/B^2$ speed, depending on altitude, latitude, and season

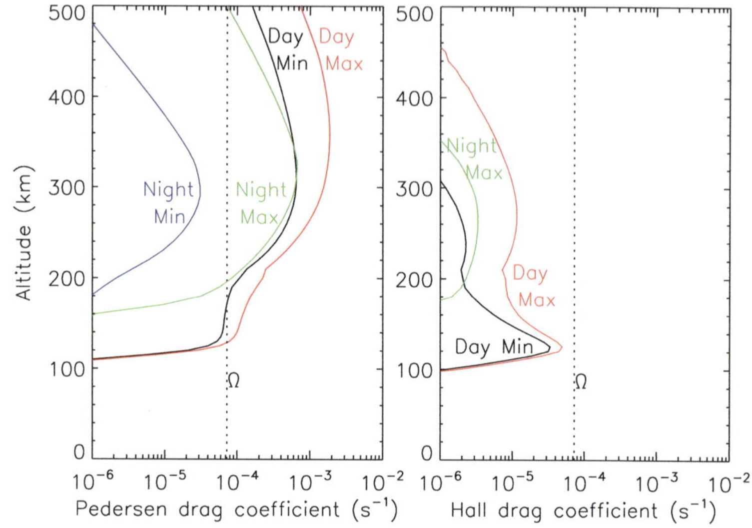

Fig. 13.4 Ion-drag coefficients $\sigma_P B^2/\rho$ (Pedersen) and $\sigma_H B^2/\rho$ (Hall), for the same conditions as in Fig. 13.1. Ω is the angular rotation rate of the Earth, which give a characteristic value both for inertial effects and for the Coriolis acceleration

(e.g., Johnson and Virdi, 1991; Killeen et al., 1992; Richmond et al., 2003a; Kwak and Richmond, 2007). These neutral winds will have a significant impact on the current, and thus on the momentum and energy transfer. The impact will not only affect the thermospheric winds and temperatures, but will also feed back on the magnetospheric dynamics that drives the plasma convection (e.g., Banks, 1972; Lyons et al., 1985; Forbes and Harel, 1989; Deng et al., 1993; Richmond, 1995a; Raeder et al., 2001; Peymirat et al., 2002; Ridley et al., 2003; Wang et al., 2007).

High-latitude Joule heating associated with magnetosphere-ionosphere electrodynamics produces a thermospheric response that also feeds back on the global electrodynamics through the disturbance-dynamo effect, especially during and following magnetic storms. At midlatitudes the disturbance winds tend to be westward, stronger at night than at day (e.g., Huang et al., 2005), and tend to produce a poleward electric field and a westward drift of the ionospheric plasma (Blanc and Richmond, 1980; Heelis and Coley, 1992). The electric field drives eastward Hall current at day that charges the dusk terminator positively and the dawn terminator negatively. The resulting daytime westward disturbance-dynamo electric field extends to low latitudes, where it tends to reduce, or even reverse, the normal low-latitude eastward electric field (Blanc and Richmond, 1980; Sastri, 1988; Fejer and Scherliess, 1995). This weakens and sometimes reverses the EEJ. When the heating stops, the global thermospheric winds largely recover toward their quiet-day patterns after a day, but full recovery can take many days (Huang et al., 2005).

13.6 Research Prospects

Electrodynamic interactions between the neutral and plasma constituents of the upper atmosphere strongly affect both constituents, and entail important feedback effects. The strength of the interaction depends on the highly variable plasma density, which itself is affected by electric fields. The complexity of the system, and its dependence on a number of inadequately known parameters, presents a challenge to researchers trying to develop accurate predictive capabilities. Complex simulation models like the Global Self-consistent Model of the Thermosphere-Ionosphere-Protonosphere (GSM-TIP) (Namgaladze et al., 1988; Klimenko et al., 2006), the TIE-GCM mentioned earlier, the Coupled Thermosphere-Ionosphere-Plasmasphere-electrodynamics (CTIPe) model (Fuller-Rowell et al., 1987; Millward et al., 2001), and the Global Coupled Thermosphere-Ionosphere-Electrodynamics Model of the Institute of Geology and Geophysics of the Chinese Academy of Sciences (GCITEM-IGGCAS) (Ren et al., 2009) are capable of representing global-scale interactions. They provide a tool for analyzing the underlying physics, but are dependent on knowledge of processes like night-time ionization rates, the character and variability of tides propagating from below, and the manner in which the magnetosphere generates electrodynamic effects and responds to thermospheric/ionospheric electrodynamic feedback. In order to be representative of the actual conditions in the thermosphere/ionosphere system, the models require wide-spread simultaneous observations of a large number of parameters that can constrain the model inputs. Formal data-assimilation procedures need to be developed and applied to this problem, in order to have a means of combining the complex models with the extensive and diverse data.

Electrical conductivity is a key element of the ion-neutral coupling, but there are often considerable difficulties in obtaining an accurate representation of the conductivities for modeling purposes. Not only do we need to have a more accurate understanding of the uncertain and variable ionization rates over the dark hemisphere and in the auroral regions, but we need to understand the correlations between spatial structure in the conductivities and electric fields in the auroral ionosphere, and we need to understand better the nonlinear influences of strong electric fields on the effective conductivities. A combination of well-designed observations, data analysis, and theory will be required to further advance our understanding of ionosphere-thermosphere electrodynamic coupling.

Acknowledgments I thank Astrid Maute for providing Fig. 13.2. The National Center for Atmospheric Research is sponsored by the National Science Foundation (NSF). This work was supported in part by NSF Award No. ATM-0836386, NASA grant NNX09AN57G, and AFOSR Contract FA9550-08-C-0046.

References

Akasofu SI, DeWitt RN (1965) Dynamo action in the ionosphere and motions of the magnetospheric plasma, 3. The Pedersen conductivity, generalized to take account of acceleration of the neutral gas. Planet Space Sci 13:737–744

Alken P, Maus S (2010a) Electric fields in the equatorial ionosphere derived from CHAMP satellite magnetic field measurements. J Atmos Solar-Terr Phys 72:319–326

Alken P, Maus S (2010b) Relationship between the ionospheric eastward electric field and the equatorial electrojet. Geophys Res Lett 37:L04104. doi:1029/2009GL041989

Anderson D, Anghel A, Chau JL, Yumoto K, Pyle R, Munakata K, Yasue S, Kato C, Akahane S, Koyama M, Fujii Z, Duldig ML, Humble JE, Silva MR, Trivedi NB, Gonzalez WD, Schuch NJ (2006) Global, low-latitude, vertical E×B drift velocities inferred from daytime magnetometer observations. Space Weather 4:S08003. doi:1029/2005SW000193

Anderson D, Anghel A, Yumoto K, Ishitsuka M, Kudeki E (2002) Estimating daytime vertical E×B drift velocities in the equatorial F-region using ground-based magnetometer observations. Geophys Res Lett 29:10.1029/2001GL014562

Axford WI, Hines CO (1961) A unifying theory of high-latitude geophysical phenomena and geomagnetic storms. Can J Phys 39:1433–1464

Baker WG, Martyn DF (1953) Electric currents in the ionosphere, I. The conductivity. Phil Trans R Soc A246: 281–294

Banks PM (1972) Magnetospheric processes and the behavior of the neutral atmosphere. Space Res 12:1051–1067

Banks PM, Yasuhara F (1978) Electric fields and conductivity in the nighttime E-region: a new magnetosphere-ionosphere-atmosphere coupling effect. Geophys Res Lett 5(12): 1047–1050

Banks PM, Schunk RW, Raitt WJ (1974) NO+ and O+ in the high latitude F-region. Geophys Res Lett 1:239–242

Bartels J, Johnston HF (1940) Geomagnetic tides in horizontal intensity at Huancayo. Terr Magn Atmos Elect 45:269

Bauske R, Noel S, Proelss GW (1997) Ionospheric storm effects in the nighttime E region caused by neutralized ring current particles. Ann Geophys 15:300–305

Bilitza D (ed) (1990) International Reference Ionosphere 1990, National Space Science Data Center publication 90–22

Blanc M, Richmond AD (1980) The ionospheric disturbance dynamo. J Geophys Res 85:1669–1686

Buchert SC, Hagfors T, McKenzie JF (2006) Effect of electrojet irregularities on DC current flow. J Geophys Res 111:A02305. doi:1029/2004JA010788

Buchert SC, Tsuda T, Fujii R, Nozawa S (2008) The Pedersen current carried by electrons: a non-linear response of the ionosphere to magnetospheric forcing. Ann Geophys 26:2837–2844

Burnside RG, Walker JCG, Behnke RA, Gonzales CA (1983) Polarization electric fields in the nighttime F layer at Arecibo. J Geophys Res 88:6259–6266

Deng W, Killeen TL, Burns AG, Roble RG, Slavin JA, Wharton LE (1993) The effects of neutral inertia on ionospheric currents in the high-latitude thermosphere following a geomagnetic storm. J Geophys Res 98:7775–7790

Denisenko VV, Biernat HK, Mezentsev AV, Shaidurov VA, Zamay SS (2008) Modification of conductivity due to acceleration of the ionospheric medium. Ann Geophys 26: 2111–2130

Dickinson RE, Ridley EC, Roble RG (1975) Meridional circulation in the thermosphere, I. Equinox conditions. J Atmos Sci 32:1737–1754

Eccles JV (1998) A simple model of low-latitude electric fields. J Geophys Res 103:26699–26708

Eccles JV (2004) The effect of gravity and pressure in the electrodynamics of the low-latitude ionosphere. J Geophys Res 109:A05304. doi:1029/2003JA010023

Emmert JT, Fejer BG, Shepherd GG, Solheim BH (2002) Altitude dependence of middle and low-latitude daytime thermospheric disturbance winds measured by WINDII, J Geophys Res 107(A12):1483. doi:1029/2002JA009646

Evans DS, Maynard NC, Trøim J, Jacobsen T, Egeland A (1977) Auroral vector electric field and particle comparisons, 2. Electrodynamics of an arc. J Geophys Res 82: 2235–2249

Fambitakoye O, Mayaud PN, Richmond AD (1976) Equatorial electrojet and regular daily variation SR, III. Comparison of observations with a physical model. J Atmos Solar-Terr Phys 38:113–121

Fang TW, Richmond AD, Liu JY, Maute A (2008) Wind dynamo effects on ground magnetic perturbations and vertical drifts. J Geophys Res 113:A11313. doi:1029/2008JA013513

Fejer BG (1993) F region plasma drifts over Arecibo: solar cycle, seasonal, and magnetic activity effects. J Geophys Res 98:13645–13652

Fejer BG, Scherliess L (1995) Time dependent response of equatorial ionospheric electric fields to magnetospheric disturbances. Geophys Res Lett 22:851–854

Fejer BG, de Paula ER, Heelis RA, Hanson WB (1995) Global equatorial ionospheric vertical plasma drifts measured by the AE-E satellite. J Geophys Res 100:5769–5776

Fejer BG, de Souza J, Santos AS, Costa Pereira AE (2005) Climatology of F region zonal plasma drifts over Jicamarca. J Geophys Res 110:A12310. doi:1029/2005JA011324

Footitt RJ, Bailey GJ, Moffett RJ (1983) Ion transport in the mid-latitude F1-region. Planet Space Sci 31:671–687

Forbes JM (1981) The equatorial electrojet. Rev Geophys Space Phys 19:469–504

Forbes JM (1995) Tidal and planetary waves. In: Johnson RM, Killeen TL (eds) The upper mesosphere and lower thermosphere. American Geophysical Union, Washington, DC, pp 67–87

Forbes JM, Harel M (1989) Magnetosphere-thermosphere coupling: an experiment in interactive modeling. J Geophys Res 94:2631–2644

Forbes JM, Vial F (1991) Semidiurnal tidal climatology of the E region. J Geophys Res 96:1147–1157

Foster JC, St-Maurice J-P, Abreu VJ (1983) Joule heating at high latitudes. J Geophys Res 88:4885–4896

Fuller-Rowell T, Evans DS (1987) Height-integrated Pedersen and Hall conductivity patterns inferred from the TIROS-NOAA satellite data. J Geophys Res 92:7606–7618

Fuller-Rowell TJ, Quegan S, Rees D, Moffett RJ, Bailey GJ (1987) Interactions between neutral thermospheric composition and the polar ionosphere using a coupled ionosphere-thermosphere model. J Geophys Res 92: 7744–7748

Gagnepain J, Crochet M, Richmond AD (1977) Comparison of equatorial electrojet models. J Atmos Solar-Terr Phys 39:1119–1124

Hagan ME, Maute A, Roble RG (2009) Tropospheric tidal effects on the middle and upper atmosphere. J Geophys Res 114:A01302. doi:1029/2008JA013637

Harper RM, Walker JCG (1977) Comparison of electrical conductivities in the E-and F-regions of the nocturnal ionosphere. Planet Space Sci 25:197–199

Heelis RA, Coley WR (1992) East-west ion drifts at mid-latitudes observed by dynamics explorer 2. J Geophys Res 97:19461–19469

Huang C-M, Richmond AD, Chen M-Q (2005) Theoretical effects of geomagnetic activity on low-latitude ionospheric electric fields. J Geophys Res 110:A05312. doi:1029/2004JA010994

Johnson RM, Virdi TS (1991) High-latitude lower thermospheric neutral winds at EISCAT and Sondrestrom during LTCS 1. J Geophys Res 96:1099–1116

Killeen TL, Roble RG (1984) An analysis of the high-latitude thermospheric wind pattern calculated by a thermospheric general circulation model, 1. Momentum forcing. J Geophys Res 89:7509–7522

Killeen TL, Roble RG (1988) Thermosphere dynamics: contributions from the first 5 years of the dynamics explorer program. Rev Geophys 26:329–367

Killeen TL, Nardi B, Purcell PN, Roble RG, Fuller-Rowell TJ, Rees D (1992) Neutral winds in the lower thermosphere from dynamics explorer 2. Geophys Res Lett 19:1093–1096

Klimenko MV, Klimenko VV, Bryukhanov VV (2006) Numerical simulation of the electric field and zonal current in the Earth's ionosphere: the dynamo field and equatorial electrojet. Geomag Aeron 46(4):457–466. doi:10.1134/S0016793206040074 (Engl. trans.)

Kwak Y-S, Richmond AD (2007) An analysis of the momentum forcing in the highlatitude lower thermosphere. J Geophys Res 112:A01306. doi:1029/2006JA011910

Lyons LR, Killeen TL, Walterscheid RL (1985) The neutral wind "flywheel" as a source of quiet time, polar-cap currents. Geophys Res Lett 12:101–104

Maruyama N, Richmond AD, Fuller-Rowell TJ, Codrescu MV, Sazykin S, Toffoletto FR, Spiro RW, Millward GH (2005) Interaction between direct penetration and disturbance dynamo electric fields in the storm-time equatorial ionosphere. Geophys Res Lett 32:L17105. doi:1029/2005GL023763

Maruyama T, Nakamura M (2007) Conditions for intense ionospheric storms expanding to lower midlatitudes. J Geophys Res 112:A05310. doi:1029/2006JA012226

Matsushita S (1967) Lunar tides in the ionosphere. In: Flügge S (ed) Handbuch der Physik, vol 49/2. Springer, Berlin, p 547

Merkin VG, Milikh G, Papadopoulos K, Lyon J, Dimant YS, Sharma AS, Goodrich C, Wiltberger M (2005) Effect of anomalous electron heating on the transpolar potential in the LFM global MHD model. Geophys Res Lett 32:L22101. doi:1029/2005GL023315

Millward GH, Müller-Wodarg ICF, Aylward AD, Fuller-Rowell TJ, Richmond AD, Moffett RJ (2001) An investigation into the influence of tidal forcing on F region equatorial vertical ion drift using a global ionosphere-thermosphere model with coupled electrodynamics. J Geophys Res 106:24,733–24,744

Namgaladze AA, Korenkov Yu N, Klimenko VV, Karpov IV, Bessarab FS, Surotkin VA, Glushchenko TA, Naumova NM (1988) Global model of the thermosphereionosphere-protonosphere system. Pure Appl Geophys 127:219–254

Pancheva D, Merzlyakov E, Mitchell NJ, Portnyagin Y, Manson AH, Jacobi C, Meek CE, Luo Y, Clark RR, Hocking WK, MacDougall J, Muller HG, Kurschner D, Jones GOL, Vincent RA, Reid IM, Singer W, Igarashi K, Fraser GI, Fahrutdinova AN, Stepanov AM, Poole LMG, Malinga SB, Kashcheyev BL, Oleynikov AN (2002) Globalscale tidal variability during the PSMOS campaign of June–August 1999: interaction with planetary waves. J Atmos Solar-Terr Phys 64:1865–1896

Pedatella NM, Forbes JM (2009) Modulation of the equatorial F-region by the quasi- 16-day planetary wave. Geophys Res Lett 36:L09105. doi:1029/2009GL037809

Pesnell WD, Omidvar K, Hoegy WR (1993) Momentum transfer collision frequency of O+–O. Geophys Res Lett 20: 1343–1346

Peymirat C, Richmond AD, Roble RG (2002) Neutral wind influence on the electrodynamic coupling between the ionosphere and the magnetosphere. J Geophys Res 107(A1). 10.1029/2001JA900106

Picone JM, Hedin AE, Drob DP, Aikin AC (2002) NRLMSISE-00 empirical model of the atmosphere: statistical comparisons and scientific issues. J Geophys Res 107(A12):1468. doi:1029/2002JA009430

Raeder J, Wang Y, Fuller-Rowell TJ (2001) Geomagnetic storm simulation with a coupled magnetosphere-ionosphere-thermosphere model. In: Song P, Singer HJ, Siscoe G (eds) Space weather: progress and challenges in research and applications. Geophysical monograph series, vol 125. American Geophysical Union, Washington, DC, pp 377–384

Raghavarao R, Sridharan R, Suhasini R (1984) The importance of vertical ion currents on the nighttime ionization in the equatorial electrojet. J Geophys Res 89:11033–11037

Rastogi RG (1989) The equatorial electrojet: magnetic and ionospheric effects. In: Jacobs JA (ed) Geomagnetism, volume 2. Academic, San Diego, CA, p 461

Reddy CA (1989) The equatorial electrojet. Pure Appl Geophys 131:485–508

Reddy CA, Devasia CV (1981) Height and latitude structure of electric fields and currents due to local east-west winds in the equatorial electrojet. J Geophys Res 86:5751–5767

Ren Z, Wan W, Liu L (2009) ITEM-IGGCAS: a new global coupled ionospherethermosphere-electrodynamics model. J Atmos Solar-Terr Phys 71:2064–2076. doi:1016/j.jastp.2009.09.015

Richmond AD (1995a) The ionospheric wind dynamo: effects of its coupling with different atmospheric regions. In: Johnson RM, Killeen TL (eds) The upper mesosphere and lower thermosphere. American Geophysical Union, Washington, DC, pp 49–65

Richmond AD (1995b) Ionospheric electrodynamics. In: Volland H (ed) Handbook of atmospheric electrodynamics, volume II. CRC Press, Boca Raton, FL, pp 249–290

Richmond AD, Lathuillère C, Vennerstroem S (2003a)Winds in the high-latitude lower thermosphere: dependence on the interplanetary magnetic field. J Geophys Res 108(A2):1006. doi:1029/2002JA009493

Richmond AD, Peymirat C, Roble RG (2003b) Long-lasting disturbances in the equatorial ionospheric electric field simulated with a coupled magnetosphere-ionospherethermosphere model. J Geophys Res 108(A3): 1118. doi:1029/2002JA009758

Richmond AD, Ridley EC, Roble RG (1992) A thermosphere/ionosphere general circulation model with coupled electrodynamics. Geophys Res Lett 19:601–604

Ridley AJ, Richmond AD, Gombosi TI, De Zeeuw DL, Clauer CR (2003) Ionospheric control of the magnetospheric configuration: Thermospheric neutral winds. J Geophys Res 108(A8):1328. doi:1029/2002JA009464

Rishbeth H (1971) Polarization fields produced by winds in the equatorial F-region. Planet Space Sci 19:357–369

Roble RG (1992) The polar lower thermosphere. Planet Space Sci 40:271–297

Roble RG, Ridley EC, Richmond AD, Dickinson RE (1988) A coupled thermosphere/ionosphere general circulation model. Geophys Res Lett 15:1325–1328

Rogister A (1971) Nonlinear theory of 'Type I' irregularities in the equatorial electrojet. J Geophys Res 76(31): 7754–7760

Ronchi C, Sudan RN, Similon PL (1990) Effect of short-scale turbulence on kilometer wavelength irregularities in the equatorial electrojet. J Geophys Res 95:189–200

Rowe JF Jr, Mathews JD (1973) Low-latitude nighttime E region conductivities. J Geophys Res 78:7461–7470

Sastri JH (1988) Equatorial electric fields of ionospheric disturbance dynamo origin. Ann Geophys 6:635–642

Singhal RP (1991) The effect of the electric field and neutral winds on E-region ion densities and conductivities at low latitudes. J Atmos Solar-Terr Phys 53:949–957

Stening RJ (1986) Inter-relations between current and electron density profiles in the equatorial electrojet and effects of neutral density changes. J Atmos Solar-Terr Phys 48: 163–170

Takeda M, Araki T (1985) Electric conductivity of the ionosphere and nocturnal currents. J Atmos Solar-Terr Phys 47:601–609

Volland H (1976a) Coupling between the neutral wind and the ionospheric dynamo current. J Geophys Res 81:1621–1628

Volland H (1976b) The atmospheric dynamo. J Atmos Solar-Terr Phys 38:869–877

Volland H (1988) Atmospheric tidal and planetary waves. Kluwer, Dordrecht

Volland H, Mayr HG (1971) Response of the thermospheric density to auroral heating during geomagnetic disturbances. J Geophys Res 76:3764–3776

Wang W, Burns AG, Wiltberger M, Solomon SC, Killeen TL (2007) An analysis of neutral wind generated currents during geomagnetic storms. J Atmos Solar-Terr Phys 69: 159–165

Yamashita S, Iyemori T (2002) Seasonal and local time dependences of the interhemispheric field-aligned currents deduced from the Ørsted satellite and the ground geomagnetic observations. J Geophys Res 107(A11):1372. doi:1029/2002JA009414

Chapter 14

Daytime Vertical E×B Drift Velocities Inferred from Ground-Based Equatorial Magnetometer Observations

David Anderson

Abstract The daytime equatorial electrojet is a narrow band of enhanced eastward current flowing in the 100–120 km altitude region within ±2° latitude of the dip equator. A unique way of determining the daytime strength of the electrojet is to observe the difference in the magnitudes of the Horizontal (H) component between a magnetometer placed directly on the magnetic equator and one displaced 6–9° away. The difference between these measured H values provides a direct measure of the daytime electrojet current, and in turn, the magnitude of the vertical E×B drift velocity in the F region ionosphere. This paper emphasizes two major topics related to the title: (1) Describes and summarizes the techniques developed for obtaining the daytime, E×B drift velocities from ground-based magnetometer observations, and (2) Describes and summarizes the equatorial, ionospheric physical transport mechanisms that have been addressed using these techniques.

14.1 Introduction

In the low latitude, F region ionosphere, the ambient ion and electron density distributions are determined through the combined physical processes of: (1) Production by solar EUV radiation; (2) Loss through charge exchange with N_2 and O_2 and; (3) Transport parallel to geomagnetic field lines by diffusion and neutral winds and perpendicular to **B** by E×B drift. In the daytime E region (90–120 km), dynamo processes generate eastward electric fields, which are transmitted to F region altitudes (150–800 km) by equipotential geomagnetic field lines, causing both ions and electrons to drift upward, perpendicular to **B** with an E×B $/B^2$ drift velocity. At the same time, forces parallel to **B** due to gravity and plasma pressure gradients act to transport plasma along the magnetic field lines. The net effect is to create crests in electron density on either side of the magnetic equator at ±15 to 18° dip latitude, known as the equatorial anomaly. Trans-equatorial neutral winds transport ionization from one hemisphere to the other causing asymmetries in both peak electron densities and peak altitudes in the equatorial anomaly (Hanson and Moffett, 1966).

The primary transport mechanism for creating the equatorial anomaly is vertical E×B drift in the F region. Fejer (1991) and Scherliess and Fejer (1999) discuss the large day-to-day variability in vertical drift velocities as measured by the Jicamarca Incoherent Scatter Radar (ISR) located at the magnetic equator in Peru. This day-to-day variability in E×B drift is responsible for the large day-to-day variability in the low latitude F region ion and electron density distributions with altitude, latitude and local time (Anderson, 1973). Because the daytime upward E×B drift is so critical, it is very important to be able to specify the drift on a day-to-day basis, since the Jicamarca ISR makes drift observations only two or three times a month. For this reason, the Jicamarca Unattended Long-term Ionosphere Atmosphere (JULIA) radar has been used to obtain the daytime E×B drift velocities utilized in this study to develop the relationships described in Section 14.2.

D. Anderson (✉)
Cooperative Institute for Research in Environmental Sciences, University of Colorado, Boulder, CO, USA; Space Weather Prediction Center, National Oceanic and Atmospheric Administration, Boulder, CO, USA
e-mail: David.Anderson@noaa.gov

M.A. Abdu, D. Pancheva (eds.), A. Bhattacharyya (Coed.), *Aeronomy of the Earth's Atmosphere and Ionosphere*, IAGA Special Sopron Book Series 2, DOI 10.1007/978-94-007-0326-1_14, © Springer Science+Business Media B.V. 2011

Rastogi and Klobuchar (1990) demonstrated that the strength of the daytime equatorial electrojet could be measured using two magnetometers, one situated on the magnetic equator and the other displaced by 6–9° away. Using this technique, Rastogi and Klobuchar (1990) were able to infer whether the daytime vertical E×B drift velocity in the F region was large or small. They compared the difference in the Horizontal (H) component values between magnetometers at Trivandrum (8.5°N, 77.0°E, 0.5°S dip latitude) and Alibag (18.5°N, 72.9°E, 13.0°N dip latitude) with the observations of Total Electron Content (TEC) measured by a chain of polarimeters as a function of latitude and local time in the Indian Sub-Continent. For their study, they incorporated one magnetometer on the magnetic equator (Trivandrum) and one at 10° N. dip latitude (Alibag). It is important that one magnetometer be within the equatorial electrojet (±3° dip latitude) and one away from the dip equator, but not too far away. In a recent paper, Manoj et al. (2006) investigated the evidence for short spatial correlation lengths of the noontime equatorial electrojet inferred from a comparison of satellite (CHAMP) and ground magnetic data. For this study they located magnetometers roughly15° away from the dip equator.

It is well known that large upward E×B drift velocities produce the equatorial anomaly with crests in the peak electron density, Nmax, and TEC at ± 15° dip latitude while the absence of E×B drift does not create the anomaly. They verified that a weak equatorial electrojet was accompanied by an absence in TEC crests, while a strong electrojet (large ΔH values) was accompanied by observed daytime crests in TEC at ± 15° dip latitude. They also found that measuring the day-to-day fluctuation in H at only one station – Trivandrum – was not a realistic measure of the strength of the equatorial electrojet. Anderson et al. (1992) subsequently carried out theoretical calculations of TEC as a function of local time and latitude and compared these with the Indian TEC observations. They found excellent agreement for both "weak" and "strong" electrojet days. Neither Rastogi and Klobuchar (1990) nor Anderson et al. (1992) presented a quantitative relationship between ΔH and E×B drift – only a qualitative one.

In a recent paper by Anderson et al. (2002) it was demonstrated that there exists *quantitative* relationships whereby the vertical E×B drift velocity in the equatorial F-region can be estimated using ground-based magnetometer observations. Such quantitative relationships were developed for the South American sector, during the Solar Maximum period, 1998–1999. This represented the first time such a unique relationship had been quantitatively established. The Jicamarca Incoherent Scatter Radar (ISR) provided the daytime, vertical E×B drift velocities in conjunction with magnetometers at Canete and Piura in Peru. However, the data sets were only available for a total of 11 days between 1998 and 1999.

This chapter emphasizes two major topics related to the title: (1) Describes and summarizes the techniques for obtaining the daytime, E×B drift velocities from ground-based magnetometer observations, and (2) Describes and summarizes the equatorial, ionospheric physical transport mechanisms that have been addressed using these techniques.

14.2 Ground-Based Magnetometer Techniques

It is well known that the effect of neutral winds together with diurnal and semi-diurnal tidal components in the atmosphere cause currents to flow in the 100–130 km altitude region. This is the so-called Sq (Solar quiet) wind dynamo current system in the E region. Resulting from this current system is an electrostatic field directed eastward from dawn to dusk at low latitudes. The strength of this electric field is about 0.5 mV/m and is responsible for the upward E×B drift velocities of ~ 20 m/s measured by the Jicamarca ISR. As a result of this electric field, within ±2° of the magnetic equator, an enhanced eastward current flows (between 100 and 110 km altitude) known as the equatorial electrojet (see Forbes, 1981; Richmond, 1989; Reddy, 1989, for in-depth reviews of the neutral wind dynamo and the equatorial electrojet).

Figure 14.1 depicts the eastward electric field (yellow arrow), the consequent vertical electric field (red arrow) and the current systems that are associated with the electrojet. The view is to the North at the magnetic equator viewing the dayside region. If an eastward electric field exists and is perpendicular to B, then a Hall current is generated in the downward direction. Because of the particular geometry at the magnetic equator where magnetic field lines are horizontal, the Hall current, carried by upward moving

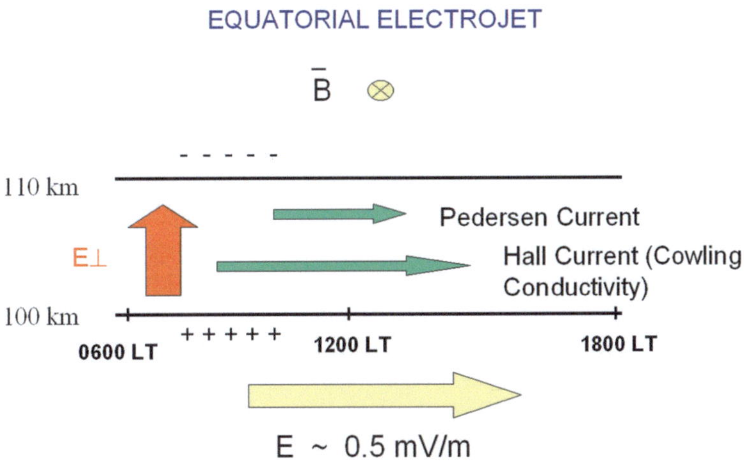

Fig. 14.1 Electric fields that generate the equatorial electrojet (see text for details)

electrons, quickly polarizes the ionospheric E layer so that an upward directed polarization electric field is produced. This electric field (red arrow) is about 5–10 times stronger than the eastward electric field (yellow arrow) that produced it. It is this vertical electric field that is responsible for the eastward equatorial electrojet current. This current produces the strong enhancement in the H component observed by magnetometers within $\pm$ 5° of the magnetic equator.

Figure 14.2 is a schematic plot of typical noontime magnetometer H component observations as a function of latitude. Note the 100 nanoTesla (nT) increase near the dip equator superimposed on the "global" Sq current magnetometer observations. When the H component observations from a magnetometer 6–9° away from the magnetic equator are subtracted from the H component values measured by a magnetometer on the magnetic equator, the difference is only related to the electrojet contribution which, in turn, is directly related to the eastward electrostatic field that created the electrojet current. Carrying out this subtraction to provide a ΔH value is necessary in order to eliminate the magnetospheric ring current component in H, resulting in a ΔH value that is only related to the ionospheric electrojet current and hence the east-west electric field. This eastward electric field might originate from the Sq Wind dynamo mechanism or could be associated with a penetration electric field from high latitudes, or both. It is emphasized that the electric field is ionospheric in origin and is not associated with the magnetospheric ring currents, or the Tail currents.

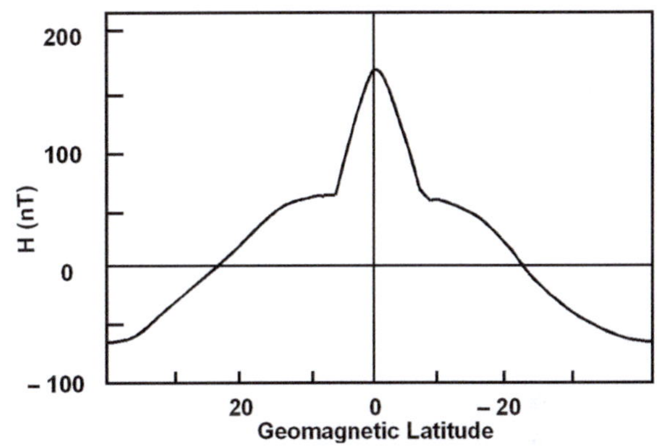

Fig. 14.2 Schematic plot of noontime magnetometer H component observations

In the Anderson et al. (2002) paper, the Jicamarca Incoherent Scatter Radar (ISR) provided the daytime, vertical E×B drift velocity. In the paper by Anderson et al. (2004), the JULIA radar at Jicamarca, Peru provided the daytime, vertical E×B drift velocities that has been related to the ground-based magnetometer observations. The JULIA radar is a low power 50 MHz coherent scatter system located at the Jicamarca Radar Observatory near Lima, Peru. In this study, JULIA observations are devoted to the daytime echo returns that occur near the 150 km altitude region.

Echoes from 150 km echoes were first observed in the early 1960s (Balsley, 1964), however their existence is still puzzling (e.g., Kudeki and Fawcett, 1993; Blanc et al., 1996; Tsunoda and Ecklund, 2000).

Nonetheless as it has been mentioned before, their Doppler velocities can be used to measure the zonal electric field in the equatorial ionosphere. Kudeki and Fawcett (1993) obtained a high correlation between the Doppler velocities from 150 km echoes and simultaneous ground magnetogram records made in Ancon. In addition, Woodman and Villanueva (1995) verified via incoherent scatter experiments that the 150 km echo phase velocities are indeed good estimates of *F* region vertical plasma drifts.

Between August 2001 and December 2003, almost 270 days of 150 km echoes with the JULIA system were obtained using one half of the Jicamarca antenna for both transmission and reception, pointing perpendicular to B with good detectability. The Doppler estimates are obtained via a spectral estimation routine similar to the one applied by Chau (1998) to process echoes from the lower atmosphere.

In order to establish the relationships between ΔH and E×B drift velocities in the Peruvian sector, Anderson et al. (2004) chose three approaches (1) A linear regression analysis; (2) A multiple regression approach and; (3) A neural network approach. The neural network method gives slightly lower RMS error values compared with the other two methods. The relationships for all three techniques were validated using an independent set of E×B drift observations, for 38 days between April 2001 and November 2003, from the Jicamarca Incoherent Scatter Radar (ISR) located at Jicamarca, Peru. They found that the neural network approach gave the lowest RMS error of the 3 approaches. Figure 14.3 displays the E×B drift velocity vs local time compared with the Jicamarca ISR observations on September 25, 2003 and April 17, 2002.

To determine whether or not realistic daytime, vertical E×B drift velocities can be obtained at other longitudes using the neural network trained in the Peruvian longitude sector, Anderson et al. (2006b) calculated quiet-time, vertical E×B drift velocities at two different longitude sectors. Magnetometer observations were obtained for the period between January 2001 and December 2004 from the magnetometers at Jicamarca (0.8 N. dip latitude) and Piura (6.8 N. dip latitude) in Peru and from Davao (1.4 S. dip latitude) and Muntinlupa (6.3 N. dip lat) in the Philippine sector. They choose only geomagnetically "quiet" days, when the three-hourly Kp value never exceeded a value of 3 over the entire day, and when the Daily Ap value was less than 10. These were "binned" into three seasons, December solstice, Equinox and June solstice periods. The neural network trained for the Peruvian sector was applied to each of the days in both the Peruvian and Philippine sectors providing ΔH-inferred vertical E×B drift velocities between 0700 and 1700 Local Time. For each season, the average E×B drift velocity curves are compared with the Scherliess-Fejer (1999), climatological E×B drift velocity curves in both the Peruvian and Philippine sectors. In the Peruvian sector, the comparisons are excellent and in the Philippine sector they are very good. They demonstrated that realistic magnetometer-inferred E×B drifts could be obtained in the Peruvian sector on a day-to-day basis and found that realistic E×B drifts could also be obtained in the Philippine sector. Figure 14.4 presents the comparisons for the Equinoctial period in the Peruvian and Philippine longitude sectors where the blue line represents the Scherliess-Fejer climatological E×B drift velocities and the red line is the average of the ΔH-inferred E×B drift velocities.

14.3 Areas of Research That Have Incorporated the ΔH Technique

A number of research efforts have been carried out, subsequent to the development of the ΔH-inferred,

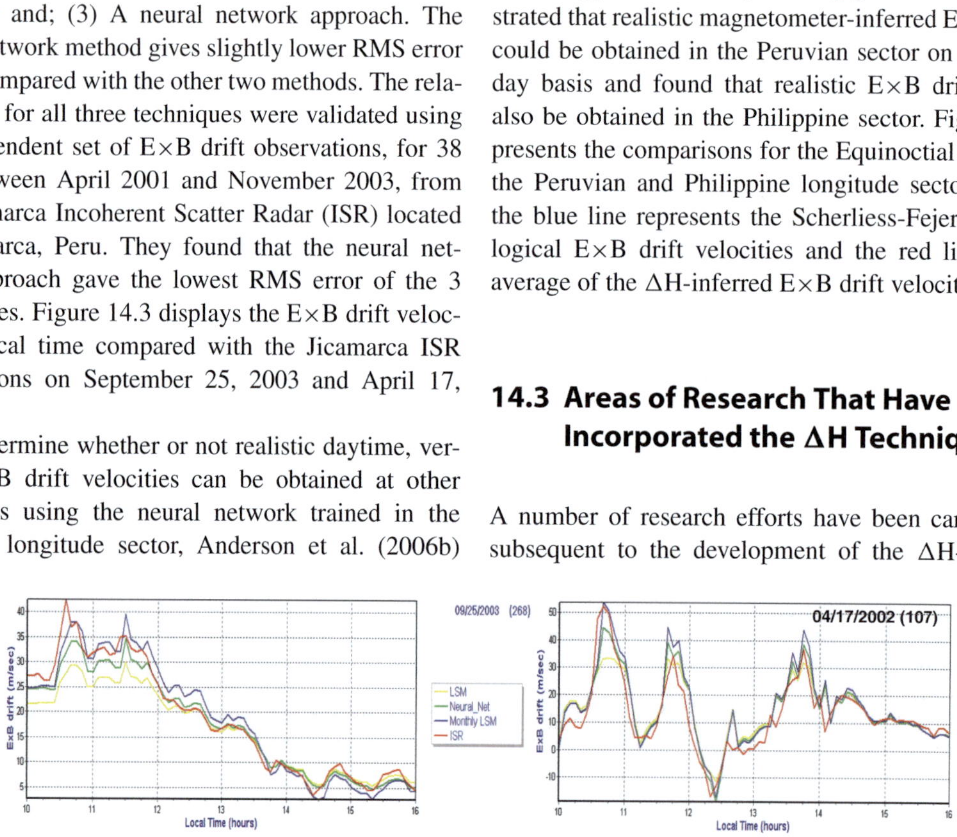

Fig. 14.3 Comparison between 3 ΔH-inferred E×B drift approaches (**a**) linear regression (monthly LSM), (**b**) multiple regression (LSM), (**c**) neural network and Jicamarca observed E×B drift (ISR) on September 25, 2003 (*left*) and April 17, 2002 (*right*)

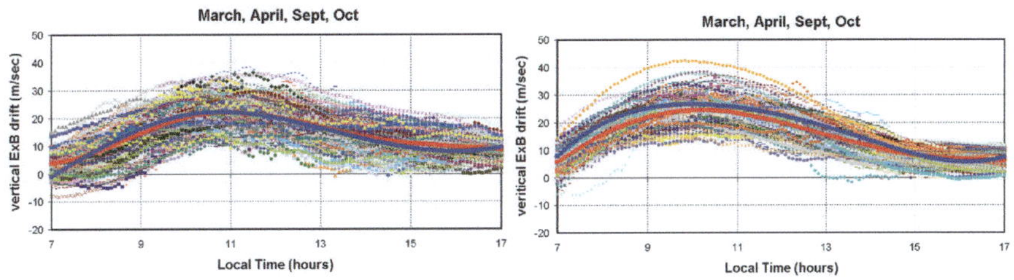

Fig. 14.4 Comparison between the average, equinoctial, quiet-day, ΔH-inferred E×B drift (*red line*) vs LT and the Scherliess-Fejer climatological model (*blue line*) in the Peruvian sector (*left*) and the Philippine sector (*right*) (see text for details)

daytime E×B drift technique. These fall into roughly three general areas (1) Prompt penetration of high latitude electric fields to the equatorial ionospheric region, (2) Modeling studies that incorporate the ΔH-inferred E×B drift velocity observations, and (3) Emerging studies that (a) Incorporate CHAMP magnetometer observations, and, (b) Compare sharp longitudinal gradients in E×B drift velocities associated with the 4-cell, non-migrating pattern seen in TOPEX observations.

14.3.1 Prompt Penetration Studies

In November 2005, a 3 day workshop was held at MIT's Haystack Observatory in Westford, MA to study "Storm-time Penetration Electric Fields and their Effects: Magnetospheric and Ionospheric Perspectives". The various papers that were presented addressed several outstanding questions: (1) What is the effect of the penetration electric field on ionospheric plasma? (2) How much of the interplanetary electric field can penetrate to the equatorial ionosphere? (3) How long can the interplanetary/magnetospheric electric field penetrate into the low latitude ionosphere? (4) How effective is the shielding process during magnetic storms? A sub-set of these presentations were submitted, peer-reviewed and published in a Special Issue of JASTP in July 2007.

Several of the papers utilized magnetometer-inferred, daytime E×B drift velocities to address several of these outstanding questions. Anghel et al. (2007) used the ΔH-inferred E×B drifts and electric fields in the Peruvian sector and found that about 10% of the interplanetary electric fields penetrate to the equator under disturbed conditions. Nicolls et al. (2007) examined the frequency characteristics between the low-latitude electric field and the interplanetary electric field and developed a linear transfer function. They found that a frequency-dependent, linear transfer function relating penetrating electric fields and the interplanetary electric fields had a broad maximum at the 2 h period. Kelley et al. (2007) applied this linear transfer function to two storm cases, in three longitude sectors, and found that the linear relationship between the interplanetary electric field and the equatorial ionospheric electric field works very well. In addition, Maruyama et al. (2007) presented simulated storm-time electric fields for the March 31, 2001 and April 17–18, 2002 geomagnetic storms and compared the calculated electric fields with low latitude observations of E×B drift velocities. The simulations highlight the possibility that penetrating electric fields may exist for several hours when the IMF is southward for a prolonged period of time. They find that the disturbance dynamo electric field becomes important at later times.

14.3.2 Modeling Studies

A number of investigations have incorporated the ΔH-inferred E×B drift velocities into low latitude, ionospheric theoretical modeling studies. Anderson et al. (2006a) studied the low-latitude ionospheric response to the Halloween geomagnetic storm on October 29 and 30, 2003 and the November 20, 2003 geomagnetic storm using a time-dependent Low Latitude Ionospheric Sector model (LLIONS) that incorporated the daytime, ΔH-inferred E×B drift velocities in the Peruvian sector. They compared their calculated values of TEC vs local time and latitude with an array of ground-based GPS receiver observations in the Peruvian longitude sector and with DMSP in-situ electron densities at 840 km altitude. On average, they found very good to excellent agreement.

Wang et al. (2008) studied the ionospheric electric field variations during the April 2–5, 2004 geomagnetic storm as simulated by a coupled magnetosphere-ionosphere-thermosphere (CMIT) model and compared their calculated, low-latitude electric fields with the ΔH-inferred electric fields in the Peruvian sector. The model captured the temporal variations seen in the ΔH-inferred E$\times$B drift observations, but overestimated the magnitude of the daytime E$\times$B drift velocities.

Two excellent theoretical studies have recently been published that (1) Model the wind dynamo effects on ground magnetic perturbations and vertical E$\times$B drifts (Fang et al., 2008a) and (2) Carry out model simulations of the equatorial electrojet in the Peruvian and Philippine longitude sectors (Fang et al., 2008b). The paper by Fang et al. (2008a) incorporates the NCAR Thermosphere-Ionosphere-Electrodynamic General Circulation Model (TIE-GCM) to demonstrate that the altitude variation in the wind velocity in the low latitude region can modify the ground magnetic perturbations a few degrees away from the magnetic equator. They find that only by combining the effects of both the EEJ and the off-equatorial wind-driven currents can the magnitude of ΔH and its relation to E$\times$B drift velocities be accurately estimated.

The paper by Fang et al. (2008b) also utilizes the TIE-GCM to simulate the EEJ current and the magnetic perturbations on the ground. They simulate the diurnal, seasonal (Equinox, June and December solstice periods) for solar cycle activity (F10.7 = 80, 140 and 200) variations in ΔH in the Peruvian and Philippine longitude sectors and the relationship between ΔH and daytime vertical E$\times$B drift velocities. The simulations show that the diurnal, seasonal and solar cycle variations are captured by the model. They find that the agreements between the simulated and observed magnitude of ΔH and the linear relationships to vertical E$\times$B drift are improved by modifying the TIE-GCM daytime E region photoionization rate in order to match observed E region electron densities.

14.3.3 Emerging Studies

A number of recent papers have incorporated the ground-based, ΔH technique for inferring the equatorial electrojet and compared these observations with satellite observations of the equatorial electrojet, most notably from the CHAMP magnetometer observations. Manoj et al. (2006) addressed the question to what extent is the highly variable noon-time current density observed by CHAMP due to temporal or spatial variations in the ionosphere. They find that the correlation coefficients between the ground and satellite data as a function of longitude between measurements vary from a high correlation when CHAMP is directly over the pair of observatories, to much lower values when the longitude separation between CHAMP and the observatories is $\pm 15°$ to the east or west. This holds for all longitude sectors.

Another investigation utilizing CHAMP observations of in-situ electron densities, the Jicamarca Unattended Long-term Ionosphere and Atmosphere (JULIA) radar measuring the daytime, vertical E$\times$B drift velocities and the ΔH (Jicamarca-Piura) observations was carried out by Stolle et al. (2008). They used electric field proxies to estimate the strength of the Equatorial Ionization anomaly (EIA). Analyzing 5 years of CHAMP electron density observations, they found high correlation coefficients (cc > 0.8) between the EIA values and both the JULIA observations and the ΔH values. They find a typical response time of the EIA to variations in the eastward electric field (JULIA observations) of 1–2 h and a response time of 2–4 h after EEJ strength variations (ΔH values).

Another area where the ΔH-inferred, daytime E$\times$B drift velocity technique has played a major and significant role relates to defining how sharp longitude gradients are in the E$\times$B vertical drift velocities at the boundaries of the non-migrating, 4-cell pattern observed by CHAMP (England et al., 2006) and TOPEX TEC observations (Scherliess et al., 2008). Anderson et al. (2009), utilizing appropriately-placed magnetometers in the Peruvian, Indian, Philippine and Indonesian longitude sectors, compared daytime, equatorial E$\times$B drift velocities and TOPEX/TEC observations associated with the 4-cell, non-migrating tidal structures. They compared the seasonal and longitude E$\times$B drift structures with TOPEX/TEC observed longitude structures. The paper demonstrates that large gradients in E$\times$B drift velocities exist at the boundary of a 4-cell pattern between the Philippine and Indonesian sectors, with a longitude gradient in vertical E$\times$B drift of 10 m/s over 15° in longitude which is responsible for the sharp gradients in TEC between these two sectors.

14.4 Summary and Future Work

This chapter has reviewed the techniques that have been developed for obtaining realistic, daytime, vertical E×B drift velocities from ground-based magnetometer observations and has demonstrated that these E×B drift velocities can be obtained wherever appropriately-placed magnetometers exist. The chapter also reviews and summarizes the subsequent papers and investigations that have utilized these techniques in the general areas of (1) Prompt penetration electric field studies, (2) Modeling equatorial electrojet studies and, (3) Emerging studies such as CHAMP-inferred electric field comparisons and studies that can account for the sharp longitude gradients in TOPEX/TEC observations at the boundaries of the 4-cell, non-migrating tidal structures. Future, near-term studies include the investigation of the unique, daytime E×B drift velocities associated sudden stratospheric warming (SSW) events (Chau et al., 2009) and lower atmospheric driving forces that affect the equatorial ionosphere (Fuller-Rowell et al., 2008).

Acknowledgements We thank Dr. Koki Chau, Director of the Jicamarca Radio Observatory, for providing the Jicamarca and Piura magnetometer data. The Jicamarca Radio Observatory is a facility of the Instituto Geofisico del Peru, Ministry of Education, and is operated with support from the NSF Cooperative agreement ATM-o432565. We also thank Prof. Kiyo Yumoto, Dept. of Earth and Planetary Sciences, Kyushu University, Fukuoka, Japan for supplying the Davao and Muntinlupa magnetometer observations.

References

Anderson DN (1973) A theoretical study of the ionospheric, F-region equatorial anomaly, I. Theory, planet. Space Sci 21:409–419

Anderson D, Anghel A, Araujo EA, Valladares C, Lin C (2006a) Theoretically modeling the low-latitude, ionospheric response to large magnetic storms. Radio Sci 41:RS5S04. doi:10.1029/2005RS003376

Anderson D, Anghel A, Chau J, Veliz O (2004) Daytime vertical E×B drift velocities inferred from ground-based magnetometer observations at low latitudes. Space Weather 2:S11001. doi:10.1029/2004SW000095

Anderson D, Anghel A, Chau J, Yumoto K (2006b) Global, low-latitude, vertical E×B drift velocities inferred from daytime magnetometer observations. Space Weather 4:S08003. doi:10.1029/2005SW000193

Anderson D, Anghel A, Yumoto K, Ishitsuka M, Kudeki E (2002) Estimating daytime vertical E×B drift velocities in the equatorial F-region using ground-based magnetometer observations. GRL 29(12):1596. doi:10.1029/2001GL014562

Anderson D, Araujo-Pradere EA, Scherliess L (2009) Comparing daytime, equatorial E×B drift velocities and TOPEX/TEC observations associated with the 4-cell, non-migrating tidal structure. Ann Geophys 7:1–7

Anderson DN, Klobuchar JA, Doherty PH, Rastogi RG (1992) A comparison of theoretical modeling of the low latitude ionosphere against TEC data from the indian longitudes during solar minimum. Int Beacon Symposium, MIT, Boston, MA

Anghel A, Anderson DN, Maruyama N, Chau J, Yumoto K, Bhattacharyya A, Alex S (2007) Interplanetary electric fields and their relationship to low-latitude electric fields under disturbed conditions. J Atmos Solar-Terr Phys 69: 1147–1159

Balsley BB (1964) Evidence of a stratified echoing regions at 150 kilometers in the vicinity of magnetic equator during daylight hours. JGR 69:1925

Blanc E, Mercandalli B, Houngninou E (1996) Kilometric irregularities in the E and F regions of the daytime equatorial ionosphere observed by a high resolution HF radar. GRL 23:645

Chau JL (1998) Examination of various techniques for measuring wind velocities using clear-air radars, with emphasis on vertical wind measurements. Ph.D Thesis, University of Colorado at Boulder

Chau JL, Fejer BG, Goncharenko LP (2009) Quiet variability of equatorial E×B drifts during sudden stratospheric warming events. Geophys Res Lett 36:L05101. doi:10.1029/2008GL36785

England SL, Maus S, Immel TJ, Mende SB (2006) Longitudinal variation of the E-region electric fields caused by atmospheric tides. Geophys Res Lett 33:L21105. doi:10.1029/2006GL027465

Fang TW, Richmond AD, Liu JY, Maute A (2008a) Wind dynamo effects on ground magnetic perturbations and vertical drifts. J Geophys Res 113:A11313. doi:10.1029/2008JA013513

Fang TW, Richmond AD, Liu JY, Maute A, Lin CH, Harper B (2008b) Model simulation of the equatorial electrojet in the Peruvian and Philippine sectors. J Atmos Solar-Terr Phys 70:2203–2211

Fejer BG (1991) Low latitude electrodynamic plasma drifts: a review. J Atmos Solar-Terr Phys 53:677–693

Forbes JM (1981) The equatorial electrojet. Rev Geophys 19:469–504

Fuller-Rowell TJR, Akmaev RA, Wu F, Anghel AF, Maruyame N, Anderson DN, Codrescu MV, Iredell M, Moorthi S, Juang H-M, Hou Y-T, Milllward G (2008) Impact of terrestrial weather on the upper atmosphere. Geophys Res Lett 35:L09808. doi:10.1029/2007GL032911

Hanson WB, Moffett RJ (1966) Ionization transport effects in the equatorial F region. J Geophys Res 71: 5559–5572

Kelley MC, Nicolls MJ, Anderson D, Anghel A, Chau JL, Sekar R, Subbaro KSV, Bhattacharyya A (2007) Multi-longitude case studies comparing the interplanetary and equatorial ionospheric electric fields using an empirical model. J Atmos Solar-Terrestr Phys 69:1174–1181

Kudeki E, Fawcett C (1993) High resolution observations of 150 km echoes at Jicamarca. GRL 18:1987

Manoj C, Luhr H, Maus S, Nagarajan N (2006) Evidence for short spatial correlation lengths of the noon-time equatorial electrojet – inferred from a comparison of satellite and ground magnetic data. J Geophys Res 111:A11312. doi:10.1029/2006JA011855

Maruyama N, Sazykin S, Spiro R, Andearson D, Anghel A, Wolf RA, Toffoletto FR, Fuller-Rowell TJ, Codrescu MV, Richmond AD, Millward G (2007) Modeling storm-time electrodynamics of the low-latitude ionosphere-thermosphere system: can long lasting disturbance electric fields be accounted for? J Atmos Solar-Terr Phys 69:1182–1199

Nicolls MJ, Kelley MJ, Chau JL, Veliz O, Anderson D, Anghel A (2007) The spectral properties of low latitude daytime electric fields inferred from magnetometer observations. J Atmos Solar-Terr Phys 69:1160–1173

Rastogi RG, Klobuchar JA (1990) Ionospheric electron content within the equatorial F2 layer anomaly belt. JGR 95: 19045–19052

Reddy CA (1989) The equatorial electrojet. PAGEOPH 131:485–508

Richmond AD (1989) Modeling the ionospheric wind dynamo: a review. PAGEOPH 131:413–435

Scherliess L, Fejer BG (1999) Radar and satellite global equatorial F region vertical drift model. JGR 104:6829–6842

Scherliess L, Thompson DC, Schunk RW (2008) Longitudinal variability of low-latitude total electron content: tidal influences. J Geophys Res 113:A01311. doi:10.1029/2007JA012480

Stolle C, Manoj C, Luhr H, Maus S Alken P (2008) Estimating the daytime equatorial ionization anomaly strength from electric field proxies. J Geophys Res 113:A09310. doi:10.1029/2007JA012781

Tsunoda RT, Ecklund WL (2000) On the nature of 150-km radar echoes over the magnetic dip equator. GRL 27: 657–660

Wang W, Lei J, Burns AG, Wiltberger M, Richmond AD, Solomon SC, Killeen TL, Talaat ER, Anderson DN (2008) Ionospheric electric field variations during a geomagnetic storm simulated by a coupled magnetosphere ionosphere thermosphere (CMIT) model. Geophys Res Lett 15:L18105. doi:10.1029/2008GL035155

Woodman RF, Villanueva F (1995) Comparisons of electric fields measured at F-region heights with 150 km – irregularity drift measurements. Paper presented at the 9th international symposium on equatorial aeronomy, Bali, Indonesia

Chapter 15

Three-Dimensional Modeling of Equatorial Spread F

J.D. Huba, G. Joyce, and J. Krall

Abstract The Naval Research Laboratory (NRL) has recently developed a three-dimensional code to study equatorial spread F (ESF). The code is based on the comprehensive NRL 3D ionosphere model SAMI3 and includes a potential equation to self-consistently solve for the electric field. The model assumes equipotential field lines so a 2D electrodynamic problem is considered. Results are presented showing the evolution of atomic and molecular ions, as well as that of the electron and ion temperatures during ESF.

15.1 Introduction

Equatorial spread F (ESF) (Haerendel, 1974; Ossakow, 1981; Hysell, 2000) is a post-sunset phenomenon in which the equatorial F-region ionosphere becomes unstable: large-scale (10's km) electron density "bubbles" can develop and rise to high altitudes ($\gtrsim$ 1000 km at times). To understand the complex and dynamic evolution of equatorial spread F, numerical simulation models are required. The majority of simulation studies have been based on two-dimensional spatial models of the nighttime F region (i.e., longitude and altitude) (Scannapieco and Ossakow, 1976; Ossakow et al., 1979; Zalesak and Ossakow, 1980; Zalesak et al., 1982; Sekar, 2003; Huba and Joyce, 2006). However, recently the Naval Research Laboratory (NRL) has developed a three-dimensional code to study equatorial spread F (ESF). The code is based on the comprehensive NRL 3D ionosphere model SAMI3 and includes a potential equation to self-consistently solve for the electric field. The model assumes equipotential field lines so a 2D electrodynamic problem is considered. In this chapter we describe the model and present results showing the evolution of atomic and molecular ions (Huba et al., 2008, 2009a), as well as electron and ion temperatures (Huba et al., 2009b) during ESF.

J.D. Huba (✉)
Naval Research Laboratory, Washington, DC, USA
e-mail: huba@ppd.nrl.navy.mil

15.2 Numerical Model

A modified version of the NRL 3D global ionosphere code SAMI3 is used in this analysis. SAMI3 is based on the 2D ionosphere model SAMI2 (Huba et al., 2000a). SAMI3 models the plasma and chemical evolution of seven ion species (H^+, He^+, N^+, O^+, $N_2{}^+$, $NO_2{}^+$ and $O_2{}^+$). The continuity and velocity equations solved are the following:

$$\frac{\partial n_i}{\partial t} + \nabla \cdot (n_i \mathbf{V}_i) = P_i - L_i n_i \qquad (15.1)$$

$$\frac{\partial \mathbf{V}_i}{\partial t} + \mathbf{V}_i \cdot \nabla \mathbf{V}_i = -\frac{1}{\rho_i}\nabla \mathbf{P}_i + \frac{e}{m_i}\mathbf{E} + \frac{e}{m_i c}\mathbf{V}_i \times \mathbf{B} + \mathbf{g} - \nu_{in}(\mathbf{V}_i - \mathbf{V}_n) - \textstyle\sum_j \nu_{ij}\left(\mathbf{V}_i - \mathbf{V}_j\right) \qquad (15.2)$$

Here, P_i are the ion production terms and L_i are the ion loss terms that involve photoionization, radiative recombination, and chemistry, ν_{in} is the ion-neutral collision frequency, ν_{ij} is the ion-ion collision frequency, and the summation is over ion species $j \neq i$. These terms are explicitly defined and discussed in detail in Huba et al. (2000a).

M.A. Abdu, D. Pancheva (eds.), A. Bhattacharyya (Coed.), *Aeronomy of the Earth's Atmosphere and Ionosphere*, IAGA Special Sopron Book Series 2, DOI 10.1007/978-94-007-0326-1_15, © Springer (outside the USA) 2011

The basic temperature equation for the α species solved is (Blanks and Kockarts, 1973; Millward et al., 1996)

$$\frac{\partial T_\alpha}{\partial t} + \mathbf{V}_\alpha \cdot \nabla T_\alpha + \frac{2}{3} T_\alpha \nabla \cdot \mathbf{V}_\alpha - \frac{2}{3} \frac{1}{n_\alpha} \nabla \cdot \kappa_\alpha \nabla T_\alpha = Q_{\alpha\beta} + Q_{\alpha n} \tag{15.3}$$

We solve the temperature equation for the electrons and three ion species: H^+, He^+, and O^+. We set the temperature of the molecular ions $N_2{}^+$, NO^+, and $O_2{}^+$ equal to the O^+ temperature. In Eq. (15.3), κ_α is the thermal conductivity, $Q_{\alpha\beta}$ is collisional heating/cooling between the α and β species, and $Q_{\alpha n}$ is frictional heating with the neutrals. There is also an electron heating source associated with photoelectrons in SAMI3/ESF not explicitly shown in Eq. (15.3). In solving Eq. (15.3) we assume that $\mathbf{V}_i = V_{is}\hat{e}_s + \mathbf{V}_\perp$ and $\mathbf{V}_e = \mathbf{V}_\perp$ where s is the direction along (i.e., parallel) to the geomagnetic field, $\mathbf{V}_\perp$ is the $E \times B$ drift velocity and we neglect the parallel electron velocity.

In addition to the plasma equations, it is necessary to solve for the electrostatic potential in order to derive the electric field to transport the plasma orthogonal to the geomagnetic field. The potential equation is derived from current conservation ($\nabla \cdot \mathbf{J} = 0$) in dipole coordinates and is given by

$$\frac{\partial}{\partial p} \Sigma_{pp} \frac{\partial \Phi}{\partial p} + \frac{\partial}{\partial \phi} \Sigma_{p\phi} \frac{\partial \Phi}{\partial \phi} = \frac{\partial F_{\phi g}}{\partial \phi} \tag{15.4}$$

where $\Sigma_{pp} = \int (p\Delta/b_s)\sigma_P \, ds$, $\Sigma_{p\phi} = \int (1/pb_s\Delta)$ $\sigma_P \, ds$, $F_{\phi g} = \int (r_E \sin^3\theta/\Delta)(B_0/c)(1/\Omega_i)\sigma_{Hi} g_p \, ds$, and σ_P is the Pedersen conductivity, σ_{Hi} is the ion component of the Hall conductivity, $\Delta = (1 + 3\cos^2\theta)^{1/2}$, $b_s = (r_E^3/r^3)\Delta$, and r_E is the radius of the earth. The field-line integration is along the entire flux-tube (i.e., s direction) with the base of the field lines at 85 km. The Hall conductivity is neglected on the LHS of Eq. (15.4); inclusion of this term only introduces a small asymmetry in the evolution of an ESF plasma bubble. Finally, we note that the ionospheric electric field driven by the neutral wind dynamo is not included in the simulation model.

The 3D simulation model is initialized using results from the two-dimensional SAMI2 code. SAMI2 is run for 48 hrs using the following geophysical conditions: F10.7 = 170, F10.7A = 170, Ap = 4, the geographic longitude is 0° so universal time and local time are the same, and the day-of-year 263 (e.g., 20 September 2002). The plasma is modeled from hemisphere to hemisphere up to ±26° magnetic latitude; the peak altitude at the magnetic equator is ~1600 km. The $E \times B$ drift is prescribed by the Fejer/Scherliess model (Fejer and Scherliess, 1995). The plasma parameters (density, temperature, and velocity) at time 1920 UT (of the second day) are used to initialize the 3D model at each magnetic longitudinal plane.

The 3D model uses a grid with magnetic apex heights from 90 to 2400 km, and a longitudinal width of 4° (e.g., $\simeq$ 460 km). The grid is $(nz, nf, nl) = (101, 202, 96)$ where nz is the number grid points along the magnetic field, nf the number in "altitude," and nl the number in longitude. This grid has a resolution of ~10 km × 5 km in altitude and longitude in the magnetic equatorial plane. The grid is periodic in longitude. In essence we are simulating a narrow "wedge" of the ionosphere in the post-sunset period. Finally, a 15% Gaussian-like perturbation in the ion density is imposed in the bottomside F region at $t = 0$.

15.3 Results

15.3.1 Ion Dynamics

Figure 15.1 shows contour plots of the electron, oxygen ion, and hydrogen ion densities (logarithmic scale) at time 2324 UT (approximately 2 h after the start of the simulation). The left panels show the densities as a function of altitude and latitude, while the right panels as a function of altitude and x (west/east direction). The velocity vectors are shown for the oxygen ions O^+ and hydrogen ions H^+ in the left panels. Here, the symbol is at the base of the vector; the maximum velocity is ~1.1 km/s. Finally, the solid line on the electron density contour plots is the H^+/O^+ transition altitude (i.e., where the oxygen ion density equals the hydrogen ion density). Note, the maximum contour level for the electrons and O^+ is 10^7 cm^{-3}, but the maximum contour level for H^+ is two orders of magnitude smaller 10^5 cm^{-3}.

At this time the H^+/O^+ transition altitude in the background ionosphere is ~ 1000 km at the magnetic equator; however, the transition altitude has increased

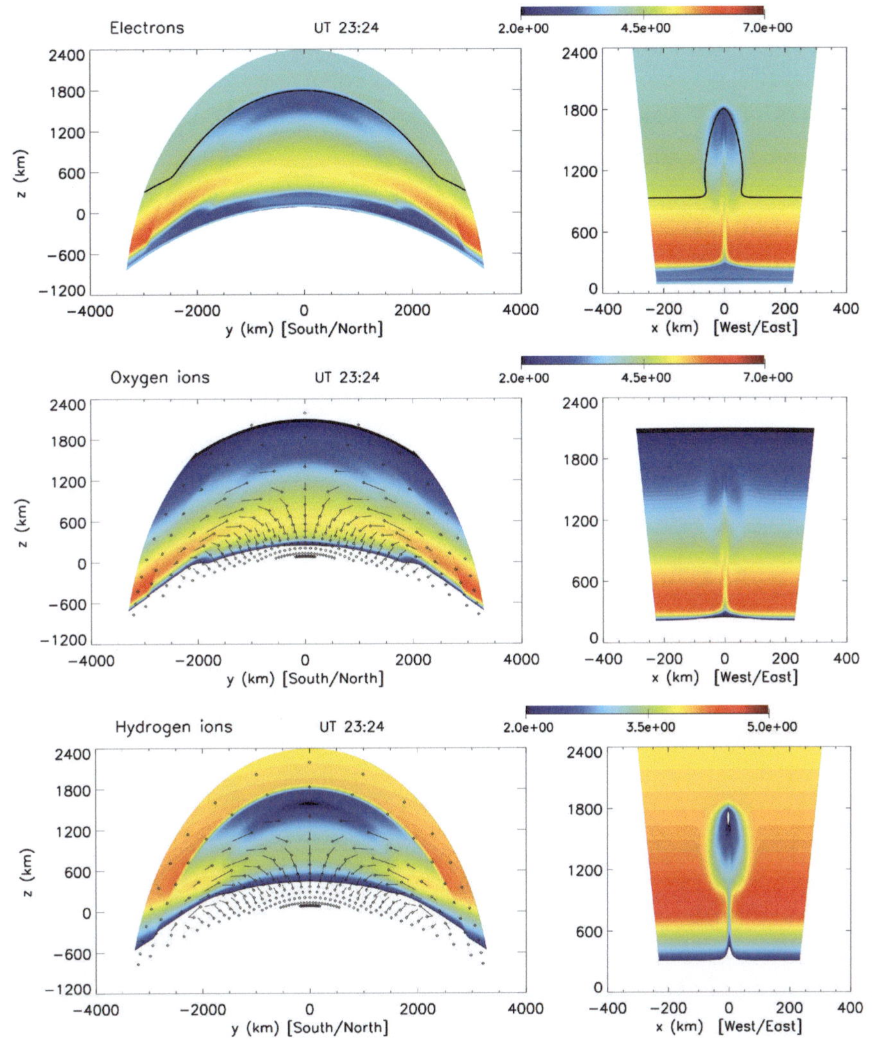

Fig. 15.1 Contour plots of the electron, oxygen ion, and hydrogen ion densities (logarithmic scale) at time 2324 UT. The *left panels* show the densities as a function of z (altitude at the magnetic equator) and y (south/north direction), while the *right panels* as a function of z and x (west/east direction). The velocity vectors are shown for the oxygen ions O^+ and hydrogen ions H^+ in the *left panels*. Here, the symbol is at the base of the vector

to the top of the plasma bubble in both latitude and longitude as seen in the top panels of Fig. 15.1. This is caused primarily by a large reduction in the H^+ density within the plasma bubble. This is shown quite vividly in the lower right panel. This large reduction of H^+ is due to the "drainage" of H^+ along the geomagnetic field associated with the downward field-aligned O^+ and H^+ flows evident in the velocity vectors shown in the middle and lower left panels. We refer to this process as a "super fountain" effect because the upflowing velocities are of order 100s m/s, substantially greater than the nominal ionospheric upwelling velocities are of order 10 s m/s. In fact, one can see the development of H^+ ionization crests forming at $y \simeq \pm 2000$ km (i.e., $\pm 18°$). The reduction of the electron density in the bubble at this time is similar to the development of a high altitude electron hole reported by Huba et al. (2000b).

In Fig. 15.2 we show contour plots similar to Fig. 15.1 but for the molecular ion NO^+. The three panels are for $t = 1921$ UT (bottom), $t = 2042$ UT (middle), and $t = 2203$ UT (top). The bottom panel is the start of the simulation. Note that the scale is logarithmic and ranges from 10^{-10} to 10^5. The middle panel

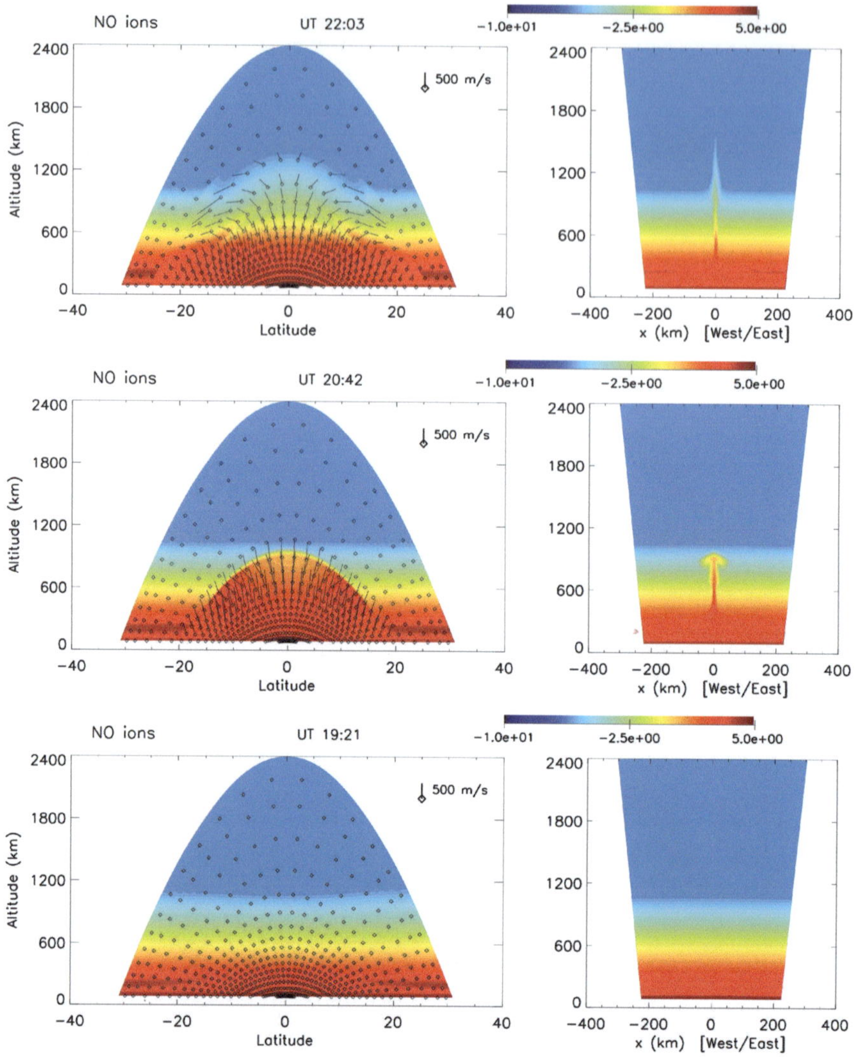

Fig. 15.2 Contour and velocity vector plots of the molecular ion NO^+. The three panels are for $t = 1921$ UT (*bottom*), $t = 2042$ UT (*middle*), and $t = 2203$ UT (*top*). Note that the scale is logarithmic and ranges from 10^{-10} to 10^5

is at $t = 2042$ UT. At this time, relatively high density NO^+ has been lifted to over 1000 km. The top panel is at time $t = 2203$ UT. By this time the NO^+ density is decreasing but there is still an small enhancement at altitudes as high as ~ 1500 km.

Hanson and Sanatani (1973) presented satellite data of O^+ and NO^+ as a function of dip latitude while passing through an equatorial spread *F* bubble. They found that the O^+ density fell from the background level of $\sim 10^6$ cm^{-3} to $\sim 10^3$ cm^{-3} inside the bubble, while the NO^+ density rose from ~ 40 cm^{-3} to ~ 500 cm^{-3}. These observations are consistent with the results shown in Figs. 15.1 and 15.2.

15.3.2 Temperature Dynamics

In Fig. 15.3 we show contour plots of the electron, oxygen ion, and hydrogen ion temperatures at time 22:19 UT (approximately 2 h after the start of the simulation). The left panels show the temperatures as a function of altitude and latitude, while the right panels as a function of altitude and *x* (west/east direction). The velocity vectors are shown for the oxygen ions O^+ and hydrogen ions H^+ in the left panels. Here, the symbol is at the base of the vector, i.e., the velocity is *away* from this point. Finally, the solid line on the electron temperature contour plots is the H^+/O^+ transition

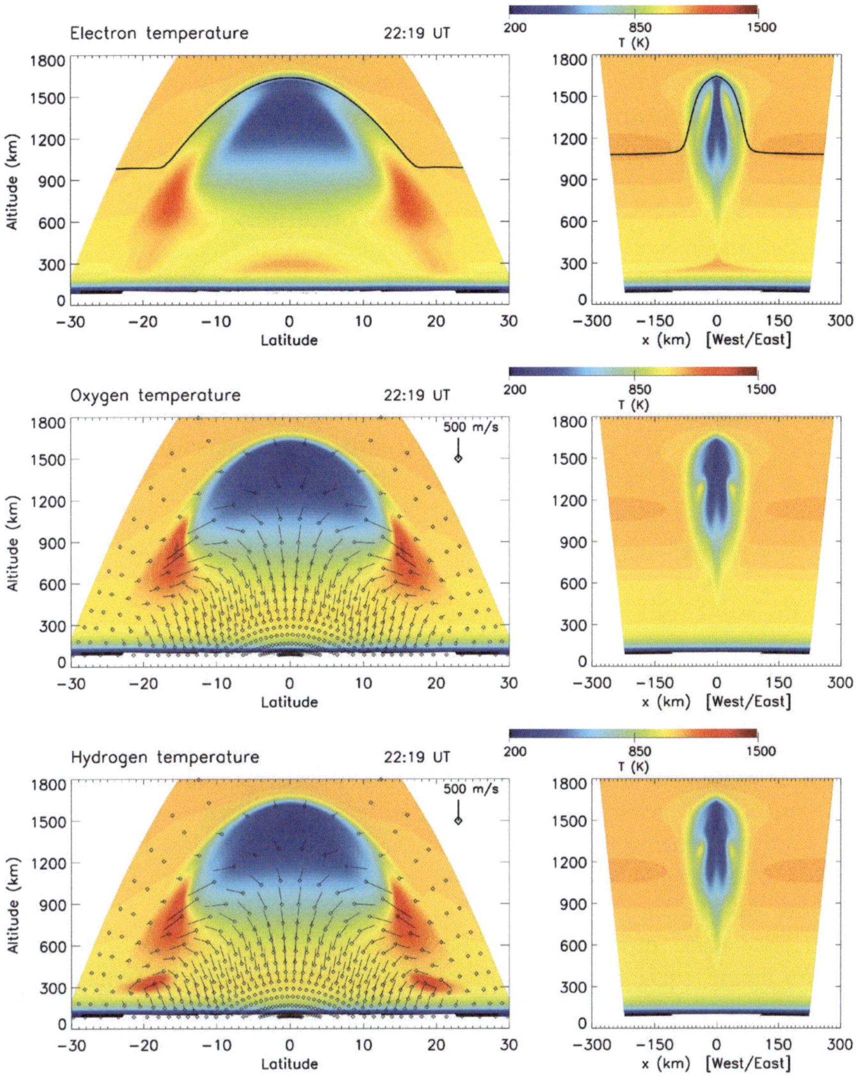

Fig. 15.3 Contour plots of the electron, oxygen ion, and hydrogen ion temperatures at time 2219 UT. The velocity vectors are shown for the oxygen ions O^+ and hydrogen ions H^+ in the left panels. Here, the symbol is at the base of the vector, i.e., the velocity is *away* from this point. The *solid line* in the electron figure is the H^+/O^+ transition altitude

altitude (i.e., where the oxygen ion density equals the hydrogen ion density).

The key points of this figure are the following. First, there is a clear transition in the temperatures along the field line that extends from $-25°$ to $+25°$ and reaches a maximum altitude of ~1600 km. This transition corresponds to the low density plasma bubble boundary (Huba et al., 2009a). Second, the plasma has cooled inside the bubble; the ion and electron temperatures decrease from ~1050 K outside the bubble to as low as ~ 400 K inside the bubble at 1500 km in the equatorial plane. The latitudinal extent of the ion cooling is $\sim \pm 15°$ while that of the electrons is $\sim \pm 10°$. The cooling is adiabatic, caused by the increase in volume of the flux tubes as the plasma rises to high altitudes (Huba et al., 2000a). Third, there is heating of the O^+ and H^+ ions in the altitude range ~ 400 – 1000 km at latitudes $\sim \pm 15°$. This is caused by the compression of the ions as they stream down the converging magnetic field lines. The ion vectors in the left panel show the "super fountain" effect described in Huba et al. (2009). The ions attain velocities exceeding 1 km/s.

The electrons are heated via collisions with the ions in this region.

The electron temperature increases along the field lines in the upper portion of the plasma bubble (above apex altitudes $\sim$ 1400 km) more than the ions. This is caused by electron heat conduction which is more effective than ion heat conduction because $\kappa_e >> \kappa_i$. We have carried a simulation in which $\kappa_e = 0$ and this high altitude electron heating did not occur.

Finally, relatively intense heating of H^+ occurs at an altitude $\sim$ 300 km at latitudes $\sim \pm 20°$. The H^+ temperature increases from 1050 K to 1400 K in this region. There is no heating of the O^+ ions but there is some collisional heating of the electrons. The low altitude heating of the H^+ ions is caused by frictional heating with the neutral atmosphere, i.e., the term Q_{H^+n} in Eq. (15.3). This term is $\propto \nu_{H^+n}|V_{H^+s} - V_{ns}|^2$. The H^+ develops a large velocity along the magnetic field in this region (i.e., $V_{H^+s} \sim 1.2$ km/s) early in the simulation. It is due to the fountain effect: the hydrogen is lifted upward because of the $E \times B$ drift and then falls along the magnetic field because of gravity and pressure gradient. This does not occur for the O^+ because its pressure gradient is different from that of hydrogen; hence, there is no O^+ heating in this region. Although the H^+ heating is interesting, the H^+ density is extremely low in this region: it drops from its ambient value 10^4 cm^{-3} to 10^{-2} cm^{-3}.

Oyama et al. (1988) made observations of the electron temperature during equatorial spread F on the Hinotori satellite at an altitude $\sim$ 600 km. They found that "the electron temperature inside the plasma bubbles is either higher or lower than that outside and can also be equal to the electron temperature outside, depending on the local time and on the place where the data were taken." In the developmental phase of ESF (e.g., around 21:00 LT) they found that the electron temperature inside the plasma bubble is 100–300 K cooler than the ambient temperature outside the bubble. Later in the evening they observe electron heating up to 1000 K hotter than the ambient electron temperature; they attribute these very hot electrons to high energy electrons in the South Atlantic anomaly region.

In Fig. 15.4 we plot the electron temperature T_e (K) as a function of latitude at an altitude of 600 km (consistent with the Oyama et al. (1988) observational altitude) at four times: 19:21 UT (start of the simulation), 20:21 UT, 21:02 UT, and 21:42 UT. At time 20:21 UT the plasma bubble has reached 600 km and the electrons have cooled by $\sim$ 100 K near the magnetic equator. At time 21:02 UT the bubble has risen higher and the latitudinal extent of the cooler plasma extends from $-18°$ to $18°$. However, at time 21:42 UT there is electron heating at the latitudes $\sim \pm 18°$. This heating is attributed to adiabatic heating of the O^+ ions that is transferred to the electrons via collisions. Thus, we find that the electron temperature (as well as ion) can be "either higher or lower than that outside temperature" consistent with the observations of Oyama et al. (1988).

Park et al. (2008) reported equatorial plasma bubbles with enhanced temperatures (BETs) in the topside ionosphere from spacecraft data (ROCSAT-1, KOMPSAT-1, and DMSP F15). A key aspect of this

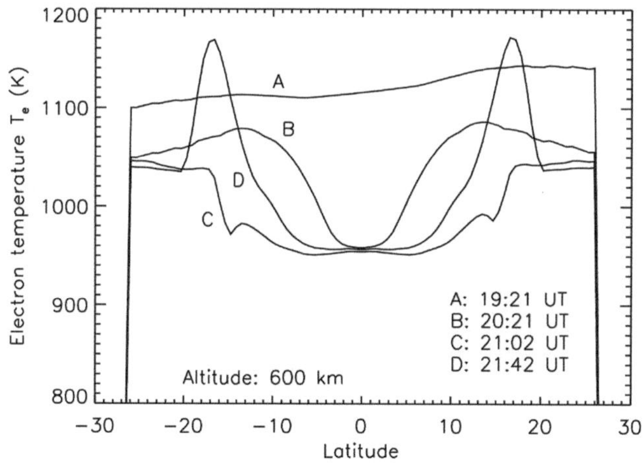

Fig. 15.4 Electron temperature T_e (K) as a function of latitude and time at an altitude 600 km

work is that the higher temperature regions were associated with very fast, field-aligned poleward flows of ions (e.g., 100 s m/s). They also report that the fraction of O^+ ions (i.e., $\gtrsim$ 80%) in BETs is similar to that of the background ionosphere. These exerimental observations are consistent with our simulation results.

15.4 Discussion

We have presented simulation results from a three-dimensional model of equatorial spread F. The model is referred to as SAMI3/ESF because it is based on the comprehensive NRL ionosphere model SAMI3. In this chapter we have highlighted the evolution of the atomic and molecular ions, as well as that of the ion and electron temperatures in the absence of a neutral wind. The key results are the following.

1. Low density atomic ions (H^+ and O^+) can be lifted to higher altitudes by the upward $E \times B$ drift generated by the instability.
2. High density molecular ions (NO^+ and ${O_2}^+$) can be lifted to higher altitudes by the upward $E \times B$ drift generated by the instability.
3. Low density plasma regions can form at high altitude (e.g., $\gtrsim$ 1000 km) because of plasma "falling" to lower altitudes along geomagnetic field lines. This is caused by a reduction of plasma pressure associated with the instability at mid-latitudes ($\gtrsim \pm 15°$) in the F layer.
4. The electron and ion temperatures decrease inside the plasma bubble because of adiabatic cooling associated with the rarfaction of the plasma.
5. The electron and ion temperatures increase in the altitude range ~400– 1000 km at latitudes $\sim \pm 15°$. This is caused by the compression of the ions as they stream down the converging magnetic field lines and collisional heat transfer from the ions to the electrons.
6. The electron temperature increases along the field lines in the upper portion of the plasma bubble (above apex altitudes $\sim$ 1400 km) because of electron heat conduction.

Additional modeling studies have been carried out not reported in this chapter. Krall et al. (2009b) investigated the impact of meridional neutral winds on ESF bubble evolution and found that these winds have a stabilizing influence on ESF. Huba et al. (2009c) studied the influence of zonal neutral winds on ESF bubbles and found that these winds distort (i.e., bend) the bubbles as a function of longitude and altitude. Krall et al. (2009a) performed an extensive analysis on the development of density *enhancements* associated with ESF bubbles. And finally, Krall et al. (2010) addressed the question: why do equatorial bubbles stop rising? They find that equatorial bubbles stop rising when the magnetic flux-tube integrated ion mass density inside the bubble equals that of the surrounding background ionosphere.

Finally, we recently completed the first global simulation study of equatorial spread F bubble evolution using a comprehensive, global 3D ionosphere model (Huba and Joyce, 2006). The NRL ionosphere model SAMI3 was modified to self-consistently model both the neutral wind dynamo electric field (e.g., pre-reversal enhancement) and the gravity-driven bubble electric field. This has been achieved by including a high resolution longitudinal grid in the pre- to post-sunset sector. Initial results indicate that ESF can be triggered by pre-sunset ionospheric density perturbations, and that an existing ESF bubble can trigger a new ESF bubble.

Acknowledgements This research has been supported by NASA and ONR.

References

Banks PM, Kockarts G (1973) Aeronomy. Academic, San Diego, CA

Fejer BG, Scherliess L (1995) Time-dependent response of equatorial ionospheric electric fields to magnetospheric disturbances. Geophys Res Lett 22(7):851

Haerendel G (1974) Theory of equatorial spread F, preprint, Max-Planck Inst. für Extraterr. Phys Munich, Germany

Hanson WB, Sanatani S (1973) Large Ne gradients below the equatorial F peak. J Geophys Res 78:1167

Huba JD, Joyce G (2006) Equatorial spread F modeling: multiple bifurcated structures, secondary instabilities, large density 'bite-outs,' and supersonice flows. Geophys Res Lett 34:L07105. doi:10.1029/2006GL028519

Huba JD, Joyce G, Fedder JA (2000a) SAMI2 (Sami2 is Another Model of the Ionosphere): a new low-latitude ionosphere model. J Geophys Res 105:23035

Huba JD, Joyce G, Fedder JA (2000b) The formation of an electron hole in the topside equatorial ionosphere. Geophys Res Lett 27:181

Huba JD, Joyce G, Krall J (2008) Three-dimensional equatorial spread F modeling. Geophys Res Lett 35:L10102. doi:10.1029/2008GL033509

Huba JD, Krall J, Joyce G (2009a) Atomic and molecular ion dynamics during equatorial spread F. Geophys Res Lett 36:L10106. doi:10.1029/2008GL033059

Huba JD, Krall J, Joyce G (2009b) Ion and electron temperature evolution during equatorial spread F. Geophys Res Lett 36:L15102. doi:10.1029/2009GL038872

Huba JD, Ossakow SL, Krall J, Joyce G, England SL (2009c) Three-dimensional equatorial spread F modeling: zonal neutral wind effects. Geophys Res Lett 36:L19106. doi:10.1029/2009GL040284

Huba JD, Joyce G (2010) Global modeling of equatorial plasma bubbles. Geophys Res. Lett. 37:L17104.doi:10.1029/2010GL044281

Hysell DL (2000) An overview and synthesis of plasma irregularities in equatorial spread F. J Atmos Solar-Terr Phys 62:1037

Krall J, Huba JD, Joyce G, Yokoyama T (2009a) Density enhancements associated with equatorial spread F. Ann Geophys

Krall J, Huba JD, Joyce G, Zalesak ST (2009b) Three-dimensional simulation of equatorial spread F with meridional wind effects. Ann Geophys 27:1821

Krall J, Huba JD, Ossakow SL, Joyce G (2010) Why do equatorial bubbles stop rising? submitted to Geophys Res Lett

Millward GH, Moffett RJ, Quegan W, Fuller-Rowell TJ (1996) A coupled thermospheric-ionospheric-plasmasphere model (CTIP). In: Schunk RW (ed) STEP: handbook of ionospheric models. Utah State University, Logan, UT, p 173

Ossakow SL (1981) Spread F theories: a review. J Atmos Solar-Terr Phys 43:437

Ossakow SL, Zalesak ST, McDonald BE, Chaturvedi PK (1979) Nonlinear spread F: dependence of altitude of F peak and bottomside background electron density gradient scale length. J Geophys Res 84:17

Oyama K-I, Schlegel K, Watanabe S (1988) Temperature structure of the plasma bubbles in the low latitude ionosphere around 600 km altitude. Planet Space Sci 36:553

Park J, Min KW, Kim VP, Kil H, Su S-Y, Chao CK, Lee J-J (2008) Equatorial plasma bubbles with enhanced ion and electron temperatures. J Geophys Res 113:A09318. doi:10.1029/2008JA013067

Scannapieco AJ, Ossakow SL (1976) Nonlinear equatorial spread F. Geophys Res Lett 3:451

Sekar R (2003) Plasma instabilities and their simulations in the equatorial F region – recent results. Space Sci Rev 107:251

Zalesak ST, Ossakow SL (1980) Nonlinear equatorial spread F: spatially large bubbles resulting from large horizontal scale initial perturbations. J Geophys Res 85:2131

Zalesak ST, Ossakow SL, Chaturvedi PK (1982) Nonlinear equatorial spread F: the effect of neutral winds and background conductivity. J Geophys Res 87:151

Chapter 16

Coupling Processes in the Equatorial Spread F/Plasma Bubble Irregularity Development

Mangalathayil Ali Abdu and E. Alam Kherani

Abstract The plasma convection pattern of the evening sector equatorial ionosphere sets the condition for the plasma structuring through instability processes leading to the Equatorial Spread F (ESF)/plasma bubble irregularity development and evolution. Vertical coupling through upward propagating atmospheric waves controls/modifies the ionosphere-thermosphere interactive processes that eventually lead to the irregularity development. Instabilities grow by the Rayleigh-Taylor mechanism at the bottom side gradient region of a "rapidly" rising post sunset F layer in the presence of precursor conditions in terms of perturbations in plasma density, convection velocity and polarization electric fields. Field line integrated conductivity controlled by thermospheric meridional/trans-equatorial winds regulates the instability growth. The day-to-day and short term variabilities in the ESF are of major concern for space application and operational systems. Our efforts to understand such variabilities and to predict the ESF occurrence pose important scientific challenges especially because of the complexity of the diverse coupling processes that control them. There is convincing new evidences that during magnetically quiet conditions, the coupling processes due to upward propagating planetary waves and/or modulated tides, and gravity waves, with their highly variable propagation conditions, energy fluxes and periodicities control the ESF variability. Penetrating electric fields and disturbance dynamo electric fields from magnetosphere-ionosphere coupling processes also cause large degree of variability. This chapter provides a review of our current understanding of the ESF development processes and its day-to-day variability originating from the different coupling processes mentioned above.

16.1 Introduction

The plasma structuring of the nighttime Equatorial ionosphere, widely known as Equatorial Spread F (ESF) irregularities, continues to pose challenging problems for scientific investigations and for a variety of operational space application systems. Discovered as range spreading echoes in the post sunset ionograms by Booker and Wells (1938) the ESF irregularities have been a subject of intensive investigations during the last few decades. As a result we now have a reasonably clear picture of their spatial and temporal distributions over the globe in wide ranging scales as well as an in-depth understanding of the processes and mechanisms that operate in their development (see for example, Farley et al., 1970; Woodman and La Hoz, 1976; Sobral et al., 1980; Rastogi, 1980; Abdu et al., 1981, 1983; Huang and Kelley, 1996; Sultan, 1996; Fejer et al., 1999; Sastri et al., 1997; Kil and Heelis, 1998; Hysell and Burcham, 1998; McClure et al., 1998; Haerendel et al., 1973; Abdu, 2001; Bhattacharyya, 2004; Fukao et al., 2006; Tsunoda, 2010, and references therein). However, we do not yet have sufficient understanding of the nature of the large degree of variability in the dynamical conditions of the background ionosphere-thermosphere system in which the ESF irregularities develop, presenting variabilities at different temporal and spatial scales. In this chapter we will

M.A. Abdu (✉)
National Institute for Space Research, São Jose dos Campos, SP, Brazil
e-mail: maabdu@dae.inpe.br

M.A. Abdu, D. Pancheva (eds.), A. Bhattacharyya (Coed.), *Aeronomy of the Earth's Atmosphere and Ionosphere*, IAGA Special Sopron Book Series 2, DOI 10.1007/978-94-007-0326-1_16,

address the ESF temporal variability, focusing on its short term regime and on the atmosphere-ionosphere-thermosphere parameters that play key roles in the coupling processes that control such variability.

The post sunset ESF occurs in sequence to a rapid rise of the F layer (by vertical plasma drift) under the action of the evening prereversal enhancement in zonal electric field (hereafter denoted as the PRE). The F layer dynamo intensification at sunset is responsible for PRE development which begins with the start of sunset at conjugate E layers that are coupled to the equatorial F layer (Abdu et al., 1981). The E-layer conductivity local time/longitude gradient and the thermospheric zonal wind (eastward in the evening) play the major roles in the electrodynamic coupling processes leading to the PRE development (Rishbeth, 1971; Heelis, et al., 1974; Farley et al., 1986; Batista et al., 1986). The steep bottom-side density gradient of the rising F layer becomes unstable to density perturbations leading to instability growth by the Rayleigh-Taylor (R-T) mechanism by which the rarified plasma of the F layer bottom side rises up to the topside ionosphere in the form of flux tube aligned plasma depletions/plasma bubbles in scale sizes of tens to hundreds of kilometers. With the non linear growth of the bubble, secondary instabilities take over in cascading process leading to irregularity structures, also magnetic field aligned, down to meter sizes (Haerendel, 1973), the entire process constituting an ESF development. VHF radar diagnostics have shown that the irregularities develop close to or before the terminator passage at the field line apex in the F region (Yokoyama et al., 2004; Fukao et al., 2006) and the structures generally move eastward.

The ESF phenomenon presents large degree of variability that can be broadly classified as: (1) long time scale variation dependent on solar activity cycle, (2) medium term variation at monthly and seasonal scales, and (3) day-to-day and shorter term variability. The long time scale variation is understood to be basically driven by a corresponding variation in the PRE that is controlled by the solar EUV flux, represented by the proxy index F10.7 (see for example, Abdu et al., 1985; Fejer et al., 1991, 1999). The medium term variations arise from a longitude dependent seasonal effects which is controlled to a first degree by the magnetic declination angle that varies with longitude (Abdu et al., 1981; Maruyama and Matuura, 1984; Tsunoda, 1985), as well as by the solar declination angle that varies with season. Abdu et al. (1981) showed that the degree of simultaneity of the sunset at the conjugate E layers (connected to the equatorial F layer bottom side), that is, the degree of alignment of the sunset terminator with the magnetic meridian, controls the intensity of the PRE to the first order so that the near alignment condition produces larger PRE intensity (Abdu et al., 1981, 1992; Batista et al., 1986). Since the ESF is largely controlled by the PRE a longitude dependent seasonal variation that is present in PRE is also present in ESF. Tropospheric convective activity, through its generation of upward propagating gravity waves, has been suggested to be a potential source of seed perturbations for the ESF instability growth based on statistical data base (McClure et al., 1998; Rottger, 1981; Tsunoda, 2010). Results from case studies based on multi-instrument data base obtained from a Spread F experimental campaign in Brazil have provided evidence on the precursor role of tropospheric gravity waves in the development of the ESF (Fritts et al., 2008; Abdu et al., 2009b; Kherani et al., 2009). Therefore the season dependent longitudinal distribution of tropospheric convective activity, reckoned in terms of the proximity to the dip equator of the inter tropical convergence zone (ITCZ), could also contribute to the longitude dependent seasonal variation of the ESF (McClure et al., 1998; Tsunoda, 2010). This component of the ESF seasonal variation is not sufficiently well understood, however. A better understanding of this aspect is essential for a more precise characterization of the seasonal and month-to-month variation in the ESF. The role of tropospheric convective activity, as a source of gravity waves for instability seeding needs to be understood better also for an assessment of the causes of the ESF day-to-day and short period variability which remains to be the least tackled and therefore the most challenging scientific problem to be addressed in this field. This class of ESF variability is governed by different types of coupling processes such as: (1) vertical coupling by upward propagating waves with associated neutral atmosphere-ionosphere interactive processes; and (2) and ionosphere-magnetosphere coupling through penetrating/disturbance electric fields. The manifestations of these coupling processes can be identified and analyzed from direct or indirect observational data on the following set of ionospheric and thermospheric parameters:

1. The F region plasma vertical drift/zonal electric field in the evening hours, that is, the evening pre-reversal enhancement in the vertical drift arising from the F layer dynamo which is controlled by the E layer conductivity local time gradient at sunset and the thermospheric zonal wind (eastward in the evening);
2. Thermospheric meridional/trans-equatorial winds that control the field line integrated conductivity of the potentially unstable flux tubes;
3. Perturbations in density and vertical drift/zonal electric field as precursor seed for the initiation of the ESF instability.

The interconnection among these parameters and the underlying coupling processes are represented in the schematic of Fig. 16.1. Although the ESF development is directly controlled by all the three factors listed above, in the way depicted in this figure, the control by the prereversal vertical drift enhancement is by far the most extensively investigated and therefore better known aspect of the ESF variability (Abdu et al., 1983, 2009; Jayachandran et al., 1993; Fejer et al., 1999; Chapagain et al., 2009). The evening F layer uplift due to the PRE is known to be a basic requirement for the spread F instability initiation at the bottom-side of the layer. The collisional interchange instability (CII) driven by the Rayleigh-Taylor mechanism is believed to be responsible for the generation of the bubble irregularity structures. The governing equation for the growth of the polarization electric field associated with the CII is given by:

$$\partial \delta E_x / \partial t - \gamma_R \delta E_x = S_x \qquad (16.1)$$

where, $\gamma_R = \left(-E_{0x}/B_0 - W_y + g/\nu_i\right)/l_0$; $S_x \approx B_0 \Delta U_{0x}/l_0 \left(\delta W_y/\kappa_i - \delta W_x\right)$; $1/l_0 = d \log n_0/dy$; γ_R is the linear growth rate for the polarization electric field; E_{0x} is the prereversal zonal electric field (positive westward); ΔU_{0x} is differential ion-neutral zonal velocity; W_y is the vertical neutral wind; $\delta W_{x,y}$ are the gravity wave associated winds; n_0 is the electron density, κ_i is the ratio of the ion gyro frequency, Ω_i , to ion-neutral collision frequency, ν_{in} (for more details, see Kherani et al., 2009; Abdu et al., 2009b). In the expression for γ_R the term $-E_{0x}/B_0$ represents the contribution from the PRE which has been (as stated above) by far the most widely investigated aspect in the ESF variability. Some new results on the ESF and PRE relationship as obtained from recent investigations, including a conjugate point equatorial experiment, will be discussed in Section 16.2. Figure 16.1 depicts the PRE as arising from the sunset electrodynamic processes under the combined effects of the E layer conductivity longitude gradient (designated as SS. $\Delta\Sigma$) and the thermospheric zonal wind (eastward in the evening) as explained by Rishbeth (1971). The presence of sporadic E layers (E_s layer) near the terminator could modify the E layer conductivity distribution and hence the intensity of the PRE. However, the PRE development could control the very Es layer formation as indicated in the figure (for details of this mechanism see Abdu et al., 2003b; Carrasco et al., 2007). In this way there exists an interactive connection between the PRE and E_s layer which will also be discussed in Section 16.2. The E layer winds, responsible for the dynamo electric field, basically driven by tidal modes can be modified by the upward propagating Planetary Waves (PWs) and the resulting modification of the E layer conductivity

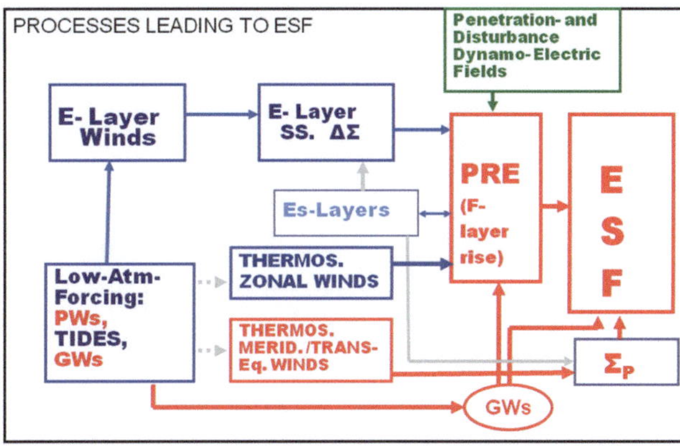

Fig. 16.1 A scheme of the coupling processes leading to ESF/plasma bubble development

longitudinal gradient in the sunset sector appears to be an important source of variability in PRE and hence ESF (Abdu et al., 2006c) which will be discussed in Section 16.3. The instability linear growth rate γ_R may lead to vertically rising flux tube aligned plasma depletion which is a nonlinear process and controlled by the integrated Pederson conductivity (Σ_P) of the unstable flux tube. The Σ_P can be enhanced by the action of meridional/trans-equatorial winds that can cause asymmetry in latitudinal distribution of the EIA (equatorial ionization anomaly) plasma. The connection between the thermospheric meridional/ trans-equatorial wind and ESF through the Σ_P is also indicated in the figure and it will be discussed in Section 16.4. Precursor signatures in the F layer bottom side in the form of perturbation in electron density or polarization electric field produced by GWs, possibly originating form tropospheric sources, can initiate the ESF instability process as also indicated in the figure. This will be discussed in Section 16.5. All these aspects of the ESF development arise from vertical coupling processes that dominate geomagnetic quiet conditions. Under geomagnetically disturbed conditions forcing from magnetosphere in the form of disturbance electric fields can drastically modify the scenario of ESF development, thereby contributing to its day-to-day variability. This aspect will be discussed in Section 16.6. A final discussion on these various aspects and some conclusions will be made in Section 16.7

16.2 The Evening Prereversal Vertical Drift and the ESF

Ionosphere–thermosphere interaction and the E- and F-region electrical and electrodynamic coupling processes drive the F layer dynamo primarily responsible for the generation of the evening enhancement in the zonal electric field (that is, the PRE). The zonal electric field contributes to the linear growth rate γ_R in two ways: (a) through the vertical drift term and (b) by raising the layer to heights where the reduced collision frequency increases the g/ν_i term (Eq. (16.1)). A larger γ_R is helpful for a more intense nonlinear growth of the polarization electric field of the CII so that it is to be expected that a more intense PRE could cause (a) higher vertical rise velocity of the bubble structure over the equator and (b) a more intense ESF structure. Through field line mapping the bubble rise velocity over the dip equator determines the onset time of the ESF at latitudes farther away from the dip equator, a higher rise velocity causing an earlier onset. This relationship is evident in the results from a conjugate point equatorial experiment (COPEX, Abdu et al., 2009a) presented in Fig. 16.2. In the four panels of this figure the mean vertical drift values (V_Z) over the equatorial station Cachimbo (9.47°S, 54.83°W, dip latitude: −2°) are shown for four groups of days sorted out on

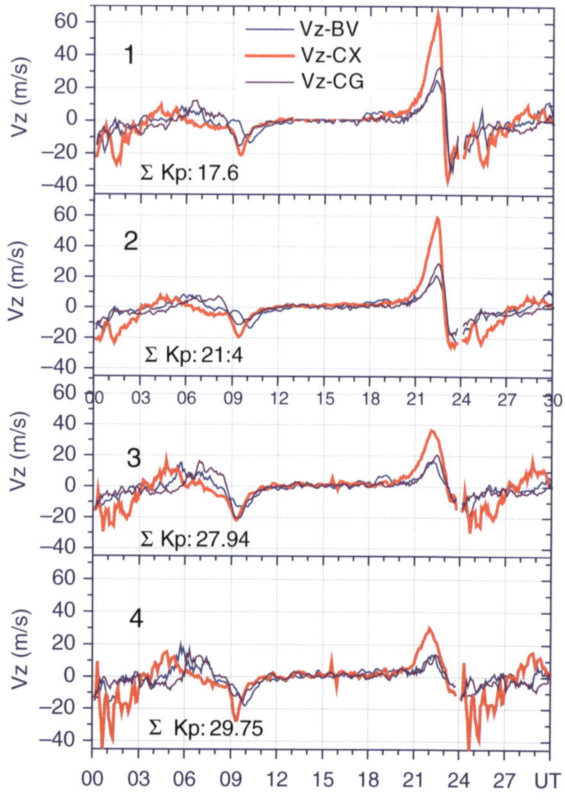

Fig. 16.2 Mean vertical drift velocities over the three COPEX sites are plotted for different groups of days: Panel (1) – for the cases of early spread F occurrence (at or before 20 LT) over Cachoeira Paulista; Panel (2) – for the group of days when SF occurred at and before 22 LT over CP; Panel (3) – for the group of days when SF did not occur before midnight over Cachoeira Paulista; and Panel (4) – for the group of days when SF did not occur over Campo Grande and CP. (ΣKp: the daily sum of the 3-hourly Kp values representing the mean value for the days considered). Station names are BV; Boa Vista; CX: Cachimbo; and CG: Campo Grande. In the x-axis the LT exceeding 24 h corresponds to a new day so that 24 should be subtracted from those values (Abdu et al., 2009b). (Reproduced with permission of the American Geophysical Union)

considerations of systematic increase in onset time of SF at locations (latitudes) farther away from the dip equator. These locations are: Campo Grande (−20.44°; 54.64°W, dip latitude: −11°) and Cachoeira Paulista (22.6°S, 45°W, dip latitude: −16°). (See also Fig. 1b of Abdu et al., 2009a). The average ΣKp values corresponding to the sorted group of days are also shown in these panels. We may point out that in the evening and early night hours (and whenever the F layer height is > 300 km, in general) the V_z values, calculated here as dh_F/dt from ionogram true heights (h_F), do represent realistic vertical drift (Bittencourt and Abdu, 1981). Based on the flux tube aligned bubble development the spread F occurrence at early post sunset local times (before 20 LT) over Cachoeira Paulista corresponds to a rather fast vertical bubble rise velocities (>100 m/s) over the equator. For this case the mean V_Z peak values (V_{zp}) over Cachimbo (near magnetic equator) in panel 1 is ~62 m/s. For a second group of SF events occurring at and before 22 LT over CP the mean V_{zp} is only slightly smaller, being 59 m/s. The results in panel 3 show the case of SF not occurring till midnight over CP, which signifies that the vertical bubble growth over the equator did not attain till midnight an apex altitude of 900 km that is field line mapped to the bottomside F region over CP. The corresponding bubble rise velocity did not exceed 40 m/s. The V_{zp} over Cachimbo for this case is around 35 m/s. The field line apex height for Campo Grande bottom-side F region is ~650 km. For the cases of post sunset spread F not occurring over Campo Grande the mean V_{zp} in panel 4 is 30 m/s which suggests that bubble vertical growth to ~650 km could be possible when the prereversal vertical drift is higher than this value. For the cases of non occurrence of SF over the equator the mean V_{zp} (not shown here) was just 22 m/s, which can be considered to be the threshold minimum below which bottom-side spread F did not occur over the equator for this observational interval. We note thus that in statistical terms the average V_{zp} presents a trend suggesting its systematic control on the spread F development, with the lowest V_{zp} (22 m/s) for the case of SF totally absent to the highest V_{zp} (~60 m/s) for a rapid vertical growth of the bubble (with rise velocity >100 m/s) over the equator. It is relevant to mention here that threshold conditions on the evening prereversal vertical drift for UHF and L-band scintillation occurrence in the Peruvian sector have been discussed by Anderson et al. (2004). In Fig. 16.2 are shown also the V_Z variations over the conjugate sites of Cachimbo (Campo Grande –20.44°, 54.64°W; and Boa Vista 02.81°N, 60.66°W, dip latitude: −11°) where the V_{zp} values are significantly smaller than over equator with a small asymmetry suggesting the presence of a generally weak trans-equatorial wind in the F layer bottom-side, which indicate that the control of the bubble rise velocity by the PRE intensity shown above is largely unaffected by trans-equatorial winds.

The degree to which the PRE controls the spread F intensity is illustrated in Fig. 16.3. These results are based on Jicamarca radar observations and show the variation of the PRE vertical drift as a function of the solar flux index (F10.7) for different degrees of SF intensity (Chapagain et al., 2009). For the case of weak spread F (generally confined to the bottom-side) as also for the plume type spread F (that extends to the topside), the vertical drift at the SF onset time, as also the PRE amplitude, increases with solar flux. Here the SF intensity is measured in terms of the height extension of the radar echoes, and it is seen that for any given solar flux value the vertical drift at the SF onset time as well as the PRE amplitude is higher for plume type SF than for weak SF (which is similar to the results of Fig. 16.2). They also present some degree of seasonal dependence. There is however, a large degree of scatter in the relationship which points to the presence of other important coupling processes operating in the ESF variability which we will be discussing below.

16.2.1 The Role of E_s Layers

Sporadic E layers present in the sunset sector can influence the development of PRE depending upon the degree to which the longitudinal gradient in the E layer conductivity (across the terminator) is modified by the conductivity distribution in the E_s layer. Assuming this to be uniform in the area of interest it could reduce the value of SS. ΔΣ resulting in a reduction in the intensity of the PRE (Carrasco et al., 2005) which develops from the E-F region electrical coupling processes mentioned before. A reduced PRE could lead to a reduced SF intensity. Further, an E_s layer, present at the feet of a flux tube of instability growth by R-T mechanism could cause, through its contribution to the flux tube integrated conductivity, a reduction of the instability

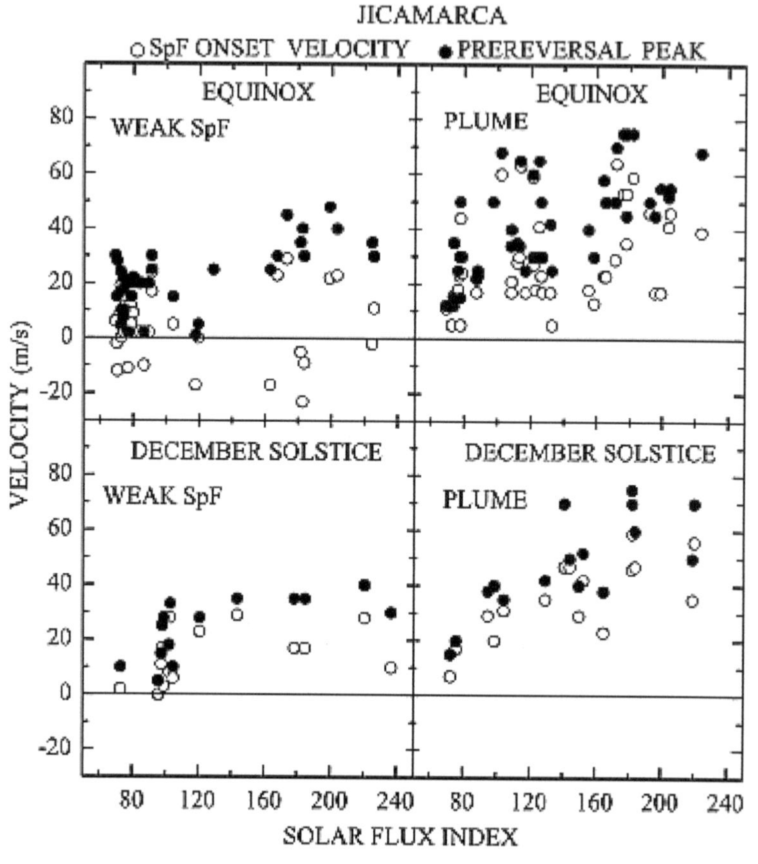

Fig. 16.3 Scatter plots of weak spread F and plume onset velocities and corresponding prereversal drift peaks as a function of solar flux index (Chapagain et al., 2009). (Reproduced with permission of the American Geophysical Union)

growth rate thereby inhibiting an SF development. However, a negative relationship between the Es layer and the SF, expected on these bases, has not been observed (see for example, Batista et al., 2008). This is because the role of Es layer in the SF generation process through the PRE development is a very complex and interactive one. Abdu et al. (2003b) showed from analysis of digisonde data over Fortaleza (3.9°S, 38.45°W, dip angle: −9°) , Brazil, that the electric field vertical structure associated with the PRE can exercise significant influence on the Es layer formation at these hours. It was shown that Es layer formation can be disrupted in the vicinity of the dip equator within about ± 12° while it can be enhanced at latitudes farther away within the low latitude limit of the F layer dynamo influence. In the post sunset plasma flow vortex pattern that describes the PRE the electric field is upward in the F layer bottom side where the plasma flow is westward and it is downward at higher up altitudes where the plasma flow is eastward, with the zero electric field at the center of the vortex flow, occurring near 300 km at around 19 LT (Kudeki and Bhattacharyya, 1999). It is near this local time that Es layer disruption/enhancement usually takes place. It was shown based on model calculation (Carrasco et al., 2007; Abdu et al., 2003b) that an upward directed vertical electric field is efficient in disrupting the ion convergence required for E_s layer formation at these hours, (whereas a downward electric field is expected to produce an opposite effect). Since the magnetic field line apex at 300 km (vortex center) maps to conjugate E layers at ± 12° latitudes the observed E_s layer disruption (enhancement) within (outside) this latitude limit being associated with the PRE development can be seen as arising from the vertical electric field structure of the post sunset plasma flow vortex (see also, Abdu and Brum, 2009). This aspect of the coupling process in the SF development needs to be investigated further for a better understanding and evaluation of the role of Es layers in the ESF variability.

16.3 Planetary Wave Effects on PRE and ESF

Atmosphere–ionosphere coupling driven by upward propagating waves (vertical coupling) is known to be an important source of variability in the ionospheric parameters. The planetary waves (PWs), in particular, have been observed to play a significant role in the widely known day-to-day variability of the atmosphere and ionosphere of the equatorial region. PWs of quasi two-day and 3- to 5-day periodicities in equatorial mesospheric winds were reported by Gurubaran et al. (2001) and Vincent (1993), respectively, and in mesospheric airglow intensity by Takahashi et al. (2005). PWs oscillations in different periodicities have been identified also in the equatorial electrojet current and ionization anomaly (Forbes and Leveroni, 1992; Abdu et al., 2006c) equatorial mesopause temperature and EEJ strength (Vineeth et al., 2007), and equatorial F layer height and vertical plasma drift (Abdu et al., 2006c; Takahashi et al., 2005; Fagundes et al., 2005). Our interest here is on the PWs modulation of the sunset electrodynamics processes leading to the development of the PRE and the post sunset F layer heights increase that basically control the spread F/plasma bubble irregularity generation and its day to day variability. Some evidence for mesosphere–thermopshere coupling through simultaneous presence of PW oscillations in mesospheric winds and in the prereversal F layer vertical drift velocity is shown in Fig. 16.4 (Abdu et al., 2006a). The top panel shows the peak evening F layer vertical drift (V_{zp}) over the equatorial and low latitude sites, Cachimbo and Campo Grande, respectively, in Brazil, as obtained during an observational Campaign (COPEX, Abdu et al., 2009a) in October–December 2002. Their respective wavelet power spectra are shown in the lower two panels. The lower most four panels present the wavelet power spectra of mesospheric zonal and meridional winds at 95 and 100 km as measured by a meteor radar at Cachoeira Paulista (22.6°S, 45°W; dip angle: –32°). We may note oscillation periods in the power spectra, ranging from ~3 to 7 days, during the interval from day 305 to day 320, that are common in the V_{zp} and mesospheric winds. None of these periodicities in the V_{zp} or in the mesospheric winds are related to magnetic or solar flux indices as was verified from wavelet power spectral analysis of theses indices (not shown here, but see Abdu et al., 2006a). Therefore the day-to-day variations in the evening prereversal electric field do seem to contain a significant contribution from planetary waves. A wave decomposition analysis for this case showed dominance of 4- to 5-day and 7-day periods in both the V_{zp} and mesospheric winds that presented downward phase propagation similar to that observed by Pancheva et al. (2003) from analysis of sporadic E layer features and mesospheric winds.

Takahashi et al. (2007) showed that Ultra Fast Kelvin (UFK) waves trapped in equatorial latitude propagating from troposphere to ionospheric heights could cause significant modulation in the mesospheric winds and post sunset F region parameters, h'F and foF2. Their findings are based on the results presented in Fig. 16.5 that show the band-pass filtered wave oscillations of 3.5-day period in foF2 and h'F over Fortaleza and in the zonal wind at 90 km over Cariri (7.4°S, 36.5°W). The ionospheric data corresponded to 21 LT whereas the zonal winds are hourly mean values. The dominant oscillations in all these parameters that can be noted around day 66 presented downward phase propagation with vertical wavelength of ~40 km. From the propagation characteristics of these waves as diagnosed from measurements at widely separated longitudes and latitude as described in their paper Takahashi et al. confirmed the UFK wave characteristics of these oscillations. It should be noted that the post sunset (21 LT) F layer height is an integrated effect of the prereversal vertical drift and therefore the connection between the 21 LT F layer height and the mesospheric winds in the results by Takahashi et al. (2007) is a corroboration of the connection between the PRE and mesospheric winds found by Abdu et al. (2006a).

These results demonstrate the existence of a strong vertical coupling through upward propagating planetary waves that results in the day-to-day oscillations in mesospheric winds, the PRE and post sunset F region parameters. Since the PWS are not known to propagate to heights beyond the dynamo region, that is the lower thermosphere, the connection between the oscillations in the mesospheric and F region parameters needs to be understood in terms of a two-stage process, that is, the dynamic coupling by the PWs propagating till the E layer heights to be followed by electrodynamic coupling between the E and F layers. As explained by Abdu et al. (2006a) the planetary wave modulation of E layer tidal winds can modify

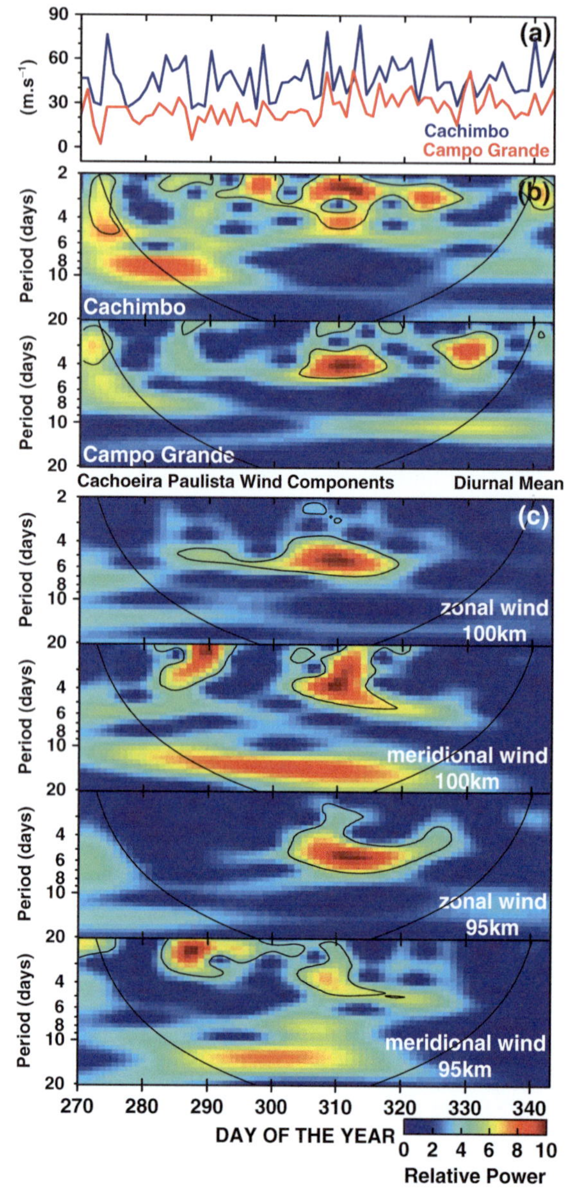

Fig. 16.4 (**a**) Vertical drift velocity (V_{zp}) variations over CX and CG for the period October–December 2002; (**b**) Morlet wavelet power spectral distribution of V_{zp} oscillations over the two stations; (**c**) wavelet spectral distribution of the daily means of mesospheric zonal and meridional winds at 100 km over Cachoeira Paulista (the *top* two plots) and at 95 km (the *bottom* two plots) (Abdu et al., 2006b). (Reproduced with permission of the American Geophysical Union)

the E layer integrated Pedersen conductivity longitudinal/local time gradient, that is present across the sunset terminator, which in turn can modulate the intensity of the PRE.

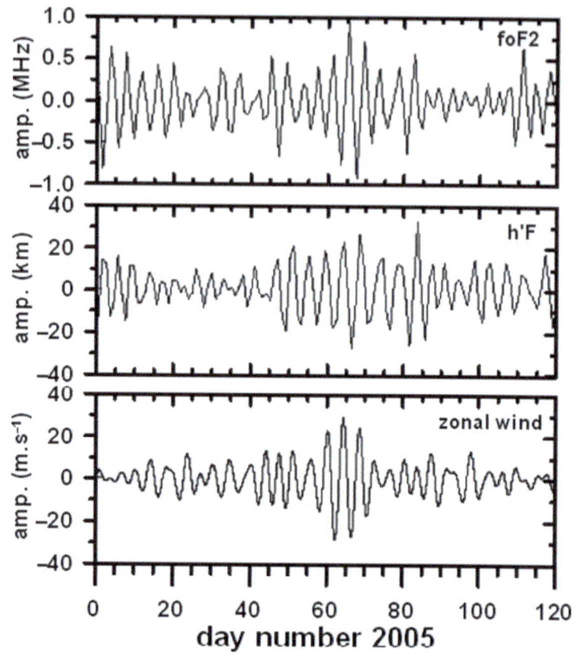

Fig. 16.5 Amplitude and phase of the 4±1-day filtered oscillations of Fortaleza *foF*2 (*top*), *h_F* (*middle*), and Cariri zonal wind at 90 m (*bottom*) during the period from January 1 to April 30, 2005 (Takahashi et al., 2007)

16.4 Thermospheric Meridional/Trans-equatorial Winds and ESF Suppression

There are two main component of the meridional wind whose coupling with the ionospheric plasma could control the spread F irregularity development. These are: (1) trans-equatorial and meridional horizontal winds (2) vertical wind that appears to be part of an equatorward converging wind. A meridional/trans-equatorial wind is efficient for interhemispheric transport of F region plasma that produces north-south asymmetry in the EIA. As a result the field line integrated conductivity becomes enhanced contributing to a reduction in the instability linear growth rate as well as retarding the instability nonlinear growth to form topside bubble structures as was shown from model calculations by Maruyama (1988). The reduction in the linear growth rate by enhanced field line integrated conductivity can be verified by expressing the growth rate in terms of flux tube integrated parameters as given by (Sultan, 1996):

$$\gamma_{FT} = \frac{\Sigma_P^F}{\Sigma_P^E + \Sigma_P^F}\left(\frac{E}{B} - U_{FT}^P + \frac{g}{\nu_{eff}}\right)\frac{1}{L_{FT}} - \beta_{FT} \qquad (16.2)$$

Here $\Sigma_P^{E,F}$ is the field line integrated conductivities for the E and F-regions segments of a field line; U_{FT}^P is the conductivity weighted flux tube integrated vertical wind; β is the recombination loss rate; ν_{eff} is the effective ion-neutral collision frequency. the subscript FT in Eq. (16.2) stands for flux tube integrated quantity. A trans-equatorial wind causes an uplift of F region plasma of the upwind hemisphere and therefore a reduction in the integrated conductivity which happens to be smaller than the increase of the integrated conductivity on the downwind hemisphere. This is due to the unequal molecular ion distribution in the conjugate F region caused by the wind as explained by maruyama (1988). Thus there is a net increase in the field line integrated conductivity, $\Sigma_P^{E,F}$ and hence a decrease in linear growth rate as per the Eq. (16.2).

Kherani et al. (2005) have derived three dimension growth rate of CII at the equator as follows:

$$\gamma_R = \frac{1}{en(1+\zeta)}\left(J_\perp \cdot \nabla_\perp \log n + J_\parallel \cdot \nabla_\parallel \log n\right) \qquad (16.3)$$

$$\zeta = 1 + \frac{\sigma_{II}}{\sigma_P}\frac{k_{II}^2}{k_\perp^2}$$

where n is the ionospheric number density, $(k_\perp, k_\parallel)$ are the perpendicular and parallel (to the Earth's magnetic field) wave number associated with density or wind perturbations, $(\sigma_P, \sigma_\parallel)$ are Pederson and parallel conductivities in the Ionosphere, $(J_\perp, J_\parallel)$ are the perpendicular and parallel current densities. Kherani et al. (2005) have shown that the explicit 3D growth rate expression (16.3) is equivalent to field-line-integrated growth rate expression (16.2). In (16.3), $J_\parallel$ is mainly driven by a trans-equatorial wind (TEW) and in that case the second term in (16.3) is always negative since the parallel density gradient caused by a TEW is always opposite to the driven parallel current. This suggests a reduction in the growth of the CII in the presence of a TEW.

Observational verification of the ESF suppression by meridional/trans-equatorial winds has met with varying degree of success (see for example, Mendillo et al., 2001; Abdu, 2001). A more convincing evidence on such effect appears to be possible from observations using ionosondes latitudinally spaced (e. g., Abdu et al., 2006b), or operated at conjugate low latitude sites (Abdu et al., 2009a; Saito and Maruyama, 2006). As an example Fig. 16.6a shows the evening vertical drift (due to the PRE) over the three COPEX sites (Abdu et al., 2009a) on three selected days during October 2002. Over Cachimbo, close to the dip equator the prereversal drift amplitudes are nearly the same (64 m/s) on the three evenings. However, plasma bubble development occurred only on two of these evening (10 and 12 October). This can be verified from the fact that on both evenings the spread F traces developed simultaneously over the conjugate sites, Campo Grande (shown in Fig. 16.6b for 10 October) and Boa Vista (not shown here), in sequence to its first appearance over the equator, Cachimbo (not shown here). On both these evenings the amplitude of the PRE at the conjugate site was nearly equal/very similar indicating that the trans-equatorial wind was almost absent in this case. In contrast to this the PRE amplitude on the evening of 23 October was markedly different at the two conjugate sites thereby indicating the presence of significant intensity of trans-equatorial wind, which was estimated to be ~25 m/s northward. As a consequence spread F traces did not develop over the two sites as can be noted in Fig. 16.6b (lower half) shown for CG with same result for BV (not shown here) (see Abdu et al., 2009a, for more details). These results do suggest that the trans-equatorial/meridional winds contribute to suppress or retard the vertical growth of plasma bubbles.

The possibility of ESF suppression by meridional wind was investigated also by determining the magnitude of such wind from hmF2 data using the servo method proposed by Rishbeth et al. (1978). The SERVO method works on the premise that the F layer peak height, in the absence of any vertical transport by winds or electric field, occurs at a balance height, h_0, where the recombination and diffusion terms balance. The measured layer peak height, *hmF2*, could differ from h_0 depending upon the intensity of the meridional wind and the zonal dynamo electric field that cause vertical transport of the F region plasma. Therefore the magnetic meridional wind U_{mm} can be calculated from the knowledge of h_0 and *hmF2*. The U_{mm} was calculated for three conditions of the spread F (SF) occurrence: (1) absence of SF at SL (Sao Luis) (2.33 S, 44.2 W, dip angle: –0.5°) an equatorial site and CP (Cachoeira Palista) a low latitude site, (2) SF over SL only, indicating limited or no vertical growth of bubble, that is, dominantly bottom side SF, and (3) presence of SF at both SL and CP indicative of well developed bubbles. The result of the study showed that the largest U_{mm} in the evening hours occurred for the case (2) of very limited or no vertical growth of the bubble over equator even when the prereversal electric field/vertical drift was the same for the cases (2) and (3) (for details, see Abdu et al., 2006b). This showed that meridional wind can indeed suppress the bubble vertical development which is in agreement with the theoretical prediction by Maruyama (1988). If the U_{mm} considered in this discussion is part of a TEW the ESF suppression will be effective independent of its direction, otherwise it needs to be directed poleward to be effective in suppressing the ESF.

Using GPS network over Brazilian sector, Muella et al. (2010) have presented detailed scintillation and TEC maps in equator-low-latitude ionosphere. They have observed no scintillation for an asymmetric TEC distribution across the equator which was considered as an indication of the presence of TEW. Using expression (16.3) in (16.1), they have shown that the presence of TEW leads to the suppression of GPS scintillation or SF provided its magnitude is sufficiently larger than PRE. In physical terms the presence of a TEW causes an asymmetry in the equatorial ionization anomaly which is accompanied by increased field line integrated Pederson conductivity. More molecular ions (O_2^+ and NO^+) created on the downwind side of the field line results in enhanced Pederson conductivity which is not balanced by the conductivity decrease in the upwind side, thus leading to a net increase of the field line integrated Pederson conductivity that can slow down the nonlinear instability growth to form ESF bubbles (Maruyama, 1988).

The vertical component at equator of a meridional circulation wind could influence the instability linear growth rate as per the Eq. (16.2), a downward (upward) directed wind increasing (decreasing) the growth rate (see for example, Sekar et al., 1994). The

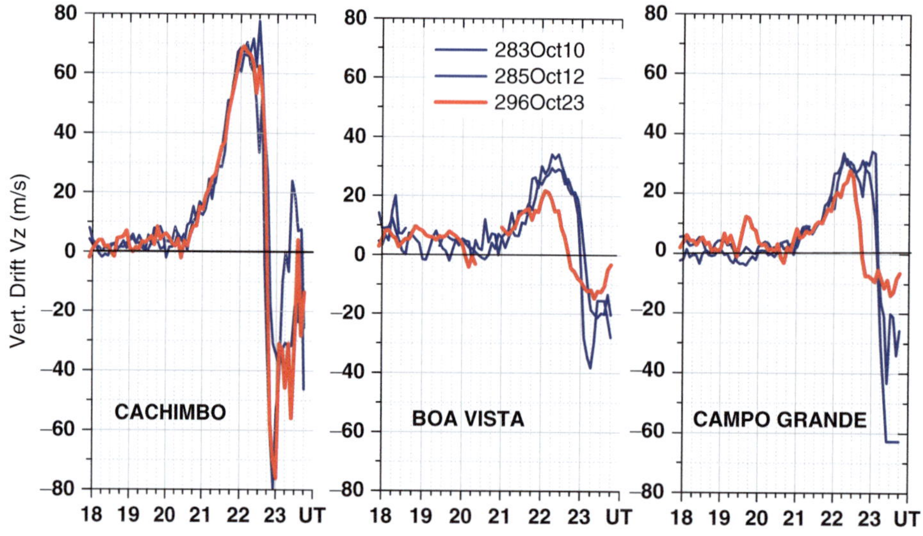

Fig. 16.6 (**a**) Vertical drift velocities over Cachimbo (*left panel*), Boa Vista (*middle panel*) and Campo Grande (*right panel*), during three quiet days October 10, 12 and 23 when they were nearly equal on all the three days during sunset time over Cachimbo, on 2 days (October 10 and 12) they were comparable/symmetric, while on one day (October 23) they were asymmetric, over the conjugate stations (Abdu et al., 2009b). (Reproduced with permission of the American Geophysical Union). (**b**) The set of 9 panels in the upper half of the figure shows ionograms at 5-min interval from, 22:35 UT (~19:00 LT) till 23:15 UT (~19:40 LT) on October 10, 2002. Development of rather intense spread F can be seen being initiated from the "off-vertical trace" in the 22:50 UT (19:10 LT) ionogram. Well developed spread F traces continued past the last ionogram presented here. The lower set of nine panels shows ionograms similar to above starting at 23:10 UT (19:30 LT) for October 23, 2002. Here no overhead spread F development is noted (Abdu et al., 2009b). (Reproduced with permission of the American Geophysical Union)

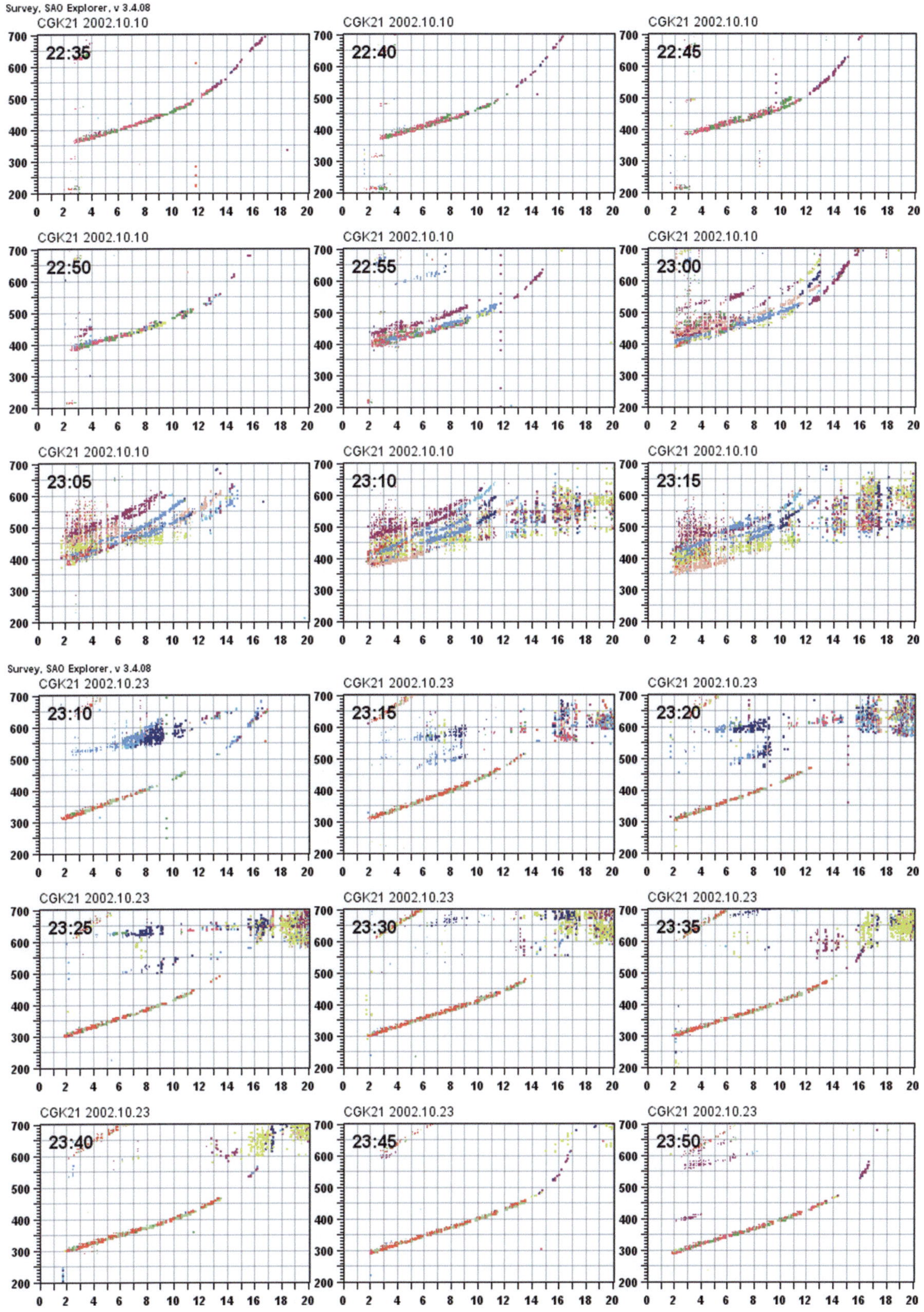

Fig. 16.6 (continued)

possible influence of vertical wind on the development of bottomside SF has been discussed by Jyoti et al. (2004).

16.5 Gravity Wave Effects on ESF Development

Atmospheric gravity waves (GWs) present at the thermospheric and F region heights can originate from different sources. Of special interest here are those originating from tropospheric convective sources that dissipate energy and momentum in E- and F regions (Fritts et al., 2009; Vadas and Liu, 2009; Chapter 10, this volume). The vertical coupling process through gravity waves is believed to contribute to, or influence, the ESF development and its day-to-day variability in different ways that can be classified as follows:

(1) Large scale GWs can modulate the post dusk sector F layer heights through large undulations in electron density iso-lines whereby the elevated bottom-side density gradient regions become unstable to GRTI (generalized Rayleigh-Taylor instability) with large instability growth rates (e.g., Kelley et al., 1981);
(2) It is well-known that the background thermospheric zonal wind that turns eastward in the evening is the major driver of the PRE. A superposition of this background zonal wind with a perturbation zonal wind due to GWs can cause large increase or decrease of the PRE amplitude depending upon the GW phase that in turn can lead to enhanced or reduced CII growth rate factor (V_Z/l_0). The consequent rise or descent of the F layer could also enhance or reduce the gravitational RTI growth rate factor ($g/\nu_i l_0$). A GW imposed zonal perturbation wind in the post sunset hours could modify also the wind driven instability growth rate such as that discussed by Kudeki et al. (2007);
(3) GWs can seed the instability by providing perturbations in the bottom side electron density in which polarization electric field grows in such a way that a further growth of a density depletion contributes to intensify the causative polarization electric field through a cycle of positive feedback, constituting the instability growth process.

We will not be discussing in this paper the GW effects defined in Items 1 and 2. Instead we will focus on the role of GWs in the direct seeding of the instability as defined in Item 3. In order to examine any possible role of GWs in the initiation of post sunset spread F/bubble irregularities it is first necessary to look for precursor signatures in the form of perturbation in the F layer density or iso-density heights during the evening hours preceding the spread F onset. We know that the PRE is the most basic requirement for the post sunset spread F to develop in such a way that its intensity needs to attain a minimum threshold limit for the instability to grow. For a given back-ground ionosphere-thermosphere conditions that threshold limit could be dependent on the intensity of the GW seed perturbation (along with other factors mentioned earlier). We will present below a case study showing the complementary roles of the GW precursor and the PRE in the instability growth leading to spread F development. The study is based on an analysis of radar and digisonde data from a recent SpreadFEx Campaign conducted in Brazil (Fritts et al., 2008; Abdu et al., 2009b). Figure 16.7a shows for 23 October 2005 over Sao Luis the F layer true heights at the plasma frequencies 5, 6, 7 and 8 MHz (obtained from digisonde SAO explorer), the plasma vertical drift, the F layer height oscillations for periods less than ~3 h (over Fortaleza), and finally the CII (Collisional Interchange Instability) growth rate factor (V_Z/l_0 Calculated from the upper two panels). The top panel is the RTI map of 5-m plasma structures, from the 30 MHz radar at Sao Luis. We may note that the irregularity plumes (indicating topside bubble structure) developed within ~30 min after the peak in the V_Z which is consistent with the peak growth rate factor of $5 \times 10^{-3} s^{-1}$ for this case. We further note that the precursor perturbations in the dhF parameter at 5–8 MHz do not clearly indicate the presence of significant gravity waves (although some estimation of the GW parameters can be obtained from them). This contrasts with the results of 24 October (Fig. 16.7b) that clearly indicate the presence of significant GWs as can be verified from the downward phase propagation in the height (dhF) oscillations around the sunset hours. On this evening topside bubbles did not develop promptly which can be attributed to the weaker PRE and lower F layer heights on this evening (see also, Abdu et al., 2009b). The radar plume that developed after 2220 UT (1920 LT) was mostly confined to bottomside. A simulation of the

CII growth by the Rayleigh-Taylor (R-T) mechanism was performed that took into account the polarization electric field development from gravity wave winds together with the background ionospheric conditions, such as the prereversal zonal electric field, the F layer heights and bottomside density gradient, to explain these results (Abdu et al., 2009b; Kherani et al, 2009). The simulation that used the back ground ionospheric conditions of 23 October together with the estimated GW winds produced an instability polarization electric linear growth that attained exponential rate within a short time (<30 min from the instability initiation) which is consistent with the topside plume structure that promptly occurred on this evening. When the same simulation was done for the background conditions of 24 October, but using the GW winds as on 23 October, the polarization electric field did not begin to grow till after one hour. However, when the GW winds of 24 October (which was around 3 times larger than that of the 23 October) was used in the calculation the growth rate attained significant values (but still at a slower rate than on 23 October), consistent with the limited plume growth that was observed. Thus this result clearly demonstrated the role of GW winds in seeding the Spread F instability process. A few other cases analyzed also confirmed these results. It should be pointed out that the amplitude of the PRE in these cases were relatively smaller (< 25 ms^{-1}) than the usual

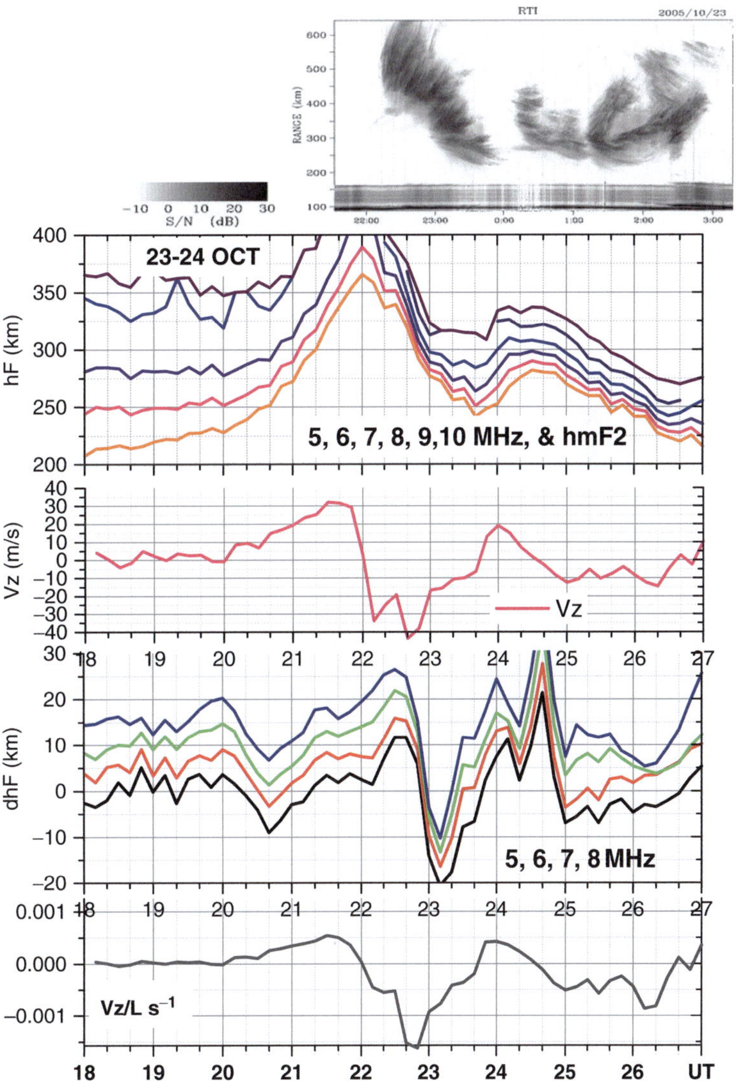

Fig. 16.7 (**a**) Plots for the 23–24 October 2005. RTI map of 5-m irregularity distribution as observed by the 30 MHz radar over Sao Luis (*top panel*); F layer true heights at plasma frequencies 5, 6, 7 and 8 MHz (panel 2 from *top*); Mean of the drift velocities calculated as d(hF)/dt at frequencies 5, 6, 7, and 8 MHz (panel 3 from *top*); Band-pass filtered (for 20-min to 3-h periods) height oscillations at frequencies 5, 6, 7, and 8 MHz (panel 4 from *top*); instability growth rate due to the term V z/L of the instability linear growth rate factor (*bottom panel*) (Abdu et al., 2009a) (**b**) Plots similar to those of Fig. 16.2a, but for the evening-night period of 24–25 October 2005

Fig. 16.7 (continued)

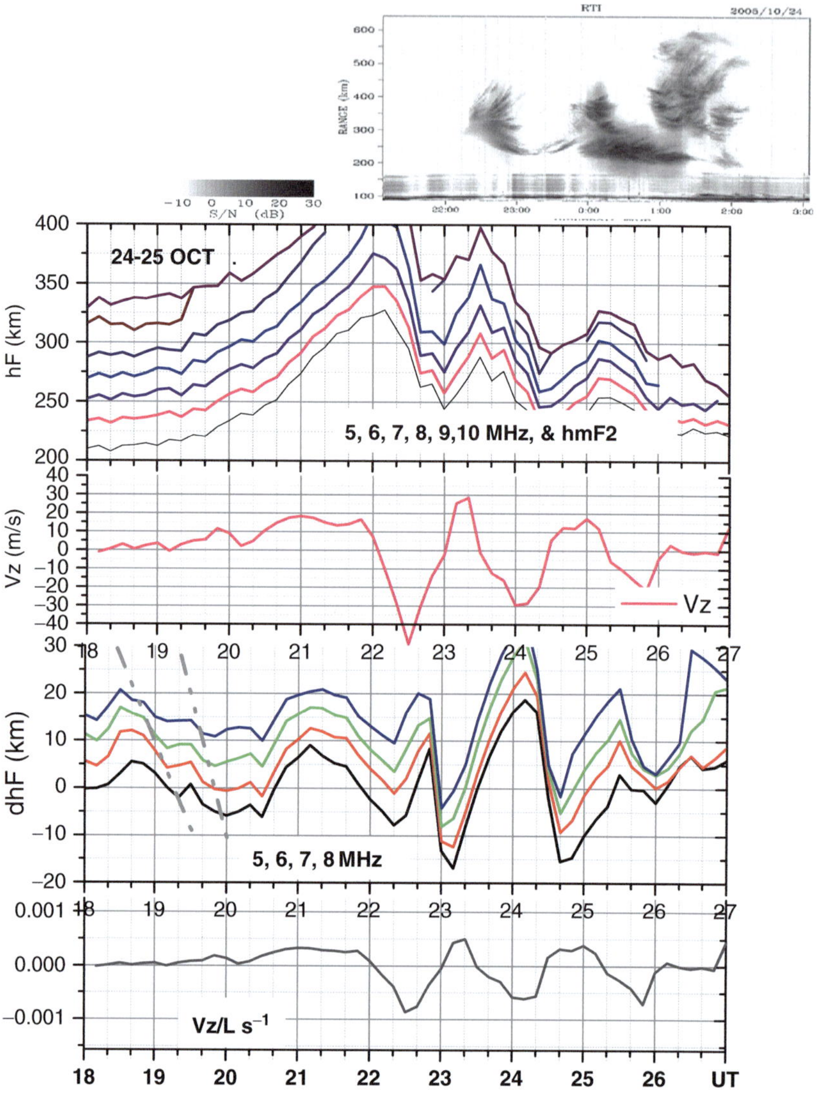

statistical threshold limit for fully developed plasma bubbles which has been found to be ~35 ms^{-1}(in a study by Abdu et al., 2006b).

16.6 Disturbance Electric Fields and the ESF

During magnetically disturbed condition the PRE and hence the ESF can undergo large variations. For example, magnetic disturbances can cause their development even during seasons when they do not occur under quiet conditions. Sastri et al. (1997) and Abdu et al. (2003a) have shown that during June solstice months when ESF do not normally develop over Brazilian longitude the electric fields due to magnetic disturbances occurring near the sunset sector can cause its development (see also, Abdu et al., 1995). On the other hand, the PRE and ESF can be suppressed during the season of their normal development, when the penetration electric field has westward polarity in the evening (Abdu et al., 2009c). They can be suppressed also by the disturbance dynamo electric field that follows the main phase of a storm occurring a few to several hours before the sunset (for example, Scherliess and Fejer, 1997; Abdu et al., 1995). Some details of these processes are explained below.

Interplanetary and magnetospheric electric fields penetrate to the ionosphere during magnetic storms. During the development phase of a storm, under the Bz south condition and AE activity intensification a "dawn-dusk" electric field promptly penetrate to equatorial latitudes. This under-shielding electric field has eastward polarity and large intensity in the dusk sector equatorial ionosphere (Fejer et al., 2008; Richmond et al., 2003). Correspondingly with the weakening, or turning north, of the Bz, marking the AE recovery phase, the over-shielding electric field that penetrate to the equatorial latitudes, has westward polarity and large amplitude in the dusk sector. Superposition of an under-shielding (eastward) electric field (originating from the storm development) with the PRE can produce large post sunset eastward electric field/vertical drift leading to topside bubble development even during the season when bubbles do not develop under normal conditions of the PRE as was shown by Abdu et al. (2003a). On the other hand the superposition of an over-shielding (westward) electric field with the PRE can result in total cancellation of the post sunset zonal electric field or even result in strong westward electric field thereby stabilizing the post sunset F layer. The total suppression of the ESF Bubble development under such conditions during the epoch of its normal occurrence has recently been demonstrated by Abdu et al. (2009c).

In Fig. 16.8 we present examples of the two types of effects, that is, the enhancement and suppression of the PRE (and ESF), that occurred on two close by days during the month of September which is in the season of ESF occurrence over Brazil. We may note that on 25 September 2001 when the Bz marginally turned south starting at ~20 UT (17 LT) the AE activity started to increase (top two panels) and the F layer vertical drift over Fortaleza, an equatorial site, (bottom panel) as measured by a digisonde (red curve) started to increase much faster than the normal prereversal vertical drift (blue curve) and attained a peak values of 75 m/s that soon resulted in spread F bubble development. Also on this evening the bubble development occurred at 2140UT/1840LT which is significantly earlier than the typical onset time for this month which is after 19 LT. In contrast to this on the evening of 23 September the Bz turned north (top panel) coinciding with the AE recovery (panel 3 from top) near 20 UT/17 LT which was followed by total suppression of the PRE and, hence, in no bubble development. This was caused by a strong over-shielding electric field of westward polarity that characterized the storm recovery in the dusk sector. However, on this day enhanced AE activity had prevailed several hours prior to sunset suggesting that the presence of a disturbance dynamo electric field of westward polarity in the evening hours might also have contributed in some degree to the observed PRE and ESF suppression. Cases of such suppressions caused exclusively by the over-shielding electric field of westward polarity have been discussed by Abdu et al. (2009c). Eastward over-shielding electric field that could cause ESF development during pre midnight hours has been reported in the Indian sector by Chakrabarty et al. (2006).

16.7 Discussion and Conclusions

We have presented here very briefly the different coupling processes that play varying roles in the development of post sunset ESF bubble irregularities. In efforts to advance our ability to predict the occurrence of these irregularities it is important to evaluate in quantitative terms the contributions arising from each of these processes in a given observed pattern of events. A synthesis of their relative role in such a context is a complex task especially when many or all of those processes happen to control simultaneously a given event. However there are situations when some of these processes will not be important contributor to the sequences leading to an event. For example, under verifiably quiet magnetic conditions the role of disturbance electric field can be neglected. Similarly during the equinoctial months and in locations of small magnetic declination angle the zonal wind dominates the equatorial evening hours so that any possible role of a meridional/trans-equatorial winds in the processes leading to ESF development can be neglected. The most important factors that are present in all situations of the instability processes leading to plasma bubble growth, irrespective of the location and geophysical conditions, are (1) the prereversal vertical drift that is needed for the post sunset rise of the F layer, and (2) a precursor perturbation in wind of right sense arising from gravity waves of appropriate orientation that could provide perturbations in plasma density and polarization electric field needed for the initiation of the instability process. The most crucial

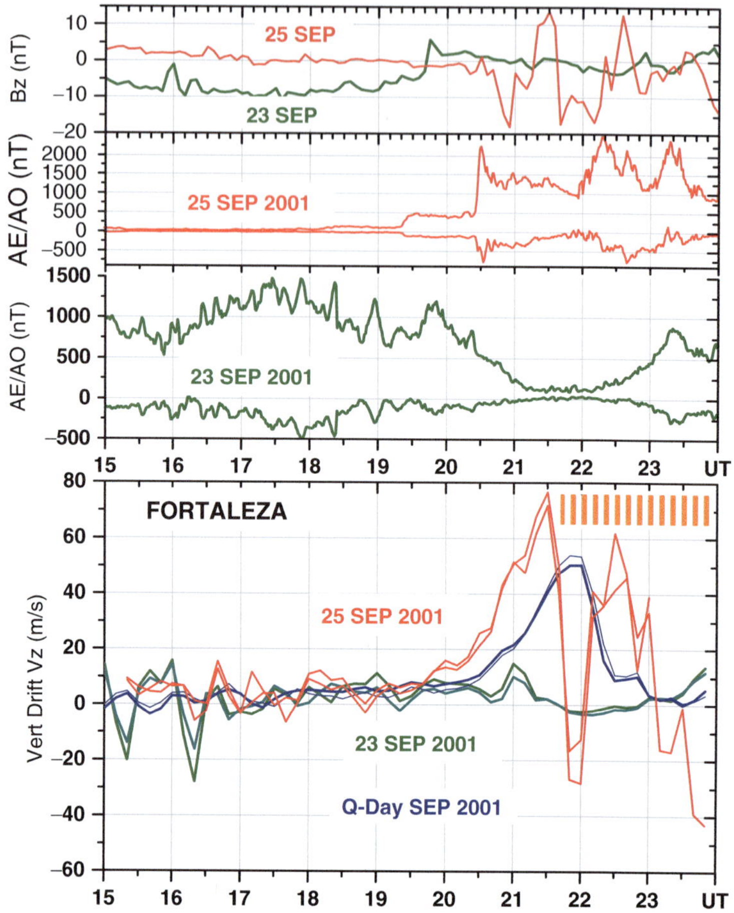

Fig. 16.8 Similar to Figs 16.1 and 16.2, but for the 2 days, 23 and 25 September 2001. The Vz values (*bottom panel*) are for Fortaleza calculated at 6 and 7 MHz. They are shown together with a quiet day reference. The ESF occurrence staring at 2140 UT is indicated by *vertical bars* (Abdu et al., 2009c). (Reproduced with permission of the American Geophysical Union)

task in synthesizing the causes that lead to an observed irregularity event therefore consists in evaluating the contributions from these two factors. In the results presented in Figs. 16.7a, b and discussed in Section 16.5 it could be noticed that when the F layer height is larger, as a results of the larger PRE amplitude, and is followed by ESF bubble development (as on 23 October) the GW amplitude and propagation features were less clear in the digisonde data, probably because the density fluctuations and the time resolution were marginally within the detection limit of the instrument (as was discussed in detail by Abdu et al., 2009b). On the other hand, when the PRE amplitude and the F layer height were smaller (as on 24 October) the precursor GW oscillations that were required to account for the CII growth rate (consistent with the irregularity plume of limited vertical extent that appeared in sequence) were indeed well above the limit of detection by the Digisonde. In fact an approximate estimate showed that the amplitude of the GW oscillations in the latter case was at least 3 times larger than that in the former case. Here as well as in other few analyzed examples (not discussed here) we have noted a tendency for the GW amplitudes as detected by the digisonde to be more pronounced for the cases of lower PRE amplitude, and therefore lower evening F layer heights, than for the cases when the PRE and heights are larger.

Tropospheric convection is now widely believed to be the source of GWs that propagate upward dissipating energy and momentum at F region heights. This belief is based on theoretical/model calculations on the upward propagation characteristics of such GWs (Vadas, 2007; Fritts et al., 2008) as well as from statistical correlation found between global distribution of tropospheric active regions, represented by the ITCZ, and the global ESF occurrence pattern (Rottger, 1981; McClure et al., 1998, Tsunoda, 2010). If this tendency for a reduced GW intensity with increasing F layer height and PRE represents a real trend (which is being verified from more analysis of the data) it would offer the following possible scenario of conditions for the GW propagation into the F layer bottomside. (1) The tropospheric GWs dissipating energy and momentum in the thermosphere begin loosing intensity with increasing height in the height region spanned by the post sunset F layer rise so that a higher (lower) F layer height would correspond to a lower (higher) GW intensity; and/or (2) The GWs loose energy by interaction with the background zonal wind as shown from model calculations by Kherani et al. (Chapter 10, this volume). The back ground zonal wind (which is eastward in the evening) is known to drive the PRE so that a larger (smaller) zonal wind responsible for a stronger (weaker) PRE could result in higher (smaller) degree of weakening of the GW amplitude. It is important to conduct more studies towards clarifying these questions.

More systematic measurements of the precursor gravity waves such as those presented by Abdu et al. (2009b) would be essential for clarifying any of the above hypotheses. Improved observational settings permitting higher temporal resolution and better detection limits by digisondes or other measurement techniques will be useful for detailed evaluation of the precursor gravity waves. It may be noted that using the C/NOFS satellite radio beacon experiment, Thampi et al. (2009) have detected GW oscillation in the TEC during the evening hours that were identified as representing the large scale wave structure earlier discussed by Tsunoda (2008) as a precondition for spread F occurrence. It is interesting to note that the scale sizes of these structures from C/NOFS satellite are comparable to the precursor gravity waves in the bottom-side F layer that were detected by the Digisonde as reported by Abdu et al. (2009b).

From the results and their implications discussed above it would appear that for that case of spread F irregularities that do not present expressive vertical development, meaning that they tend to be dominating lower F layer heights (that is, the predominantly bottom-side events), the GW presence tends to be more visible at least for diagnostic by digisondes. More recently cases of possible GW association with bottom-side irregularity structures have been diagnosed also from optical imaging observations (Takahashi et al., 2010). On the other hand the well developed bubble type ESF events are those for which the PRE amplitude /F layer heights are expressively large. Statistically it has been shown that the PRE vertical drift of at least 35 m/s could lead to well developed topside bubbles (Abdu et al., 2006b) for the specific ambient ionospheric conditions studied. These are also the cases for which the meridional/trans-equatorial winds, if present, could play important roles for suppressing their development as was discussed in Section 4 above. An evaluation of such effect was possible from latitudinally spaced or conjugate point, observations by Digisondes as was discussed by Abdu et al. (2009a). Here again there is need to make quantitative assessment of the importance of such winds in suppressing a bubble instability development against the role of the prereversl vertical drift to enhance its growth. In other words, it is important to determine the magnitude of the meridional/trans-equatorial wind that could inhibit/limit a given instability growth process from becoming a well developed bubbles that could otherwise have resulted from an optimum combinations the PRE intensity and GW winds.

Under magnetically disturbed conditions a synthesis of the contributions to the ESF instability growth from the different sources becomes less tractable, obviously so because of the complexity in the background ionospheric responses to such disturbances. These responses are characterized by drastic changes in the zonal electric field and in zonal, as well as meridional/trans-equatorial, winds, with probably little changes in terms of the GW fluxes. At this stage there exist only some case studies of isolate events of spread F variability under disturbed conditions such as those already considered above (see also, Basu et al., 2010; Li et al., 2010; Tulasi Ram et al., 2008).

The main conclusions from the results and discussion presented above are the following: The nature of the different coupling processes that control the ionosphere–thermosphere interaction sequence leading to the Spread F bubble irregularity development is reasonably known. What is lacking in our understanding concerns the relative dominance of the different elements of the processes for varying degrees of the irregularity/instability growth. The main dynamic/electrodynamic factors as precursors that determine the development of the ESF/Bubble irregularities are (1) the prereversal vertical drift and post sunset F layer rise, (2) a seed perturbation in the form of gravity waves, (3) integrated conductivity of the potentially unstable flux tube, and (4) vertical as well as horizontal density gradients in the F layer bottomside that are not discussed here. Enhancement of the flux tube integrated conductivity to limit the spread F instability nonlinear growth can be caused by meridional or trans-equatorial winds based on analysis of Digisonde data. The possibility of enhanced conductivity due to sporadic E layers contributing to such effects (that is, limiting the ESF growth) cannot be easily verified due to the interactive processes operating in the developments of the PRE and Es layer formation (the Es layer formation is disrupted at the time of the PRE development of significant amplitude). On a statistical basis the PRE controls in a systematic way the degree of the vertical growth, and hence the latitudinal extent, of the spread F structure. Planetary waves propagating upward to the dynamo region produces large day-to-day oscillations in the amplitude of the PRE and hence in the intensity of the associated ESF events. Large intensity of evening vertical drift (PRE) (with vertical drift > 35 m/s) is necessary for well developed spread F/bubble events to occur. Such cases appear to require only relatively weaker GW contribution as seed perturbation. In order for the PRE with relatively smaller amplitude to lead to developing bubbles the instability growth would need to be enhanced by significant amplitude of GW seed perturbation such as that are normally detected by a digisonde. Disturbance electric fields during magnetically active conditions are an important source of ESF variability. Further studies on all aspects of the background conditions controlling the ESF variability need to be pursued for a better understanding of the related coupling processes, key for improving predictive capability in this field.

Acknowledgements MAA wishes to acknowledge the supports received from the CNPq (Conselho Nacional de Pesquisa e Desenvolvimento) through the grant: 300883/2008-00. EAK wish to acknowledge the support from FAPESP through the grant 07/00104-0.

References

Abdu MA (2001) Outstanding problems in the equatorial ionosphere thermosphere system relevant to spread F. J Atmos Solar-Terr Phys 63:869–884

Abdu MA, Batista PP, Batista IS, Brum CGM, Carrasco AJ (2006a) Planetary wave oscillations in mesospheric winds, equatorial evening prereversal electric field and spread F. Geophys Res Lett 33:L07107. doi:10.1029/2005GL024837

Abdu MA, Batista IS, Reinisch BW, de Souza JR, Sobral JHA, Pedersen TR, Medeiros AF, Schuch NJ, de Paula ER, Groves KM (2009a) Conjugate point equatorial experiment (COPEX) campaign in Brazil: electrodynamics highlights on spread F development conditions and day-to-day variability. J Geophys Res 114:A04308. doi:10.1029/2008JA013749

Abdu MA, Batista IS, Takahashi H, MacDougall J, Sobral JH, Medeiros AF, Trivedi NB (2003a) Magnetospheric disturbance induced equatorial plasma bubble development and dynamics: a case study in Brazilian sector. J Geophys Res 108(A12):1449. doi:10.1029/2002JA009721

Abdu MA, Batista IS, Walker GO, Sorel JHA, Thrived NB, de Paula ER (1995) Equatorial ionosphere electric field during magnetospheric disturbances: local time/longitude dependences from recent EITS campaigns. J Atmos Solar-Terr Phys 57:1065–1083

Abdu MA, Batista IS, Sobral JHA (1992) A new aspects of magnetic declination control on equatorial spread F and F region dynamo. J Geophys Res 97(A10):14897–14904

Abdu MA, Bittencourt JA, Batista IS (1981) Magnetic declination control of the equatorial F region dynamo electric field development and spread F. J Geophys Res 86: 11443–11446

Abdu MA, Brum CGM (2009) Electrodynamics of the vertical coupling processes in the atmosphere-ionosphere system of the low latitude region. Earth Planets Space 61:385–395

Abdu MA, de Medeiros RT, Bittencourt JA, Batista IS (1983)Vertical ionization drift velocities and range spread F in the evening equatorial ionosphere. J Geophys Res 88:399–402. doi:10.1029/ JA088iA01p00399

Abdu MA, Iyer KN, de Medeiros RT, Batista IS, Sobral, JHA (2006b) Thermospheric meridional wind control of equatorial spread F and evening prereversal electric field. Geophys Res Lett 33(L07106):1–4

Abdu MA, Kherani EA, Batista IS, de Paula ER, Fritts DC, Sobral JHA (2009b) Gravity wave initiation of equatorial spread F/plasma bubble irregularities based on observational data from the SpreadFEx campaign. Ann Geophys 27:1–16

Abdu MA, Kherani EA, Batista IS, Sobral JHA (2009c) Equatorial evening prereversal vertical drift and spread F suppression by disturbance penetration electric fields. Geophys Res Lett 36: L19103. doi:10.1029/2009GL039919

Abdu MA, MacDougall JW, Batista IS, Sobral JHA, Jayachandran PT (2003b) Equatorial evening prereversal electric field enhancement and sporadic E layer disruption: a manifestation of E and F region coupling. J Geophys Res 108(A6):1254. doi:10.1029/2002JA009285

Abdu MA, Ramkumar TK, Batista IS, Brum CGM, Takahashi H, Reinisch BW, Sobral JHA (2006c) Planetary wave signatures in the equatorial atmosphere-ionosphere system, and mesosphere- E- and F- region coupling. J Atmos Solar-Terrest Phys 68:509–522

Abdu MA, Sobral JHA, Nelson OR, Batista IS (1985) Solar cycle related range type spread F occurrence characteristics over equatorial and low latitude stations in Brazil. J Atmos Solar-Terr Phys 47:901–905

Anderson DN, Reinisch BW, Valladares C, Chau J, Veliz O (2004) Forecasting the occurrence of ionospheric scintillation activity in the equatorial ionosphere on a day-to-day basis. J Atmos Solar-Terr Phys 66:1567–1572. doi:10.1016/j.jastp.2004.07.010

Basu S, MacKenzie E, Bridgwood C, Valladares CE, Groves KM, Carrano C (2010) Specification of the occurrence of equatorial ionospheric scintillations during the main phase of large magnetic storms within solar cycle 23. Radio Sci 45:RS5009. doi:10.1029/2009RS004343

Batista IS, Abdu MA, Bittencourt JA (1986) Equatorial F-region vertical plasma drifts: seasonal and longitudinal asymmetries in the American sector. J Geophys Res 91: 12055–12064

Batista IS, Abdu MA, Carrasco AJ, Reinisch BW, de Paula ER, Schuch NJ, Bertoni F (2008) Equatorial spread F and sporadic E-layer connections during the Brazilian Conjugate Point Equatorial Experiment (COPEX). J Atmos Solar-Terr Phys 70:1133–1143

Bhattacharyya A (2004) Role of E region conductivity in the development of equatorial ionospheric plasma bubbles. Geophys Res Lett 31:L06806. doi:10.1029/ 2003GL018960

Bittencourt JA, Abdu MA (1981) A theoretical comparison between apparent and real vertical ionization drift velocities in the equatorial F-region. J Geophys Res 86: 2451–2455.

Booker HG, Wells HW (1938) Scattering of radio waves in the F region of ionosphere. Terr Magn Atmos Electr 43:249

Carrasco AJ, Batista IS, Abdu MA (2005) The pre-reversal enhancement in the vertical drift for Fortaleza and the sporadic E layer. J Atmos Solar-Terr Phys 67: 1610–1617

Carrasco AJ, Batista IS, Abdu MA (2007) Simulation of the sporadic E layer response to prereversal associated evening vertical electric field enhancement near dip equator. J Geophys Res 112:A06324. doi:10.1029/2006JA012143

Chakrabarty D, Sekar R, Narayanan R, Patra AK, Devasia CV (2006) Effects of interplanetary electric field on the development of an equatorial spread F event. J Geophys Res 111:A12316. doi:10.1029/2006JA011884

Chapagain NP, Fejer BG, Chau JL (2009) Climatology of post sunset equatorial spread F over Jicamarca. J Geophys Res 114:A07307. doi:10.1029/2008JA013911

Fagundes PR, Pillat VG, Bolzan MJA, Sahai Y, Becker-Guedes F, Abalde JR, Aranha SL, Bittencourt JA (2005) Observations of F layer electron density profiles modulated by planetary wave type oscillations in the equatorial ionospheric anomaly region. J Geophys Res 110:A12302. doi:10.1029/2005JA011115

Farley DT, Balsley BB, Woodman RF, McClure JP (1970) Equatorial spread F, implications of VHF radar observations. J Geophys Res 75:7199–7216

Farley DT, Bonelli E, Fejer BG, Larsen MF (1986) The prereversal enhancement of the zonal electric 4eld in the equatorial ionosphere. J Geophys Res 91:13723–13728

Fejer BG, de Paula ER, Gonzalez SA, Woodman RF (1991) Average vertical and zonal F-region plasma drifts over Jicamarca. J Geophys Res 96:13901–13906

Fejer BG, Jensen JW, and Su S-Y (2008) Seasonal and longitudinal dependence of equatorial disturbance vertical plasma drifts. Geophys Res Lett 35:L20106. doi:10.1029/2008GL035584

Fejer BG, Scherliess L, de Paula ER (1999) Effects of the vertical plasma drift velocity on the generation and evolution of equatorial spread F. J Geophys Res 104:19854–19869

Forbes JM, Leveroni S (1992) Quasi 16-day oscillations in the ionosphere. Geophys Res Lett 19:981–984

Fritts DC, Vadas SL, Riggin DM, Abdu MA, Batista IS, Takahashi H, Medeiros A, Kamalabadi F, Liu H-L, Fejer BG, Taylor MJ (2008) Gravity wave and tidal influences on equatorial spread F based on observations during the Spread F Experiment (SpreadFEx). Ann Geophys 26:3235–3252

Fukao S, Yokoyama T, Tayama T, et al (2006) Eastward traverse of equatorial plasma plumes observed with the equatorial atmosphere radar in Indonesia. Ann Geophys 24(5): 1411–1418

Gurubaran S, Sridharan S, Ramkumar TK, Rajaram R (2001) The mesospheric quasi 2-day wave over Tirunelveli. J Atmos Solar-Terr Phys 63:975–985

Haerendel G (1973) Theory of equatorial spread F. Report, Maxplanck-Institut fur Extraterre. Phys Garching, Germany

Heelis RA, Kendall PC, Moffet RJ, Windle DW, Rishbeth H (1974) Electrical coupling of the E- and F- region and its effects on the F-region drifts and winds. Planet Space Sci 22:743–756

Huang C-S, Kelley MC (1996) Nonlinear evolution of equatorial spread-F. 4. Gravity waves, velocity shear, and day-to-day variability. J Geophys Res 101:24523

Hysell DL, Burcham JD (1998) JULIA radar studies of equatorial spread F. J Geophys Res 103:29155–29167

Jayachandran B, Balan N, Rao PB, Sastri JH, Bailey GJ (1993) HF doppler and ionosonde observations on the onset conditions of equatorial spread F. J Geophys Res 98:13741–13750

Jyoti N, Devasia CV, Sridharan R, Tiwari D (2004) Threshold height (h0F)c for the meridional wind to play a deterministic role in the bottom side equatorial spread F and its dependence on solar activity. Geophys Res Lett 31:L12809. doi: 10.1029/2004GL019455

Kelley MC, Larsen MF, La Hoz C (1981) Gravity wave initiation of equatorial spread-F: a case study. J Geophys Res 86: 9087–9100

Kherani EA, Abdu MA, de Paula ER, Fritts DC, Sobral JHA, de Meneses FC Jr (2009) The impact of gravity waves rising from convection in the lower atmosphere on the generation and nonlinear evolution of equatorial bubble. Ann Geophys 27:1657–1668

Kherani AE, Mascarenhas M, De Paula ER, Sobral JHA, Bertoni F (2005) A three-dimensional simulation of collisional-interchange-instability in the equatorial-low-latitude ionosphere. 121(1–4), November. doi:10.1007/s11214-006-6158-x

Kil H, Heelis RA (1998) Global distribution of density irregularities in the equatorial ionosphere. J Geophys Res 103: 407–417

Kudeki E, Akgiray A, Milla M, Chau JL, Hysell DL (2007) Equatorial spread-F initiation: post-sunset vortex, thermospheric winds, gravity waves. J Atmos Solar-Terr Phys 69:2416–2427

Kudeki E, Bhattacharya S (1999) Post sunset vortex in equatorial F region plasma drifts and implications for bottomside spread-F. J Geophys Res 104(A12):28163–28170

Li G, Ning B, Hu L, Liu L, Yue X, Wan W, Zhao B, Igarashi K, Kubota M, Otsuka Y, Xu JS, Liu JY (2010) Longitudinal development of low-latitude ionospheric irregularities during the geomagnetic storms of July 2004. J Geophys Res 115:A04304. doi:10.1029/2009JA014830

Maruyama T (1988) A diagnostic model for equatorial spread F 1. Model description and application to electric field and neutral wind effects. J Geophys Res 93:14611–4622

Maruyama T, Matuura N (1984) Longitudinal variability of annual changes in activity of equatorial Spread F and plasma bubbles. J Geophys Res 89(A12):10903–10912

McClure JP, Sing S, Bamgboye DK, Johnson FS, Kil H (1998) Occurrence of equatorial F region irregularities: evidence for tropospheric seeding. J Geophys Res 103: 29119–29135

Mendillo M, Meriwether J, Biondi M (2001) Testing the thermospheric neutral wind suppression mechanism for the day-to-day variability of equatorial spread F. J Geophys Res 106:3655– 3663

Muella MTAH, Kherani EA, de Paula ER, Cerruti AP, Kintner PM, Kantor IJ, Mitchell CN, Batista IS, Abdu MA (2010) Scintillation-producing Fresnel-scale irregularities associated with the regions of steepest TEC gradients adjacent to the equatorial ionization anomaly. J Geophys Res 115:A03301. doi:10.1029/2009JA014788

Pancheva D, Houldoupis C, Meek CE, Manson AH, Mitchell NJ (2003) Evidence of a role for modulated atmospheric tides in the dependence of sporadic E layers on planetary waves. J Geophys Res 108(A5):1176. doi:101029/2002JA009788

Rastogi RG (1980) Seasonal variation of equatorial spread F in the American and Indian zones. J Geophys Res 85:22

Richmond AD, Peymirat C, Roble RG (2003) Long-lasting disturbances in the equatorial ionospheric electric field simulated with a coupled magnetosphere-ionosphere-thermosphere model. J Geophys Res 108(A3):1118. doi:10.1029/2002JA009758, 2003

Rishbeth H (1971) Polarization fields produced by winds in the equatorial F region. Planet Space Sci 19:357–369

Rishbeth H, Ganguly S, Walker JCG (1978) Field-aligned and field-perpendicular velocities in the ionospheric F2 layer. J Atmos Terr Phys 40:767–784

Rottger J (1981) Equatorial spread F by electric 4elds and atmospheric gravity waves generated by thunderstorms. J Atmos Solar-Terr Phys 43:453–462

Saito S, Maruyama T (2006) Ionospheric height variations observed by ionosondes along magnetic meridian and plasma bubble onsets. Ann Geophys 24:2991–2996

Sastri JH, Abdu MA, Batista IS, Sobral JHA (1997) Onset conditions of equatorial (range) spread F at Fortaleza, Brazil, during the June solstice. J Geophys Res 102(A11): 24013–24021

Scherliess L, Fejer BG (1997) Storm time dependence of equatorial disturbance dynamo zonal electric field, J Geophys Res 1022(A11):2403–24046

Sekar R, Suhasini R, Raghavarao R (1994) Effects of vertical winds and electric fields in the nonlinear evolution of equatorial spread F. J Geophys Res 99(A2):2205–2213

Sobral JHA, Abdu MA, Zamlutti CJ, Batista IS (1980) Association between plasma bubble irregularities and airglow disturbances over Brazilian low latitudes. Geophys Res Lett 1:980–982

Sultan PJ (1996) Linear theory and modeling of the Rayleigh-Taylor instability leading to the occurrence of equatorial spread F. J Geophys Res 101:26875–26891

Takahashi H, Lima LM, Wrasse CM, Abdu MA, Batista IS, Gobbi D, Buriti RA, Tsuda T (2005) Evidence on 2–4 day modulation of the equatorial ionosphere h'F and mesospheric airglow emission. Geophys Res Let 32:L12102. doi:10.1029/2004GL022318

Takahashi H, Wrasse CM, Fechine J, Pancheva D, Abdu MA, Batista IS, Lima LM, Batista PP, Clemesha BR, Schuch NJ, Shiokawa K, Gobbi D, Mlynczak MG, Russell JM (2007) Signatures of ultra fast Kelvin waves in the equatorial middle atmosphere and ionosphere. Geophys Res Lett 34:L11108. doi:10.1029/2007GL029612, 2007

Takahashi H et al (2010) Equatorial ionosphere bottom-type spread F observed by OI 630.0 nm airglow imaging. Geophys Res Lett 37:L03102. doi:10.1029/2009GL041802

Thampi SV, Yamamoto M, Tsunoda RT, Otsuka Y, Tsugawa T, Uemoto J, Ishii M (2009) First observations of large-scale wave structure and equatorial spread F using CERTO radio beacon on the C/NOFS satellite. Geophys Res Lett 36:L18111. doi:10.1029/2009GL039887

Tsunoda RT (1985) Control of the seasonal and longitudinal occurrence of equatorial scintillation by longitudinal gradient in integrated Pedersen conductivity. J Geophys Res 90:447–456

Tsunoda RT (2008) Satellite traces: an ionogram signature for large scale wave structure and a precursor for equatorial spread F. Geophys Res Lett 35:L20110. doi:10.1029/2008GL035706

Tsunoda RT (2010) On seeding equatorial spread F during solstices. Geophys Res Lett 37:L05102. doi:10.1029/2010GL042576

Tsunoda RT, White BR (1981) On the generation and growth of equatorial backscatter plumes, 1- Wave structure in the bottomside F layer. J Geophys Res 86:3610–1981

Tulasi Ram S, Rama Rao PVS, Prasad DSVVD, Niranjan K, Gopi Krishna S, Sridharan R, Ravindran S (2008) Local time dependent response of postsunset ESF during geomagnetic storms. J Geophys Res 113:A07310. doi:10.1029/2007JA012922

Vadas SL, Liu H-L (2009) The generation of large-scale gravity waves and neutral winds in the thermosphere from the dissipation of convectively-generated gravity waves. J Geophys Res 114:A10310. doi:10.1029/2009JA014108

Vincent RA (1993) Long-period motions in the equatorial mesosphere. J Atmos Solar-Terr Phys 55:1067–1080

Vineeth C, Pant TK, Devasia CV et al (2007) Atmosphere-ionosphere coupling observed over the dip equatorial MLTI region through the quasi 16-day wave. Geophys Res Lett 34(12):L12102, June 16 2007

Woodman RF, La Hoz C (1976) Radar observations of F region equatorial irregularities. J Geophys Res 81:5447

Yokoyama T, Fukao S, Yamamoto M (2004) Relationship of the onset of equatorial F region irregularities with the sunset terminator observed with the Equatorial Atmosphere Radar. Geophys Res Lett 31:L24804. doi:10.1029/2004GL021529

Chapter 17

Influences on the Development of Equatorial Plasma Bubbles: Insights from a Long-Term Optical Dataset

Jonathan J. Makela and Ethan S. Miller

Abstract We present an analysis of an eight-year dataset of field-aligned optical images of the low-latitude, nighttime ionosphere. The duration of this dataset allows us to study the climatology of ionospheric structures, such as equatorial plasma bubbles (EPBs) and medium-scale traveling ionospheric disturbances (MSTIDs) during both high and low solar activity. In this chapter, we concentrate on the climatology of EPBs during both high and low solar activity and a possible linkage between EPBs and MSTIDs during low solar activity. In contrast to climatologies derived from forward-scattering (scintillation) measurement techniques and satellite-based in-situ measurements, we do not find a reduction in the occurrence of EPBs at low solar activities. However, we do find that EPBs typically occur later in the evening at low solar activity levels. We also present statistics for the occurrence of MSTIDs seen to the south of Hawaii during low solar activity, which indicate that the increase in post-midnight, quiet-time EPB occurrence during low solar activity is accompanied by an increase in MSTID occurrence.

17.1 Introduction

The Earth's ionosphere is known to play host to a wide variety of irregularity processes. These irregularities can be detrimental to radio wave propagation, as they can cause rapid fluctuations in both signal phase and amplitude. For this reason, understanding when and where ionospheric irregularities occur has long been a focus of the Space Weather community. Different physical processes dominate at different local times, levels of solar and geomagnetic activity, and locations, making the study of ionospheric irregularities a rich field of investigation.

Perhaps the irregularity process with the longest history, dating back to the 1930s (Booker and Wells, 1938), is the one that has come to be known as equatorial spread-F. The first observations were made by ionosondes, but many other observational modalities have now been brought to bear on studying this class of irregularities. Through the collected data, much has been learned about the conditions conducive to the development of nighttime equatorial irregularities. In short, due to the horizontal nature of the magnetic field and the rapid recombination of the bottomside of the F layer after sunset, the equatorial ionosphere is subject to irregularity development through the gravitational Rayleigh-Taylor instability (RTI), a mechanism first proposed by Dungey (1956). Although much progress has been made over the decades, a fundamental understanding of the physics governing the day-to-day variability of this phenomenon is still lacking, and the study of nighttime equatorial irregularities is still an active research priority of the Space Weather community.

One of the complicating factors in understanding the day-to-day variability has been the lack of comprehensive, long-term measurements. The need for long-term measurements arises from the fact that the irregularity development process depends on factors that vary as a function of season and solar cycle.

J.J. Makela (✉)
Department of Electrical and Computer Engineering, University of Illinois at Urbana-Champaign, Urbana, IL, USA
e-mail: jmakela@illinois.edu

M.A. Abdu, D. Pancheva (eds.), A. Bhattacharyya (Coed.), *Aeronomy of the Earth's Atmosphere and Ionosphere*, IAGA Special Sopron Book Series 2, DOI 10.1007/978-94-007-0326-1_17, © Springer Science+Business Media B.V. 2011

Additionally, different facets of the physical properties of the irregularities affect different measurement techniques. As a result, specific features of equatorial irregularity climatologies depend on the measurement technique used to create the climatology. Comparisons of these differing climatologies can lead to seemingly contradictory conclusions.

Poleward of the region of the Earth's ionosphere in which equatorial spread-F occurs, a second class of ionospheric structures commonly referred to as medium-scale traveling ionospheric disturbances (MSTIDs), can develop. The Perkins' instability (Perkins, 1973) is often cited as the cause of the development of MSTIDs, although it does have its shortcomings in describing all of the observed properties of MSTIDs. Similar to the perturbations that develop at the magnetic equator, MSTIDs have been observed by a variety of observational modalities, and a general consensus has arisen as to the structures' phenomenology and occurrence climatologies (e.g., Garcia et al., 2000; Shiokawa et al., 2003). However, the fundamental mechanisms responsible for their generation and propagation have not been fully clarified.

In this chapter, we contribute an analysis of a nearly continuous eight-year optical dataset obtained using a field-aligned imaging system in the mid-Pacific. From this dataset, we infer the climatology of the occurrence of equatorial plasma bubbles (EPBs), the optical manifestation of equatorial spread-F. To our knowledge, this is the most comprehensive dataset of field-aligned images of EPBs collected to date. Based on the climatology, we suggest and provide evidence of the influence MSTIDs can have on the development of EPBs during low solar activity levels. Finally, we discuss how the climatology of EPBs derived here differs from previous climatologies constructed from other observing modalities. These differences highlight the strengths and limitations of the wide variety of measurement techniques that have been used to study this phenomenon.

17.2 Data

In order to study differing influences on the development of equatorial plasma bubbles (EPBs) during solar minimum and maximum conditions, we consider an optical dataset collected in the mid-Pacific beginning in 2002. The data consist of images of the naturally occurring 630.0-nm nightglow emission taken with the Cornell Narrow Field Imager (CNFI). This instrument was installed at the Maui Space Surveillance Site on the Haleakala Volcano on Maui, Hawaii (geographic: 20.7°N, 203.7°E; geomagnetic: 21.3°N, 271.4°E) in late 2001 by Cornell University and continues to be operated by Cornell University and the University of Illinois at Urbana-Champaign.

As described in more detail in Kelley et al. (2002), CNFI observes towards the southern horizon with a 47° field of view in a magnetic field aligned viewing geometry first proposed by Tinsley (1982). This geometry allows for the observation of EPBs with high spatial resolution, as lines of sight from the imager are nearly tangent to the magnetic field lines at the altitude of the peak in the 630.0-nm emission intensity. Images are obtained using 90-s exposures through a narrowband interference filter (2.0-nm full width, half maximum) with a sampling time varying from 3 to 9 min, depending on the number of additional filters used by CNFI in a given observation mode. As such, CNFI provides excellent observations of the spatial and temporal dynamics associated with the development, propagation, and decay of EPBs.

In this study, we make use of the entire dataset of 630.0-nm images obtained by CNFI from 1 January 2002 through 31 December 2009. CNFI operated near continuously during this period, obtaining images from astronomical sunset through astronomical sunrise on a nightly basis. Unlike typical wide-angle imaging systems, images were obtained when the moon was above the horizon, as in most seasons the moon is not in the direct field of view of CNFI. The resulting images during moon-up periods have increased background intensities from scattered moonlight, but the presence, or lack, of ionospheric structure can still be discerned. Although the dataset provided by CNFI is nearly continuous, there are several extended periods during which the instrument was not operational due to maintenance issues. However, this dataset represents the largest collection of field-aligned optical images for studying EPBs obtained to date.

EPBs are, by far, the most prevalent ionospheric feature seen in the CNFI dataset. Although EPBs can take on many different, complex forms, their general morphology is well documented (see Makela 2006 and references therein). The typical EPB appears as latitudinally-elongated dark regions, extending from the magnetic equator at the south of the imager's field of view, towards the north. Oftentimes, the EPBs extend out of the northern field of view of the imager. During geomagnetically quiet periods, EPBs observed with CNFI drift from west to east at a velocity on the order of 100 m/s (Yao and Makela, 2007). This drift velocity can reduce, or even reverse, during periods of geomagnetic activity (Sahai et al., 2009). Additionally, EPBs are aligned with the magnetic meridians, but eventually tilt to the west with increasing latitude as the evening progresses (Makela and Kelley, 2003).

Although not as common as EPBs, medium-scale traveling ionospheric disturbances (MSTIDs) are also seen in the CNFI dataset. In contrast to the predominately field-aligned EPBs emanating from the magnetic equator and drifting to the east, MSTIDs are aligned at a significant angle from the magnetic meridians and drift to the southwest. They typically drift into the field of view of CNFI from the north and have been observed in the CNFI dataset to reach the magnetic equator in contrast to results from the Japanese sector, which suggest a propagation limit of 18° in magnetic latitude (Shiokawa et al., 2002). Due to their distinct characteristics during geomagnetically quiet periods, EPBs and MSTIDs are easily distinguished from one another, as seen in Fig. 17.1. During geomagnetic activity, this distinction becomes a bit more difficult as the morphology of both types of structures can significantly deviate from their quiet-time patterns (e.g., Makela et al., 2009).

17.3 Analysis and Results

In order to reduce the large dataset collected by CNFI, we perform a manual analysis of each night of data. The images are labeled as one of six categories: (1) clear; (2) cloudy; (3) EPB; (4) MSTID; (5) simultaneous MSTID and EPB; or (6) unusable data. Note that a clear image refers to an image in which a majority of the field of view is free from obscuring clouds and neither EPB nor MSTID structure is observed. An image is determined to be cloudy if clouds obscure enough of the sky that it would be difficult to discern whether or not an EPB or MSTID was present. Unusable data are typically caused by equipment malfunction (e.g., the camera shutter not opening) or light contamination in the environment surrounding CNFI.

Once the entire dataset is labeled in this way, the labels are further reduced into one-hour bins for each night of observation. When studying EPBs, if any of the images taken during a given hour are labeled EPB or simultaneous MSTID and EPB, the entire bin is labeled as EPB. If no EPBs are observed in a given hour, the bin is labeled as clear if more than 25% of the

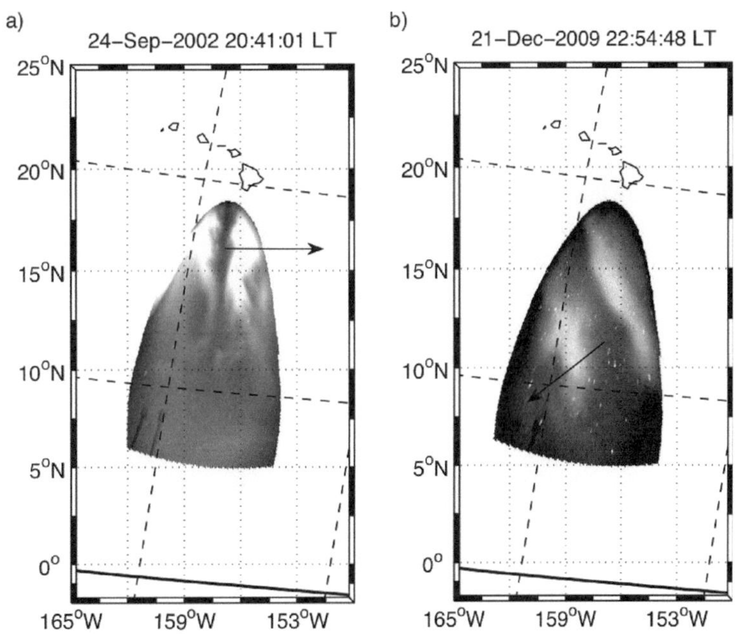

Fig. 17.1 Example images obtained by CNFI to study (**a**) EPBs and (**b**) MSTIDs. The *dashed lines* indicate magnetic latitude and longitude. The *arrows* indicate the direction of structure propagation

images in that hour were individually labeled as clear (in the case of studying EPBs, an image showing only an MSTID is considered to be clear); otherwise, the hour is labeled as cloudy. Although the clear/cloudy threshold of 25% is somewhat arbitrary, given the field of view of CNFI and the typical drift velocity of the ionospheric structures being considered, we are reasonably confident that, if structure were present in a given hour, we would observe it given this threshold. At this point, an hourly bin on a given night is labeled as clear, cloudy, or EPB. When studying MSTIDs, the same procedure is followed, with images labeled as MSTID or simultaneous MSTID and EPB contributing to the MSTID bins, and EPB-only images considered to be clear.

We further reduce the labeled and binned data into a climatology of EPB or MSTID occurrence in two ways. The procedure is the same for both EPBs and MSTIDs and follows the methodology previously used in Makela et al. (2004); for brevity, we will only describe the procedure in terms of calculating the climatology of EPBs. The first procedure entails performing a 20-day running filter on the hourly-binned data, counting the number of EPB bins at a given hour in the 20 days surrounding a given date and dividing by the sum of the number of EPB and clear bins at that hour in the 20-day window. The choice of a 20-day window was made to balance the conflicting desires of having a window long enough to have a large number of bins being considered without being too long that meaningful trends (e.g., monthly, seasonal) are smoothed out. The result is an estimate of the occurrence probability of EPBs on that date and hour based upon the occurrence rate of EPBs in the surrounding dates.

The climatology results compiled in this way are presented in Fig. 17.2. Confirming the results of our initial study examining the first 2 years of this data (Makela et al., 2004), we see a preponderance of EPB activity during the June through October period, with a maximum occurrence rate in September. A secondary maximum is seen in the April through early May period. This is in general agreement with the established understanding of the seasonal/longitudinal EPB climatologies (e.g., Tsunoda, 1985).

Also, as discussed in Makela et al. (2004), the EPBs observed from June through August occur later in the evening and, as seen in individual images (not shown), many drift into the field of view fully formed, suggesting that they developed to the west of Hawaii. Finally, it is quite clear from Fig. 17.2 that EPBs are often observed into the early morning hours, oftentimes until local sunrise.

In order to study the solar cycle dependence of the occurrence of EPBs, we reduce the dataset using a second method that results in monthly occurrence statistics. Again, treating each labeled and binned hour as a data point, we count the number of EPB bins in a given month and divide by the sum of the number of EPB and clear bins at that hour in that month. While this methodology results in a monthly climatology that is, perhaps, easier to compare to monthly climatologies derived from other measurement techniques (e.g., Aarons, 1993), the results are a bit cruder in terms of being able to temporally resolve the "spread-F" seasons at the Hawaiian longitude.

To study the influences of the solar cycle on EPB development, we further divide the monthly datasets into periods of high solar activity (HSA; 2002–2004, inclusive; mean $f_{10.7} = 138.1 \times 10^{-22}$ W/m^2/Hz) and low solar activity (LSA; 2007–2009, inclusive; mean $f_{10.7} = 70.8 \times 10^{-22}$ W/m^2/Hz). Figure 17.3 presents the monthly EPB occurrence rate separated into pre-midnight (Fig. 17.3a; 1900–2300 LT) and post-midnight (Fig. 17.3b; 0000–0400 LT) time periods. The bottom panel of each plot gives the total number of data points (clear or EPB) considered in each bin.

Examining the data from the pre-midnight period presented in Fig. 17.3a, we note that there is a slight reduction in EPB occurrence during LSA (15% yearly occurrence rate) as compared to HSA (23% yearly occurrence rate). This result is in agreement with previous studies in the American sector (Sahai et al., 2000). However, during the main "spread-F season", there is still a high incidence of EPBs during LSA (42% in September). The overall trend in the occurrence pattern remains the same in the pre-midnight hours during both HSA and LSA, suggesting that the overarching physical control determining a given longitude's spread-F season remains the same under a wide range of solar activity levels.

Interestingly, the post-midnight data presented in Fig. 17.3b show an opposite solar cycle effect. Occurrence rates increase from a 31% yearly occurrence during HSA to 43% during LSA. The increase

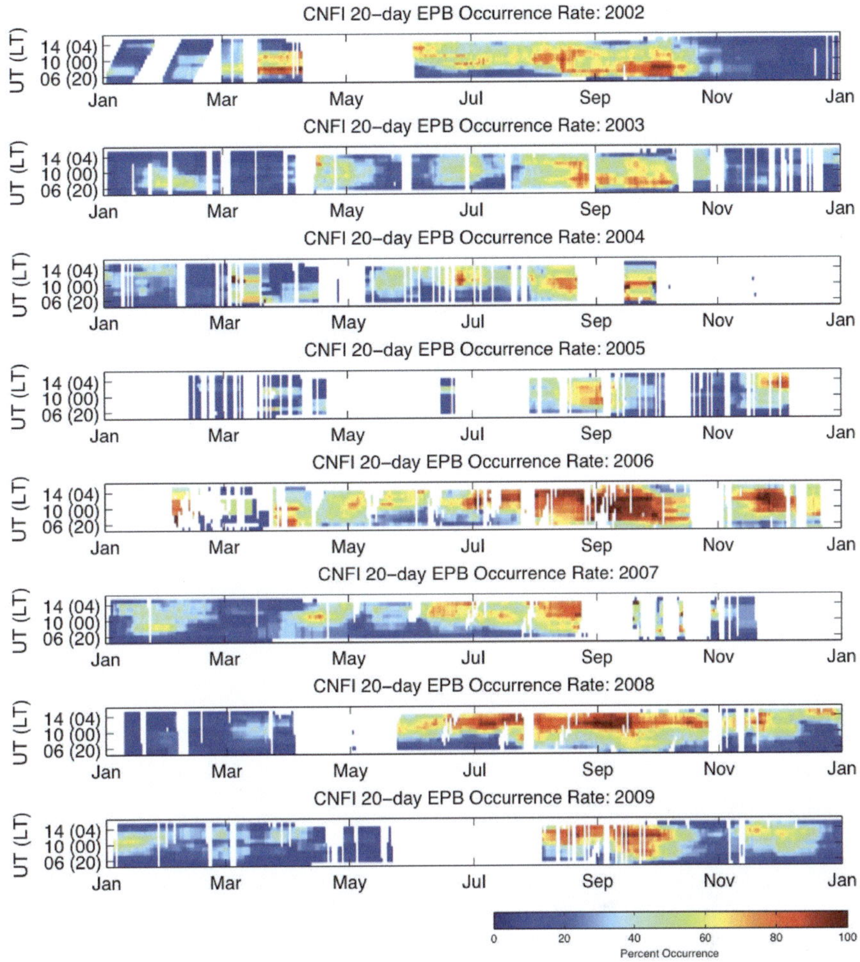

Fig. 17.2 20-day occurrence statistics for EPBs observed by CNFI from 2002 to 2009

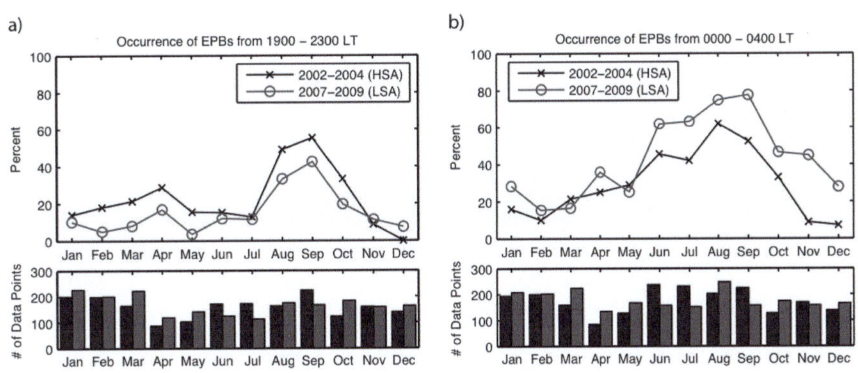

Fig. 17.3 Monthly occurrence statistics for (**a**) the evening and (**b**) post-midnight time periods. Statistics from HSA (*black line*, "x" marks) and LSA (*gray line*, "o" marks) are shown. The *bottom panels* give the number of hourly-binned data points that are included in each monthly calculation (*black*: HSA; *gray*: LSA)

is most notable in the November through January solstice period as EPBs are a fairly rare occurrence during HSA (11%) as compared to LSA (33%), although this increase is more broadly seen from June through January.

Additional insight into the differences between EPB occurrence in LSA and HSA can be gleaned from examining the monthly plots presented in Fig. 17.4. Here, we have separated the data into hourly bins for each month and calculated the occurrence rate of EPBs

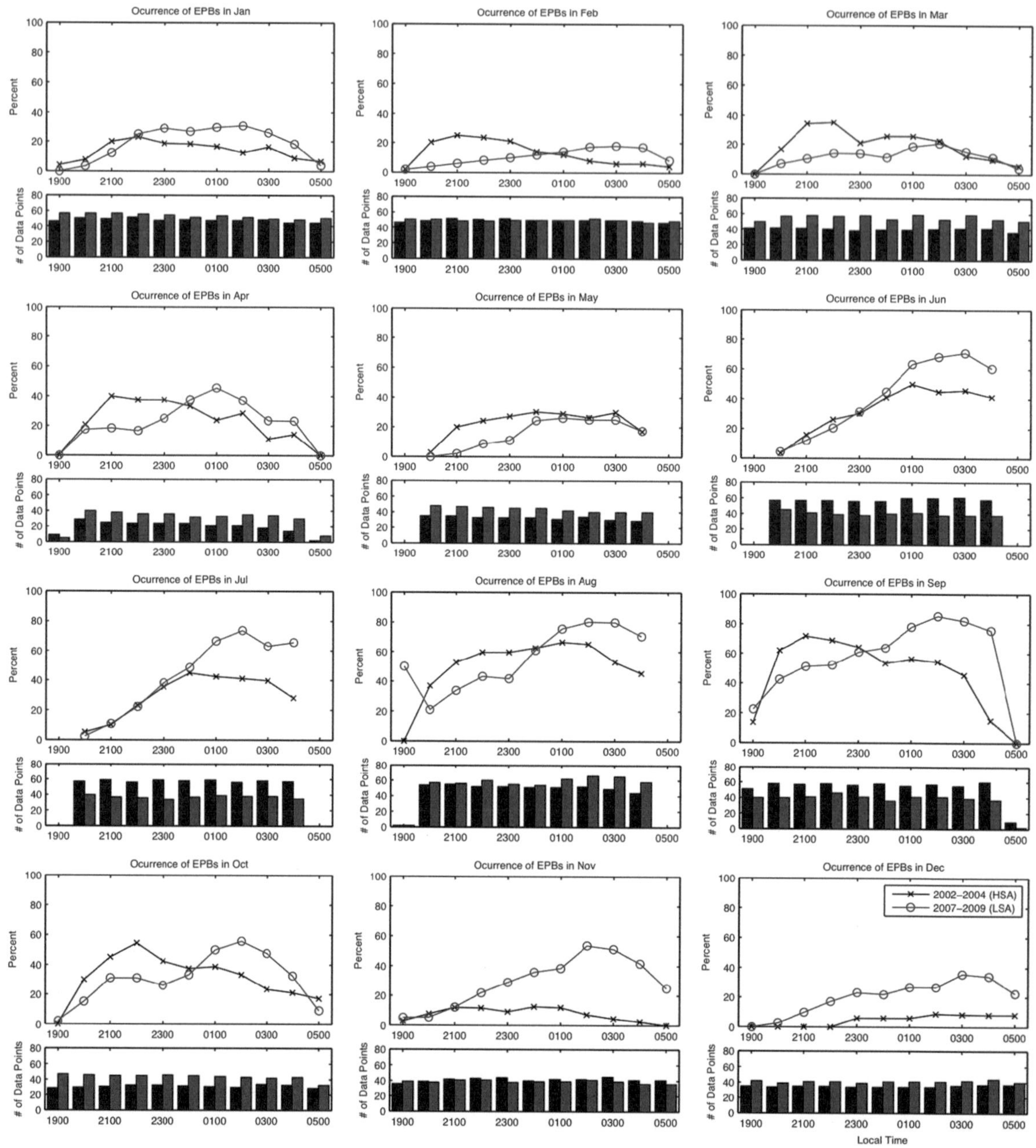

Fig. 17.4 Same as Fig. 17.3, except for hourly occurrence statistics organized by months

for LSA and HSA. In months corresponding to the spread-F season (April; August through October), we see that during HSA, EPB occurrence peaks early in the evening, by 2100 or 2200 LT, with EPBs becoming less likely as time goes on. However, during LSA, the peak in occurrence shifts dramatically, with the maximum occurring after local midnight (0100–0200 LT). Note that the occurrence rates during these months are roughly comparable for each month in HSA and LSA, but the peak has shifted 4 h later.

A slightly different trend is seen in June and July, months in which EPBs were prevalent in the post-midnight hours during HSA. A similar trend is seen in these months during LSA, with relatively few EPBs observed pre-midnight but many observed post-midnight. However, the occurrence rate for post-midnight EPBs during these months is higher during LSA.

In November and December, and to a lesser extent January, months showing the largest increase in EPB occurrence from HSA to LSA, we see that this increase is for all times after about 2100 LT, although the increase is most dramatic in the post-midnight hours.

17.4 Discussion

Several climatological features, as presented above, deserve additional recognition and discussion. The first is the transition toward EPB activity later in local time toward solar minimum. The second is new, recurring (annually, possibly semiannually) peaks in EPB activity in solstice months after local midnight during LSA. The third is the departure of the optical EPB climatology from the collocated L-band scintillation climatology (not shown) as well as satellite-derived climatologies (e.g., Gentile et al., 2006), which show very little EPB activity during LSA.

Consider the local time distributions of EPBs presented in Fig. 17.2. In 2002–2003, the onset of EPBs is soon after local sunset. In 2007–2009, it is somewhat delayed. Miller et al. (2010), performed a similar analysis of the CNFI dataset by developing in parallel a climatology of radar backscatter from the Christmas Island 50-MHz coherent scatter radar (CXI), which shares a common magnetic field-of-view with CNFI. The CXI radar data clearly show a trend of backscatter beginning later in local time with decreasing solar activity. This feature is probably due to reductions in the instability growth rate on account of lower F-region densities and weaker background vertical drifts, shifting the local time of EPB generation to later local times.

Another feature that emerges in Fig. 17.2 is the dramatic increase in post-midnight EPB activity toward LSA, particularly around the December solstice. Note that these are not the typical post-midnight events studied previously (e.g., Aarons, 1993), as they are not associated with increased geomagnetic activity. Study of the individual images collected by CNFI during this period indicates that many of these post-midnight events were plasma depletions whose initiation coincided with the passage of an MSTID (Miller et al., 2009). They hypothesized that, under low solar activity conditions, the polarization electric field associated with MSTIDs would drive an $\mathbf{E}\times\mathbf{B}$ drift which is vertical in the geomagnetic apex frame. This is similar to the hypothesis of Tsunoda (2007) that polarization fields associated with sporadic-E layers drive an $\mathbf{E}\times\mathbf{B}$ drift at the apex. As the MSTID approaches the magnetic equator, the apex altitude affected by this vertical drift approaches the bottomside and can cause a perturbation to be formed, initiating the Rayleigh-Taylor instability. The results of Miller et al. (2010) show that these post-midnight, quiet-time EPBs are associated with the generation of fresh 3-m irregularities, causing the backscatter observed by the CXI coherent scatter radar system, rather than "fossilized" EPBs drifting into the field of view of the instruments.

To examine this in more detail, Fig. 17.5 presents a monthly breakdown of the occurrence of MSTIDs during LSA. Figure 17.5a shows that pre-midnight MSTID occurrence is relatively rare (less than 12% yearly occurrence) in the region south of Hawaii observed by CNFI. December (21%) and January (19%) show the highest occurrence rates of pre-midnight MSTIDs. In the post-midnight timeframe, illustrated in Fig. 17.5b, MSTID occurrence increases (31% yearly occurrence). December (57%) and January (48%) remain the months with the largest occurrence rate for MSTIDs. However, broad maximum in the occurrence rates are seen between October through February (42%) and June through August (34%).

Previous studies have shown that MSTIDs have seasonal occurrence patterns, with maximum occurrence during local summer with a smaller peak in local winter over Japan (Shiokawa et al., 2003) and both local winter (Garcia et al., 2000) and summer (Martinis et al., 2010) over the Caribbean. The negative correlation of nighttime MSTID occurrence with solar activity has also been reported by Kotake et al. (2006). All of this provides more evidence that the increase in post-midnight, quiet-time EPBs to the south of Hawaii during LSA is related to the increased occurrence in MSTIDs in this region.

Figure 17.6 shows two typical examples of the low solar activity relationship between MSTIDs and EPBs in the post-midnight period during quiet geomagnetic conditions. In Fig. 17.6a, from 21 January 2009

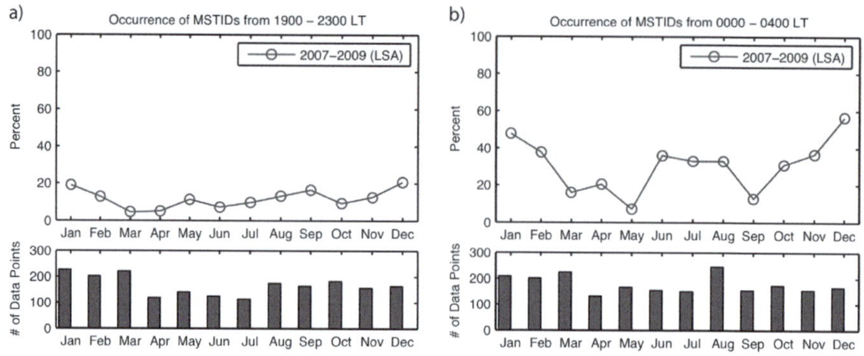

Fig. 17.5 Same as Fig. 17.3 but for the monthly occurrence rate of MSTIDs. Only LSA is shown

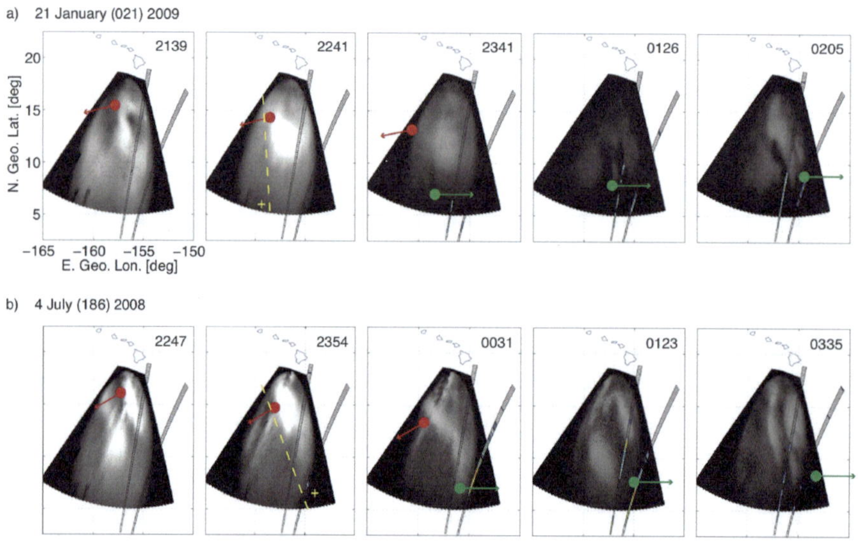

Fig. 17.6 Two cases of EPBs initiated by MSTIDs in the central Pacific during quiet geomagnetic conditions. The CXI radar beams have been geomagnetically mapped to the same 250-km altitude as the images of the 630.0-nm airglow emission. Typical examples are shown from (**a**) the December solstice period, 21 January (021) 2009, and (**b**) the June solstice period, 4 July (186) 2009. *Dots* and *arrows* indicate the approximate propagation of points on the MSTIDs (*red*) and EPBs (*green*). Notice that the MSTIDs are not associated with FAI backscatter in the radar until depletions form. See text for discussion

(Kp < 2), the MSTID structure appears propagating toward the west-southwest at 2139 LT. As the phase fronts of the MSTID reach the geomagnetic equator, EPBs grow (a nascent EPB is indicated with the + mark at 2241 LT) and drift toward the east. As these EPBs drift across the north and east beams of the Christmas Island radar (shown in these figures mapped along the geomagnetic field lines to the airglow layer) from 2341 to 0205 LT, backscatter is received from 3-m field-aligned irregularities (FAI) within the EPBs. Note also the poleward expansion of the EPB between 2341 and 0128 LT; this is clearly an actively developing EPB.

Figure 17.6b illustrates a similar example, except from the opposite solstice period, 4 July 2008 (Kp < 2), suggesting that this behavior is not restricted to the December solstice. Of particular interest to distinguishing MSTIDs from EPBs, notice that not only does the MSTID drift to the southwest, it also is not associated with radar backscatter at 2247 LT although the undulations occur on the same field lines observed by the radar. In this example, an EPB forms within the east beam of the radar, as suggested by the thin backscatter layer marked by + at 2354 LT. Again, the MSTID drifts to the southwest as the EPBs drift to the east. A more atypical example that highlights extreme interaction between MSTID and EPB structures that appear to drive 3-m FAI until local sunrise was presented by Miller et al. (2009).

Finally, we turn to the overall occurrence of EPBs at high and low solar activity levels. At HSA, the climatology derived here for the Pacific sector agrees quite well with the climatologies derived by other methods. However, this is not the case for LSA. Previous studies based upon forward-scattering (scintillation) measurements (e.g., Aarons, 1993) have shown a dramatic decrease in the occurrence of scintillations, especially at higher frequencies, as solar activity levels decrease. For the 2002–2009 time period over Hawaii, this is confirmed with collocated GPS L1 scintillation measurements (not shown). Similarly, measurements from the Defense Meteorological Satellite Program (DMSP) satellites between 1989 and 2004 show a drastic decrease in the number of EPBs encountered at 840 km (Gentile et al., 2006). These two results seem to be in stark contrast to the results of EPB occurrence rates obtained from the ground-based optical observations presented here, which show very similar overall occurrence rates of EPBs at HSA and LSA.

Resolving the discrepancy with climatologies based upon forward scattering is straightforward. It is well understood that amplitude scintillation strength depends on the magnitude of electron density gradients of irregularities at the Fresnel scale (e.g., Yeh and Liu, 1982) and, thus, strong scintillations require large background electron densities. During LSA, the background electron densities are decreased leading to a reduction of the overall level of activity for scintillations. Thus, as our climatology presented here shows, although the underlying EPB structure is almost equally likely to be present during HSA and LSA, scintillations will only occur when the electron density gradients are large enough, a condition more likely to be satisfied during HSA.

Unlike the forward-scattering measurements, which rely upon having large background electron densities, the in-situ measurement made by the DMSP satellites has no such constraint. The in-situ density probe should register electron density reductions associated with EPBs regardless of the underlying mean electron density (assuming it is above the sensitivity threshold of the instrument). In their study, Gentile et al. (2006) explain the drastic reduction in the number of EPBs detected by the DMSP satellites during LSA by suggesting that the differing ionospheric conditions at LSA limit the number of EPBs that reach DMSP altitudes at the appropriate local times. The DMSP satellites fly in a near sun-synchronous orbit and the satellites used in the Gentile et al. study crossed the magnetic equator in the early evening hours (between 1900 and 2200 LT). Assuming that EPBs are always generated near local sunset, this leads to the conclusion that the vertical growth velocity of EPBs is reduced at LSA, and thus do not reach the DMSP altitudes at the correct local times. Alternately, one could conclude that EPBs simply do not reach high altitudes during periods of LSA, as has been reported using radar (Hysell and Burcham, 2002) and imaging (Sahai et al., 2000) data.

Using the climatology derived here from ground-based optical observations, along with the information on the spatial/temporal dynamics provided by the individual images collected by CNFI, we can add new insight to the discussion of EPB occurrence during LSA. Our climatology shows that the reduction in bubbles observed by the DMSP satellites is most likely due primarily to a shift in the local time of their occurrence, rather than a slowing of the vertical growth velocity or an overall reduction in their maximum altitude, although these factors likely also play a contributing role. The field of view of CNFI maps to apex altitudes from approximately 300 to 1000 km. In our analysis of the images, any image showing an EPB was counted as an EPB, regardless of its maximum altitude. If the vertical growth of EPBs was simply slower during LSA, but they still occurred early in the evening, we would observe them during their slowed growth phase. The data presented in Figs. 17.3 and 17.4 show this is not the case. As discussed above, the number of EPBs (at any altitude) was reduced at LSA before local midnight, as compared to HSA. After midnight, the number of EPBs increased for LSA. What this suggests is that the entire generation mechanism for EPBs is delayed during LSA, and the lack of bubbles reaching the DMSP altitude is not just due to a reduction in the vertical growth velocity of the bubbles themselves.

This interpretation is confirmed by the 50-MHz backscatter climatology presented in Miller et al. (2010) showing that the seasonal peak in the back scatter occurrence rate associated with EPBs shifts from about 2030 LT in 2003 to 0130 LT in 2008. As with the optical observations, the radar observations would still measure coherent backscatter if slowly growing EPBs were present at lower altitudes earlier in the evening. The fact that this is not the case suggests that there is a relationship between the generation time of EPBs and solar activity level, most likely due to the reduction in the RTI growth rate caused by the reduction in electron density gradients and vertical drift velocities at low solar activities. This solar cycle dependence of EPB generation time has also been observed in radar data collected in the S. American sector (Fejer et al., 1999; Hysell and Burcham, 2002).

We have not yet performed an analysis of the altitude obtained by individual bubbles for the entire dataset presented here, and cannot conclusively state if there is a solar-cycle dependence of the altitudes obtained by EPBs. This will be the subject of future work. However, from the data analysis performed to create the climatology presented here, we can state anecdotally that EPBs do indeed reach high altitudes (> 900 km) during LSA. We, therefore, conclude that the small number of bubbles observed by the DMSP satellites in the Gentile et al. (2006) study is a result of the local times of the satellites and the shift to later local times in the occurrence of EPBs at LSA.

17.5 Conclusions

We have presented an analysis of an eight-year dataset of field-aligned optical images of the low-latitude, nighttime ionosphere. From the climatology derived from these observations we have shown that, for the Hawaiian sector, equatorial plasma bubbles (EPBs) are most likely to occur from August through October and, to a lesser extent, April. In contrast to climatologies derived from forward-scattering (scintillation) measurement techniques and satellite-based in-situ measurements, we do not find a drastic reduction in the occurrence of EPBs at low solar activities. The reduced occurrence of scintillations with solar activity is expected, based upon the reduce background electron density. We find that the EPBs typically occur later in the evening at low solar activity levels, explaining the discrepancy with the satellite-derived climatologies.

In addition, we have presented statistics for the occurrence of medium-scale traveling ionospheric disturbances (MSTIDs) seen to the south of Hawaii during low solar activity. We further develop the hypothesis presented by Miller et al. (2009) that the increase in post-midnight, quiet-time EPB occurrence during low solar activity is due to coupling of the electric fields internal to MSTIDs by showing that the increase in the post-midnight, quiet-time EPB occurrence is accompanied by an increase in the number of MSTID occurrences. Further work, especially through modeling studies, is needed to confirm this hypothesis.

Acknowledgements Airglow observations on Mount Haleakala are supported by the Air Force Office of Scientific Research under Contract FA9550-05-0160494 to Cornell University. The continued operation of the instrument would not be possible without the continued on-site support of Jacob Burger. We thank Professor Michael C. Kelley for making the CNFI data available for our use. Work at the University of Illinois is supported by the National Science Foundation through grant ATM-06-044654. Ethan Miller acknowledges further support from a National Science Foundation CEDAR postdoctoral researcher award to JHU/APL.

References

Aarons J (1993) The longitudinal morphology of equatorial F-layer irregularities relevant to their occurrence. Space Sci Rev 63:209–243

Booker HG, Wells HW (1938) Scatter of radio waves by the F-region of the ionosphere. Terr Mag Atmos Elec 43(3): 249–256

Dungey JW (1956) Convective diffusion in the equatorial F-region. J Atmos Solar-Terr Phys 9(5–6):304–310

Fejer BG, Scherliess L, de Paula ER (1999) Effects of the vertical plasma drift velocity on the generation and evolution of equatorial spread F. J Geophys Res 104(A9): 859–869. doi:10.1029/1999JA900271

Garcia FJ, Kelley MC, Makela JJ, Huang C-S (2000) Airglow observations of mesoscale low-velocity traveling ionospheric disturbances at midlatitudes. J Geophys Res 105(A8): 18407–18415

Gentile LC, Burke WJ, Rich FJ (2006) A climatology of equatorial plasma bubbles from DMSP 1989–2004. Radio Sci 41: RS5S21. doi:10.1029/2005RS003340

Hysell DL, Burcham JD (2002) Long term studies of equatorial spread F using the JULIA radar at Jicamarca. J Atmos Solar-Terr Phys 64:1531–1543

Kelley MC, Makela JJ, Ledvina BM, Kintner PM (2002) Observations of equatorial spread-F from Haleakala, Hawaii. Geophys Res Lett 29(20):2003. doi:10.1029/2002GL015509

Kotake N, Otsuka Y, Tsugawa T, Ogawa T, Saito A (2006) Climatological study of GPS total electron content variations caused by medium-scale traveling ionospheric disturbances. J Geophys Res 111:A04306. doi:10.1029/2005JA011418

Makela JJ (2006) A review of imaging low-latitude ionospheric irregularity processes. J Atmos Solar-Terr Phys 68(13):1441–1458

Makela JJ, Kelley MC (2003) Field-aligned 777.4-nm composite airglow images of equatorial plasma depletions. Geophys Res Lett 30(8):1442. doi:10.1029/2003GL017106

Makela JJ, Kelley MC, Tsunoda RT (2009) Observations of midlatitude ionospheric instabilities generating meter-scale waves at the magnetic equator. J Geophys Res 114:A01307. doi:10.1029/2007JA012946

Makela JJ, Ledvina BM, Kelley MC, Kintner PM (2004) Analysis of the seasonal variations of equatorial plasma bubble occurrence observed from Haleakala, Hawaii. Ann Geophys 22(9):3109–3121

Martinis C, Baumgardner J, Wroten J, Mendillo M (2010, in press). Seasonal dependence of MSTIDs obtained from 630.0 nm airglow imaging at Arecibo. Geophys Res Lett 37: L11103. doi:10.1029/2010GL043569

Miller ES, Makela JJ, Groves KM, Kelley MC, Tsunoda RT (2010, in press). Coordinated study of coherent radar backscatter and optical airglow depletions in the central pacific. J Geophys Res 115:A06307. doi:10.1029/2009JA014946

Miller ES, Makela JJ, Kelley MC (2009) Seeding of equatorial plasma depletions by polarization electric fields from middle latitudes: experimental evidence. Geophys Res Lett 36(18):L18105. doi:10.1029/2009GL039695

Perkins F (1973) Spread F and ionospheric currents. J Geophys Res 78(1):218–226

Sahai Y, Fagundes PR, Bittencourt JA (2000) Transequatorial F-region ionospheric plasma bubbles: solar cycle effects. J Atmos Solar-Terr Phys 62:1377–1383

Sahai Y et al (2009) Observations of the F-region ionospheric irregularities in the South American sector during the October 2003 "Halloween Storms". Ann Geophys 27: 4463–4477. doi:10.5194/angeo-27-4463-2009

Shiokawa K, Ihara C, Otsuka Y, Ogawa T (2003). Statistical study of nighttime medium-scale traveling ionospheric disturbances using mid-latitude airglow images. J Geophys Res 108(A1):1052. doi:10.1029/2002JA009491

Shiokawa K, Otsuka Y, Ejiri MK, Sahai Y, Kadota T, Ihara C, Ogawa T, Igarashi K, Miyazaki S, Saiti A (2002). Imaging observations of the equatorward limit of midlatitude traveling ionospheric disturbances. Earth Planets Space 54: 57–62

Tinsley BA (1982) Field aligned airglow observations of transequatorial bubbles in the tropical F-region. J Atmos Solar-Terr Phys 44(6):547–557

Tsunoda RT (1985) Control of the seasonal and longitudinal occurrence of equatorial scintillations by the longitudinal gradient in integrated E region Pedersen conductivity. J Geophys Res 90(A1):447–456.

Tsunoda RT (2007) Seeding of equatorial plasma bubbles with electric fields from an Es-layer instability. J Geophys Res 112:A06304

Yao D, Makela JJ (2007) Analysis of equatorial plasma bubble zonal drift velocities in the Pacific sector by imaging techniques. Ann Geophys 25(3):701–709

Yeh KC, Liu CH (1982) Radio wave scintillations in the ionosphere. Proc IEEE 70(4):324–360

Chapter 18

A Review of the Recent Advances in the Investigation of Equatorial Spread F and Space Weather Effects over Indian Sector Using Optical and Other Techniques

R. Sekar and D. Chakrabarty

Abstract Observations using a narrow spectral band (0.3 nm) airglow photometer with a narrow field-of-view (3°) developed recently enabled to capture the impact of space weather events on neutral airglow emission over low latitudes and also opened up new vistas to characterize different plasma structures associated with Equatorial Spread F (ESF). Plasma enhancement structures in buoyancy-dominated altitude region, as predicted by earlier numerical simulation investigation, were identified based on simultaneous VHF radar and airglow observations. A few other varieties of plasma structures that include confined structure in a limited altitude region and fossil plasma bubbles that became active at a later time, had been observed. Moreover, it was also shown that the prompt penetration of interplanetary electric field (IEF) can trigger ESF during post-sunset hours when ionospheric condition is not suitable for the occurrence of ESF. It was also shown that over-shielding effect of IEF can resurrect a plasma plume from a plasma structure confined in a narrow altitude region even during pre-midnight hours. Numerical simulation investigation confirmed that eastward electric field associated with over-shielding electric field is necessary but not sufficient condition for pre-midnight plume event. The importance of pre-seed to pre-midnight plume event in the guise of confined ESF structure developed during post-sunset hours is discussed. An important finding came up in the form of identifying the response of thermospheric OI 630.0 nm airglow emission during nighttime over low latitude corresponding to IEF variations. It was shown that quasi-periodic fluctuations (~0.5 and 1.0 h) in IEF can affect thermospheric airglow emission over low latitudes. The importance of narrow spectral band airglow variations in bringing out these fluctuations is discussed. Interestingly, these periodicities were also found in the north-south component of nocturnal magnetic field (H) over the entire Indian sub-continent indicating the current responsible to be of magnetospheric origin. It was also shown during another event that while the fast fluctuations (~40–45 min) in the vertical drift are governed by IEF during a storm that encompassed multiple substorms, the slowly varying component of ionospheric height variations corroborate with the slowly varying component of the variation in the auroral electrojet index, AE. The results discussed in this article highlight the important coupling aspects of ionosphere and thermosphere during quiet and disturbed space weather conditions.

R. Sekar (✉)
Space and Atmospheric Science Division, Physical Research Laboratory, Ahmedabad, India
e-mail: rsekar@prl.res.in

18.1 Introduction

Earth's ionosphere and thermosphere act as a mutually coupled system and some of the features in one of these have imprint on the other. This coupling is decided by the time constant for transfer of momentum and energy. The time constant is usually larger for the transfer of momentum from ions to neutral as the ion densities in the ionosphere are at least less by three orders of magnitude compared to neutral densities. In addition to this coupling, the low-latitude ionosphere–thermosphere system (ITS) is susceptible to forcing from the magnetosphere via the high-latitude region

M.A. Abdu, D. Pancheva (eds.), A. Bhattacharyya (Coed.), *Aeronomy of the Earth's Atmosphere and Ionosphere*, IAGA Special Sopron Book Series 2, DOI 10.1007/978-94-007-0326-1_18, © Springer Science+Business Media B.V. 2011

and also to forcing from the lower atmospheric regions by means of waves and winds. The characteristic phenomena of the low latitude ITS like the plasma fountain, equatorial Spread F (ESF) etc manifest the underlying complex coupling processes (e.g. see Abdu, 1997). It is, therefore, imperative that observations by multiple techniques are needed from various vantage points to address critical scientific issues comprehensively. A number of important results have recently been obtained from Indian sector that aptly elicit the critical coupling pertaining to ITS over low latitude region under magnetically quiet condition and coupling between high and low latitude regions during disturbed space weather conditions. Keeping in mind the enormity of the literature available on the coupling issues of ITS and for the sake of brevity, attempts are made here to highlight only the results obtained over Indian zone using multiple techniques.

As mentioned, low latitude ITS exhibit various characteristic processes during day and nighttime. The post-sunset equatorial ionospheric electrodynamics is particularly significant as the plasma irregularities generated by ESF has many crucial ramifications in the communication, navigation etc. It is now known that pre-reversal enhancement (PRE) of zonal equatorial electric field arising due to the F region dynamo mechanism (Rishbeth, 1971) at the sunset terminator, plays a very crucial role in the generation of ESF (Sekar and Kelley, 1998; Fejer et al., 1999). A well-developed PRE lifts the equatorial F layer to higher altitudes and in the presence of rapidly depleting E layer during post-sunset hours provides the favorable condition for the seed perturbations at the bottomside to grow nonlinearly and convect upward and eventually cascading into plasma irregularity structures of wide-range (from a few hundred kilometers to a few centimeters) of spatial scale sizes. This gives rise to what is known as ESF because of the spread in the echoes when the probing radio wave passes through the F region of the ionosphere. The background conditions that include the variability in PRE, electron density gradient, peak height of F layer, neutral winds, E region conductivity etc. make the occurrence and development of ESF highly variable in space and time. Moreover, other factors like prompt penetration and overshielding of interplanetary electric fields (IEF), disturbance dynamo effects etc. also contribute to the occurrence variability and the development of ESF phenomenon by modulating the low latitude ionospheric electric field.

The electric fields due to the prompt penetration and overshielding processes primarily arise owing to the imbalances between the region 1 (R1) and region 2 (R2) field aligned currents (FAC). R1 FAC lies poleward of the auroral oval and is responsible in mapping the polar cap voltage drop (the voltage drop IEF generates across the polar cap region) immediately to the nightside outer magnetosphere. Under dawn-to-dusk IEF, the earthward plasma convection is enhanced in the nightside outer magnetosphere and the magnetospheric hot plasma gets deflected by the gradient-curvature drift resulting in the formation of the ring current. Strong pressure gradient in the ring current, thereafter, forms R2 FAC connecting the inner edge of the ring current with the equatorward edge of the auroral oval. In a steady state, the R2 FAC tends to minimize the electric field at low latitudes producing the shielding effect. However, during intervals when IEF suddenly turns to dawn-to-dusk direction, the ionosphere over low latitudes responds directly to IEF as the shielding electric field at the inner magnetosphere gets temporally incapacitated. This is because of the fact that the shielding effect produced by R2 FAC tends to lag behind the polar cap electric field variations as convection requires time to redistribute the magnetospheric plasma. This is prompt penetration effect (e.g. Nishida, 1968; Vasyliunas, 1970; Wolf, 1970, Reddy et al., 1979; Spiro et al., 1988; Sastri et al., 1992, 1993; Kikuchi et al., 2000; Chakrabarty et al., 2005; Kelley et al., 2007). In the overshielding scenario (e.g. Rastogi and Patel, 1975; Kelley et al., 1979; Chakrabarty et al., 2006; Ebihara et al., 2008), IEF suddenly reduces or changes polarity after a sustained dawn-to-dusk condition. This makes the residual inner magnetospheric shielding field dominant for some time. The magnitude of the effects of prompt penetration and overshielding processes crucially depends on the local ionospheric conditions. Moreover, as the shielding mechanism is associated with the plasma convection and subsequent injection of energetic magnetospheric particles in the ring current, both storm and substorm can contribute to the electric field perturbations over low latitudes. While the substorm-related electric fields can persist for 2–3 h (e.g. Sastri, 2002) and overshielding electric field decays substantially after half an hour (e.g. Peymirat et al., 2000), prompt penetration events lasting several hours (e.g. Huang et al., 2005) are reported. Theoretically predicted shielding time constant of the order ~20 min (e.g. Senior and Blanc,

1984) deviates considerably, on many occasions, with the observations. While the prompt-penetration/over-shielding effects are almost simultaneously felt over low latitudes on many occasions, disturbance dynamo effects are delayed as this effect is associated with the auroral heating and subsequently modified wind system (Blanc and Richmond, 1980).

As many of the coupling processes are still not comprehensively understood, it is difficult to have an all-inclusive view of the low latitude ionosphere in the presence and absence of ESF phenomenon. Moreover, considering the enormity of the work done on the low latitude ionosphere, it is difficult to cover all the work in this brief article. Therefore, attempts are made to narrow down the scope to only the recent multi-technique investigations of nighttime low latitude ITS in the presence and absence of ESF. Many important investigations have been carried out from the Indian region to understand different aspects of ESF under magnetically quiet (e.g. Raghavarao et al., 1987; Sridharan et al., 1997; Patra et al., 1995; Devasia et al., 2002) and disturbed (e.g. Chakrabarty et al., 2006; Kakad et al., 2007; Tulasi Ram et al., 2008) space weather conditions. A detailed review of the important works on ESF in the recent past from India is provided in Sekar and Chakrabarty (2008a). However, using multiple techniques, investigations on nighttime ITS are far from comprehensive. In this chapter, the recent investigations over Indian zone using VHF radar, ionosonde, magnetometer and satellite measurements along with thermospheric airglow intensity in a narrow spectral band are highlighted.

18.2 Motivation for the Co-ordinated Optical and Radar Investigations

The motivation for the co-ordinated optical and radar measurements came from the previous attempt to interpret the VHF radar observation of ESF structures using nonlinear simulation model (Sekar et al., 2001). A peculiar pattern of VHF return echoes revealed ESF structure with the bottomside envelope descending down to upper E region (Patra et al., 1997). In order to understand this peculiarity, a non-linear numerical simulation model investigation was carried out that revealed the importance of more than one mode as seed perturbations. Results from this simulation investigation also revealed the presence of down-drafting plasma enhancements in the altitude region well beyond 350 km. In order to verify the prediction of this simulation work and to experimentally establish the presence of plasma enhancement unequivocally, VHF radar observation alone was not sufficient as the signal-to-noise ratio of the return echo is proportional to the square of the fluctuations in electron densities$(\Delta N_e)^2$ at half the radar wavelength. The indications for plasma depletion or enhancement can be obtained from the Doppler velocities of plasma structures as the depletions, in general, move upward while the plasma enhancement draft downward. However, these indications, on occasions, can be misleading as the depleted structures in decaying phase of ESF were shown to move downward (Sekar et al., 2007). Thus the Doppler velocities inside the plasma structures cannot be used to identify plasma depletion/enhancement unambiguously. In this connection, the optical observations on OI airglow emissions (630.0 and 777.4 nm) are useful. The charge exchange of O^+ with O_2 followed by the dissociative recombination of the resultant O_2^+ with ambient electron is the principal source for the nighttime 630.0 nm airglow whose centroid of emission lies between 250 and 300 km. On the other hand, the radiative recombination of O^+ is responsible for 777.4 nm airglow emission and thus the centroid of this emission lies at the peak altitude of F-region of the ionosphere. Thus the intensity of 777.4 nm is proportional to N_e^2 assuming O^+ is the dominant ion at the peak of the F-layer while the intensity of 630.0 nm is proportional to N_e. Therefore, the temporal fluctuations in the intensities of these emission lines ($\Delta I_{630.0}$ and $\Delta I_{777.4}$) are proportional to electron density fluctuations (ΔN_e). Thus optical observations of airglow emissions were conducted in campaign mode from Gadanki (13.5°N, 79.2°E dip lat 6.4°N) in conjunction with VHF radar during January–March for past 5 years (2003–2007) and also during April 2006.

18.3 Experimental Details

The Indian VHF radar, operating at a frequency of 53 MHz, is located at Gadanki. The beam width of the radar is about 3° and a beam transverse to geomagnetic field is available to probe field-aligned ionospheric

Table 18.1 Radar specifications for the coordinated ESF and airglow experiments from Gadanki, India

Location	Gadanki (13.5°N, 79.2°E, dip lat. 6.4°N)
Frequency	53 MHz
Beam width (3 dB)	~3°
Beam direction	14.8°N (for F region studies)
Peak power-aperture product	3×10^{10} Wm^2
Antenna	32 × 32, 3 element orthogonal Yagi arrays
Inter-pulse period (IPP)	4 ms
Pulse-width	32 μs
No. of FFT points	256
Range coverage	84–588 km
Range resolution	4.8 km
Velocity range	± 354 ms^{-1}
Velocity resolution	2.76 ms^{-1}

irregularities. This VHF radar is coherent pulsed radar whose maximum peak power-aperture product can go up to 3×10^{10} Wm^2. The descriptions of this VHF radar system are discussed in detail by Rao et al. (1995). The operational specifications of the radar for the equatorial spread F experiments are given in Table 18.1.

A multi-wavelength airglow photometer was designed and fabricated at Physical Research Laboratory, Ahmedabad. The field of view of this photometer (3°) is chosen to coincide with the beam width of VHF radar. The field-of-view of the photometer being narrow, plasma structures are expected to stand out in the intensity variations more conspicuously. In addition, a temperature-tuned narrow band (0.3 nm) interference filter is used in this photometer. As the typical night sky background intensity closer to 630.0 nm is ~10 R/nm (Broadfoot and Kendall, 1968), this photometer enables detection of intensity fluctuations greater than 3R (1 Rayleigh, R = 10^6 photons cm^{-2}s^{-1}ster^{-1}). Therefore, this photometer is capable of measuring much less intensity fluctuations than the typical broad-band photometers wherein filters with band width >1 nm are employed. The integration time is in general 10 s for most of the airglow observations and on occasions, observations with high temporal resolution of 1 s are also obtained. By operating the photometer in a bi-directional mode along zonal and vertical directions and using identifiable features in the plasma structures, the zonal plasma drifts were determined. Experimental campaigns were conducted during January–March 2003–2007 and in April 2006 as a special campaign mode by operating simultaneously airglow photometer and VHF radar in ionospheric mode.

18.4 Results and Discussion

Figure 18.1 depicts composite results obtained from VHF radar and photometer on an ESF night (left panel) and non-ESF night (right panel). The figure consists of range-time-intensity (RTI) maps in Fig. 18.1a, e, range-time-velocity (RTV) maps in Fig. 18.1b, f that are obtained from VHF radar observations on an ESF and a non-ESF night. The HF Radar (ionosonde) data to monitor the ionospheric layer movement are also overlaid in one of the subplots of both the panels (Fig. 18.1c, g). In addition, variations in the vertical columnar intensities of 630.0 nm (Fig. 18.1c, g) and 777.4 nm (Figs. 18.1d, h) airglow emissions are also depicted in Fig. 18.1. Temporal variations in 630.0 and 777.4 nm airglow emission intensities reveal small (~15 min) and large (~a few hours) scale fluctuations. The small-scale variations in airglow are absent on non-ESF nights while the large-scale variations are present on both ESF and non-ESF nights. The large-scale variations correspond to the ionospheric layer movement while the small-scale variations are identified to be due to plasma structures associated with ESF. In addition to the commonly observed plasma depletions (D) with upward movement, the observation also reveals the presence of plasma enhancements (E) that move downwards (Sekar et al., 2004). Some of the plume structures at higher altitudes (~500 km) are found to be bifurcated similar to the results obtained by the simulation investigations of Huba and Joyce (2007). A thin bottomside spread F layer was noticed at 1915–2000 IST and during 2100–2200 IST in Fig. 18.1a. It is verified based on the bi-directional observations (zenith and

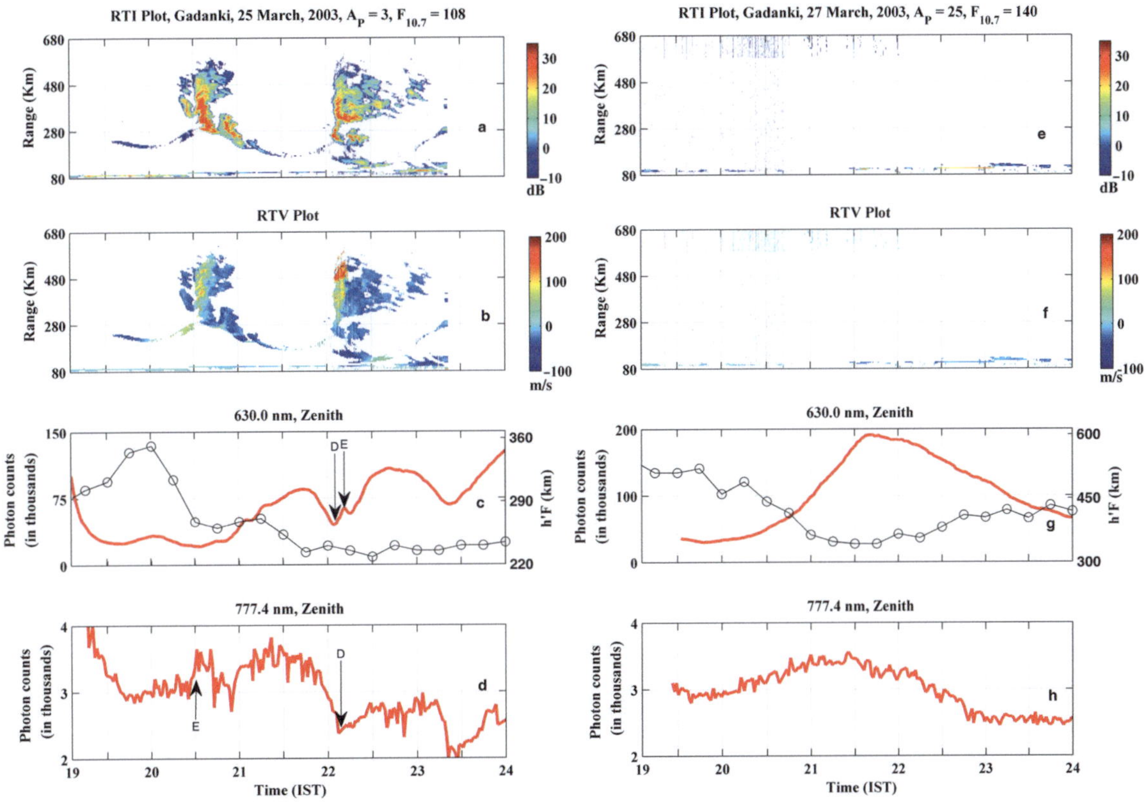

Fig. 18.1 Range-time-intensity (RTI) and range-time-velocity (RTV) maps on equatorial spread F (ESF) night (*left panel*) and non-ESF night (*right panel*) along with the corresponding emission intensities in 630.0 and 777.4 nm airglow lines (red curves). Large time scale variations are found to be anti-correlated with ionosphere layer height (h'F) variation shown by circles. Plasma depletions (D) and plasma enhancements (E) are identified using coordinated measurements (based on Sekar et al., 2004)

45° east) of 630.0 nm airglow intensities on this night that was reported earlier (Sekar et al., 2008), the zonal plasma drift was eastward at the 630.0 nm emission altitude region. However, as the F-region was devoid of ESF structures during these time intervals on this night, characteristic intensity signatures corresponding to ESF structures were not obtained in 777.4 nm airglow emission. Therefore, the association of this thin confined ESF structure with shear is not clear based on the present dataset. Moreover, plasma enhancement structures in the present case are found to ride over the plasma depletions. However, isolated structure, predominantly moving downward barring a small region wherein the velocities are upward, is also observed (Sekar et al., 2008). The generation of such isolated enhancement structure calls for detailed theoretical investigation.

Figure 18.2 depicts a few examples (Left panel: February 11, 2004 and Right panel: February 07, 2005) of ESF structures with wave-like bottom side structures in addition to the case discussed in Fig. 18.1. The plumes are found to ride over the bottom side structures (Fig. 18.2a, e) in these cases. The velocities (Fig. 18.2b, f) inside the bottom side structures are predominantly downward and over the plumes are predominantly upward except at a few times on March 25, 2003 (in Fig. 18.1) and February 11, 2004 (in Fig. 18.2). The optical measurements corresponding to these structures are depicted in the lower subplots (Fig. 18.2c, d, g, h) which are useful to identify nature of plasma structure. As the VHF radar acts as a slit camera (Woodman and La Hoz, 1976), temporally varying wave-like bottom side structures are the manifestation of spatial variation of plasma structures moving into the field of view of VHF radar from zonal direction. By making use of the estimation of zonal velocity obtained from the bi-directional optical observations, the zonal wavelength of the bottom side

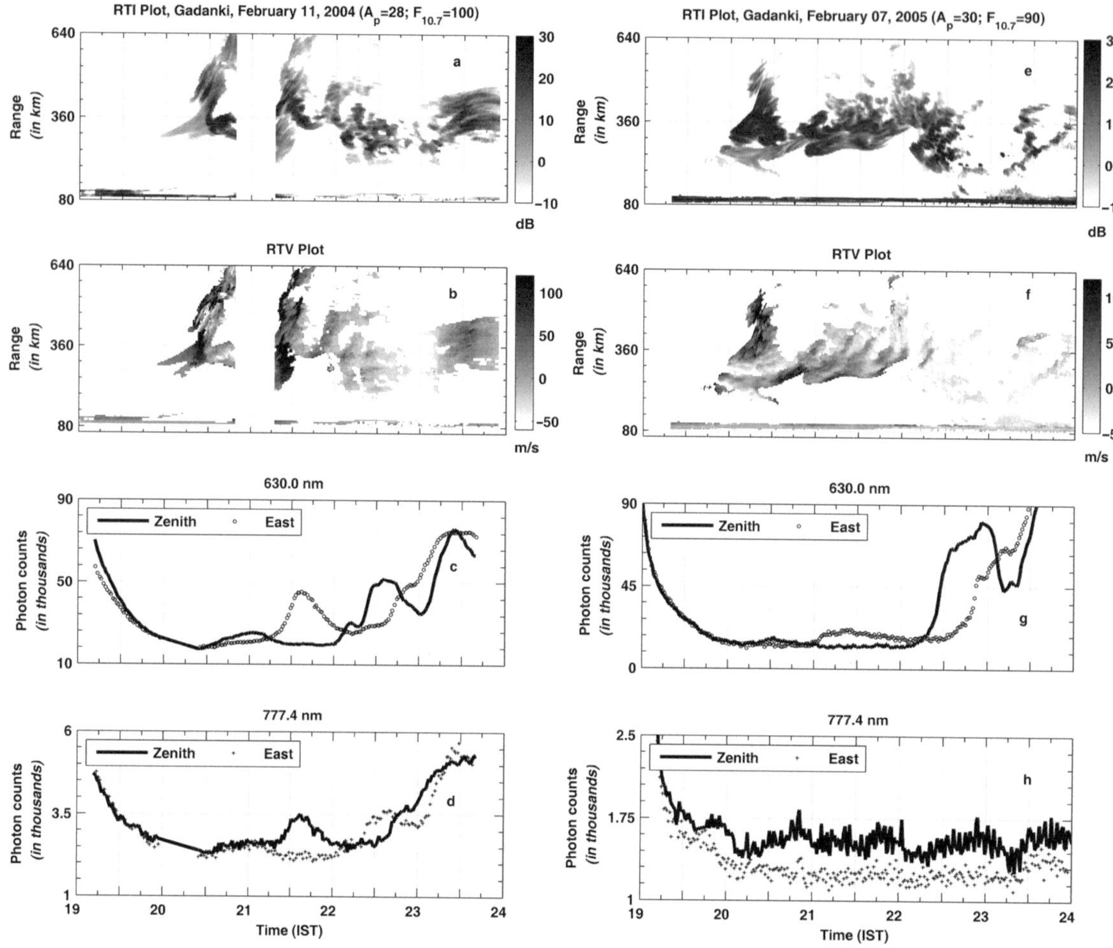

Fig. 18.2 Plasma structures and velocities along with airglow emission intensities are depicted in a sequence similar to Fig. 18.1. Plumes are found to ride over large scale bottomside structure (based on Sekar et al., 2008)

structure observed on March 25, 2003 was estimated to be around 600 km. As pointed out by Kelley et al. (1981), such large bottomside ESF structures are not uncommon and gravity waves are the probable candidates for seeding such a large bottom side structures. In contrast to the bottom side wave-like structure extended over a large altitude region, an undulated structure confined to a limited altitude (250–300 km) region without any plume structure was observed (Fig. 18.3) on March 19, 2004. The first two subplots in Fig. 18.3 depict the RTI (Fig. 18.3a) and RTV maps (Fig. 18.3b) on March 19, 2004. The temporal variations in the 630.0 nm airglow emission intensities over eastward direction reveal abrupt change around 2015 IST while a similar change occurs only at 2045 over zenith direction (Fig. 18.3c). This indicates that the confined structures around 250–300 km altitude region moved westward. However, the zonal plasma drift above 250–300 km altitude region is principally governed by the F-region dynamo during nighttime which drives the plasma eastward. This is due to the fact that the F-region dynamo during nighttime is driven by eastward wind determined by the thermal gradient in the thermosphere which is directed only eastward during nighttime. Thus the inference of westward drift around the emission altitude of 630.0 nm clearly indicates a vertical shear in the zonal plasma drift. This observation is consistent with the need for the vertical shear in the zonal plasma drift for the confinement of ESF structures as shown by Sekar and Kelley (1998).

The background conditions of the ionosphere during post-sunset hours are, in general, conducive for

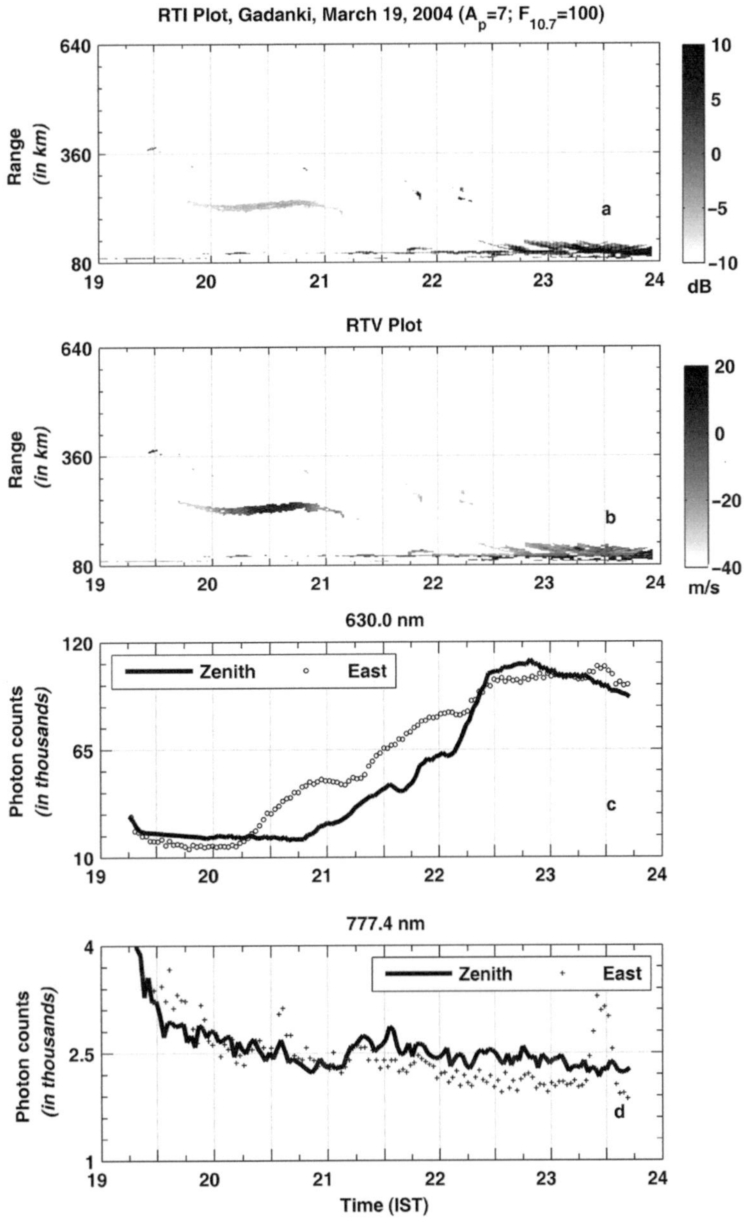

Fig. 18.3 ESF structure confined to lower altitude region along with airglow observations. This bi-directional OI 630.0 nm airglow observation paved way to identify shear in zonal plasma flow (refer text for detail, after Sekar et al., 2008)

the generation of plasma irregularity structures associated with ESF. A seed perturbation in electron density under the post- sunset background condition develops non-linearly into plasma bubbles that manifest as plasma plumes in VHF radar observations of meter scale-size plasma irregularity structures. The structures subsequently decay by midnight hours and remain weak structure known as "fossil bubbles" which cannot support meter scale-size irregularities in general. As the structures are weak they move with the ambient plasma drift in the eastward (e.g. Bhattacharyya et al., 2001) and downward directions. The scattering of radio waves by these "fossil bubbles" was recorded by scintillation receivers (Basu et al., 1978) and not recorded by VHF radar. The neutral wind drives secondary instability structures on the western wall of these "fossil bubbles" and makes them active with a weak turbulent structure. The presence of such "active fossil bubbles" was identified during an occasion (Sekar et al., 2007) using a co-ordinated optical and VHF radar observations. Figure 18.4 depicts RTI and RTV maps as well as range-time-width (RTW) map (in units of velocity) on April 26–27, 2006 in Fig. 18.4a–c respectively. Figure 18.4d, e depicts the variation in 630.0 nm airglow intensity observed from the radar site at Gadanki and variation in h'F over Thumba. The ESF structures were weak and less turbulent (as indicated by the RTI and RTW maps

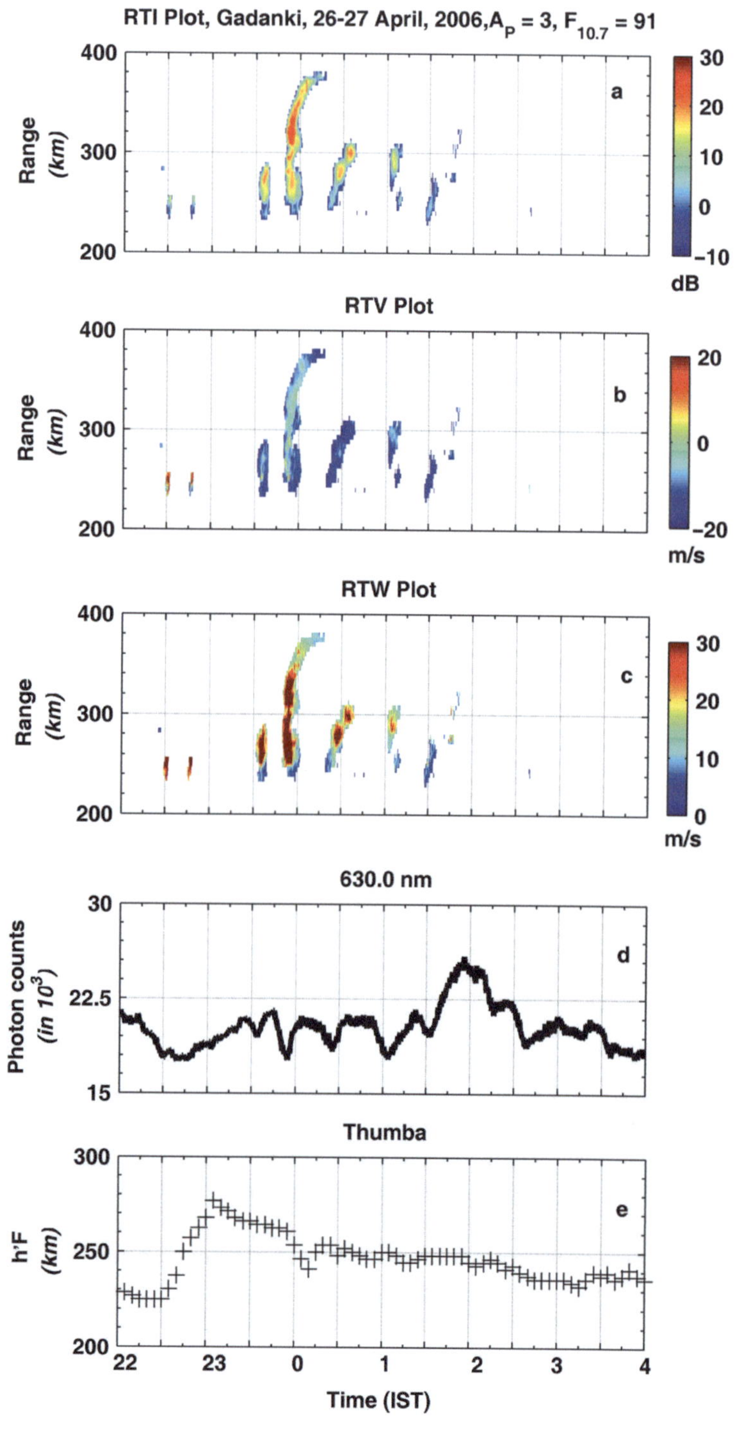

Fig. 18.4 The plasma depleted structures moving downward during the decaying phase of an ESF event that are identified as "active fossil bubbles". The range-time-width (RTW) map enables to determine the degree of turbulence (after Sekar et al., 2007)

respectively) and moved into the vicinity of the VHF radar field of view during midnight. A few depletions in airglow intensities (Fig. 18.4d) are unambiguously identified as plasma depletions as ionospheric layer height did not change significantly. The optical observations on this occasion clearly revealed that those structures are depleted structures and are located on the western walls of the bubbles. Unlike the normal plasma bubble, these structures moved predominantly downward (Fig. 18.4b) similar to the "fossil bubbles".

As mentioned earlier, the observations were conducted in a campaign mode from Gadanki using airglow photometer along with VHF and HF radars. During these observational periods, most of the nights were magnetically quiet and a few of them were magnetically disturbed. Further, most of the nights, the observations were marked with presence of ESF while a few nights without the presence of ESF structures (for example, February 12 and 20, 2004 in the right and left panels of Fig. 18.5 respectively). A comparison of optical observations made during two non-ESF nights led to an important result. It is to be noted here that the VHF radar provides only the structure and dynamics of E-region irregularities on non-ESF nights. Comparison of 630.0 nm airglow emission intensities on a magnetically disturbed non-ESF night (12 February 2004) with magnetically quiet non-ESF night (20 February 2004) revealed the presence of small (~30 min–1.0 h) scale-size fluctuations (Fig. 18.5a) during the disturbed night. The residual fluctuations are found to correlate well with the variations in the F-layer height on the same night (Fig. 18.5c). In order to find the reason for the airglow fluctuations, interplanetary electric field (IEF) values are derived from the observations of Advanced Composition Explorer (ACE) satellite. Propagation lags are calculated for each datum point (Chakrabarty

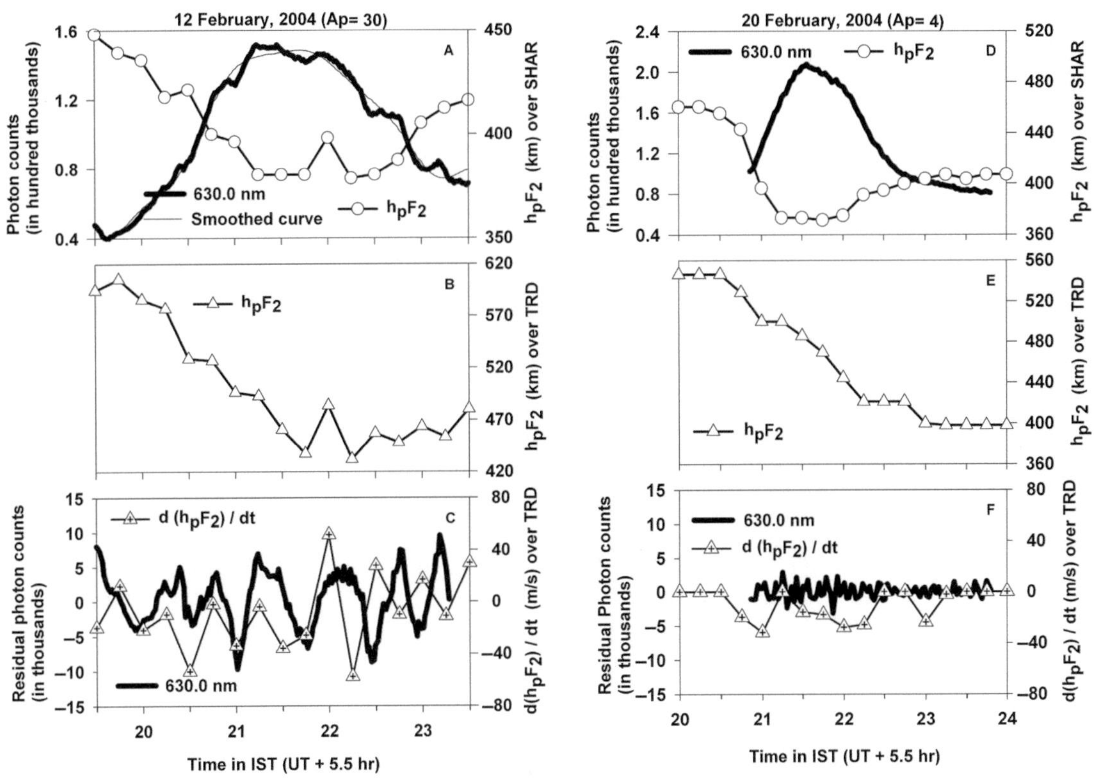

Fig. 18.5 The airglow observations on 630.0 nm emission and ionospheric layer height variations (h_pF_2) on a magnetically disturbed night (*left panel*) in comparison with magnetically quiet night (*right panel*). Note that ESF (*left panel*) was not present on both the nights (based on Chakrabarty et al., 2005)

et al., 2005) and variation in the lag-corrected dawn-to-dusk component of IEF (IEFy) is plotted in Fig. 18.6. A detailed correlation analysis was carried out using this propagation lag corrected IEF (Fig. 18.6a), ionosphere layer height variation (Fig. 18.6b) and the fluctuations in 630.0 nm airglow intensities (Fig. 18.6c). Harmonic analyses (Fig. 18.6d) and cross-spectrum analyses (Fig. 18.6e) revealed that certain (~0.5 and 1.0 h) periodic components in the IEF affect low latitude ionosphere and subsequently affect the airglow variations (Chakrabarty et al., 2005). In order to ensure that these fluctuations originated from magnetosphere, variations in the horizontal component of magnetic field from four magnetometer stations in the Indian sector were also examined (Fig. 18.7) for February 12, 2004 (Fig. 18.7a) and February 20, 2004 (Fig. 18.7b). Same periodicities are also found to be present in the horizontal magnetic field variations from equator to a location well beyond the Sq focus on the magnetically disturbed night in comparison with the quiet night (Fig. 18.7c). This clearly indicates that the current responsible for the fluctuations in horizontal

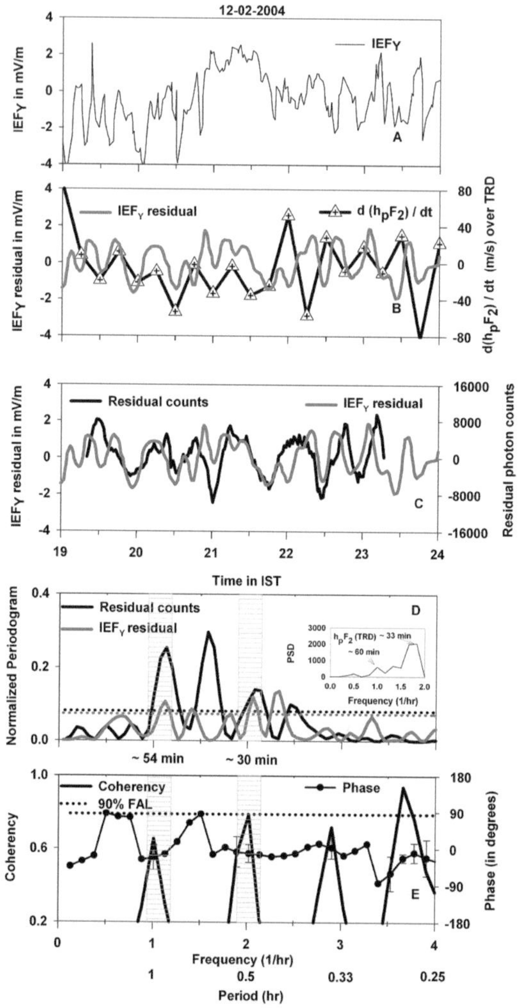

Fig. 18.6 The Y (dawn-to-dusk component in Geocentric Solar Ecliptic coordinate system) component of interplanetary electric field (IEF) as deduced from Advanced Composition Explorer (ACE) satellite measurement on a magnetically disturbed night and the similarities in the residual IEFy with the residuals in ionospheric layer height variations and 630.0 nm airglow emission intensities. Cross-spectrum analysis between IEFy and airglow intensities reveals the imprint of 30 and 60 min components of IEFy on 630.0 nm airglow over low latitudes (after Chakrabarty et al., 2005)

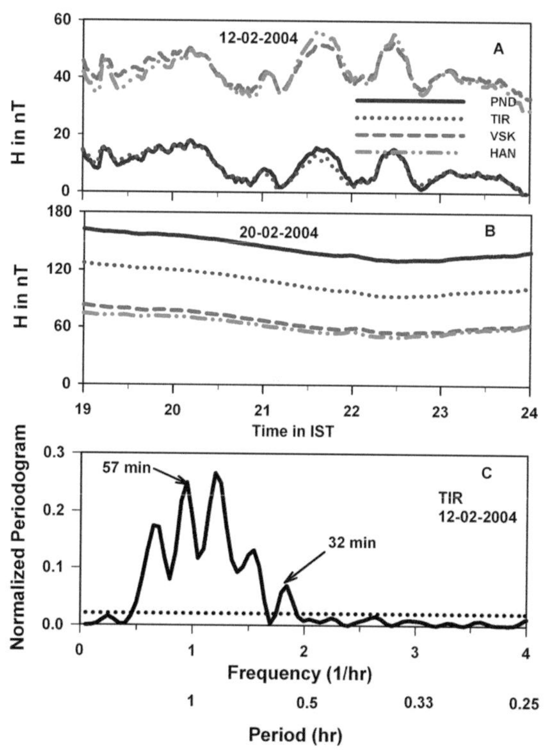

Fig. 18.7 The instantaneous values of horizontal component (H) of magnetic field over India on a magnetically disturbed night (12 February, 2004) and quiet night (20 February, 2004). Spectral components with periodicites 30 min and 60 similar to IEFy and airglow variations are observed. This observations of H at the dip equatorial station (Trivandrum, TIR), the intermediate stations (Visakhapatanm, VSK and Pondicherry, PND) and at a station well beyond Sq focus region (Hanle, HAN), reveal that the current responsible is magnetospheric in origin (after Chakrabarty et al., 2005)

component of magnetic field is of magnetospheric origin. Thus the results from three independent techniques confirmed the effect of IEF on the neutral airglow emission. Further, Chakrabarty et al. (2005) estimated the amplitude of these airglow fluctuations to be 5–10R for a typical nighttime intensity level of 50–100R. The narrow spectral band airglow photometer employed in the present investigations is capable of detecting intensities greater than 3R unlike the widely employed broad-band photometers (>1 nm) which can detect intensities more than 10R. Moreover, as the field-of-view of the photometer is narrow, small features in airglow intensity were not smeared out. Thus the importance of narrow spectral-band and narrow field-of-view observations of 630.0 nm airglow is evident in recording the effects of space weather events on neutral OI 630.0 nm airglow emission.

The implications of IEF over low-latitude ionosphere are multi-fold. Another event, when the prompt penetration of IEF into the equatorial ionosphere triggered the development of ESF, was captured by the radars and airglow photometer (Right panel of Fig. 18.8). The layer movement of the ionosphere on January 7, 2005 was found to be downward between 1900 and 2000 IST (Fig. 18.8R2). This condition normally stabilizes the perturbations in the bottom side of the ionosphere and therefore, not conducive for the occurrence of ESF on this night. At that time, the IEF was shown (Chakrabarty et al., 2006) to penetrate promptly into the low latitude ionosphere (Fig. 18.8R1) and the polarity of the zonal electric field reversed that made the equatorial ionosphere to move upward (Fig. 18.8R2). As a consequence, the ESF was found to get triggered around 2040 IST (Fig. 18.8R3). The linear growth rate calculations were performed by Chakrabarty et al. (2006) and it was shown that the growth rate was negative during 1900–2000 IST which became positive at 2030 IST with the growth time of about 10 min. The development of ESF, as recorded by the VHF radar, is consistent with these calculations. Further, a reversal in the electric field of the ionosphere was also noticed around 2100 IST (Fig. 18.8R2). The 630.0 nm airglow observations (Fig. 18.8R4) were found to corroborate with the ESF structures. Bi-directional observations of 630.0 nm (Fig. 18.8R4), as discussed earlier, indicated the presence of shear in the altitude region of ~300 km. Thus the ESF structure during 2040–2140 IST remained confined to a narrow altitude region without any plume structure during this time interval. The zonal electric field over equatorial ionosphere reversed once again around 2200 IST owing to the over-shielding effects of IEF. A plasma plume was found to get resurrected under the influence of the eastward electric field associated with the over-shielding effects of IEF. The optical observations and the effects of prompt penetration of IEF in triggering the development of ESF and subsequent development of plasma plume under the influence of over-shielding electric fields are described in detail by Chakrabarty et al. (2006).

It is generally observed that penetration of IEF into Earth's ionosphere initiates geomagnetic storm. It was initially believed that the occurrence of ESF is inhibited during magnetic storm. Later, by segregating the ESF events based on their occurrence time, the post-midnight ESF was found to be triggered by storm (e.g. Aarons et al., 1980) while the post-sunset ESF events are found to be correlated with the effects of magnetic storm on some occasions and not so on other occasions. Abdu (1997) had suggested that the triggering or inhibition of ESF depends on the phase of the magnetic storm. Two contrasting events exhibiting variabilities in the occurrence of ESF during magnetic storms are encountered during the coordinated optical and radar campaign periods from Gadanki. Comparison of these two storm events (Fig. 18.8) reveals that the ionospheric electric fields (Fig. 18.8L2 and R2) reversed its polarity from westward to eastward around the same local time of 2145 on both the occasions due to the effects of over-shielding electric field (Fig. 18.8L1 and R1). Further, the magnitude and the duration of reversal (~0.5 h) to eastward direction remained same for both the events. However, during one of the events, the development of pre-midnight plume was observed after ~0.5 h of the reversal of the electric field from westward to eastward (Fig. 18.8R3), while ESF was found to be absent during the other event (Fig. 18.8L3). The important difference between the two cases is the reversal of the polarity of the zonal electric field just after sunset (2000 h) on the ESF night compared to the non-ESF event (Fig. 18.8L2 and R2). On a non-ESF night, the prompt penetration of certain frequency components of IEF in the equatorial ionosphere is observed without any eastward excursion of the electric field during 1930–2100 h. The development of plasma plume during nighttime condition, when the peak height of the layer is around 350 km, is highly unlikely as revealed

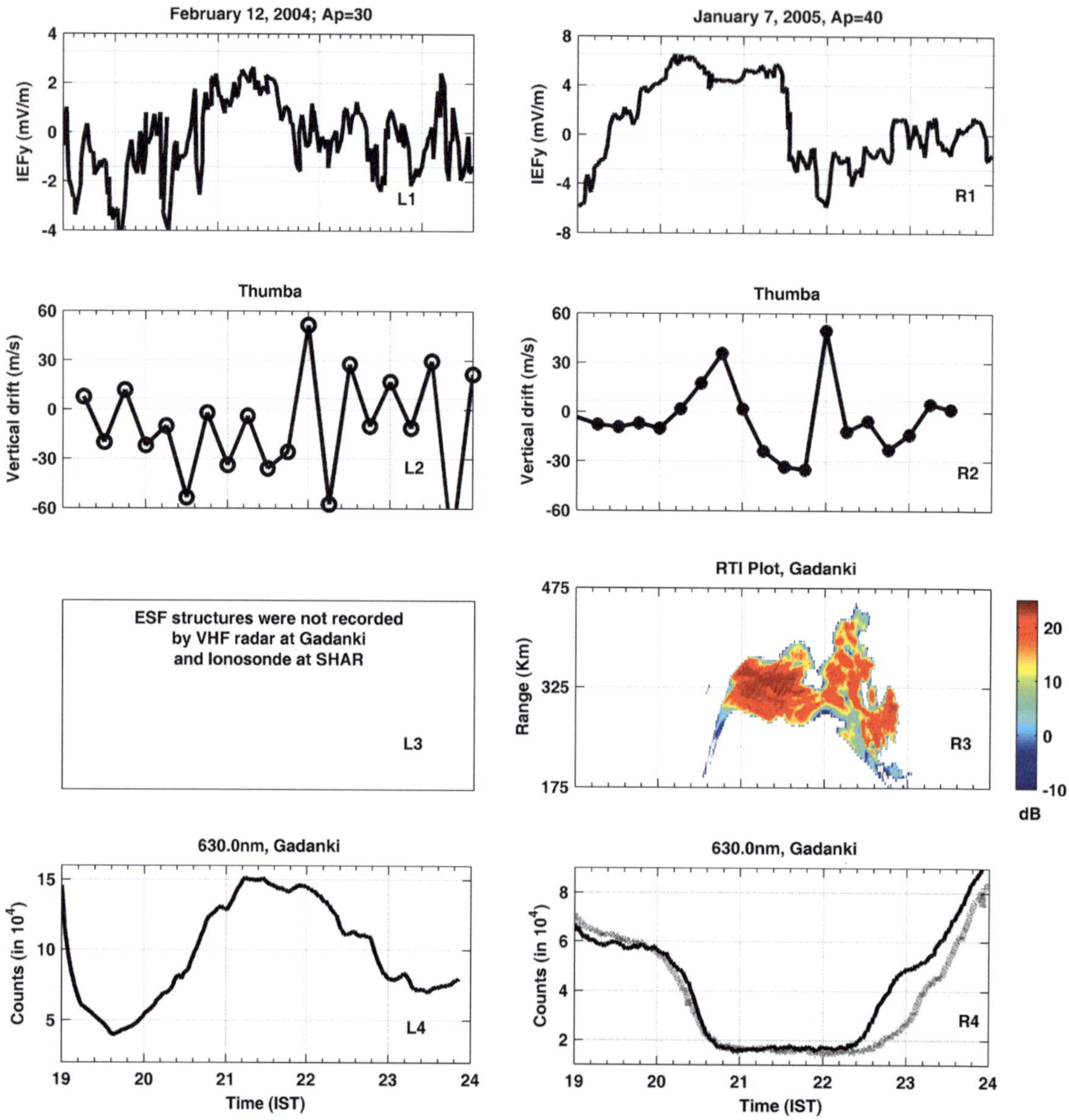

Fig. 18.8 Comparison of interplanetary, ionospheric electric fields, development of ESF and 630.0 nm airglow on two magnetically disturbed nights. The ESF was found to trigger on January 7, 2005 due to prompt penetration of IEFy at 2000 IST over dip equator and a generation of a plume event after 2200 IST was observed. None of these was recorded on 12 February 2004 (after Sekar and Chakrabarty, 2008b)

by earlier simulations (Ossakow et al., 1979; Ossakow, 1981). In order to understand the development of late night plume event on one of those occasions, simulations were carried out using a two dimensional non linear simulation model of ESF (Raghavarao et al., 1992; Sekar et al., 1994). The model describes the evolution of plasma bubble under perturbed conditions by solving the basic plasma fluid equations under F region approximations which reduces them into two coupled partial differential equations.

$$\nabla.\,(\upsilon_{\text{in}} N \nabla \Phi) = B\left[-g + W_y \upsilon_{\text{in}} + \left(\frac{E_{xo}}{B}\upsilon_{\text{in}}\right)\right]\frac{\partial N}{\partial x} \tag{18.1}$$

$$\frac{\partial N}{\partial t} - \frac{\partial}{\partial x}\left[\left(\frac{N}{B}\right)\left(E_{yo} + \frac{\partial \Phi}{\partial y}\right)\right] + \frac{\partial}{\partial y}\left[\left(\frac{N}{B}\right)\left(-E_{xo} + \frac{\partial \Phi}{\partial x}\right)\right] = -\upsilon_R N \tag{18.2}$$

In the above equations, N, Φ, g, ν_{in}, E_{xo}, E_{yo}, ν_R, B and W_y represent electron number density, perturbation potential, acceleration due to gravity, ion-neutral collision frequency, x-component (positive westward) of background electric field, y-component (positive upward) of background electric field, recombination rate, geomagnetic field strength and vertical wind. The detailed descriptions of this model and the scientific results obtained from this model are available in the literature (Sekar et al., 1994; Sekar, 2003; Sekar and Chakrabarty, 2008b). Thus initial perturbations with superposition of two modes are taken as follows.

$$N(x,y) = N_o(y) \left[1 - \left[a_1 \cos\left(\frac{2\pi}{\lambda_1}x\right) + a_2 \cos\left(\frac{2\pi}{\lambda_2}x\right)\right]\right] \tag{18.3}$$

A larger (λ_1) wavelength mode of 240 km and another mode with wavelength (λ_2) of 60 km without any phase difference between them are used in the present investigation. The amplitude of the second wavelength mode (a_2) is chosen to be ~5% of the background electron density and assumed to remain constant over an entire altitude region. However, the amplitude of the longer wavelength mode (a_1) is varied over a limited altitude region where steep gradient (small plasma scale length, L) exists to support the presence of large-scale irregularities developed during post-sunset hours. In the remaining altitude regions, the amplitudes are kept constant at a value of 5% of the background electron density. These aspects are discussed in detail in Sekar and Chakrabarty (2008b). As the investigation pertains to pre-midnight (2130–2200) hours, the altitude profile of electron density was chosen (Fig. 18.9a) to represent pre-midnight condition based on the morphological studies of true height analyses carried out (Chandra et al., 1973) over Indian zone. Further, to represent the electric field conditions with and without the effects of over-shielding, the vertical plasma drifts as depicted in Fig. 18.9b are used as inputs to the model. Figure 18.10 depicts the simulation outputs for three different initial conditions.

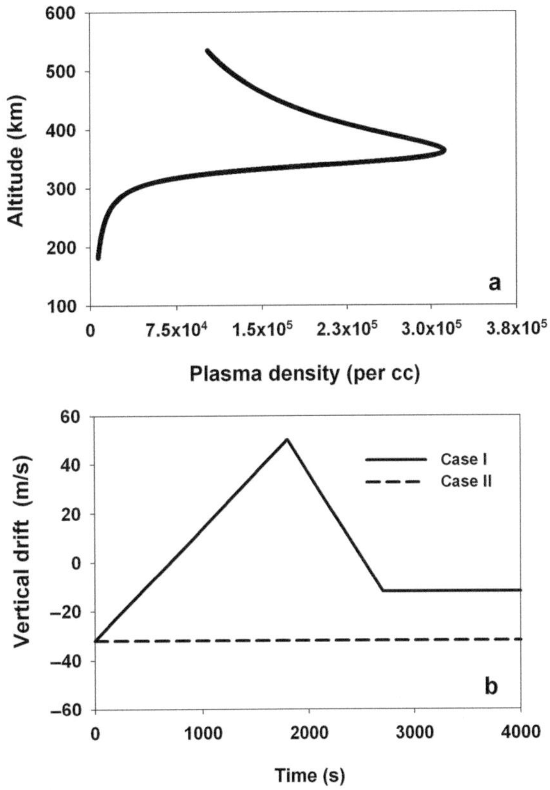

Fig. 18.9 A typical electron density profile during pre-midnight hours over Indian zone is used as input to the simulation model. The vertical drifts corresponding to different zonal electric field conditions are represented in case-I and case-II. Case-I represents the presence of the eastward electric field due to over-shielding effect and Case-II represents the absence of such effect during pre-midnight hours (based on Sekar and Chakrabarty, 2008b)

Each initial condition is defined by the plasma scale length (L), Height of the F-layer peak, the amplitudes of both the wavelengths of perturbation (a_1 and a_2) and the background electric field (E). Iso-electron density contours in zonal and vertical plane 2000s after the initiation of instability process are plotted in each subplot of Fig. 18.10 for three different initial conditions. The development of plasma bubble during pre-midnight condition penetrating up to an altitude of 450 km within 2000s, similar to the observation, is not noticed when the amplitude of large scale perturbation is only 5% of the ambient plasma density along with storm-induced eastward electric field (Fig. 18.10a). A second simulation (Fig. 18.10b) was carried out with 35% amplitude of the ambient plasma density as initial perturbation for larger wavelength mode and the other

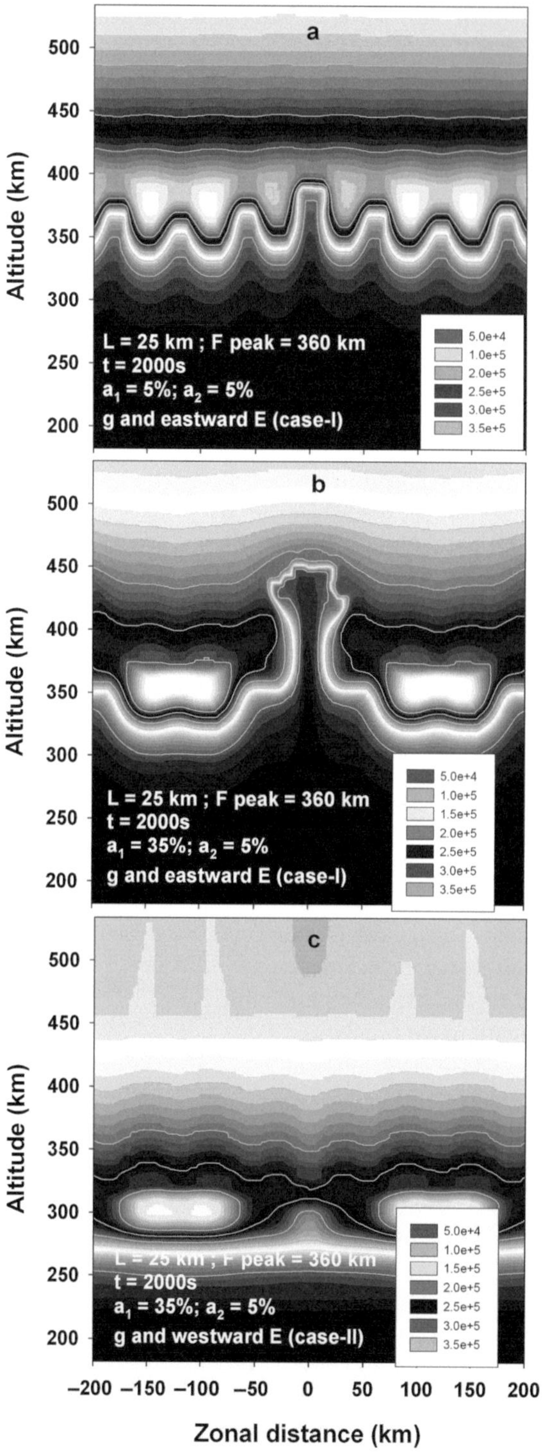

Fig. 18.10 Non-linear numerical simulation output in the form of iso-electron density contours for the development of plasma bubble for initial perturbations of 5, 35, and 35% and in typical pre-midnight ionospheric conditions. In the first two cases, eastward electric field (Case-I of Fig. 18.9) is present and in the last case (Case-II of Fig. 18.9), typical nighttime westward electric field is present (based on Sekar and Chakrabarty, 2008b)

conditions remaining the same as that of previous one. The result reveals the development of plasma bubble extending up to 450 km within 2000s similar to that of the observations. Thus these results of the simulations emphasize the need for larger amplitude of initial perturbation in the form of pre-seed for the development of plasma plume during pre-midnight hours. A third simulation (Fig. 18.10c) was carried out with the same conditions as that of second simulation except without the effect of over-shielding electric field into equatorial ionosphere (i.e. usual nighttime westward electric field as shown in Fig. 18.9). This simulation reveals that the initial perturbations are mitigated without the formation of plasma bubble. Further multiple simulations were carried out to quantify the amplitudes of the required amount of pre-seed under varying initial altitude profiles of electron densities that have been observed from one night to another. Thus these simulations reveal that the storm-induced eastward electric field during nighttime over-dip-equator region is a necessary but not a sufficient condition for the development of pre-midnight plume structure. In addition to the storm-induced eastward electric field, a pre-seed in varying degrees in a localized altitude region and a suitable background ionospheric condition is needed for the development of late night plume event. The confined ESF irregularities developed in post-sunset hours are suggested to provide such large amplitude perturbations during pre-midnight hours.

As already mentioned, PRE plays a very crucial role in the generation of ESF. In one peculiar case (Fig. 18.11), it is found that equatorial F layer continued its ascent to very high altitude till 2130 h

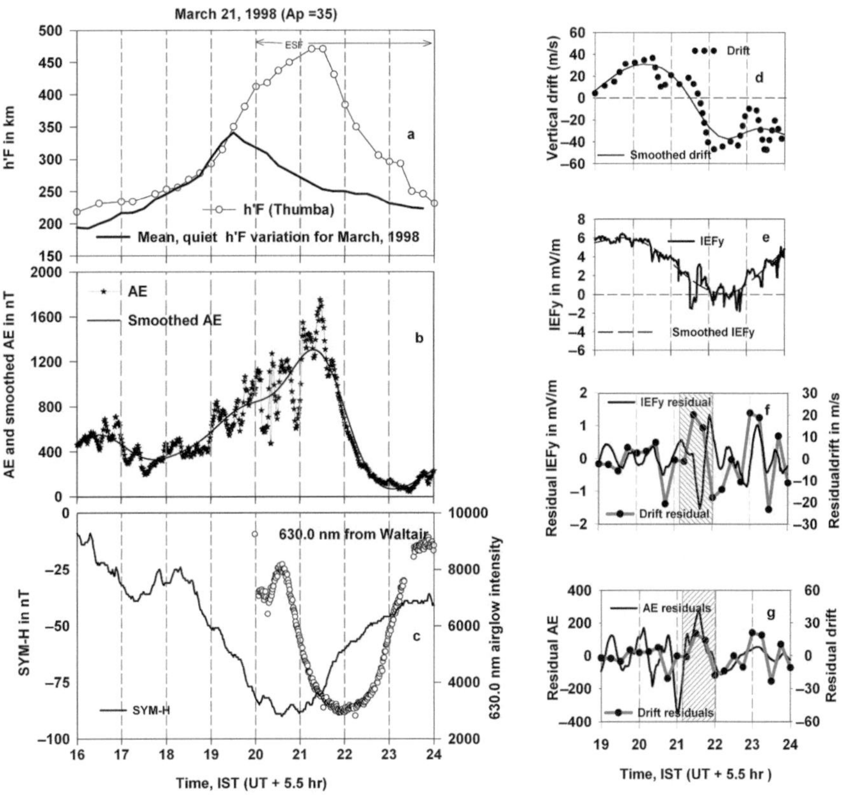

Fig. 18.11 A case of unusual post-sunset ascent of equatorial F layer that vary in tandem with the slowly varying component of auroral electrojet index, AE. The fast fluctuations are found to vary in accordance with the fast fluctuations in IEFy except during the onset of a substorm expansion phase (marked by *hatched box*) when drift residuals follow AE residuals. The variation in the ring current and the 630.0 nm airglow intensity variation from Waltair (dip lat 10.09°N) are also shown (based on Chakrabarty et al., 2008)

instead of coming down as expected by typical variation (Fig. 18.11a). Critical inspection revealed that although the duration is marked by the main phase of a geomagnetic storm (Fig. 18.11c), the slowly varying component of ionospheric F layer over dip equator followed the slowly varying component in the variation of the auroral electrojet current (marked by AE indices) as revealed by Fig. 18.11b. It is interesting to note that the 630.0 nm airglow intensity variation from a low latitude station also indicated the ascent of F layer during this time (Fig. 18.11c). Interestingly, it is found that the small scale fluctuations (~40–45 min fluctuations) in the vertical plasma drift (Fig. 18.11d) over dip equator (derived from the temporal changes in the ionospheric height) were controlled by IEF during most of the time (Fig. 18.11e) in this interval except when a substorm is triggered with the reversal in IEF polarity (over-shielding) as shown in Fig. 18.11f, g. During the interval (~30–40 min) when substorm is triggered, the fluctuations in the ionospheric zonal electric field were found to be governed by the substorm-related transient electric field (Chakrabarty et al., 2008). The investigation brought out the distinctive effects of storm and substorm on the equatorial ionosphere during post-sunset hours.

18.5 Summary

Co-ordinated multi-technique observations including narrow-band airglow observations from the low latitude regions in India as well as numerical simulation investigations bring out the following salient results.

- Experimental evidences are obtained for the presence of plasma enhancement structures during ESF as predicted by numerical simulation model.
- "Active Fossil Bubbles" are identified during post-midnight hours.
- It is shown that IEF affects thermospheric airglow emission even over low latitude region. Quasi-periodic variations (0.5 and 1.0 h) in IEF are found to affect the OI 630.0 nm nighttime airglow emission over low latitude.
- Similar quasi-periodic (0.5 and 1.0 h) variations are also found in the nighttime northward magnetic field variations over the whole Indian subcontinent implying the currents responsible for this variation magnetospheric are of origin.
- Prompt penetration of IEF is shown to trigger ESF during post-sunset hours. Linear growth rate analysis supports the observations.
- Over-shielding effects of IEF is found to be responsible for the resurrection of plasma plume during pre-midnight hours.
- Numerical simulations elicit that storm-time eastward over-shielding electric field is a necessary but not a sufficient condition for the pre-midnight resurrection of plasma plume. A large amplitude pre-seed perturbation is needed. Post-sunset ESF irregularities in a limited altitude region are suggested as pre-seed perturbation.
- Distinctive effects of IEF and substorm on equatorial ionosphere are identified. While the quasi-periodic (~40–45 min) fluctuations in the vertical drift are found to be controlled by IEF, the slowly varying component of ionospheric height over dip equator follows the slowly varying component of auroral electrojet index, AE.

Acknowledgements One of the authors (RS) thanks Dr. Archana Bhattacharyya for the invitation to present this work in the 11th Scientific Assembly of IAGA (2009) held at Sopron, Hungary. The authors acknowledge Dr. M. A. Abdu for the invitation to be a part of the IAGA Special Sopron Book Series. The authors also acknowledge the data obtained from National Atmospheric Research Laboratory (Gadanki), Space Physics Laboratory (Thiruvananthapuram), Indian Institute of Geomagnetism (Navi Mumbai) and Indian Institute of Astrophysics (Bangalore) as well as ACE Science Center, WDC-C2 (Kyoto), the 210 MM magnetometer chain. The inputs from Dr. S. P. Gupta during the preparation of this article and the discussions with Dr. J. H. Sastri on substorm physics are duly acknowledged. This work is supported by Department of Space, Government of India.

References

Aarons J et al (1980) Seasonal and geomagnetic control of equatorial scintillations in two longitudinal sectors. J Atmos Solar-Terr Phys 42:861–866

Abdu M (1997) Major phenomena of the equatorial ionosphere thermosphere system under disturbed conditions. J Atmos Solar-Terr Phys 59:1505–1519

Basu S, Basu S, Aarons J, McClure JP, Cousins MD (1978) On the coexistence of kilometer- and meter-scale irregularities in the nighttime equatorial F region. J Geophys Res 83: 4219–4226

Bhattacharyya A, Basu S, Groves KM, Valladares CE, Sheehan R (2001) Dynamics of equatorial F region irregularities from spaced receiver scintillation observations. Geophys Res Lett 28:119–122

Blanc M, Richmond AD (1980) The ionospheric disturbance dynamo. J Geophys Res 85:1669–1686

Broadfoot LA, Kendall KR (1968) The airglow spectrum, 3100-10,000A. J Geophys Res 73(1)426–428

Chakrabarty D, Sekar R, Narayanan R, Devasia CV, Pathan BM (2005) Evidence for the interplanetary electric field effect on the OI 630.0 nm airglow over low latitude. J Geophys Res 110:A11301. doi:10.1029/2005JA011221

Chakrabarty D, Sekar R, Narayanan R, Patra AK, Devasia CV (2006) Effects of interplanetary electric field on the development of an equatorial spread F event. J Geophys Res 111:A12316. doi: 10.1029/2006JA011884

Chakrabarty D, Sekar R, Sastri JH, Ravindran S (2008) Distinctive effects of interplanetary electric field and substorm on nighttime equatorial F layer: a case study. Geophys Res Lett 35:L19108. doi:10.1029/2008GL035415

Chandra H, Rajaram G, Rastogi RG (1973) Electron density distribution over the magnetic equator. Ind J Radio Space Phys 2:243–250

Devasia CV, Jyoti N, Subbarao KSV, Viswanathan KS, Tiwari D, Sridharan R (2002) On the plausible linkage of thermospheric meridional winds with the equatorial spread F. J Atmos Solar-Terr Phys 64:1–12

Ebihara Y, Nishitani N, Kikuchi T, Ogawa T, Hosokawa K, Fok M-C (2008) Two-dimensional observations of overshielding during a magnetic storm by the Super Dual Auroral Radar Network (SuperDARN) Hokkaido radar. J Geophys Res 113:A01213. doi:10.1029/2007JA012641

Fejer B, Scherliess L, de Paula ER (1999) Effects of the vertical plasma drift velocity on the generation and evolution of equatorial spread F. J Geophys Res 104: 19859–19869

Huang C-S, Foster JC, Kelley MC (2005) Long-duration penetration of the interplanetary electric field to the low-latitude ionosphere during the main phase of magnetic storms. J Geophys Res 110:A11309. doi:10.1029/2005JA011202

Huba JD, Joyce G (2007) Equatorial spread F modeling: multiple bifurcated structures, secondary instabilities, large density 'bite-outs,' and supersonic flows. Geophys Res Lett 34:L07105. doi:10.1029/ 2006GL028519

Kakad B, Jeeva K, Nair KU, Bhattacharyya A (2007) Magnetic activity linked generation of nighttime equatorial spread F irregularities. J Geophys Res 112:A07311. doi:10.1029/2006JA012021

Kelley MC, Fejer BG, Gonzales CA (1979) An explanation for anomalous equatorial ionospheric electric fields associated with a northward turning of the interplanetary magnetic field. Geophys Res Lett 6:301–304

Kelley MC, Larsen MF, LaHoz C, McClure JP (1981) Gravity wave initiation of equatorial Spread F: a case study. J Geophys Res 86:9087–9100

Kelley MC, Nicolls MJ, Anderson D, Anghel A, Chau JL, Sekar R, Subbarao KSV, Bhattacharyya A (2007) Multi-longitude case studies comparing the interplanetary and equatorial ionospheric electric fields using an empirical model. J Atmos Solar-Terr Phys 69: 1174–1181

Kikuchi T, Lühr H, Schlegel K, Tachihara H, Shinohara M, Kitamura T-I (2000) Penetration of auroral electric fields to the equator during a substorm. J Geophys Res 105:23251–23261

Nishida A (1968) Coherence of geomagnetic DP 2 fluctuations with interplanetary magnetic variations. J Geophys Res 73:5549–5559

Ossakow SL (1981) Spread-F theories-A review. J Atmos Solar-Terr Phys 43:437–452

Ossakow SL, Zalesak ST, McDonald BE, Chaturvedi PK (1979) Nonlinear equatorial spread F: Dependence of altitude of F peak and bottomside background electron density gradient scale length. J Geophys Res 84:17–29

Patra AK, Anandan VK, Rao PB, Jain AR (1995) First observations of equatorial spread F from Indian MST radar. Radio Sci 30(4):1159–1165

Patra AK, Rao PB, Anandan VK, Jain AR (1997) Radar observations of 2.8 m equatorial spread-F irregularities. J Atmos Solar-Terr Phys 59(13):1633–1641

Peymirat C, Richmond AD, Kobea AT (2000) Electrodynamic coupling of high and low latitudes: simulations of shielding/overshielding effects. J Geophys Res 105:22991–23003

Raghavarao R, Gupta SP, Sekar R, Narayanan R, Desai JN, Sridharan R, Babu VV, Sudhakar V (1987) *In-situ* measurements of winds, electric fields and electron densities at the onset of equatorial spread-F. J Atmos Solar-Terr Phys 49:485–492

Raghavarao R, Sekar R, Suhasini R (1992)Nonlinear numerical simulation of equatorial spread-F: effects of winds and electric fields. Adv Space Res 12:227–230

Rao PB, Jain AR, Kishore P, Balmuralidhar P, Damle SH, Vishwanathan G (1995) Indian MST radar, system description and sample vector wind measurements in ST mode. Radio Sci 30:1125–1138

Rastogi RG, Patel VL (1975) Effect of interplanetary magnetic field on ionosphere over the magnetic equator. Proc Indian Acad Sci 82A:121

Reddy CA, Somayajulu VV, Devasia CV (1979) Global scale electrodynamic coupling of the auroral and equatorial dynamo regions. J Atmos Solar-Terr Phys 41:189–201

Rishbeth H (1971) Polarization fields produced by winds in the equatorial F region. Planet Space Sci 19:357–369

Sastri JH (2002) Equatorial geomagnetic and ionospheric effects of substorms. Ind J Radio Space Phys 31:309–320

Sastri JH, Ramesh KB, Karunakaran D (1992) On the nature of substorm-related transient electric field disturbances in the equatorial ionosphere. Planet Space Sci 40:95–103

Sastri JH, Rao JVSV, Ramesh KB (1993) Penetration of polar electric fields to the night side dip equator at times of geomagnetic sudden commencements. J Geophys Res 98:17517–17523

Sekar R (2003) Plasma instabilities and their simulations in the equatorial F region – recent results. Space Sci Rev 107: 251–262

Sekar R, Chakrabarty D (2008a) Equatorial spread-F research in India: a brief review. Ind J Radio Space Phys 37:7–27

Sekar R, Chakrabarty D (2008b) Role of overshielding electric field on the development of pre-midnight plume event: simulation results. J Atmos Solar-Terr Phys 70:2212–2221

Sekar R, Chakrabarty D, Narayanan R, Patra AK (2008) Thermospheric airglow intensity variations corresponding to

various ESF structures revealed by VHF radar maps. Ann Geophys 26:3863–3873
Sekar R, Chakrabarty D, Narayanan R, Sripathi S, Patra AK, Subbarao KSV (2004) Characterization of VHF radar observations associated with equatorial spread F by narrow-band optical measurements. Ann Geophys 22:3129–3136
Sekar R, Chakrabarty D, Sarkhel S, Patra AK, Devasia CV, Kelley MC (2007) Identification of active fossil bubbles based on coordinated VHF radar and airglow measurements. Ann Geophys 25:2099–2102
Sekar R, Kelley MC (1998) On the combined effects of vertical shear and zonal electric field patterns on nonlinear equatorial spread F evolution. J Geophys Res 103: 20735–20747
Sekar R, Kherani EA, Rao PB, Patra AK (2001) Interaction of two long wavelength modes in the nonlinear numerical simulation model of equatorial spread F. J Geophys Res 106:24765–24775
Sekar R, Suhasini R, Raghavarao R (1994) Effects of vertical winds and electric fields in the nonlinear evolution of equatorial spread-F. J Geophys Res 99:2205–2213
Senior C, Blanc M (1984) On the control of magnetospheric convection by the spatial distribution of ionospheric conductivities. J Geophys Res 89:261–284
Spiro RW, Wolf RA, Fejer BG (1988) Penetration of high latitude electric field effects to low latitudes during SUNDIAL 1984. Ann Geophys 6:39–50
Sridharan R, Chandra H, Das SR, Sekar R, Sinha HSS, Pallam Raju D, Narayanan R, Raizada S, Misra RN, Raghavarao RN, Vyas GD, Rao PB, Ramarao PVS, Somayajulu VV, Babu VV, Danilov AD (1997) Ionization hole campaign – a co-ordinated rocket and ground based study at the onset of equatorial spread -F: first results. J Atmos Solar-Terr Phys 59:2051–2067
Tulasi Ram S, Rama Rao PVS, Prasad DSVVD, Niranjan K, Gopi Krishna S, Sridharan R, Ravindran S (2008) Local time dependent response of postsunset ESF during geomagnetic storms. J Geophys Res 113:A07310. doi:10.1029/2007JA012922
Vasyliunas VM (1970) Mathematical models of magnetospheric convection and its coupling to the ionosphere. In: McCormac BM (ed) Particles and fields in the magnetosphere. Springer, New York, NY pp 60–71
Wolf RA (1970) Effects of ionospheric conductivity on convective flow of plasma in the magnetosphere. J Geophys Res 75:4677–4698
Woodman RF, La Hoz C (1976) Radar observations of F region equatorial irregularities. J Geophys Res 81:5447–5466

Chapter 19

Radar and Optical Observations of Irregular Midlatitude Sporadic E Layers Beneath MSTIDs

David L. Hysell, Tatsuhiro Yokoyama, Elnana Nossa, Russell B. Hedden, Miguel F. Larsen, John Munro, Steven Smith, Michael P. Sulzer, and Sixto A. González

Abstract An irregular sporadic E ionization layer was observed by the Arecibo incoherent scatter radar and a coherent scatter radar imager located on St. Croix during the passage of an MSTID observed by the Boston University all-sky camera in 630 nm imagery. The MSTID in question was not very intense and was barely detectable in the vertical F region plasma drifts measured by Arecibo. The intensity of the coherent scatter from small-scale irregularities in the sporadic E layer appeared nonetheless to be modulated by the MSTID and was strongest in the F region airglow crests, mapped along magnetic field lines to the E region volume being observed. The coherent scatter Doppler shifts were highly correlated with altitude displacements in the sporadic E layer, and the sign of the correlation was controlled by the sign of the background zonal electric field. The MSTID did not appear to modulate the morphology of the sporadic E layer irregularities themselves, which took the form of convective rolls and which drifted with the ambient neutral wind.

19.1 Introduction

Midlatitude sporadic E ionization layers have been affecting human activity since the earliest days of radio and remain an important means of long-path communications in the HF and VHF bands (e.g, see reviews by Whitehead, 1972, 1989; Mathews, 1998). Interest in the electrodynamics of the layers grew in the 1990s after Riggin et al. (1986) and Yamamoto et al. (1991, 1992) found that they could be populated by intense, small-scale field-aligned irregularities (FAIs) and probed using coherent scatter radar. The echoes have been termed "quasiperiodic" or QP because of apparent periodicities in range-time-intensity representations of the observations. The echoes have since been observed throughout the Asian and American sectors and over a wide span of latitudes. Rocket experiments have shown that they are associated with irregular sporadic E layers and accompanying, strong polarization electric fields and also with strong neutral wind shear (e.g. Fukao et al., 1998, Yamamoto et al., 2005 and references therein). Studies using radar interferometry and imaging have shown that the echoes are frequently organized spatially along frontal structures extending in oblique directions, separated by a few tens of kilometers, and propagating at speeds of a few tens of meters per second (e.g. Chu and Wang, 1997; Hysell and Burcham, 2000, Hysell et al., 2002b, Saito et al., 2006). More recent studies incorporating incoherent scatter observations have also been able to associate the QP echoes with billowy density irregularities of the kind first observed at Arecibo by Miller and Smith (1978) and Smith and Miller (1980) (e.g. Hysell et al., 2004; Larsen et al., 2007; Hysell et al., 2009) and attributed to neutral dynamical instability.

The E region structures are also reminiscent of frontal structures observed in the midlatitude F region associated with medium scale traveling ionospheric disturbances (MSTIDs). These structures are aligned northwest-to-southeast in the northern hemisphere (southwest-to-northeast in the southern). While they

D.L. Hysell (✉)
Department of Earth and Atmospheric Sciences, Cornell University, Ithaca, NY 14853, USA
e-mail: dlh37@cornell.edu

M.A. Abdu, D. Pancheva (eds.), A. Bhattacharyya (Coed.), *Aeronomy of the Earth's Atmosphere and Ionosphere*, 269
IAGA Special Sopron Book Series 2, DOI 10.1007/978-94-007-0326-1_19, © Springer Science+Business Media B.V. 2011

have been observed with incoherent scatter radar (Behnke, 1979) and all-sky imagers (Miller et al., 1997; Garcia et al , 2000; Otsuka et al 2004), they are especially vivid in GPS-TEC maps (e.g. Saito et al., 1998; Tsugawa et al., 2007). Field-aligned irregularities (FAIs) are also now known to be embedded in the frontal structures (Fukao et al., 1991; Saito et al., 2002, 2008).

What, if any, is the relationship between patchy sporadic E layers and QP echoes and MSTIDs and the F region plasma irregularities within them? While there are some superficial morphological similarities, the wavelengths and scale sizes of the E and F region phenomena are dissimilar, and the occurrence phenomenology is also different.

Experimentally, irregularities in sporadic E layers have been found to take the form of patchy, elongated, roll-like deformations. Irregularity fronts appear to propagate with periods of a few minutes, wavelengths of a few tens of km, and directions preferentially toward the southwest or northeast in the northern hemisphere, although direction can vary considerably from event to event or even within an event. The fronts are strongly polarized, presenting electric fields often large enough to excite Farley-Buneman instability (Haldoupis and Schlegel, 1994). FAIs exist throughout the patchy layers even when the condition for Farley Buneman instability is not met. It is typical for the irregularities to come in bursts.

MSTIDs meanwhile exhibit periods of the order of an hour, wavelengths of hundreds of km, and propagation speeds of several hundred m/s. They propagate strictly southwestward in the northern hemisphere. They are also polarized. Small-scale field-aligned irregularities have been detected in the troughs of MSTIDs observed in all-sky imagery (Otsuka et al., 2009). Observations show E region FAIs occurring on the same geomagnetic field lines as MSTID wavefronts observed in airglow and by GPS-TEC (Otsuka et al., 2007; Saito et al., 2007). Not only the intensity of the coherent scatter from the FAIs but also the Doppler shifts have been found to be controlled by the phase of the MSTID on common magnetic field lines (Otsuka et al., 2007.) Other, particularly tantalizing experiments have been performed recently in Japan, where a portable radar was temporarily situated to observe E region FAIs on the same flux tubes as F region FAIs seen simultaneously by the MU radar (personal communication, T. Yokoyama, 2009.) In preliminary observations, the F region FAIs that appeared generally formed on geomagnetic field lines with strong E region FAIs. The reverse was not true. Polarization electric fields in patchy sporadic E layers were evidently strong enough to create not only E region but also F region FAIs in the most opportune phase of an ongoing, strong MSTID. The E and F regions were the source and load, respectively. The role of any kind of feedback was not obvious in this preliminary work.

Analysis of MSTIDs often invoke the mechanism due to Perkins (1973) (see also Miller, 1996; Hamza, 1999). In this instability, field-aligned slabs of plasma displaced upward (downward) have reduced (increased) flux tube integrated Pedersen conductivities due to the height variation of the ion-neutral collision frequency. Reduced (increased) Pedersen currents flow in the slabs as a result. An induced polarization electric field normal to the slabs maintains a continuous current in that direction. The currents parallel to the slabs are unmatched, however, and so the $\mathbf{J}\times\mathbf{B}$ force on the slabs is different in different phases of the perturbation. The vertical components of the unmatched forces, $((\Sigma_{\pm}\mathbf{E}_{\parallel}' \times \mathbf{B}) \cdot \hat{z}$ contributes to the force balance that maintains the heights of the slabs. Here, $\Sigma_{\pm}$ is the flux tube integrated Pedersen conductivity in upward and downward displaced slabs, $\mathbf{E}_{\parallel}{}'$ is the component of the effective electric field parallel to the slab, $\mathbf{E}' = \mathbf{E} + \mathbf{u} \times \mathbf{B}$, $\mathbf{u}$ is the neutral wind velocity, and $\hat{z}$ is the vertical direction. The tendency is for upward (downward) displaced slabs to be forced upward (downward) and for instability. The linear, local growth rate of the instability is proportional to the force mismatch:

$$\gamma = \frac{|E'| \cos I}{BH} \sin(\theta - \alpha) \sin(\alpha)$$

where θ and α are the angles the effective electric field and the wave normal make to the eastward direction, respectively, in the plane perpendicular to $\mathbf{B}$, I is the magnetic inclination, and H is the neutral scale height. The growth rate is intuitive, since $(\sin(\theta - \alpha)$ selects the component of the effective electric field parallel to the slabs and $\sin(\alpha)$ selects the vertical component of the associated $\mathbf{J} \times \mathbf{B}$ force. Growth is fastest when $\alpha = \theta/2$.

The strength of the Perkins theory is that it predicts favored propagation directions. For predominantly zonal F region winds, the growth rate is maximized for southwestward propagation in the northern hemisphere. The weakness is that the predicted growth time is very long – hours or more for realistic conditions. Absent extraordinary forcing, Perkins would seem to be irrelevant. C. E. Seyler (personal communication) performed an eigen-analysis of electrostatic waves in the midlatitude ionosphere and reinterpreted Perkin's theory this way: while there are no intrinsic, fast-growing solutions in the presence of ordinary background forcing, waves which are not NW-SE aligned experience rapid damping, regardless of the driving mechanism.

While some experimental results point to neutral instability as the source of irregularities in sporadic E layers, Cosgrove and Tsunoda (2002, 2004) have attributed them to wind shear-driven E_S plasma instabilities. Their mechanism may be viewed as an E region analog of the Perkins instability. In this case, slabs of ionization displaced vertically in the vicinity of the wind shear node will experience different wind forcing, and the slabs will become polarized to the extent the current forcing is otherwise divergent. The vertical component of the polarization electric field will then drive vertical ion drifts in phase with the original displacements, and instability results. Tsunoda (2006) derived an expression for the linear, local growth rate of the E_S layer instability which is formally similar to that of the Perkins instability, only with the scale length of the vertical shear replacing the neutral atmospheric scale height. For zonal wind forcing, this growth rate is proportional to $(\cos(\alpha)\sin(\alpha)$, the former term selecting the components of the wind parallel to the wavefronts (which alone cause polarization), and the latter term selecting the vertical component of the polarization electric field (which alone leads to wave growth). Like the Perkins instability, the E_S instability has a favored propagation direction ($\alpha = \pi/4$), although this arises from the assumption of predominantly zonal E region winds, which is typically not what we observe. Cosgrove (2007a) considered meridional winds, finite layer thickness, and wavelength-dependent F region loading in a followup analysis of the instability.

Cosgrove and Tsunoda (2004) have argued that the growth rate of the Perkins and E_S instabilities acting together is larger than that of either one acting alone. The hypothesis has been tested in a number of numerical investigations (Cosgrove 2007b; Yokoyama et al., 2008, 2009). The last of these included rather exhaustive and well documented runs of a fully three-dimensional initial value simulation code. The results are summarized as follows: (1) Strong frontal E region irregularities (no matter what their source) can most easily couple to and deform the F region when the wavefronts are aligned NW-SE. The F region deformations can grow much more rapidly than would be possible under Perkins instability. However, Perkins can then sustain and amplify the deformation, once created. (2) Rotational wind shear in the E region can cause the E region frontal structures to remain in phase, or nearly so, with the F region MSTIDs, thus increasing the coupling efficiency. (3) Large-scale deformations in the F region conductivity can then spatially modulate, to some degree, the occurrence of E region irregularities, potentially explaining QP-echo clustering and burst timescales. Overall, however, the E region is the source, and the F region is the load.

19.2 Observations

Observations of the midlatitude ionosphere were made between July 17–26, 2009, concentrated in the first few hours after local sunset. The Arecibo incoherent scatter radar, lidars, Boston University all-sky imager, and sounder were involved in the study. Also involved was the St. Croix 30 MHz coherent scatter radar imager, which operates remotely and observes field-aligned plasma density irregularities in the E region over Puerto Rico (Hysell et al., 2009). The intent of the study was to identify causal relationships between large-, intermediate-, and small-scale plasma density irregularities occurring together in the E and F region.

Strong sporadic E ionization layers were observed on the evenings of July 17/18, 19/20, and 22/23. The last of these events was the most long-lived and intense and was also accompanied by favorable optical conditions, and this Chapter concentrates on that event. The E_S layer in question was already visible in ISR observations commencing at 1600 LT but intensified at 1800 LT, when the blanketing frequency rose from 3 to over 6 MHz. Between 2000 and 2030 LT, the blanketing frequency fell abruptly from 6 to below 3 MHz, and the layer became patchy. The layer maintained a top frequency of 6 MHz until local midnight and dissipated completely by 0200 LT on July 23. By comparison, foF2 decreased from about 4.8 to 4.0 MHz between 2030 and 2200 LT.

The Boston University all-sky imager captured red- and green-line imagery serially from 2000 LT through 0200 LT. The integration time for each line was 90 s, and the overall cycle time was 212 s. The red-line observations were dominated by a medium-scale traveling ionospheric disturbance (MSTID) which was present at the start of observations and persisted throughout the entire period. A representative image frame is shown in Fig. 19.1. This image has been unwarped and projected to a reference altitude of 250 km. The period of the MSTID at this time was about 20 min, and the wavelength about 275 km, implying a phase speed of about 230 m/s in the Earth-fixed frame. The propagation direction varied over time but was always approximately to the southwest.

Figure 19.2 shows Arecibo incoherent scatter data from a time preceding the E_S layer emergence through 2200 LT, when ISR observations ceased. The data are plotted here to emphasize the characteristics of the F region. The top (bottom) two panels show data from the linefeed (Gregorian) feed systems, which were run simultaneously and directed vertically (at 15° zenith angle). Data were acquired using the coded long-pulse mode, which ran for 40 s every minute, and the MRACF seven-frequency mode, which ran for the remaining 20 s. The incoherent integration time applied to data from both modes was 10 s. (See Sulzer, (1986a, b) for a description of these modes.) Grayscales are a proxy for the relative electron density, constructed from the first lag of the coded long-pulse autocorrelation function, corrected for range, normalized to the noise power level, and plotted on an arbitrary scale. This proxy measurement is most useful at night, when the electron and ion temperatures are equal and the F region temperature is approximately uniform. Plotted symbols with error bars represent line-of-sight plasma drifts averaged over the F region. The plotting convention here is that positive values denote drifts away from the radar.

The linefeed data in Fig. 19.2 depict a disturbed F region bottomside, with variations in layer height tracking the corresponding variations in the corresponding vertical velocity measurement. There is considerable fine-structure in the vertical drifts, with

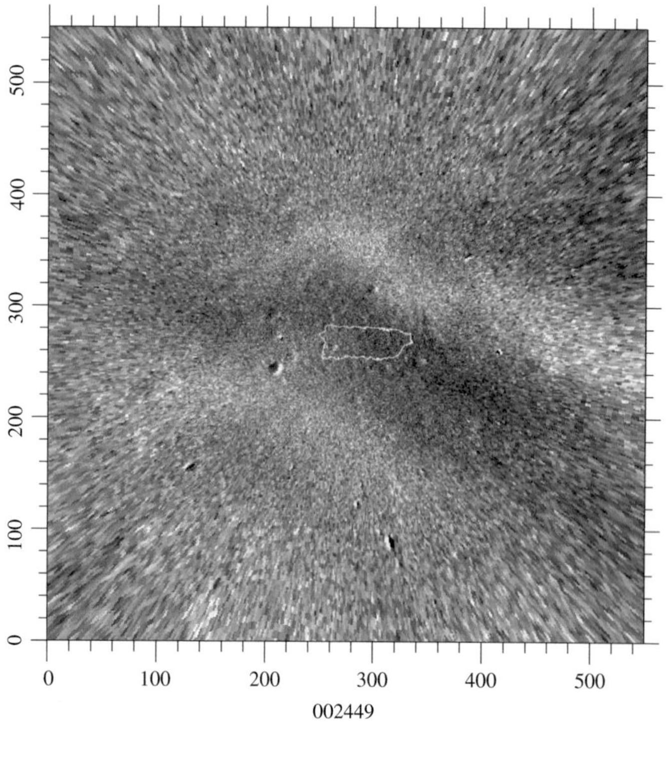

Fig. 19.1 630 nm image of an MSTID over Arecibo on 23 July, 00:24:49 UT (UT = LT + 4 hr). The dimensions of the image are (1050 km)^2 at a reference altitude of 250 km

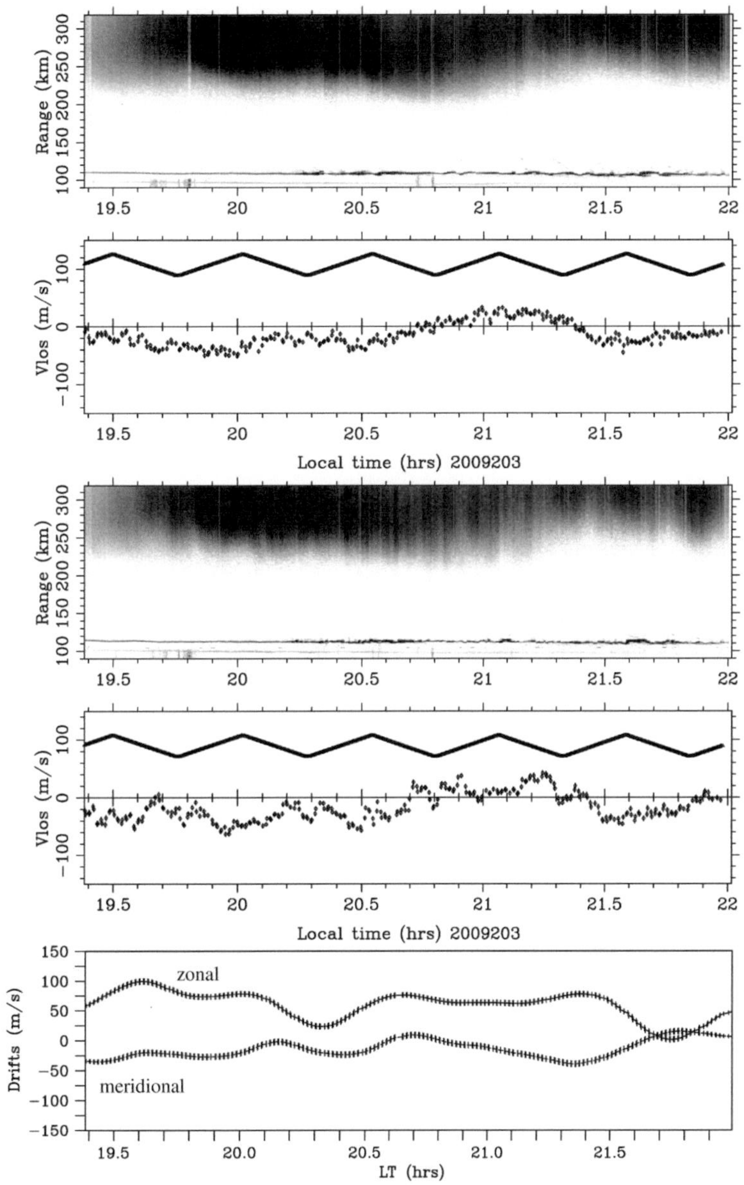

Fig. 19.2 Arecibo radar observations of an irregular sporadic E ionization layer beneath and F region disturbed by an MSTID. *Top panel*: relative plasma density versus range and time as observed by the vertically-looking linefeed system at Arecibo. *Second panel*: average line-of-sight F region plasma drifts observed by the linefeed. *Third panel*: relative plasma density versus range and time as observed by the Gregorian feed system, which was pointed at 15° zenith angle. *Fourth panel*: average line-of-sight F region plasma drifts observed by the Gregorian feed system. The pointing azimuth is also indicated here. *Fifth panel*: estimates of the zonal and meridional F region plasma drift speeds derived from a regularization procedure

periods between 5 and 60 min. evident. The fine structure precedes the onset of and is not obviously correlated with the occurrence of E_S layer irregularities. Similar remarks hold for the Gregorian system data, which are more structured still. This is evidence of significant horizontal heterogeneity in the ionosphere. Note that the symbols in the fourth panel in the figure represent the azimuth angle of the Gregorian beam, which varied between 0 and 360° here, such that the beam pointed northward when the symbols are at either extreme excursion in the plot.

The bottom panel of Fig. 19.2 shows estimates of the zonal and meridional horizontal F region plasma drift speed, derived from the linefeed and Gregorian datasets using a regularization procedure (see Sulzer et al. (2005) for details.) These curves reflect zonal and meridional plasma drift estimates which are consistent with available data and also maximally smooth in the sense of the L2 norm of their second temporal derivatives. They are calculated assuming spatial homogeneity, which is a necessary if unjustified assumption. Because of inhomogeneity and the need for significant damping in the regularization, fine structure is unresolved by the procedure, and only broad trends are recovered. The curves suggest mainly southeastward plasma drifts, which are consistent with climatology for this season and local time.

The 30 MHz St. Croix coherent scatter radar imager was used to monitor field-aligned plasma density irregularities (FAIs) in the E region over Arecibo throughout the experiments. Three-dimensional images of the radar backscatter from nonthermal 5-m FAIs in the ionosphere can be constructed with a time resolution of a few Seconds. Details about the radar and the way in which spaced-receiver data are processed to produce the images can be found in Hysell et al. (2009).

Figure 19.3 shows two representative examples of the results, projected here into two-dimensional plan view. Patches of color indicate the regions of space from which backscatter is received. The white contours are the loci of perpendicularity for the St. Croix radar site and indicate the altitudes where the Bragg scattering condition for FAIs is satisfied. Refraction and finite aspect sensitivity effects permit finite deviations about the loci. Most of the backscatter shown here arrived from regions close to the 110 km contour.

Numerous previous measurements of this kind have shown a tendency for the backscatter to be organized into wavelike bands propagating most often to the northeast or southwest. Both the power and the Doppler shifts have been so organized, suggesting the propagation of polarized waves in the background ionization. The period and wavelengths of the waves has been typically 10 min and 30 km, respectively, with some variability (See for example Hysell et al., 2004 and Larsen et al., 2007). This phenomenology is all but absent in the dataset in question, however. The upper panel in Fig. 19.3 shows a single example when clear wavelike behavior was briefly evident. The 630 nm airglow imagery at this time showed the crest of an intense, southwest-propagating MSTID overhead in the F-region volume connected to the E region by magnetic field lines. Much more common was the appearance of disconnected, drifting scattering

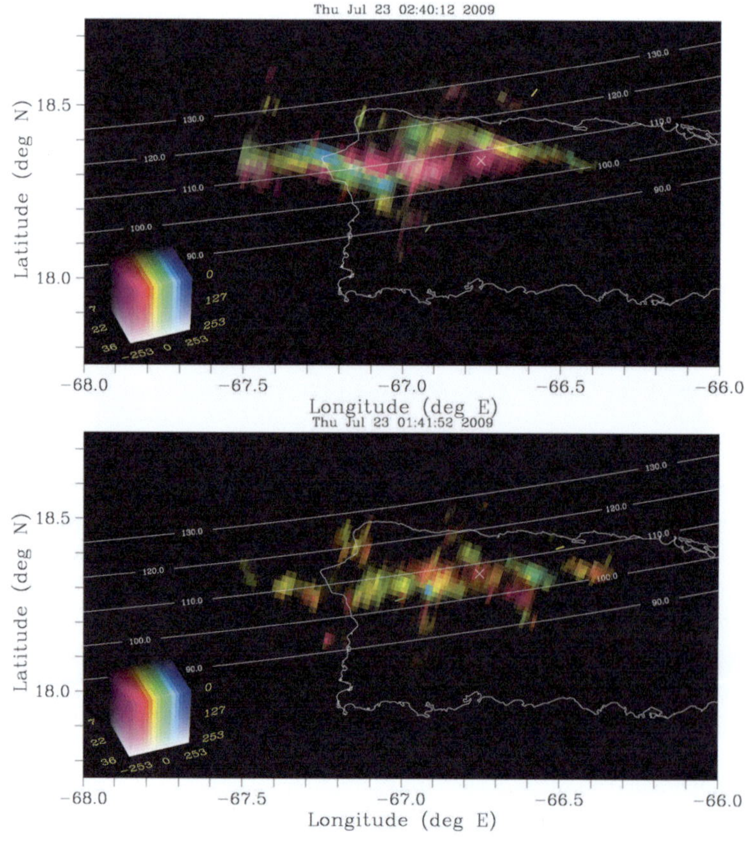

Fig. 19.3 Radar images from 0240 and 0142 UT, July 23, 2009, respectively (UT = LT + 4 hr.) The brightness, hue, and saturation of the colors in the images represent signal-to-noise ratio on a dB scale, Doppler velocity, and spectral width, respectively. Positive (negative) velocities imply blue (red) shifts and are color-coded accordingly

centers like those shown in the lower panel of Fig. 19.3. Scattering centers such as these predominated between 2015 and 2200 LT in July 22, exhibiting southeastward drifts before 2100 LT and southwestward drifts afterward. The lifetimes of the scattering centers was no more than a few minutes before about 2100 LT. After 2100 LT, the lifetimes became considerably longer, with some sites persisting in the radar imagery for the entire time it took them to drift from the eastern to the western edges of the images.

Notice that some of the Doppler shifts measured approached 250 m/s. This speed is comparable to the ion acoustic speed for iron ions, and such large Doppler shifts are therefore possibly indicative of the presence of Farley Buneman waves in the E_S layers.

19.3 Synthesis

Here, we begin to synthesize the results from the ISR, coherent scatter radar, and 630 nm imager. Figure 19.4 combines 630 nm imager data in the form of a keogram with a range-time-intensity (RTI) plot of the coherent scatter radar data, plotted on a common time scale. The keogram was calculated by integrating the optical images along diagonal cuts running from northwest to southeast. There are 1100 distinct diagonal cuts, and the integrated irradiance for each cut versus time represents a different row of the keogram. The vertical extent of the keogram spans 1485 km. MSTIDs propagating to the southwest are clearly revealed as diagonal traces. The horizontal red line demarks the cut that passes through the F region volume at 250 km altitude that maps along magnetic field lines to the E region directly over Arecibo.

The lower panel of Fig. 19.4 shows a RTI representation of the coherent scatter radar data. The color scheme used for this plot is the same as was used for the radar images in Fig. 19.3. The range from the St. Croix radar to the E region over Arecibo is about 240 km and is indicated by the red horizontal line.

The combined datasets indicate a relationship between the F region MSTIDs and the E region coherent scatter. Specifically, the 20 min MSTID periodicity appears to be present also in the bursts of coherent scatter. The cyan vertical lines superimposed on the figure indicate the times when MSTID nodes intersect the horizontal red line in the keogram, with reduced (enhanced) airglow falling on the left (right) side of the lines by convention. Tracing the lines down to the horizontal red line through the RTI diagram, we can see that coherent scatter over Arecibo generally occurs to the right of the intersection point, i.e. in the enhanced airglow phases of the MSTIDs. The rule has exceptions. No MSTID signatures accompany the coherent scatter observed between 0115 and 0145 UT (2115 and 2145 LT). From Fig. 19.2, however, it is clear that the F region ascended in this period such that the there was little ionization left at the 630 nm emission height, and the keogram consequently shows weak emissions at this time. Strong backscatter was also observed over Arecibo at about 0245 UT in the reduced phase of an MSTID. Otherwise, the correspondence between the coherent scatter over Arecibo and the F region airglow on common magnetic field lines is striking.

The phase speeds of the MSTIDs seen in the airglow imagery were between 200 and 300 m/s to the southwest or south-southwest. Meanwhile, the scattering sites in the coherent scatter radar images move with velocities of a few tens of m/s at most. They appear to drift to the southeast before about 0100 UT and to the southwest or west afterward. They do not themselves keep pace or remain in phase with the MSTIDs. What Fig. 19.4 implies is that the condition for FAI generation in the E region is locked to the phase of the F region MSTIDs overhead. The E region scattering sites themselves are not.

Figure 19.5 shows Arecibo linefeed data, focusing here on the E region. The grayscales are a proxy measurement for the E region plasma density. Vertically extended, patchy sporadic E layer irregularities of the kind first reported by Miller amd Smith (1978) and Smith Miller (1978) are evident. Their morphology is suggests convective rolls.

The solid black line shows the coherent scatter detected by the St. Croix radar at the range to the E region over Arecibo. The coherent scatter occurred in six bursts separated by roughly 15–20 min each. The coherent scatter was clearly initiated with the appearance of the irregular sporadic layers. While a few of the coherent scatter peaks seem to be associated with deformations in the sporadic layer, the details in the coherent scatter do not precisely match the fine structure of the sporadic E layer. Some convective rolls appear to give rise to coherent scatter, whereas others do not. There is no clear 15–20 min periodicity in the sporadic E layer density morphology, and we cannot

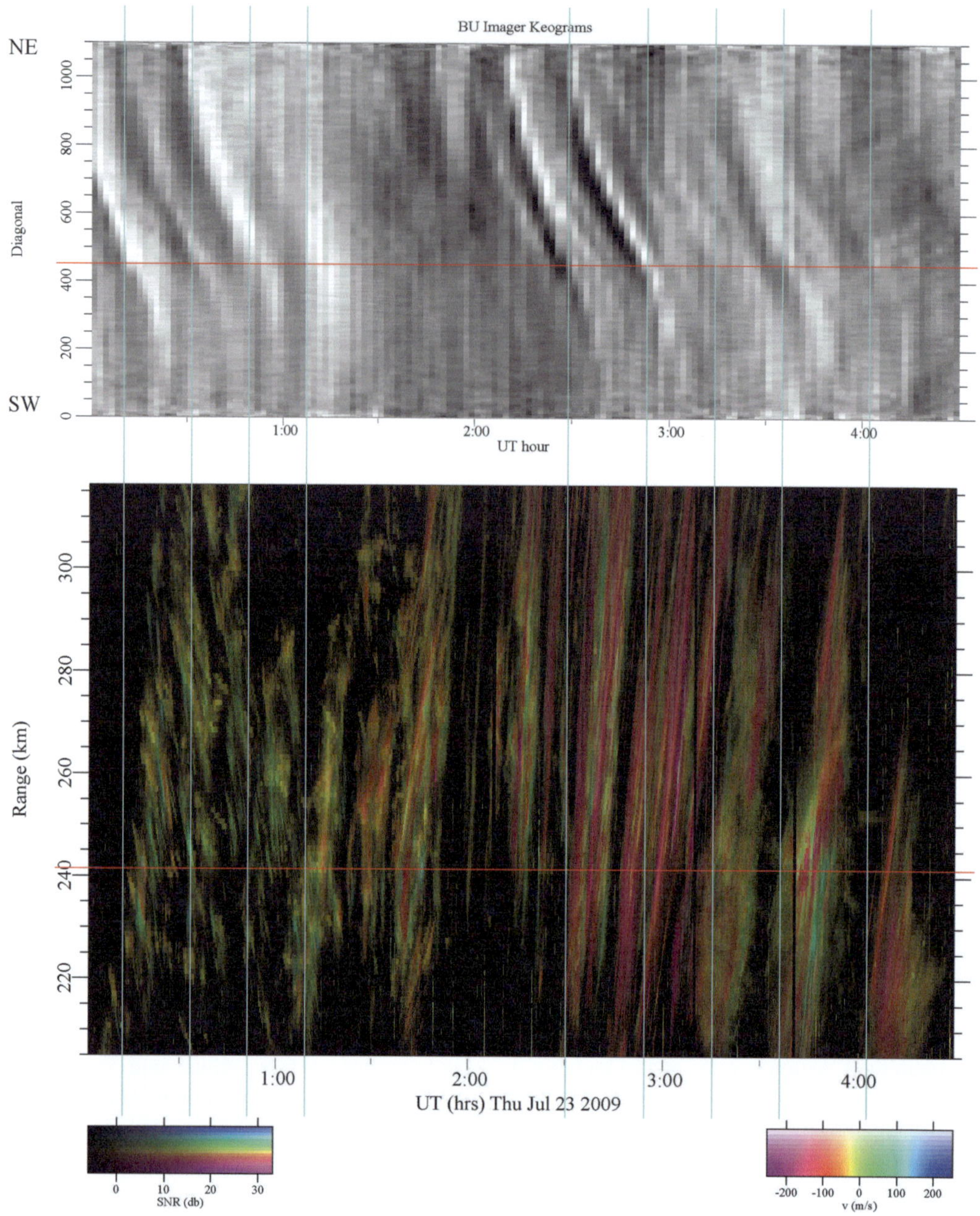

Fig. 19.4 Coherent scatter range-time-intensity (RTI) map plotted beneath 630 nm imager keogram. The color scales adopted for the RTI plot are the same as for the radar images in the previous figure. The *horizontal red lines* are specific to Arecibo's location (see text)

attribute that periodicity in the coherent scatter observations to the spatio/temporal structure of the layers themselves.

It is noteworthy that the MSTIDs were present over Arecibo by 2000 LT, whereas the irregular sporadic E layers did not appear until about 2015 LT, which was about the time that the crest of an MSTID propagated into the F region volume sharing magnetic field lines with the E layer over Arecibo.

The Doppler shifts of the coherent scatter echoes turn out to be more closely related to the sporadic E layer morphology than were the echo intensities

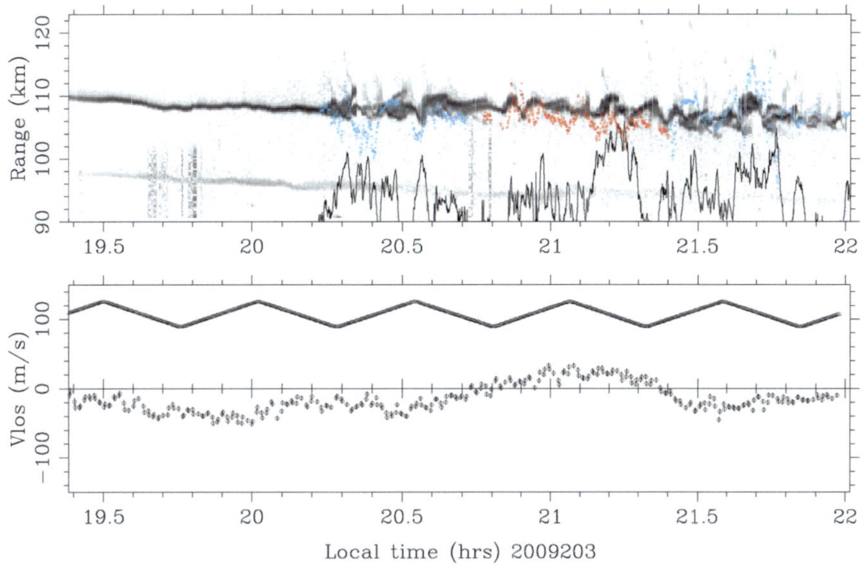

Fig. 19.5 (Combined incoherent and coherent scatter observations. *Top panel*: The grayscales represent plasma density in an irregular sporadic E layer. The *solid line* shows the coherent scatter signal-to-noise ratio for the scattering volume directly over Arecibo detected by the St. Croix radar. SNR is plotted on a logarithmic scale spanning -4–46 dB. The *colored symbols* represent the Doppler shifts of the coherent scatter echoes on a ±200 m/s scale (see text). *Bottom panel*: vertical F region plasma drifts derived from incoherent scatter

themselves. The Doppler shifts of the coherent echoes received from the scattering volume over St. Croix are plotted as colored symbols in Fig. 19.5. Red and blue symbols were plotted according to two different sign conventions. Where a red (blue) symbol is used, a positive value implies a red (blue) Doppler shift in the St. Croix radar data. The former (latter) convention was applied when the F region was ascending (descending), according to the data in the lower panel of Fig. 19.5. The associated zonal ionospheric electric field is eastward (westward) in each case. Using this convention, the deflections of the plotted points become highly correlated with the vertical displacements of the sporadic E layer. The coherent scatter line-of-sight Doppler shifts are always highly correlated with the vertical E_S layer displacements, but the sign of the correlation depends on the sign of the background zonal electric field.

Finally, the Doppler shifts also manifest the 15–20 min. periodicity seen in the coherent scatter intensity in the sense that weak echoes always have small Doppler shifts. The Doppler shift time history is effectively modulated by an envelope that has a small amplitude wherever the backscatter intensity itself becomes small. The modulation is not particularly clear in the Doppler data, which are also modulated by vertical displacements in the E_S layer, the background electric field, and almost certainly by variations in the E region wind forcing. The modulation is necessarily present, however.

Figure 19.5 presents a comprehensive analysis of the incoherent scatter data obtained from the E_S layer using the Gregorian feed system at Arecibo, which was directed at a 15° zenith angle and at a time-varying azimuth angle. The figure concentrates on the period between 20 and 22 LT when the E_S layer showed the most vertical development. The layer looks somewhat more structured here than in Fig. 19.5, but this is just a consequence of azimuth swinging and the additional conflation of spatio-temporal variations in the layer.

Linefeed data confirm that the vertical drifts in the layer were negligible. A weighted-damped least squares regularization technique was applied to the Gregorian line-of-sight drift profile measurements in an attempt to infer the complete zonal (U) and meridional (V) drift profile components. The procedure was described by Hysell et al. (2009). It finds the model profiles that are minimally structured while remaining consistent with the line-of-sight measurements in terms of the χ-squared parameter, taking into account the experimental error variances. Structure here is measured by the global L2 norm of the combined second derivatives in space and time. The regularization parameter is set at the threshold above which χ-squared begins to increase sharply. The object is to produce find U and V profiles with no obvious artifacts due to beam swinging.

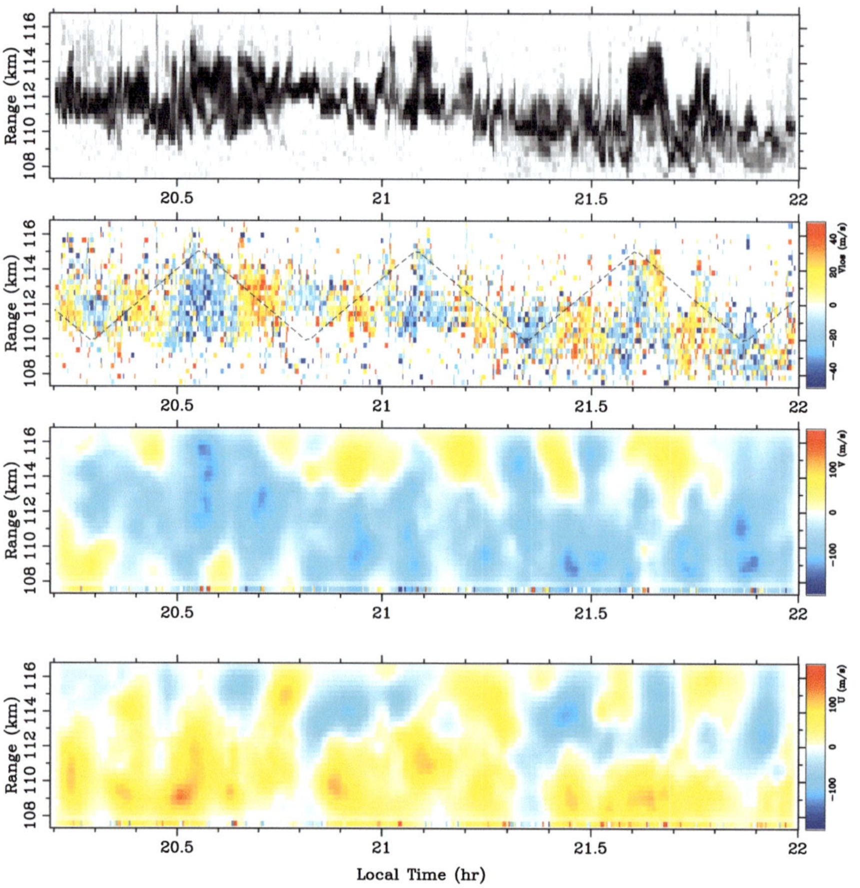

Fig. 19.6 Incoherent scatter observations of the sporadic E layer obtained with the Gregorian feed system at Arecibo. *Top panel*: relative electron density. *Second panel*: line-of-sight ion drifts. *The dashed* line indicates the pointing azimuth of the feed. *Third panel*: meridional drift profiles obtained through a regularization procedure. *Fourth panel*: zonal drift profile. Note that the data are plotted against range rather than altitude

The third and fourth panels of Fig. 19.6 imply that the ion drifts in the E_s layer were mainly southward, switching from southeastward at 2030 LT to southwestward at 2200 LT. This is consistent with the apparent drift of the scattering sites seen in the coherent scatter radar imagery. The drift profiles also imply anticyclonic flow with strong shears and a gradual downward phase propagation with time. As the ions are still unmagnetized at the altitudes in question, we attribute these effects to neutral forcing and neutral wind shear. Neutral shear and dynamical instability is a strong candidate for the source of free energy responsible for producing the E_s layer irregularities at the scale sizes evident in Fig. 19.6.

19.4 Summary and Analysis

Irregular, patchy E_s layers formed over Arecibo at about 2015 LT after MSTIDs had been in clear evidence in red-line imagery for at least 15 min. The MSTIDs might have been present even earlier, although we cannot know for certain without daytime imaging capability. Coherent scatter from sporadic E layers was observed after the layers ceased to be blanketing and instead became structured, showing vertical development and what appear to have been convective rolls. Coherent scatter radar imagery suggests that features in the irregular sporadic E layers drifted with the lower thermospheric neutral wind. The MSTIDs propagated to the southwest while the background F region plasma drifted to the east or southeast, and so the former were not merely frozen into the background flow.

The coherent backscatter intensity was modulated by the passage of the MSTIDs and was strongest mainly when the airglow in the F region on common field lines with the E region scattering volume was a maximum. The coherent backscatter intensity also exhibited periodicity that matched the period of the MSTIDs. The large-scale irregular density features in the E_S layers meanwhile did not exhibit similar periodicity and did not appear to be modulated or by the MSTIDs. Similarly, the scattering sites seen by radar imagery did not follow in their motions the phase propagation of the MSTIDs. Instead, only the condition for FAI generation appeared to track the MSTIDs. Likewise, the wavelengths and periods of wavelike features sometimes seen in radar images of the E_s layers do not match those of the MSTIDs.

The Doppler shifts of the coherent scatter observed from St. Croix were correlated with the vertical displacement of the E_S layers, and the sign of the correlation coefficient depended on the background zonal electric field in the ionosphere. The intensity of the coherent scatter, meanwhile, was not so clearly correlated with individual features in the E_S layer ionization; some features gave rise to stronger coherent scatter than others. Strong, rotational wind shears accompany the irregular sporadic E layers as a rule.

Measurements from rockets and earlier radar measurements (see, e.g. Larsen, 2000), including those from the St. Croix and Arecibo radars (Hysell et al., 2004; Larsen et al., 2007; Hysell et al., 2009), suggest that the neutral wind profiles are either unstable or close to instability at the time when the QP echoes occur. These earlier studies have furthermore shown evidence for strong coupling between structures in the neutral flow and associated large-scale structure in the plasma irregularities. Specifically, the same horizontal structure was found in green-line optical imager data and in the coherent scatter radar images obtained with the St. Croix radar during a time when the mesospheric emission layer height and the sporadic E layer height overlapped (and when no noteworthy features were present in red-line imagery) (Larsen et al., 2007). Resonant lidar measurements have also shown evidence for the development of neutral uplift at the altitude of the sporadic E layer prior to the development of plasma irregularities generated by that uplift. In general, the development of the coherent scatter echoes have been found to be strongly linked to uplift in sporadic E layers, viz., coherent scatter occurs only when and where sporadic E layers exhibit uplift and deformation, as exemplified by Fig. 19.5. The time sequence of high vertical resolution wind profile estimates from the Arecibo ISR described here suggest that the neutral flow was either unstable or close to the instability condition in the Richardson number sense throughout the time when QP structures were present in the E region.

Irregular features in E_S layers induced by neutral uplift would behave like dielectric patches in the ionosphere. Background electric fields would induce electric dipoles in the features, causing them to polarize. Where the conductivity is anisotropic, the tendency can be for strong polarization electric fields to form in the direction normal to the magnetic field and the applied background field (i.e. Shalimov et al., 1998; St.-Maurice and Hamza, 2001; Hysell and Drexler, 2006; Dimant et al., 2009). The polarization electric field, in turn, can drive strong Hall current parallel (antiparallel) to the original background electric field, assuming that the underlying density irregularity is a local enhancement (depletion), and assuming that this current can close somehow (see below). This process is responsible for the equatorial electrojet. Generally speaking, coherent radar echoes from small-scale irregularities embedded in the medium would exhibit Doppler shifts consistent with the direction of the flow of electrons carrying the current.

Given an eastward background zonal electric field, a coherent scatter radar looking westward should observe red-shifted echoes from patches of plasma with locally enhanced Hall conductivities and blue-shifted echoes from local depletions. The situation reverses for a westward background field. In view of the fact that the Hall conductivity increases with decreasing layer height at 110 km altitude, positive and negative layer height displacements constitute negative and positive Hall conductivity perturbations, respectively. This is consistent with the Doppler shift behavior shown in Fig. 19.5. We do not expect the correspondence to be perfect; E region winds also drive background currents and polarization electric fields and complicate the picture. Nevertheless, the correspondence is striking, and we argue that the irregular E_S layer behaved like a corrugated sheet dielectric.

However, the analysis by Hysell et al. (2002a) showed that the aforementioned process is arrested in the case of electrically isolated E region conductivity irregularities in which the Hall current cannot close. In simulation, current closure through the F region was necessary for large polarization electric fields to form (larger than the background field) and for large Hall currents carried by rapidly-moving charge carriers to flow.

The July, 2009 Arecibo and St. Croix data suggest that the principle coupling between the MSTIDs and the irregular sporadic E layers was through the modification of the F region flux-tube-integrated conductivity by the former and the effect this had on polarization electric field production efficiency in the latter. The sporadic E layer did not create the MSTID, and the MSTID did not greatly influence the morphology, growth, or motion of the sporadic E layers. While electric field perturbations associated with the MSTID must have had some influence on the E region, these were evidently small compared to the background field, which clearly dominates the E region drift data in Fig. 19.2. The fine structure in the E layer polarization electric fields, which influences the coherent scatter Doppler shifts, is mainly controlled by the E layer conductivity and the background electric field. However, the polarization electric fields are broadly modulated by variations in the F layer conductivity associated with the MSTID. We see this most clearly not in the Doppler shifts themselves but in the coherent scatter echo power. Coherent echoes come from meter-scale waves driven by strong currents in the E_S layers. These currents and the large polarization electric fields that drive them can only be maintained when the currents are allowed to close in the F layer.

Acknowledgments This work was supported by awards ATM-0541526 and ATM-0541593 from the National Science Foundation to Cornell University, Clemson University, and the University of the Virgin Islands. The Arecibo Observatory is part of the National Astronomy and Ionosphere Center, which is operated by Cornell University under a cooperative agreement with the National Science Foundation. We appreciate the assistance of the USDA in hosting the St. Croix radar.

References

Behnke RA (1979) F layer height bands in the nocturnal ionosphere over Arecibo. J Geophys Res 84:974

Chu YH, Wang CY (1997) Interferometry observations of 3-dimensional spatial structures of sporadic E using the Chung-Li VHF radar. Radio Sci 32(2):817–832

Cosgrove RB (2007a) Wavelength dependence of the linear growth rate of the Es layer instability. Ann Geophys 25:1311–1322

Cosgrove RB (2007b) Generation of mesoscale F layer structure and electric fields by the combined Perkins and Es layer instabilities, in simulations. Ann Geophys 25: 1579–1601

Cosgrove RB, Tsunoda RT (2002) A direction-dependent instability of sporadic- E layers in the nighttime midlatitude ionosphere. Geophys Res Lett 29(18):1864. doi:10.1029/2002JA009,728

Cosgrove RB, Tsunoda RT (2004) Instability of the E–F coupled nighttime midlatitude ionosphere. J Geophys Res 109.doi:10.1029/2003JA010,243

Dimant YS, Oppenheim MM, Milikh GM (2009) Meteor plasma trails: effects of external electric field. Ann Geophys 27: 279–296

Fukao S, Kelley MC, Shirakawa T, Takami T, Yamamoto M, Tsuda T, Kato S (1991) Turbulent upwelling of the midlatitude ionosphere, 1. Observational results by the MU radar. J Geophys Res 96:3725–3746

Fukao S, Yamamoto M, Tsunoda RT, Hayakawa H, Mukai T (1998) The SEEK (Sporadic-E Experiment over Kyushu) campaign. Geophys Res Lett 25:1761

Garcia FJ, Kelley MC, Makela JJ, Huang CS (2000) Airglow observations of mesoscale low-velocity traveling ionospheric disturbances at midlatitudes. J Geophys Res 105:18407–18415

Haldoupis C, Schlegel K (1994) Observation of the modified two-stream plasma instability in the midlatitude E region ionosphere. J Geophys Res 99:6219

Hamza A (1999) Perkins instability revisited. J Geophys Res 104:22567–22575

Hysell DL, Burcham J (2000) The 30 MHz radar interferometer studies of midlatitude E region irregularities. J Geophys Res 105:12797

Hysell DL, Drexler J (2006) Polarization of E region plasma irregularities. Radio Sci 41:RS4015. doi:10.1029/2005RS003,424

Hysell DL, Larsen MF, Zhou QH (2004) Common volume coherent and incoherent scatter radar observations of midlatitude sporadic E-layers and QP echoes. Ann Geophys 22:3277–3290

Hysell DL, Nossa E, Larsen MF, Munro J, Sulzer MP, Aponte N, González SA (2009) Sporadic E layer observations over Arecibo using coherent and incoherent scatter radar: assessing dynamic stability in the lower thermosphere. J Geophys Res 114:A12303. doi:10.1029/2009JA014,403

Hysell DL, Yamamoto M, Fukao S (2002a) Simulations of plasma clouds in the midlatitude E region ionosphere with implications for type I and type II quasiperiodic echoes. J Geophys Res 107:1313

Hysell DL, Yamamoto M, Fukao S (2002b) Imaging radar observations and theory of type I and type II quasiperiodic echoes. J Geophys Res 107:1360

Larsen MF (2000) A shear instability seeding mechanism for quasi-periodic radar echoes. J Geophys Res 105(A11):24931–24940

Larsen MF, Hysell DL, Zhou QH, Smith SM, Friedman J, Bishop RL (2007) Imaging coherent scatter radar, incoherent scatter radar, and optical observations of quasiperiodic structures associated with sporadic E layers. J Geophys Res 112:A06321. doi:10.1029/2006JA012,051

Mathews JD (1998) Sporadic E: Current views and recent progress. J Atmos Solar-Terr Phys 60:413

Miller CA (1996) On gravity waves and the electrodynamics of the mid-latitude ionosphere. Ph.D. thesis, Cornell University, Ithaca, NY

Miller CA, Swartz WE, Kelley MC, Mendillo M, Notingham D, Scali J, Reinisch B (1997) Electrodynamics of midlatitude spread F, 1. Observations of unstable gravity wave-induced ionospheric electric fields at tropical latitudes. J Geophys Res 102:11521–11532

Miller KL, Smith LG (1978) Incoherent scatter radar observations of irregular structure in mid-latitude sporadic E layers. J Geophys Res 83:3761

Otsuka Y, Onoma F, Shiokawa K, Ogawa T, Yamamoto M, Fukao S (2007) Simultaneous observations of nighttime medium-scale traveling ionospheric disturbances and E region field-aligned irregularities at midlatitudes. J Geophys Res 112:A06317. doi:10.1029/2005JA011,548

Otsuka Y, Shiokawa K, Ogawa T, Yokoyama T, Yamamoto M (2009) Spatial relationship of nighttime medium-scale traveling ionospheric disturbances and F region field-aligned irregularities observed with two spaced all-sky airglow imagers and the middle and upper atmosphere radar. J Geophys Res 114:A05302. doi:10.1029/2008JA013,902

Otsuka Y, Shiokawa K, Ogawa T, Yokoyama T, Yamamoto M, Fukao S (2004) Spatial relationship of equatorial plasma bubbles and field-aligned irregularities observed with an all-sky airglow imager and the equatorial atmosphere radar. Geophys Res Lett 31:L20802. doi:10.1029/2004GL020,869

Perkins F (1973) Spread F and ionospheric currents. J Geophys Res 78:218

Riggin D, Swartz WE, Providakes J, Farley DT (1986) Radar studies of long-wavelength waves associated with mid-latitude sporadic E layers. J Geophys Res 91:8011

Saito A, Fukao S, Miyazaki S (1998) High resolution mapping of TEC perturbations with the GSI GPS network over Japan. Geophys Res Lett 25:3079

Saito A, Nishimura M, Yamamoto M, Fukao S, Tsugawa T, Otsuka Y, Miyazaki S, Kelley MC (2002) Observations of traveling ionospheric disturbances and 3-m scale irregularities in the nighttime F-region ionosphere with the MU radar and a GPS network. Earth Planets Space 54:31–44

Saito S, Yamamoto M, Hashiguchi H (2008) Imaging observations of nighttime mid-latitude F region field-aligned irregularities by an MU radar ultra-multi-channel system. Ann Geophys 26:2345–2352

Saito S, Yamamoto M, Hashiguchi H, Maegawa A (2006) Observation of threedimensional signatures of quasi-periodic echoes associated with mid-latitude sporadic-E layers by MU radar ultra-multi-channel system. Geophys Res Lett 33:L14109. doi:10.1029/2005GL025,526

Saito S, Yamamoto M, Hashiguchi H, Maegawa A, Saito A (2007) Observational evidence of coupling between quasi-periodic echoes and medium-scale traveling ionospheric disturbances. Ann Geophys 25:2185–2194

Shalimov S, Haldoupis C, Schlegel K (1998) Large polarization electric fields associated with midlatitude sporadic E. J Geophys Res 103:11617

Smith LG, Miller KL (1980) Sporadic-layers and unstable wind shears. J Atmos Solar-Terr Phys 42:45

St.-Maurice JP, Hamza AM (2001) A new nonlinear approach to the theory of E region irregularities. J Geophys Res 106:1751

Sulzer MP (1986a) A phase modulation technique for a sevenfold statistical improvement in incoherent scatter data-taking. Radio Sci 21:737

Sulzer MP (1986b) A radar technique for high range resolution incoherent scatter autocorrelation function measurements utilizing the full average power of klystron radars. Radio Sci 21:1033–1040

Sulzer MP, Aponte N, González SA (2005) Application of linear regularization methods to Arecibo vector velocities. J Geophys Res 110(A10). doi:10.1029/2005JA011,042

Tsugawa T, Otsuka Y, Coster AJ, Saito A (2007) Medium-scale traveling ionospheric disturbances detected with dense and wide TEC maps over North America. Geophys Res Lett 34:L22101. doi:10.1029/2007GL031,663

Tsunoda RT (2006) On the coupling of layer instabilities in the nighttime midlatitude ionosphere. J Geophys Res 111:A11304. doi:10.1029/JA011,630

Whitehead JD (1972) The structure of sporadic E from a radio experiment. Radio Sci 7:355

Whitehead JD (1989) Recent work on mid-latitude and equatorial sporadic E. J Atmos Solar-Terr Phys 51:401

Yamamoto M, Fukao S, Ogawa T, Tsuda T, Kato S (1992) A morphological study of midlatitude E-region field-aligned irregularities observed with the MU radar. J Atmos Solar-Terr Phys 54:769

Yamamoto M, Fukao S, Tsunoda RT, Pfaff R, Hayakawa H (2005) SEEK-2 (Sporadic-E Experiment over Kyushu 2) – Project outline. Ann Geophys 23:2295–2305

Yamamoto M, Fukao S, Woodman RF, Ogawa T, Tsuda T, Kato K (1991) Mid-latitude E region field-aligned irregularities observed with the MU radar. J Geophys Res 96:15943

Yokoyama T, Otsuka Y, Ogawa T, Yamamoto M, Hysell DL (2008) First three-dimensional simulation of the Perkins instability in the nighttime midlatitude ionosphere. Geophys Res Lett 35:L03101. doi:10.1029/2007GL032,496

Yokoyama T, Otsuka Y, Ogawa T, Yamamoto M, Hysell DL (2009) Three-dimensional simulation of the coupled Perkins and Es layer instabilities in the nighttime midlatitude ionosphere. J Geophys Res 114:A03308. doi:10.1029/2008JA013,789

Chapter 20

Instabilities in the Midlatitude Ionosphere in Terms of *E–F* Coupling

Tatsuhiro Yokoyama

Abstract There are two intriguing phenomena in the midlatitude ionosphere: quasi-periodic (QP) radar echoes associated with a sporadic-*E* (E_s) layer in the *E* region and medium-scale traveling ionospheric disturbances (MSTIDs) in the *F* region, both of which often show northwest–southeast alignment and propagate southwestward in the Northern Hemisphere. The Perkins instability in the *F* region and the E_s-layer instability in the *E* region can produce the tilted structure, but the growth rate of the Perkins instability alone cannot account for the observed MSTIDs. The electrodynamical coupling between these instabilities is now believed to be an important factor for the rapid growth of MSTIDs. From recent numerical studies, it is found that the polarization process in the *E* region is essential for the full development of MSTIDs as well as for the seeding of NW-SE perturbation in the *F* region. Further observations to identify the cause and effect of the coupling process are required in the future.

20.1 Introduction

Midlatitude ionosphere has attracted considerable attention for the past two decades since strong meter-scale irregularities were detected by the MU radar (Fukao et al., 1991; Yamamoto et al., 1991). One of the intriguing phenomena which still remains unresolved in the nighttime midlatitude ionosphere is so-called medium-scale traveling ionospheric disturbances (MSTIDs). MSTID structure is elongated from northwest to southeast (from northeast to southwest) and propagates southwestward (northwestward) in the Northern (Southern) Hemisphere with a wavelength of a few hundred kilometers and with a velocity of approximately 100 m s^{-1}. MSTIDs are mainly caused by density perturbation in the bottomside *F* region, and the horizontal structures have been studied by all-sky imagers (e.g., Miller et al., 1997) and GPS-TEC maps (e.g., Saito et al., 1998). Another interesting phenomenon in the *E* region is structured coherent backscatter radar echoes, so-called quasi-periodic (QP) echoes (Yamamoto et al., 1991). From rocket and ground-based observations, it has been found out that the horizontal inhomogeneity of a sporadic-*E* (E_s) layer is important as a source of QP echoes (e.g., Yamamoto et al., 2005). However, what causes the inhomogeneity of an E_s layer is not yet clear. Recently, electrodynamical coupling between the *E* and *F* regions has been considered to be an important factor of the generation mechanism of MSTIDs and QP echoes. It is controlled by electric field mapping along geomagnetic field lines so that the electric field generated in one region can map to the other region instantaneously. The efficiency of the mapping is determined by the conductivity ratio, $(\sigma_0/\sigma_P)^{1/2}$, where σ_0 is the specific or parallel conductivity and σ_P is the Pedersen conductivity (Farley, 1959). Because the ratio is about 100(1000) in the *E*(*F*) region, electric fields with a few-kilometer scale size in the *E* region can map along the magnetic field lines and easily reach the *F*-region altitude (Kelley, 1989). In this chapter, observational, theoretical, and numerical studies of ionospheric irregularities are reviewed in terms of the coupling between the *E* and *F* regions (*E–F* coupling).

T. Yokoyama (✉)
NASA Goddard Space Flight Center, Greenbelt, MD, USA
e-mail: tatsuhiro.yokoyama@gsfc.nasa.gov

M.A. Abdu, D. Pancheva (eds.), A. Bhattacharyya (Coed.), *Aeronomy of the Earth's Atmosphere and Ionosphere*,
IAGA Special Sopron Book Series 2, DOI 10.1007/978-94-007-0326-1_20,

20.2 Observations

One clear evidence that the electrodynamical coupling along the magnetic field lines is important is shown in Fig. 20.1 (Otsuka et al., 2004). MSTIDs are observed in geomagnetic conjugate hemispheres simultaneously, and their phases match each other almost perfectly. Therefore, it is natural to think that the coupling between the *E* and *F* regions should work in some way because the distance along the magnetic field lines between the two regions (a few hundred kilometers) is much shorter than that between the hemispheres (several thousand kilometers).

It has become a popular topic since several observational evidences in 1960s and 1970s were revisited by Tsunoda and Cosgrove (2001). Meanwhile, observations aiming at an evidence of the *E–F* coupling were conducted, and a hypothesis was proposed that polarization electric fields are set up within E_S-layer patches and map along the magnetic field lines up to the *F* region to excite spread *F* (Haldoupis et al., 2003; Kelley et al., 2003). Since then, a theory in terms of the *E–F* coupling has been established by Cosgrove and Tsunoda (2004), which is described in the following section, and several observational evidences have been presented (Saito et al., 2007; Otsuka et al., 2007, 2008; Swartz et al., 2009). Figure 20.2 shows two-dimensional radar interferometric images of QP echoes observed with the MU radar (34.9°N, 136.1°E) and two-dimensional GPS-TEC perturbation maps with the superposed images of the QP echoes

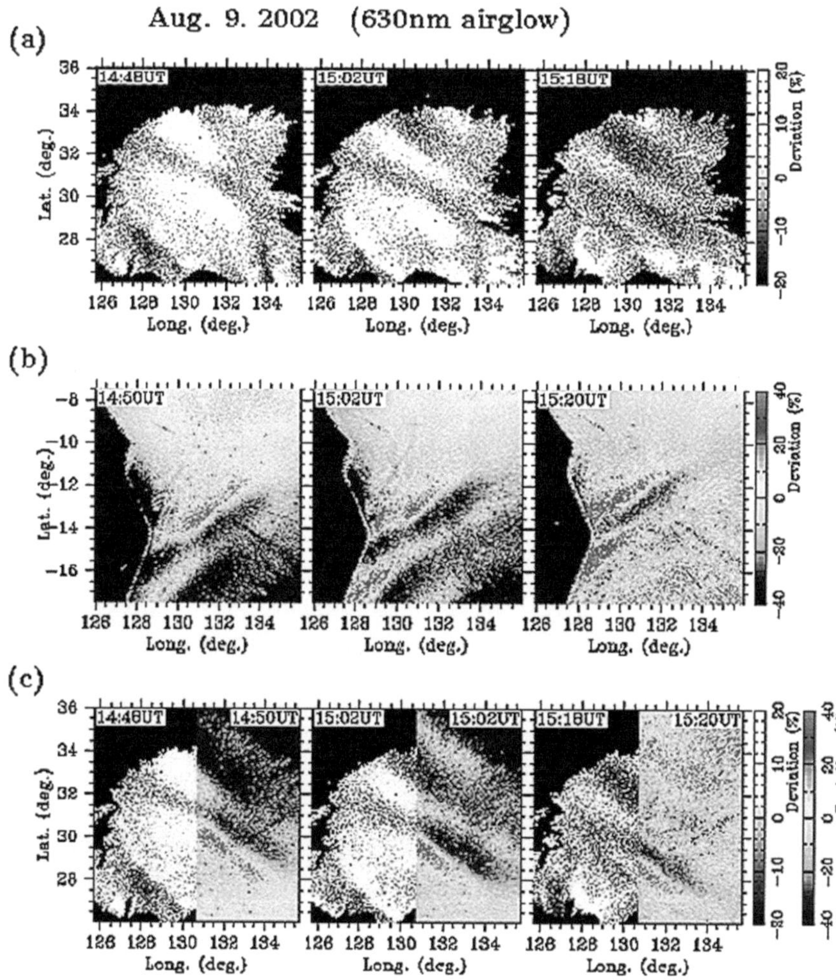

Fig. 20.1 Two-dimensional maps of 630-nm airglow intensity observed at Sata and Darwin that are geomagnetic conjugate locations [After Otsuka et al. (2004)]. (**a**) Sata. (**b**) Darwin. (**c**) Sata (*left*) and conjugate of Darwin (*right*)

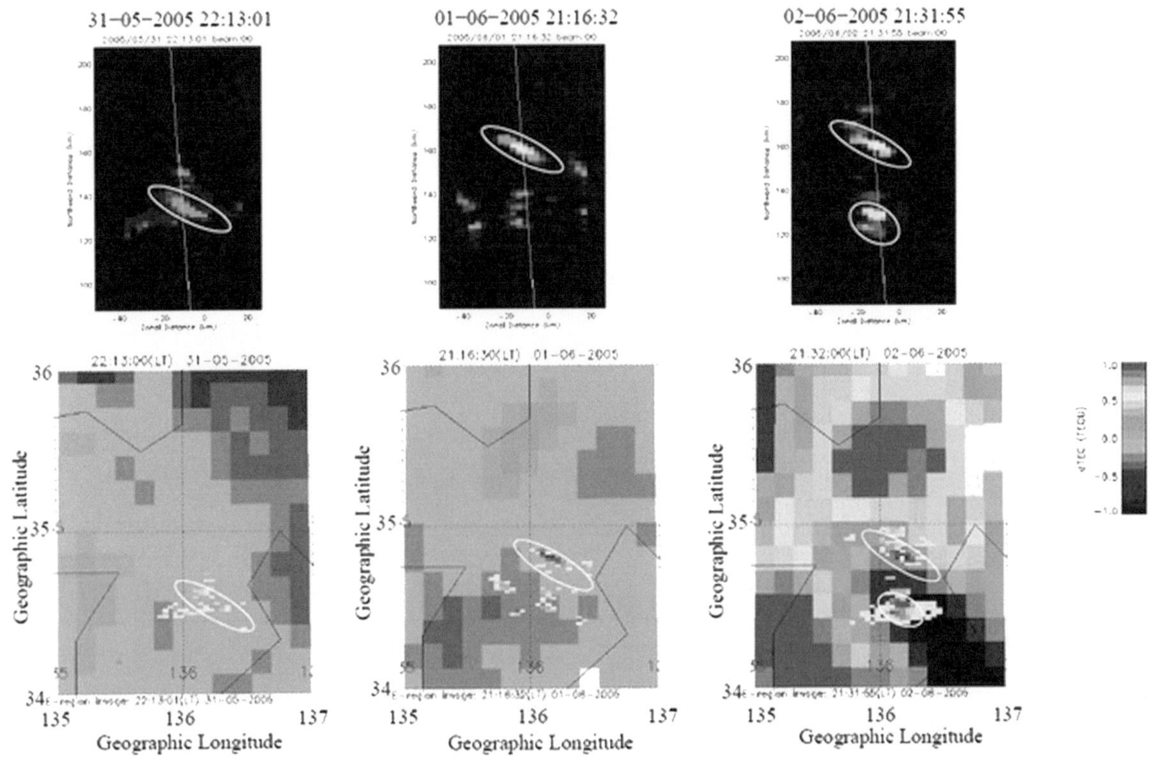

Fig. 20.2 Two-dimensional radar interferometric images of QP echoes observed with the MU radar (*top*) and two-dimensional GPS-TEC perturbation maps with the superposed images of the QP echoes mapped to 300 km altitude along the magnetic field lines (*bottom*). The *vertical axis* of the top panels shows northward distance from the MU radar (from 90 to 205 km), and the *horizontal axis* shows zonal distance (from −50 to 25 km) [After Saito et al. (2007)]

mapped to 300 km altitude along the magnetic field lines (Saito et al., 2007). The alignment of the QP echoes is similar to that of the MSTIDs, although the horizontal scale of the QP echoes is much smaller than that of MSTIDs. Using the Arecibo incoherent scatter radar, a patchy E_S layer and the structured bottomside *F* region were observed simultaneously (Swartz et al., 2009). It is also shown that the time variation of the Doppler velocity of QP echoes is correlated with the MSTID phase pattern (Saito et al., 2007; Otsuka et al., 2007).

Generally, both MSTID and E_S-layer activities are the highest around summer solstice, but MSTID activity has a second peak around winter solstice (Kotake et al., 2006). The correlation between MSTID and E_S-layer activities in summer is shown in Fig. 20.3 (Otsuka et al., 2008). The MSTID activity clearly shows positive correlation with a critical frequency of E_S layer (f_oE_S) and the difference between a critical and blanket frequency of E_S layer ($f_oE_S - f_bE_S$), which represents inhomogeneity of E_S layer. If an E_S layer in the summer hemisphere is assumed to be the source of MSTIDs, the *F* region in the winter hemisphere could also be modulated through the coupling process, which could produce the second peak of MSTIDs. However, a conclusive observational evidence has yet to be obtained to identify the cause and effect of the coupling process.

20.3 Theory

The generation of MSTIDs has been attributed to the Perkins instability (Perkins, 1973) because it has been the only mechanism to explain the peculiar northwest–southeast (NW–SE) orientation of MSTIDs, which are not aligned with the magnetic meridian as equatorial plasma bubbles are. The key idea of the Perkins instability is as follows. When a current flows along NW-SE wave fronts, a zonal component of current **J** produces

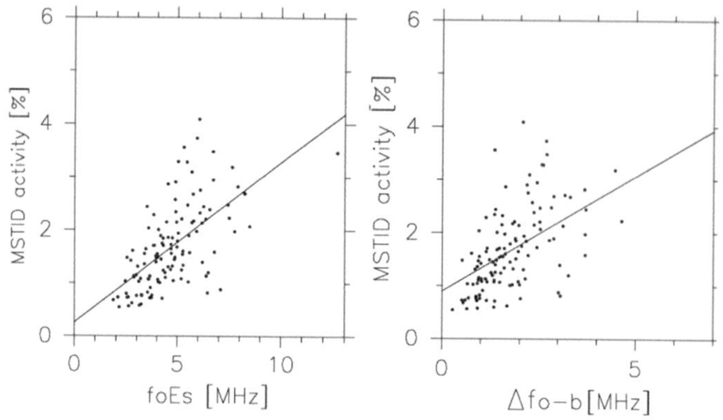

Fig. 20.3 Scatter plots of MSTID activity against f_oE_s (*left*) and $f_oE_s - f_bE_s$ (*right*) over the same area during May–August 2001. The MSTID activity is defined as the ratio of the standard deviation of TEC for 1 h within an area bounded by 33.75–37.80°N and 137.50–141.55°E to the average background TEC value ($\delta I/\bar{I}$) [After Otsuka et al. (2008)]

a vertical $\mathbf{J} \times \mathbf{B}$ force. The $\mathbf{J} \times \mathbf{B}$ force either intensifies or damps the perturbation depending on the angles between a background electric field and a $\mathbf{k}$ vector of the perturbation. That is, if the current flows northwestward along the wave fronts, the perturbation with larger conductivity (lower altitude) meets a downward $\mathbf{J} \times \mathbf{B}$ force which further intensifies the conductivity perturbation. However, as the growth rate of the Perkins instability is very small, another mechanism to reinforce the instability, such as gravity wave seeding (Kelley and Fukao, 1991), is required to account for the observations. A similar mechanism, called E_s-layer instability, is also proposed where an E_s layer located at a zonal wind shear node is unstable for NW-SE perturbation (Cosgrove and Tsunoda, 2002). As the ion-neutral collision frequency is larger than the ion gyrofrequency in the lower E region (90–120 km), ions are considered as unmagnetized plasma. Under the unstable condition, an upward component of polarization electric field is set up in the raised part of the E_s layer, where ions are forced to drift further upward by the electric field. Because of the similarity of the Perkins and E_s-layer instabilities, the coupling process between the E and F regions is proposed, and it is shown that the coupled system including both instabilities gives a larger growth rate than either alone (Cosgrove and Tsunoda, 2004).

In what follows, the growth rates of these instabilities derived by Tsunoda (2006) are presented, and how the coupling process effectively works on the seeding of MSTIDs is discussed. See Tsunoda (2006) for the detailed derivation of the growth rates. The growth rates of the Perkins (γ_P) and E_s-layer (γ_E) instabilities are written as

$$\gamma_P = \frac{|\mathbf{E}_0 + \mathbf{U}^F \times \mathbf{B}| \cos I}{BH} \sin(\theta - \alpha) \sin \alpha \quad (20.1)$$

where $\mathbf{E}_0$ is the background electric field, $\mathbf{U}^F$ is the F-region neutral wind, $\mathbf{B}$ is the geomagnetic field, θ is the angle between the total effective electric field $\mathbf{E}_0 + \mathbf{U}^F \times \mathbf{B}$ and the east direction, I is the magnetic inclination angle, α is the angle between the direction normal to the frontal structure and the east direction, and H is the atmospheric scale height and

$$\gamma_E = \frac{U_x^E \cos I}{H_U \rho_i} \left(\frac{\Sigma_H^{E_s}}{\Sigma_P^{E_s} + \Sigma_P^E + \Sigma_P^F} \sin \alpha \cos \alpha - 1 \right) \quad (20.2)$$

where U_x^E is the zonal wind velocity in the E region, $H_U \equiv U_x^E (\partial U_x^E / \partial z)^{-1}$ is the scale length of the vertical shear of U_x^E, $\rho_i = \nu_{in}/\Omega_i$ where ν_{in} is the ion-neutral collision frequency and Ω_i is the ion angular gyrofrequency, and Σ_H and Σ_P are the field-line integrated Hall and Pedersen conductivities in the designated region, respectively. θ and α are given in the plane perpendicular to the geomagnetic field. Both instabilities have directional preference, that is, γ_P depends on $\sin(\theta - \alpha) \sin \alpha$ with maximum occurring where $\alpha = \theta/2$ and γ_E depends on $\sin \alpha \cos \alpha$ with maximum occurring where $\alpha = 45°$. Including the vertical drift velocity driven by polarization electric fields generated in the coupled region, the growth rates of the coupled instabilities ($\gamma_{\rm P}^{\rm c}$ and $\gamma_{\rm E}^{\rm c}$) become

$$\gamma_{\rm P}^{\rm c} = \gamma_{\rm P} + \frac{\Sigma_{\rm H}^{E_{\rm s}}}{\Sigma_{\rm P}^{E_{\rm s}} + \Sigma_{\rm P}^{E} + \Sigma_{\rm P}^{F}} \frac{U_x^E \cos I}{H_U} \cos^2\alpha \quad (20.3)$$

$$\gamma_{\rm E}^{\rm c} = \gamma_{\rm E} + \frac{|\mathbf{E}_0 + \mathbf{U}^F \times \mathbf{B}| \cos I}{\rho_{\rm i} B H} \cos(\theta - \alpha) \sin\alpha \quad (20.4)$$

The second term in Eq. (20.3) comes from the $\mathbf{E} \times \mathbf{B}$ drift in the F region produced by the E-region polarization electric field which is responsible for γ_E. This term depends on $\cos^2\alpha$ and is much larger than $\gamma_{\rm P}$ because $H_U \ll H$. However, it cannot reach its maximum because γ_E is negative with $\alpha = 0°$ so that the $E_{\rm s}$-layer instability would be damped. The second term in Eq. (20.4) comes from the Pedersen drift in the E region produced by the F-region polarization electric field which is responsible for $\gamma_{\rm P}$. This term is small compared to γ_E because $H \gg H_U$, so it does not affect the directional preference of γ_E. The derivation of Eqs. (20.3) and (20.4) is not exactly based on the coupled dynamics but the simple addition of the terms in the other region. For example, the second term of Eq. (20.3) does not include the gravity balance at a new altitude which must be considered in the derivation of $\gamma_{\rm P}$. The discrepancy between $\gamma_{\rm P}^{\rm c}$ and an actual growth rate observed in numerical results is shown in the following section.

20.4 Numerical Studies

The formation of NW-SE structure from random density perturbation due to the Perkins instability has been shown by numerical simulations (Kelley and Miller 1997; Zhou et al., 2005; Yokoyama et al., 2008). Now a full three-dimensional model which can simulate the coupling process self-consistently is available (Yokoyama et al., 2009). Figure 20.4 shows the time variation of the F-region Pedersen conductance perturbation (integrated along **B**) in the cases of the Perkins instability alone and the coupled instability. Plotted data are taken from the previous studies (Yokoyama et al., 2008, 2009). The growth of the Perkins instability alone is very slow as expected from the linear growth rate and tends to saturate when the amplitude reaches 3–4%. Including the coupling process,

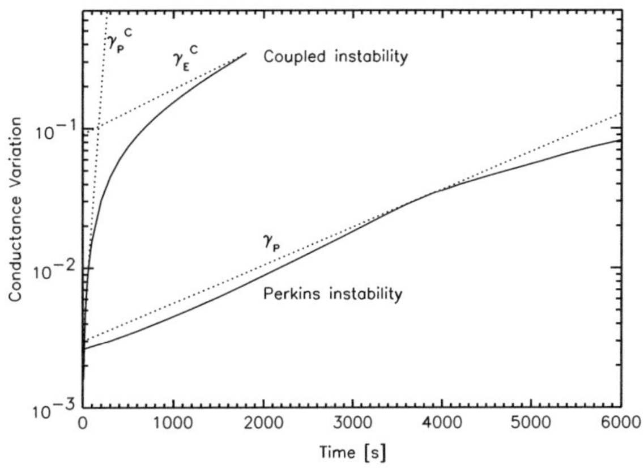

Fig. 20.4 Time variation of the F-region Pedersen conductance perturbation (integrated along **B**) in the cases of the Perkins instability alone and the coupled instability. The *dotted lines* represent designated linear growth rates [plotted data are after Yokoyama et al. (2008, 2009)]

the perturbation grows very rapidly following $\gamma_{\rm P}^{\rm c}$ in the first 100 s because of the polarization electric field mapped from the E region as predicted by Tsunoda (2006). However, it starts to saturate and approach $\gamma_E^{\rm c}$ after 100 s. $\gamma_{\rm P}^{\rm c}$ is no longer available after the F region is modulated to a certain degree because the second term of Eq. (20.3) does not include the gravity balance at a new altitude; the F region raised by the E-region electric field also undergoes a faster downward motion due to a lower collision frequency. Nevertheless, the growth speed of the coupled instability is much faster than that of the Perkins instability alone and comparable to the observed time scale of the appearance of MSTIDs.

Following the study by Yokoyama et al. (2009), a new numerical model named Midlatitude Ionosphere Electrodynamics COupling (MIECO) has recently been developed in which dipole magnetic field lines are adopted to study latitudinal dependence of MSTIDs in a large simulation domain (Yokoyama and Hysell, 2010). As an example of simulation by MIECO, a rotational wind shear, which has a southward component at the center of the E_s layer, is applied, and random density perturbation is applied on an E_s layer. Figure 20.5 shows Hall (Pedersen) conductance variation integrated along **B** over the E(F) region and projected onto an altitude of 102 km (300 km) at $t = 3600$ and 7200 s. The NW-SE structure in both the E and the F regions is clearly formed and propagates southward (southwestward as apparent phase propagation) with increasing amplitude. Note that the

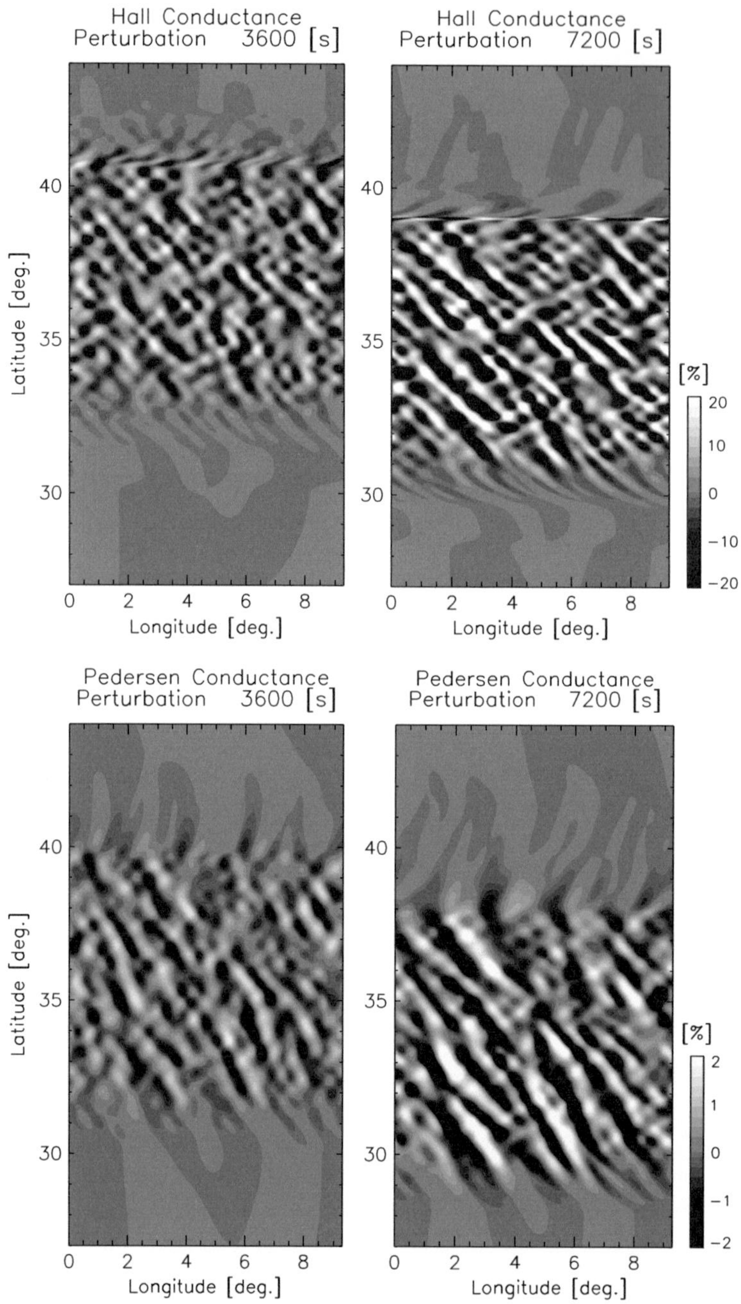

Fig. 20.5 Simulation results by MIECO when the rotational wind shear is applied in the E region. Hall (Pedersen) conductance variation integrated along **B** over the E(F) region and projected onto an altitude of 102 km (300 km) at t = 3600 and 7200 s [After Yokoyama and Hysell (2010)]

wake of the propagating NW-SE structure in the F region tends to be damped after it is decoupled from the E_s layer, even though the unstable condition for NW-SE perturbation is always satisfied in terms of the Perkins instability. This suggests that the polarization process in the E region is essential for the full development of MSTIDs as well as for the seeding of NW-SE perturbation in the F region.

20.5 Summary

In this chapter, the electrodynamical coupling process between the *E* and *F* regions in the midlatitude ionosphere is presented. Several observations imply the importance of the coupling process for both QP echoes in the *E* region and MSTIDs in the *F* region. The coupled Perkins and E_S-layer instabilities is the most plausible mechanism of the rapid growth of MSTIDs. From recent numerical studies, it is found that the polarization process in the *E* region is essential for the full development of MSTIDs as well as for the seeding of NW-SE perturbation in the *F* region. The *E–F* coupling may also be important in the equatorial and low-latitude ionosphere in terms of the seeding of equatorial plasma bubbles. Further observations to identify the cause and effect of the coupling process are required in the future.

Acknowledgment The author was supported by Japan Society for the Promotion of Science (JSPS) Postdoctoral Fellowships for Research Abroad for the present work at Cornell University, USA.

References

Cosgrove RB, Tsunoda RT (2002) A direction-dependent instability of sporadic-*E* layers in the nighttime midlatitude ionosphere. Geophys Res Lett 29(18):1864. doi:10.1029/2002GL014669

Cosgrove RB, Tsunoda RT (2004) Instability of the *E-F* coupled nighttime midlatitudeionosphere. J Geophys Res 109:A04305. doi:10.1029/2003JA010243

Farley DT Jr (1959) A theory of electrostatic fields in a horizontally stratified ionospheresubject to a vertical magnetic field. J Geophys Res 64:1225–1233

Fukao S, Kelley MC, Shirakawa T, Takami T, Yamamoto M, Tsuda T, Kato S (1991) Turbulent upwelling of the midlatitude ionosphere, 1. Observational results by the MU radar. J Geophys Res 96:3725–3746

Haldoupis C, Kelley MC, Hussey GC, Shalimov S (2003) Role of unstablesporadic-*E* layers in the generation of midlatitude spread *F*. J Geophys Res 108(A12):446. doi:10.1029/2003JA009956

Kelley MC (1989) The Earth's Ionosphere: plasma physics and electrodynamics. International Geophysics, vol. 43. Academic, San Diego, CA

Kelley MC, Fukao S (1991) Turbulent upwelling of the midlatitude ionosphere, 2. Theoretical framework. J Geophys Res 96:3747–3753

Kelley MC, Miller CA (1997) Electrodynamics of midlatitude spread *F* 3. Electrohydrodynamicwaves? A new look at the role of electric fields in thermospheric wave dynamics. J Geophys Res 102:11,539–11,547

Kelley MC, Haldoupis C, Nicolls MJ, Makela JJ, Belehaki A, Shalimov S, Wong VK (2003) Case studies of coupling between the *E* and *F* regions during unstable sporadic-*E* conditions. J Geophys Res 108(A12):1447. doi:10.1029/2003JA009955

Kotake N, Otsuka Y, Tsugawa T, Ogawa T, Saito A (2006) Climatological study ofGPS total electron content variations caused by medium-scale traveling ionospheric disturbances. J Geophys Res 111:A04306. doi:10.1029/2005JA011418

Miller CA, Swartz WE, Kelley MC, Mendillo M, Nottingham D, Scali J, Reinisch B (1997) Electrodynamics of midlatitude spread *F*, 1. Observations of unstable,gravity wave-induced ionospheric electric fields at tropical latitudes. J Geophys Res 102:11,521–11,532

Otsuka Y, Onoma F, Shiokawa K, Ogawa T, Yamamoto M, Fukao S (2007) Simultaneous observations of nighttime medium-scale traveling ionospheric disturbancesand *E* region field-aligned irregularities at midlatitude. J Geophys Res 112:A06317. doi:10.1029/2005JA011548

Otsuka Y, Shiokawa K, Ogawa T, Wilkinson P (2004) Geomagnetic conjugate observationsof medium-scale traveling ionospheric disturbances at midlatitude using all-skyairglow imagers. Geophys Res Lett 31:L15803. doi:10.1029/2004GL020262

Otsuka Y, Tani T, Tsugawa T, Ogawa T, Saito A (2008) Statistical study of relationshipbetween medium-scale traveling ionospheric disturbance and sporadic *E* layer activitiesin summer night over Japan. J Atmos Solar-Terr Phys 70:2196–2202

Perkins F (1973) Spread *F* and ionospheric currents. J Geophys Res 78:218–226

Saito A, Fukao S, Miyazaki S (1998) High resolution mapping of TEC perturbationswith the GSI GPS network over Japan. Geophys Res Lett 25:3079–3082

Saito S, Yamamoto M, Hashiguchi H, Maegawa A, Saito A (2007) Observationalevidence of coupling between quasi-periodic echoes and medium scale traveling ionosphericdisturbances. Ann Geophys 25:2185–2194

Swartz WE, Kelley MC, Aponte N (2009) E- and F-region coupling between anintense sporadic E layer and a mesoscale traveling ionospheric disturbance. Ann Geophys 27:2475–2482

Tsunoda RT (2006) On the coupling of layer instabilities in the nighttime midlatitudeionosphere. J Geophys Res 111:A11304. doi:10.1029/2006JA011630

Tsunoda RT, Cosgrove RB (2001) Coupled electrodynamics in the nighttime midlatitudeionosphere. Geophys Res Lett 28:4171–4174

Yamamoto M, Fukao S, Woodman RF, Ogawa T, Tsuda T, Kato S (1991) Midlatitude*E* region field-aligned irregularities observed with the MU radar. J Geophys Res 96:15,943–15,949

Yamamoto M, Fukao S, Tsunoda RT, Pfaff R, Hayakawa H (2005) SEEK-2(Sporadic-*E* Experiment over Kyushu 2) – project outline, and significance. Ann Geophys 23:2295–2305

Yokoyama T, Hysell DL (2010) A new midlatitude ionosphere electrodynamics couplingmodel (MIECO): Latitudinal dependence and propagation of medium-scale travelingionospheric disturbances. Geophys Res Lett 37:L08105. doi:10.1029/2010GL042598

Yokoyama T, Hysell DL, Otsuka Y, Yamamoto M (2009) Three-dimensional simulationof the coupled Perkins and *E*s layer instabilities in the nighttime midlatitude ionosphere. J Geophys Res 114:A03308. doi:10.1029/2008JA013789

Yokoyama T, Otsuka Y, Ogawa T, Yamamoto M, Hysell DL (2008) First threedimensionalsimulation of the Perkins instability in the nighttime midlatitude ionosphere. Geophys Res Lett 35:L03101. doi:10.1029/2007GL032496

Zhou Q, Mathews JD, Du Q, Miller CA (2005) A preliminary investigation of thepseudo-spectral method numerical solution of the Perkins instability equations in the homogeneousTEC case. J Atmos Solar-Terr Phys 67:325–335

Chapter 21

Statistical Study of Medium-Scale Traveling Ionospheric Disturbances Observed with a GPS Receiver Network in Japan

Y. Otsuka, N. Kotake, K. Shiokawa, T. Ogawa, T. Tsugawa, and A. Saito

Abstract Using densely spaced GPS receivers in Japan, we investigated two-dimensional maps of total electron content (TEC) perturbations with a spatial resolution of $0.15 \times 0.15°$ in longitude and latitude to reveal the statistical characteristics of medium-scale traveling ionospheric disturbances (MSTIDs). We found that MSTIDs can be categorized into three groups: daytime, nighttime, and dawn and dusk MSTIDs. Daytime MSTIDs frequently occur in winter and tend to propagate southeastward. We speculate that daytime MSTIDs are caused by atmospheric gravity waves in the thermosphere. Nighttime MSTIDs frequently occur in summer and propagate southwestward. This propagation direction supports the notion that polarization electric fields could play an important role in generating nighttime MSTIDs. Dawn and dusk MSTIDs frequently occur in summer and propagate eastward and north-northwestward, respectively.

21.1 Introduction

A medium-scale traveling ionospheric disturbance (MSTID) is a wavy structure of the plasma density in the F region, with horizontal scale sizes of 100–500 km (Hunsucker, 1982). MSTIDs have been observed by using various techniques such as ionosondes, HF Doppler sounders, satellite beacons, and incoherent scatter radars (Hunsucker, 1982; Hocke and Schlegel, 1996). The observational results suggest that MSTIDs are caused by atmospheric gravity waves that propagate upward from the lower atmosphere, or are created in conjunction with auroral activity (Hines, 1960; Hooke, 1968). Recent novel observations using highly sensitive CCD cameras and dense GPS receiver networks revealed the two-dimensional horizontal structure of the MSTIDs (e.g., Mendillo et al., 1997; Saito et al., 1998). By observing 630-nm airglow with CCD cameras, Garcia et al. (2000) and Shiokawa et al. (2003a) statistically investigated the period, phase velocity, and horizontal wavelength of nighttime MSTIDs over Arecibo (18.3°N, 66.7°W) in Puerto Rico and Rikubetsu (43.5°N, 143.8°E) and Shigaraki (34.8°N, 136.1°E) in Japan, respectively. They reported that most nighttime MSTIDs propagate southwestward. Miller et al. (1997) and Kelley and Miller (1997) pointed out that this preferred propagation direction cannot be explained by the classical theory of atmospheric gravity waves alone. Furthermore, they suggested that the nighttime MSTIDs could be generated by electrodynamical forces. The Perkins instability is one of the possible mechanisms generating the nighttime MSTIDs, through electro-dynamical processes involving polarization electric fields (Perkins, 1973). The preferred alignment (NW-SE) of the MSITD can be explained by this instability, but not by gravity wave theories.

Saito et al. (1998) first showed two-dimensional maps of total electron content (TEC) perturbations caused by MSTIDs over Japan using a dense global positioning system (GPS) network, which consisted of about 1,000 GPS receivers. Kotake et al. (2007) applied this method to the TEC data obtained from

Y. Otsuka (✉)
Solar-Terrestrial Environment Laboratory, Nagoya University, Nagoya, Japan
e-mail: otsuka@stelab.nagoya-u.ac.jp

M.A. Abdu, D. Pancheva (eds.), A. Bhattacharyya (Coed.), *Aeronomy of the Earth's Atmosphere and Ionosphere*,
IAGA Special Sopron Book Series 2, DOI 10.1007/978-94-007-0326-1_21,

350 GPS receivers in Southern California in 2002, and reported statistical characteristics of the MSTIDs. The present paper reports, for the first time, statistical characteristics of the MSTIDs observed with the GPS network in Japan.

21.2 Data and Method of Analysis

More than 1000 dual-frequency (1.57542 and 1.22760 GHz) GPS receivers have been installed in Japan by the Geographical Survey Institute. They provide data of dual frequency GPS signals, including phase advances and group delays (P-code pseudoranges), every 30 s. From the carrier phase advances, the TEC along a ray path between the GPS satellite and receiver is accurately obtained, although the level of the TEC is unknown because of the unknown initialization constant in phase measurements. This ambiguity in phase measurements was removed by using the measured pseudoranges. We obtained the perturbation component of the TEC, which could be caused by MSTID, by subtracting the 1-h running average (average over ±30 min centered on the corresponding data) from the original TEC time series for each pair of satellites and receivers. The perturbation component of the TEC includes both temporal and spatial variations. The TEC data with elevation angles less than 35° were excluded in this study in order to reduce the effects of the cycle slips and the uncertainty of location where the TEC is mapped. The TEC perturbations so obtained were multiplied by a slant factor to convert the slant TEC to vertical TEC. The slant factor S is defined as τ_0/τ_1, where τ_1 is the length of the ray path between altitudes of 250 and 450 km and τ_0 is the thickness of the ionosphere (200 km) for the zenith path. S varies in a rage between 1 and 1.6 when the satellite elevation angle is larger than 35°. TECs were mapped on the ionospheric shell at a 300 km altitude with a horizontal cell of $0.15° \times 0.15°$ in latitude and longitude in order to distinguish temporal and spatial variations of TEC. The TEC values within each horizontal cell were averaged. This method was the same as that described by Saito et al. (1998). Figure 21.1 shows a two-dimensional map of the TEC perturbations which occurred over Japan at 2250 LT on August 6, 2002. MSTIDs with phase fronts aligned from NW to SE can be seen on the map.

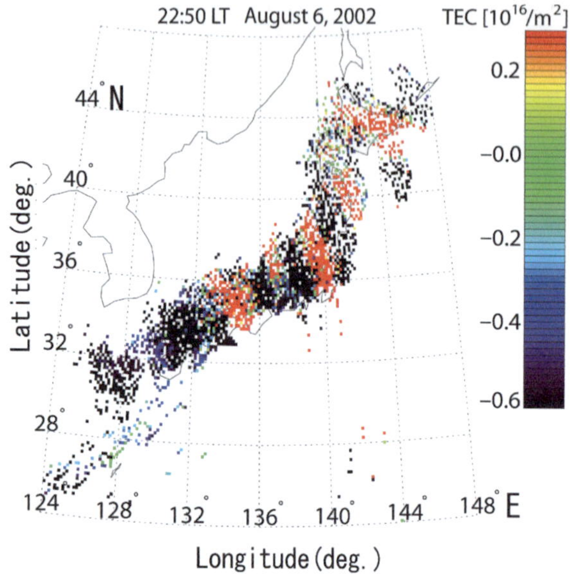

Fig. 21.1 Two-dimensional map of TEC perturbations observed with GPS network in Japan at 2250 LT on August 6, 2002

We adopted the method of Kotake et al. (2007) to identify the TEC perturbations as MSTIDs by using time sequences of two-dimensional TEC perturbation maps. Briefly, the direction of the phase front of the TEC perturbations is determined by eye on the map of the TEC perturbations every 10 min. Assuming that the propagation direction of the MSTIDs is perpendicular to the phase front of the TEC perturbations, we estimate the horizontal wavelength and period of the MSTIDs using plots of horizontal distance and time cross-section of the TEC perturbations. The horizontal phase velocity (V) of the MSTIDs is obtained from $V = \lambda/T$, where λ is the horizontal wavelength and T is the period of the MSTIDs. MSTID parameters (propagation direction, period, horizontal wavelength, phase velocity, and amplitude) are measured every 10 min. The MSTIDs are defined as the TEC perturbations that satisfy the following four criteria: (1) the TEC perturbation has amplitude exceeding 0.2 TECU (1 TECU $= 10^{16}$ electrons/m^2); (2) the horizontal wavelength of the TEC perturbations is shorter than 1000 km; (3) the period of the TEC perturbations is less than 60 min; and (4) the TEC perturbation has more than two phase fronts and propagates on the map.

21.3 Results

We analyzed the TEC data obtained in Japan in 2002. In order to investigate seasonal variations in the MSTID characteristics, the 1-year interval was divided into three seasons, namely, summer (May–August), winter (November–February), and equinox (March, April, September, and October).

Figure 21.2 shows local time and seasonal variations of the MSTID occurrence rate in monthly and hourly bins. We defined the occurrence rate as the ratio of the time duration of the MSTIDs to the monthly and hourly observation intervals. From the figure, we find that the MSTID occurrence rate strongly depends on the season and local time. A high occurrence rate can be seen at four regions: dawn (0500–0700 LT) in summer, daytime (0800–1200 LT) in winter, dusk (1700–2000 LT) in summer, and nighttime (2100–0300 LT) in summer. The occurrence rate in the winter daytime and summer nighttime is the highest and exceeds 40%, which is approximately twice that at dawn in summer. For the nighttime MSTID occurrence, a minor peak can also be seen between November and December.

Figure 21.3 shows the local time variation of the occurrence rate of the MSTID horizontal propagation direction over Japan for (a) all seasons, (b) equinoxes, (c) summer, and (d) winter. The occurrence rate was defined as the ratio of the time period for the MSTIDs to the entire observation interval. The azimuth of the propagation direction was defined as the angle clockwise from geographic north. We classified all the data into 30°-interval bins and calculated the occurrence rate in each azimuthal bin. From the figure, we find that the azimuth of the MSTID propagation direction is a function of local time. The propagation direction of the dawn MSTIDs, which frequently appear in summer, is distributed from north-northeast to east-southeast (from 30° to 120° in azimuth). The daytime MSTIDs, which are observed frequently in winter, propagate mainly southward and southeastward (from 90° to 120°). At dusk, the MSTIDs appear frequently in summer and propagate mostly north-northwestward (from 300° to 360°). During the nighttime, most MSTIDs propagate southwestward (between 210° and 270°). The nighttime MSTIDs are most frequently observed in summer.

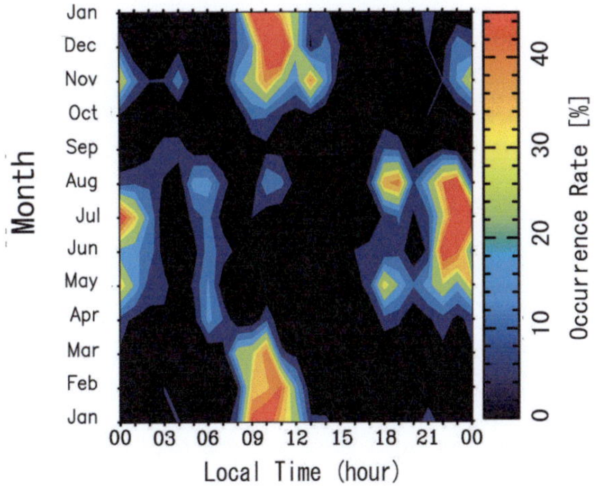

Fig. 21.2 Local time and seasonal variations in MSTID occurrence rate over Japan in 2002

Figure 21.4 shows the occurrence rate of the horizontal wavelength of the MSTIDs observed over Japan for (a) all seasons, (b) equinoxes, (c) summer, and (d) winter, respectively. Horizontal wavelengths are divided into 50-km bins. Most of the horizontal wavelengths are between 250 and 450 km in the summer nighttime, whereas they are between 100 and 350 km in the winter daytime. On the other hand, the wavelengths at dawn in summer are distributed widely, from 200 to 450 km. The wavelengths at dusk in summer are between 150 and 300 km.

Figure 21.5 shows the local time variation of the occurrence rate for period for the MSTIDs observed over Japan for (a) all seasons, (b) equinoxes, (c) summer, and (d) winter, respectively. All the period data were divided into 5-min interval bins. Most of the periods of the nighttime MSTIDs are between 40 and 60 min, which is longer than those of the daytime MSTIDs, the periods of which are between 20 and 45 min. In summer, the periods of the MSTIDs tend to be longer at dawn, between 30 and 45 min, than at dusk, between 25 and 40 min.

Figure 21.6 shows the local time variation of the occurrence rate of the horizontal phase velocity for (a) all seasons, (b) equinoxes, (c) summer, and (d) winter, respectively. The data were divided into 30-m/s bins. The main distribution of phase velocity is between 60 and 140 m/s in summer nighttime and between 80 and 160 m/s in winter daytime. The phase velocity is found to be slightly faster in the daytime than nighttime. The phase velocities of the MSTIDs at dawn are distributed widely between 100 and 180 m/s, and those at dusk are distributed between 100 and 120 m/s.

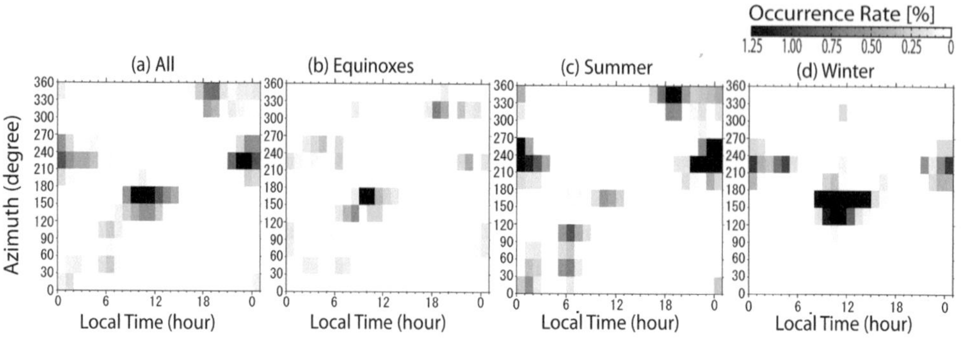

Fig. 21.3 Local time variation of horizontal propagation direction of MSTIDs observed in Japan for (**a**) all seasons, (**b**) equinoxes, (**c**) summer, and (**d**) winter

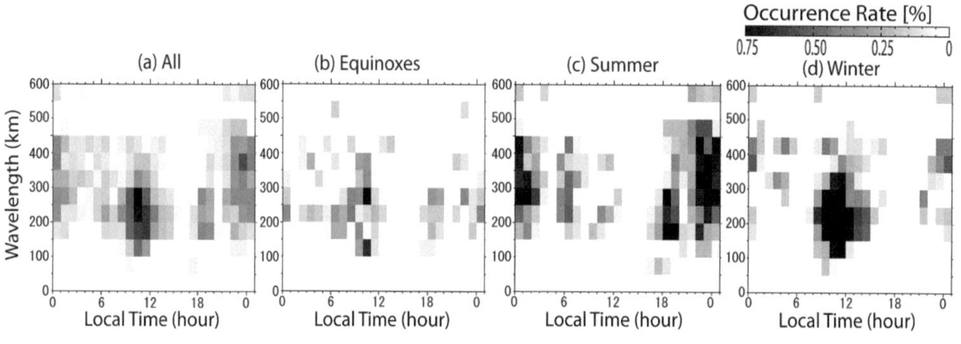

Fig. 21.4 Local time variation of horizontal wavelength of MSTIDs observed in Japan for (**a**) all seasons, (**b**) equinoxes, (**c**) summer, and (**d**) winter

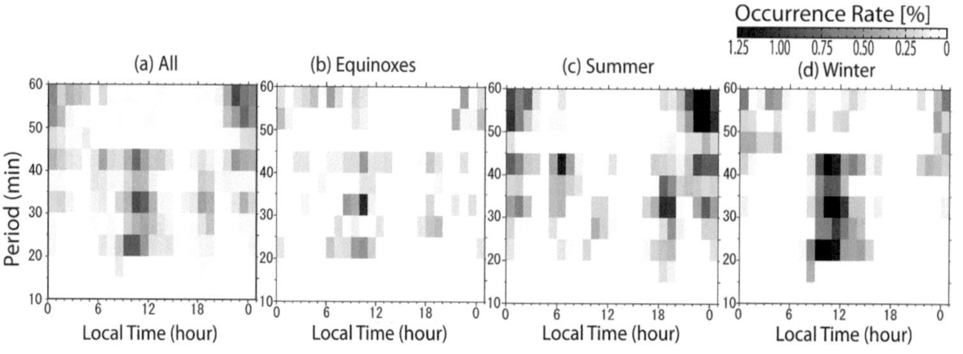

Fig. 21.5 Local time variation of period of MSTIDs observed in Japan for (**a**) all seasons, (**b**) equinoxes, (**c**) summer, and (**d**) winter

21.4 Discussion

By analyzing the GPS-TEC data within the area of 124–148°E and 24–48°N, we have shown the seasonal and local time variations of the MSTID characteristics, such as occurrence rate, propagation direction, period, horizontal wavelength, and phase velocity. On the basis of these characteristics, we found that the MSTIDs over Japan can be classified into

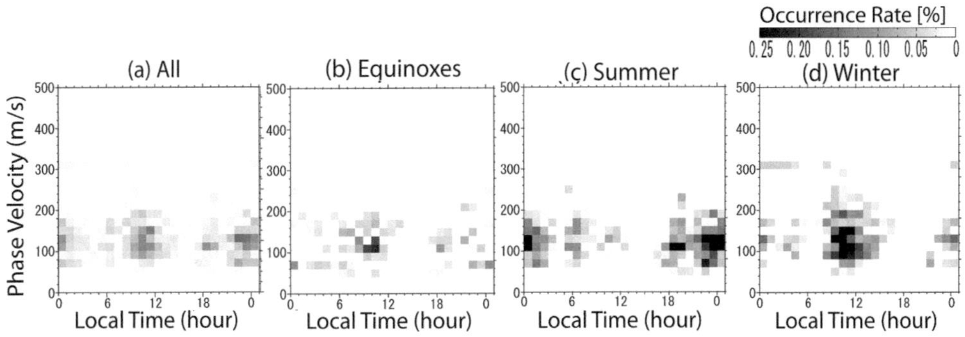

Fig. 21.6 Local time variation of horizontal phase velocity of MSTIDs observed in Japan for (**a**) all seasons, (**b**) equinoxes, (**c**) summer, and (**d**) winter

the following three types: 1) daytime MSTIDs, which propagate southeastward and more frequently occur in winter; 2) nighttime MSTIDs, which propagate southwestward and frequently occur in summer; and 3) dawn and dusk MSTIDs, which propagate eastward and northwestward, respectively, and frequently occur in summer. These characteristics are almost the same as those observed in Southern California (Kotake et al., 2007). Previous studies also have noted that the occurrence rate and propagation direction of the MSTIDs depend on season and local time (Jacobson et al., 1995; Mercier, 1996; Kelley and Miller, 1997). The local time difference of the MSTID characteristics could arise from a corresponding difference in the mechanisms causing MSTIDs. In the following subsections, we discuss the mechanisms causing the MSTIDs in daytime, nighttime, and dawn/dusk, respectively.

21.4.1 Daytime MSTIDs

Daytime MSTIDs over Japan have a wavelength of 100–350 km and occur frequently in winter. The daytime MSTIDs mostly propagate south-southeastward. The seasonal variation of the MSTID occurrence rate and the horizontal wavelengths are consistent with the findings from previous studies (Kotake et al., 2006; Evans et al., 1983; Oliver et al., 1997).

MSTIDs have been considered to be a manifestation of atmospheric gravity waves (Hines, 1960; Hooke, 1968). The gravity wave is an oscillation of neutral gases. Through the neutral-ion collisions, ions in the F region move along the geomagnetic field lines (B), whereas ion motion across the magnetic field line is restricted. This directivity of ion mobility causes directivity in the response of the plasma density variations to the neutral motion due to the gravity waves. Hooke (1968) first formulated a linear theory that provides a reasonable description of the relationship between plasma density perturbations and neutral wind oscillation caused by gravity waves. According to his theory, gravity waves propagating equatorward could cause larger amplitude of TEC perturbations than gravity waves propagating toward other directions because a neutral wind oscillation parallel to the geomagnetic field is larger for gravity waves propagating equatorward than those propagating in other directions. Such a directivity in the response of the plasma to the gravity waves could be responsible for the southward tendency in the propagation direction of the daytime MSTIDs. Furthermore, an anisotropic frictional ion drag force could also contribute to the southward tendency of the daytime MSTID propagation direction (Liu and Yeh, 1969; Kelley and Miller, 1997).

On the basis of these theories, most MSTIDs are expected to show a tendency to propagate equatorward along the magnetic field lines. However, our statistical results show that MSTIDs propagate south-southeastward. The daytime MSTIDs observed over Southern California also tend to propagate southeastward (Kotake et al., 2007). Further investigations are needed to explain such an eastward shift of the MSTID propagation direction.

The present results show that daytime MSTIDs appear frequently in winter. Kotake et al. (2006), who investigated the TEC perturbations caused by MSTIDs using GPS networks across the world, showed that the daytime MSTIDs are active in winter in both the northern and southern hemispheres. Bristow et al.

(1996) showed that the occurrence rate of the MSTIDs observed with the Super Dual Auroral Radar Network (Super-DARN) is highest in winter and suggested that this tendency of the MSTIDs to occur in winter can be attributed to the reflection of gravity waves near the mesopause. The altitude gradient of the neutral temperature is steep, especially at high latitudes and the neutral temperature is low near the mesopause in summer, such that some of the gravity waves cannot propagate upward from the middle atmosphere through the mesopause region. Our results suggest that this theory could also be applicable to mid-latitudes.

21.4.2 Nighttime MSTIDs

The present study shows that most nighttime MSTIDs over Japan propagate southwestward. This southwestward tendency is consistent with that obtained from previous observations in the northern hemisphere using the GPS-TEC maps over Southern California (Kotake et al., 2007) and the 630-nm airglow imaging technique (e.g., Miller et al., 1997; Garcia et al., 2000; Shiokawa et al., 2003a). In the southern hemisphere, the nighttime MSTIDs have phase fronts aligned in NE to SW direction (Otsuka et al., 2004; Shiokawa et al., 2005). The nighttime MSTIDs in both hemispheres have mirrored structures connected by the geomagnetic fields.

According to Otsuka et al. (2007, 2009), the NW-SE phase fronts of the nighttime MSTIDs in the northern hemisphere can be explained as follows. The field-line-integrated electric current in the F region, **J**, is written as $\mathbf{J} = \Sigma(\mathbf{E} + \mathbf{U} \times \mathbf{B})$, where Σ is the field-line-integrated Pedersen conductivity, **E** is the background electric field, and **U** is the neutral wind velocity. During the nighttime, **E** is expected to be smaller and anti-parallel to $\mathbf{U}\times\mathbf{B}$, because **E** is generated by the F-region dynamo mechanism (Rishbeth, 1971). Consequently, **J** probably flows in almost the same direction as $\mathbf{U}\times\mathbf{B}$. Since **U** blows southeastward during the nighttime at mid-latitudes (Hedin et al., 1996), **J** is northeastward. When a perturbation of Σ ($\delta\Sigma$) has a structure elongated from NW to SE, **J** traverses the $\delta\Sigma$. In such case, to maintain a divergence-free **J**, a polarization electric field ($\boldsymbol{\delta E}$) should be generated perpendicular to the MSTID phase fronts and the current continuity in the perpendicular to the wave vector requires an equation of $\Sigma(\mathbf{E} + \mathbf{U} \times \mathbf{B}) \cdot \mathbf{k} = (\Sigma - \delta\Sigma)$ $(\mathbf{E} + \boldsymbol{\delta E} + \mathbf{U} \times \mathbf{B})\ \mathbf{k}$, where **k** is a unit vector perpendicular to the phase fronts. Therefore, The $\boldsymbol{\delta E}$ is northeastward (southwestward) in the regions of low (high) Σ. The $\boldsymbol{\delta E}$ moves the plasma upward (downward) by $\mathrm{E}\times\mathrm{B}$ drift, causing the plasma density perturbations. This is a possible mechanism for generating the nighttime MSTIDs with phase fronts elongated from NW to SE in the northern hemisphere. Magnitude of the $\boldsymbol{\delta E}$ is comparable to or lager than that of the background electric field **E** (Shiokawa et al., 2003b; Otsuka et al., 2007). In particular, under the condition that the wave vector of the $\delta\Sigma$ lies between the direction of **J** and the east, the perturbations in both plasma density and electric field grow with time through the Perkins instability (Perkins, 1973). In the daytime, the $\boldsymbol{\delta E}$ could be short-circuited by high conductivity in the E region such that the above mechanism does not work during the daytime. This might explain why the NW-SE alignment of MSTID is seen only in the nighttime.

The above scenario accounts for the NW-SE alignment of the MSTID, but cannot explain southwestward propagation. Garcia et al. (2000) have pointed out that the propagation velocity of the MSTIDs is not consistent with the background plasma velocity because the background plasma velocity, which could be caused by ExB drift, is southeastward and smaller than the propagation velocity of the MSTIDs. Yokoyama et al. (2009), who performed three-dimensional simulations of E and F region instabilities, showed that the F region structures follow the E region drift such that the southward component of the rotational wind shear in the Es layer produces the southwestward propagation of the NW-SE structures in both the E and F regions. The importance of the electromagnetic coupling processes between the E and F regions for the generation of the E and F region structures was identified theoretically by Tsunoda and Cosgrove (2001) and Cosgrove and Tsunoda (2004). Simultaneous observations of the MSTID and E region irregularities were performed using airglow imagers, VHF/HF radars, and a GPS network (Saito et al., 2007; Otsuka et al., 2007; Ogawa et al., 2009). Furthermore, by statistically analyzing GPS-TEC and ionosonde data in Japan, Otsuka et al. (2008) showed that the MSTID activity is closely correlated with two Es parameters (*foEs* and *foEs – fbEs*). These observational results also indicate that MSTID and Es structures could be generated by an electro-dynamical coupling process between the E and F regions.

21.4.3 Dawn and Dusk MSTIDs

Dawn and dusk MSTIDs over Japan occur frequently in summer. The mean propagation direction of the dawn and dusk MSTIDs is eastward and north-northwestward, respectively. These features are consistent with those of the MSTIDs over Southern California (Kotake et al., 2007). As neutral winds in the thermosphere are driven by the pressure gradient of the neutral atmosphere, the neutral winds blow from the dayside to the nightside, and then westward at dawn and eastward at dusk. The zonal component of the horizontal propagation velocity of the observed dawn and dusk MSTIDs tends to be opposite to the direction of the neutral winds. This relationship between the MSTID propagation direction and background neutral winds suggests that the wind filtering mechanism could be operating effectively (Cowling et al., 1971; Waldock and Jones, 1984). When the horizontal phase velocity of the gravity waves is close to the background neutral wind velocity (a critical level), the gravity waves cannot propagate through the critical level (Bretherton, 1967). The favored azimuth of the MSTIDs propagation could counter the background wind. Consequently, the preferred azimuth of the MSTIDs propagation direction could be eastward at dawn and westward at dusk.

However, as described above, equatorward-propagating gravity waves could cause a larger amplitude of TEC perturbations than gravity waves propagating in other directions (Hooke, 1970). This is because the amplitude of the electron density perturbations is proportional to the neutral wind velocity in the direction parallel to the geomagnetic field lines. Therefore, gravity waves with a large amplitude of neutral wind perturbations could be needed to produce the dawn and dusk MSTIDs. Mechanisms generating such gravity waves with large neutral wind perturbations have not been clarified as yet.

By analyzing the GPS-TEC data for Japan and its geomagnetically conjugate point, Australia, Afraimovich et al. (2009) have pointed out that the dusk MSTIDs over Japan appear 1.5–3 h before the evening solar terminator over Japan and when the solar terminator passes over Australia. On the basis of this result, they have suggested that MSTIDs are caused by slow magnetosonic waves. Recently, terminator waves of neutral density, wind and plasma temperature have been observed in the morning and evening by the CHAMP satellite (Forbes et al., 2008; Liu et al., 2009). These waves have a horizontal wavelength of ~1,000 km, whereas dawn and dusk MSTIDs observed by GPS-TEC in Japan have horizontal wavelengths of 150–450 km. The relationship between the terminator waves observed by CHAMP and dawn and dusk MSTIDs is under investigation.

21.5 Summary

We have investigated the statistical characteristics of the MSTIDs observed over Japan with densely spaced GPS receivers, and found that the characteristics of MSTIDs are different at daytime, nighttime, and dawn and dusk. The observed characteristics can be summarized as follows:

1. The occurrence rate of daytime MSTIDs is high in winter and during equinoxes. Most of the daytime MSTIDs propagate southeastward. Since neutral wind oscillation parallel to the geomagnetic field line is larger for gravity waves propagating equatorward than for gravity waves propagating in other directions, equatorward-propagating gravity waves could cause a larger amplitude of TEC perturbations. Such a directivity in response of the F region plasma to the gravity waves could be responsible for the observed southward tendency in the propagation direction. Dumping of the gravity waves due to ion drag could also contribute to this directivity.
2. The occurrence rate of nighttime MSTIDs has a peak in summer. Most of the nighttime MSTIDs propagate southwestward. This result suggests that electrodynamical forces could play an important role in generating the nighttime MSTIDs.
3. Dawn and dusk MSTIDs occur frequently in summer. On average, MSTIDs propagate eastward at dawn and north-northwestward at dusk. The zonal component of the horizontal propagation velocity of these MSTIDs tends to be opposite to the direction of the background neutral winds.

This result indicates that these MSTIDs could be caused by atmospheric gravity waves and that the wind filtering mechanism could be operating effectively.

The characteristics of MSTIDs obtained in this study are consistent with previous observational results. These findings suggest that the mechanism for generating MSTIDs depends on the local time.

Acknowledgements GEONET GPS data were provided by the Geographical Survey Institute of Japan.

References

Afraimovich I, Edemskiy K, Leonovich AS, Leonovich LA, Voeykov SV, Yasyukevich YV (2009) MHD nature of night-time MSTIDs excited by the solar terminator. Geophys Res Lett 36:L15106. doi:10.1029/2009GL039803

Bretherton FP (1967) The propagation of groups of gravity waves in a shear flow. Quant J R Meteoroll Soc 92:466–480

Bristow WA, Greenwald RA, Villain JP (1996) On the seasonal dependence of medium-scale atmospheric gravity waves in the upper atmosphere at high latitudes. J Geophys Res 101:685–699

Cosgrove RB, Tsunoda RT (2004) Instability of the E-F coupled nighttime midlatitude ionosphere. J Geophys Res 109:A04305. doi:10.1029/2003JA010243

Cowling DH, Webb HD, Yeh KC (1971) Group rays of internal gravity waves in a wind stratified atmosphere. J Geophys Res 76:213–220

Evans JV, Holt JM, Wand RH (1983) A differential-Doppler study of traveling ionospheric disturbances from Millstone Hill. Radio Sci 18:435–451

Forbes JM, Bruinsma SL, Miyoshi Y, Fujiwara H (2008) A solar terminator wave in thermosphere neutral densities measured by the CHAMP satellite. Geophys Res Lett 35:L14802. doi:10.1029/2008GL034075

Garcia FJ, Kelley MC, Makela JJ, Huang CS (2000) Airglow observations of mesoscale low-velocity traveling ionospheric disturbances at midlatitudes. J Geophys Res 105: 18407–18415

Hedin AE, Fleming EL, Manson AH, Scmidlin FJ, Avery SK, Clark RR, Franke SJ, Fraser GJ, Tsunda T, Vial F, Vincent RA (1996) Empirical wind model for the upper, middle, and lower atmosphere. J Atmos Terr Phys 58: 1421–1447

Hines CO (1960) Internal atmospheric gravity waves at ionospheric heights. Can J Phys 38:1441–1481

Hocke, K, Schlegel K (1996) A review of atmospheric gravity waves and traveling ionospheric disturbances: 1982–1995. Ann Geophys 14:5917–5940

Hooke WH (1968) Ionospheric irregularities produced by internal atmospheric gravity waves. J Atmos Solar-Terr Phys 30:795–823

Hooke WH (1970) Ionospheric response to internal gravity waves. 1 The F2 region response, J Geophys Res 75(28):5535–5544

Hunsucker R (1982) Atmospheric gravity waves generated in the highlatitude ionosphere: a review. Rev Geophys 20:293–315

Jacobson AR, Carlos RC, Massey RS, Wu G (1995) Observations of traveling ionospheric disturbances with a satellite-beacon radio interferometer: seasonal and local time behavior. J Geophys Res 100:1653–1665

Kelley MC, Miller CA (1997) Electrodynamics of midlatitude spread F, 3 Electrohydrodynamic waves? A new look at the role of electric fields in thermospheric wave dynamics. J Geophys Res 102:11539–11547

Kotake N, Otsuka Y, Tsugawa T, Ogawa T, Saito A (2006) Climatological study of GPS total electron content variations caused by medium-scale traveling ionospheric disturbances. J Geophys Res 111:A04306. doi:10.1029/2005JA011418

Kotake N, Otsuka Y, Tsugawa T, Ogawa T, Saito A (2007) Statistical study of medium-scale traveling ionospheric disturbances observed with the GPS networks in Southern California. Earth Planets Space 59:95–102

Liu H, Lühr H, Watanabe S (2009) A solar terminator wave in thermospheric wind and density simultaneously observed by CHAMP. Geophys Res Lett 36:L10109. doi:10.1029/2009GL038165

Liu CH, Yeh KC (1969) Effect of ion drag on propagation of acoustic-gravity waves in the atmospheric F region. J Geophys Res 74:2248–2255

Mendillo M, Baumgardner J, Nottingham D, Aarons J, Reinisch B, Scali J, Kelley M (1997) Investigations of thermospheric-ionospheric dynamics with 6300°A images from the Arecibo Observatory. J Geophys Res 102: 7331–7343

Mercier C (1996) Some characteristics of atmospheric gravity waves observed by radio-interferometry. Ann Gephys 14: 42–58

Miller CA, Swartz WE, Kelley MC, Mendillo M, Nottingham D, Scali J, Reinisch B (1997) Electrodynamics of midlatitude spread F, 1 Observations of unstable, gravity wave-induced ionospheric electric fields at tropical latitudes. J Geophys Res 102:11,521–11,532

Ogawa T, Nishitani N, Otsuka Y, Shiokawa K, Tsugawa T (2009) Medium-scale traveling ionospheric disturbances observed with the SuperDARN Hokkaido radar, all-sky imager and GPS network, and their relation to concurrent sporadic-E irregularities. J Geophys Res 114:A03316. doi:10.1029/2008JA013893

Oliver WL, Otsuka Y, Sato M, Takami T, Fukao S (1997) A climatology of F region gravity wave propagation over the middle and upper atmosphere radar. J Geophys Res 102:14499–14512

Otsuka Y, Onoma F, Shiokawa K, Ogawa T, Yamamoto M, Fukao S (2007) Simultaneous observations of nighttime medium-scale traveling ionospheric disturbances and E-region field-aligned irregularities at midlatitude. J Geophys. Res 112:A06317. doi:10.1029/2005JA011548

Otsuka Y, Shiokawa K, Ogawa T, Wilkinson P (2004) Geomagnetic conjugate observations of medium-scale traveling ionospheric disturbances at midlatitude using all-sky airglow imagers. Geophys Res Lett 31:15803. doi:101029/2004GL020262

Otsuka Y, Shiokawa K, Ogawa T, Yokoyama T, Yamamoto M (2009) Spatial relationship of nighttime medium-scale traveling ionospheric disturbances and F region field-aligned irregularities observed with two spaced all-sky airglow imagers and the middle and upper atmosphere radar. J Geophys Res 114:A05302. doi:10.1029/2008JA013902

Otsuka Y, Tani T, Tsugawa T, Ogawa T (2008) Statistical study of relationship between medium-scale traveling ionospheric disturbance and sporadic E layer activities in summer night over Japan. J Atmos Solar-Terr Phys 70: 2196–2202

Perkins F (1973) Spread F and ionospheric currents. J Geophys Res 78:218-226

Rishbeth H (1971) The F-layer dynamo. Planet Space Sci 19:263–267

Saito A, Miyazaki S, Fukao S (1998) High resolution mapping of TEC perturbations with the GSI GPS network over Japan. Geophys Res Lett 25:3079–3082

Saito S, Yamamoto M, Hashiguchi H, Maegawa A, Saito A (2007) Observational evidence of coupling between quasi-periodic echoes and medium scale traveling ionospheric disturbances. Ann. Geophys 25:2185–2194

Shiokawa K, Ihara C, Otsuka Y, Ogawa T (2003a) Statisitical study of nighttime medium-scale traveling ionospheric disturbances using midlatitude airglow images. J Geophys Res 108:1052. doi:101029/2002JA009491

Shiokawa K, Otsuka Y, Ihara C, Ogawa T, Rich FJ (2003b) Ground and satellite observations of nighttime medium-scale traveling ionospheric disturbance at midlatitude. J Geophys Res 108(A4):1145. doi:10.1029/2002JA009639

Shiokawa K, Otsuka Y, Tsugawa T, Ogawa T, Saito A, Ohshima K, Kubota M, Maruyama T, Nakamura T, Yamamoto M, Wilkinson P (2005) Geomagnetic conjugate observation of nighttime medium-scale and largescale traveling ionospheric disturbances: FRONT3 campaign. J Geophys Res 110:A05303. doi:10.1029/2004JA010845

Tsunoda RT, Cosgrove RB (2001) Coupled electrodynamics in the nighttime midlatitude ionosphere. Geophys Res Lett 28:4171–4174

Waldock JA, Jones TB (1984) The effects of neutral winds on the propagation of medium-scale atmospheric gravity wave at mid-latitudes. J Atmos Solar-Terr Phys 46: 217–231

Yokoyama T, Hysell DL, Otsuka, Y, Yamamoto M (2009) Three-dimensional simulation of the coupled Perkins and layer instabilities in the nighttime midlatitude ionosphere. J Geophys Res 114:A03308. doi:10.1029/2008JA013789

Part IV
Thermosphere–Ionosphere Coupling, Dynamics and Trends

Chapter 22

New Aspects of the Coupling Between Thermosphere and Ionosphere, with Special regards to CHAMP Mission Results

Hermann Lühr, Huixin Liu, Jeahueng Park, and Sevim Müller

Abstract The recent years have brought enormous progress in our understanding of the characteristics and the variability of the thermosphere. This is to a good part based on the high-resolution measurements of the accelerometer on board the CHAMP satellite. It sampled the air drag from the solar maximum in 2001 to the minimum in 2009. The thermosphere reacts quite differently to solar and magnetic forcing. This is in particular evident when comparing the mass density on the day and night sides at low latitudes. We were able to delineate the variations due to solar flux, season, and magnetic activity. Another important finding is the close relation of thermospheric features to the geomagnetic field geometry. Examples of this are the equatorial mass density anomaly, the cusp density anomaly, and the high-speed wind channel along the dip equator. None of these phenomena is so far considered in any of the atmospheric models. Thermospheric wind is a prime driver for ionospheric dynamics. As an example we present the F-region dynamo current system and its variation with local time. This wind-driven dynamo over the equator is assumed to cause special ionospheric effects after sunset. Typical post-sunset features at low latitude are the equatorial plasma bubbles. Here it is shown that some of these bubbles are accompanied by depletions in neutral density. Neutral depletions can form only when the ion-neutral friction is high enough. In the pressure balance equation, the neutral depletions make up a significant part. The comprehensive set of CHAMP neutral and plasma measurements enabled us to interpret the ionosphere–thermosphere as a coupled system.

H. Lühr (✉)
Helmholtz Centre Potsdam, GFZ German Research Centre for Geosciences, 14473 Potsdam, Germany
e-mail: hluehr@gfz-potsdam.de

22.1 Introduction

The high atmospheric layer between about 100 and 1000 km altitude, termed thermosphere, is characterized by strong changes in response to solar and geomagnetic activity. In particular, during geomagnetic storms, thermospheric density and composition experience dramatic changes which are still not fully understood (see Prölss, 1997). During non-storm times also, significant variations are observed that await explanation. Traditionally the atmospheric density was deduced from orbit changes of spacecraft due to air drag. But this provided only an average value along the orbit. It is difficult to derive regional variations from these data. Later on, satellites carrying mass spectrometers (e.g. ESRO4, AE-E, DE2) provided information about the thermospheric composition (e.g. Trinks and von Zahn, 1975). In principle, mass density can be derived from it, but the sampling is quite sparse. New opportunities for detailed thermospheric studies arose with the advent of recent satellites such as CHAMP and GRACE carrying sensitive tri-axial accelerometers. The continuous and homogeneous data set over almost one solar cycle enables the investigation of various aspects of high atmospheric properties.

One important aim is to find a suitable mathematical formulation that is able to describe the climatological properties of the atmosphere. There are different families of empirical models (e.g. Jacchia, MSIS, JB) which are based on fitting a set of base functions to available observations. These models are updated and

M.A. Abdu, D. Pancheva (eds.), A. Bhattacharyya (Coed.), *Aeronomy of the Earth's Atmosphere and Ionosphere*, IAGA Special Sopron Book Series 2, DOI 10.1007/978-94-007-0326-1_22, © Springer Science+Business Media B.V. 2011

improved, as additional observations become available. A pending question is how well the base functions describe the relation between the controlling parameters and the thermospheric response. The controlling parameters are, for example, the forcing terms such as solar EUV flux and geomagnetic activity. Also location, local time, and season have to be considered properly. Long-duration mass density data sets may be used for delineating the various dependences.

Recent satellite observations have provided convincing evidence that certain thermospheric features are controlled by the geomagnetic field. There are, for example, the equatorial mass density anomaly (Liu et al., 2005, 2007) and the cusp-related density anomaly (Lühr et al., 2004; Rentz and Lühr, 2008). These and other magnetic field-related phenomena are not considered by any of the present atmospheric models. Their properties and generation mechanisms have to be determined, in order to enable the design of suitable base functions for implementation in the models.

The ionosphere is embedded in the thermosphere. In certain height regions, a strong interaction takes place between the neutral and ionized particles. This makes the thermosphere–ionosphere a coupled system. Many studies in the past have suffered from lack of simultaneous observations of plasma and neutral dynamics. Instructive examples of ion-neutral coupling are the F-region dynamo (Lühr and Maus, 2006) and the neutral density depletions accompanying equatorial plasma bubbles (Park et al., 2010).

In the sections to follow we will first introduce the CHAMP satellite and then describe the approach used for deriving thermospheric density and wind from the accelerometer data. Subsequently, quantitative expressions for the dependence of the mass density at 400 km altitude on external drivers are offered. In Section 22.4 we present examples of thermospheric phenomena controlled by the magnetic field geometry. It is followed by a description of the typical properties of the F-region dynamo and of the density depletions accompanying plasma bubbles. Conclusions round off this overview article.

22.2 Observation and Data Processing

New opportunities for thermospheric research evolved from the accelerometer measurements performed by the satellite CHAMP and GRACE. The long duration and global coverage of these missions revealed several new features.

22.2.1 CHAMP Satellite

The CHAMP satellite, launched on 15 July 2000, cycles the Earth on a near-polar (inclination 87.25°) and almost circular orbit at an altitude of about 400 km (Reigber et al., 2002). The orbit has decayed from 456 km at the beginning of the mission to below 300 km in 2010. The orbital plane precesses through 1 h of local time in 11 days. Thus, it takes 131 days for CHAMP, when combining ascending and descending orbital arcs, to cover all local times.

Figure 22.1 shows the arrangement of instruments on the spacecraft. For the investigation of thermospheric properties we have made intensive use of the readings taken by the STAR (Space Three-Axes Accelerometer for Research missions) accelerometer (ACC) during the years 2002–2005. This instrument is located at the spacecraft centre of gravity (CoG) in order to sense properly the non-gravitational forces acting on the satellite. The insets depict the accelerometer sensor and to the left the proof mass which is kept freely floating in the cage by electrostatic forces. Ionospheric properties are derived from the planar Langmuir probe (PLP). This probe provides readings of the electron density and the electron temperature. Electric currents can be sensed by the magnetometers on board CHAMP. In particular, the fluxgate magnetometer (FGM), accommodated on the boom in order to reduce the influence of the spacecraft noise, is employed for estimating current density and direction.

22.2.2 Deriving Thermospheric Mass Density and Wind

The data considered are the pre-processed level-2 ACC data. These scaled and cleaned data are averages over 10 s, which correspond to a sampling distance of 76 km. Another preparation step is the removal of the acceleration due to solar radiation pressure.

The deceleration exerted on a satellite when it travels through the atmosphere can be expressed as

$$\boldsymbol{a} = -\frac{1}{2}\rho\frac{C_\mathrm{d}}{m}A_\mathrm{eff}V^2\boldsymbol{v} \qquad (22.1)$$

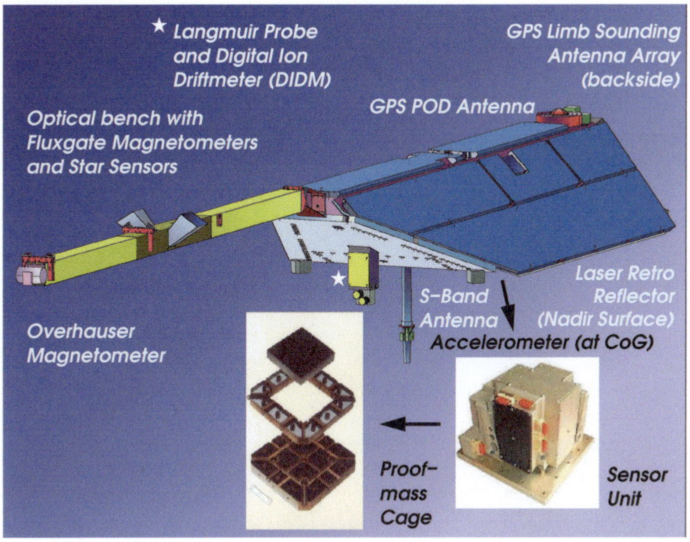

Fig. 22.1 Schematic illustration of the CHAMP satellite and the allocation of the scientific instruments

where $\boldsymbol{a}$ is the total acceleration due to air drag, ρ is the local mass density, C_d is the drag coefficient, A_{eff} is the effective cross-sectional area in ram direction, m is the satellite mass, V is the total velocity relative to the atmosphere, and $\boldsymbol{v}$ is the unit vector of the velocity in ram direction. (For further details about the quantities entering Eq. (22.1), see Liu et al., 2005, 2006). Since all quantities in Eq. (22.1) are known or measured, it can be solved for ρ, thus providing the thermospheric mass density along the orbit.

For the estimation of the thermospheric winds, we make use of the property such that the acceleration vector $\boldsymbol{a}$ is aligned with the velocity $\boldsymbol{v}$ as reflected in Eq. (22.1). Under this assumption the components of the acceleration vector are related to the velocity as follows (Liu et al., 2006):

$$\frac{a_y}{a_x} = \frac{v_y}{v_x} \tag{22.2}$$

where the x component is aligned with the spacecraft axis pointing nominally into the along-track direction and the y component with the cross-track spacecraft axis. The z component contribution can be ignored since the accelerations in vertical direction are known to be very small. From Eq. (22.2) we can deduce the zonal wind velocity u, since for CHAMP on its near-polar orbit the cross-track velocity is well aligned with the east/west direction:

$$u = v_x \frac{a_y}{a_x} - v_{cor} \tag{22.3}$$

where v_x is the orbital velocity (7.6 km/s) and v_{cor} is the corotation velocity of the atmosphere at CHAMP's position. The resulting zonal wind velocity u is thus given in the frame of an observer at the Earth's surface.

22.3 Parameterization of Thermospheric Mass Density

The long and homogeneous series of CHAMP accelerometer data provides the opportunity to investigate the dependence of the mass density at an altitude of around 400 km on the various kinds of drivers. In a dedicated study, Müller et al. (2009) tried to identify quantitative relations for the effects on the air densities. They concentrated their effort on the low-latitude region (±30° lat.) since this provides the largest contribution to satellite air drag. It is known that the solar extreme ultraviolet (EUV) and the geomagnetic activity are the main causes for mass density variations. The sensitivity to these drivers, however, may be dependent on local time. For this reason the authors performed the investigations for day- and night-side data sets separately. As time interval representing the daytime most suitably, they identified the 10:30 to 16:30 local time (LT) sector and correspondingly for night-time, they selected 22:30 to 04:30 LT. The displacement of these two intervals by 1.5 h from noon and midnight is consistent with the delayed reaction of the thermosphere to the diurnal variation of solar insulation.

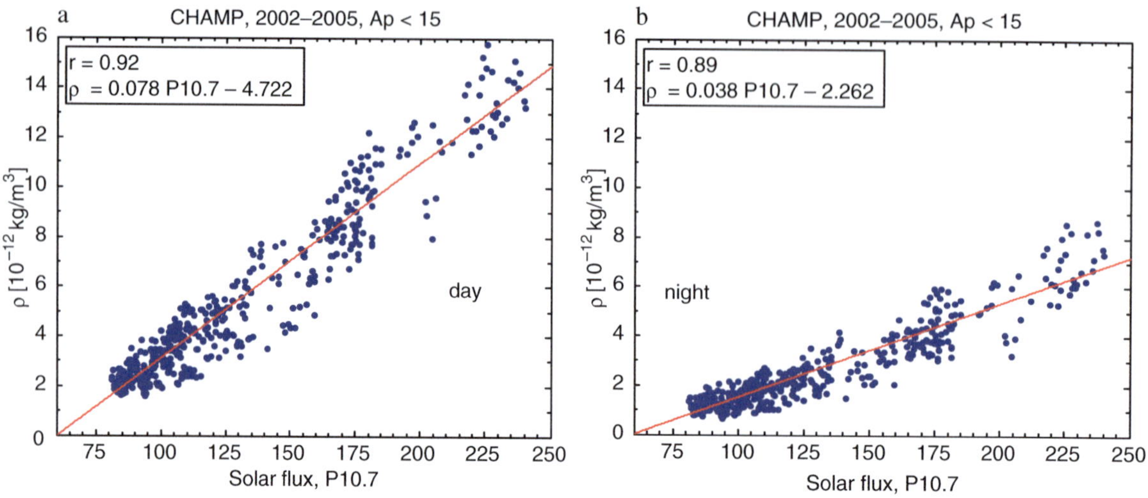

Fig. 22.2 Correlation between mass densities at 400 km altitude and the solar flux level (P10.7) separately for day and nightsides. The density readings are longitudinally averaged daily means (adopted from Müller et al., 2009)

As a first step the dependence of the mass density on the solar EUV flux was determined. The time interval considered was the years 2002–2005, the declining phase of solar cycle 23. In order not to confuse the solar with the geomagnetic effect, only data from magnetically quiet days with an activity index $A_p < 15$ have been considered. In Fig. 22.2 the relation between the solar flux, represented by the index P10.7,[1] and air density is shown separately for daytime and night-time. As can be seen, excellent correlations emerge between the two quantities over the whole covered solar flux range, P10.7 = 80–240 sfu, for both time sectors. Motivated by the high correlation coefficient of 0.9, linear regression lines were fitted. From the difference in slope of the regression line, as given in the frames of Fig. 22.2, we may conclude that the sensitivity of the thermospheric density to solar flux variation is twice as high on the day side as on the night side. This ratio seems to be independent of the intensity of solar flux. In the morning and evening sectors we find intermediate slopes. All this means is that the response of the thermosphere to solar flux is local time dependent (and of course also latitude dependent). For these reasons it may be questioned whether it is a good idea to make use of density values obtained from orbital averages, regardless of local time and latitude range, for determining the thermospheric dependence on solar flux (e.g. Gou et al., 2008).

Consequently, Müller et al. (2009) investigated also the dependence of the thermospheric density on geomagnetic activity. For this purpose, density readings from active days with $A_p > 15$ were selected. In order to remove the effect of solar flux variations, all density values were first normalized to a constant flux level of P10.7 = 130 sfu by making use of the equations given in the frames of Fig. 22.2. In addition, the authors allowed for a delayed reaction of the low-latitude thermosphere to magnetic activity of 3 and 4.5 h for the day side and night side, respectively. As controlling parameters the two versions of the 3-hourly magnetic activity indices A_p and a_m were tested. The index A_p is the linear equivalent to the planetary activity index K_p, while a_m is an alternative version of A_p. For more details, see Menvielle and Marchaudon (2007). It turned out that both indices are suitable, but slightly better results are obtained when using the a_m index.

In the case of magnetic activity, Müller et al. (2009) considered the absolute increase of mass density added on top of the background. For the correlation the increases in air density readings were first sorted into logarithmically spaced a_m bins. Then the median value of each bin was used in the analysis. As can be read from Fig. 22.3, excellent linear correlations emerge and the regression lines for both day side and night

[1]P10.7 = 0.5 (F10.7 + F10.7a), where F10.7a is the 81-day running average of the daily F10.7.

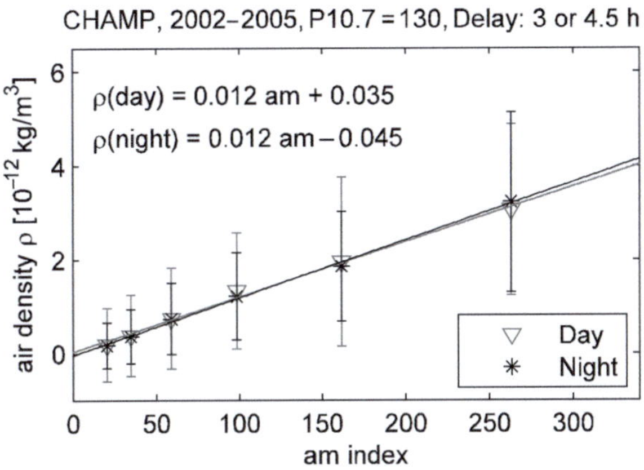

Fig. 22.3 Dependence of the low-latitude (±30° lat.) mass density on magnetic activity, as represented by the index a_m. A linear fit to the medians of each bin was applied (*solid lines*). The error bars indicate the standard deviation of the values within a bin (adopted from Müller et al., 2009)

side are practically identical. Numerical expressions for the fitted lines are given in Fig. 22.3. Both start as expected at the origin. Although the spread of the individual density readings is considerable, on average we find a linear additive increase in the air density with magnetic activity, independent of local time.

Another characteristic investigated is the seasonal variation of the thermospheric density. In order to delineate the various influences, Müller et al. (2009) selected only data from magnetically quiet days ($A_p < 15$) and normalized the mass density readings to a constant solar flux level of $P = 130$ sfu employing again the equations from Fig. 22.2. These pre-processed CHAMP mass density data of all 4 years were plotted versus day of year (DoY).

As can be seen in Fig. 22.4, a clear seasonal variation of the normalized density readings emerges for both the day side and the night side. Lowest densities are recorded around June solstice, and peak densities are found around spring equinox. Data from all 4 years homogeneously support this intra-annual modulation, although the solar and magnetic activities varied considerable from year to year. Annually harmonic functions were fitted to the data. Resulting coefficients are listed in Table 22.1. The seasonal variations are very similar in shape on the day side and night side, as can be deduced from the phases. The amplitudes are, as expected, about twice as large on the day side as on the night side.

We may conclude from the results obtained by Müller et al. (2009) that there is a linear dependence of thermospheric mass density on solar EUV flux. The coefficient relating these two quantities varies, however, with local time. At low latitudes it is twice as high on the day side as on the night side. In contrast, magnetic activity causes an additive enhancement of the mass density on top of the quiet-time density.

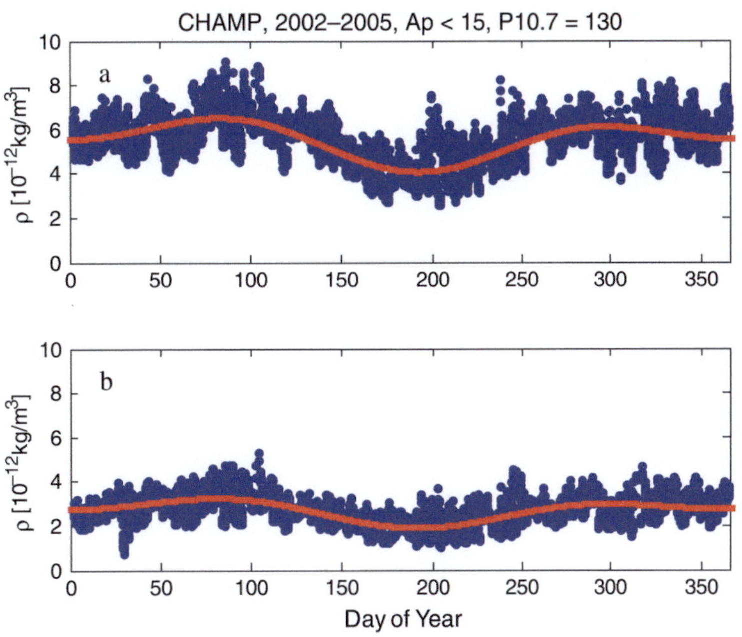

Fig. 22.4 Seasonal variation of the mass density in the daytime (**a**) and night-time (**b**) sectors. The *solid curves* represent the fitted annual harmonics functions (cf. Table 22.1) (adopted from Müller et al., 2009)

Table 22.1 Coefficients of the annually harmonic functions. Prevailing conditions are A_p<15, P10.7=130, h=400 km. The uncertainty within a 95% confidence is about 2.4×10^{-14} and 1.6×10^{-14} kg/m^3 for the amplitudes of all the harmonics on the day side and night side, respectively. The uncertainty of the phases (day of peak amplitude) is about ±5 days (from Müller et al., 2009)

	Day side		Night side	
	Amplitude (10^{-12} kg/m^3)	Phase (DoY)	Amplitude (10^{-12} kg/m^3)	Phase (DoY)
Annual	0.77	23	0.45	25
Semi-annual	0.70	97	0.37	96
Ter-annual	0.24	52	0.11	38
Constant	5.57	–	2.72	–
RMSE	0.75	–	0.50	–

The amount of additional density is proportional to the activity index A_p or a_m and is the same at all local times. But the thermospheric response occurs after slightly different delay times in the different time sectors. The seasonal variation of the mass density can be described sufficiently well by the annual and semi-annual harmonics. The amplitude of the ter-annual does not exceed the uncertainty range. Highest densities are reached during spring equinox (DoY 83) and lowest values around June solstice (DoY 192). The ratio between the extremes amounts to 1.63. Both annual and semi-annual variations, having approximately the same amplitude, exhibit a minimum around the middle of the year, thus causing together the deep July minimum.

An important conclusion from this study is that the investigated forcings of the thermosphere can be treated by a linear combination of the different influences. No non-linear interaction between the drivers has been identified.

22.4 Magnetic Forcing of the Thermosphere

In this section the spatial distribution of thermospheric density and winds is presented when geophysical conditions are constant. Dominating features are the equatorial mass density anomaly (EMA) and the fast wind jet at the dip equator instead of at the geographic equator.

A survey of the mass density distribution based on CHAMP measurements was performed by Liu et al. (2005). It revealed the main characteristics of the thermosphere, as is shown in Fig. 22.5. For comparison, the lower frame presents the prediction of the MSIS atmospheric model for the same observational conditions. Although the gross features agree between observations and model, there are significant differences. CHAMP does not detect the largest densities at sub-solar latitudes but some 20° north and south of it. This effect is best visible when the data are binned by magnetic latitude. In a dedicated study, Liu et al. (2007) have investigated the typical characteristics of this equatorial mass density anomaly (EMA). As shown in Fig. 22.6, two bands of enhanced density track closely the latitude variations of the magnetic dip equator. This demonstrates the influence of the geomagnetic field on the neutral air distribution. The EMA is predominantly a daytime phenomenon building up after 09 LT and ceasing shortly before 20 LT. A hemispheric symmetric distribution of the two density bulges is observed during equinox months. During solstices the bulge in the summer hemisphere strongly dominates and the whole system is displaced by ~10° in latitude towards the summer hemisphere. An interesting feature that suggests a link between the mass density anomaly and the ionospheric dynamics is the dependence of the crest-to-trough ratio on the solar EUV flux level. The basic idea for the generation mechanism, offered by Liu et al. (2005), is that the geomagnetic field acts as guiding frame for plasma motion. Its geometry therefore determines where ion energy is deposited in the thermosphere. A suitable process proposed earlier by Fuller-Rowell et al. (1997) is chemical heating fuelled by charge exchange between O^+ and O_2^+ at E region altitudes. The source for an enhanced O^+ flux could be the equatorial ionization anomaly. A final verification of this suggestion is, however, still pending. Also note that the neutral density in Fig. 22.6 is not uniformly distributed in longitudes. Instead, it maximizes at four longitudes around –120°, –20°, 60°, and 140°, forming a wave-4 structure in the zonal direction. Though out of the

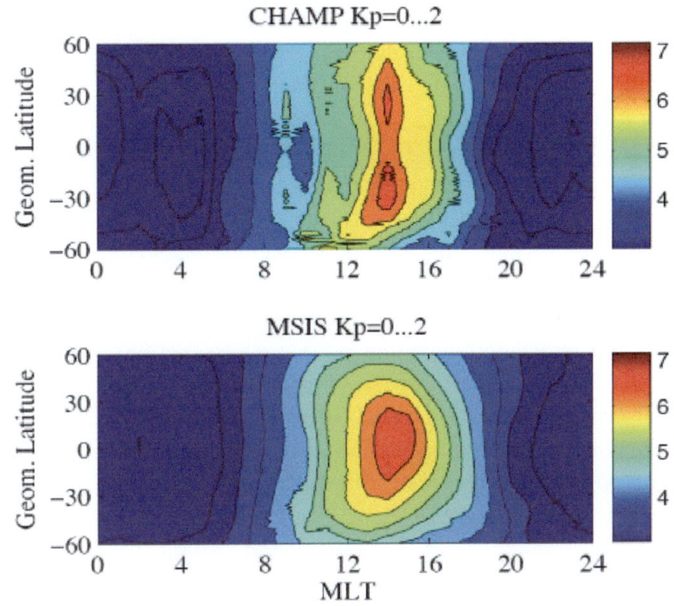

Fig. 22.5 Latitudinal distribution of the thermospheric density (in 10^{-12} kg/m^3) versus local time, (*top panel*) CHAMP observations and (*bottom panel*) model predictions. The data have been sorted by magnetic latitude (after Fig. 2 of Liu et al., 2005)

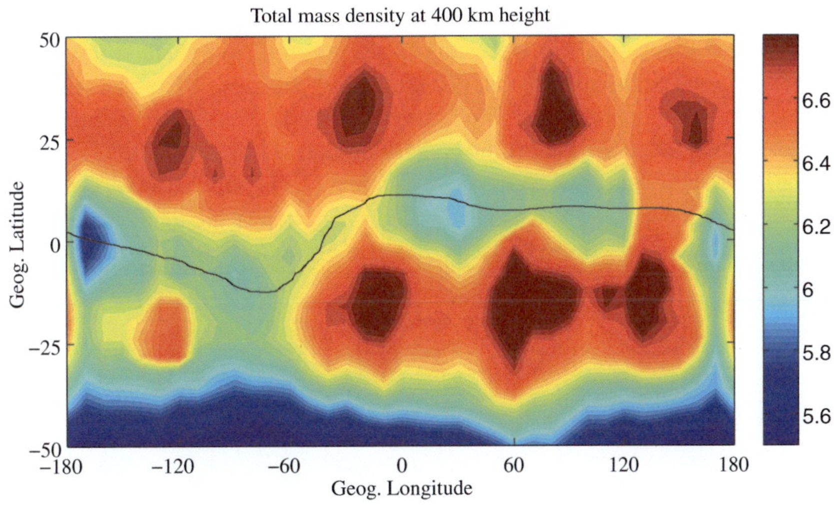

Fig. 22.6 Longitude distribution of the mass density (in 10^{-12} kg/m^3) on the day side. The two bands of enhanced density track closely the latitude variation of the magnetic dip equator (adjusted from Fig. 2 in Liu et al., 2007)

scope of the present paper, this zonal structure is an important feature implying the influence of the upper thermosphere by upward propagating tidal waves as discussed in detail in Liu et al. (2009b).

Another example of geomagnetic field-controlled thermospheric modification is the polar cusp-related density anomaly (Lühr et al., 2004). Localized density enhancements are observed rather regularly around noon at cusp latitudes. The associated upwelling of air is expected to be caused by a combined action of soft particle precipitation and small-scale field-aligned currents. A detailed study of the cusp density anomalies was performed by Rentz and Lühr (2008).

The other thermospheric quantity that can be derived from CHAMP accelerometer measurements is the zonal wind. In a first global study, Liu et al. (2006) determined from CHAMP data the diurnal variation of the wind speed at equatorial latitudes and its dependence on solar flux level and magnetic activity. A comparison with the horizontal wind model (HWM93) revealed good agreements during night-time but significant differences during daylight hours. Explanations for that are still pending.

In a follow-up study, Liu et al. (2009a) considered the zonal wind distribution at low and mid-latitudes. Figure 22.7 illustrates the zonal wind distribution in the frame of magnetic dip latitude versus magnetic local time in quasi-dipole coordinates. Although some differences exist (e.g. absolute wind speed and reversal time, which will be discussed in the next section), both CHAMP and DE-2 observations reveal fairly similar wind patterns. The wind at equatorial latitudes blows eastwards during night and westwards before afternoon. Towards higher latitude the morning wind reversal occurs progressively earlier, leading to a pronounced triangle shape in the 2D wind pattern. On the night side, the wind maximizes at the dip equator, forming a fast wind jet.

It is important to note that the net wind, a 24-h average of the wind, is not zero as shown in the right column of Fig. 22.7. At equatorial and low latitudes, the net wind is positive. This is the atmospheric superrotation, as discussed in Liu et al. (2006). Due to the magnetic control of the neutral wind, the atmospheric super-rotation maximizes at the dip equator instead of at the geographic equator, which is quite different from that on other planets like Venus.

The fast wind jet is examined from a different perspective, as shown in Fig. 22.8. This figure presents the seasonally averaged zonal wind distribution during night in the geographic versus geomagnetic frame. It is seen that the eastward zonal wind blows strongest along the dip equator, as indicated by the solid line.

This wind jet along the Earth's dip equator instead of at the geographic equator again demonstrates strong magnetic control of the thermospheric dynamics. As pointed out in Liu et al. (2009a), the ion drag is the main factor that shifts the fast wind from the geographic equator to the dip equator. With the EIA structure prevailing in the equatorial region, the ion drag minimizes at the equator, facilitating faster wind to flow there.

22.5 The F-Region Dynamo

In this section we present an example where the thermosphere is the prime driver for an ionospheric process. The concept of a wind-driven dynamo at F-region altitudes was proposed already several decades ago by Rishbeth (1971). The first observational evidence of the generated currents was deduced from Magsat magnetic field data (Maeda et al., 1982), but limited to the

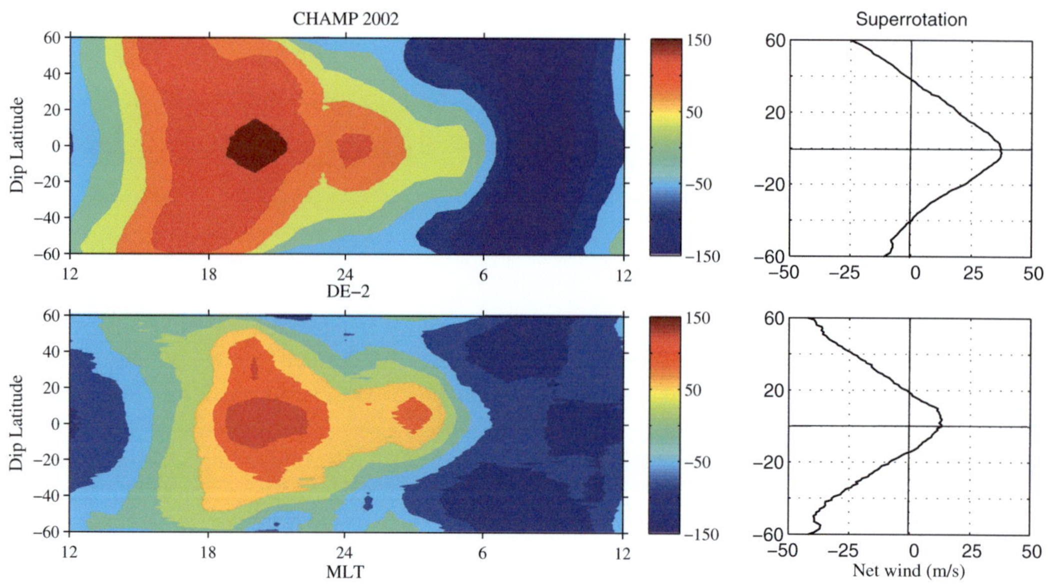

Fig. 22.7 Latitudinal distribution of the thermospheric zonal wind (in m/s) versus local time: (**a**) CHAMP observations, (**b**) Dynamics Explorer 2 observations. The data have been sorted by magnetic latitude (adjusted from Fig. 2 in Liu et al., 2009a)

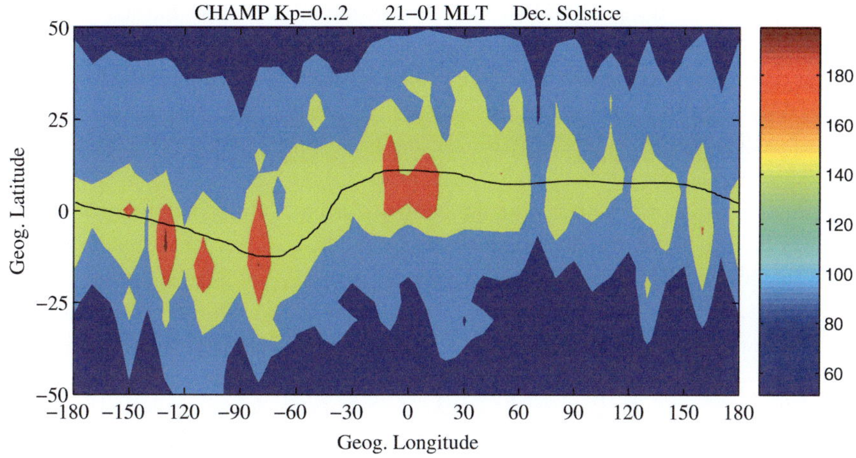

Fig. 22.8 Longitudinal distribution of the zonal wind speed (in m/s) in the 18–24 LT sector. The channel of enhanced velocity closely tracks the latitude variation of the magnetic dip equator (adjusted from Fig. 1 in Liu et al., 2009a)

17–18 LT sector. Only recently, Lühr and Maus (2006) could deduce the average current distribution for all local times from CHAMP data.

The principle concept starts off from zonal winds at F-region heights. The neutral particles try to push the plasma, due to ion drag, in the same direction. The Lorentz force deflects ions and electrons into opposite directions causing at low magnetic latitudes, where the field lines are almost parallel to the Earth's surface, a vertical electric field. This electric field is mapped along the field lines onto the ionospheric E layers in both hemispheres driving Hall and Pedersen currents there. A schematic drawing of the F-region dynamo concept is shown in Fig. 22.9.

The strength of the electric field $\boldsymbol{E}$ is controlled by the ratio between the conductances of the F and E regions. It can be expressed as follows:

$$\boldsymbol{E} = \frac{\Sigma_{\mathrm{P}}^{\mathrm{F}}}{\Sigma_{\mathrm{P}}^{\mathrm{E,N}} + \Sigma_{\mathrm{P}}^{\mathrm{E,S}} + \Sigma_{\mathrm{P}}^{\mathrm{F}}} \boldsymbol{u} \times \boldsymbol{B} \qquad (22.4)$$

where ${\Sigma_{\mathrm{P}}}^{\mathrm{E,N}}$ and ${\Sigma_{\mathrm{P}}}^{\mathrm{E,S}}$ are the E region conductances of the northern and southern hemispheres, ${\Sigma_{\mathrm{P}}}^{\mathrm{F}}$ is the flux tube-integrated conductivity in the F region, $\boldsymbol{u}$ is the zonal wind velocity, and $\boldsymbol{B}$ is the ambient magnetic field. Highest F-region conductances are expected on flux tubes containing the equatorial ionisation anomaly (EIA). A typical location of the EIA at 400 km altitude is $\pm 15°$ MLat. These flux tubes intersect the E layer at about $\pm 20°$ MLat.

For the Pedersen current J_{P} in the E layer, we can write

$$J_{\mathrm{P}} = \Sigma_{\mathrm{P}}^{E} E \qquad (22.5)$$

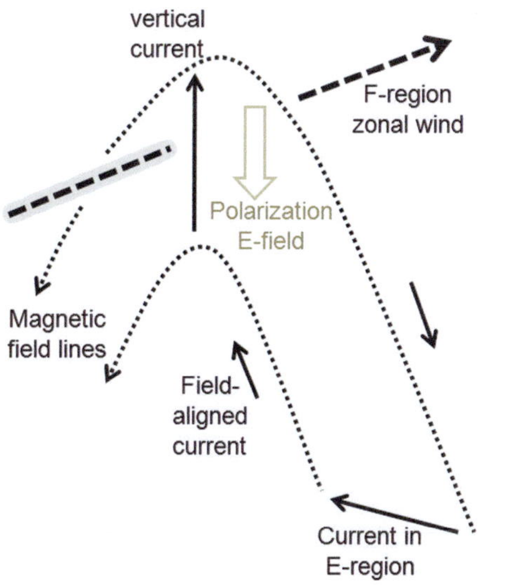

Fig. 22.9 Principle concept of the F-region dynamo. Shown is the configuration for eastward zonal wind (evening, nighttime sector)

where E is the sum of E- and F-region electric fields. When inserting Eq. (22.4) into (22.5) and assuming the same conductances in both E regions and that the sum of these conductances is much larger than the flux tube-integrated conductivity in the F region, which is reasonable during daytime, we obtain

$$J_{\mathrm{P}} = {}^{1}\!/_{2}\, \Sigma_{\mathrm{P}}^{\mathrm{F}} [uB]^{\mathrm{F}} - \Sigma_{\mathrm{P}}^{\mathrm{E}} \left[uB_z \right]^{\mathrm{E}} \qquad (22.6)$$

From Eq. (22.6) we see that the sheet current density in the E region depends on the zonal winds in the E and F regions. Around noon the zonal winds at the two layers have, according to the wind model HWM07, opposite directions, and the product (uB) is about five times stronger in the F region than in the E layer. Thus, J_P is the sum of two almost equally strong contributions from both layers. However, the F-region dynamo current is only marginally influenced due to the small E-region field. Therefore the vertical current J_Z over the equator is primarily proportional to the flux tube-integrated conductivity in the F region.

$$J_Z \approx \Sigma_P^F \left\{ [uB]^F - \left[uB_z\right]^E \right\} \qquad (22.7)$$

Since the E-region contribution is five times smaller, the F-region dynamo current is practically independent of E-layer conditions during day time. The vertical sheet current density J_Z can be deduced from the peak-to-peak variations of the magnetic field east/west component ΔB_y when CHAMP crosses the dip equator:

$$J_Z = \Delta B_y / \mu_0 \qquad (22.8)$$

where μ_0 is the permeability of free space. When inserting Eq. (22.8) into (22.7), all the quantities for determining the F-region conductance are observed by CHAMP. A typical value derived around noon is $\Sigma_P{}^F = 2S$.

The diurnal variation of the vertical current driven by the F-region dynamo has been deduced from typical bipolar magnetic field B_y deflections observed by CHAMP, as presented in Fig. 1 of Lühr and Maus (2006). The data considered are averages over the period August 2000 through October 2004. Only quiet-time measurements ($K_p < 2.3$) have been used. This interval spans the maximum of solar cycle 23 with an average solar flux level of F10.7=153.

Figure 22.10 shows the vertical current density as derived when applying Eq. (22.8) to the CHAMP average magnetic field variations. During the hours around noon, we find downward-directed currents. This is consistent with a westward zonal wind in that time sector. The current density goes through a maximum shortly before 12 LT. This implies that the product, wind speed times F-region conductance, peaks at that time. Conversely, maximum upward currents are deduced after 18 LT coinciding well with peak eastward wind speeds at that time (cf. Fig. 22.7). Towards later local times the ionospheric conductance seems to go down dominating more and more the current density decay. After midnight, no current can be sensed anymore. In particular, the conductance of the E region reduces faster after sunset than that of the F region. Then our assumption of Eq. (22.6) is no longer valid. In that case, Eq. (22.4) approaches the relation $\boldsymbol{E} = \boldsymbol{u} \times \boldsymbol{B}$. This means, the F-region dynamo can be regarded as a voltage generator during night-time.

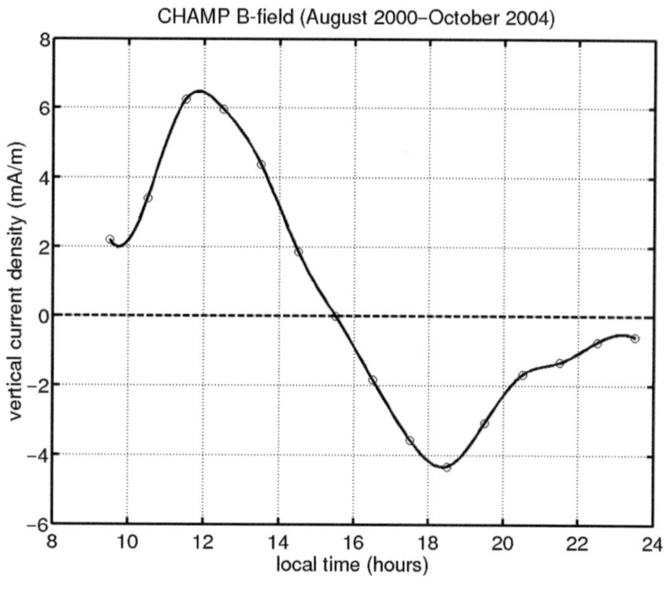

Fig. 22.10 Diurnal variation of the vertical currents driven by the F-region dynamo. Plotted is the sheet current density. Positive values represent downward flowing currents

A particular interesting point is the zero crossing of the vertical current. This can uniquely be related to the switch in zonal wind direction from westwards to eastwards. According to Fig. 22.10, this occurs shortly after 15 LT. For comparison with wind observations, we may have a look at Fig. 22.7. The switch in zonal wind direction is found in the CHAMP data at about 13 and 16 LT in the DE-2 data. This indicates a certain bias of the CHAMP-derived winds towards a too early start of the eastward wind, while DE-2 data seem to favour a later switch. As a consequence of this finding, it is recommended that the CHAMP-derived winds are recalibrated and studies performed with the old data set are revisited. As a benchmark test for the new winds, the F-region dynamo currents can be used.

22.6 Mass Density Depletions Accompanying Plasma Bubbles

Another example of close ionosphere–thermosphere interaction is the synchronous appearance of neutral density depletions (NDDs) with equatorial plasma bubbles (EPBs) and magnetic field strength enhancements. Illés-Almár et al. (1998) reported first about NDDs with scale sizes of 100–200 km. The authors attributed the occurrence of NDD to the plasma bubble generation process. This suggestion was supported by Bencze et al. (2000), who presented three examples of NDDs collocated with EPBs. This phenomenon has been revisited by Park et al. (2010) in a statistical study making use of CHAMP observations over the years 2002–2005. This revealed the characteristic properties of NDDs.

An example of NDD occurrences is shown in Fig. 22.11. The two events displayed are bracketed by dashed lines. In the top panel, displaying the neutral density variations, we see in two cases localized density depletions. Although not prominent, they are significant. These events are accompanied by pronounced plasma density depletions (middle panel) and distinct enhancements of the magnetic field strength (bottom panel). The relation between plasma density and magnetic field variations has earlier been investigated (e.g. Lühr et al., 2003; Stolle et al., 2006). Here we focus on the relation between NDDs and EPBs. Within the 4 years under study, we detected 131 events. The global distribution of the event occurrence rate is shown in Fig. 22.12 separately for the seasons.

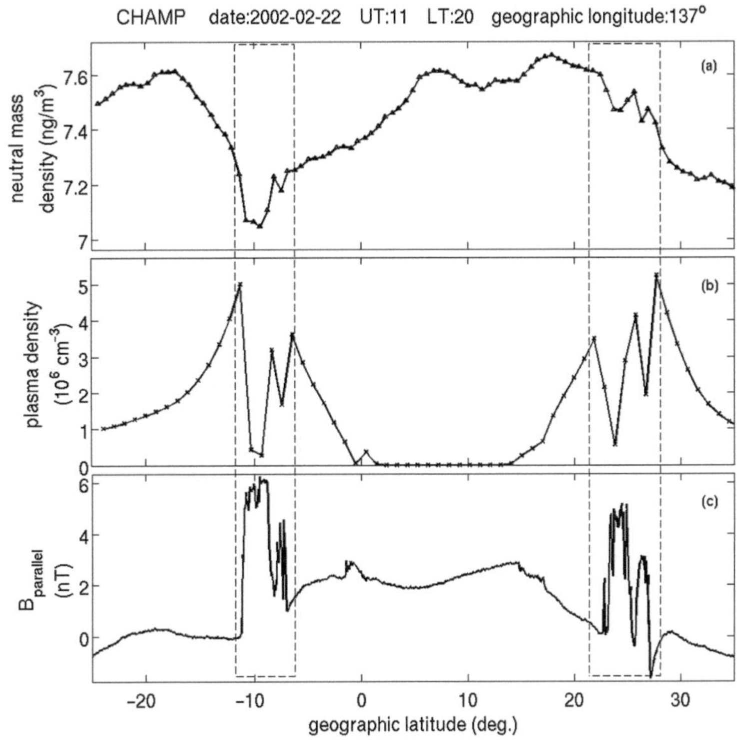

Fig. 22.11 Neutral density depletions accompanying equatorial plasma bubbles: (**a**) neutral mass density variations, (**b**) plasma density measured by planar Langmuir probe, (**c**) residual magnetic field strength variations. The two examples are bracketed by dashed lines (after Park et al., 2010)

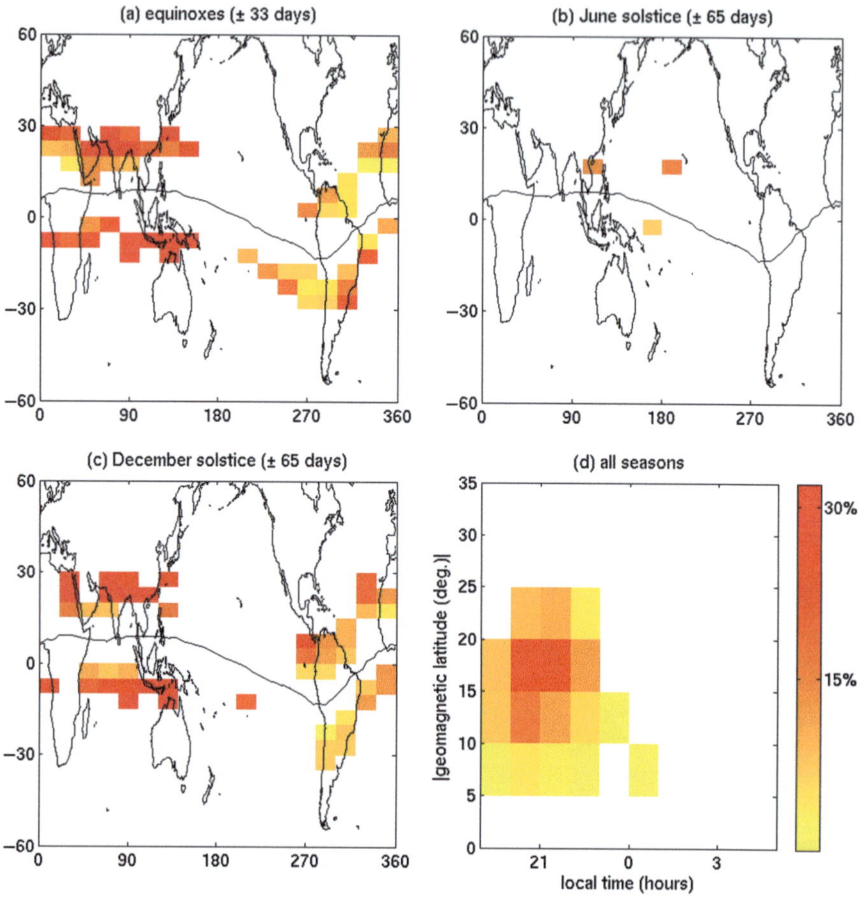

Fig. 22.12 Global distribution of NDD-to-EPB ratio separately for the different seasons (**a, b, c**). Panel (**d**) shows the NDD-to-EPB ratio as a function of geomagnetic latitude and local time (all seasons) (after Park et al., 2010)

Typically, less than 30% of the EPBs are accompanied by NDDs. It is obvious that equinoxes and December solstice are favourite seasons for the formation of NDDs. During the months around June solstice, we detected only three events. Also the longitudinal distribution is not even. During equinox and December solstice months, large parts of the Pacific Ocean are void of events, while around June solstice, NDDs seem to occur preferably there. From the local time versus latitude graph we learn that NDDs have highest occurrence rates around 21 LT. We find in the CHAMP data (~400 km) no secondary occurrence peak past midnight, as reported by Bencze et al. (2000). The highest count rates of NDDs are enCountrytered at magnetic latitudes between 15° and 20°. This is somewhat poleward of the EPB occurrence peak at 9° reported by Stolle et al. (2006) also based on CHAMP data.

For the generation of the neutral density depletions, Park et al. (2010) suggest the ion-neutral friction as the primary coupling process. In the frame of neutral particles, the plasma momentum equation can be described by

$$0 = \mathbf{j} \times \mathbf{B} + \nabla n_{\mathrm{i}}(T_{\mathrm{e}} + T_{\mathrm{i}})k_{\mathrm{B}} - n_{\mathrm{i}} m_{\mathrm{i}} \nu_{\mathrm{in}} \boldsymbol{v}_{\mathrm{i}} \qquad (22.9)$$

where $\boldsymbol{j}$ is the current density, $\boldsymbol{B}$ is the ambient magnetic field, n_{i} is the ion number density, m_{i} is the ion mass, T_{e} and T_{i} are electron and ion temperatures, respectively, k_{B} is the Boltzmann constant, ν_{in} is the ion/neutral collision frequency, and $\boldsymbol{v}_{\mathrm{i}}$ is the ion velocity. The first term on the right side represents the magnetic force term, the second the plasma pressure gradient and the last the ion-neutral friction. Here the effect of the gravitational acceleration is neglected. Similarly, the neutral particle momentum balance is

$$0 = \nabla n_{\mathrm{n}} T_{\mathrm{n}} k_{\mathrm{B}} + n_{\mathrm{i}} m_{\mathrm{i}} \nu_{\mathrm{in}} \boldsymbol{v}_{\mathrm{i}} \qquad (22.10)$$

Here the first term on the right side is the neutral pressure gradient and the second term is the ion-neutral friction as above. Merging Eqs. (22.9) and (22.10) implies that the sum of the three different types of pressure variations (magnetic, plasma, and neutral particle pressure) vanishes:

$$\Delta P_m + \Delta P_p + \Delta P_n = 0 \quad (22.11)$$

Park et al. (2010) have shown that the undulations of the neutral density are caused by the plasma drift, and the coupling efficiency is determined by the ion-neutral friction, $n_i m_i \nu_{in}$. The ion-neutral collision frequency can be estimated according to Schunk and Nagy (2000), $\nu_{in} = 4.45 \times 10^{-17} n_n \times T_i$ (s^{-1}), when assuming the dominant interaction between oxygen ions and atoms, which is reasonable at CHAMP altitude. The detection threshold for NDDs in CHAMP data is a mass density of 2.5×10^{-14} kg/m^3. Based on that value Park et al. (2010) reported that detectable NDDs can only form when the product of electron density and neutral mass density, $n_i \rho > 15$ kg/m^6, that detectable NDDs accompanying EPBs can form. In Section 22.3, we have shown that the mass density attains its minimum close to June solstice, and also the F-region electron density reaches its annual minimum in the middle of the year (e.g. Rishbeth and Müller-Wondrag, 2006). For these reasons it is no surprise that we hardly detected any NDD around June solstice.

Another interesting finding is that NDDs contribute a significant part to the pressure balance. Park et al. (2010) showed for the two events presented in Fig. 22.11 that the tiny mass density reduction contributes almost the same amount to the total particle pressure decrease as the substantial electron density depletion. This means that we can interpret the magnetic field increase within plasma bubbles correctly only when both the plasma and neutral depletions are considered. This is of particular interest for modellers of the core and crustal magnetic field who have to remove the ionospheric influences from satellite magnetic field data.

22.7 Conclusions

In this chapter, we have given some insight into the contributions the CHAMP satellite mission has made for improving our understanding of the ionosphere–thermosphere coupling. All the topics addressed show that the processes in the high atmosphere can be understood properly only when we consider the plasma and neutral particle dynamics simultaneously. This statement is not new, but a proper consideration was mostly hampered by the lack of a complete set of plasma and neutral quantities observation. The common approach for mitigating this problem was to make assumptions about the quantities not observed or to use models for their determination. A good step towards a good and simultaneous observational coverage of various ionosphere–thermosphere parameters was achieved with the CHAMP mission. As expected, a number of surprises emerged. Among them are the thermospheric density anomalies which are controlled by the geometry of the geomagnetic field, e.g., the cusp-related density anomaly and the equatorial mass density anomaly. Furthermore, long predicted concepts, such as the F-region dynamo, have now experimentally been confirmed. The data set is available for a detailed investigation of the F-region dynamo climatology. This again may yield unexpected results.

For future research, ESA's Swarm mission is now in preparation. With the scheduled launch in 2012, this multi-satellite mission will provide new and improved observational opportunities. With its suite of complementary particle and field instruments and with the specific constellation of three-satellite orbits so far, untouched ionosphere–thermosphere phenomena can be investigated. Further details about the Swarm mission can be found at the ESA web site (http://www.esa.int/esaLP/LPswarm.html).

Acknowledgements We thank Claudia Stolle and Stefan Maus for fruitful discussions about the ionosphere–thermosphere results presented here. We also thank W. Köhler for processing the CHAMP accelerometer data. The CHAMP mission is sponsored by the Space Agency of the German Aerospace Center (DLR) through funds of the Federal Ministry of Economics and Technology, following a decision of the German Federal Parliament (grant code 50EE0944). The data retrieval and operation of the CHAMP satellite by the German Space Operations Center (GSOC) is acknowledged.

References

Bencze P, Almár I, Illés-Almár E (2000) Further results referring to the neutral density depletions attributed to plasma bubbles. J Atmos Solar-Terr Phys 62:1339

Fuller-Rowell, TJ, Codrescu MV, Fejer BG, Borer W, Marcos F, Anderson DN (1997) Dynamics of the low-latitude thermosphere: quiet and disturbed conditions. J Atmos Solar-Terr Phys 61:1533–1540

Guo J, Wan W, Forbes JM, Sutton E, Nerem RS, Woods TN, Bruinsma S, Liu L (2007) Effects of solar variability on thermosphere density from CHAMP accelerometer data. J Geophys Res 112:A10308. doi:10.1029/2007JA012409

Illés-Almár E, Almár I, Bencze P (1998) Neutral density depletions attributed to plasma bubbles. J Geophys Res 103(A3):4115

Liu H, Lühr H, Henize V, Köhler W (2005) Global distribution of the thermospheric total mass density derived from CHAMP. J Geophys Res 110:A04301. doi:10.1029/2004JA010741

Liu H, Lühr H, Watanabe S (2007) Climatology of the equatorial thermospheric mass density anomaly. J Geophys Res 112:A05305. doi:10.1029/2006JA012199

Liu H, Lühr H, Watanabe S, Köhler W, Henize V, Visser P (2006) Zonal winds in the equatorial upper thermosphere: decomposing the solar flux, geomagnetic activity, and seasonal dependencies. J Geophys Res 111:A07307. doi:10.1029/2005JA011415

Liu H, Watanabe S, Kondo T (2009a) Fast thermospheric wind jet at the Earth's dip equator. Geophys Res Lett 36:L08103. doi:10.1029/2009037377

Liu H, Yamamoto M, Lühr H (2009b) Wave-4 pattern of the equatorial mass density anomaly- A thermospheric signature of tropical deep convection. Geophys Res Lett 36:L18104. doi:10.1029/2009GL039865

Lühr H, Maus S (2006) Direct observation of the F region dynamo currents and the spatial structure of the EEJ by CHAMP. Geophys Res Lett 33:L24102. doi:10.1029/2006GL028374

Lühr H, Rother M, Maus S, Mai W, Cooke D (2003) The diamagnetic effect of the equatorial Appleton anomaly: its characteristics and impact on geomagnetic field modelling. Geophys Res Lett 30(17):1906. doi:10.1029/2003GL017407

Lühr H, Rother M, Köhler W, Ritter P, Grunwaldt L (2004) Thermospheric up-welling in the cusp region, evidence from CHAMP observations. Geophys Res Lett 31:L06805. doi:10.1029/2003GL019314

Maeda H, Iyemori T, Araki T, Kamei T (1982) New evidence of a meridional current system in the equatorial ionosphere. Geophys Res Lett 9:337–340

Menvielle M, Marchaudon A (2007) Geomagnetic indices in solar-terrestrial physics and space weather, Space weather, research toward applications in Europe. In: Lilensten J (ed) Astrophysics and space science library. Springer, Berlin, pp 277–288

Müller S, Lühr H, Rentz S (2009) Solar and magnetospheric forcing of the low latitude thermospheric mass density as observed by CHAMP. Ann Geophys 27:2087–2099

Park J, Lühr H, Min KW (2010) Neutral density depletions associated with equatorial plasma bubbles as observed by the CHAMP satellite. J Atmos Solar-Terr Phys 72:157–163. doi:10.1016/j.jastp.2009.11.003

Prölss GW (1997) Magnetic storm associated perturbations of the upper atmosphere. In: Tsurutani BT, Gonzale WD, Kamide Y, Arballo JK (eds) Magnetic storms. Geophysical monograph series, vol 98. AGU, Washington, DC, p 227

Reigber C, Lühr H, Schwintzer P (2002) CHAMP mission status. Adv Space Res 30(2):129–134

Rentz S, Lühr H (2008) Climatology of the cusp-related thermospheric mass density anomaly, as derived from CHAMP observations. Ann Geophys 26:2807–2823

Rishbeth H (1971) The F-layer dynamo. Planet Space Sci 19:263–267

Rishbeth H, Müller-Wodarg ICF (2006) Why is there more ionosphere in January than in July? The annual asymmetry of the F-layer. Ann Geophys 24:3293–3311

Stolle C, Lühr H, Rother M, Balasis G (2006) Magnetic signatures of equatorial spread-F, as observed by the CHAMP satellite. J Geophys Res 111:A02304. doi:10.1029/2005JA011184

Trinks H, von Zahn U (1975) The ESRO4 gas analyzer. Rev Sci Instrum 46:213–217

Chapter 23

Equatorial Ionization Anomaly: The Role of Thermospheric Winds and the Effects of the Geomagnetic Field Secular Variation

Inez S. Batista, Erica M. Diogo, Jonas R. Souza, Mangalathayil Ali Abdu, and Graham J. Bailey

Abstract The vertical plasma drift is the well known driver of the equatorial ionization anomaly (EIA). The latitudinal distribution of ionization in the EIA is determined also by thermospheric meridional wind whose precise role can only be evaluated through the use of theoretical models because it depends not only upon the local configuration of the wind, but is a complex function of its distribution along the entire magnetic field line. Besides, in the Brazilian region, the magnetic field secular variations are fast enough for their effects on the ionosphere to be observed in the time span of the order of a solar cycle. In this work we use the Sheffield University Plasmasphere-Ionosphere Model (SUPIM) to investigate the role of the vertical plasma drift, thermospheric meridional wind and of the magnetic field secular variations in the changing trend of the EIA over the Brazilian region.

23.1 Introduction

The equatorial ionization anomaly (EIA) is one of the important features of the equatorial and low-latitude ionosphere. It is generated by the well known fountain effect, produced by the upward vertical ExB/B^2 plasma drift that elevates the F-region ionosphere plasma to higher altitudes over the magnetic equator, followed by diffusion along the geomagnetic field lines, that moves the plasma down and away from the equator, forming ionization peaks/crests in the subtropics on both sides of the magnetic equator and an ionization trough over the dip equator. This configuration can be significantly modified by the thermospheric meridional wind. The latitudinal plasma distribution that characterizes the EIA (a through at the magnetic equator and two crests at approximately $\pm 15°$ latitude) is well reproduced by many theoretical models (Hanson and Moffett, 1966; Anderson, 1973a, b; Balan and Bailey, 1996; Bailey and Balan, 1996; Bittencourt et al., 2007).

In this work we use the Sheffield University Plasmasphere-Ionosphere Model (SUPIM) (Bailey and Selek, 1990; Bailey et al., 1997) to investigate the EIA response to changes in zonal electric fields and thermospheric winds, and also to investigate the long term trends in the EIA over the Brazilian region in terms of how it is connected to the rather rapid secular variation of the geomagnetic field in the Brazilian sector as manifested in the peculiar northwestward movement of the geomagnetic equator over northeast Brazil (Rangarajan and Muniz Barreto, 2000).

The Sheffield University Plasmasphere Ionosphere Model (Bailey et al., 1993; Bailey and Balan, 1996) solves the coupled time-dependent equations of the continuity, momentum, and energy balance along the geomagnetic field lines and calculates the density, momentum flux and temperature for electrons and for the ions O^+, He^+, N_2^+, O_2^+, NO^+ and N^+ (see also Souza et al., 2000a, b) for applications of the model to low latitudes over Brazil). In the present version of the SUPIM, the geomagnetic field is represented by a tilted centered dipole with the angle of tilt and the magnetic declination angle given by the International Geomagnetic Reference Field (IGRF) model (Maus et al., 2005). The solar flux values used as input to the SUPIM are derived from the Extreme

I.S. Batista (✉)
National Institute for Space Research, São José dos Campos, SP, Brazil
e-mail: inez@dae.inpe.br

M.A. Abdu, D. Pancheva (eds.), A. Bhattacharyya (Coed.), *Aeronomy of the Earth's Atmosphere and Ionosphere*,
IAGA Special Sopron Book Series 2, DOI 10.1007/978-94-007-0326-1_23,

Ultra Violet flux model for Aeronomic Calculations (EUVAC) (Richards et al., 1994) that gives the EUV flux at 37 wavelength groups between 50 and 1050 Å. The flux for the X-rays and Lyman-α (1216 Å) were taken from the model SOLAR2000 (Tobiska et al., 2000). The neutral atmosphere parameters were taken from the model NRLMSISE-00 (Picone et al., 2002) that gives density and temperature for the neutral constituents He, O, N_2, O_2, Ar, H, and N. The number density for the neutral constituent NO was calculated using the empirical expression given by Mitra (1968)

$$n(\mathrm{NO}) = 4 \times 10^{-1} \exp\left(\frac{-3700}{T_n}\right) n(\mathrm{O_2}) + 5 \times 10^{-7} n(\mathrm{O}).$$

The neutral wind used as input in the SUPIM model is given by the empirical model HWM93 (Horizontal Wind Model) (Hedin et al., 1995) and the vertical drift at the equatorial F-region is given by the empirical model from Scherliess and Fejer (1999).

23.2 Effect of the Vertical Drift on the EIA Development

The vertical ExB plasma drift over the geomagnetic equator is the main driver of the EIA. The drift is upward during the day and downward during the night. Before its evening reversal to downward the drift undergoes an enhancement, the pre-reversal enhancement (PRE) (see for example Fejer et al., 1989, 1991) that is responsible for the intensification of the EIA after sunset and during pre-midnight hours. The degree of development of the post sunset EIA is directly related to the amplitude of the PRE. Depending upon the amplitude of the PRE the effect on the EIA may last for extended durations, however. In order to quantify the effect of the PRE amplitude in the EIA development we have used two drift models, shown in Fig. 23.1, as inputs to the SUPIM model. The main difference between the two drift models is observed around sunset and at nighttime. The pre-reversal enhancement amplitude in drift model D1 is much higher than in model D2. Also the enhancement starts much earlier and the reversal occurs later in model D1 than in model D2. Those differences in the PRE behavior are reflected in the EIA development as shown in Fig. 23.2.

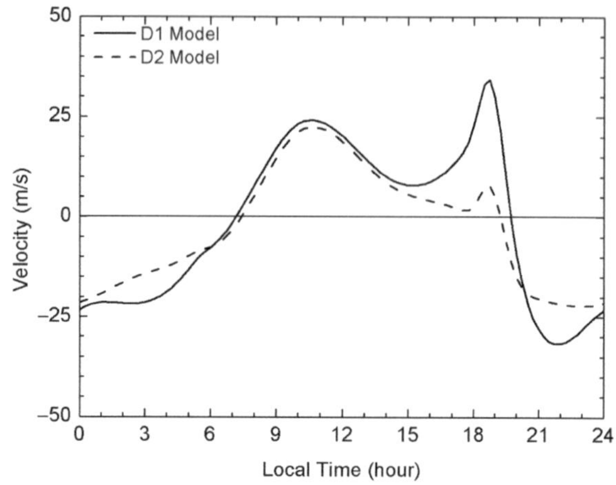

Fig. 23.1 Vertical drift models used as input in the simulations. Model D1 is representative of high solar activity and Model D2 is representative of low solar activity

Figure 23.2 shows the latitude versus height distribution of the F layer plasma frequencies at 2000 LT (local time) for the September equinox of high solar activity (left panels) and low solar activity (right panels). All the results were obtained using the SUPIM model run for the Brazilian region (45°W). Figure 23.2a, c were obtained using the drift model D1 as input to the SUPIM and Fig. 23.2b, d were obtained using drift model D2 (note that the plots are in geographic latitude). At the 45°W longitude the magnetic equator is now located at ~3°S. The results for both high and low solar activity (HSA and LSA) periods using the drift model D1 (Fig. 23.2a, c) show a well developed EIA with clear and pronounced density peaks to the north and south of the equator and the trough close to the magnetic equator. The peaks are formed at similar latitudes during HSA and LSA. The lower intensity of the EIA crest/peak densities during the LSA period is due to the low background ionization density that results from lower solar radiation intensity. Comparing the latitudinal distribution of the F-layer plasma frequencies obtained from the SUPIM using drift model D1 and D2 it can be noticed that the reduction of the pre-reversal enhancement in model D2 by a factor of approximately 4 times (in relation to the PRE amplitude of model D1) resulted in a reduction by ~25 and ~35% in the average (of the

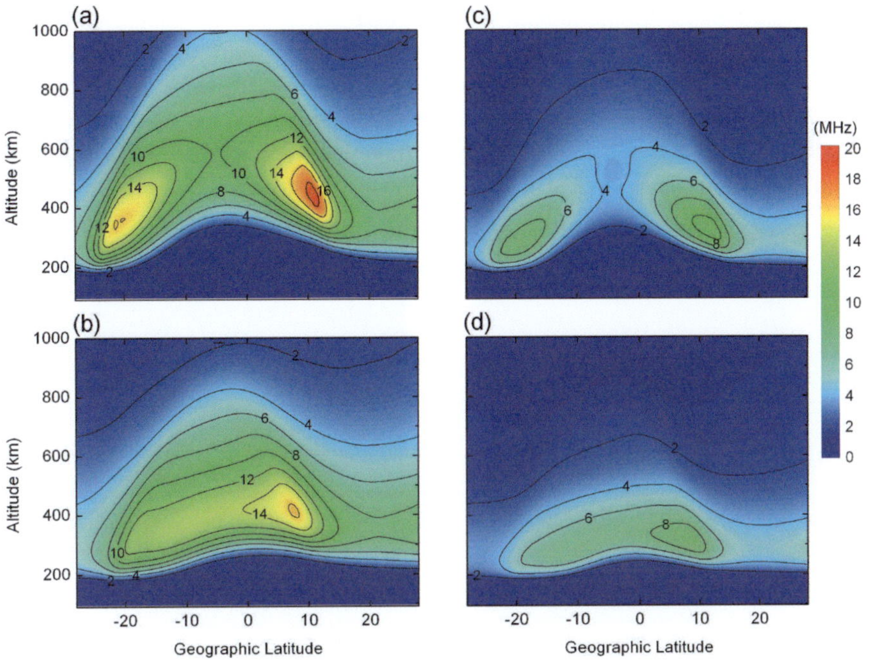

Fig. 23.2 Contour maps of the F layer plasma frequency as a function of geographic latitude and altitude. The results are for the 45°W meridian, September equinox, at 2000 LT. Panels (**a**) and (**b**) are the results for high solar activity and panels (**c**) and (**d**) are for low solar activity. Note that the drift model D1 was used as input for results in panels (**a**) and (**c**) and that the drift model D2 was used for results in panels (**b**) and (**d**)

northern and southern crests) peak electron density[1] for HSA and LSA, respectively. Correspondingly the position of the anomaly crest is shifted equatorwards by around 6°, the crest-to-trough ratio is significantly reduced and the vertical extension of the equatorial anomaly is reduced by approximately 200 km.

23.3 Effect of the Meridional Wind in the EIA Development

The thermospheric neutral wind has a very important effect in the final configuration of the EIA. The initial latitudinal distribution is modified by the neutral-ion collisional drag that transports the ionization along the magnetic field lines at the same velocity as that of the wind component along it. This plasma movement will result in inter-hemispheric transport of ionization. Also the ionization moves upwards in the upwind hemisphere and downwards in the downwind hemisphere of the magnetic field line. This upward (downward) movement will bring ionization to regions of lower (higher) atmospheric density and recombination rates. As a result the EIA will present asymmetric crests with respect to the magnetic dip equator with unequal values of electron densities and heights for the F-region peak. In order to evaluate the effect of the meridional wind in the equatorial anomaly development we have run the SUPIM model for September equinox conditions of HSA and LSA. The vertical drift models D1 and D2 (Fig. 23.1) were used as input for the SUPIM runs for HSA and LSA respectively. The results of the plasma frequencies outputs from the SUPIM runs at 45°W longitude for the input winds from HWM are shown for different local times in Figs. 23.4 and 23.6. It is important to stress that, in a region with high magnetic declination as in the present case (magnetic declination ~20°W), the effective meridional wind,

[1] The relation between peak electron density and frequency in given by

$$N(\text{cm}^{-3}) = 1.24 \times 10^4 (f(\text{MHz}))^2$$

which is the same as the magnetic meridional wind along the magnetic field line, has contributions of both the meridional and the zonal components of the wind. The effective winds, presented in Figs. 23.3 and 23.5 were calculated using the expression

$$U_{eff} = (U_\phi \cos D + U_\theta \sin D) \cos I$$

where U_ϕ is the meridional component of the wind (positive northward), U_θ is the zonal component of the wind (positive eastward), D is the magnetic declination angle and I is the dip angle. According to the above expression, it can be seen that U_{eff} is the projection of the magnetic meridional wind along the magnetic field line.

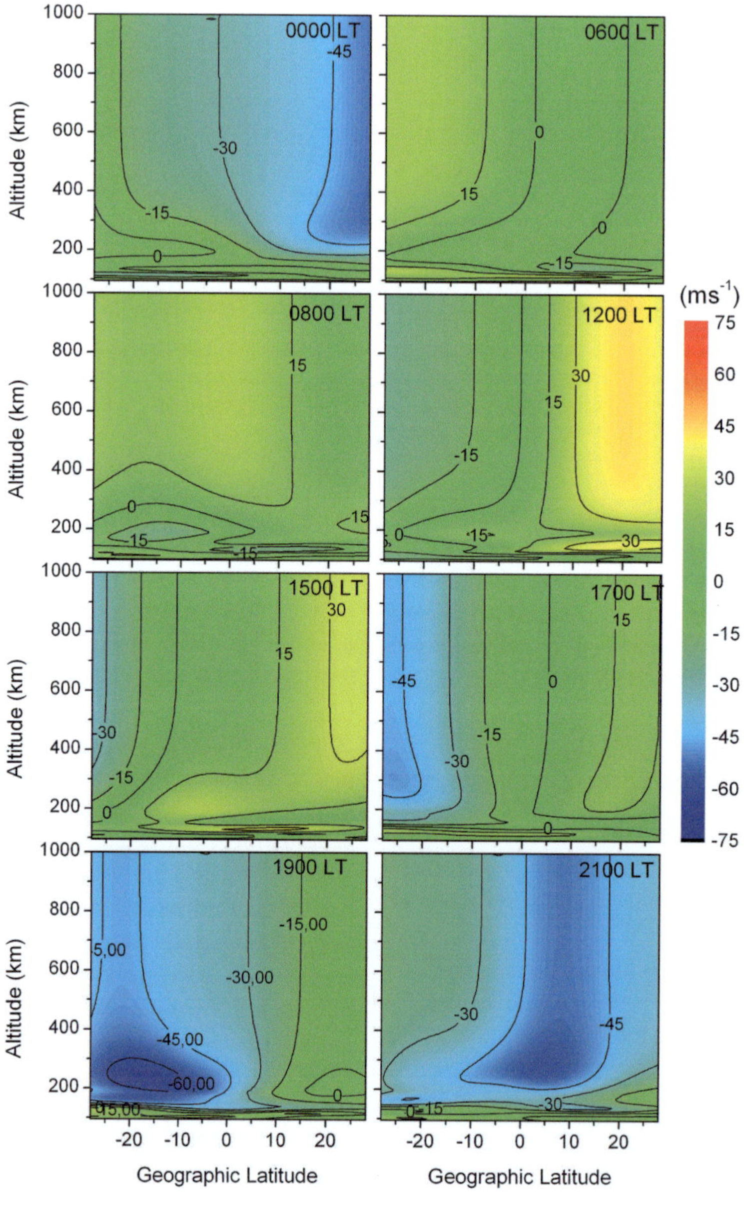

Fig. 23.3 Contour maps of effective meridional wind as a function of geographic latitude and altitude at different local times. The results are for the 45°W meridian, September equinox, and high solar activity. Positive winds are northward

For the HSA period (Fig. 23.4) we can observe that the EIA is well developed during all day hours and that it is intensified after sunset due to the vertical drift intensification (pre-reversal enhancement). For the LSA period (Fig. 23.6) no post-sunset enhancement in EIA is observed due to the much less significant prereversal enhancement in the vertical drift. In fact, during the LSA the anomaly intensity steadily decreases after 1500 LT.

The 1200 LT panel on Fig. 23.4 shows an almost symmetrical development of the EIA. At the same local time the effective meridional wind from Fig. 23.3 is northward in the northern hemisphere and southward (but weak) in the southern hemisphere (diverging wind at the equator) which results in a downward movement of the ionization (along the magnetic field line) in both hemispheres. The much higher wind intensity at the northern hemisphere will imply in a more

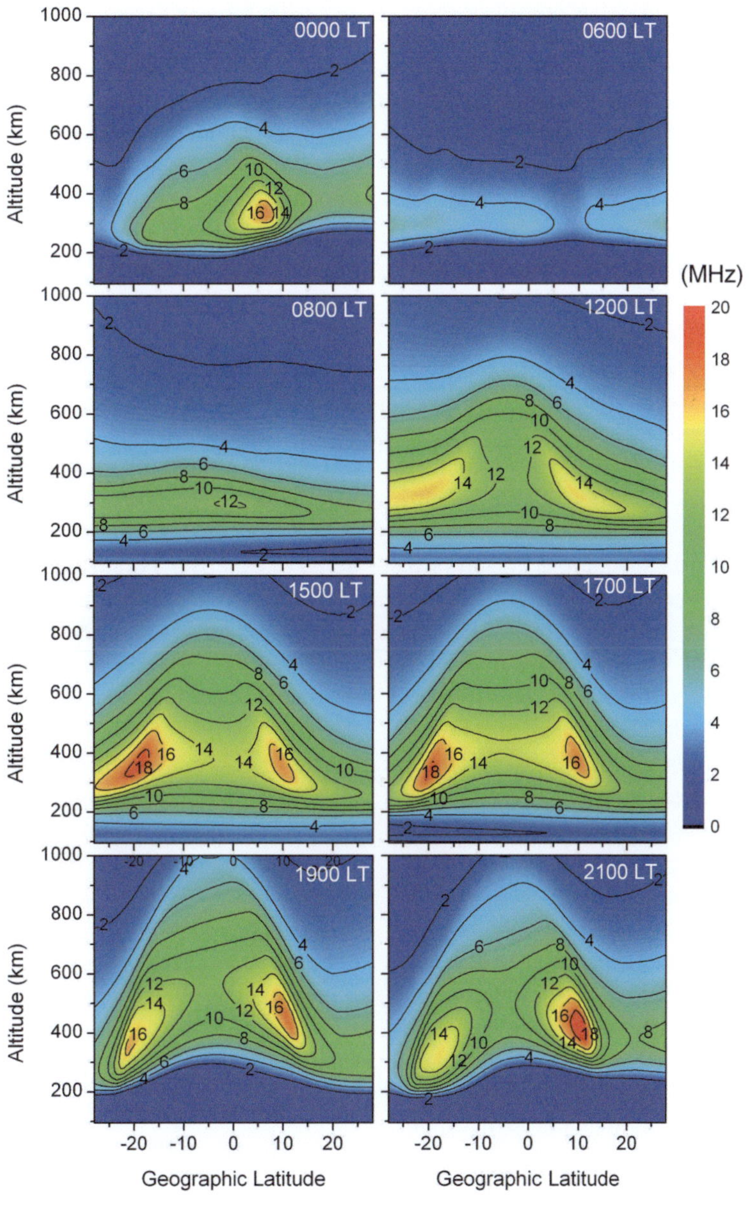

Fig. 23.4 Contour maps of F layer plasma frequency as a function of geographic latitude and altitude at different local times. The results are for the 45°W meridian, September equinox, and high solar activity. Drift model D1 was used as input

effective downward movement of ionization in that hemisphere, and a consequent asymmetry in the two crests of the EIA. This effect will only be seen approximately 3 h later (as seen in the 1500 LT plot) when the southern crest is more intense (higher density) and also at slightly higher altitude than the northern crest. At 1700 LT the wind pattern is again divergent at the equator, but now the southern hemisphere component is a bit more intense than the component at the northern hemisphere. Approximately 2 h later (as seen in the 1900 LT plot) the relative behavior observed before for the crests will reverse, with the northern crest more intense and at higher altitudes than the southern crest. The persistent southward wind that is observed later is the responsible for the strengthening of the asymmetry observed at later times (probably due to the recombination effect under the relatively more persistent southward wind, and hence the downward plasma transport, in the southern hemisphere). The EIA response to the wind during LSA (Figs. 23.5 and 23.6) is quite similar to that during the HSA. The asymmetrically divergent wind (stronger at one hemisphere) will affect the latitudinal distribution of ionization approximately 2–3 h later.

In order to evaluate the effect of the wind on the ionization distribution and in producing asymmetric EIA crests it is necessary to investigate its behavior along the entire field line. Figure 23.7 shows the net wind integrated between 33°N and 33°S latitude (in arbitrary units) as a function of local time. In the same figure we show the asymmetry of the EIA measured as the relative difference in peak electron density and the layer peak height at the anomaly crests. The *density asymmetry* is calculated as the difference between the peak electron density in the northern hemisphere crest ($NmF2_{NH}$) and the peak electron density in the southern hemisphere crest ($NmF2_{SH}$), divided by ($NmF2_{NH}$) that is, *Density Asymmetry* = ($NmF2_{NH}$ - $NmF2_{SH}$)/$NmF2_{NH}$. The *height asymmetry* is the difference between the peak electron density height in the northern hemisphere crest ($hmF2_{NH}$) and that at the southern hemisphere crest ($hmF2_{SH}$), that is, *Height Asymmetry* = ($hmF2_{NH}$ - $hmF2_{SH}$). As discussed before, a northward wind will push the F layer down in the northern hemisphere and up in the southern hemisphere causing negative *height asymmetry*. This is clearly seen in Fig. 23.7 which shows that the *height asymmetry* parameter has opposite phase to the net integrated wind. The response of the F layer height to the meridional wind is very prompt as can be seen from Fig. 23.7, in which the zeros at the two parameters are very close in time. On the other hand, the density changes in response to the meridional wind are seen after 2–3 h, as already pointed out in the previous discussion. From Fig. 23.7 we can see that the afternoon reversal of the net integrated wind from positive (northward) to negative (southward) occurs at 1600 LT, but reversal on the peak *density asymmetry* is observed only 2 h later (1800 LT). The maximum negative (southward) integrated wind occurs at 2100 LT while the maximum *density asymmetry* occurs around midnight. Although most of the time a net transequatorial wind (net positive or negative integrated wind) is present, its effect on the electron density latitudinal distribution is not straightforward as the height response. The *density asymmetry* response presents a time delay which could be caused by the plasma diffusion along the field line that plays an important role in establishing a new peak density (foF2) in response to a change in the wind. On the other hand the plasma transport by meridional/trans-equatorial wind occurs through (height dependent) collisional effect and therefore the F layer height responds promptly to a change in the wind intensity. Thus the asymmetry in foF2 responds to a change in the meridional/trans-equatorial wind with a time delay of the order of 2–3 h as seen in Fig. 23.7 which is comparable with the observational results over Fortaleza for F layer peak density response to changes in zonal electric field as reported earlier by Abdu et al. (1990). On the other hand the hmF2 presents a prompt/fast response to a change in the meridional/trans-equatorial wind clearly seen in the results of Fig. 23.7. Despite of all the aspects discussed above, the *height asymmetry* appears to have a significant effect in the development of the *density asymmetry* (with some time delay that depends on the recombination time constant), at least under the equinox wind configuration used in the present simulation, in which a net positive (negative) integrated wind does not necessarily means that a transequatorial wind is present at the magnetic equator.

23.4 Secular Variation of the Geomagnetic Field and Its Effect on the EIA

Over the Brazilian region magnetic field secular variations are fast enough so that their ionospheric signatures can be observed in the time interval of one

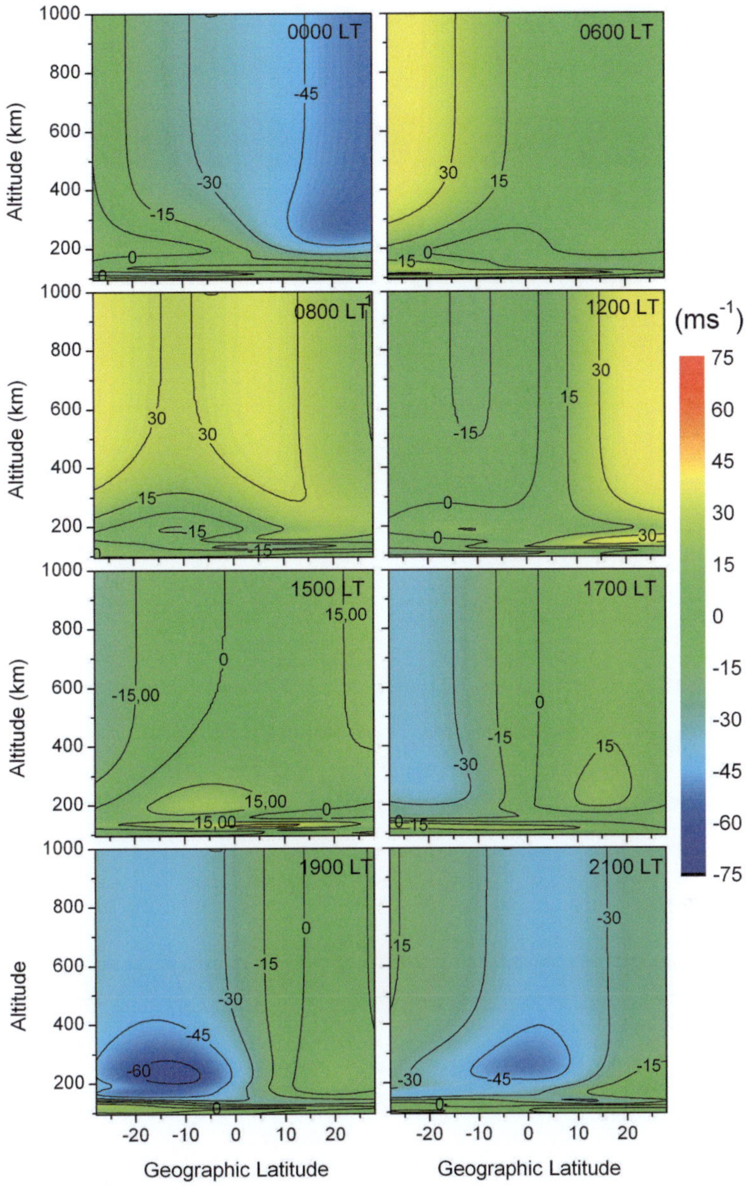

Fig. 23.5 Contour maps of effective meridional wind as a function of geographic latitude and altitude. The results are for the 45°W meridian, September equinox, and low solar activity. Positive winds are northward

or more solar cycles (Abdu et al., 1996). According to the International Geomagnetic Reference Field (IGRF) Model (Maus et al., 2005) (the online version can be found at http://ngdc.noaa.gov/geomagmodels/struts/calcGridIGRF) the magnetic inclination over northeast Brazil varies at a rate of 20' per year, corresponding to an apparent northwestward movement of the magnetic equator. This peculiarity of the magnetic equator over Brazil allowed studying long term trends of the sporadic E (Es) layer behavior (Abdu et al., 1996) and of the F3 layer occurrence (Batista et al., 2002) over the Brazilian station Fortaleza (2.8 to 38). In this work we have used the SUPIM model in order to investigate to what extent this movement of the geomagnetic equator affects the EIA position over the Brazilian region.

Figures. 23.8 and 23.9 show the distribution of F region plasma frequency as a function of geographic latitude and altitude at 1800 LT for the September equinox of high and low solar activity, respectively,

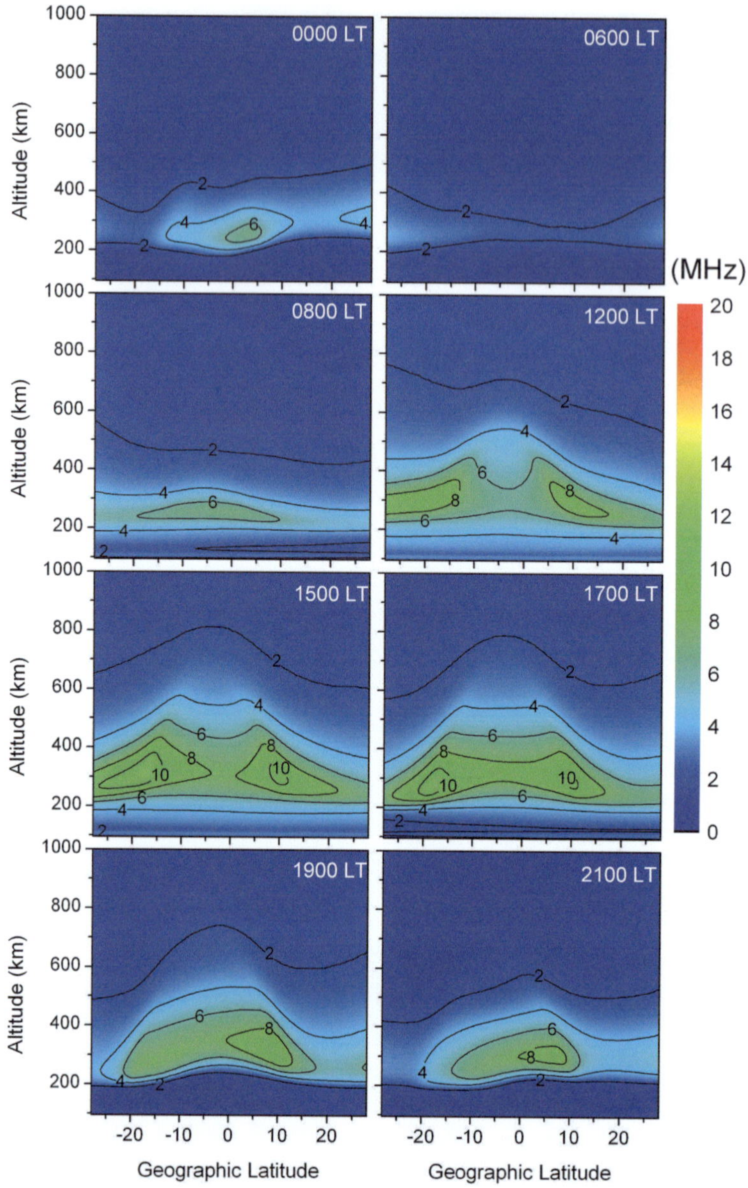

Fig. 23.6 Contour maps of F layer plasma frequency as a function of geographic latitude and altitude. The results are for the 45°W meridian, September equinox, and low solar activity. Drift model D2 was used as input

at 45°W longitude. Results in Fig. 23.8 (HSA) were obtained using the drift model D1 as input for the SUPIM and those in Fig. 23.9 (LSA) were obtained using drift model D2.

Results in Fig. 23.8a–c were obtained running the SUPIM with IGRF inputs for the high solar activity years 1978, 1989 and 2002, respectively. In order to ensure that the changes observed from 1 year to the other were due to the secular variation in the geomagnetic field alone, all the other input parameters (such as vertical drift, solar flux, neutral wind, magnetic activity, etc.) were kept unchanged from one run to the other. Figure 23.9a–c are similar to Fig. 23.8a–c but for the low solar activity years of 1986, 1996 and 2006, respectively.

As we can see from Fig. 23.8, at the 45°W longitude the crests of the EIA are moving northward (for reference vertical lines are drawn at the latitudes

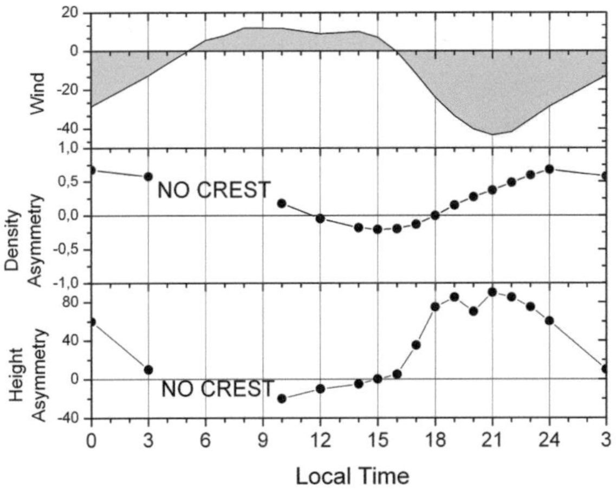

Fig. 23.7 Net wind integrated along the magnetic meridian (*upper panel*). EIA crests density asymmetry calculated as ($NmF2_{NH} - NmF2_{SH}$)/$NmF2_{NH}$ (*middle*) and height asymmetry calculated as ($hmF2_{NH} - hmF2_{SH}$) (*lower panel*)

22.5°S and 8°N). In the southern hemisphere the position of the EIA crest has moved from approximately 25°S to ~21°S between 1978 and 2002 (Fig. 23.8a, c). Similar variation is also observed in the position of the northern hemisphere crest, which moved from ~5.5°N to ~10°N. Figure 23.9 shows the same steady movement of the EIA crest at both hemispheres as observed in Fig. 23.8, but for low solar activity. From the results presented in Figs. 23.8 and 23.9 it was possible to calculate the rate of change of the geographic latitude of the EIA crest at the 45°W longitude meridian. These results are shown in Fig. 23.10 as a linear fit to the data points corresponding to the EIA crest position at the southern hemisphere, for high and low solar activity. The geographic latitude of the EIA crest varies at a rate of 10'/year and 9.5'/year for the high and low solar activity periods, respectively. These rates are very close to the rate of change of the dip equator at the same meridian that is equal to 11.6'/year, according to IGRF results. The linear fit to the data points corresponding to the dip equator position (geographic latitude) at the 45°W longitude meridian, for years 1978, 1984, 1990, 1996, 2002 and 2006, is also plotted in Fig. 23.10.

23.5 Summary

The effect of the vertical plasma drift and thermospheric wind over the equatorial ionization anomaly was investigated using the Sheffield University Plasmasphere-Ionosphere Model. Furthermore, the effect of the magnetic field secular variations on the EIA location over the Brazilian region was also investigated using the SUPIM.

The simulation results show that reducing the PRE amplitude by a factor of 4 implies in a decrease by ~25 and ~35% in the F layer peak density at the EIA crest during high and low solar activity, respectively. The position of the anomaly crest is shifted equatorwards by around 6°, the crest-to-trough ratio is significantly reduced and the vertical extension of the equatorial anomaly is reduced by approximately 200 km.

Thermospheric meridional wind can introduce asymmetries in height and density of the EIA crests. The response of the F layer height to the meridional wind is very fast. A net wind along the magnetic field line produces an asymmetry in the EIA crest height. If the net wind is northward (southward), the southern hemisphere crest will be observed at higher (lower) altitude than the crest at the opposite hemisphere. On the other hand, the density changes in response to the meridional wind are not straightforward. They result of a more complicated process that involves diffusion and recombination. They seem to occur with a time delay of 2–3 h (which is similar to the response delay to electric field changes shown from ionosonde measurements by Abdu et al., 1990). When the F layer peak heights at EIA crests are at different heights in the opposite hemispheres, the height dependent collision factors (together with the diffusion and recombination processes) could cause density asymmetries at the crests. This could be the main cause of the delayed response of the density asymmetry at the EIA crests. At least under the equinox wind configuration used in the present simulation, in which a net positive (negative) integrated wind does not necessarily means that a transequatorial wind is present at the magnetic equator, the *height asymmetry* appears to have a significant effect in the development of the *density asymmetry*.

There are some regions of the globe where geomagnetic field secular variations are fast enough so that their ionospheric signatures can be observed in the time interval of a solar cycle. This is the case with the northeast region over Brazil where the rate of secular variation is the largest over the low latitude regions of the globe. Over the 45°W meridian (Brazilian region) the EIA crest position undergoes a steady northward excursion, closely following the magnetic equator

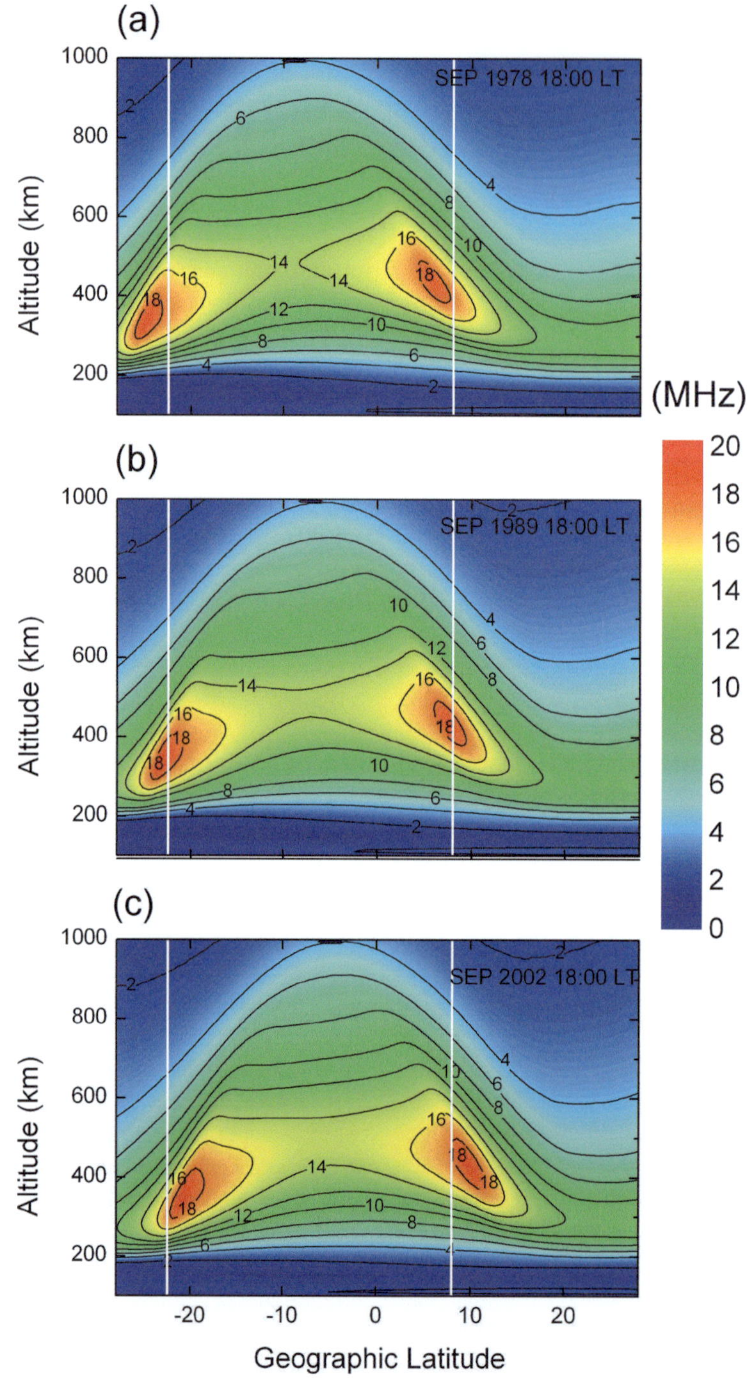

Fig. 23.8 Contour maps of the F layer plasma frequency as a function of geographic latitude and altitude. The results are for the 45°W meridian, September equinox, high solar activity, at 1800 LT. Panels (**a**), (**b**) and (**c**) were obtained using IGRF for the years 1978, 1989 and 2002, respectively. For reference vertical lines are drawn at 22.5°S and 8°N

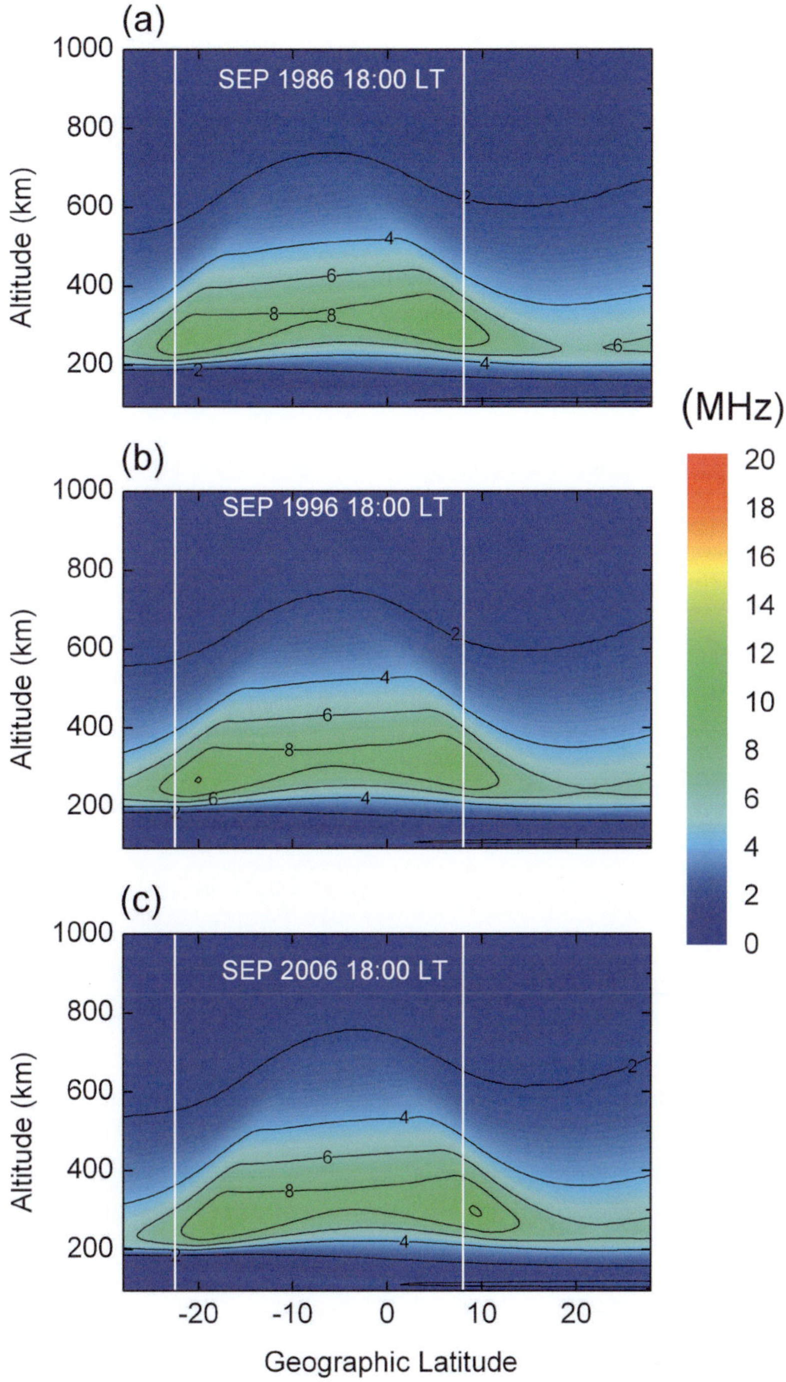

Fig. 23.9 Contour maps of the F layer plasma frequency as a function of geographic latitude and altitude. The results are for the 45°W meridian, September equinox, low solar activity, at 1800 LT. Panels (**a**), (**b**) and (**c**) were obtained using IGRF for the years 1976, 1986 and 2006, respectively. For reference vertical lines were drawn at 22.5°S and 8°N

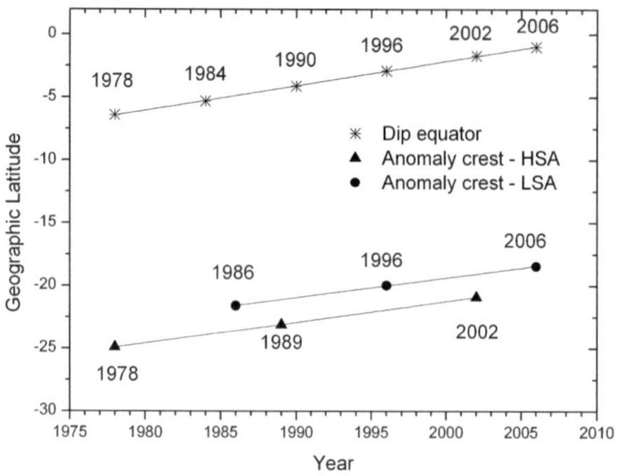

Fig. 23.10 Time variation of the geographic latitude of the EIA crest at the southern hemisphere, over the 45°W meridian high solar activity (*triangles*) and low solar activity (*circles*). The time variation of the geomagnetic equator position at the same longitude is also shown (*stars*)

displacement. This should be taken into account when long term trend studies are undertaken using data from this region.

Acknowledgments This work was partially supported by CNPq under grant 301643/2009-1.

References

Abdu MA, Batista IS, Muralikrishna P, Sobral JHA (1996) Long term trends in sporadic E-layer and electric fields over Fortaleza, Brazil. Geophys Res Lett 23:757–760

Abdu MA, Walker GO, Reddy BM et al (1990) Electric field versus neutral wind control of the equatorial anomaly under quiet and disturbed conditions: a global perspective. Ann Geophys 8:419–430

Anderson DN (1973a) Theoretical study of the ionospheric F region equatorial anomaly – 1. Theory. Planet Space Sci 21:409–419

Anderson DN (1973b) Theoretical study of the ionospheric F region equatorial anomaly – II. Results in the American and Asian sectors. Planet Space Sci 21:421–442

Bailey GJ, Balan N (1996) Some modelling studies of the equatorial ionosphere using the Sheffield University Plasmasphere Ionosphere Model. Adv Space Res 18:59–68

Bailey GJ, Balan N, Su YZ (1997) The Sheffield University Ionosphere-Plasmasphere Model – a review. J Atmos Solar-Terr Phys 59:1541–1552

Bailey GJ, Sellek R (1990) A mathematical model of the Earths plasmasphere and its application in a study of He+ at L=3. Ann Geophys 8:171–189

Bailey GJ, Sellek R, Rippeth Y (1993) A modeling study of the equatorial topside ionosphere. Ann Geophys 11: 263–272

Balan N, Bailey GJ (1996) Modeling studies of equatorial plasma fountain and equatorial anomaly. Adv Space Res 18:107–116

Batista IS, Abdu MA, MacDougall J, Souza JR (2002) Long term trend in the frequency of occurrence of the F3 layer over Fortaleza, Brazil. J Atmos Solar-Terr Phys 64: 409–1412

Bittencourt JA, Pillat VG, Fagundes PR et al (2007) LION: a dynamic computer model for the low-latitude ionosphere. Ann Geophys 25:2375–2392

Fejer BG, de Paula ER, Batista IS et al (1989) Equatorial F-region vertical plasma drifts during solar maxima. J Geophys Res 94:12049–12054

Fejer BG, de Paula, ER, Gonzalez, SA, Woodman RF (1991) Average vertical and zonal F-region plasma drift over Jicamarca. J Geophys Res 96:13901–13906

Hanson WB, Moffett RJ (1966) Ionization transport effects in the equatorial F region. J Geophys Res 71:5559–5572

Hedin AE, Fleming EL, Manson AH et al (1995) Empirical wind model for the upper, middle and lower atmosphere. J Atmos Solar-Terr Phys 58:1421–1447

Maus S, Macmillan S, Chernova T et al (2005) The 10th generation international geomagnetic reference field. Phys Earth Planet Interiors 151:320–322

Mitra AP (1968) A review of D-region processes in non-polar latitudes. J Atmos Solar-Terr Phys 30:1065–1114

Picone JM, Hedin AE, Drod DP (2002) NRLMSISE-00 empirical model of the atmosphere: statistical comparisons and scientific issues. J Geophys Res 107:1468, 2002

Rangarajan GK, Muniz Barreto L (2000) Secular change in the location of the magnetic dip equator in the twentieth century. Geofis Intern 39:323–336

Richards PG, Fennelly JA, Torr DG (1994) EUVAC: a solar EUV flux model for aeronomic calculations. J Geophys Res 99:8981–8992

Scherliess L, Fejer BG (1999) Radar and satellite global equatorial *F* region vertical drift model. J Geophys Res 104: 6829–6842

Souza JR, Abdu MA, Batista IS, Bailey GJ (2000a) Determination of vertical plasma drift and meridional wind using the Sheffield University Plasmasphere Ionosphere Model and ionospheric data at equatorial and low latitudes in Brazil: summer solar minimum and maximum conditions. J Geophys Res 105:12813–12821

Souza JR, Bailey GJ, Abdu MA, Batista IS (2000b) Ionospheric modelling at low latitudes over Brazil during summer solar minimum. Adv Space Res 25:133–138

Tobiska WK, Woods T, Eparvier F et al (2000) The SOLAR2000 empirical solar irradiance model and forecast tool. J Atmos Solar-Terr Phys 62:1233–1250

Chapter 24

Characteristics of Temperature and Density Structures in the Equatorial Thermosphere Simulated by a Whole Atmosphere GCM

Hitoshi Fujiwara, Yasunobu Miyoshi, Hidekatsu Jin, Hiroyuki Shinagawa, and Kaori Terada

Abstract Numerical simulations of the thermospheric temperature and mass density variations have been performed with a whole atmosphere general circulation model (GCM) which includes all the atmospheric regions: troposphere, stratosphere, mesosphere, and thermosphere. The GCM simulations represent hour-to-hour variations of the thermosphere due to effects from the lower atmosphere. The GCM also reproduces some characteristics of the thermospheric temperature and density structures: e.g., the midnight temperature maximum (MTM) and midnight density maximum (MDM) in the equatorial upper thermosphere. The MTMs and MDMs simulated here have the maximum amplitudes of 73 K and 19%, respectively, which are consistent with previous observations. The MTMs and MDMs simulated by the GCM also vary from hour to hour. The amplitude and location of the MTM depend on UT or longitude. In the dayside of the low-latitude region, the double-hump structure of the mass density is also seen in the GCM results. The lower atmospheric effects on the thermosphere would be important for generation of the mass density structure.

24.1 Introduction

In the equatorial thermosphere and ionosphere, there are peculiar phenomena which result from coupling between the lower and upper atmospheres through propagating atmospheric waves and the low inclination of the geomagnetic field. The midnight temperature maximum (MTM) and the midnight density maximum (MDM) are such phenomena in the equatorial thermosphere. The MTM and a related enhancement of 630 nm airglow emission (e.g., midnight brightness wave: MBW) have been investigated since the early 1970s or 1960s using photometers, Fabry-Perot interferometers (FPIs), all-sky camera systems, radars, and satellites as mentioned by Herrero et al. (1993), Colerico and Mendillo (2002), Faivre et al. (2006), and Meriwether et al. (2008).

The MTM is an enhancement in neutral temperature near local midnight at low latitudes in the upper thermosphere. The previous observations reported that the amplitude of the MTM was between 40–200 K in the American sector (e.g., Colerico and Mendillo, 2002, Faivre et al., 2006) and 80–570 K in the Indian sector (Rao and Sastri, 1994). The AE-E satellite observations showed that the MTM amplitude was larger in summer than in winter and that the MTM occurred earlier in summer than in winter (Herrero et al., 1983, 1993). These suggest longitudinal and seasonal variations of the MTM. The solar cycle dependence of the MTM amplitude seems to be weak (Faivre et al., 2006), while the MDM observations suggested solar activity effects on the relative phases of the MDM in solstice and equinox (Arduini et al., 1997).

The generation of the MTM is considered to be a result from tidal mode coupling. Mayr et al. (1979) and Hererro et al. (1983) ascribed the principal importance for generating the MTM to ion-neutral coupling and the solar EUV-driven neutral wind which produce semidiurnal variations of the thermospheric parameters. Fesen et al. (1986) showed importance of the propagating semidiurnal tide from the lower atmosphere in the generation of the MTM using

H. Fujiwara (✉)
Department of Geophysics, Tohoku University, Sendai, Japan
e-mail: fujiwara@pat.gp.tohoku.ac.jp

M.A. Abdu, D. Pancheva (eds.), A. Bhattacharyya (Coed.), *Aeronomy of the Earth's Atmosphere and Ionosphere*, IAGA Special Sopron Book Series 2, DOI 10.1007/978-94-007-0326-1_24,

the Thermosphere Ionosphere General Circulation Model (TIGCM) developed at the National Center for Atmospheric Research (NCAR). In addition, Fesen (1996) concluded that the upward propagating tidal waves were more important than the ion-neutral coupling for generating the MTM. Although the NCAR models, which included effects from the upward propagating semidiurnal tides, successfully reproduced the MTM in simulation results, the amplitudes of the MTM simulated by the models ($\sim$20 K) were much smaller than observed ones. This should indicate importance of additional tidal modes for reproducing the observed MTM as suggested by Mayr et al. (1979) and Hererro et al. (1983); namely, the terdiurnal component may be important. From the Fourier analysis of the neutral density data of the Sun Marco 3 and 5 satellites, Arduini et al. (1997) found that the third and fourth order harmonics were required to reproduce the MDM observed. Colerico et al. (2006) and Faivre et al. (2006) also pointed out the importance of terdiurnal mode to explain the observed MTM amplitude and night-to-night variability.

General circulation models (GCMs), which include all the atmospheric regions, have been developed (e.g., Miyoshi and Fujiwara, 2003, 2006, 2008; Akmaev et al., 2009). The atmospheric tides with various modes are generated within the models. Recently, Akmaev et al. (2009) and Fujiwara and Miyoshi (2010) successfully simulated the MTM with amplitudes of 50–150 K. Akmaev et al. (2009) showed UT and longitudinal dependence of the MTM and contributions of tidal components with wave numbers 2–5 to generation of the MTM, particularly terdiurnal and semidiurnal ones. They also reproduced some MTM features; the tropical MTM peaks earlier and attains larger magnitudes in summer than in winter. Akmaev et al. (2009) and Fujiwara and Miyoshi (2010) showed night-to-night variations of the MTM at low- to mid-latitudes. In addition, Fujiwara and Miyoshi (2010) discussed relation between the MTM and the neutral wind convergence and abatement of the meridional wind.

In the present study, we investigate temperature and mass density structures in the equatorial thermosphere, e.g., the MTM and MDM, by using a whole atmosphere GCM which covers all the atmospheric regions. In addition, we briefly describe an update and information regarding our GCM to suggest some challenges with a way forward for the future.

24.2 A Whole Atmosphere GCM

A whole atmosphere GCM has been developed as an extension of the middle atmosphere GCM developed at Kyushu University (Miyahara et al., 1993; Miyoshi, 1999). The GCM was originally developed as a tropospheric GCM at the Japan Meteorological Agency (Kanamitsu et al., 1983), and developed as a community climate model at the University of Tokyo. The GCM was extended to include the processes of the middle atmosphere, and then the thermosphere.

The whole atmosphere GCM solves the full nonlinear primitive equations for momentum, thermodynamics, continuity, and hydrostatics. The continuity equation of mass mixing ratio for the major species (N_2, O_2, and O) is also solved taking into account the photo-dissociation of O_2, oxygen chemistry, and molecular diffusion. This GCM is a global spectral model with spatial resolution versions of T21, T42, and T85 (Miyoshi and Fujiwara, 2008). In this study, we use the GCM with triangular truncation of T21 which is equivalent to a spatial resolution of 5.6° for latitude and longitude. The GCM covers 75 vertical pressure levels (vertical resolution of 0.4 scale height above the tropopause) and contains all the atmospheric regions from the ground to exobase. The time step for integrating the equations is 100 s. The corrected centred dipole is assumed to describe the Earth's magnetic field. The magnetospheric convection electric field modeled by Volland (1975) and empirical ionosphere based on Chiu (1975), Fuller-Rowell and Evans (1987) are used in the model. The effects of auroral particle precipitation on heating the neutral gases are evaluated by using an analytical prescription into the auroral oval developed by Roble and Ridley (1987). The magnetospheric convection electric field and empirical ionosphere are used for calculating Joule heating and ion-drag force. The details of the GCM are described in Miyoshi and Fujiwara (2003, 2006, 2008), Fujiwara and Miyoshi (2006), and references therein.

24.3 Results

In this section, we show some results from simulations with the whole atmosphere GCM. We focus our attention on temperature and density variations in solstice (December) and equinox (March), and their UT/longitudinal dependences. The solar minimum and geomagnetically quiet condition is assumed in this study; we perform numerical simulations in cases of March 1–31 and December 1–31, assuming that the cross polar cap potential is 30 kV and the F10.7 index is 70×10^{-22} W/m^2/Hz.

First, we show examples of GCM calculations. Figure 24.1 represents global distributions of temperature and horizontal wind at 01:00 UT on December 15 on constant pressure surfaces of about (a) 12, (b) 50, (c) 89, (d) 134, (e) 180, and (f) 306 km altitudes. The characteristic features of wind patterns and temperature structures are seen at each pressure (or altitude)

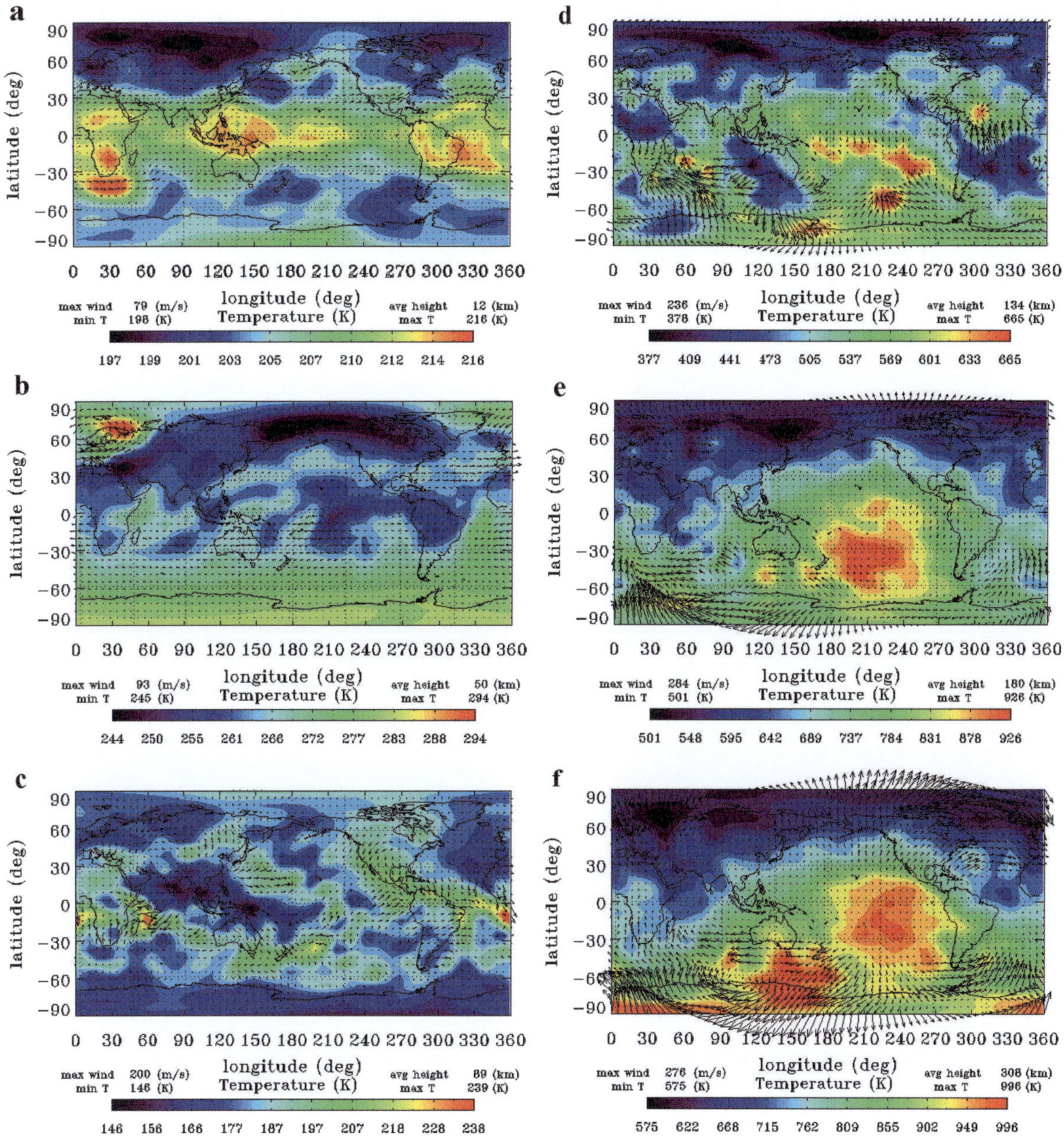

Fig. 24.1 Global distributions of temperature and horizontal wind at 01:00 UT on December 15 on constant pressure surfaces of about (**a**) 12, (**b**) 50, (**c**) 89, (**d**) 134, (**e**) 180, and (**f**) 306 km altitudes

level. Since we focus our attention on the atmospheric structures/variations in the upper thermosphere, we show GCM results in the altitude region above about 300 km in what follows.

Figure 24.2 shows global distributions of temperature and horizontal wind at 01:00 UT in December: (a) monthly average and (b) snapshot on December 15 on a constant pressure surface of about 300 km altitude. Figure 24.2b is the same as Fig. 24.1f except for the color scale. As mentioned by Fujiwara and Miyoshi (2010), who showed a 5-day average of GCM results, the monthly averaged temperature and wind show smoothed structure, while the snapshot of temperature and wind distributions have localized structures. In addition, the minimum and maximum temperatures are lower (by 26 K) and higher (by 52 K), respectively, in the snapshot case than in the monthly-average case. Since the amplitude of the diurnal temperature variation is larger in the snapshot case than in the monthly-average case, the larger amplitude of the MTM appears in the snapshot case. As Fujiwara and Miyoshi (2010) showed the MTM amplitudes of 50–150 K, which are in agreement with previous observations, the MTM amplitude seen in Fig. 24.2b is 73 K.

Figure 24.3 is the same as Fig. 24.2 except for March. The minimum and maximum temperatures are lower (by 7 K) and higher (by 16 K), respectively, in the snapshot case than in the monthly-average case. Localized structures are seen in the equatorial region in the snapshot case although the amplitudes of the localized structures are small. The temperature structures in the equatorial region are simpler in March than in December. The MTM amplitude in March (Fig. 24.3b) is smaller that that in December (Fig. 24.2b); namely, the MTM amplitude in the equatorial region for the March case is 45 K.

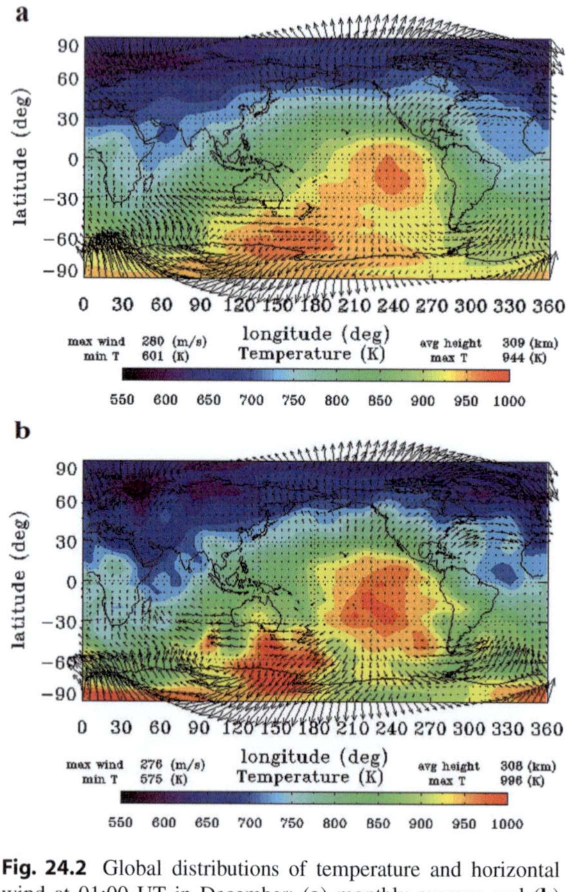

Fig. 24.2 Global distributions of temperature and horizontal wind at 01:00 UT in December: (**a**) monthly average and (**b**) snapshot on December 15 on a constant pressure surface of about 300 km altitude

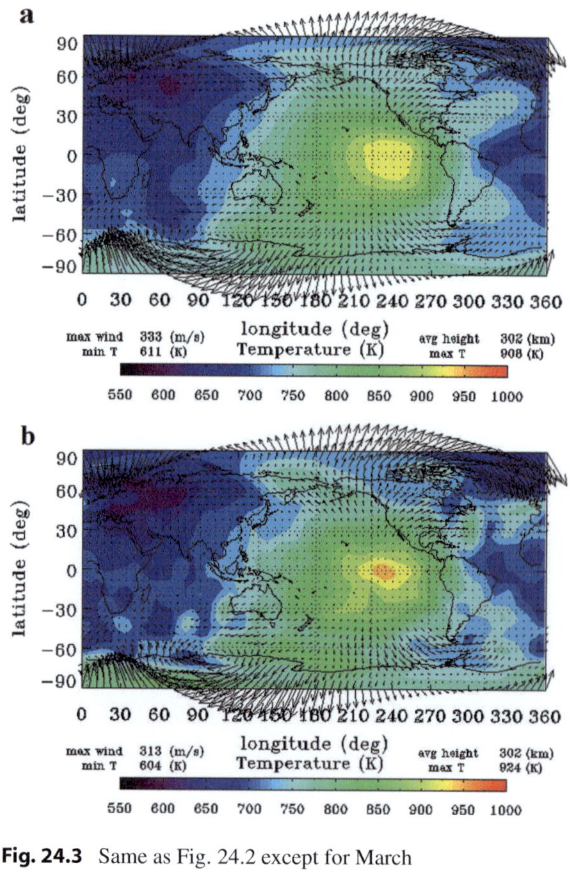

Fig. 24.3 Same as Fig. 24.2 except for March

Global distributions of temperature and horizontal wind in the northern winter solstice at (a) 06:00 UT on December 22, (b) 12:00 UT on December 22, (c) 18:00 UT on December 22, and (d) 00:00 UT on December 23 are shown in Fig. 24.4. The wind and temperature in Fig. 24.4 are calculated on a constant pressure surface of about 309 km altitude. Temperature enhancements in the night side of the low-latitude region are seen in each panel. In particular, the MTM is clearly seen at 12:00 UT (Fig. 24.4b) in this case. Although the maximum wind velocities are not so different among the four cases, the temperature ranges are various; e.g., 506–1004 K at 12:00 (Fig. 24.4b) and 576–972 K at 00:00 UT (Fig. 24.4d).

Figure 24.5 shows global distribution of the neutral mass density at 01:00 UT in December: (a) monthly average and (b) snapshot on December 15 at 400 km altitude. Although the atmospheric parameters are calculated on constant pressure surfaces because of assumption of the hydrostatics, we can infer the parameters at a specific height from the output data with the spline interpolation. As seen in Figs. 24.2 and 24.3, localized structures are seen in the snapshot case (Fig. 24.5b) while the structures are smoothed out in the monthly average case (Fig 24.5a). The largest neutral mass density is seen in the dayside of the low-latitude region. In Fig. 24.5b, the double-hump structure of the neutral mass density appear at around –30° (southern hemisphere) and the equator (northern hemisphere). In addition, enhancement of the neutral mass density is seen in the night side of the equatorial region (Fig. 24.5b).

Figure 24.6 is the same as Fig. 24.5 except for March. The density enhancement in the night side of the low-latitude region is not clearly seen in March compared to December. The density distribution shows more symmetric structure about the equator comparing to temperature. The double-hump structure

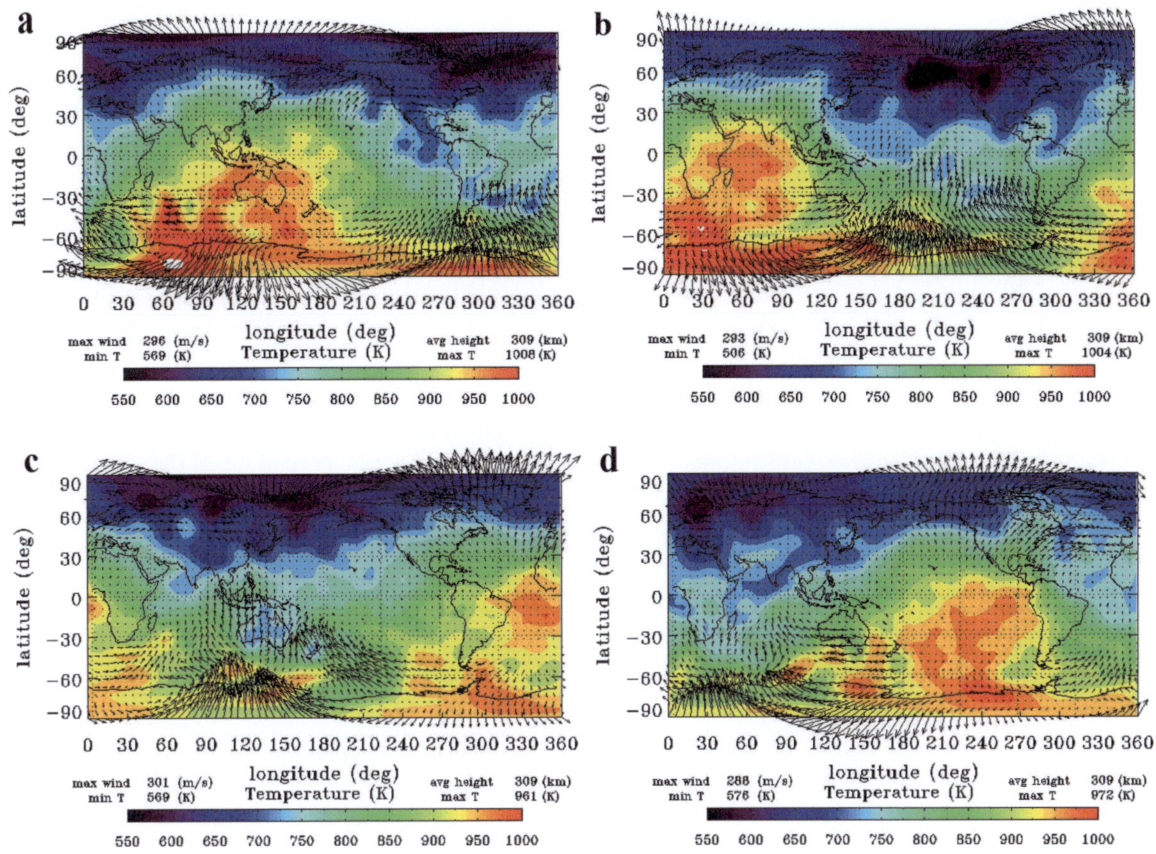

Fig. 24.4 Global distributions of temperature and horizontal wind in the northern winter solstice at (**a**) 06:00 UT on December 22, (**b**) 12:00 UT on December 22, (**c**) 18:00 UT on December 22, and (**d**) 00:00 UT on December 23

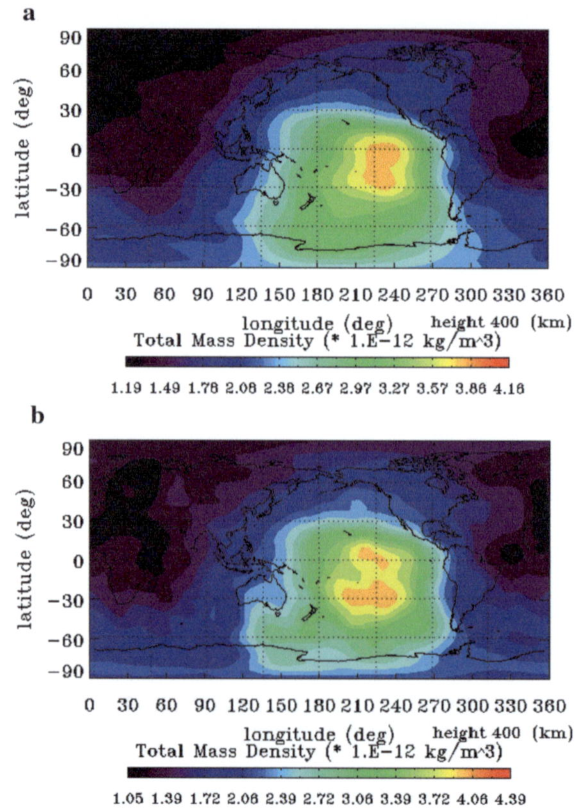

Fig. 24.5 Global distribution of the neutral mass density at 01:00 UT in December: (**a**) monthly average and (**b**) snapshot on December 15 at 400 km altitude. Note that the neutral mass densities at specific heights are obtained from the simulation output data with the spline interpolation

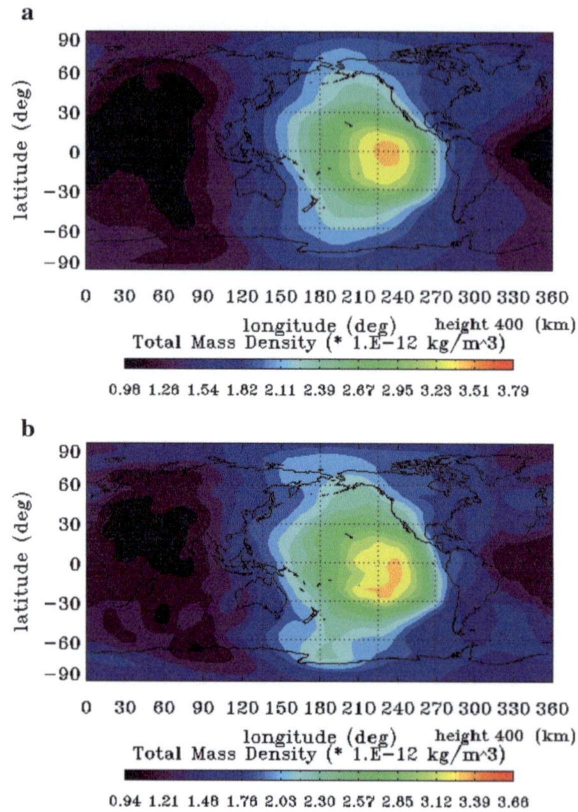

Fig. 24.6 Same as Fig. 24.5 except for March

is not clearly seen in March, although the double-hump structure (or density structure which would be related to the double- hump one) is slightly seen in Fig. 24.6b. The global structures of the neutral mass density are almost the same as those of temperature (see Figs. 24.2, 24.5, 24.3, and 24.6). The MDM amplitudes are 13 and 19% in the March (Fig. 24.6b) and December (Fig. 24.5b) cases, respectively; the night-time minimum densities are 1.04×10^{-12} and 1.15×10^{-12} kg/m^3 in March and December, respectively, and the night-time maximum densities are 1.17×10^{-12} and 1.37 $\times 10^{-12}$ kg/m^3 in March and December, respectively at 2.8° latitude.

24.4 Discussion

The MTM and MDM, which are ones of the characteristic features of temperature and neutral mass density in the low-latitude upper thermosphere, are reproduced from simulations with a whole atmosphere GCM for a low solar and geomagnetically quiet condition. Since the MTM shows significant variation from hour to hour, the amplitude of the MTM is larger in a snapshot case than in the monthly average case. In addition, UT or longitudinal dependence of the MTM appearance is shown in Fig. 24.4. In December, the MTM seems to appear clearly during the periods of 14:00–16:00 UT in the GCM results (not shown here). As shown in Fig. 24.4, the range of temperature variation (or day-night temperature difference) changes significantly in the four cases (00:00, 06:00, 12:00, and 18:00 UT). Change in the day-night temperature difference and/or change in baseline of the midnight temperature, which would be caused by effects from the lower atmosphere, would contribute to causing the large amplitudes of the MTM as well as heating due to convergence of winds and advection.

We obtain global distributions of the neutral mass density at specific heights from simulation outputs with the spline interpolation. The values of the neutral mass density obtained are in the range of previous observations, e.g., the neutral mass density derived from the CHAMP observations (Liu et al., 2005).

Enhancement of the neutral mass density is calculated at around the midnight in the low-latitude region in December (Fig. 24.5b). Arduini et al. (1997) reported that amplitudes of the midnight density maximum (MDM), which was obtained from the Sun Marco 3 and 5 satellite observations, were about 10% in solstice and about 20% in equinox at 380 km altitude. The amplitudes of the MDM in December (Fig. 24.5b) and March (Fig. 24.6b) are about 19 and 13%, respectively, at –2.8° latitude at 400 km altitude. The values of the MDM amplitude seem to be almost consistent with those by Arduini et al. (1997); however, GCM results show the larger MDM amplitude in December than in March although results by Arduini et al. (1997) indicated the larger MDM amplitude in equinox than in solstice. The features of the MDM including this discrepancy should be investigated in more detail in the future.

Anomalies of the neutral species in the equatorial thermosphere were first shown by Philbrick and McIsaac (1972) and Hedin and Mayr (1973). Liu et al. (2005) discovered the equatorial neutral mass density anomaly (or double-hump structure of the neural mass density) from the CHAMP satellite observations. The present GCM results also show the double-hump structure clearly in the December case although the structure is not clear in the March case (Figs. 24.5b and 24.6b), while Liu et al. (2007) showed that the double-hump structure is most prominent around equinox during moderate solar activity periods. Investigation of this discrepancy between the present GCM simulations and the CHAMP observations should be one of the most important challenges for understanding the fundamental structure of the equatorial thermosphere. Liu et al. (2007) pointed out that the large-scale meridional wind circulation may influence the equatorial mass density anomaly (or double-hump structure) superposed on the ion drag and chemical heating effects. In addition to effects of the wind circulation, ion drag, and chemical heating, the present GCM results show importance of effects from the lower atmosphere to produce the double-hump structure. In order to understand generation mechanism of the double-hump structure, we should perform more GCM simulations under the conditions of strong ion drag and meridional wind. In addition, we also need a comprehensive modeling study which can describe contributions of the above effects, e.g., meridional wind, ion drag, lower atmospheric effects, simultaneously. We are now developing a coupled model of the whole atmosphere, ionosphere, and ionospheric electrodynamics. As the first step of the work, we investigated low-latitude ionospheric variations (e.g., wavenumber-4 structure) caused by the atmospheric tidal forcings by using an ionospheric dynamo model with use of the GCM output data (Jin et al., 2008). The contributions of the meridional wind, ion-drag force, and the lower atmospheric effects to formation of the double-hump structure will be investigated with the coupled model.

24.5 Conclusions

We have performed numerical simulations with a whole atmosphere general circulation model (GCM) which includes all the atmospheric regions for low solar and geomagnetically quiet condition. The GCM reproduces characteristic features of wind patterns and temperature structures from the troposphere to thermosphere.

The midnight temperature maximum (MTM) is seen in the GCM results. Since the amplitude of the diurnal temperature variation is larger in the snapshot case than in the monthly-average case, the larger amplitude of the MTM appears in the snapshot case. The amplitude of the MTM depends on UT or longitude; the maximum amplitude of the MTM is calculated during 14:00–16:00 UT in the December solstice case. GCM simulations show that the day-night temperature difference would vary from hour to hour. This suggests that the baseline of the midnight temperature would vary from hour to hour and that change in the day-night temperature difference would contribute to causing large amplitude MTMs as well as heating due to convergence of winds and heat advection.

Global distributions of the neutral mass density at specific heights are obtained from simulation outputs with the spline interpolation. The amplitudes of the midnight density maximum (MDM) (13–19%) are consistent with those shown by previous observations; however, GCM results show the larger MDM amplitude in December than in March although results by Arduini et al. (1997) indicated the larger MDM amplitude in equinox than in solstice.

In the dayside of the low-latitude region, the double-hump structure of the mass density is also seen in the GCM results; the double-hump structure is clearly seen in December although the structure is not clear in March. This GCM result is different from the CHAMP observations; the double-hump structure is most prominent around equinox during moderate solar activity periods (Liu et al., 2007). We will investigate this discrepancy between the present GCM simulations and the CHAMP observations, which should be one of the most important challenges for understanding the fundamental structure of the thermospheric density. The present GCM simulations show that the large-scale structure of the mass density is similar to that of temperature. The variation from hour to hour of the double-hump structure (amplitude and location) would be also caused by the significant variation of the temperature which strongly depends on the lower atmospheric effects.

Acknowledgments This work was supported in part by Grant-in-Aid for Scientific Research C (20540435) and Scientific Research on Innovative Areas (20200047), and the Global COE program "Global Education and Research Center for Earth and Planetary Dynamics" at Tohoku University by the Ministry of Education, Science, Sports and Culture, Japan. A part of this work was also supported by the joint research program of the Solar-Terrestrial Environment Laboratory, Nagoya University.

References

Akmaev RA, Wu F, Fuller-Rowell TJ, Wang H (2009) Midnight temperature maximum (MTM) in Whole Atmosphere Model (WAM) simulations. Geophys Res Lett 36:L07108. doi:10.1029/2009GL037759

Arduini C, Laneve G, Herrero FA (1997) Local time and altitude variation of equatorial thermosphere midnight density maximum (MDM): San Marco drag balance measurements. Geophys Res Lett 24:377–380

Colerico MJ, Mendillo M (2002) The current state of investigations regarding the thermospheric midnight temperature maximum (MTM). J Atmos Solar-Terr Phys 64:1361–1369

Faivre M, Meriwether J, Fesen CG, Biondi MA (2006) Climatology of the midnight temperature maximum phenomenon at Arequipa, Peru. J Geophys Res 111:A06302. doi:10.1029/2005JA011321

Fesen CG (1996) Simulation of the low-latitude midnight temperature maximum. J Geophys Res 101:26,863–26,874

Fesen CG, Dickinson RE, Roble RG (1986) Simulation of the thermospheric tides at equinox with the National Center for Atmospheric Research Thermospheric General Circulation Model. J Geophys Res 91:4471–4489

Fujiwara H, Miyoshi Y (2006) Characteristics of the large-scale traveling atmospheric disturbances during geomagnetically quiet and disturbed periods simulated by a whole atmosphere general circulation model. Geophys Res Lett 33:L20108. doi:10.1029/2006GL027103

Fujiwara H, Miyoshi Y (2010) Morphological features and variations of temperature in the upper thermosphere simulated by a whole atmosphere GCM. Ann Geophys 25: 427–437

Fuller-Rowell TJ, Evans DS (1987) Height-integrated Pedersen and Hall conductivity patterns inferred from the TIROS-NOAA satellite data. J Geophys Res 92:7606–7618

Hedin AE, Mayr HG (1973) Magnetic control of the near equatorial neutral thermosphere. J Geophys Res 78:1688–1691

Herrero FA, Mayr HG, Spencer NW (1983) Latitudinal (seasonal) variation in the thermospheric midnight temperature maximum: a tidal analysis. J Geophys Res 88:7225–7235

Herrero FA, Spencer NW, Mayr HG (1993) Thermosphere and F-region plasma dynamics in the equatorial region. Adv Space Res 13:201–220

Jin H, Miyoshi Y, Fujiwara H, Shinagawa H (2008) Electrodynamics of the formation of ionospheric wave number 4 longitudinal structure. J Geophys Res 113:A09307. doi:10.1029/2008JA013301

Kanamitsu M, Tada K, Kudo T, Sato N, Isa S (1983) Description of the JMA operational spectral model. J Meteorol Soc Japan 61:812–828

Meriwether J, Faivre M, Fesen C, Sherwood P, Veliz O (2008) New results on equatorial thermospheric winds and the midnight temperature maximum. Ann Geophys 26: 447–466

Liu H, Lühr H, Henize V, Köhler W (2005) Global distribution of the thermospheric total mass density derived from CHAMP. J Geophys Res 110:A04301. doi:10.1029/2004JA010741

Liu H, Lühr H, Watanabe S (2007) Climatology of the equatorial thermospheric mass density anomaly. J Geophys Res 112:A05305. doi:10.1029/2006JA012199

Mayr HG, Harris I, Spencer NW, Hedin AE, Wharton LE, Porter HS, Walker JCG, Carlson CH Jr (1979) Tides and the midnight temperature anomaly in the thermosphere. Geophys Res Lett 6:447–450

Miyahara S, Yoshida Y, Miyoshi Y (1993) Dynamic coupling between the lower and upper atmosphere by tides and gravity waves. J Atmos Solar-Terr Phys 55: 1039–1053

Miyoshi Y (1999) Numerical simulation of the 5-day and 16-day waves in the mesopause region. Earth Planets Space 51: 763–772

Miyoshi Y, Fujiwara H (2003) Day-to-day variations of migrating diurnal tide simulated by a GCM from the ground surface to the exobase. Geophys Res Lett 30:1789. doi:10.1029/2003GL017695

Miyoshi Y, Fujiwara H (2006) Excitation mechanism of intraseasonal oscillation in the equatorial mesosphere and lower thermosphere. J Geophys Res 111:D14108. doi:10.1029/2005JD006993

Miyoshi Y, Fujiwara H (2008) Gravity waves in the thermosphere simulated by a general circulation model. J Geophys Res 113:D01101. doi:10.1029/2007JD008874
Philbrick CR, McIsaac JP (1972) Measurements of atmospheric composition near 400 km. Space Res 12:743–750
Rao HNR, Sastri JH (1994) Characteristics of the equatorial midnight temperature maximum in the Indian sector. Ann Geophys 12:276–278
Roble RG, Ridley EC (1987) An auroral model for the NCAR thermospheric general circulation model (TGCM). Ann Geophys 54:369–382
Volland H (1975) Models of the global electric fields within the magnetosphere. Ann Geophys 31:159–173

Chapter 25

Longitudinal Variations of the Thermospheric Zonal Wind Induced by Nonmigrating Tides as Observed by CHAMP

Kathrin Häusler and Hermann Lühr

Abstract In July 2000 the very successful German mini-satellite mission CHAMP, an acronym for Challenging Minisatellite Payload, was launched. One of the scientific instruments on board is an accelerometer that allows us to derive the zonal wind at CHAMP's altitude ($\sim$ 400 km). Previous to its launch, continuous and globally distributed measurements of the upper thermospheric wind have been rather sparse. With the launch of the CHAMP satellite we are now able to investigate the upper thermospheric zonal wind, and in particular its longitudinal variability, in a climatological sense. This capability has led to exciting and unanticipated findings such as the coupling from the troposphere to the thermosphere via nonmigrating tides. In this chapter we talk about the longitudinal variations of the CHAMP zonal wind at equatorial latitudes. Further, we present the nonmigrating tidal spectra embedded in the CHAMP zonal wind with special emphasis on the eastward propagating diurnal tide with zonal wavenumber 3 (DE3).

25.1 Introduction

Recently, satellite-borne observations of the geospace environment revealed exciting results not anticipated before. A prominent four-peaked longitudinal structure was identified in various upper atmospheric parameters such as the electron density (e.g., Lin et al., 2007; Lühr et al., 2007), electric field (e.g., Hartman and Heelis, 2007; Kil et al., 2007), equatorial electrojet (e.g., England et al., 2006; Lühr et al., 2008) as well as in ionospheric emissions (Sagawa et al., 2005; Immel et al., 2006) or in neutral mesosphere lower thermosphere (MLT) temperature (Forbes et al., 2006) and nitric oxide (Oberheide and Forbes, 2008b). Immel et al. (2006) were the first to relate nonmigrating tides of tropospheric origin to the longitudinal observations aloft.

Within this chapter we report on longitudinal variations of the upper thermospheric neutral zonal wind and trace back the likewise observed wave-4 structure in the zonal wind to the influence of nonmigrating tides and in particular to the eastward propagating diurnal tide with zonal wavenumber 3 (DE3).

Our research is based on measurements performed on board the German mini-satellite CHAMP (Challenging Minisatellite Payload). In July 2000, the satellite was launched into a quasi circular, near-polar orbit with an inclination of 87.3° and an altitude of about 450 km. CHAMP precesses through 1 h of local time (LT) every 11 days, thus taking its daily measurements at a quasi constant local time which is crucial for our analysis (cf. Section 25.5). Combining the ascending (northward bound) and descending (southward bound) orbital arcs, a 24 h local time coverage is obtained after 131 days. For detailed information on the CHAMP mission the reader is referred to Reigber et al. (2003, 2005). An illustration of the satellite can be found in Chapter 22 by Lühr et al. within this volume (their Fig. 22.1).

The findings of this chapter will bring forth the influence of atmospheric tides on upper atmospheric dynamics. Atmospheric tides are persistent global-scale oscillations in temperature, wind, and density

K. Häusler (✉)
GFZ German Research Centre for Geosciences, Telegrafenberg, Potsdam, Germany
e-mail: kathrin@gfz-potsdam.de

M.A. Abdu, D. Pancheva (eds.), A. Bhattacharyya (Coed.), *Aeronomy of the Earth's Atmosphere and Ionosphere*,
IAGA Special Sopron Book Series 2, DOI 10.1007/978-94-007-0326-1_25,

with periods that are harmonics of a solar day. One distinguishes between migrating and nonmigrating tides (e.g., Chapman and Lindzen, 1970; Forbes, 1995). Migrating tides have a zonal wavenumber, s, equal to their frequency in cycles per day. They propagate westward with a phase speed matching the apparent motion of the sun and are thus sun-synchronous and longitude independent. The absorption of solar energy by tropospheric water vapor as well as stratospheric ozone are the primary sources of migrating tides in the lower and middle atmosphere (Oberheide et al., 2002). Unlike migrating tides, nonmigrating tides are non-sun-synchronous. The zonal wavenumber, s, does not equal their frequency in cycles per day and they can either propagate eastward ($s < 0$), westward ($s > 0$) or remain standing ($s = 0$). Nonmigrating tides are excited for instance by zonal asymmetries, i.e. topography, land-sea differences, longitude dependences in absorbing species (Forbes et al., 2003), or by nonlinear interactions between the migrating diurnal tide and planetary waves (Hagan and Roble, 2001) or gravity waves (McLandress and Ward, 1994). Latent heat release in the tropical troposphere is another important source for nonmigrating tides and denotative for the observed wave-4 structure.

Atmospheric tides can be expressed, in general, in the form

$$A_{n,s} \cos(n\Omega t + s\lambda - \phi_{n,s}), \qquad (25.1)$$

where $A_{n,s}$ is the amplitude, n denotes the harmonics of a solar day, Ω the angular rotation rate of Earth, t is the universal time, s is the above mentioned zonal wavenumber, λ the longitude, and $\phi_{n,s}$ is the phase (Forbes et al., 2006).

Oscillations with periods of 24 and 12 h are referred to as diurnal (D) and semidiurnal (S) tides and correspond to $n = 1$, 2, respectively. Throughout this Chapter we will use the notation DWs and DEs to describe westward (W) or eastward (E) propagating diurnal tides with zonal wavenumber s. For semidiurnal tides D will be replaced by S and the standing oscillations are termed D0 and S0. Stationary planetary waves with zonal wavenumber m are denoted as SPWm. The phase is defined as the universal time (UT) when the wave maximum passes the Greenwich meridian.

The CHAMP satellite, as already mentioned before, performs its measurements in a quasi constant local time frame. Hence, Expression (25.1) has to be converted from universal time to a local time frame. This is realized by inserting $t = t_{\mathrm{LT}} - \lambda/\Omega$ into Expression (25.1) changing it to

$$A_{n,s} \cos(n\Omega t_{\mathrm{LT}} + (s - n)\lambda - \phi_{n,s}). \qquad (25.2)$$

Examining Expression (25.2), we find that a wave-4 structure in satellite data cannot be related unambiguously to a certain tidal component. Any values of s and n leading to $|s - n| = 4$ are able to generate it. In other words, the observed wave-4 pattern can be triggered by a diurnal tide ($n = 1$) with $s = -3$ or $s = +5$, a semidiurnal tide ($n = 2$) with $s = -2$ or $s = +6$, or by a stationary planetary wave ($n = 0$) with $s = 4$. Table 25.1 summarizes the individual contributions to the observed wave structures in satellite data for wavenumbers $m = 1, 2, 3, 4$. The individual contributions to the observed wave-4 pattern can, however, be discriminated by their local time evolution. Within 24 h LT, a wave-4 structure caused by DW5 propagates 90° to the west while DE3 propagates 90° to the east; SW6 displays a phase shift of 180° to the west while SE2 moves 180° to the east; SPW4 shows no phase shift at all. In order to quantify the oscillations causing the wave-4 structure, we have to isolate the individual contributions. This can be done by performing a two-dimensional (2-D) Fourier transform on the data which is described in more detail in Section 25.5.

This chapter is organized as follows. Section 25.2 overviews the zonal wind determination from accelerometer measurements. Section 25.3 presents the observed longitudinal variations. Section 25.4 reports on the prominent wave-4. Section 25.5 depicts the tidal spectra and compares the CHAMP results to TIME-GCM (thermosphere ionosphere mesosphere electrodynamics general circulation model) simulations and HME (Hough Mode Extension) analysis. Section 25.6 contains the conclusions.

Table 25.1 The listed diurnal (D) and semidiurnal (S) oscillations propagating either eastward (E) or westward (W) with a given zonal wavenumber, s, or standing (0) as well as stationary planetary waves (SPWm) can account for the observed wave-m structures in satellite data taken at quasi-constant local time. Adapted from Häusler and Lühr (2009)

Observed		Contributing oscillations				
Wave-1	→	DW2,	D0,	SW3,	SW1,	SPW1
Wave-2	→	DW3,	DE1,	SW4,	S0,	SPW2
Wave-3	→	DW4,	DE2,	SW5,	SE1,	SPW3
Wave-4	→	DW5,	DE3,	SW6,	SE2,	SPW4

25.2 Deriving the Zonal Wind

We use CHAMP accelerometer measurements to calculate the upper thermospheric zonal wind. The raw accelerometer data have to be carefully preprocessed before using them for the zonal wind determination. The preprocessing steps include removing other non-gravitational forces (i.e., the solar radiation pressure), corrections using bias and scale factors, and rotation of the accelerometer measurements into a suitable coordinate system. It was not possible to calibrate the CHAMP accelerometer to sufficient precision on ground. Thus, bias and scale factors were determined in orbit (Visser and IJssel, 2003). The used coordinate system, also referred to as a Velocity-Nadir (VN) coordinate system, has its origin at the satellite's center of mass. The x-axis is aligned with the satellites velocity vector along the nominal orbit, the z-axis points downward and y is perpendicular to the orbital plane, completing the right-hand triad. Liu et al. (2006) describe how to derive the zonal wind velocity, u_{zonal} (positive towards east), from CHAMP accelerometer measurements. Thereby, the governing equation is

$$u_{\text{zonal}} = -\frac{a_y}{a_x} \cdot v_x - v_c \tag{25.3}$$

in which a_x and a_y are the preprocessed horizontal acceleration components. We neglect the meridional wind component and approximate the along-track velocity component, v_x, with the satellite's orbital velocity (7.6 km/s). The approximation is justified considering the large orbital velocity of the satellite. The velocity component, v_c, is the corotation depending on height and latitude ($v_c = \Omega(R_E + h)cos\theta$, where Ω is the angular rotation rate of Earth, R_E the radius of the Earth, h the height of CHAMP, and θ the geographic latitude). After calculating the zonal wind using Eq. (25.3), the wind readings for each orbit are averaged within two latitude bands of ± 10 deg centered around the magnetic or geographic equator, respectively. Due to the fact that the equators are crossed twice with each orbit, we obtain two averaged local wind readings, u_n, per orbit which are separated in local time by 12 h. For our research we use zonal wind residuals that are free of zonal mean. These residuals, Δu_n, are obtained by subtracting a zonal mean average over a day, u_m, from the local wind reading, u_n. The obtained mean-free zonal wind component is used in our statistical analysis. The readings are sorted for each month into 3 h local time bins and 24 overlapping longitude bins of 30° width. An estimate of the achieved precision of a single wind velocity reading derived from CHAMP measurements is given in the appendix of Liu et al. (2006). It amounts to about 20 m/s.

25.3 Longitudinal Variation of the Equatorial Zonal Wind

In this section we will discuss the observed longitudinal variations of the upper thermospheric zonal wind at magnetic equator latitudes. For our analysis we used 4 years (corresponding to ~45250 equatorial overflights) of CHAMP measurements taken between 2002–2005. As described in Section 25.2, we use mean-free zonal winds that were sorted for every month into local time and longitude bins. Figure 25.1 shows the longitudinal distribution of the zonal delta wind for the months of March, June, September, and December for two different local times, namely 06 LT and 18 LT. The bars attached to the data points give the standard deviation of the samples within a longitude bin. They do not reflect the measurements uncertainty but the variability of the prevailing wind. Large longitudinal variations are observed in the month of December at 06 LT with eastward directed zonal winds from 30°W–165°E (peaking at 90°E with 42 m/s) and westward directed zonal wind residuals for the remaining longitudes. At 75°W we find a pronounced westward wind maximum of −69 m/s. In June we observe a reversed longitudinal variation of the wind at 06 LT. While the wind is directed eastward from 150°W–90°W, it is directed westward for the other longitudes interrupted by some brief eastward turning. For the pronounced western hemispheric wind maximum we find that it is now directed eastward (57 m/s) and located at 120°W. These large scale features in the morning hours are almost absent during the equinox months. These months exhibit more or less the switch in polarity between the two solstices as was already pointed out by Häusler et al (2007). However, when looking at Fig. 25.1 for the month of September, we find the zonal wind varying harmonically with longitude exhibiting a four-peaked structure. This four-peaked longitudinal structure will be discussed in more detail in the following section. The observed large-scale longitudinal variations in June and December are also absent in the evening hours.

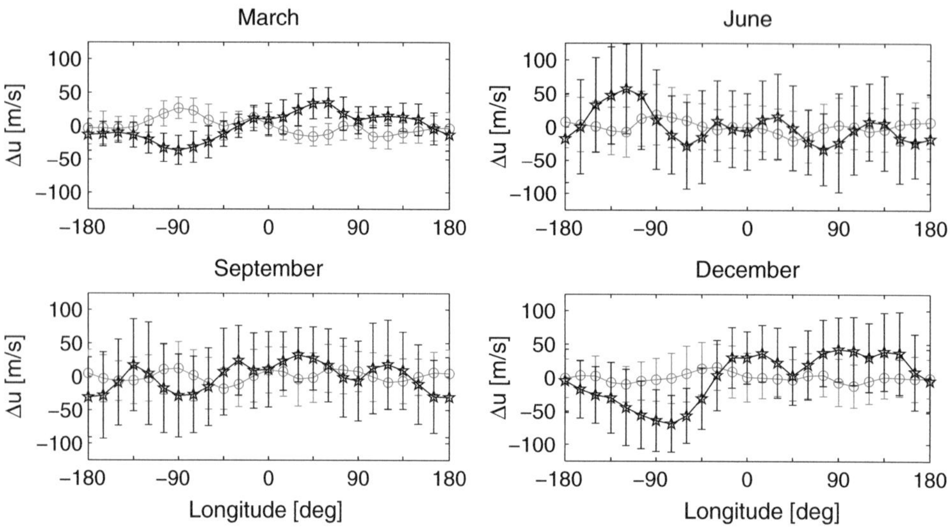

Fig. 25.1 Longitudinal variation of the zonal delta wind at 400 km altitude along the magnetic equator for March, June, September, and December. The mean zonal wind velocity is subtracted from each curve. Observations at two different local times (LT) are shown; *black curve*: 06 LT, *gray curve*: 18 LT

The longitudinal variation of the zonal wind for these 2 months at 18 LT is relatively small (cf. Fig. 25.1).

25.4 The Prominent Wave-4

In Section 25.3, we found that the zonal wind is featuring a four-peaked longitudinal structure which is also termed wave-4. In the course of this section we will identify the possible reason for this longitudinal pattern. We isolate the single harmonic structure embedded in the zonal wind by performing a Fourier transform on the longitude distribution of the zonal wind residuals obtained in Section 25.2. In order to calculate the observed wave form of the zonal wind for a given wavenumber, m, the determined Fourier coefficients a_m and b_m are employed:

$$\text{wave}\text{-}m(\lambda) = a_m \cos(m\lambda) + b_m \sin(m\lambda), \qquad (25.4)$$

with $m = 1, 2, 3, 4$. Using Eq. (25.4) for calculating the longitudinal variation of the wavenumber 4 for each local time hour, we get the time evolution of the wave-4 signal over the course of 24 h of local time. Häusler and Lühr (2009) have investigated the annual variation of the wave-4 structure observed in the CHAMP zonal wind. Figure 25.2 demonstrates the longitude vs. local time distribution of the observed wave-4 structure in the zonal wind for the months of September and December. While in September the wave-4

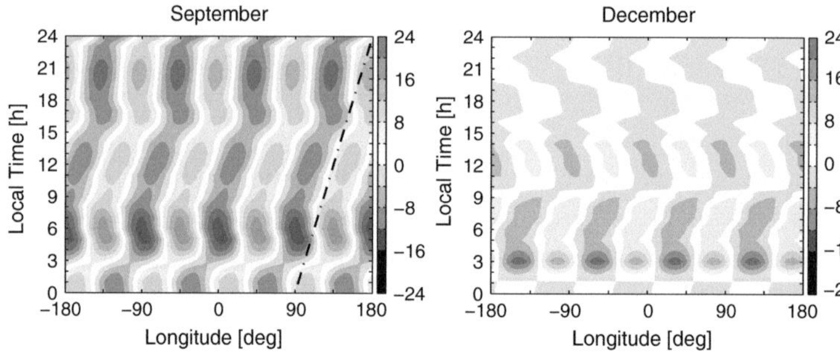

Fig. 25.2 Longitude vs. local time distribution of the observed wave-4 structure in the zonal wind [m/s] for September and December. The *black guiding line* is indicating a phase propagation of 90° to the east within 24 h LT. Modified version of Fig. 2 in Häusler and Lühr (2009)

structure shifts 90° to the east within 24 h LT, the month of December does not show a clear phase propagation at all. The clear 90° shift to the east within 24 h LT already highlights the DE3 tide as the likely driver of the wave-4 pattern (cf. Section 25.1). The annual variation (not shown) exhibits a maximum in wave-4 amplitudes for the months of July through September and a smaller secondary maximum in the months of March and April (Häusler and Lühr, 2009) suggesting that DE3 changes through the seasons.

25.5 Nonmigrating Tidal Spectra

As already discussed in Section 25.1, the wavenumber, m, observed by a satellite moving slowly in local time can be caused by several tidal components. This ambiguity can be expressed mathematical (Lühr et al, 2008) as

$$\begin{aligned} A_m \cos(m\lambda + \varphi) = a_{n-m} \cos\left\{m\lambda - \frac{\pi n}{12}(\mathrm{LT} - t_{n-m})\right\} \\ + a_m \cos\left\{m\lambda - \frac{\pi n}{12} t_m\right\} \\ + a_{n+m} \cos\left\{m\lambda + \frac{\pi n}{12}(\mathrm{LT} - t_{n+m})\right\}, \end{aligned} \quad (25.5)$$

where A_m is the amplitude of the total m-th harmonic, λ is the longitude, φ a local time dependent phase shift, n $(= 1,2)$ denotes harmonics of a day, a_{n-m}, a_m, a_{n+m} are the amplitudes of the various tidal components and stationary planetary waves, and t_{n-m}, t_m, t_{n+m} are the corresponding phases, i.e. the universal time at which the maximum is passing the Greenwich meridian. For the stationary planetary waves they denote the longitude of the maximum. In cases where the signal is available from all longitudes and local times, Eq. (25.5) can be solved analytically for a_{n-m}, a_m, a_{n+m} and t_{n-m}, t_m, t_{n+m} by performing a second Fourier transform for every combination of n and m. Thus Eq. (25.5) is used to identify all the tidal and stationary planetary wave contributions that may be present in the CHAMP zonal wind and which appear differently in a representation such as depicted in Fig. 25.2. A 24 h LT coverage is a mandatory requirement in order to perform the 2-D Fourier transform. This demand is fairly satisfied for each month combining 4 years of CHAMP measurements.

Figure 25.3 shows the annual variation of the individual tides adding to the wave-4 along the magnetic equator according to Table 25.1, namely DW5, DE3, SW6, SE2, and SPW4. We see that DE3 is the most striking component throughout the year, and thus we conclude it is the prime cause for the observed wave-4 structure as already presumed in Section 25.4. In correspondence with the seasonal wave-4 variability discussed in Section 25.4, DE3 exhibits two maxima, one from July through September and a smaller one in March and April. The biggest amplitude is obtained in July with a value of 10.5 m/s followed by 9.6 m/s in August and September. March and April amplitudes reach values of 6.7 m/s and 7.2 m/s, respectively. From December through February, the amplitudes of DE3 do not exceed 4 m/s and they are comparable in strength to the other wave-4 constituents. Hence it explains why during this time of the year no clear phase propagation of the wave-4 is recognizable (cf. Fig. 25.2). It has to be noticed that the binning of the zonal wind residuals into longitude bins of 30° width and 3 h LT composites

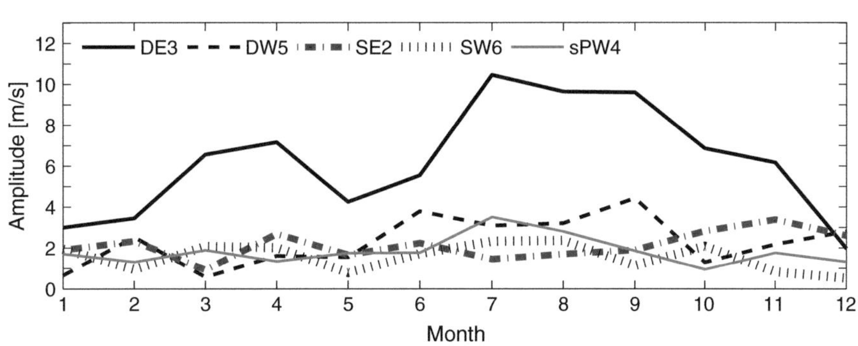

Fig. 25.3 Amplitude of the diurnal (DE3, DW5), semidiurnal (SE2, SW6) and stationary (SPW4) components contributing to the observed wave-4 structure in the zonal wind along the magnetic equator versus month of the year. DE3 is the most prominent component throughout the year and thus responsible for the observed wave-4 structure. Modified version of Fig. 3 in Häusler and Lühr (2009)

induces a damping on the tidal amplitudes (Häusler and Lühr, 2009). Running our analyzing routines with test data sets revealed that the damping is following the well-known function $x^{-1} \sin x$. Considering the damping due to the longitude bins, we find that $x = \pi/12, \pi/6, \pi/4, \pi/3$ for the wave-1, wave-2, wave-3, wave-4, respectively. This leads to an underestimation of the corresponding tides due to the longitude bins by approximately 1%, 4%, 10%, 17% observed as wave-1, wave-2, wave-3, and wave-4. Regarding the local time binning, $x = \pi/8$ and $\pi/4$ for the diurnal and semidiurnal tides yielding to an undervaluation of the diurnal and semidiurnal tides by 3% and 10%. The total damping is the product between the two contributions. Therefore, the diurnal tides observed as wave-1, wave-2, wave-3, and wave-4 are underestimated by our approach by 4%, 7%, 13%, and 20%. The semidiurnal tides observed as wave-1, wave-2, wave-3, and wave-4 are underestimated by 11%, 14%, 19%, and 26%. Owing to the fact that the stationary planetary waves are local time independent, we do not have to consider the damping due to the LT binning. Hence, the stationary planetary waves with zonal wavenumber $m =$ 1,2,3,4 are reduced by 1%, 4%, 10%, 17% considering our analysis approach. The amplitudes reported herein have been corrected for this effect. Comparing the annual variation of DE3 at 400 km altitude with the one for the zonal wind at 100 km altitude (Pedatella et al., 2008), we find a big resemblance between the two altitudes giving us confidence in the CHAMP results. How the tidal signature propagates upward will be discussed later on in this section.

Considering all combinations of $m (= 1, 2, 3, 4)$ and $n (= 1, 2)$ in Eq. (25.5) and performing the 2-D Fourier transform yields the diurnal and semidiurnal spectrum embedded in the zonal wind at CHAMP altitude. For illustration, the diurnal tidal spectrum of the zonal wind along the magnetic equator and along the geographic equator is shown in Fig. 25.4. Notably, the migrating tide (DW1) is filtered out and hence does not appear in Fig. 25.4 due to the pre-processing applied to the wind data. When looking at the diurnal tidal spectra in Fig. 25.4, it may be noted that overall larger amplitudes are observed along the magnetic equator (top panel). We find the strongest amplitudes of the tidal spectrum in December/January for DW2 with an amplitude of 12.0 m/s and 14.3 m/s, respectively. Along the geographic equator (bottom panel), the amplitudes of DW2 for December/January are reduced to 5.3 m/s and 7.7 m/s, respectively. Another strong

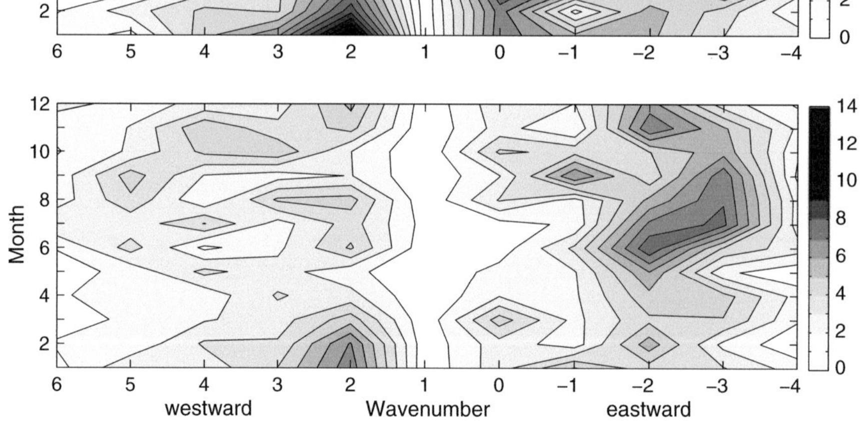

Fig. 25.4 Diurnal tidal spectra of the zonal wind [m/s] along the magnetic equator (*top panel*) and the geographic equator (*bottom panel*) observed by CHAMP. Positive (negative) wavenumbers correspond to westward (eastward) propagating tides. Modified version of Fig. 4 in Häusler and Lühr (2009)

tidal component observed along the magnetic equator is the standing diurnal oscillation D0 peaking in December with an amplitude of 11.3 m/s which almost vanishes when considered in geographic coordinates. Currently we do not know whether DW2 and D0 are generated in situ or whether they propagate upward from lower altitudes. A puzzling result is revealed for the annual behavior of the DE2 and DE3 tides within both coordinate systems. While generally DE3 exhibits stronger amplitudes when observed in the magnetic frame, DE2 displays maximum amplitudes in the geographic reference frame. The generation mechanism is the same for both tides.

25.5.1 CHAMP/TIME-GCM Tidal Comparison

For exploring and understanding the open issues of dynamical coupling between the lower and the upper atmosphere, we performed a comparison between the TIME-GCM model and in situ CHAMP observations concerning tidal signatures. TIME-GCM is a three-dimensional, time-dependent global grid point model developed at the National Center for Atmospheric Research (NCAR). Detailed information on the model is given by Roble (1995, 1996), Roble and Ridley (1994), and references therein.

Important to us is the fact that TIME-GCM inherently calculates the atmospheric tides that are excited by the absorption of ultraviolet and EUV radiation in the middle and upper atmosphere. Because the model does not extend to the ground, it cannot account for tides excited in the troposphere and lower stratosphere propagating upward into the model domain (Hagan et al., 2007). For this reason, the TIME-GCM lower boundary is perturbed with results of the global scale wave model (GSWM) (Hagan and Roble, 2001) which can account for tides driven outside the model domain (e.g. Hagan and Forbes, 2002, 2003).

Modeling studies performed by Hagan et al. (2007, 2009) with the TIME-GCM model revealed that DE3 is capable of propagating well into the upper atmosphere. However, an open question was how well TIME-GCM is performing regarding tidal signatures in the upper thermosphere. For the comparison between TIME-GCM and CHAMP, we ran the model for the 15th day of March, July, September, and December and let it settle to a diurnally reproducible state. In the beginning, solar minimum conditions were applied, thus the F10.7 value was set to 75 sfu (sfu = $10^{-22}\text{m}^{-2}\text{Hz}^{-1}$). Here, we want to focus only on DE3 and DE2 since they are excited by latent heat release in the tropical troposphere. The full comparison can be found in Häusler et al. (2010).

Figure 25.5 shows the intra-annual variation of the DE3 and the DE2 tide as observed by CHAMP and predicted by TIME-GCM. For the CHAMP DE3 and DE2 tide it is the same intra-annual course already depicted in Fig. 25.4, lower panel. It is visible that TIME-GCM exhibits a weakness in simulating the observed intra-annual variation for both tides. TIME-GCM predicts DE3 amplitudes of 13.3 m/s for March and December, 7.2 m/s for June, and the largest amplitude of 17.3 m/s is obtained in September. In contrast, the corresponding CHAMP DE3 amplitudes are much smaller. They equal 3.9 m/s in March, 3.8 m/s in June, 7.7 m/s in September, and 2.5 m/s in December. However, while TIME-GCM is overestimating the strength of DE3 at least by a factor of two in all simulated months, it shows hardly any DE2 contributions, except for December where TIME-GCM discloses an amplitude of 9.2 m/s. For all the other months the DE2 amplitudes are below 1.6 m/s. Yet, DE2 dominates the CHAMP tidal spectrum in the month of June with an amplitude of 8.8 m/s.

The discrepancies for DE2 between the model and observations can be attributed to insufficient lower boundary forcing. The GSWM DE2 amplitudes reported by Hagan and Forbes (2002) are also comparatively weak. Inaccuracies in the model dissipation schemes could be a factor that leads to the high DE3

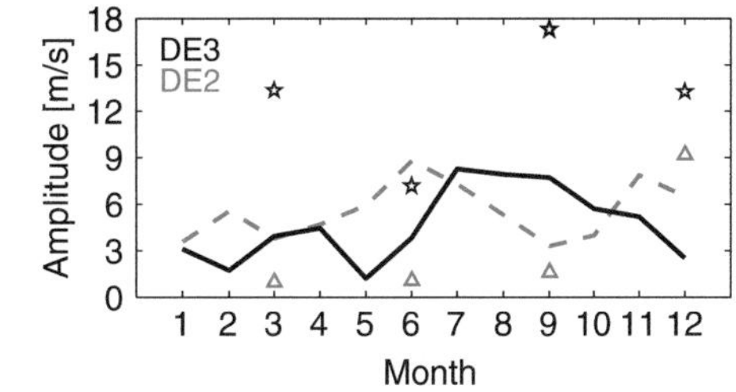

Fig. 25.5 Comparison between the CHAMP DE3 (*solid*) and DE2 (*dashed*) tidal amplitudes with the TIME-GCM DE3 (*pentagram*) and DE2 (*triangle*) tidal amplitudes obtained from the March, June, September, and December model runs under solar quiet conditions (F10.7 = 75 sfu)

amplitudes in TIME-GCM (Hagan et al., 2009). In order to overcome these discrepancies, it is planned to update the TIME-GCM lower boundary conditions. This will be achieved with new GSWM forcing that is currently in development.

25.5.2 Troposphere–Thermosphere Coupling by Nonmigrating Tides

TIME-GCM supports direct DE3 tidal upward propagation from the troposphere into the upper thermosphere (Hagan et al., 2007, 2009). Oberheide and Forbes (2008a) report that at least 50% of the CHAMP DE3 tidal signal in the zonal wind can be attributed to troposphere forcing. For their study, the authors use a physics-based empirical fit model based on Hough Mode Extension fits to TIMED (Thermosphere Ionosphere Mesosphere Energetics and Dynamics) tides. A classical Hough mode itself represents the solution of Laplace's tidal equation in an idealized, i.e. isothermal, wind- and dissipationless, atmosphere (e.g., Forbes, 1995). Using their model, Oberheide and Forbes (2008a) are able to extend TIMED tidal observations into the upper thermosphere, hence closing the gap between the MLT and measurements taken at about 400 km altitude. Revisiting the subject in a more recent study, Oberheide et al., (2009) report that within the limits of uncertainties, the DE3 tidal winds at CHAMP altitude are fully attributable to troposphere forcing. Figure 25.6 shows the good agreement between CHAMP DE3 zonal wind amplitudes and HME analysis. However, ascribing the DE3 tidal signal in the zonal wind only to direct tidal upward propagation raises the question why we observe stronger DE3 amplitudes in the zonal wind along the magnetic equator compared to the analysis along the geographic equator. This challenging result indicates additional electrodynamical coupling which still has to be resolved.

Figure 25.6 indicates another interesting result. The lower panel of Fig. 25.6 depicts the strength of the solar flux within the 4 years of CHAMP measurements that coincide with the declining phase of solar cycle 23. We observe stronger DE3 amplitudes in CHAMP altitude when solar activity is reduced. A dependence of the upper thermospheric tidal wind signatures on solar activity was also predicted by TIME-GCM (Häusler et al., 2010). According to Oberheide et al. (2009), the increase of DE3 amplitudes with decreasing solar flux is due to reduced dissipation in the lower thermosphere during solar minimum conditions.

25.6 Conclusions

The CHAMP mission has contributed significantly to the understanding of upper thermospheric dynamics. Due to the paucity of available zonal wind measurements, the unique CHAMP data set has proven

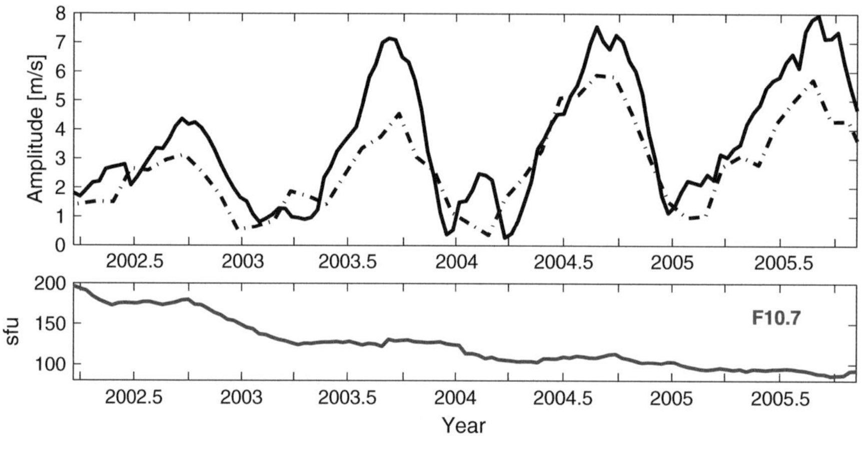

Fig. 25.6 DE3 zonal wind amplitudes from CHAMP (*solid*) and from the HME analysis (*dash-dotted*), averaged between 5°S and 5°N. *Bottom panel* shows the strength of the solar flux within the 4 years of measurements. Modified version of Fig. 14 in Oberheide et al., (2009)

to be a valuable tool for numerous studies. This chapter aims to highlight the findings accumulated in the works of Häusler et al. (2007, 2010) and Häusler and Lühr (2009) concerning the longitudinal variation of the upper thermospheric zonal wind as observed by the CHAMP satellite. We show that the observed four-peaked longitudinal structure is caused by the nonmigrating DE3 tide propagating from the troposphere into the upper thermosphere. The discovery that tropospheric weather systems in the tropics have an effect on upper atmospheric dynamics via troposphere-thermosphere coupling by nonmigrating tides is an important new consideration for aeronomy in these altitudes. For the proper interpretation of nonmigrating tidal signals in the CHAMP observations, numerical simulations are needed. The initial comparison between TIME-GCM predictions and in situ CHAMP observations is therewith an essential beginning. Recently, the CHAMP accelerometer data have been recalibrated and an extended data set of wind readings is now available starting from June 2001 to August 2010 allowing us to investigate in more detail the open questions. The upcoming multi-satellite SWARM mission will likely provide further beneficial contributions to our understanding of upper thermospheric dynamics.

Acknowledgements We thank Jens Oberheide, Maura Hagan, and Astrid Maute for fruitful discussions on the topic. K.H. is supported by the Deutsche Forschungsgemeinschaft (DFG) through its Priority Program CAWSES (SPP1176).

References

Chapman S, Lindzen RS (1970) Atmospheric tides: thermal and gravitational. D. Reidel Publishing Company, Dordrecht, Holland

England SL, Maus S, Immel TJ, Mende SB (2006) Longitude variation of the E-region electric fields caused by atmospheric tides. Geophys Res Lett 33:L21105. doi:10.1029/2006GL027465

Forbes JM (1995) Tidal and planetary waves. In: Johnson RM, Killeen TL (eds) The upper mesosphere and lower thermosphere: a review of experiment and theory. Geophysical Monograph Series, vol 87. AGU, Washington, DC

Forbes JM, Russell J, Miyahara S, Zhang X, Palo S, Mlynczak M, Mertens CJ, Hagan ME (2006) Troposphere-thermosphere tidal coupling as measured by the SABER instrument on TIMED during July–September 2002. J Geophys Res 111:A10S06. doi:10.1029/2005JA011492

Forbes JM, Zhang X, Talaat ER, Ward W (2003) Nonmigrating diurnal tides in the thermosphere. J Geophys Res 108(A1):1033. doi:10.1029/2002JA009262

Hagan ME, Forbes JM (2002) Migrating and nonmigrating diurnal tides in the middle and upper atmosphere excited by tropospheric latent heat release. J Geophys Res 107(D24):4754. doi:10.1029/2002JD001236

Hagan ME, Forbes JM (2003) Migrating and nonmigrating semidiurnal tides in the upper atmosphere excited by tropospheric latent heat release. J Geophys Res 108(A2):1062. doi:10.1029/2002JA009466

Hagan ME, Maute A, Roble RG (2009) Tropospheric tidal effects on the middle and upper atmosphere. J Geophys Res 114:A01302. doi:10.1029/2008JA013637

Hagan ME, Maute A, Roble RG, Richmond AD, Immel TJ, England SL (2007) Connections between deep tropical clouds and the Earth's ionosphere. Geophys Res Lett 34:L20109. doi:10.1029/2007GL030142

Hagan ME, Roble RG (2001) Modeling the diurnal tidal variability with the National Center for Atmospheric Research thermosphere-ionosphere-mesosphere-electrodynamics general circulation model. J Geophys Res 106(A11): 24869–24882

Hartman WA, Heelis RA (2007) Longitudinal variations in the equatorial vertical drift in the topside ionosphere. J Geophys Res 112:A03305. doi:10.1029/2006JA011773

Häusler K, Lühr H (2009) Nonmigrating tidal signals in the upper thermospheric zonal wind at equatorial latitudes as observed by CHAMP. Ann Geophys 27:2643–2652. URL www.ann-geophys.net/27/2643/2009/

Häusler K, Lühr H, Hagan ME, Maute A, Roble RG (2010) Comparison of CHAMP and TIME-GCM nonmigrating tidal signals in the thermospheric zonal wind. J Geophys Res 115:D00I08. doi:10.1029/2009JD012394

Häusler K, Lühr H, Rentz S, Köhler W (2007) A statistical analysis of longitudinal dependences of upper thermospheric zonal winds at dip equator latitudes derived from CHAMP. J Atmos Solar-Terr Phys 69:1419–1430. doi:10.1016/j.jastp.2007.04.004

Immel TJ, Sagawa E, England SL, Henderson SB, Hagan ME, Mende SB, Frey HU, Swenson CM, Paxton LJ (2006) Control of equatorial ionospheric morphology by atmospheric tides. Geophys Res Lett 33:L15108. doi:10.1029/2006GL026161

Kil H, Oh SJ, Kelley MC, Paxton LJ, England SL, Talaat ER, Min KW, Su SY (2007) Longitudinal structure of the vertical ExB drift and ion density seen from ROCSAT-1. Geophys Res Lett 34:L14110. doi:10.1029/2007GL030018

Lin CH, Wang W, Hagan ME, Hsiao CC, Immel TJ, Hsu ML, Liu JY, Paxton LJ, Fang TW, Liu CH (2007) Plausible effect of atmospheric tides on the equatorial ionosphere observed by the FORMOSAT-3/COSMIC: three-dimensional electron density structures. Geophys Res Lett 34:L11112. doi:10.1029/2007GL029265

Liu H, Lühr H, Watanabe S, Köhler W, Henize V, Visser P (2006) Zonal winds in the equatorial upper thermosphere: decomposing the solar flux, geomagnetic activity, and seasonal dependencies. J Geophys Res 111:A07307. doi:10.1029/2005JA011415

Lühr H, Häusler K, Stolle C (2007) Longitudinal variation of F region electron density and thermospheric zonal wind caused by atmospheric tides. Geophys Res Lett 34:L16102. doi:10.1029/2007GL030639

Lühr H, Rother M, Häusler K, Alken P, Maus S (2008) The influence of nonmigrating tides on the longitudinal variation of the equatorial electrojet. J Geophys Res 113:A08313. doi:10.1029/2008JA013064

McLandress C, Ward WE (1994) Tidal/gravity wave interactions and their influence on the large-scale dynamics of the middle atmosphere: model results. J Geophys Res 99:8139–8155

Oberheide J, Forbes JM (2008a) Tidal propagation of deep tropical cloud signatures into the thermosphere from TIMED observations. Geophys Res Lett 35:L04816. doi:10.1029/2007GL032397

Oberheide J, Forbes JM (2008b) Thermospheric nitric oxide variability induced by nonmigrating tides. Geophys Res Lett 35:L16814. doi:10.1029/2008GL034825

Oberheide J, Forbes JM, Häusler K, Wu Q, Bruinsma SL (2009) Tropospheric tides from 80 to 400 km: propagation, interannual variability and solar cycle effects. J Geophys Res 114:D00I05. doi:10.1029/2009JD012388

Oberheide J, Hagan ME, Roble RG, Offermann D (2002) Sources of nonmigrating tides in the tropical middle atmosphere. J Geophys Res 107(D21):4567. doi:10.1029/2002JD002220

Pedatella NM, Forbes JM, Oberheide J (2008) Intra-annual variability of the low-latitude ionosphere due to nonmigrating tides. Geophys Res Lett 35:L18104. doi:10.1029/2008GL035332

Reigber C, Lühr H, Schwintzer P (eds) (2003) First CHAMP Mission results for gravity, magnetic and atmospheric studies. Springer, London

Reigber C, Lühr H, Schwintzer P, Wickert J (eds) (2005) Earth observation with CHAMP – results from three years in orbit. Springer, London

Roble RG (1995) Energetics of the mesosphere and thermosphere. In: Johnson RM, Killeen TL (eds) The upper mesosphere and lower thermosphere: a review of experiment and theory. Geophysical Monograph Series, vol 87. AGU, Washington, DC, pp 1–21

Roble RG (1996) The NCAR thermosphere-ionosphere-mesosphere-electrodynamics general circulation model (TIME-GCM). In: Schunk RW (ed) STEP handbook on ionospheric models. Utah State University, Logan, UT, pp 281–288

Roble RG, Ridley EC (1994) A thermosphere-ionosphere-mesosphere-electrodynamics general circulation model (time-GCM): equinox solar cycle minimum simulations (30–500 km). Geophys Res Lett 21:417–420. doi:10.1029/93GL03391

Sagawa E, Immel TJ, Frey HU, Mende SB (2005) Longitudinal structure of the equatorial anomaly observed by IMAGE/FUV. J Geophys Res 110:A11302. doi:10.1029/2004JA010848

Visser PNAM, IJssel J van den (2003) Verification of CHAMP accelerometer observations. Adv Space Res 31:1905–1910. doi:10.1016/S0273-1177(03)00165-0

Chapter 26

Causal Link of Longitudinal Plasma Density Structure to Vertical Plasma Drift and Atmospheric Tides – A Review

Hyosub Kil and Larry J. Paxton

Abstract This chapter reviews recent advances in our understanding of the characteristics and driving mechanisms of the longitudinal plasma density structure in the low-latitude *F* region. Various ionospheric observations have shown the development of a longitudinal wave-like pattern in plasma density. Typically, the wave number-4 (wave-4) pattern is pronounced during July–September, and the wave-3 pattern is pronounced during December–January. Variation of the longitudinal plasma density pattern with local time and season is causally linked to vertical plasma drift in the *F* region (or *E*-region dynamo electric fields). The wave-4 pattern is of special interest to the ionosphere-thermosphere community because this phenomenon is closely associated with the diurnal eastward propagating zonal wave number 3 tide (DE3). This idea is supported by observations of annual variation of the DE3 amplitude and diurnal variation of the DE3 phase that are consistent with the annual and diurnal variations of the wave-4 patterns in plasma density and dynamo electric fields. The connection of the ionospheric wave-3 pattern presumably to the diurnal eastward propagating zonal wave number 2 tide (DE2) further demonstrates the significant role of atmospheric tides in the formation of large-scale ionospheric structures. Formation of the large-scale longitudinal structures of the ionosphere is attributed to the modulation of the *E*-region dynamo electric fields by atmospheric tides, but recent studies indicate that the ionosphere and thermosphere can be directly modulated by the penetration of atmospheric tides into the *F*-region height.

H. Kil (✉)
Applied Physics Laboratory, The Johns Hopkins University, Laurel, MD 20723, USA
e-mail: hyosub.kil@jhuapl.edu

26.1 Introduction

Electric fields and neutral winds are the major drivers of the low-latitude ionospheric morphology (Kelley, 1989). The effects of neutral winds and electric fields on the ionosphere are sensitive to the geomagnetic field configuration: magnetic field strength, magnetic declination, and displacement of the geomagnetic equator from the geographic equator. Therefore, longitudinal variation of the plasma density is often understood in the context of geomagnetic field configuration (e.g., West and Heelis, 1996; Su et al., 1997; Thuillier et al., 2002; Kil et al., 2006; Ren et al., 2009a). However, the longitudinal wave number three (wave-3) or wave number four (wave-4) patterns that have been extensively investigated in recent years are not explained by the geomagnetic field configuration. Atmospheric tides are recognized as a new source that controls the large-scale longitudinal variation of the *F*-region morphology in low latitudes (Sagawa et al., 2005; Immel et al., 2006).

This chapter reviews recent progress in the study of longitudinal plasma density structure, especially, wave-4 pattern. In Section 26.2, characteristics of the longitudinal plasma density structure are described. In Section 26.3, we present observations and model simulation results that demonstrate the causal link of the longitudinal plasma density structure to the *E*-region dynamo electric fields. In Section 26.4, connections between the dynamo electric fields and atmospheric

M.A. Abdu, D. Pancheva (eds.), A. Bhattacharyya (Coed.), *Aeronomy of the Earth's Atmosphere and Ionosphere*, IAGA Special Sopron Book Series 2, DOI 10.1007/978-94-007-0326-1_26, © Springer Science+Business Media B.V. 2011

tides are described. In Section 26.5, we describe the generation mechanism of E-region dynamo electric fields and explain how tidal winds can longitudinally modulate the dynamo electric fields. Conclusions are given in Section 26.6.

26.2 Longitudinal Plasma Density Structure

The wave-like longitudinal plasma density pattern was first reported by topside soundings from the Russian satellite Intercosmos-19 (Kochenova, 1987, 1988; Karpachev, 1988; Benkova et al., 1990). Intercosmos-19 observed the formation of four peaks in plasma density (wave-4 pattern in longitude at a fixed local time) at the magnetic equator (Benkova et al., 1990). The authors deduced the vertical $\mathbf{E} \times \mathbf{B}$ drift from the separation of the northern and southern equatorial ionization anomalies (EIAs) and showed that the equatorial plasma density peaks at the locations where the vertical $\mathbf{E} \times \mathbf{B}$ drift is minimal. Formation of the longitudinal wave-4 pattern was also identified in nightglow observations in late 1960s from the Orbital Geophysical Observatory (OGO) 4 and 6 (see references cited by Thuillier et al., 2002), but the focus of the Thuillier et al. was on the hemispheric asymmetry induced by winds.

Global-scale ionospheric and thermospheric observations from multiple satellites in the 2000s advanced our understanding of the characteristics of the longitudinal plasma density structure and its driving mechanisms. Study of the ionospheric wave structures in the 2000s may have arisen from the investigation of the wave-like longitudinal structure in the equatorial electrojet (EEJ) by using the magnetometer data from the Ørsted satellite (Jadhav et al., 2002; Ivers et al., 2003). Jadhav et al. (2002) suggested atmospheric tides as one of the possible generation mechanisms of the longitudinal wave-like structure in the EEJ. The EEJ is an enhanced eastward current during the daytime in the equatorial region that is driven by dynamo electric fields in the E region (e.g., Kelley, 1989). Therefore, the behavior of the E-region electric fields can be deduced from the observation of the EEJ. Because the dynamo electric fields in the E region play a dominant role in the F-region ionospheric morphology, the EEJ can be used as an indicator of F-region plasma density structure.

Sagawa et al. (2005) elaborated on the connection between the ionospheric wave-4 structure and atmospheric tides. They identified the formation of the wave-4 pattern in the F-region plasma density from the optical observation of the nighttime ionosphere by far ultraviolet imager (FUV) onboard the Imager for Magnetopause-to-Aurora Global Exploration (IMAGE) satellite. They suggested that non-migrating atmospheric tides that propagate from lower altitudes to the E region can modulate the E-region dynamo electric field and produce the observed wave-4 plasma density pattern in the F region. Immel et al. (2006) and England et al. (2006a, b) specified that the diurnal eastward propagating zonal wave number 3 tide (DE3) is responsible for the generation of the ionospheric wave-4 plasma density structure; DE3 longitudinally modulates the E-region dynamo and creates the wave-4 plasma density structure (Immel et al., 2006). DE3 was suggested as the source on the basis of the coincidence of the crest longitudes of Global Scale Wave Model (GSWM) (Hagan and Forbes, 2002) during the spring equinox in 2002 and the observed wave-4 plasma density patterns by IMAGE/FUV and Global Ultraviolet Imager (GUVI) onboard the Thermosphere Ionosphere Mesosphere Energetics and Dynamics (TIMED) satellite. The GSWM DE3 shows a good agreement with wave-4 plasma density pattern in the spring equinox, but not in other seasons. In later studies, the connection of wave-4 plasma density structure with the DE3 was identified by observations of winds and temperature in the lower thermosphere.

The wave number in the longitudinal plasma density structure describes the wave pattern at a fixed local time. On the other hand, the phase of atmospheric tides is expressed by the zonal wave number, s (=... −2, −1, 0, 1, 2, ...), and oscillation frequency in terms of cycles per solar day, n (= 1 (diurnal), 2 (semidiurnal), 3 (terdiurnal), ...) (e.g., Forbes et al., 2006). A non-migrating tide appears as a wave number $|s-n|$ in longitude when it is viewed at a fixed local time. As an example, DE3 ($s = -3$, $n = 1$) has a wave-4 pattern ($|s-n| = 4$) at a fixed local time. Note that the wave-4 pattern can be produced by various tides. Häusler and Lühr (2009) tabulated the tides that contribute to various ionospheric longitudinal wave patterns. As we describe later, the dominant tides that affect the ionospheric longitudinal wave structure can be deduced by examining the amplitudes of tides at the dynamo region (*E* region) and the seasonal and diurnal variations of the tidal amplitude and phase.

The characteristics of the longitudinal plasma density structure have been identified from various observations: TIMED/GUVI (Henderson et al., 2005; Kil et al., 2008; Immel et al., 2006, 2009; England et al., 2009), IMAGE (Immel et al., 2006), the first Republic of China satellite (ROCSAT-1) (Kil et al., 2007, 2008; Oh et al., 2008), the CHAllenging Minisatellite Payload (CHAMP) satellite (Lühr et al., 2007; Liu and Watanabe, 2008; Pedatella et al., 2008), Defense Meteorological Satellite Program (DMSP) satellites (Kil et al., 2008; Ren et al., 2008), the global positioning system (GPS) occultation experiment (GOX) onboard the Formosa Satellite Mission #3/Constellation Observing System for Meteorology, Ionosphere and Climate (FORMOSAT-3/COSMIC or F3/C) (Lin et al., 2007a, b), GPS total electron content (TEC) maps (Wan et al., 2008), and TEC data derived from the ocean TOPography EXperiment (TOPEX) (Scherliess et al., 2008). Figure 26.1 presents the ROCSAT-1 observations of the ionospheric plasma density in January–February (a) and September–October (b). The local times are divided into 0800–0900, 0900–1000, 1000–1100, 1200–1400, 1600–1800, and 2000–2200 LT from top to bottom. For the plots, ROCSAT-1 data are processed during the solar maximum period (1999–2002) for Kp $\leq 3^{+}$. ROCSAT-1's altitude was 600 km. The longitudinal variation of the plasma density emerges in the 0800–0900 LT plots and is fully developed at 1200–1400 LT. The peak-to-trough variation of the longitudinal wave structure of the average plasma density within $\pm 20°$ magnetic latitude is about 40~50% at 1200–1400 LT in both seasons. Scherliess et al. (2008) reported about 20~30% peak-to-trough variation in TEC. Because the amplitude of the variation of the plasma density depends on the selection of the sampling height, latitude, local time, etc., the variation of the ROCSAT-1 plasma density is not directly comparable to the variation of TEC. The daytime plasma distribution at low latitudes is symmetric with respect to the magnetic equator. The seasonal difference in the longitudinal plasma density structure is clear; the wave-3 pattern is pronounced

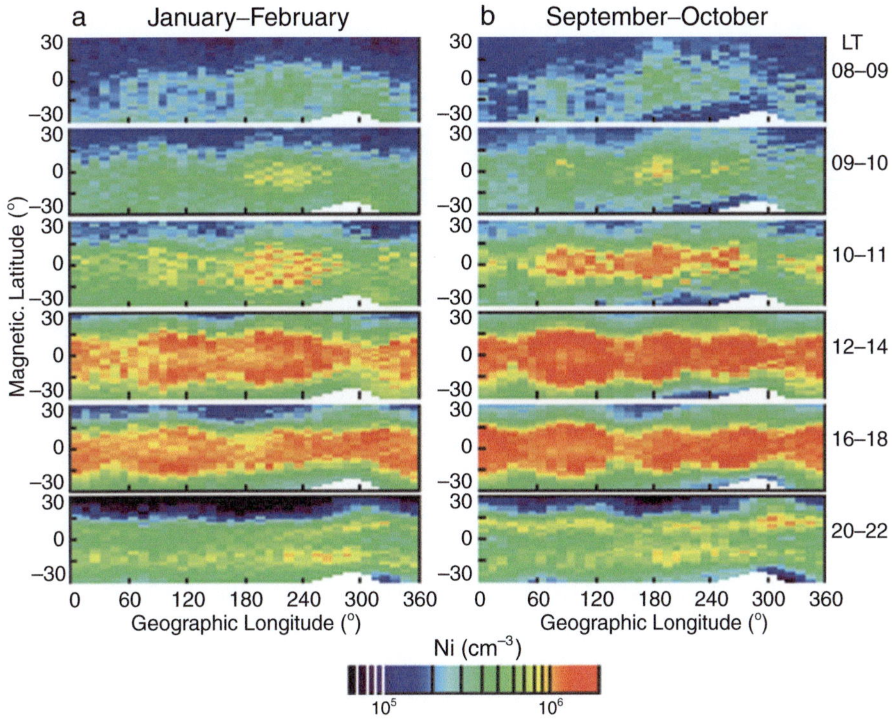

Fig. 26.1 Local time variations of the topside plasma distribution during January–February (**a**) and September–October (**b**). The density maps are produced by using ROCSAT-1 data during 1999–2002 under the condition Kp $\leq 3^{+}$(Kil et al., 2008)

in January–February, and the wave-4 pattern is pronounced in September–October. We note a slight eastward shift of the longitudes of the plasma density crests (or troughs) with an increase of local time in Fig. 26.1b and this will be discussed later. The seasonal and diurnal variations of the longitudinal plasma density structure observed by ROCSAT-1 are consistent with the TEC observations from TOPEX (Scherliess et al., 2008) and the ground GPS network (Wan et al., 2008). The wave structure formed during the day affects the nighttime plasma density structure, and we can still trace the wave features at night. The connection among plasma density, dynamo electric field, and atmospheric tides may be determined more accurately by using observations during daytime than at night: (1) the longitudinal morphology of the pre-reversal enhancement (PRE) is different from that of daytime vertical $\mathbf{E} \times \mathbf{B}$ drift, especially during solstices (Fejer et al., 2008; Su et al., 2008; Kil et al., 2009a); (2) the nighttime plasma density structure is modified from the daytime plasma density structure by PRE; (3) the hemispheric component of neutral wind modifies the nighttime plasma density structure.

The longitudinal wave pattern is variable with the season. Here division of the seasons into three (equinox, June solstice, and December solstice) or four (spring equinox, fall equinox, June solstice, and December solstice) does not precisely describe the annual variation of the wave pattern. Although the wave-3 pattern is more pronounced than the wave-4 pattern in some months (December–January), the investigation of the characteristics of longitudinal plasma density structure has focused on the wave-4 pattern. The major reason may be because we have a better understanding of the source of the wave-4 pattern than that of the wave-3 pattern.

Wan et al. (2008) investigated the annual and local time variations of the wave-4 longitudinal structure by using the GPS TEC map data provided by Jet Propulsion Laboratory. They extracted the wave-4 component from latitudinal mean of the TEC data within $\pm 30°$ magnetic latitudes during 1998–2007. Figure 26.2 shows the diurnal and seasonal variations of the amplitude of the wave-4 component for three F10.7 conditions: F10.7 < 120 (top), 120 < F10.7 < 160 (middle), and F10.7 > 160 (bottom). Because the Fourier-filtered wave-4 component repeats in longitude, the figure shows the result at one longitude segment (from 45°W to 45°E). In the figure, the wave-4 structure is most pronounced in July–September and is least developed during December–February. The wave-4 amplitude shows only a minor dependence on the solar flux. The annual variation of the wave-4 amplitude is consistent with ROCSAT-1 observations (Kil et al., 2008). We note the eastward shift (rightward tilt) of the wave-4 pattern with local time. The eastward shift of the wave-4 plasma density pattern is also observed by ROCSAT-1 (Fang et al., 2009; Ren et al., 2009b). The annual variation of the wave-4 amplitude and the eastward shift of the wave-4 phase with local time provide useful clues to identifying the connection of plasma density structure with atmospheric tides.

26.3 Causal Link of the Longitudinal Density Structure to *E*-region Dynamo Electric Field

The climatological model of the equatorial electrojet (EEJM-1) was derived from Ørsted, CHAMP, and SAC-C (Satélite de Aplicaciones Científicas-C) satellite measurements (Alken and Maus, 2007). By using the EEJM-1 model, Lühr et al. (2008) investigated the seasonal and local time variations of the wave-4 component in the EEJ. Figure 26.3 shows the EEJ patterns for three seasons: June (left column), September (middle column), and December (right column). The top panels show the EEJ peak current density; the middle panels show the wave-4 component in the EEJ (or DE3 signal); and the bottom panels show the remaining EEJ after removal of the wave-4 component. The wave-4 pattern is pronounced in September but is not well developed in December. The phase of the wave-4 pattern is shifted eastward (tilted rightward) with local time. The seasonal and diurnal behavior of the wave-4 pattern in the EEJ is consistent with that observed in the wave-4 pattern in the ionospheric plasma density. After removal of the wave-4 component in the EEJ (bottom panels), the wave-3 pattern is dominant in June and December, and the wave-2 pattern is dominant in September. Therefore, wave-2, wave-3, and wave-4 patterns are the dominant features in the EEJ.

The long-term (March 1999–June 2004) observations of ion velocity at low latitudes from ROCSAT-1 (altitude: 600 km; orbital inclination: 35°) provided a

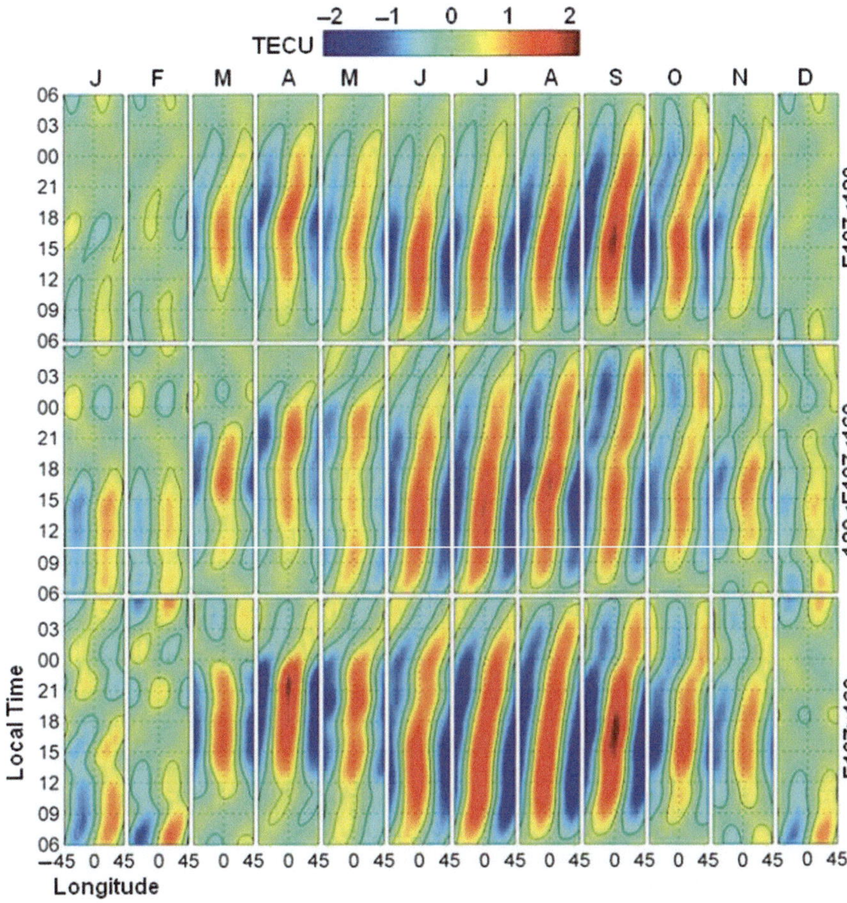

Fig. 26.2 Seasonal and local time variations of the wave-4 component in the latitudinal mean of the GPS TEC for three F10.7 conditions (Wan et al., 2008)

unique data source for the development of a high temporal and spatial resolution vertical **E** × **B** drift model. The empirical vertical **E** × **B** drift models based on the ROCSAT-1 data were developed for three seasons – equinox and solstices (Fejer et al., 2008; Kil et al., 2008). Development of wave-3 or wave-4 plasma density structure in the morning and their temporal evolution show a causal link to the empirical vertical **E** × **B** drift models. As was described above, the annual variation of the amplitude of the wave-4 plasma density pattern is not precisely divided into three or four seasons. For example, the wave-4 plasma density pattern is more pronounced in the fall equinox than in the spring equinox. Kil et al. (2009b) and Ren et al. (2009b) developed a monthly vertical **E** × **B** drift model based on the ROCSAT-1 data and showed that the annual variation of the wave-4 amplitude in vertical **E** × **B** drift is consistent with the annual variation of the wave-4 amplitude in plasma density. DMSP (altitude: 840 km) observations at 0930 LT by Hartman and Heelis (2007) also identified the development of the most pronounced wave-4 pattern in vertical **E** × **B** drift in September.

The causal link of the longitudinal plasma density structures to vertical **E** × **B** drift is further evidenced by model simulations. Oh et al. (2008) produced the observed wave-4 pattern in plasma density by using the ROCSAT-1 vertical **E** × **B** drift data as an input to the SAMI2 (Sami2 is Another Model of the Ionosphere (Huba et al., 2000)) model. Fang et al. (2009) tested the connection of the annual and diurnal variations of the wave-4 plasma density structure with vertical **E** × **B** drift by carrying out the global ionosphere plasmasphere (GIP) model (Millward et al., 2001) simulations. Figure 26.4 shows the bimonthly wave-4 patterns extracted from the ROCSAT-1 vertical **E** × **B**

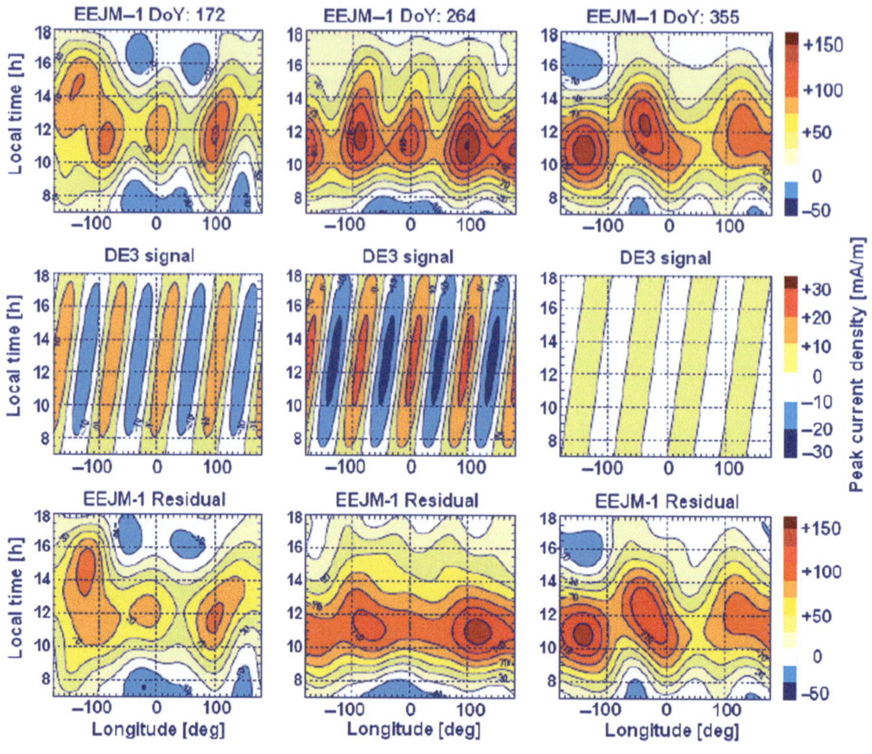

Fig. 26.3 Annual and local time variations of the EEJ in June (*left column*), September (*middle column*), and December (*right column*). The *top panels* show the peak EEJ current density. The *middle panels* show the wave-4 pattern extracted from the EEJ peak current density. The *bottom panels* are the residual of EEJ current density after removal of the wave-4 component (Lühr et al., 2008)

drift (top), GIP model plasma density (middle), and ROCSAT-1 plasma density (bottom). The ROCSAT-1 vertical **E** × **B** drift data were used as an input to the GIP model. The mean plasma densities between –20°N and 20°N magnetic latitude at 600-km altitude are used for the wave-4 component in the middle and bottom panels. Because the wave-4 pattern is periodic in longitude, one cycle of data between 60°E and 150°E longitude is presented. The wave-4 component in the vertical **E** × **B** drift shows maximum amplitude in July–October and eastward shift with local time. Ren et al. (2009b) obtained similar results from the analysis of the ROCSAT-1 vertical ion drift data. Similar annual and local time variations of the wave-4 component are identified in the GIP model (middle panel) and ROCSAT-1 (bottom panel) plasma densities, which provides strong evidence that the vertical **E** × **B** drift (or *E*-region dynamo electric field) during the daytime is responsible for the formation of the longitudinal wave-4 structure in plasma density.

26.4 Connection Between Ionosphere and Atmospheric Tides

Atmospheric tides are global-scale oscillations in atmosphere temperature, wind, and density. The main tidal source is daily absorption of solar radiation by water vapor and ozone in the troposphere and stratosphere (Lindzen, 1967). Tides propagate into the lower thermosphere and induce dynamo electric fields in the *E* region (Chapman and Lindzen, 1970). Tidal modes are distinguished by zonal wave number, oscillation frequency in cycles per day (diurnal, semi-diurnal, etc.), propagation in zonal direction (eastward or westward). A diurnal migrating (sun-synchronous) tide generates the dawn-to-dusk electric field in the *E* region (e.g., Kelley, 1989; Heelis, 2004). The dawn-to-dusk electric field drives the diurnal variation of the *F*-region plasma drift: upward drift during the day and downward drift at night (Fejer et al., 1991).

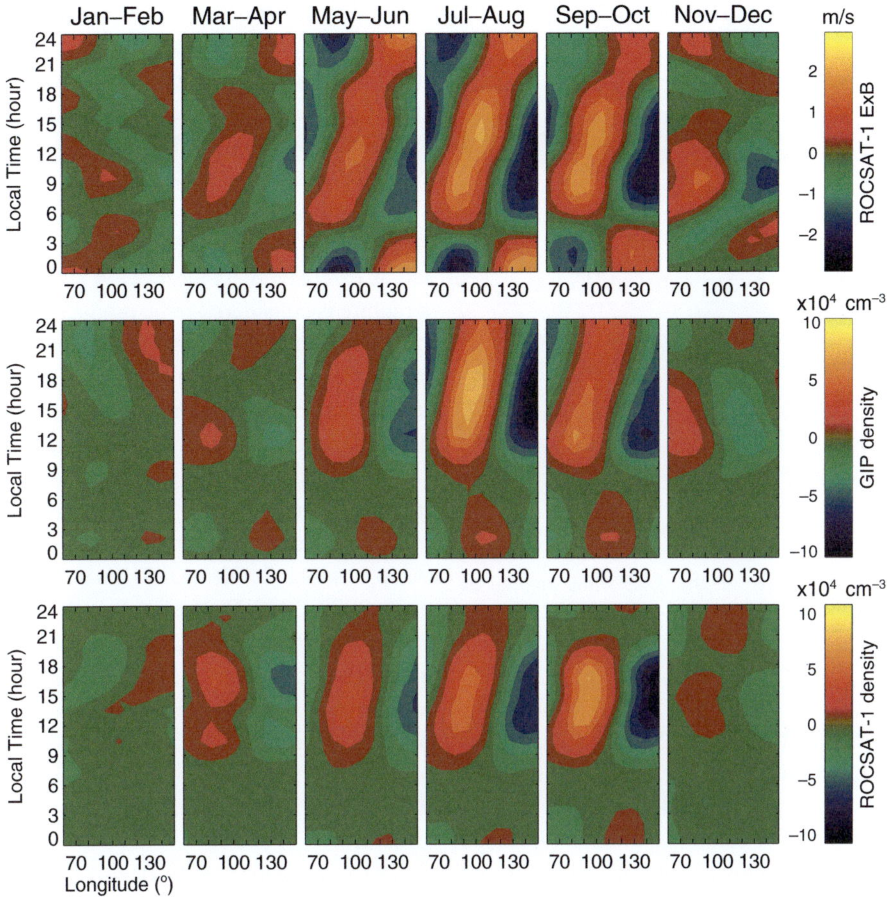

Fig. 26.4 Annual and local time variations of the wave-4 patterns in ROCSAT-1 vertical **E** × **B** (*top*), GIP model ionospheric density (*middle*), and ROCSAT-1 ionospheric density (*bottom*) (Fang et al., 2009)

Non-migrating tides are excited by longitudinal variations in water vapor or ozone due to land-sea differences or topography (Kato et al., 1982; Forbes and Groves, 1987; Kato, 1989; Forbes et al., 2003). Non-migrating tides modulate the dawn-to-dusk electric field driven by the diurnal migrating tide and can produce the longitudinal wave-like structure in plasma density or in the *E*-region dynamo electric field. DE3, driven by latent heat release associated with tropical convection in the troposphere (Hagan and Forbes, 2002), is the most likely source of the wave-4 longitudinal modulation of the dynamo electric field because its amplitude is pronounced in the dynamo region (*E* region), peaks near the equator, is symmetric in latitude, and is seasonally variable (Talaat and Lieberman, 1999; Forbes et al., 2006, 2008; Oberheide et al., 2006; Oberheide and Forbes, 2008a, b).

The annual variation of the amplitude of the DE3 has been investigated by using the following observation data: horizontal wind measurements from the High Resolution Doppler Imager (HRDI) and the Wind Imaging Interferometer (WINDII) onboard the Upper Atmosphere Research Satellite (UARS) (Talaat and Lieberman, 1999; Forbes et al., 2003) and from the TIMED Doppler Interferometer (TIDI) onboard the TIMED satellite (Oberheide et al., 2006; Oberheide and Forbes, 2008a; Ren et al., 2010); measurements of neutral temperature from the Sounding of the Atmosphere using Broadband Emission Radiometry (SABER) instrument onboard the TIMED satellite (Forbes et al., 2008, 2009; Kil et al., 2008; Oberheide and Forbes, 2008a; England et al., 2009); and neutral composition data derived from the observations of the TIMED and Student Nitric Oxide Explorer (SNOE)

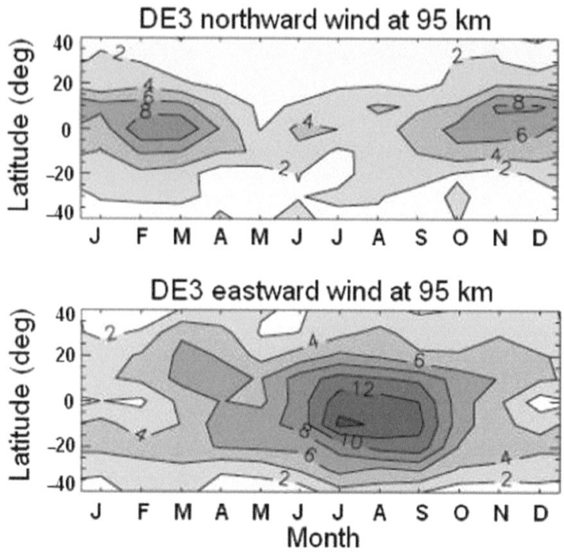

Fig. 26.5 Annual variations of the DE3-wind amplitudes observed by HRDI and WINDII (Forbes et al., 2003)

satellites (Oberheide and Forbes, 2008b). Figure 26.5 shows the northward (top) and the eastward (bottom) components of DE3 winds at 95-km altitude derived from HRDI and WINDII data. The annual variations of the two wind components are largely different; the amplitude peaks occur near March and November for the northward wind and during July–September for eastward wind. Similar annual variations of the meridional and zonal components of the DE3 winds are observed by TIMED/TIDI (Oberheide et al., 2006; Oberheide and Forbes, 2008a).

In addition to the annual variation of the DE3 amplitude, the diurnal variation of the DE3 phase is a useful indicator of the connection between wave-4 plasma density structure and DE3. Figure 26.6 shows residual temperature at 110-km altitude in the equatorial region extracted from the TIMED/SABER data. TIMED precesses at a rate such that the local time of the ascending node covers all local times every 120 days, and the local time (given on the right axis) varies with day of year. The development of the wave-4 pattern and its temporal evolution are dominated by the DE3 component. This is determined based on the fact that as local time progresses (downward) the wave-4 pattern shifts eastward at a rate of ~90 deg/24 h LT which is, by definition, the phase shift of a wave-4 structure associated with DE3. The annual and local time variations in the DE3 component identified from the measurements of winds and temperature agree with the observations of the wave-4 patterns in plasma density, EEJ, and vertical **E** × **B** drift. This is why the ionospheric wave-4 structure is attributed to DE3 and not, for example, stationary planetary wave-4 ($s = 4, n = 0, |s - n| = 4$) that shows no phase shift.

Although the wave-3 structure has not been studied as extensively as the wave-4 structure, the wave-3 structure is considered to be produced by a similar process: modulation of *E*-region dynamo electric field by atmospheric tides. But the wave-3 and wave-4 longitudinal structures of the ionosphere are produced by different tides. The wave-3 structures in plasma density, EEJ, and vertical **E** × **B** drift are most pronounced in December– January. The observation of the peak amplitude of the diurnal eastward propagating zonal wave number 2 tide (DE2) near December (Forbes et al., 2008, 2009; Pedatella et al., 2008; England et al., 2009) indicates that the DE2 is the likely driving mechanism behind the wave-3 longitudinal structures of the ionosphere. Further supporting evidence of the coupling of the ionospheric wave-3 longitudinal structure to the DE2 may be found by investigating the local time evolution of the phase in the ionospheric wave-3 structure.

The investigation of the day-to-day variability of the longitudinal wave structures in the ionosphere is challenging because a high temporal (a few-hour interval) and spatial (about 10° longitudinal interval) resolution data set is required on a global scale on a daily basis. Immel et al. (2009) reported that the ionospheric wave-4 structure varies over a 30-day period and has a 6-day periodicity from the analysis of the TIMED/GUVI data during the 2002 spring equinox. Because the ionospheric plasma density from GUVI is available at night and the TIMED orbit precesses, the authors extracted the periodicity under the assumption that the daytime ionospheric wave-4 structure is persistent throughout the night. Liu et al. (2010) reported a 5-day periodicity of the wave-4 structure in the *F*-peak height from the analysis of the COSMIC data during August-September 2008. Liu et al. (2010) attributed the 5-day periodicity to the interaction of DE3 with a 5-day planetary wave. These recent studies indicate that some part of the short-term variability of the ionosphere can be understood in terms of atmospheric tides. It remains an open question, whether this behavior is observed at other times.

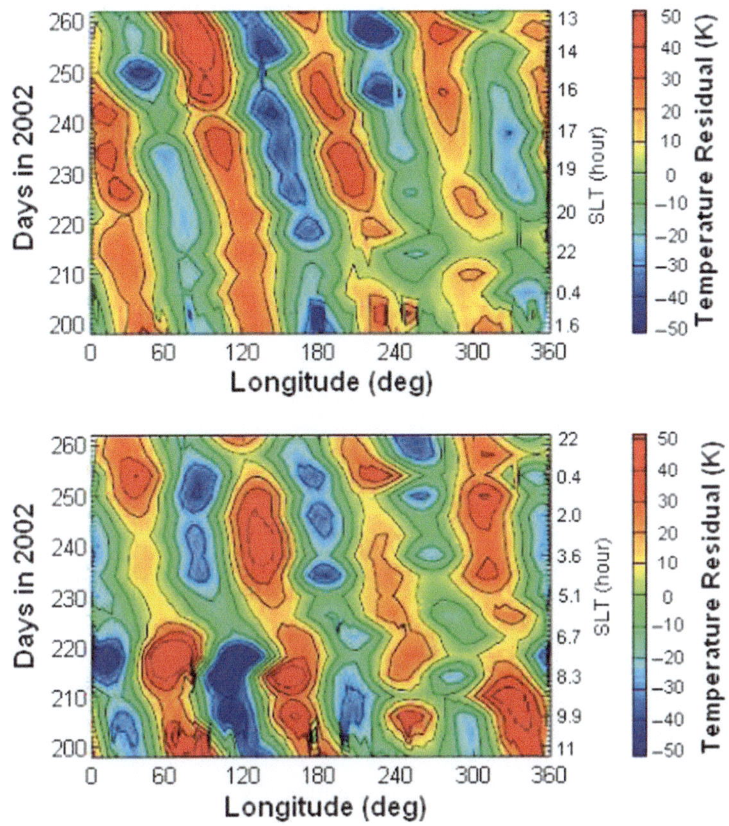

Fig. 26.6 Temperature residuals from 5-day means at 110-km altitude in the equatorial region for the ascending (*top*) and descending (*bottom*) portions of the orbit (Forbes et al., 2006)

26.5 Generation of Dynamo Electric Fields by Tidal Winds

In this section, we explain how dynamo electric fields in the *E* region are generated by tidal winds. Dynamo electric fields in the *E* region are mostly inferred from the measurements of plasma motion in the *F* region. In the *F* region, the gyration frequencies of ions and electrons are much larger than their collision frequencies with neutral particles. Under these conditions, the plasma (ions and electrons together) moves approximately in the **E** × **B** direction, perpendicular to the applied electric field (**E**) and magnetic field (**B**), with the magnitude of *E*/*B*. Because the electrical conductivity in the *E* region is much larger than that in the *F* region during the daytime, electric fields in the *E* region control the plasma motion in the *F* region and can be deduced from the measurement of plasma motions in the *F* region (Kelley, 1989).

Electric fields in the low-latitude ionosphere are driven by collisions of the plasma with neutral particles (neutral wind dynamo) (Kelley, 1989; Heelis, 2004). Ions and electrons move in different directions after collision because of the difference of the gyrofrequency to collision-frequency ratios between ions and electrons. Then electric currents are generated. Any divergence in currents induced by temporal and spatial variations of wind and/or conductivity creates electric fields. The Earth's ionosphere is a good electrical conductor. That is, charge neutrality is maintained on a time scale of $\tau \approx \varepsilon_o/\sigma <\sim 10^{-6}$s ($\varepsilon_o$: permittivity in free space; σ: electrical conductivity). The time variation of the total charge density (ρ_c) can be

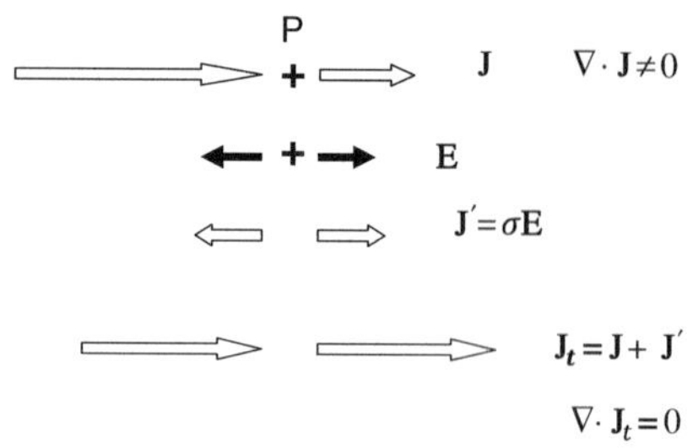

Fig. 26.7 Illustration of the generation of electric field in the ionosphere by the divergence in current density

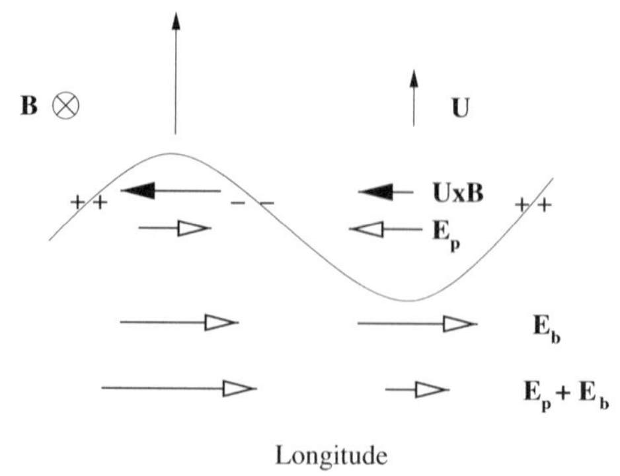

Fig. 26.8 Schematic diagram for the generation of the perturbation electric field by meridional winds (Kil et al., 2007)

ignored, on a time scale > ~10^{-6} s, in the following charge conservation equation:

$$\frac{\partial \rho_c}{\partial t} + \nabla \cdot \mathbf{J} = 0 \qquad (26.1)$$

The electric field in the low-latitude ionosphere is deduced from the requirement $\nabla \cdot \mathbf{J} = 0$. Figure 26.7 schematically explains this process. The arrows in the top indicate that the current density (**J**) is not divergence free ($\nabla \cdot \mathbf{J} \neq 0$). The gradient in current density accumulates positive charges at point *P* and induces an electric field (**E**). The currents driven by the electric field decrease and increase the current density on the left and right sides of *P*, respectively, so that the total current density ($\mathrm{J_t} = \mathrm{J} + \sigma\,\mathrm{E}$) is divergence free ($\nabla \cdot \mathrm{J_t} = 0$). In a simplified model, we ignore the field-aligned currents (or *F* region conductivity) because the conductivity in the *E* region is much greater than that in the *F* region during daytime.

The wind-driven current density in the direction perpendicular to the magnetic field is expressed as follows (Heelis, 2004):

$$\begin{aligned} \mathrm{J} &= \sigma_{\mathrm{P}}\mathrm{U} \times \mathrm{B} + \sigma_{\mathrm{H}}\mathrm{B} \times (\mathrm{U} \times \mathrm{B})/B \\ &= \sigma_{\mathrm{P}}\mathrm{U} \times \mathrm{B} + \sigma_{\mathrm{H}} B \mathrm{U}_{\perp} \end{aligned} \qquad (26.2)$$

Here **U** is the neutral wind velocity, σ_{P} is the Pedersen conductivity, and σ_{H} is the Hall conductivity. $\mathrm{U}_{\perp}$ is the component of **U** perpendicular to the magnetic field. The wind velocity and conductivity vary spatially and temporarily and, in general, $\nabla \cdot \mathbf{J} \neq 0$; an electric field must build up to make $\nabla \cdot \mathbf{J} = 0$. Kil et al. (2007) explained the longitudinal modulation of the *E*-region dynamo electric field by the longitudinal variation of the meridional wind. The diagram in Figure 26.8 shows a snapshot of the daytime *E* region at northern low-middle latitudes. The sinusoidal curve shows the modulation of the northward wind (**U**) by tides and the vertical arrows indicate the wind amplitudes. The magnetic field is directed into the paper. The northward wind drives the Pedersen current in the U × B direction (westward or dusk-to-dawn) as indicated by filled horizontal arrows. The current density is proportional to the wind velocity and positive and negative charges pile up as shown. The charge accumulation induces the opposite polarization electric fields ($\mathbf{E}_{\mathrm{p}}$) at the crests and troughs of the wind field. Addition of $\mathbf{E}_{\mathrm{p}}$ to the uniform background eastward (dawn-to-dusk) electric field ($\mathbf{E}_{\mathrm{b}}$) results in the longitudinally modulated net electric field $\mathbf{E}_{\mathrm{p}} + \mathbf{E}_{\mathrm{b}}$.

The mechanism described above may not play a major role in the creation of the ionospheric wave-4 structures because, as we can infer from the annual variations, the ionospheric wave-4 structures are more closely connected to DE3 zonal wind than to DE3 meridional wind (see Fig. 26.5). Jin et al. (2008) explained the modulation of the *E*-region dynamo electric field in terms of the DE3 zonal wind. The diagram in Fig. 26.9 schematically explains this process. The sinusoidal curve indicates the modulation of the tidal winds in the zonal direction in the *E* region. As described in Eq. (2), Hall currents ($\Delta\, \mathrm{J^{dyn}}$) flow in the wind direction ($\Delta\, u_n$). Positive and negative charges accumulate in the regions where the tidal wind converges and diverges, respectively. The polarization electric fields ($\Delta\, \mathbf{E}$) develop and longitudinally modulate the background electric field ($\mathbf{E}_{\mathrm{o}}$).

Model simulations further support the causal link between DE3 zonal wind and wave-4 density structure. Hagan et al. (2007) used the GSWM tidal winds

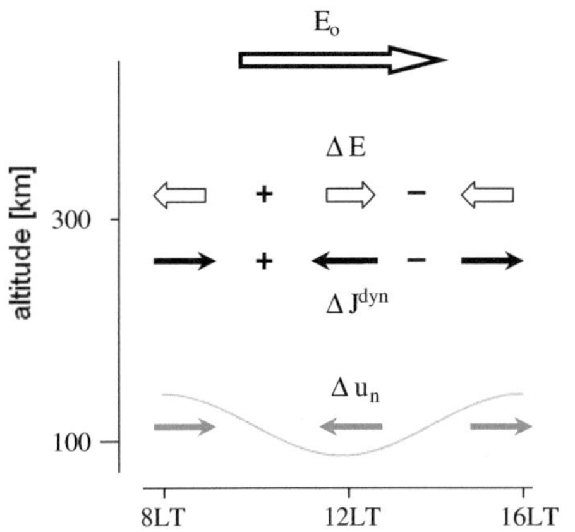

Fig. 26.9 Schematic diagram for the generation of the perturbation electric field by zonal winds (Jin et al., 2008)

as inputs to the thermosphere-ionosphere-mesosphere-electrodynamics general circulation model (TIME-GCM) and identified the DE3 zonal wind as the cause of the formation of wave-4 density structure. Using their own electrodynamics model, Jin et al. (2008) demonstrated that the wave-4 density structure and its eastward shift with local time arose from the effects of DE3 zonal wind. Ren et al. (2010) carried out the Theoretical Ionospheric Dynamo Model, Institute of Geology and Geophysics, Chinese Academy of Sciences, version II (TIDM–IGGCAS–II) model simulations by using the TIMED/TIDI tidal winds as inputs and showed that the symmetric component of the DE3 zonal wind is responsible for the ionospheric wave-4 structure.

26.6 Conclusions

The longitudinal structure of the plasma density in the low-latitude *F* region has been extensively investigated during the last 5 years by using the IMAGE, TIMED/GUVI, ROCSAT-1, FORMOSAT-3/COSMIC, CHAMP, TOPEX, DMSP, and GPS TEC data. The wave-4 structure in plasma density is dominant in July–September, and the wave-3 structure is dominant in December–January. The annual and local time variations of the ionospheric wave-4 pattern and atmospheric tides demonstrate the causal link between plasma density, daytime vertical $\mathbf{E} \times \mathbf{B}$ drift (or dynamo electric field in the *E* region), and DE3. The annual variation of the amplitudes of the ionospheric wave-3 pattern and DE2 indicates their association. Further supporting evidence may be found by investigating the local time variation of the phase of the ionospheric wave-3 pattern.

The effect of atmospheric tides on the ionosphere has been focused on the modulation of the *E*-region dynamo electric fields, but the modulation of the ionosphere and thermosphere by the direct penetration of atmospheric tides should be taken into account. The observations of the longitudinal modulation of the thermosphere (neutral wind and total neutral mass density) at *F*-region heights is interpreted as evidence for the direct penetration of atmospheric tides (Lühr et al., 2007; Forbes et al., 2009; Häusler and Lühr, 2009; Liu et al., 2009). Talaat and Lieberman (2010) showed the propagation of DE3 winds to the *F*-region height (250 km) from the measurements of neutral winds by UARS/WINDII. An important unanswered question is whether the thermospheric longitudinal structure is produced by the direct penetration of atmospheric tides (Forbes et al., 2009; Liu et al., 2009) or by the pre-existing ionospheric plasma density structure. Model simulations will provide insight into the role of the penetration of atmospheric tides and ion drag in the formation of thermospheric longitudinal structures.

Acknowledgements H. Kil and L. J. Paxton acknowledge support from NASA Grant NNX08AQ12G and NASA TIMED Program GUVI Grant NAG5-11412.

References

Alken P, Maus S (2007) Spatio-temporal characterization of the equatorial electrojet from CHAMP, Ørsted, and SAC-C satellite measurements. J Geophys Res 112:A09305. doi:10.1029/2007JA012524

Benkova NP, Deminov MG, Karpachev AT, Kochenova NA, Kusnerevsky YuV, Migulin VV, Pulinets SA, Fligel MD (1990) Longitude features shown by topside sounder data and their importance in ionospheric mapping. Adv Space Res 10:857–866

Chapman S, Lindzen RS (1970) Atmospheric tides. Gordon and Breach, New York, NY

England SL, Immel TJ, Sagawa E, Henderson SB, Hagan ME, Mende SB, Frey HU, Swenson CM, Paxton LJ (2006a) The effect of atmospheric tides on the morphology of the quiet-time post-sunset equatorial ionospheric anomaly. J Geophys Res 111:A10S19. doi:10.1029/2006JA011795

England SL, Maus S, Immel TJ, Mende SB (2006b) Longitudinal variation of the *E*-region electric fields caused by atmospheric tides. Geophys Res Lett 33:L21105. doi:10.1029/2006GL027465

England SL, Zhang X, Immel TJ, Forbes JM, DeMajistre R (2009) The effect of non-migrating tides on the morphology of the equatorial ionospheric anomaly: seasonal variability. Earth Planets Space 61:493–503

Fang T-W, Kil H, Millward G, Richmond AD, Liu JY, Oh S-J (2009) Causal link of the wave-4 structures in plasma density and vertical plasma drift in the low-latitude ionosphere. J Geophys Res 114:A10315. doi:10.1029/2009JA014460

Fejer BG, de Paula ER, Gonzalez SA, Woodman RF (1991) Average vertical and zonal *F* region plasma drifts over Jicamarca. J Geophys Res 96;13,901–13,906

Fejer BG, Jensen JW, Su S-Y (2008) Quiet time equatorial *F* region vertical plasma drift model derived from ROCSAT-1 observations. J Geophys Res 113:A05304. doi:10.1029/2007JA012801

Forbes JM, Bruinsma SL, Zhang X, Oberheide J (2009) Surface-exosphere coupling due to thermal tides. Geophys Res Lett 36:L15812. doi:10.1029/2009GL038748

Forbes JM, Groves GV (1987) Diurnal propagating tides in the low-latitude middle atmosphere. J Atmos Solar-Terr Phys 49:153–164

Forbes JM, Russell MJ, Miyahara S, Zhang X, Palo S, Mlynczak M, Mertens CJ (2006) Troposphere-thermosphere tidal coupling as measured by the SABER instrument on TIMED during July–September 2002. J Geophys Res 111:A10S06. doi:10.1029/2005JA011492

Forbes JM, Zhang X, Palo S, Russell J, Mertens CJ, Mlynczak M (2008) Tidal variability in the ionospheric dynamo region. J Geophys Res 113:A02310. doi:10.1029/2007JA012737

Forbes JM, Zhang X, Talaat ER, Ward W (2003) Nonmigrating diurnal tides in the thermosphere. J Geophys Res 108(A1):1033. doi:10.1029/2002JA009262

Hagan ME, Forbes JM (2002) Migrating and nonmigrating diurnal tides in the middle and upper atmosphere excited by tropospheric latent heat release. J Geophys Res 107(D24):4754. doi:10.1029/2001JD001236

Hagan ME, Maute A, Roble RG, Richmond AD, Immel TJ, England SL (2007) Connections between deep tropical clouds and the Earth's ionosphere. Geophys Res Lett 34:L20109. doi:10.1029/2007GL030142

Hartman WA, Heelis RA (2007) Longitudinal variations in the equatorial vertical drift in the topside ionosphere. J Geophys Res 112:A03305. doi:10.1029/2006JA011773

Häusler K, Lühr H (2009) Nonmigrating tidal signals in the upper thermospheric zonal wind at equatorial latitudes as observed by CHAMP. Ann Geophys 27:2643–2652

Heelis RA (2004) Electrodynamics in the low and middle latitude ionosphere: a tutorial. J Atmos Solar-Terr Phys 66:825–838

Henderson SB, Swenson CM, Christensen AB, Paxton LJ (2005) Morphology of the equatorial anomaly and equatorial plasma bubbles using image subspace analysis of Global Ultraviolet Imager data. J Geophys Res 110:A11306. doi:10.1029/2005JA011080

Huba JD, Joyce DG, Fedder JA (2000) Sami2 is another model of the ionosphere (SAMI2): a new low-latitude ionosphere model. J Geophys Res 105:23,035–23,053

Immel TJ, England S, Zhang X, Forbes JM, DeMajistre R (2009) Upward propagating tidal effects across the *E*- and *F*-regions of the ionosphere. Earth Planets Space 61:505–512

Immel TJ, Sagawa E, England SL, Henderson SB, Hagan ME, Mende SB, Frey HU, Swenson CM, Paxton LJ (2006) Control of equatorial ionospheric morphology by atmospheric tides. Geophys Res Lett 33:L15108. doi:10.1029/2006GL026161

Ivers D, Stening R, Turner J, Winch D (2003) Equatorial electrojet from Ørsted scalar magnetic field observations. J Geophys Res 108(A2):1061. doi:10.1029/2002JA009310

Jadhav G, Rajaram M, Rajaram R (2002) A detailed study of equatorial electrojet phenomenon using Ørsted satellite observations. J Geophys Res 107(A8):1175. doi:10.1029/2001JA000183

Jin H, Miyoshi Y, Fujiwara H, Shinagawa H (2008) Electrodynamics of the formation of ionospheric wave number 4 longitudinal structure. J Geophys Res 113:A09307. doi:10.1029/2008JA013301

Karpachev AT (1988) Characteristics of the global longitudinal effect in the night-time equatorial anomaly. Geomagn Aeronomy 28(1):46–49

Kato S (1989) Non-migrating tides. J Atmos Solar-Terr Phys 51:673–682

Kato S, Tsuda T, Watanabe F (1982) Thermal excitation of nonmigrating tides. J Atmos Solar-Terr Phys 44:131–146

Kelley MC (1989) The Earth's ionosphere. Academic, San Diego, CA

Kil H, DeMajistre R, Paxton LJ, Zhang Y (2006) Nighttime F-region morphology in the low and middle latitudes seen from DMSP F15 and TIMED/GUVI. J Atmos Solar-Terr Phys 68:1672–1681

Kil H, Oh S-J, Kelley MC, Paxton LJ, England SL, Talaat ER, Min K-W, Su S-Y (2007) Longitudinal structure of the vertical $\mathbf{E} \times \mathbf{B}$ drift and ion density seen from ROCSAT-1. Geophys Res Lett 34:L14110. doi:10.1029/2007GL030018

Kil H, Oh S-J, Paxton LJ, Fang T-W (2009b) High-resolution vertical drift model driven from the ROCSAT-1 data. J Geophys Res 114:A10. doi:10.1029/2009JA014324

Kil H, Paxton LJ, Oh S-J (2009a) Global bubble distribution seen from ROCSAT-1 and its association with the pre-reversal enhancement. J Geophys Res 114:A06307. doi:10.1029/2008JA013672

Kil H, Talaat ER, Oh S-J, Paxton LJ, England SL, Su S-Y (2008) The wave structures of the plasma density and vertical $\mathbf{E} \times \mathbf{B}$ drift in low-latitude F region. J Geophys Res 113:A09312. doi:10.1029/2008JA013106

Kochenova NA (1987) Longitudinal variations of the equatorial ionosphere according to Intercosmos-19 data. Geomagn Aeronomy 21(1):142–144

Kochenova NA (1988) Longitudinal variations of N(h) profiles at the magnetic equator. Geomagn Aeronomy 28(1): 144–146

Lin CH, Wang W, Hagan ME, Hsiao CC, Immel TJ, Hsu ML, Liu JY, Paxton LJ, Fang TW, Liu CH (2007a) Plausible effect of atmospheric tides on the equatorial ionosphere observed by the FORMOSAT-3/COSMIC: three-dimensional electron density structures. Geophys Res Lett 34:L11112. doi:10.1029/2007GL029265

Lin CH, Hsiao CC, Liu JY, Liu CH (2007b) Longitudinal structure of the equatorial ionosphere: time evolution of the four-peaked EIA structure. J Geophys Res 112:A12305. doi:10.1029/2007JA012455

Lindzen RS (1967) Thermally driven diurnal tide in the atmosphere. Q J R Met Soc 93:18–42

Liu G, Immel TJ, England SL, Kumar KK, Ramkumar R (2010) Temporal modulations of the longitudinal structure in *F*2 peak height in the equatorial ionosphere as observed by COSMIC. J Geophys Res 115:A04303. doi:10.1029/2009JA014829

Liu H, Watanabe S (2008) Seasonal variation of the longitudinal structure of the equatorial ionosphere: does it reflect tidal influences from below? J Geophys Res 113:A08315. doi:10.1029/2008JA013027

Liu H, Yamamoto M, Lühr H (2009) Wave-4 pattern of the equatorial mass density anomaly: a thermospheric signature of tropical deep convection. Geophys Res Lett 36:L18104. doi:10.1029/2009GL039865

Lühr H, Häusler K, Stolle C (2007) Longitudinal variation of *F* region electron density and thermospheric zonal wind caused by atmospheric tides. Geophys Res Lett 34:L16102. doi:10.1029/2007GL030639

Lühr H, Rother M, Häusler K, Alken P, Maus S (2008) Influence of nonmigrating tides on the longitudinal variation of the equatorial electrojet. J Geophys Res 113:A08313. doi:10.1029/2008JA013064

Millward GH, Muller-Wodarg ICF, Aylward AD, Fuller-Rowell TJ, Richmond AD, Moffett RJ (2001) An investigation into the influence of tidal forcing on *F*-region equatorial vertical ion drift using a global ionosphere-thermosphere model with coupled electrodynamics. J Geophys Res 106:24, 733–24,744

Oberheide J, Forbes JM (2008a) Tidal propagation of deep tropical cloud signatures into the thermosphere. Geophys Res Lett 35:L04816. doi:10.1029/2007GL032397

Oberheide J, Forbes JM (2008b) Thermospheric nitric oxide variability induced by nonmigrating tides. Geophys Res Lett 35:L16814. doi:10.1029/2008GL034825

Oberheide J, Wu Q, Killeen TL, Hagan ME, Roble RG (2006) Diurnal nonmigrating tides from TIMED Doppler Interferometer wind data: monthly climatologies and seasonal variations. J Geophys Res 111:A10S03. doi:10.1029/2005JA011491

Oh S-J, Kil H, Kim W-T, Paxton LJ, Kim YH (2008) The role of the vertical **E** × **B** drift for the formation of the longitudinal plasma density structure in the low-latitude *F* region. Ann Geophys 26:2061–2067

Pedatella NM, Forbes JM, Oberheide J (2008) Intra-annual variability of the low-latitude ionosphere due to nonmigrating tides. Geophys Res Lett 35:L18104. doi:10.1029/2008GL035332

Ren Z, Wan W, Liu L, Heelis RA, Zhao B, Wei Y, Yue X (2009a) Influences of geomagnetic fields on longitudinal variations of vertical plasma drifts in the presunset equatorial topside ionosphere. J Geophys Res 114:A03305. doi:10.1029/2008JA013675

Ren Z, Wan W, Liu L, Xiong J (2009b) Intra-annual variation of wavenumber-4 structure of vertical **E** × **B** drifts in the equatorial ionosphere seen from ROCSAT-1. J Geophys Res 114:A5. doi:10.1029/2009JA014060

Ren Z, Wan W, Liu L, Zhao B, Wei Y, Yue X, Heelis RA (2008) Longitudinal variations of electron temperature and total ion density in the sunset equatorial topside ionosphere. Geophys Res Lett 35:L05108. doi:10.1029/2007GL032998

Ren Z, Wan W, Xiong J, Liu L (2010) Simulated wave number 4 structure in equatorial *F*-region vertical plasma drifts. J Geophys Res 115:A05301. doi:10.1029/2009JA014746

Sagawa E, Immel TJ, Frey HU, Mende SB (2005) Longitudinal structure of the equatorial anomaly in the nighttime ionosphere observed by IMAGE/FUV. J Geophys Res 110:A11302. doi:10.1029/2004JA010848

Scherliess L, Thompson DC, Schunk RW (2008) Longitudinal variability of low-latitude total electron content: tidal influences. J Geophys Res 113:A01311. doi:10.1029/2007JA012480

Su YZ, Bailey GJ, Oyama KI, Balan N (1997) A modeling study of the longitudinal variations in the north-south asymmetries of the ionospheric equatorial anomaly. J Atmos Solar-Terr Phys 59:1299–1310

Su S-Y, Chao CK, Liu CH (2008) On monthly/seasonal/longitudinal variations of equatorial irregularity occurrences and their relationship with the post-sunset vertical drift velocities. J Geophys Res 113:A05307. doi:10.1029/2007JA012809

Talaat ER, Lieberman RS (1999) Nonmigrating diurnal tides in the mesosphere and lower thermosphere. J Atmos Sci 56:4073–4087

Talaat ER, Lieberman RS (2010) Direct observations of nonmigrating diurnal tides in the equatorial thermosphere. Geophys Res Lett 37:L04803. doi:10.1029/2009GL041845

Thuillier G, Wiens JRH, Shepered GG, Roble RG (2002) Photochemistry and dynamics in thermospheric intertropical arcs measure by the WIND imaging interferometer on board UARS: a comparison with TIE-GCM simulations. J Atmos Solar-Terr Phys 64:405–415

Wan W, Liu L, Pi X, Zhang M-L, Ning B, Xiong J, Ding F (2008) Wavenumber-4 patterns of the total electron content over the low latitude ionosphere. Geophys Res Lett 35:L12104. doi:10.1029/2008GL033755

West KH, Heelis RA (1996) Longitude variations in ion composition in the morning and evening topside equatorial ionosphere near solar minimum. J Geophys Res 101:7951–7960

Chapter 27

Longitudinal Structure of the Mid- and Low-Latitude Ionosphere Observed by Space-borne GPS Receivers

C.H. Lin, C.H. Chen, H.F. Tsai, C.H. Liu, J.Y. Liu, and Y. Kakinami

Abstract This study presents longitudinal structures of the mid- and low-latitude ionosphere using the GPS radio occultation observation on board the COSMIC satellite mission. The longitudinal structure seen in the equatorial and low-latitude ionospheric regions results from modification of the daily dynamo electric field by upward propagating atmospheric tides that are generated by latent heat release of the tropical rainstorms. Changes of the dynamo electric field modify the equatorial plasma fountain and thereby enhance the equatorial ionization anomaly (EIA). With capability of three-dimensional global ionospheric observation, altitudinal, local time, and monthly variations of this recent discovered fascinating feature are obtained for further understanding of the underlying physical mechanism. Through comparison between electron densities at various altitudes, the longitudinal structure is prominently seen at upper part of the ionosphere. Additionally, COSMIC observations provide three-dimensional structure of the Weddell Sea anomaly which is featured by the greater nighttime electron density than daytime. Not only occurring at the southern hemisphere near the Weddell Sea region of the Antarctica, a similar nighttime density enhancement feature is also found in the northern hemisphere during local summer by COSMIC observations. The anomalous signatures in both hemispheres share very similar characteristics in electron density structure, latitudinal distribution, and appearance time. They are, therefore, categorized as the mid-latitude summer nighttime anomaly (MSNA).

C.H. Lin (✉)
Department of Earth Science, National Cheng Kung University, Tainan, Taiwan; Earth Dynamic System Research Center, National Cheng Kung University, Tainan, Taiwan
e-mail: charles@mail.ncku.edu.tw

27.1 Introduction

To study the longitudinal variation of the ionosphere, it is desirable to have global observations over oceans and continents of limited ground-based observation. The radio occultation (RO) experiment (cf. Hajj et al., 2000; Schreiner et al., 1999, 2007) of the Constellation Observing System for Meteorology, Ionosphere, and Climate (or COSMIC) mission has served the purpose by providing a great opportunity to monitor the ionospheric variations climatologically after binning the monthly average of the global electron densities (cf. Lei et al., 2007; Lin et al., 2007a). The constellation was launched into a circular low-Earth orbit at 01:40 UTC on 15 April 2006 and has collected more than a million ionospheric electron density profiles around globe since then. The six microsatellites are close to each other at the initial parking orbit at a 72° inclination angle orbital plane with an altitude of around 500 km after launch. The constellation has now been deployed to its final mission orbit, having 30° separation in longitude between each of microsatellites at around 750–850 km altitude.

In this chapter, observations in 2007 are used to construct monthly ionospheric maps for studying the longitudinal structures. Since some satellites were still in the orbital lifting process and each of the six micro-satellites locates at different altitude in 2007, the ionospheric maps presented here are constructed up to 500 km altitude only. Additionally, numerical simulation of the radio occultation inversion has indicated

M.A. Abdu, D. Pancheva (eds.), A. Bhattacharyya (Coed.), *Aeronomy of the Earth's Atmosphere and Ionosphere*,
IAGA Special Sopron Book Series 2, DOI 10.1007/978-94-007-0326-1_27, © Springer Science+Business Media B.V. 2011

that the spherical symmetry assumption of the ionosphere made in the RO inversion process would result in artificial electron density structure in the E-region below 200 km altitude (Liu et al., 2010; Yue et al., 2010). To avoid such RO inversion error to contaminate the ionospheric maps, electron densities below 200 km altitude are truncated before data binning. The three-dimensional electron density maps are constructed by averaging observations during around 40 magnetically quiescent days, i.e. 5 days prior to and 5 days after each month, same as the process described in Lin et al. (2007a). The maps are binned in the format of global constant local time in order to study the longitudinal variations of the ionosphere. The advantage of the COSMIC observation is that the global electron density map can be constructed by binning 30∼40-day data, thereby suitable to study the monthly variations of the ionosphere.

Two types of the longitudinal variations are demonstrated by the radio occultation observations. One of which is the longitudinal variation of the equatorial ionosphere resulting from the upward propagating atmospheric tide. The other is the recently defined mid-latitude summer nighttime anomaly (MSNA). Both structures have drawn attentions and been studied recently by using various ground-based and space-borne observations. This chapter demonstrates major features of the two longitudinal structures based on three-dimensional map constructed by COSMIC observations in the following sections. The altitude-latitude information of the electron densities provided by the COSMIC is helpful for understanding their underlying physics.

27.2 Atmospheric Tidal Effect to the Equatorial Ionosphere

Recent satellite observations of the global ionosphere have shown that the low-latitude and equatorial ionosphere reveal a unique longitudinal structure produced by the atmospheric tides when viewing it in a global constant local time aspect (Sagawa et al., 2005; Immel et al., 2006; England et al., 2006; Lin et al., 2007b; Hagan et al., 2007; Lühr et al., 2008). The latent heating due to deep tropical convection is considered an important source for most of the atmospheric tides that result in longitudinal variability into the upper atmosphere and low and equatorial regions (Oberheide et al., 2002; Forbes et al., 2006, 2008; Oberheide and Forbes, 2008). Modulations of the neutral winds affect the E-region ionospheric dynamo and subsequently enhance the equatorial plasma fountain leading to the stronger equatorial ionization anomaly (EIA).

The most prominent feature of the longitudinal variation is the wave-4 structure of the EIA crests seen during equinoctial seasons when viewing the electron density map in the global constant local time aspect. Figure 27.1 shows the electron density and total electron content (TEC) maps, constructed from COSMIC observations, at global constant local time of 1600 LT during September 2007. The wave-4 signature is identified in electron density maps at 300, 350 and 400 km altitudes as well as the TEC integrated from 100 to 500 km altitude. The four electron density/TEC enhancement regions are consistent with airglow observation of the wave-4 structure (Sagawa et al., 2005; Immel et al., 2006) located around South America (–108°E to –52°E geographic longitude), Central Africa (–37°E to 45°E), Southeast Asia (70°E to 135°E) and Central Pacific regions (167°E to –135°E). The latitude-altitude slices of the electron density indicate expanded and contracted EIA crests. The wave-4 structure of the EIA may result from the diurnal eastward propagation of wave-3 (DE3) nonmigrating tide or other modes of the nonmigrating tide, such as semidiurnal eastward wave-2 (SE2) and diurnal westward wave-5 (DW5) tides (see Forbes et al., 2008 for details). Among these nonmigrating tides, the DE3 component has larger amplitude than other tidal modes (Forbes et al., 2008).

From the latitude-longitude slices shown in Fig. 27.1, the wave-4 longitudinal structure of the EIA becomes more discernible as altitude increase. The structure is better identified at 350 km than that at 300 km altitude. It is due to that the electron density at the EIA region is not only affected by the strength of the equatorial plasma fountain effect but also the meridional neutral wind and the longitudinal effect due to offset of the magnetic and geographic equator. Since the plasma fountain can all reach the F-layer height during most of daytime hours, the resulting EIA strengths are not well distinguished at F-layer height during daytime. These combined effects make it difficult to identify the wave-4 structure from the NmF_2 or electron density around F-layer height of

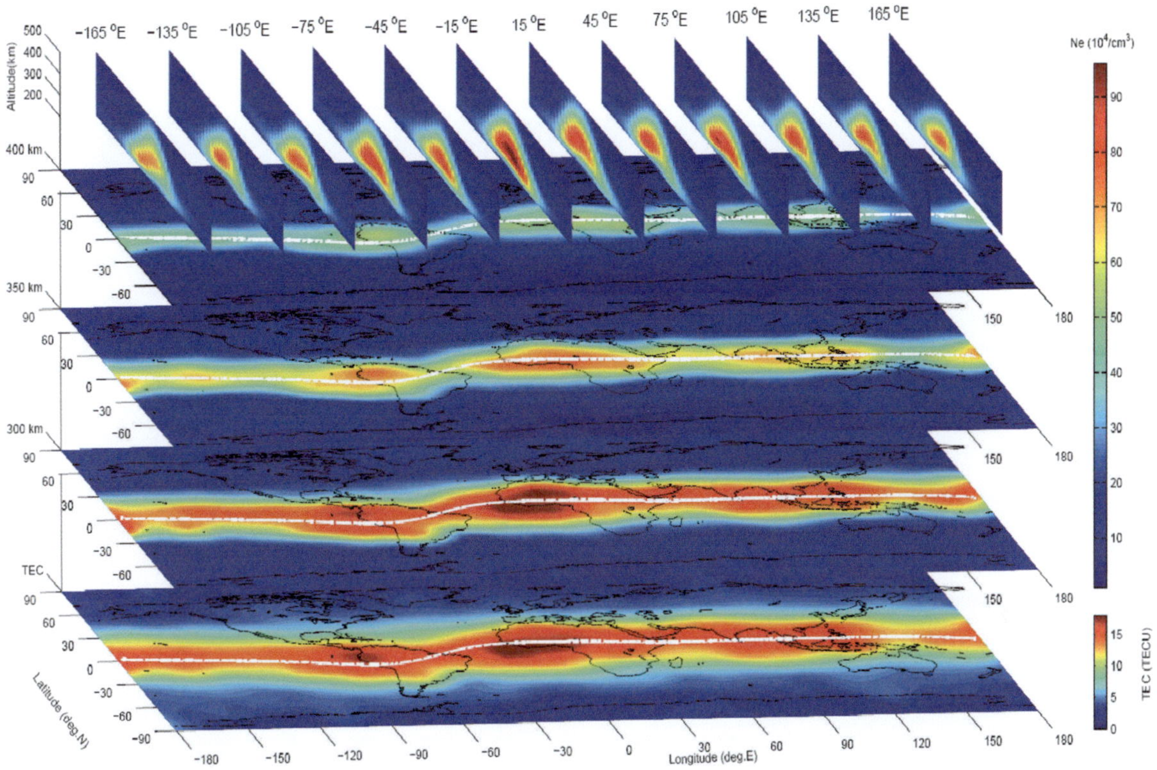

Fig. 27.1 Longitudinal wave-4 structure of the equatorial ionosphere observed by the FORMOSAT-3/COSMIC in the three-dimensional global constant local time map at 1600 LT of September 2007

300 km in every local time. At higher altitude, on the other hand, the effect of enhanced plasma fountain is better identified. A good example of this effect can be found by comparing the altitude-latitude electron density slices at 15°E and 45°E longitude (Fig. 27.1). The EIAs at 15°E and 45°E are located at stronger and weaker EIA regions, respectively. Although the EIA feature at 15°E is greater than that at 45°E, the F_2-peak density in the northern hemisphere of the EIA at 45°E is slightly greater than that at 15°E. On the other hand, the electron densities above 400 km altitude show stronger magnitude at 15°E than at 45°E clearly. The comparison suggests that the strength of the plasma fountain and the EIA may not simply be represented by magnitude of NmF_2 or electron density around F-peak. Thus, it is more convenient to compare densities above 400 km to identify the longitudinal structure. In this chapter, integrate electron content between 400 and 450 km altitudes, referred as TEC(400–450), is used to study the longitudinal structure produced by the tidal effects.

27.2.1 Monthly Variation of the Tidal Effect

Figure 27.2 shows the monthly variation of TEC(400–450) at global constant local time of 1400 LT during year 2007. From Fig. 27.2, the 4-peaked structure becomes clearly identified during June-October. After October, the 4-peaked structure evolves into 3-peaked structure and reveals prominently during November-February, although it is less distinguishable in January. Unlike the clear wave-4 structure shown during the March equinox of the TOPEX observation in Scherliess et al. (2008), the longitudinal structure is blurred during March and April in Fig. 27.2, followed by a well defined 3-peaked structure in May. According to the spectrum analysis of the nonmigrating diurnal tide in neutral temperature of Forbes et al. (2008), the diurnal eastward wave-2 (DE2) nonmigrating tide, corresponding the 3-peaked structure viewing in global constant local time, often shows

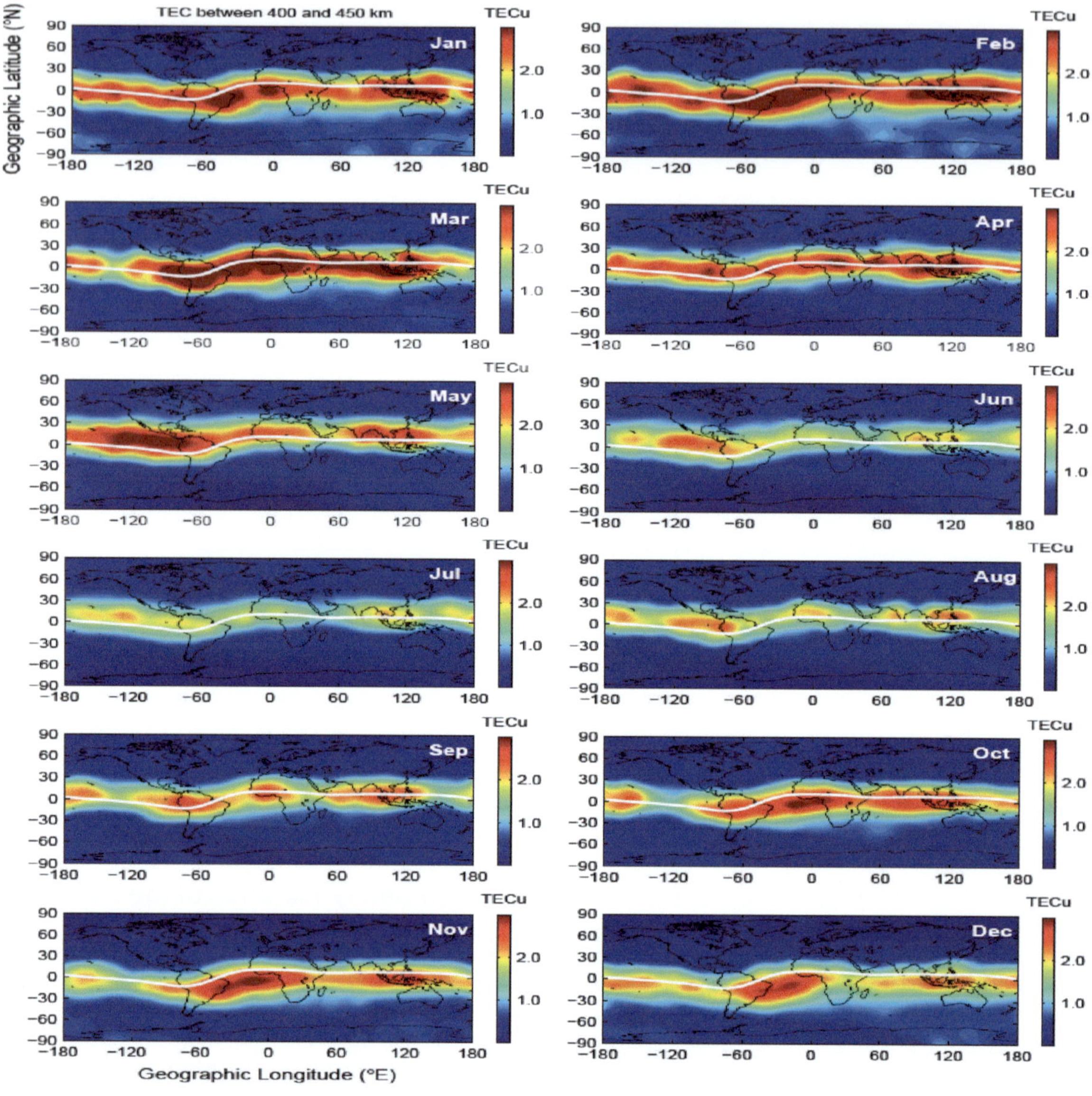

Fig. 27.2 Monthly variation of the longitudinal variation of the equatorial ionosphere in 2007. Each global constant local time map at 1400 LT shows the electron content integrated between 400 and 450 km altitudes

stronger amplitude during November-February. The diurnal eastward wave-3 (DE3) nonmigrating tide, corresponding to the 4-peaked structure, is most clearly seen during June-October period. The monthly variations of nonmigrating tidal signature in the neutral temperature shown in Forbes et al. (2008) generally agree with the COSMIC electron density observations in Fig. 27.2. It is also shown in Forbes et al. (2008) that the major dominated tides in March and April are mixture of the DE1, DE2 and DE3 nonmigrating tides, while the DE2 tide is sometimes the dominated tide in May. These dominated nonmigrating tides again agree with the electron densities during March-May. The less identifiably longitudinal structure in March and April may result from mixture of the various tides, while the 3-peaked structure shown in May might come from the dominated DE2 tide.

A desirable way to analyze the COSMIC observations for comparison with the neutral temperature observation of TIMED-SABER satellite (e.g. Forbes et al., 2008) is to perform the spectrum analysis on the electron densities. However, such comparison may encounter some difficulties as the low-latitude and equatorial ionospheric densities are not only affected by the plasma fountain effect but also other effects and the associated diurnal and seasonal effects. Nevertheless, a spectral analysis of the COSMIC electron density at 300 km altitude in year 2007 has been carried out by Wu et al. (2009) showing the day-to-day strength of the DE3 tide. The electron density enhancements result from the DE3 tide in 2007 are stronger during day-of-year (DOY) 0–50 (~January to middle March), 90–180 (~April to June) and 260–315 (~September to middle November) in Wu et al. (2009). However, the TEC(400–450) shown in this chapter reveals that the 4-peaked structure is manifest during July-October. The disagreements between results here and Wu et al. (2008) again suggest that the ionospheric density may not simply reflect the atmospheric tidal effects, especially at F-layer height of around 300 km altitude. Therefore, the direct spectral analysis of COSMIC observations may not fully explain the longitudinal structure.

27.2.2 *Diurnal Variation of the Tidal Effect*

The most prominent longitudinal variations shown in Fig. 27.2 are the wave-4 structure seen in September and wave-3 structure seen in December. Figures 27.3 and 27.4 show the diurnal variation of the TEC(400–450) in September and December, respectively. Figure 27.3 indicates that the wave-4 structure is seen from 1000 LT to 2200 LT and reveals an eastward movement similar to that described in Lin et al. (2007c), while the wave-3 structure is clearly identified during 1000–2000 LT in Fig. 27.4. The three peaks shown in Fig. 27.4 are located at South America-Africa (–75°E to 15°E), India-South Asia (50°E to 140°E), and Central Pacific (160°E to 135°E) regions. Unlike the eastward movement seen in Fig. 27.3, the three peaks shown in Fig. 27.4 remain their location mostly during their appearance. Both the wave-4 and wave-3 structures appear during similar local time period while the equatorial plasma fountain effect is active. This result indicates that the longitudinal structure is formed by modification of the plasma fountain strength by the upward propagating nonmigrating tide in various modes. The COSMIC observed electron densities provide opportunities for further detailed comparisons between the ionospheric density structure and the tidal signatures in the neutral atmosphere to understand their local time dependences.

In addition to the wave-3 structure, the electron density enhancements seen between South America and Antarctica during nighttime hours of 2200–0400 LT reveal as the most prominent ionospheric structure. It is the so called Ionospheric Weddell Sea anomaly which is categorized as part of the mid-latitude summer nighttime anomaly (MSNA) by Thampi et al. (2009) and Lin et al. (2010) and will be described with more details in the following section. It is the other type of longitudinal structure that occurs in the mid-latitude region being observed in detail by the COSMIC mission.

27.3 The Mid-Latitude Summer Nighttime Anomaly (MSNA)

An anomalous ionospheric feature which occurs around the Antarctica featured by a greater nighttime (18:00 LT~02:00 LT) electron density than daytime (08:00~18:00 LT) during local summer (around December solstice) was discovered in late 1950s. This feature observed by an ionosondes was coined as the Weddell Sea anomaly (WSA) by the location of its discovery (cf. Bellchambers and Piggott, 1958; Penndorf, 1965;, Dudeney and Piggott, 1978). With increasing satellite observations nowadays, Horvath and Essex (2003) observed the spatial extent of the WSA by TEC observation above oceans using the TOPEX/Poseidon altimeter soundings. Their result shows that the WSA is generally situated west of Antarctic Peninsula over the Bellinghausen Sea, covering a much larger area than originally expected to confine nearby the Weddell Sea area.

In addition to two-dimensional TEC observation of this peculiar nighttime electron density enhancement, the F-layer parameters of the anomaly were further revealed by using the radio occultation observation of the COSMIC satellite constellation (Burns et al., 2008). Later, Lin et al. (2009) showed the three-dimensional electron density structure of the WSA during December solstice which is helpful for

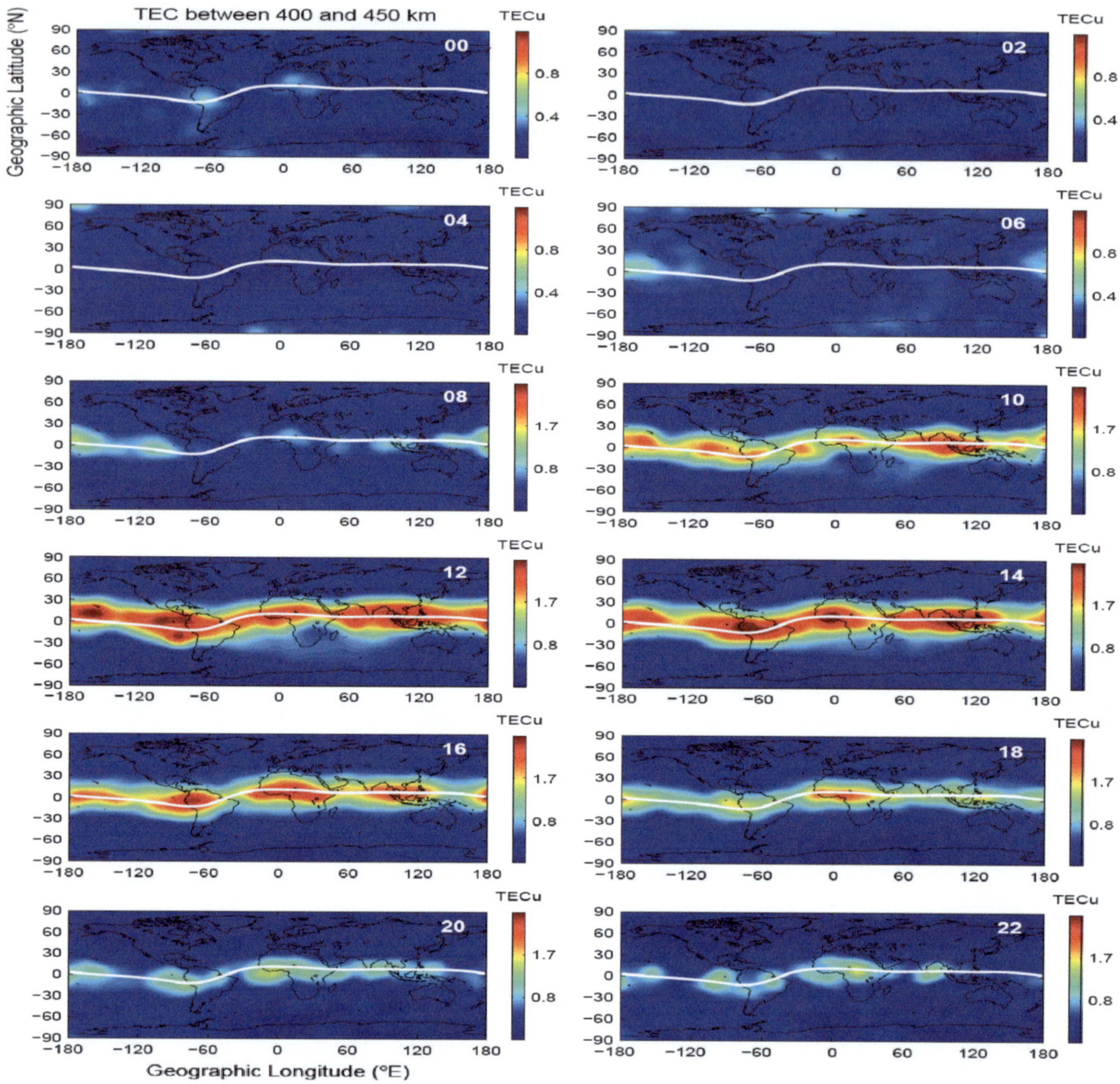

Fig. 27.3 Diurnal variation of wave-4 structure in September 2007. Each global constant local time map shows the electron content integrated between 400 and 450 km altitudes

understanding of its formation mechanism. The nighttime (2000~0400 LT) density enhancements shown in Fig. 27.4 are signatures of the WSA. It starts to develop at 1800 LT south of the EIA region and becomes well-developed around midnight. Not only occurring in the southern hemisphere, Lin et al. (2009) suggests that a similar-to-WSA-like anomalous electron density structure, featured by greater nighttime electron density than that in daytime, also occurs at the northern hemisphere around June solstice for the first time. The existence of the anomaly in the northern hemisphere is further studied by regional radar and beacon receivers and global satellite observations (Thampi et al., 2009; Liu et al., 2010). They demonstrate that the anomalous density structures in both northern and southern hemispheres share a very similar development and characteristics during evening hours. They suggest that the anomaly seen in the northern hemisphere during June solstice and the WSA are basically the same type of the nighttime density enhancement anomaly and coin the two anomalies as the mid-latitude summer nighttime anomaly (MSNA).

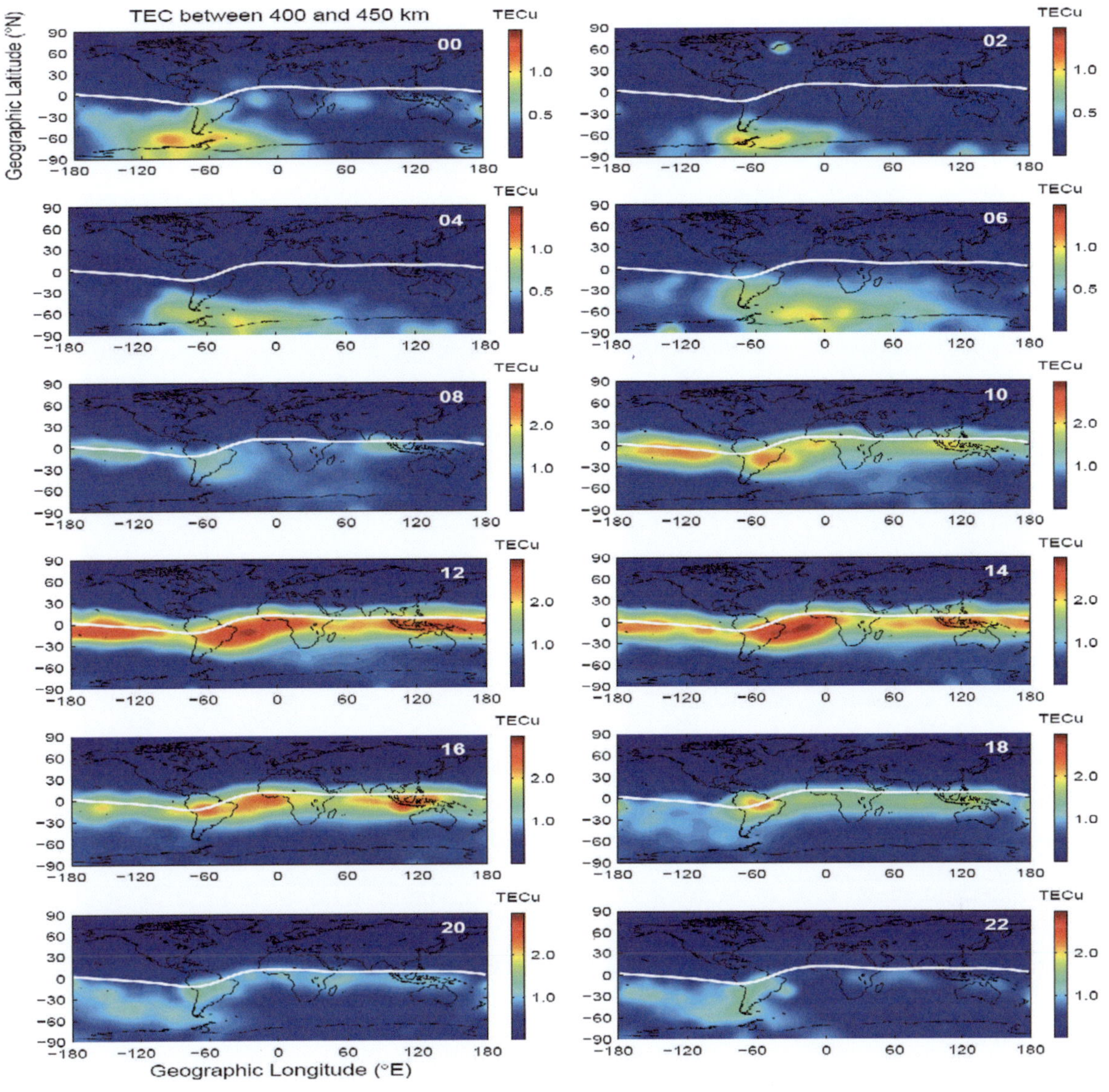

Fig. 27.4 Diurnal variation of wave-4 structure in December 2007. Each global constant local time map shows the electron content integrated between 400 and 450 km altitudes

Figure 27.5 shows the MSNA signatures in both northern and southern hemispheres by subtracting the electron densities at 1400 LT from those at 2200 LT. The MSNA signature, greater nighttime density, in the northern hemisphere occurs around east coast of North America-Europe (–60°E∼60°E) and Central Asia-Alaska (90°E to –120°E), while it occurs around South Pacific-South Atlantic oceans (–180°E to 0°E) in the southern hemisphere. The MSNA is likely to be formed due to the combined effect of the longer summer sunlight hours and the day-to-night/summer-to-winter equatorward neutral winds during evening hours (Dudeney and Piggott, 1978; Horvath and Essex, 2003). The longer sunlight and photo-ionization time in high geographic latitude region during summer time produce more electron density and it is later sustained at higher altitudes by the nighttime equatorward wind. The latitudes of the MSNA regions are mainly around ±40°N to ±60°N, ranging around 45°∼75° dip angles where the neutral wind effect in moving the plasma

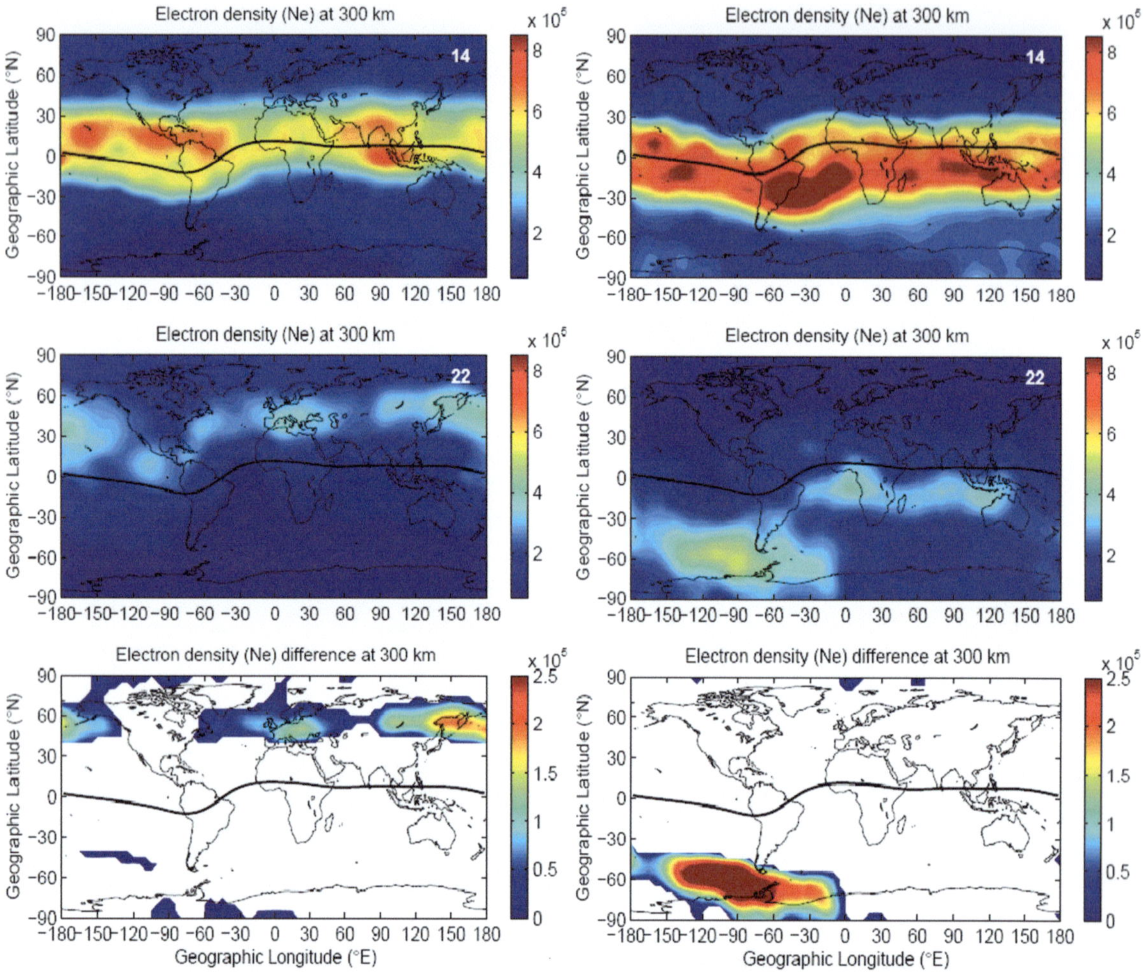

Fig. 27.5 The MSNA signatures in the northern (*Left*) and southern (*Right*) hemispheres by subtracting the electron densities at 1400 LT from those at 2200 LT

upward along the magnetic field line is still effective (cf. Kohl and King, 1967; Rishbeth and Garriot, 1969). Such effect produces an electron density enhancement in the mid-latitude region around ±40°N to ±60°N after 2000 LT as seen in Fig. 27.6.

Based on above reasons, MSNA mainly occurs at longitudes where the magnetic equator shifts toward poleward of the summer hemisphere in respect to the geographic equator. The magnetic declination angle also plays a role by projecting the geographical zonal wind into equatorward wind along the field lines (cf. Jee et al., 2009; Liu et al., 2010; Lin et al., 2010 and references therein). This effect is most prominently seen in the southern MSNA region where the eastward declination of the magnetic field lines provides an additional northward/equatorward wind component in the magnetic meridional plane projected from the nighttime eastward winds at F-layer height. It results in a much stronger southern MSNA signature between South Pacific, South America and Antarctica than other longitudes of the northern MSNA (Fig. 27.5).

The development of the MSNA in Fig. 27.6 shows not only the electron density enhancement in mid-latitude after 2000 LT hour but also the uplift of the ionospheric layer indicating the equatorward wind effect. Figure 27.6 also shows that the enhancement occurred mainly between 250 and 350 km altitudes where the recombination loss proceeds much slower than at the lower altitude. To further investigate monthly variations of the MSNA, Figs. 27.7

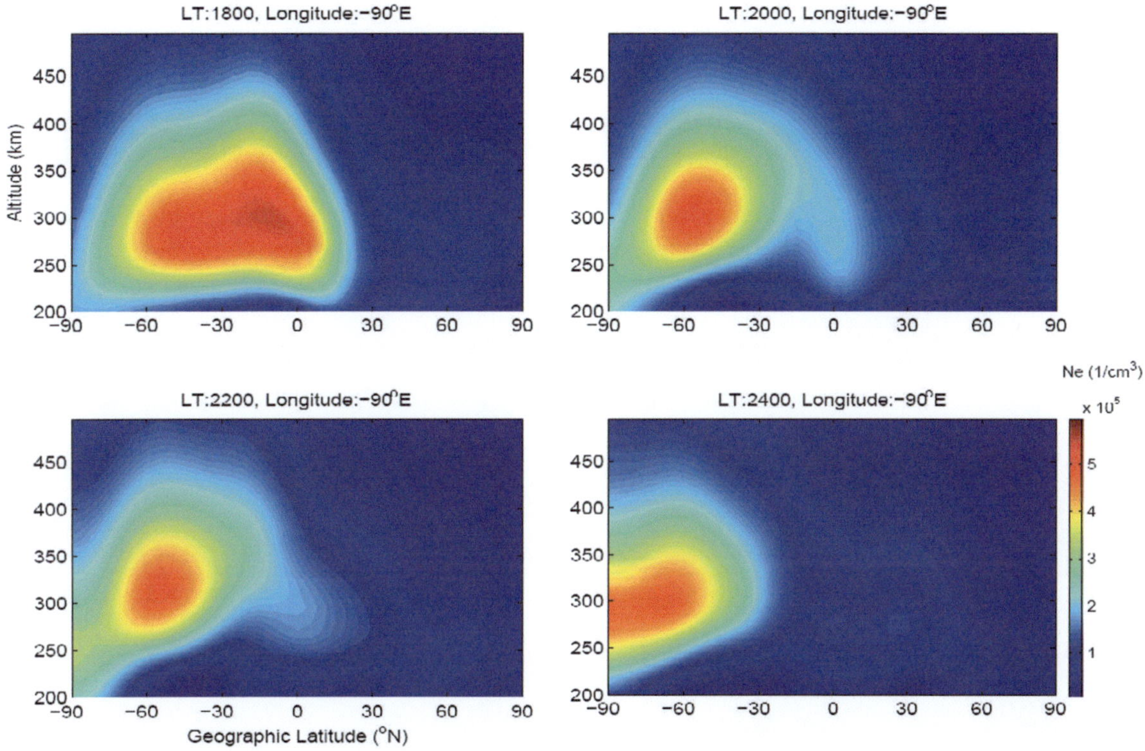

Fig. 27.6 Two-dimensional electron density slices at –90°E geographic longitude from 1800 to 2400 LT in December 2007 showing developments of the MSNA in the southern hemisphere (or Weddell Sea anomaly). The magnetic equator locates at approximately –11°N geographic latitude. This figure is taken from Lin et al. (2009)

and 27.8 show the local-time plots of the electron density at locations where the northern (52°N, 104°E in geographic coordinate) and southern (–60°N, –90°E in geographic coordinate) MSNA features are seen clearly. The MSNA signature appears as low as 250 km altitude and becomes most prominent at 300 km. The northern MSNA signature is seen during April-September (Fig. 27.7), while the southern MSNA signature appears during September-February (Fig. 27.8). The MSNA signatures in the both hemispheres exist up to 450 km altitude according to Figs. 27.7 and 27.8. The monthly variations of the MSNA indicate that the previously unexpected nighttime density enhancement is a general existing mid-latitude feature occurring during local summer nighttime.

27.4 Summary

In summary, this chapter demonstrates two types of the ionospheric longitudinal structures in low- and mid-latitude regions. The longitudinal structure in the low-latitude region results from the upward propagating atmospheric tides that modify the ionospheric electrodynamics and affect the equatorial plasma fountain consequently. The monthly and diurnal observations provided by the COSMIC ionospheric soundings are useful to cross compare with the neutral atmospheric observation of the tidal effects. Additionally, the longitudinal structure seen in the mid-latitude region, the MSNA, is now known to exist not only in the southern hemisphere but also the northern hemisphere after utilizing the COSMIC observations. Results shown in Figs. 27.7 and 27.8 indicate that the MSNA is a general existing mid-latitude ionospheric feature occurring at local summer nighttime. Furthermore, the altitude-latitude electron density maps derived from COSMIC soundings shown in Fig. 27.6 are helpful to understand the associated formation mechanism.

Acknowledgements This work is partially supported by the Taiwan National Science Council under NSC NSC 98-2111-M-006-003-MY2 and by NSPO under 98-NSPO(B)-IC-FA07-01(L) and 98-NSPO(B)-IC-FA07-01(V).

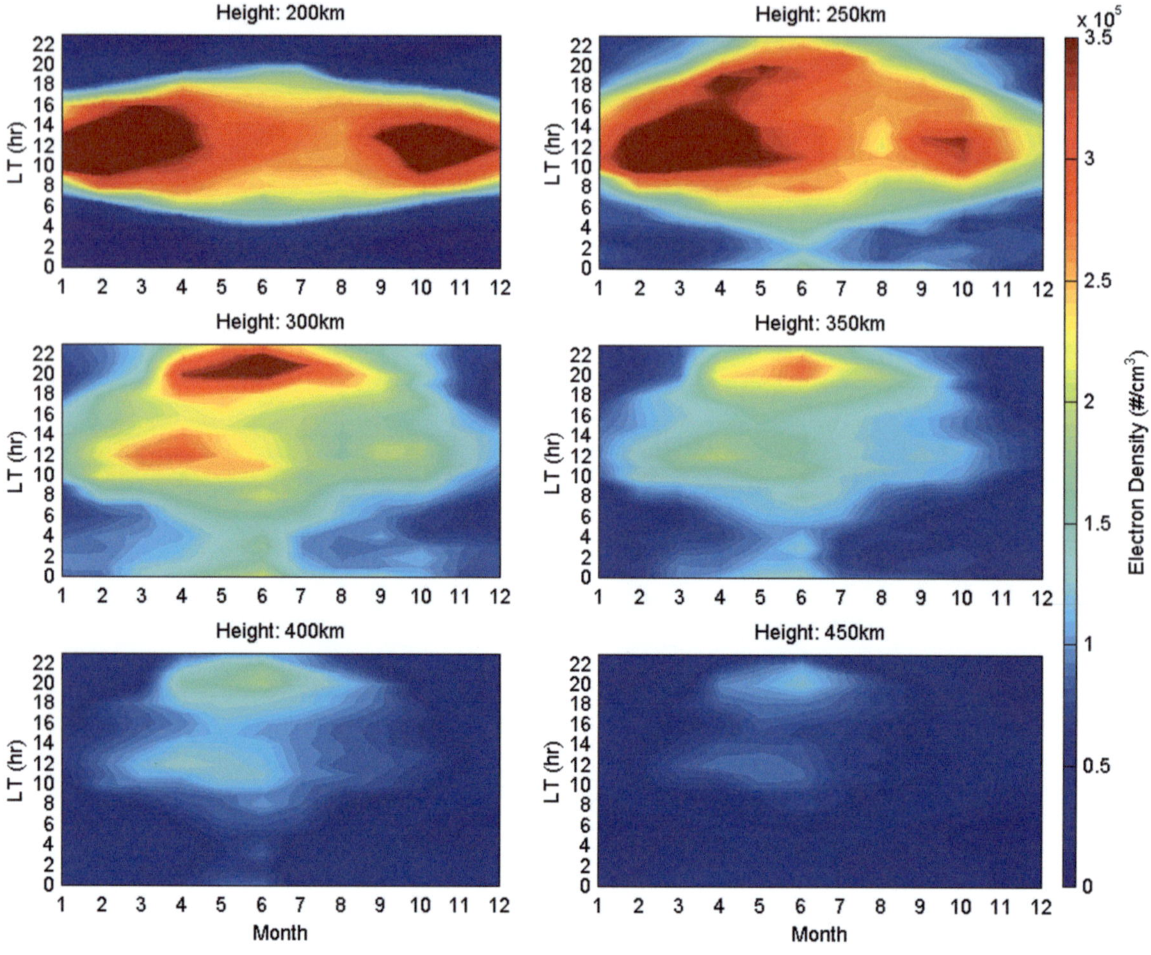

Fig. 27.7 Local time variations of the electron density at (52°N, 104°E in geographic coordinate) at various altitudes in each month of 2007. This figure indicates that the northern MSNA occurs from April to September

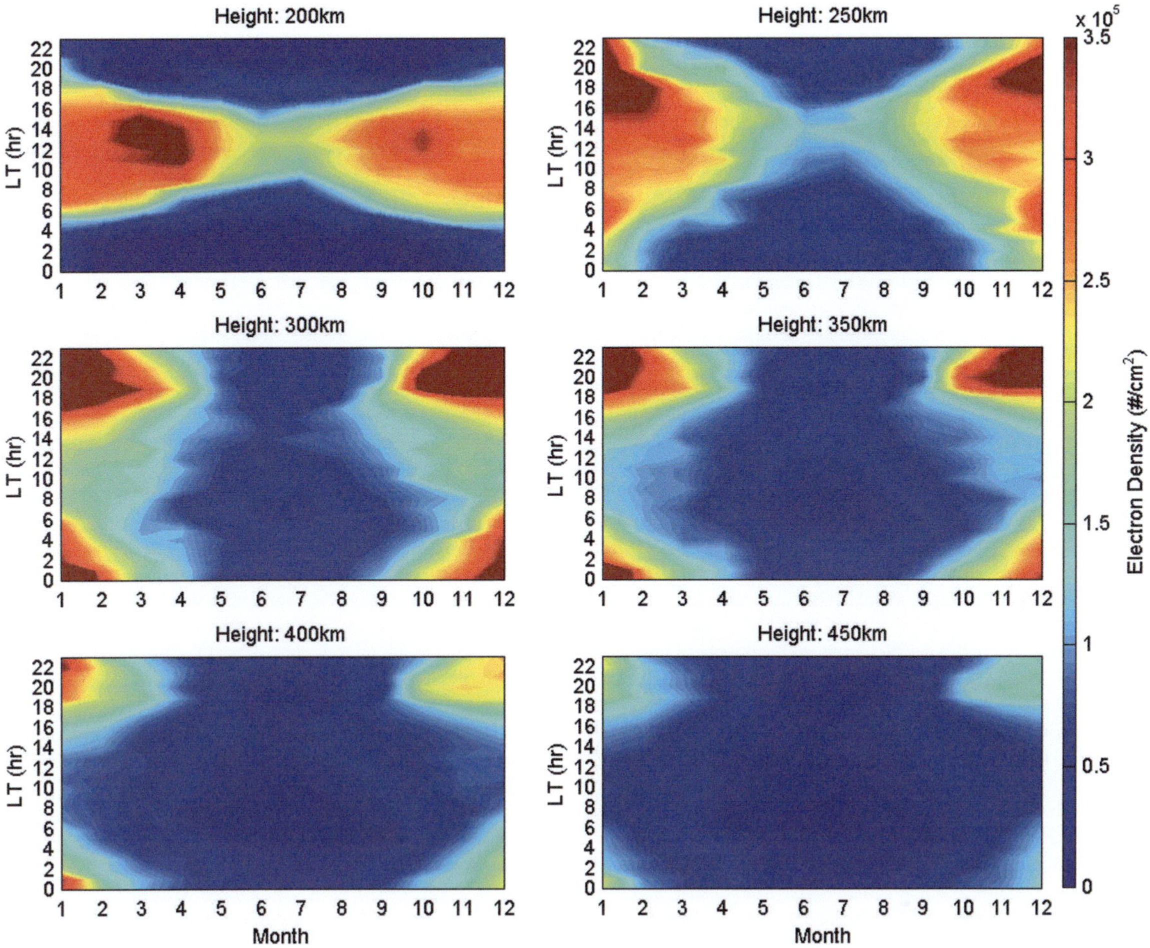

Fig. 27.8 Local time variations of the electron density at (–60°N, –90°E in geographic coordinate) at various altitudes in each month of 2007. This figure indicates that the southern MSNA (or the Weddell Sea anomaly) occurs from September to February

References

Bellchambers WH, Piggott WR (1958) Ionospheric measurements made at Halley Bay. Nature 188:1596–1597

Burns AG, Zeng Z, Wang W, Lei J, Solomon SC, Richmond AD, Killen TL, Kuo Y-H (2008) The behavior of the F_2 peak ionosphere over the South Pacific at dusk during quiet summer condition from COSMIC data. J Geophys Res 113:A12305. doi:10.1029/2008JA013308

Dudeney JR, Piggot WR (1978) Antarctic ionospheric research. In: Lanzerotti LJ, Park CG (eds) Upper atmosphere research in Antarctica. Antarctic Research Series, Washington, DC, pp 200–235

England SL, Maus S, Immel TJ, Mende SB (2006) Longitudinal variation of the E-region electric fields caused by atmospheric tides. Geophys Res Lett 33:L21105. doi:10.1029/2006GL027465

Forbes JM, Russell J, Miyahara S, Zhang X, Palo S, Mlynczak M, Mertens CJ, Hagan ME (2006) Troposphere-thermosphere tidal coupling as measured by the SABER instrument on TIMED during July –September 2002. J Geophys Res 111:A10S06. doi:10.1029/2005JA011492

Forbes JM, Zhang X, Palo S, Russell J, Mertens CJ, Mlynczak M (2008) Tidal variability in the ionospheric dynamo region. J Geophys Res 113:A02310. doi:10.1029/2007JA012737

Hagan ME, Maute A, Roble RG, Richmond AD, Immel TJ, England SL (2007) Connections between deep tropical clouds and the Earth's ionosphere. Geophys Res Lett 34:L20109. doi:10.1029/2007GL030142

Hajj GA, Lee LC, Pi X, Romans LJ, Schreiner WS, Straus PR, Wang CM (2000) COSMIC GPS ionospheric sensing and space weather. Terr Atmos Oceanic Sci 11(1): 235–273

Horvath I, Essex EA (2003) The Weddell sea anomaly observed with the Topex satellite data. J Atmos Solar-Terr Phys 65;693–706

Immel TJ, Sagawa E, England SL, Henderson SB, Hagan ME, Mende SB, Frey HU, Swenson CM, Paxton LJ (2006) Control of equatorial ionospheric morphology by

atmospheric tides. Geophys Res Lett 33:L15108. doi:10.1029/2006GL026161
Jee G, Burns AG, Kim Y-H, Wang W (2009) Seasonal and solar activity variations of the Weddell Sea Anomaly observed in the TOPEX total electron content measurements. J Geophys Res 114:A04307. doi:10.1029/2008JA013801
Kohl H, King JW (1967) Atmospheric winds between 100 and 700 km and their effects on the ionosphere. J Atmos Solar-Terr Phys 29:1045–1062
Lei J, Syndergaard S, Burns AG, Solomon SC, Wang W, Roble RG, Wu Q et al (2007) Comparison of COSMIC ionospheric measurements with ground-based observations and model predictions: preliminary results. J Geophys Res 112:A07308. doi:10.1029/2006JA012240
Lin CH, Liu JY, Cheng CZ, Chen CH, Liu CH, Wang W, Burns AG, Lei J (2009) Three-dimensional ionospheric electron density structure of the Weddell Sea Anomaly. J Geophys Res 114:A02312. doi:10.1029/2008JA013455
Lin CH, Liu JY, Fang TW, Chang PY, Tsai HF, Chen CH, Hsiao CC (2007a) Motions of the equatorial ionization anomaly crests imaged by FORMOSAT-3/COSMIC. Geophys Res Lett 34:L19101. doi:10.1029/2007GL030741
Lin CH, Wang WB, Hagan ME, Hsiao CC, Immel TJ, Hsu ML, Liu JY, Paxton LJ, Fang TW, Liu CH (2007b) Plausible effect of atmospheric tides on the equatorial ionosphere observed by the FORMOSAT-3/COSMIC: three-dimensional electron density structures. Geophys Res Lett 34:L11112. doi:10.1029/2007GL029265
Lin CH, Hsiao CC, Liu JY, Liu CH (2007c) Longitudinal structure of the equatorial ionosphere: time evolution of the four-peaked EIA structure. J Geophys Res 112:A12305. doi:10.1029/2007JA012455
Lin CH, Liu JY, Cheng CZ, Chen CH, Liu CH, Wang W, Burns AG, Lei J (2009) Three-dimensional ionospheric electron density structure of the Weddell Sea Anomaly. Geophys Res Lett 114:A02312. doi:10.1029/2008JA013455
Lin CH, Liu CH, Liu JY, Chen CH, Burns AG, Wang W (2010) Midlatitude summer nighttime anomaly of the ionospheric electron density observed by FORMOSAT-3/COSMIC. J Geophys Res 115:A03308. doi:10.1029/2009JA014084
Liu JY, Lin CY, Lin CH, Tsai HF, Solomon SC, Sun YY, Lee IT, Schreiner W, Kuo YH (2010) Artificial plasma cave in the low-latitude ionosphere results from the radio occultation inversion of the FORMOSAT-3/COSMIC. J Geophys Res 115:A07319. doi:10.1029/2009JA015079
Liu H, Thampi SV, Yamamoto M (2010) Anomalous phase reversal of the diurnal cycle in the mid-latitude ionosphere. J Geophys Res 115:A01305. doi:10.1029/2009JA014689
Lühr H, Rother M, Ha¨usler K, Alken P, Maus S (2008) The influence of nonmigrating tides on the longitudinal variation of the equatorial electrojet. J Geophys Res 113:A08313. doi:10.1029/2008JA013064
Oberheide J, Forbes JM (2008) Tidal propagation of deep tropical cloud signatures into the thermosphere from TIMED observations. Geophys Res Lett 35:L04816. doi:10.1029/2007GL032397
Oberheide J, Hagan ME, Roble RG, Offermann D (2002) Sources of nonmigratingtides in the tropical middle atmosphere.J Geophys Res 107(D21):4567. doi:10.1029/2002JD002220
Penndorf R (1965) The average ionospheric conditions over the Antarctic. In: Waynick AH (ed) Geomagnetism and aeronomy, vol 4. Antarctic Research Series. AGU, Washington, DC, [pp 1–45
Rishbeth H, Garriott OK (1969) Introduction to ionospheric physics. Academic, New York, NY
Sagawa E, Immel TJ, Frey HU, Mende SB (2005), Longitudinal structure of the equatorial anomaly in the nighttime ionosphere observed by IMAGE/FUV. Geophys Res Lett 110:A11302. doi:10.1029/2004JA010848
Scherliess L, Thompson DC, Schunk RW (2008) Longitudinal variability of low-latitude total electron content: tidal influences. J Geophys Res 113:A01311. doi:10.1029/2007JA012480
Schreiner W, Rocken C, Sokolovskiy S, Syndergaard S, Hunt D (2007) Estimates of the precision of GPS radio occultations from the COSMIC/FORMOSAT-3 mission. Geophys Res Lett 34:L04808. doi:10.1029/2006GL027557
Schreiner WS, Sokolovskiy SV, Rocken C, Hunt DC (1999) Analysis and validation of GPS/MET radio occultation data in the ionosphere. Radio Science 34:949–966
Thampi SV, Lin CH, Liu H, Yamamoto M (2009) First tomographic observations of the Mid-latitude Summer Nighttime Anomaly (MSNA) over Japan. J Geophys Res 114:A10318. doi:10.1029/2009JA014439
Wu Q, Solomon SC, Kuo Y-H, Killeen TL, Xu J (2009) Spectral analysis of ionospheric electron density and mesospheric neutral wind diurnal nonmigrating tides observed by COSMIC and TIMED satellites. Geophys Res Lett 36:L14102. doi:10.1029/2009GL038933
Yue X, Schreiner WS, Lei J, Sokolovskiy SV, Rocken C, Hunt DC, Kuo Y-H (2010) Error analysis of Abel retrieved electron density profiles from radio occultation measurements. Ann Geophys 28:217–222

Chapter 28

Ionosphere–Thermosphere Coupling in the Low-Latitude Region

Shigeto Watanabe and Tsutomu Kondo

Abstract The zonal neutral wind in the thermosphere flows strongly at the Earth's magnetic dip equator instead of the geographic equator around 20 magnetic local time (MLT). On the other hand, the fast zonal plasma drift occurs in the low-latitude F region of evening ionosphere, but the velocity decreases at the magnetic dip equator. Therefore, the fast plasma drift velocity structure forms an arch in the frame of latitude and altitude in the evening. The fast zonal neutral wind occurs inside of the arch. Since the fast zonal plasma drift is strongly associated with equatorial ionization anomaly (EIA), we suggest that the ionosphere–thermosphere coupling associated with F-region dynamo is significantly important in the phenomenology of the low-latitude, F-region ionosphere and thermosphere. The fast neutral wind occurring in the evening F region over the magnetic dip equator may result in atmospheric super-rotation in the low-latitude thermosphere.

28.1 Introduction

Though the ionization rate is less than 1% in the region of low-latitude thermosphere, the dynamics of neutral atmosphere is strongly controlled by the plasma. The coupling process between the neutral atmosphere and the plasma has been investigated by satellites such as DE-2 and CHAMP. The observations showed similar local time variations of plasma drift velocity and neutral wind (Rishbeth, 1972; Heelis et al., 1974), atmospheric super-rotation in thermosphere (Coley and Heelis, 1989), and equatorial temperature and wind anomaly (ETWA; Raghavarao et al., 1991). From CHAMP satellite data, Liu et al. (2006, 2007, 2009) presented results showing that the equatorial zonal winds of daytime and nighttime upper atmosphere at ~400 km altitude are significantly influenced by the solar flux and that the winds show atmospheric super-rotation in thermosphere depending on the season and solar flux. From ground-based imager observations of 630 nm airglow by Otsuka et al. (2005), a meso-scale structure generated by gravity wave was found at the same time at conjugate points of northern and southern hemispheres. The result suggested that the electro-magnetic coupling along magnetic field lines is significantly important in the thermosphere and the ionosphere. The NCAR TIE-GCM results indicated that the coupling between ionospheric plasma and the eastward neutral wind is strong in the evening at magnetic equator (Richmond et al., 1992). Fuller-Rowell et al. (1997) suggested the importance of chemical heating in the ETWA region. Maruyama et al. (2003) indicated the effect of ion drag in processes controlling the ETWA. The coupling between the neutral atmosphere and the ionospheric plasma is a key process to understand the dynamics and the structure of the thermosphere and the ionosphere, but we do not have enough information on the coupling process because the direct simultaneous observations of the neutral atmosphere/plasma and their comparison with modeling have not yet been performed in detail.

In this chapter, the fast equatorial zonal neutral winds as measured by the DE-2 satellites are presented. The winds are compared with plasma drifts and

S. Watanabe (✉)
Department of Earth and Planetary Sciences, Hokkaido University, Hokkaido, Japan
e-mail: shw@ep.sci.hokudai.ac.jp

M.A. Abdu, D. Pancheva (eds.), A. Bhattacharyya (Coed.), *Aeronomy of the Earth's Atmosphere and Ionosphere*, IAGA Special Sopron Book Series 2, DOI 10.1007/978-94-007-0326-1_28,

electron densities also measured onboard by the DE-2 satellite. The latitudinal structures of the winds and plasma drifts are closely associated with the equatorial ionization anomaly (EIA) (Namba and Maeda, 1939; Appleton, 1946).

28.2 Data

The inclination of the DE-2 satellite is 90°. It samples through all local times every 6 months. The thermospheric zonal neutral wind, the zonal plasma drift, and the electron density were measured by the WATS (wind and temperature spectrometer), the RPA (retarding potential analyzer), and the LANG (Langmuir probe) instruments, respectively, onboard the DE-2 satellite. Note that the detailed descriptions of these instruments have been given by Spencer et al. (1981), Hanson et al. (1981), and Krehbiel et al. (1981). The data presented here covers the altitude range between 300 and 600 km. Since the data are limited in local time, altitude, latitude, and longitude, we did not investigate the seasonal dependence in them. Spencer et al. (1981) have described in detail the derivation procedure and the related errors in the neutral wind data. We have selected the DE-2 data for the period Aug. 1981–Feb. 1983 during which the average solar radio flux value is F10.7=180. Very magnetically active periods ($K_p \geq 5$) were excluded from the data analyzed in this study.

28.3 Results

The seasonally averaged zonal neutral wind distributions in the frame of geographic latitude and longitude are presented in Fig. 28.1 for the magnetic local time (MLT) intervals 00–04, 04–08, 08–12, 12–16, 16–20, and 20–24 MLT. The solid line shows the magnetic dip equator. The data in each bin are averaged in 5° longitude/latitude and in height range between 300 and 600 km. The CHAMP satellite results also show similar neutral wind distribution (Liu et al., 2009, Luhr et al., Chapter 22, this volume). The fast zonal neutral wind flows eastward after the evening as can be noted in the panels (a), (e), and (f). The wind velocity peaks in the equatorial region and decreases toward higher latitudes. It is to be noted that the maximum wind velocity region is aligned with the Earth's magnetic dip equator, instead of the geographic equator. The wind velocity is ~150 m/s. Both the CHAMP and DE-2 observations reveal nearly identical latitudinal pattern. The neutral wind flows westward before noon as can be noted in the panels (c). The zonal neutral wind at magnetic equator averaged in local time

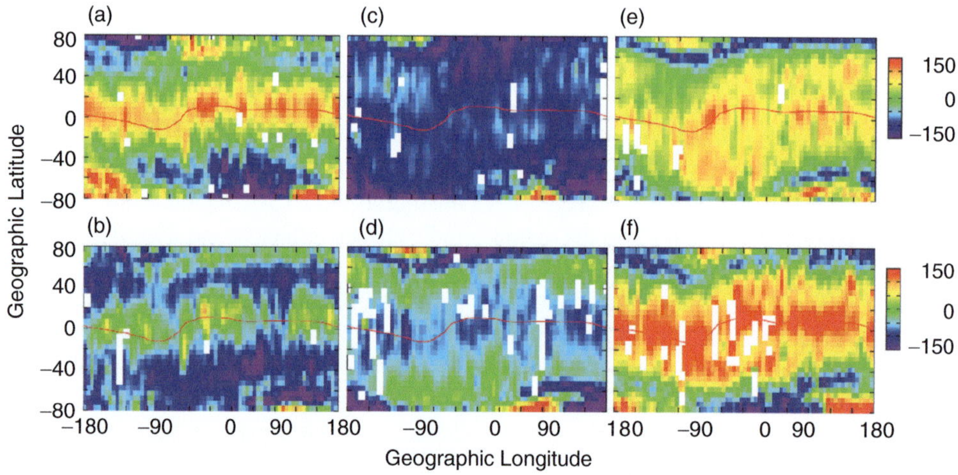

Fig. 28.1 Distribution of the seasonally averaged zonal neutral wind velocity (in unit of ms^{-1}) in the frame of geographic longitude and geographic latitude during periods of (**a**) 00–04 MLT, (**b**) 04–08 MLT, (**c**) 08–12 MLT, (**d**) 12–16 MLT, (**e**) 16–20 MLT, and (**f**) 20–24 MLT. Positive is eastward. The solid line indicates the magnetic dip equator. Note the banded structure along the magnetic dip equator, where the fastest neutral wind flows after sunset until morning

flows eastward. It corresponds to the super-rotation of atmosphere at the thermospheric heights as also shown by Liu et al. (2006).

Liu et al. (2009) and Luhr et al. (Chapter 22, this volume) have presented a comparison of the zonal neutral wind distributions in the frame of magnetic dip latitude and magnetic local time as obtained by the CHAMP and the DE-2 satellites. Although some differences exist in the mean values of the neutral wind and also in the local time of its second maximum, both the CHAMP and the DE-2 observations reveal similar neutral wind patterns. As stated before, the neutral wind flows eastward after noon to morning and westward before noon in the low-latitude region. Toward higher latitude, the morning wind reversal from eastward to westward occurs at earlier local times (also see Fig. 28.1b and c). The reversal time is ~2 MLT near ±30° latitude. On the night side, the latitudinal variation of the wind exhibits a maximum at the magnetic dip equator. This fast neutral wind continues throughout the time of eastward wind. During 08–16 MLT, both observations show a minimum westward flow at the magnetic dip equator (also see Fig. 28.1c, d). There is a good agreement in the features revealed by CHAMP and DE-2 observations.

Figure 28.2 shows the zonal plasma drifts obtained by the DE-2 satellite in the frame of geomagnetic local time and latitude (left panel), and in the frame of geographic longitude and latitude (right panel). Data in each bin are averaged in 1 h and 1° geomagnetic longitude/latitude. The averaged data correspond to the altitude range between 300 and 600 km. The zonal plasma drift velocity indicates symmetric pattern on either side of the magnetic dip equator. It flows strongly in the low-latitude region. The zonal plasma drift changes the direction from westward to eastward at around 16 MLT. It presents variation similar to that of the zonal neutral wind. Both zonal neutral wind and zonal plasma drift velocities increase eastward in the period between 18 and 20 MLT. The plasma drift velocity reaches ~100 m/s. However, the fast zonal neutral wind at around 20 MLT reaches about 150 m/s. On the other hand, the zonal plasma drift forms a velocity anomaly from 18 to 20 MLT so that the velocity trough occurs at magnetic dip equator and two peaks occur at ±10° magnetic latitudes (see right panel). The zonal plasma drift velocity at the magnetic dip equator is ~50 m/s smaller than its value at ±10° magnetic latitudes. The drift velocity anomaly disappears at around 22 MLT, and then the plasma flows strongly at magnetic dip equator until the morning.

Figure 28.3 shows the zonal neutral wind (left), the zonal plasma drift (middle), and the electron density (right) observed by the DE-2 satellite in the frame of magnetic latitude and altitude. The data are also shown for the local time intervals of 15–18, 18–21, and 21–24 MLT. The data are averaged in bin of 10 km in altitude and 1° in latitude. The zonal neutral wind flows strongly at magnetic dip equator, and the velocity increases with increasing altitude at low and middle latitudes. There is no data beyond 600 km altitude because of the lack of sensitivity of the WATS (wind

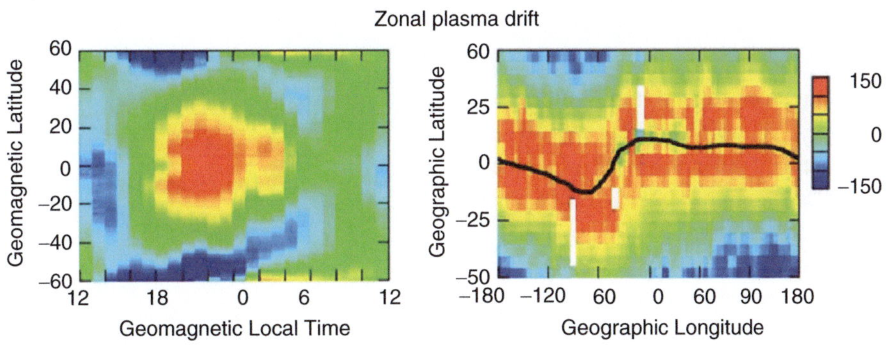

Fig. 28.2 Distributions of zonal plasma drift velocity (*left*) in the frame of geomagnetic local time and geomagnetic latitude and zonal plasma drift velocity (*right*) in the frame of geographic longitude and latitude. (The velocities are in unit of m/s.) The zonal plasma drift data averaged at altitudes between 300 and 600 km are used. The drift velocity in the right panel is for 18–24 MLT. *Black solid* line represents magnetic dip equator. Note that the zonal plasma drift anomaly (PDA) occurs near magnetic dip equator during 18–21 MLT

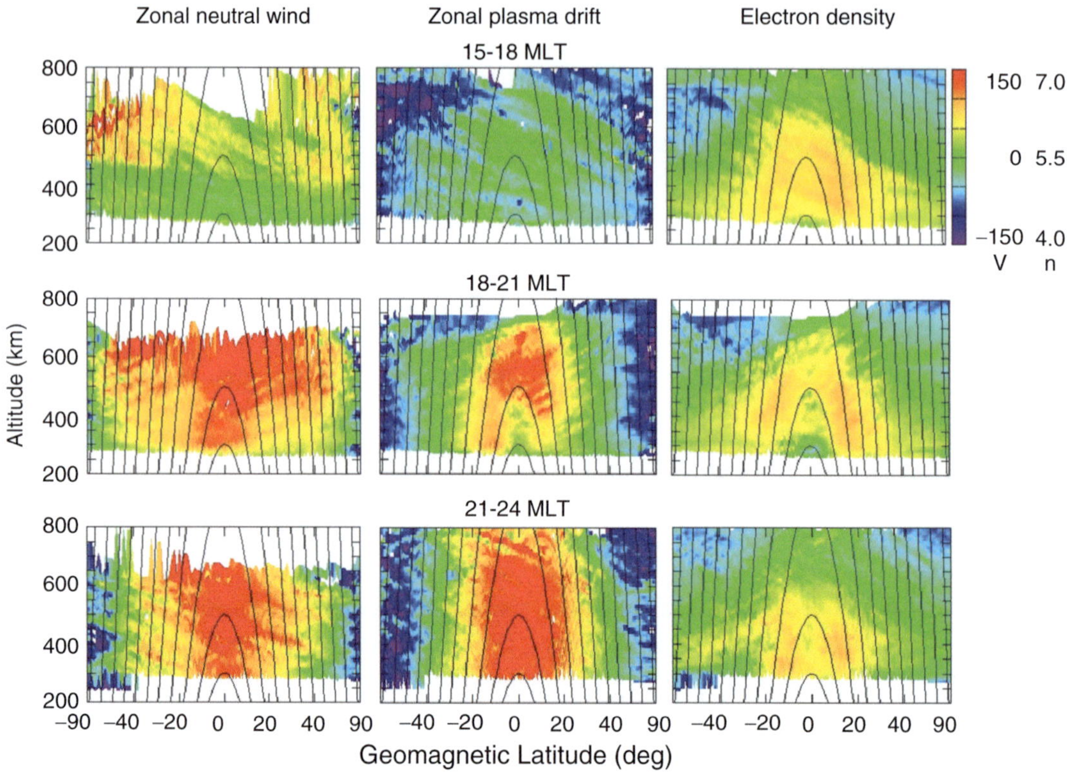

Fig. 28.3 Zonal neutral wind, zonal plasma drift, and electron density distributions in the frame of magnetic latitude and altitude during periods of 15–18, 18–21, and 21–24 MLT. The velocity and density units are in m/s and in log scale of cm^{-3}. *Solid lines* represent geomagnetic field lines. Note that the zonal plasma drift anomaly (PDA) forms an arch structure during 18–21 MLT. The PDA occurs just inside of equatorial ionization anomaly (EIA)

and temperature spectrometer) instrument in the low neutral density regions. Both the zonal neutral wind velocity and the zonal plasma drift velocity increase with increasing magnetic local time in the evening to midnight. However, the shape of fast zonal plasma drift velocity is seen as an arch in 18–21 MLT. We call this feature as zonal plasma drift velocity anomaly (PDA). Such an anomalous feature does not appear at the other local times. At altitudes between 300 and 500 km over the magnetic dip equator, the plasma drift velocity is small. The maximum plasma drift velocity of ~100 m/s occurs at ~600 km altitude over the dip equator. The latitudinal distribution of the fast zonal plasma drift is similar to that of the equatorial ionization anomaly (EIA), but the faster zonal plasma drift occurs at lower altitudes than that of EIA, i.e., PDA is inside of the EIA. Note that the shear in the zonal plasma drift velocity occurs below 600 km altitude over the dip equator.

28.4 Discussion

From the above analysis of the DE-2 data, the latitudinal structure of the zonal neutral wind has revealed a fast zonal neutral wind over the Earth's magnetic dip equator which is similar to the results reported from the CHAMP observations. It is remarkable how similar this structure is in the two independent datasets obtained two decades apart with totally different instruments. The CHAMP satellite probes the in situ neutral wind with a tri-axis accelerometer, while the DE-2 measures the wind with a wind and temperature spectrometer (WATS). The principles of these instruments are completely different as described in Liu et al. (2006) and Spencer et al. (1981). Furthermore, the neutral wind varies significantly by location, season, and solar and geomagnetic conditions (Liu et al., 2004, 2006; Oberheide et al., 2006). The consistency between the latitudinal structures revealed in the two datasets is striking. The CHAMP and the DE-2 measurements strongly confirm the existence of a belt of fast neutral wind and its stable location at the magnetic dip equator. This fast neutral wind along the Earth's magnetic dip equator instead of the geographic equator demonstrates a strong magnetic control of the thermospheric dynamics. In the upper atmosphere at low latitudes, the atmospheric pressure gradient is the primary driver of the neutral wind, with the ion drag being an important impeding force (Rishbeth, 1972). With the development of the equatorial ionization anomaly (EIA) in the equatorial ionosphere usually starting at ~09 MLT, the plasma density forms a trough at the magnetic dip equator (Balan et al., 1994; Liu and Wan, 2001). This trough offers lower ion drag, which consequently facilitates faster neutral wind to flow over the magnetic dip equator (Maruyama et al., 2003). During the period of 05–08 MLT, however, the EIA structure disappears and a peak of the plasma density peak instead of trough forms at the magnetic dip equator (Lin et al., 2007). This causes the ion drag to peak at the magnetic dip equator as well, which slows down the zonal neutral wind considerably. As a result, the magnetic dip equator becomes a channel of slow flow instead of fast flow as shown in Fig. 28.1c, d. The fast neutral wind occurs over the magnetic dip equator during 16–24 MLT and after 12 MLT (see Fig. 28.1a, b, e, and f). These observations strongly suggest the ion drag being the principal cause for shifting the fast zonal neutral wind from the geographic equator to the magnetic dip equator. Besides the similar latitudinal structure revealed by the CHAMP and the DE-2, we note that an apparent difference is seen in the occurrence time of the second wind maximum after midnight (see Liu et al., 2009), Luhr et al. Chapter 22, this volume). Raghavarao et al. (1991) examined the latitudinal variation of the wind orbit by orbit instead of in a statistical manner as we do here. As shown in Fig. 1 of Raghavarao et al. (1991), for instance, the neutral wind peak at the dip equator is very prominent and broad, with a width of about 20° in latitude. But the neutral wind trough near ±25° latitude is not clear as shown in Fig. 28.1e. Furthermore, the location of the neutral wind trough is expected to be highly variable with local time and season, following that of the EIA crests. Therefore, it is quite likely that this trough structure with shallow magnitude has been smeared out in statistical analysis, as a consequence of combining the measurements in different season, longitudes, and local times. The statistical analysis of the DE-2 by Coley et al. (1994) has revealed no band of slow neutral wind.

The zonal plasma drift forms a trough at magnetic dip equator and crests in ±10° magnetic latitudes (see Fig. 28.2), which was already reported by Aggson et al. (1987). From Jicamarca radar data, Fejer et al. (1985) showed that the zonal plasma drift increases with increasing altitude in the F region over the magnetic dip equator. However, a 3D structure of the zonal plasma drift has not been understood in detail. Figure 28.3 shows for the first time that the fast zonal neutral wind occurs inside of the region of the fast zonal plasma drift. The shape of fast zonal plasma drift in latitude and altitude frames is similar to equatorial ionization anomaly (EIA). We may call this feature as the zonal plasma drift velocity anomaly (PDA). The fast zonal plasma drift is strongly associated with the EIA development after sunset till midnight. We suggest that F-region dynamo also contributes to the generation of the fast zonal neutral wind over the magnetic dip equator, and the ionosphere–thermosphere coupling associated with F-region dynamo is significantly important in the phenomenology of the low-latitude, F-region ionosphere and thermosphere. The fast zonal neutral wind occurring in the evening equatorial F region over the magnetic dip equator may constitute a source of atmospheric super-rotation in the low-latitude thermosphere.

In summary, the DE-2 results presented here which is in good agreement with the published CHAMP results reveal a fast zonal neutral wind at the Earth's magnetic dip equator instead of the geographic equator, demonstrating the strong magnetic control of the neutral dynamics via ion drag. The fast zonal neutral wind occurs inside the zonal plasma drift velocity anomaly (PDA) in the evening hours, and such latitudinal structures in the zonal wind and plasma drifts are associated with equatorial ionization anomaly (EIA). This is a strong manifestation of the ionosphere–thermosphere coupling processes over the low latitude.

Acknowledgments The author thanks H. Liu, H. Luhr, and M.A. Abdu for discussing data and its analyses. The author thanks the use of the DE-2 data in NSSDC of NASA.

References

Aggson TL, Maynard NC, Herrero FA, Mayr HG, Brace LH, Liebrecht MC (1987) Geomagnetic equatorial anomaly in zonal plasma flow. J Geophys Res 92:311–315

Appleton EV (1946) Two anomalies in the ionosphere. Nature 157:691

Balan N, Bailey GJ, Moffett RJ (1994) Modeling studies of ionospheric variations during an intense solar cycle. J Geophys Res 99:17,467–17,475

Coley WR, Heelis RA, Spencer NW (1994) Comparison of low-latitude ion and neutral zonal drift using DE2 data. J Geophys Res 99:341–348

Fejer BG, Kudeki E, Farley DT (1985) Equatorial F region zonal plasma drift. J Geophys Res 90:12249–12255

Fuller-Rowell TJ, Codrescu MV, Fejer BG, Borer W, Marcos F, Anderson DN (1997) Dynamics of the low-latitude thermosphere: quiet and disturbed conditions. J Atmos Solar-Terr Phys 61:1533–1540

Hanson WB, Heelis RA, Power RA, Lippincott CR, Zucarro DR, Holt BJ, Harmon LH, Sanatani S (1981) The retarding potential analyzer for Dynamics Explorer-B. Space Sci Instrum 5:503

Heelis RA, Kendall PC, Moffett RJ, Rishbeth H, Windle DW (1974) Electrical coupling of the E and F regions and its effect on F region drifts and winds. Planet Space Sci 22:743

Krehbiel JP, Brace LH, Theis RF, Pinkus WH, Kaplan RB (1981) The Dynamics Explorer Langmuir probe instrument. Space Sci Instr 5:493–502

Lin CH, Liu JY, Fang TW, Chang PY, Tsai HF, Chen CH, Hsiao CC (2007) Motions of the equatorial ionization anomaly crests imaged by FORMOSAT-3/COSMIC. Geophys Res Lett 34:L19101. doi:10.1029/2007GL030741

Liu H, Luhr H, Watanabe S, Kohler W, Henize V, Visser P (2006) Zonal winds in the equatorial upper thermosphere: decomposing the solar flux, geomagnetic activity, and seasonal dependencies. J Geophys Res 111:A09S29. doi:10.1029/2005JA011415

Liu H, Luhr H, Watanabe S (2007) Climatology of the equatorial mass density anomaly. J Geophys Res 112:A05305. doi:10.1029/2006JA012199

Liu H, Watanabe S, Kondo T (2009) Fast thermospheric wind jet at the earth's dip equator. Geophys Res Lett 36:L08103. doi:10.1029/2009GL037377

Liu L, Wan W (2001) The evolution of equatorial trough of ionospheric f-region ionization. Terr Atmos Oceanic Sci 12:559–565

Liu L, Luan X, Wan W, Lei J, Ning B (2004) Solar activity variations of equivalent winds derived from global ionosonde data. J Geophys Res A12305. doi:169 10.1029/2004JA010574

Maruyama N, Watanabe S, Fuller-Rowell TJ (2003) Dynamic and energetic coupling in the equatorial ionosphere and thermosphere. J Geophys Res 108(A11):1396. doi:10.1029/2002JA009599

Namba S, Maeda K-I (1939) Radio wave propagation. Corona, Tokyo, 86pp

Oberheide J, Wu Q, Killeen TL, Hagan ME, Roble RG (2006) Diurnal nonmigrating tides from TIMED Doppler Interferometer wind data: monthly climatologies and seasonal variations. J Geophys Res 111:A10S03. doi:10.1029/2005JA011491

Otsuka Y, Shiokawa K, Ogawa T, Wilkinson P (2005) Geomagnetic conjugate observations of medium-scale traveling ionospheric disturbances at midlatitude using all-sky airglow imagers. Geophys Res Lett 31:CiteID L15803

Raghavarao R, Wharton LE, Spencer NW, Mayr HG, Brace LH (1991) An equatorial temperature and wind anomaly (ETWA). Geophys Res Lett 18(9):1193–1196

Richmond AD, Ridley EC, Roble RG (1992) A thermosphere/ionosphere general circulation model with coupled electrodynamics. Geophys Res Lett 19:601–604

Rishbeth H (1972) Thermospheric winds and the F-region: a review. J Atmos Solar-Terr Phys 183(34):1–47

Spencer NW, Wharton LE, Niemann HB, Hedin AE, Carignan GR, Maurer JC (1981) The Dynamics Explorer wind and temperature spectrometer. Space Sci Instrum 5:417–428

Chapter 29

A Tutorial Review on Sporadic *E* Layers

Christos Haldoupis

Abstract The sporadic *E* layers (*Es*) form in the dynamo region of the ionosphere when metallic ions of meteoric origin are converged vertically in a wind shear. This paper provides a comprehensive update on sporadic *E,* a topic that has been studied for many years. The aim is to render useful information and physical understanding for both the general and specialized reader, and construct an integrated picture of sporadic *E* through a critical synthesis of recent findings. The basic aspects of the layer windshear theory are reviewed and then selected observations are presented which are tested against the theoretical predictions. The emphasis is placed on the tidal wind control of the diurnal and semidiurnal variability and altitude descent of sporadic *E* layers. There is now enough evidence to suggest that mid- and low-latitude sporadic *E* is not as "sporadic" as the name implies but a regularly occurring ionospheric phenomenon. This suggests that *E* layer physics could also be incorporated in existing atmosphere-ionosphere coupling models. Furthermore, the present review summarizes recent findings which provide physical insight into long-going problems and questions about the seasonal dependence and the global occurrence of *Es*. The experimental results, which are in favor of the windshear theory, imply that the key agents controlling sporadic E are: tidal wind atmospheric dynamics, the Earth's horizontal magnetic field component, and the meteoric deposition of metallic material in the lower thermosphere.

C. Haldoupis (✉)
Physics Department, University of Crete, Heraklion, Crete, Greece
e-mail: chald@physics.uoc.gr

29.1 Introduction

The name "sporadic *E*" and its abbreviation "*Es*" refer to a legendary ionospheric phenomenon which is known since the recording of the first primitive ionograms in the early nineteen thirties. It is a generic term used for the thin layers of enhanced metallic ionization that form in the *E* region ionosphere, mostly between about 95 and 120 km. These can at times become denser than the normal *E* layer or even the peak *F* layer, thus they may affect HF radio propagation and *F* region ionosonde recordings severely and therefore can be of relevance to space weather.

Sporadic *E* has been investigated extensively, both experimentally and theoretically. Albeit the great majority of the observational studies have been performed with ionosondes, there are also a considerable number of investigations carried out with incoherent and coherent scatter radars as well as through probing in-situ with rockets. It is not the purpose of this paper to critically summarize all these numerous results and the interested reader should consult past review papers by Whitehead (1989) and Mathews (1998) and the references cited therein, as well as more recent publications quoted in the present paper.

The scope here is to provide a tutorial review on Sporadic *E* that summarizes our present understanding. Instead of providing a detailed treatise of the phenomenon and its complexities, we focus here on basic physical principles and key observational characteristics. An objective is to emphasize a fact that has been overlooked, that is, sporadic *E* is not as "sporadic" as its name implies but rather a regularly occurring phenomenon over a large range of latitudes from a few degrees off the magnetic equator to the auroral zones.

M.A. Abdu, D. Pancheva (eds.), A. Bhattacharyya (Coed.), *Aeronomy of the Earth's Atmosphere and Ionosphere*, IAGA Special Sopron Book Series 2, DOI 10.1007/978-94-007-0326-1_29,

In the following, key theoretical elements which form the basis of sporadic E understanding are presented, followed by the presentation and discussion of selected experimental properties.

29.2 Basic *Es* Theory and Processes

Here, basics of the midlatitude sporadic E theory are presented, which are pertinent to the scope of the present paper and useful in obtaining a clear physical understanding. The physics of sporadic E layer formation is described through the *Windshear Theory*, first proposed and formulated in the early sixties by Whitehead (1961), and Axford (1963), and since developed further by more authors (e.g., see Whitehead, 1989). It relies on the idea that thin layers of ionization can form in the dynamo region of the ionosphere by vertical ion convergence driven by vertical shears in the horizontal neutral wind.

This layer-forming process is controlled fully by ion dynamics, which can be adequately expressed through a simplified version of the ion momentum equation. Following Chimonas and Axford (1968) and neglecting pressure gradient (diffusion) forces at E region heights, gravity, as well as electric field forces at middle and low latitudes, the equation of ion motion includes at steady state only the ion-neutral collisional and geomagnetic Lorentz forces:

$$m_i \, \nu_i(\mathbf{v}_i - \mathbf{U}_n) - e\,\mathbf{v}_i \times \mathbf{B} = 0, \qquad (29.1)$$

where m_i and ν_i are the ion mass and ion-neutral collision frequency, $\mathbf{v}_i$ and $\mathbf{U}_n$ are the ion drift and neutral wind velocities respectively, e is the electronic charge, and $\mathbf{B}$ the geomagnetic field vector. By adopting a geomagnetic south, geomagnetic east and vertically up Cartesian (x,y,z) coordinate system for the northern hemisphere, and using the notations of Mathews (1998) for the vectors $\mathbf{v}_i$ (u, v, w), $\mathbf{U}_n$ (U, V, W) and $\mathbf{B}$ $(-B\cos I, 0, -B\sin I)$, Eq. (29.1) can be solved for the (positive upwards) vertical ion drift:

$$w = \frac{(\nu_i/\omega_i)\cos I}{1+(\nu_i/\omega_i)^2} V + \frac{\cos I \sin I}{1+(\nu_i/\omega_i)^2} U = f_{zn} V + f_{mr}\, U. \qquad (29.2)$$

Here, I denotes the magnetic dip angle while ν_i/ω_i is the ratio of ion-neutral collision frequency to ion gyrofrequency, which introduces an altitude dependence through the decrease with altitude of the ion-neutral collision frequency. The dimensionless parameters f_{zn} and f_{mr} represent the zonal and meridional "ion drift factors". In deriving Eq. (29.2), it is assumed that the vertical wind component is negligible, that is, $W \approx 0$, an assumption that is fairly reasonable.

29.2.1 *Windshear Ion Convergence Mechanisms*

The first and second terms in the right hand side of Eq. (29.2) define two processes of vertical ion convergence which associate with the vertical shears in zonal (V) and meridional wind (U), respectively. The zonal wind shear mechanism is illustrated in the upper part of Fig. 29.1. It involves the horizontal component of the magnetic field $B_H = B\cos I$ and a vertical wind shear characterized by a westward wind above and an eastward wind below. As the ions drift with the wind to the west above (east below) with V_{west} (V_{east}) they are also Lorentz-forced $eV_{\text{west}}B_H$ ($eV_{\text{east}}B_H$) to drift downwards (upwards), therefore converging at an angle to the wind shear null, where $V = 0$, to accumulate and form a layer. The lower part of Fig. 29.1 sketches the meridional wind shear mechanism which requires (for the northern hemisphere) a northward wind above and a southward wind below. Here, the ions are wind-driven horizontally while at the same time they are constrained by the Lorentz force to gyrate about the inclined magnetic field lines. As a result, the ions finally move in the direction of the magnetic field with the wind velocity component $U_{\text{north}}\cos I$ ($U_{\text{south}}\cos I$) and therefore converge to the wind shear null at $U = 0$ to form a layer. Since the electrons are strongly magnetized ($\omega_e >> \nu_e$), they are not affected directly by neutral winds. Therefore, in the ion convergence processes under discussion, the electrons are Coulomb-forced to follow the ions, moving along the magnetic field lines to maintain plasma neutrality.

Note that none of the two ion-convergence mechanisms will work well near the magnetic equator where the magnetic field is fairly horizontal. In the case of a meridional wind shear, the wind simply moves the ions along the field lines thus the Lorentz force is zero and the ions do not move vertically. In the zonal wind

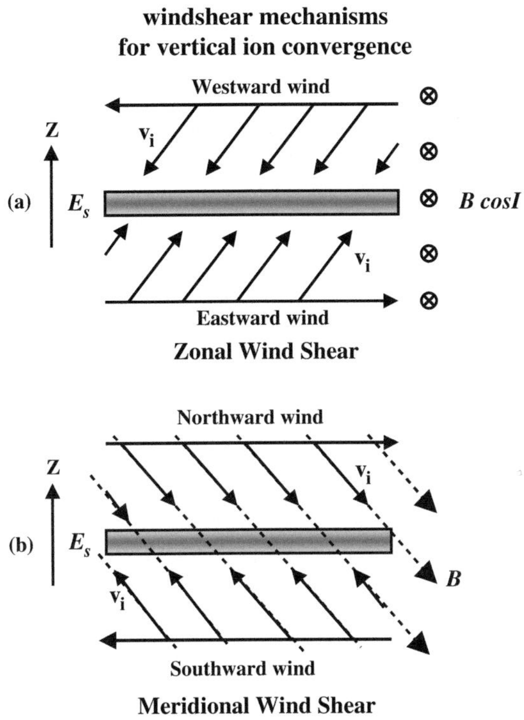

Fig. 29.1 Exemplifying sketches of the zonal (*top*) and meridional (*bottom*) wind shear mechanisms for vertical ion-convergence into a thin ionization layer forming at the wind shear velocity null. More details on the two mechanisms are given in the text

shear process the ions are strongly Lorentz-forced to move vertically, however they cannot converge into a layer easily because they are kept near a fixed magnetic field line by the strongly magnetized electrons, because the plasma must remain neutral. Also, both windshear mechanisms do not work efficiently at high magnetic latitudes (auroral zones). There, the horizontal magnetic field component which is involved in the zonal windshear process is small, whereas for the meridional windshear process the driving wind component along the magnetic field is also small. As a result of the aforementioned geometrical constraints, the strongest and more frequent layers occur at midlatitudes. We note however that the sporadic *E* layers seen occasionally at high latitudes are attributed mostly to the relatively large auroral electric fields which in this case do enter also in Eq. (29.1).

Es layers live long times (from a few to many hours), implying that the windshear ion convergence process cannot rely on ambient molecular ions which live short times, because they neutralize quickly (in a few minutes) through dissociative recombination. To get around this problem, it was suggested that *Es* layers are due to metallic (monoatomic) ions of meteoric origin undergoing slow radiative recombination which requires a three body collision. This fact has been confirmed by rocket and incoherent scatter radar observations (e.g. see Whitehead, 1989 and Mathews, 1998). Note that the metallic ion lifetimes range widely with altitude, from a few days at ~120 km to a few hours at ~95 km (MacDougall et al., 2000).

29.2.2 Ion-Convergence Times

The layer-forming efficiency of the zonal and meridional windshear mechanisms at a given altitude differ because their ion drift factor dependence on ν_i/ω_i is different. This is illustrated in Fig. 29.2, which presents the zonal and meridional ion drift factors for a typical midlatitude location as a function of altitude. The ν_i/ω_i altitude profile used here was taken

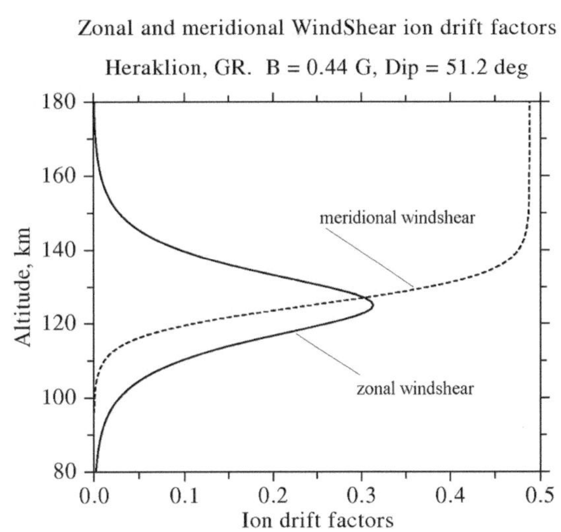

Fig. 29.2 Altitude variation of the zonal and meridional windshear ion drift factors (see Eq. (29.2)) for a typical midlatitude location. They determine the mechanisms' layer forming efficiency at different altitudes. The zonal windshear ion-convergence dominates below 115 km, at altitudes where the majority of sporadic *E* layers are situated

from Bishop and Earle (2003) and refers to metallic ion plasma with a mean ionic mass 40 AMU. As seen, both mechanisms are equally effective at around 125 km where $\nu_i \sim \omega_i$, while at lower (upper) altitudes the zonal (meridional) windshear mechanism becomes dominant. At upper heights the ion-neutral collision frequency is reduced which makes the ions more magnetized, therefore the ability of a zonal wind to move them across the magnetic field is weakened. On the other hand, the meridional windshear process, which moves the ions along the inclined magnetic field direction, works better if the ions are strongly magnetized therefore it dominates at upper heights where $\omega_i \gg \nu_i$.

At lower *E* region heights where collisions are more frequent, the ions become less magnetized, which reduces the action of Lorentz force. This affects both windshear mechanisms but the effect is more severe in the meridional rather than the zonal windshear process. Figure 29.2 shows that the meridional windshear effects on ion vertical motion become minimal below ~115 km, thus ion convergence and *Es* layer forming at altitudes below 115 km is governed by the vertical shears in the zonal wind. The latter can still affect the layers down to altitudes 90 and even 85 km, as suggested by incoherent scatter radar observations of weak lower altitude *Es*.

Equation (29.2) can be used to estimate the time required for ion convergence to cover a vertical distance Δz, as a function of altitude (through the ion-neutral collision frequency) and neutral wind velocity for both the zonal and meridional wind shear mechanisms:

$$t_{zn} = \frac{1 + (\nu_i/\omega_i)^2}{V(\nu_i/\omega_i)\cos I}\Delta z \text{ and } t_{mr} = \frac{1 + (\nu_i/\omega_i)^2}{U \cos I \sin I}\Delta z. \tag{29.3}$$

Such estimates are useful in understanding the layer altitude descent with time, as it will be discussed later. Figure 29.3 shows estimates of vertical ion convergence times for the same midlatitude location and the ion-neutral collision frequency profile used in Fig. 29.2. The ion convergence times were computed for $\Delta z = 1$ km, separately for a strong (weak) zonal and meridional wind components of 100 m/s (10 m/s). As one may infer, a typical meridional wind of 50 m/s can force the ions to drift 1.0 km vertically in a few minutes at altitudes above 120 km, whereas below

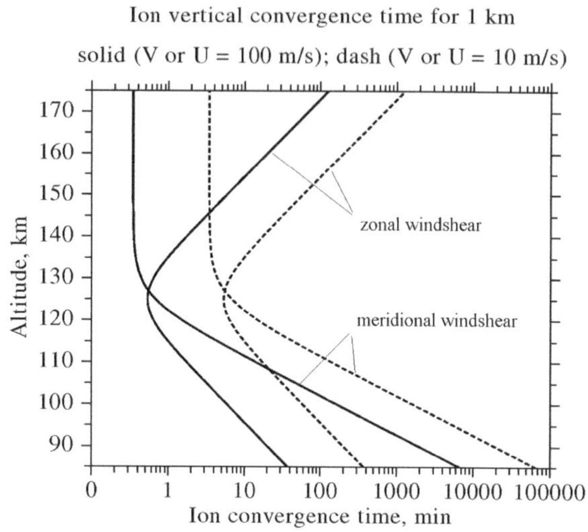

Fig. 29.3 Ion-convergence time estimates as a function of altitude for the zonal and meridional wind shear mechanisms, computed for strong (*solid*) and weak (*dash*) wind driving conditions. See text for details

100 km this requires times of the order of hours. On the other hand, a zonal wind is much more efficient in forcing the ions to move vertically faster at lower altitudes compared to an equally strong meridional wind there. For example, at 110 km the ions need ~10 min to move one kilometer vertically under the forcing of a 50 m/s zonal wind, but ~100 min for a meridional wind of the same magnitude.

It is important to realize that vertical ion-convergence becomes vertical ion-divergence for a windshear polarity opposite of that forming a layer. In this way an existing layer can be broadened and possibly be dissolved by a vertical windshear acting to move the ions apart, for instance in the case of a wind shear in which the wind is eastward above and westward below. Expectedly, the times required for windshear *de-layering* are about the same as for layer forming, which means that the layers can de-form faster (slower) at the altitudes where they form faster (slower). This is the reason why *Es* layers remain fairly stable for many hours at lower heights, especially below 100 km where ion-convergence, or divergence, times are long. On the other hand, at upper heights near 125 to 130 km, layers are more variable and shorter-lived since ion-convergence (divergence) is much faster there.

29.2.3 Plasma Diffusion

A process that acts against ion convergence is ambipolar plasma diffusion. Diffusion is commonly ignored in the sporadic *E* forming process because the ambipolar diffusion coefficient $D_a = k_B(T_i + T_e)/m_i\nu_i$ is small at *E* region altitudes. The role of plasma diffusion becomes increasingly important at upper *E* region altitudes where D_a increases because ion-neutral collisions decrease (e.g., D_a is ~1200 m^2/s at 150 km as compared to ~50 m^2/s at 100 km). Ambipolar plasma diffusion is the physical reason why sporadic *E* layers appear to be wider at higher E region altitudes, despite that meridional wind shear convergence there is more effective in layer formation (e.g., see Fig. 29.3). This also explains why strong wind shears in the *F* region meridional wind cannot form there narrow layers.

To estimate the importance of plasma diffusion in the layering process, we follow Axford (1963) and Kelley (1989) and take a sinusoidal form for a tidal zonal $V = V_0 \sin(k_z z)$, or meridional $U = U_0 \sin(k_z z)$, wind profile where V_0, U_0 and k_z $(=2\pi/\lambda_z)$ are the characteristic wind speed amplitudes and the vertical tidal wavenumber, respectively. Then, plasma diffusion prevents a layer from forming when the characteristic diffusion time $\tau_D = 1/(k_z^2 D_a)$ is smaller than the typical zonal (meridional) windshear plasma convergence time taken from Eq. (29.3), that is:

$$\tau_D \leq \frac{1 + (\nu_i/\omega_i)^2}{k_z V_0 (\nu_i/\omega_i) \sin I}; \quad \tau_D \leq \frac{1 + (\nu_i/\omega_i)^2}{k_z U_0 \cos I \sin I}. \tag{29.4}$$

Figure 29.4 exemplifies the zonal and meridional wind ion convergence and plasma diffusion times computed as a function of altitude for the same parameters used in the previous figures, and for wind speed amplitudes $V_0 = U_0 = 25$ m/s and a tidal vertical wavelength of $\lambda_z = 33$ km, which corresponds to the semidiurnal S(2,6) tidal mode. In this case, plasma diffusion is effective in opposing layer forming above ~140 km and ~180 km for the zonal and meridional windshear processes, respectively. On the other hand, diffusion times are fairly large at lower altitudes relative to convergence times there; therefore, correctly, plasma diffusion becomes unimportant in layer forming at lower heights, say below 130 to 120 km, where narrow sporadic *E* layers are mostly located.

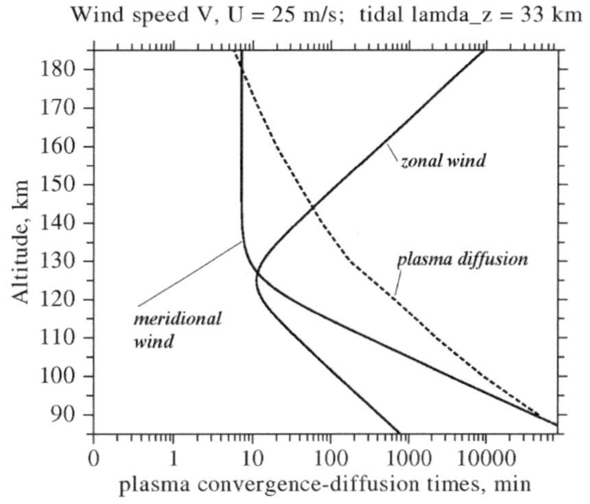

Fig. 29.4 Examples of vertical ion-convergence (or layer-forming) time estimates at different altitudes for a zonal and meridional wind (*solid lines*), as compared to plasma diffusion times (*dashed line*) estimated for a semidiurnal tidal wind of vertical wavelength $\lambda_z = 33$ km. Plasma diffusion has little effect in *Es* layer forming below about 130 km. See text for more details

29.3 *Es* Observational Properties and Interpretations

Here, selected observations that characterize key properties of sporadic E layers are presented. The emphasis is placed on incoherent scatter radar measurements, which show *Es* to be largely a non-sporadic phenomenon. The processes contributing to short-term *Es* variability, and thus to the layer sporadic character, are discussed only in brief. Recent developments on two important *Es* properties, that is, the pronounced seasonal dependence and the global distribution of sporadic *E* layer occurrence are also summarized. The results included here are interpreted in the framework of the windshear theory predictions.

29.3.1 The Non-sporadic Nature of Sporadic E

The windshear theory suggests that sporadic *E* is closely linked to atmospheric dynamics because its formation requires the presence of sheared winds in the lower thermosphere. Inherently, this implies that *Es* possesses a variable but regular character that reflects both, the complexity and repeatability of atmospheric wind and wave dynamics in the MLT (mesosphere – lower thermosphere) region. This has been confirmed by incoherent scatter radar (ISR) studies at Arecibo (Geog. Lat. ~18° N; Magnetic dip ~50°). Besides its advantage in measuring electron density profiles accurately with good time and altitude resolution, it is the superb sensitivity of ~500 electrons per cm^3 of the Arecibo ISR which reveals that sporadic *E* is present nearly all times, that is, it exhibits a non-sporadic behavior, in general. The Arecibo ISR studies showed that there is a well defined tidal variability in sporadic *E*, which prompted Mathews (1998) to introduce the term "tidal ion layers" (TIL) as more appropriate to "sporadic *Es* ".

As discussed by Mathews (1998), the Arecibo ISR observations revealed that the diurnal and semidiurnal tides are the main agents that control the formation and altitude descent of sporadic *E* layers. This was recognized to be also true at higher midlatitudes by Haldoupis et al. (2006) who used a novel method to identify tidal variations and altitude descent in ionogram data. On the average, the tidal winds may dominate and thus govern the diurnal and sub-diurnal variability and descent of the layers through their vertical wind shears which form and drag the layers along as they phase speed propagate downwards. The close connection between *Es* and tides is to be anticipated because, as shown for example by Chapman and Lindzen (1970) and Forbes et al. (2007), the neutral winds in the E region are dominated by solar tides. This setup agrees fairly well with windshear theory, which, as discussed previously, favors metallic ion layer formation at vertical shear convergence nulls. However, atmospheric dynamics in the *E* region can occasionally become rather complex, which may lead to departures from the predominant tidal wind pattern (Larsen, 2002).

The term "sporadic" had been labeled at times as "convenient" and "improper" well before the Arecibo ISR observations. This happened because it was realized that this term was adopted to a large extent as a result of an instrumental limitation rather than of a prominent physical property. The limitation, which made a variable ionospheric phenomenon appear more sporadic in occurrence than it actually is, relates with the fact that the ionosonde electron density measurements are inevitably subject to a sharp lower cutoff near 2.5×10^4 cm^{-3} caused by the instrument's lowest transmitted frequency of ~1.0 to 1.5 MHz. Thus, whenever the peak electron density in a layer would decrease below (increase above) the minimum electron density detected by ionosonde, the layer would disappear (appear), therefore making its occurrence to appear sporadic.

Figures 29.5 and 29.6 show Arecibo ISR observations, presented here in order for the reader to appreciate that *Es* follows a predictably repeating pattern, determined by the tidal modes in the lower thermosphere. Figure 29.5 illustrates a prevailing *Es* tidal variability both in time and altitude, observed during 6 full days of continuous Arecibo ISR operation. Each panel corresponds to a 24-h local day and represents a height-time-intensity (HTI) plot, where the "intensity" here refers to the logarithm of the (positive) vertical electron density gradient, *log(dNe/dz)*. This type of display is quite useful for detecting the altitude location of *Es* layers quite accurately during the course of a 24-h local day, and thus is suitable in identifying the tidal variability and descent of layers with time. As seen in Fig. 29.5, there is a daily trace pattern that exhibits iterating characteristics dominated by tidal-like periodicities. Figure 29.6 presents more typical Arecibo observations obtained from various radar runs, each lasting at least two 24-h days. Shown there are height-time-*log(dNe/dz)* intensity 24-h local day plots which illustrate a repeating pattern in sporadic layer formation and descent.

Inspection of the HTI plots, in both Figs. 29.5 and 29.6, shows that for a given 24-h local day there is usually a set of 3 different *Es* traces. These include a diurnal trace at lower altitudes below about 110 km and two semidiurnal-like upper altitude traces, a daytime trace and a less frequent nighttime trace, which appear at about 140 km prior to noon and midnight, respectively. The lower altitude diurnal layer trace is descending with a speed ~1.0 km/h which agrees reasonably well with the phase velocity of the theoretical diurnal tidal mode S(1,1) which has a vertical wavelength of ~28 km and is known to be dominant below 110 km at lower midlatitudes (e.g., see Harper, 1977). The upper altitude semidiurnal-like layers descend with speeds of about 2–4 km/h towards lower heights to often merge there with the slowly descending diurnal layer below. This repeating *Es* trace pattern shows that sporadic *E* exhibits a well-defined regularity in both formation and altitude descent, in response to tidal windshear dynamics.

As mentioned, Figs. 29.5 and 29.6 show a noticeable difference between the upper altitude semidi-

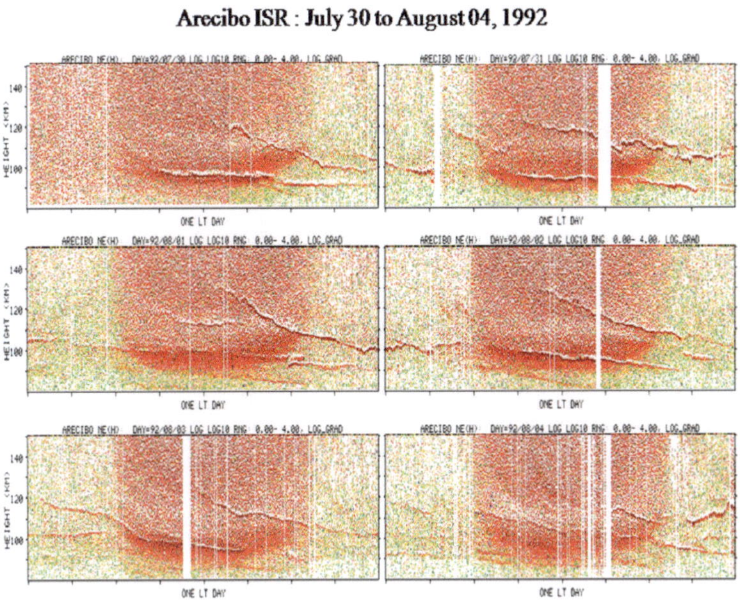

Fig. 29.5 Arecibo incoherent scatter radar *log(dNe/dz)* height-time-intensity plots measured for six consecutive days, showing a regular and repeating pattern of sporadic *E* layer occurrence. This is characterized by lower and upper *E* region altitude layers, which descend in altitude during the course of the day. The lower altitude layers appear to follow a well-defined diurnal variation which is apparently controlled by the diurnal tide. The horizontal axis in each single day plot denotes local time ranging from 0 to 24 hours. The vertical axis in each plot represents height ranging from 80 to 150 km

Fig. 29.6 Same as Fig. 29.5. It presents Arecibo ISR observations of daily sporadic *E* trace activity, which shows a repeating pattern of diurnal and semidiurnal periodicities in layer formation and altitude descent controlled by the diurnal and semidiurnal tides. The vertical height-axis in each plot ranges from 80 to 180 km

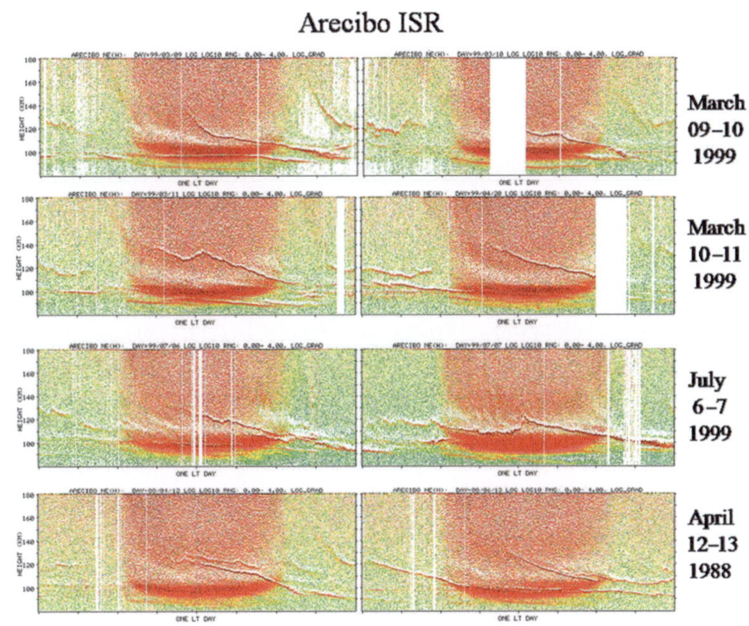

urnal layers. That is, the nighttime layer is often much weaker and less frequent than the daytime one. This is likely because the tidal windshear associated with the formation of the daytime semidiurnal layer collects the metallic ion population produced by the enhanced meteoric influx during past-midnight to morning hours and the subsequent photoionization of metallic atom deposition. In this way, the medium is depleted of metallic ions later in the day, which implies that it may become difficult for

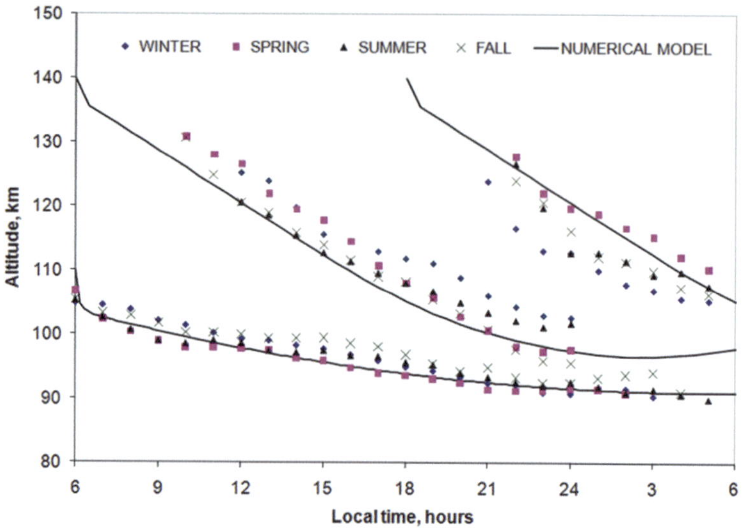

Fig. 29.7 The three sets of points represent seasonal averages for the low altitude diurnal and the upper altitude semidiurnal sporadic *E* layer daily traces as observed by the incoherent scatter radar at Arecibo. The superimposed lines are windshear theory numerical simulations for a diurnal S(1,1) tidal wind mode with a vertical wavelength $\lambda_z = 28$ km (*lower altitude line*) and a semidiurnal S(2,6) zonal tidal wind with $\lambda_z = 33$ km (*upper altitude lines*)

the upper nighttime layer to reach always detectable levels.

The seasonal tidal variability and descent of *Es* layers at Arecibo were studied statistically by Christakis et al. (2009), using for analysis a large ISR data set of 140 days of radar operation, distributed over several years and covering evenly all seasons. The key findings are summarized in Fig. 29.7, which shows the hourly mean altitude positions of the prevailing diurnal and semidiurnal *Es* layers. As seen, there is a three layer daily pattern prevailing, with some differences, in all seasons. The solid lines superimposed on top of the three sets of mean seasonal *Es* trace points represent numerical simulations computed form windshear theory (by solving Eq. (29.2)), using a simplified diurnal or semidiurnal zonal tidal wind of a given amplitude and vertical wavelength λ_z. It was found that the S(1,1) 24-h tide with a vertical wavelength $\lambda_z \sim 25$ km controls fully the formation and descent of metallic *Es* layers at low altitudes below ~110 km. The higher altitude layers and their altitude descent was approximately accounted for by using the semidiurnal tidal mode S(2,6) with $\lambda_z \sim 35$ km.

29.3.2 Physical Interpretation

Taking into consideration the observations and theoretical predictions discussed previously, a physical picture emerges which is summarized as follows. First, let's take the upper altitude daytime and nighttime descending layers that are controlled by a semidiurnal tide. These are associated with the so called intermediate descending layers which often detach from the *F* region bottomside (Mathews, 1998). They form at shear convergence nulls at higher *E* region altitudes (say, 150–180 km) but they become narrow only below about 135 km because plasma diffusion, which counteracts ion-convergence and plasma accumulation, becomes now ineffective below ~140 km (see Fig. 29.4).

According to the windshear theory, a layer remains in a shear convergence null only if it forms fast enough compared to the time required for the null to phase-propagate downward a distance equal to the layer's width. In the upper *E* region the layers form rapidly (see Fig. 29.3) thus they tend to "stick" at a wind shear convergence null as it moves downwards with the vertical tidal phase speed. This is manifested by the steady negative slopes seen in *Es* traces above about 115 to 120 km (see Figs. 29.5 and 29.6), while below these heights the situation changes gradually because ion-neutral collisions become increasingly effective in opposing/delaying ion convergence. As a result, layer descent slows down (trace slopes in Fig. 29.5 start curving) because ion convergence time becomes increasingly larger (see Fig. 29.3), therefore a layer cannot form fast enough to remain inside a tidal convergence null as it does at upper *E* region heights. In fact, the layers lag steadily behind the downward propagating tidal convergence null, therefore they keep descending at rates progressively smaller than the vertical tidal phase velocity. This slow descent continues till a divergent tidal node that follows ion convergence catches up with *Es* to impose a possible de-layering effect because the layered ions tend now to disperse vertically. Depending upon the amplitude and phase velocity of the semidiurnal tide, there is finally a lower altitude near 110 km at which the ion neutral collision frequency is high enough to not allow the 12-h tide to affect the layer. At these altitudes and below, the semidiurnal tides and their associated ion convergent (or divergent) wind shears propagate through a stagnating layer without impacting on it.

The upper altitude semidiurnal tidal layers may merge with lower altitude metallic ionization to contribute in the formation of the prominent 24-h sporadic *E* layer that is controlled by the diurnal S(1,1) tide. Relative to the semidiurnal tides, the S(1,1) 24-h tide has larger amplitudes (e.g., see Harper, 1977) and a shorter vertical wavelength so that it phase-propagates downwards slower than the 12-h tides, therefore it provides the time needed to fully control *Es* layer formation and descent from about 110 km down to 90 km (see Figs. 29.5 and 29.6). Although, the *Es* picture can at a given day become more complicated (see such days in Figs. 29.5 and 29.6), because of the complexity of atmospheric dynamics and the confluence of coexisting tidal modes, as well as gravity wave and possible neutral wind instability effects, the overall pattern of sporadic *E* formation and descent remains fairly well defined and predictable.

29.3.3 Non-tidal Es Variability

Close inspection of the daily plots in Figs. 29.5 and 29.6, show that *Es* traces undergo at times a shorter term variability, which appears to be more prominent during nighttime, as manifested by trace (layer) multiplicity, disruptions and altitudinal distortions, all constituting departures from a clear diurnal/semidiurnal picture. This is because of various causes and physical processes which also enter into action to affect in a rather unpredicted way the *Es* layers, contributing therefore to a "sporadic" character of the phenomenon. One key reason for this is the complexity of atmospheric wave dynamics due to non-linear interactions that result to additional tidal modes. For example, Arecibo ISR (e.g. see Mathews, 1998) and ionosonde studies (e.g., see Haldoupis and Pancheva, 2006) revealed at times also a role for shorter period tidal modes, such as the quarterdiurnal and terdiurnal tides, on *Es* formation and descent. Naturally, it is the confluence of co-existing tidal modes, having different amplitudes and phases, which may create at times significant discrepancies from the dominant and regular diurnal and semidiurnal *Es* pattern.

Ionosonde studies over the last 10 years established that, in addition to the tides, planetary waves (PW) also play a role on *Es* generation which imposes long-term PW periodicities in the occurrence and intensity of sporadic *E* layers (e.g., see Haldoupis and Pancheva, 2002 and more references therein). Further studies, which involved careful analysis of simultaneous ionosonde and MLT neutral wind measurements, showed that the effect of PW on *Es* is impacted indirectly rather than directly. This seems to be done through the diurnal and semidiurnal tides which are modulated by planetary waves, apparently through a nonlinear interaction process at altitudes below 100 km (for details, see Pancheva et al., 2003 and Haldoupis et al., 2003).

At shorter scales, there exist a spectrum of gravity waves in the lower thermosphere which are also expected at times to have modulating effects on *Es* occurrence and intensity. Djuth et al. (2004) showed that gravity wave-like perturbations are imbedded in thermospheric electron density, and that "sets" of waves separated by 20–60 min can be present propagating rapidly downwards with speeds higher than say 100 km/h. These short period, large vertical wavelength gravity waves are weak ($\sim$1–3%) and require special data filtering methods in order to be identified, thus very rarely are visible in the type of displays shown in Figs. 29.5 and 29.6. A rare example of such a gravity wave set was observed during daytime before local noon in July 7, 1999, but their layering effect was weak and thus hardly visible in Fig. 29.5. Traces of such waves show up more easily in the upper *E* region plasma (see Djuth et al., 2004) and have a negative slope due to downward phase propagation which curves the traces towards increasing time as the phase front approaches lower *E* region heights where ion-neutral collisions become much more frequent. The wind shears associated with such waves can affect at times upper altitude sporadic *E* layers (e.g., see Mathews, 1998). Their effect is small below about

115 km apparently because the waves propagate fast through a layer, thus there is no time for their wind shears to impact much of vertical ion convergence or divergence effects on it.

Finally, there are also other forms of complex variability in sporadic *E* which are attributed to neutral air density and/or plasma instabilities as well as localized electrodynamic processes. Neutral instability mechanisms include the wind shear or Kelvin-Helmholtz and/or neutral convective instabilities which are responsible for short scale overturning structures such a Kelvin-Helmholtz billows characterized by vertical scales of a few kilometers and times scales of hours (e.g., see Larsen, 2000; Larsen et al., 2004). Also plasma instabilities, like the two-stream and gradient-drift instabilities (e.g., see Hysell et al., 2002) can lead to the generation of medium to short scale electrostatic plasma irregularities which are field aligned and can cause strong backscatter of radar signals incident perpendicular to the earth's magnetic field. This is an important topic of sporadic *Es* plasma physics and electrodynamics that relates to the turbulent or unstable state of *Es* plasma, which, however, is beyond the scope of, and the available space for, the present paper.

29.3.4 Seasonal Variability

An important property of long-term *Es* variability is a well-defined annual *Es* dependence which is marked by a pronounced summer maximum. There exist several statistical studies (e.g., see Whitehead, 1989) which show that *Es* occurrence and intensity is marked by a conspicuous maximum which forms during June–July and December–January for the northern and southern hemispheres respectively, that is, around the northern and southern hemisphere summer solstices. This seasonal *Es* morphology is inexplicable from the windshear theory and constitutes for it an serious weakness. Although there might be some seasonal differences in the tidal modes involved in *Es* formation, there exists no evidence to suggest that something dramatic is happening in the driving tides during summer which can account for the conspicuous summer maximum in occurrence and intensity. The annual *Es* dependence attracted considerable attention through the many years of *Es* research but, despite the efforts, no comprehensive explanation emerged until recently, when long term measurements of the meteoric, and thus metallic material, deposition in the atmosphere became available.

The role of metallic ions is recognized as an essential constituent in layer forming, and often it has been invoked as one of the reasons contributing to the sporadic nature of the phenomenon. Therefore, among the possibilities proposed for the explanation of *Es* summer maximum was also the increase of metal ion content (Whitehead, 1989). This was logical to consider since sporadic *E* is due to metallic ions provided by the atmospheric ablation of meteoroids, therefore the layer mean electron density (intensity) and occurrence are expected to depend directly on metallic material deposition in the lower thermosphere. The metallic ion variability as the cause for the *Es* summer maximum has been excluded however in earlier studies on the basis that the meteoric influx was sporadic and that no evidence existed in favour of a strong seasonal dependence.

The option of metallic ion seasonal dependence and its effect on sporadic *E* occurrence and intensity was brought up again recently after the publication of a series of meteor radar measurements in both the northern and southern hemispheres. These studies revealed a strong seasonal dependence for the daily meteor count rates which, as in sporadic *E*, it was marked by a pronounced summer maximum (e.g., see Singer et al., 2004; Janches et al., 2004; Lau et al., 2006). This variability was attributed to the fact that the sporadic meteor radiants are not randomly distributed in the sky but arrive from well-defined sources located near the ecliptic plane.

By using some of these meteor measurements, a recent study by Haldoupis et al. (2007) established a close correlation between the annual variation of ionosonde sporadic *E* layer intensities and meteoric deposition rates. Figure 29.8, illustrates the good correlation that exists between the mean annual variation of daily meteor counts measured in northern Europe over a period of 6 years, and simultaneous *foEs* daily means taken from an ionosonde station in the European midlatitude sector. The quantity *foEs* is the layer critical ionosonde frequency which relates approximately to the maximum electron density N_{em} through $foEs = 9.0\,(N_{em})^{1/2}$ (where N_{em} is measured in m^{-3} and *foEs* in Hz). Since the occurrence and strength of sporadic *E* layers depends directly on the metal ion content,

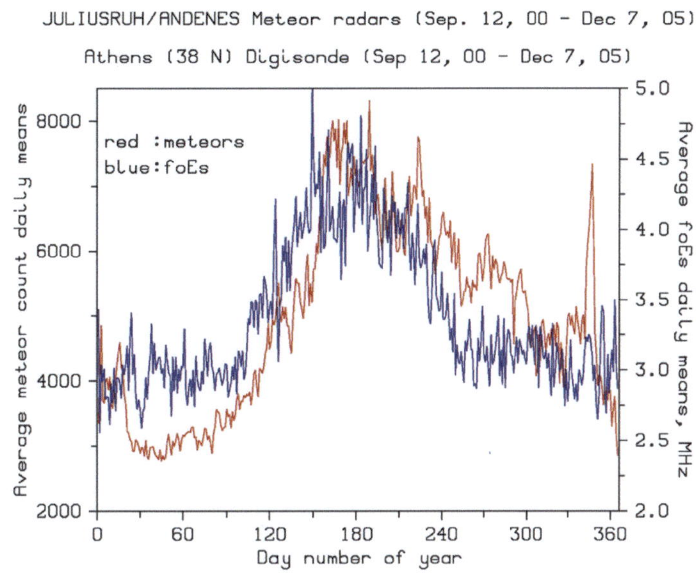

Fig. 29.8 Comparison of the mean annual variation of daily mean meteor counts measured partly in Juliusruh and partly in Andenes (*red line*), and ionosonde *foEs* statistics measured by the Athens ionosonde (38.0°N) for the period from September 2000 to December 2005. As seen, the correlation between meteor count rates and sporadic *E* intensity is fairly good, suggesting a likely cause and effect relationship

which apparently is determined by meteoric deposition, the findings of Haldoupis et al. (2007) offered for the first time a likely cause-and-effect explanation for the long-going mystery of sporadic *E* layer seasonal dependence. This agrees well with the established importance of metallic ions in the formation of sporadic *E*.

29.3.5 Global Es Occurrence and Variability

There are several studies on the world-wide distribution of Sporadic *E* made in the 70's and 80's, which were based on routine ionosonde observations from as many locations as possible (see Whitehead, 1989 for a discussion and references). Although most of these studies agreed that the strongest and most frequent layers occur at midlatitudes, their findings differed widely, a fact that might have been anticipated, given the few reliable ionosondes that were available at that time plus their uneven global distribution. Several of the early studies suggested as a reason for the world-wide occurrence and intensity of *Es* the global variability of the horizontal magnetic field component, *H*, whereas others found a small *H* dependence, or attributed the *Es* global distribution to the world-wide thunderstorm activity. Also there have been findings which contradicted the wind shear theory, for example reports on strong blanketing sporadic *E* layers near the magnetic equator where layer formation is, according to theory, inhibited. Although there are valuable information in all these old studies, which are characterized by great scientific intuition and thoroughness, overall they have been inconclusive and failed to provide an acceptable picture on the global distribution of sporadic *E* and the physical reasons behind it.

The answer to the problem of *Es* global distribution came only recently with the use of a new methodology which involved LEO (low earth orbiting) satellite GPS (global positioning system) occultation measurements that have the advantage of good global coverage (e.g., see Hocke and Tsuda, 2001). The GPS signals received at LEO satellites are modified by refractive index changes in the atmosphere and ionosphere and their analysis can provide information on various atmospheric parameters including electron density fluctuations. The method has been particularly suitable for observing GPS signal occultations caused by electron density perturbations in relation with sporadic *E* layers. The analysis of LEO occultation data yields information on the occurrence of sporadic *E* with good spatial (geographic) and altitude resolution. Thus, the method has been used for measuring the distribution of sporadic *E* world-wide (e.g., Hocke et al., 1981; Wu et al., 2005; Wu, 2006; Arras et al., 2008)

The most complete study to date on the global distribution of Sporadic *E* occurrence, which is based on

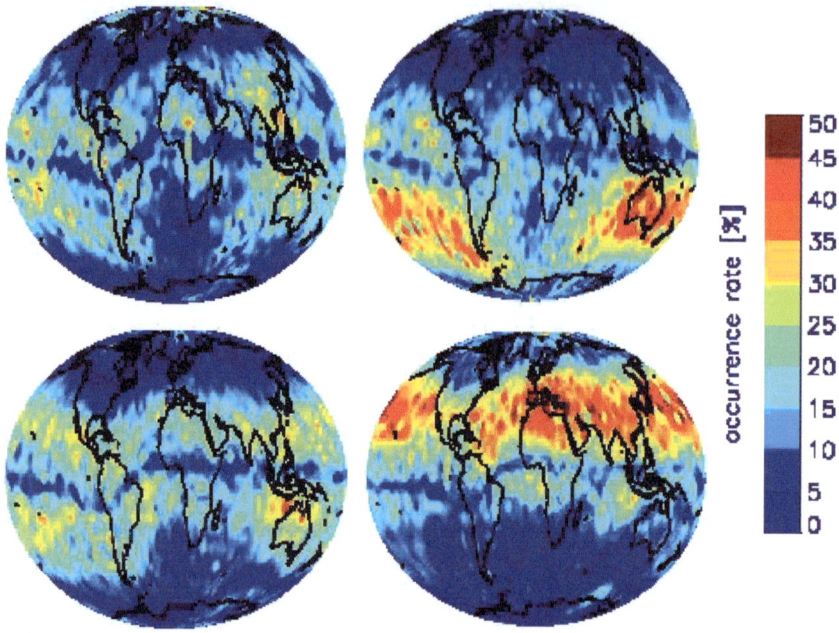

Fig. 29.9 Global distribution of sporadic *E* layer occurrence during the (north hemisphere) four seasons of 1 year (2006–2007): fall (*top left*), winter (*top right*), spring (*bottom left*) and summer (*bottom right*), as measured by GPS radio occultation methods. See more text for details. (Figure is published in Arras et al. (2008) and is kindly provided by Christina Arras)

a large data base of GPS radio occultations obtained with LEO satellites, was made by Arras et al. (2008). Figure 29.9 (courtesy of *Christina Arras*) shows the global distribution of sporadic *E* during all 4 seasons, based on a single year of occultation data from three different LEO satellites, taken between September 2006 and August 2007. As seen, the world-wide distribution characteristics of sporadic *E* occurrence have been depicted quite clearly for all four seasons. The strong seasonal variability agrees very much with that discussed in the previous section, that is, sporadic *E* exhibits a pronounced summer maximum (right panels in Fig. 29.9).

As noted by Arras et al. (2008), Fig. 29.9 shows considerable structure in the global distribution of *Es* occurrence characterized by well-defined regions of maxima and minima. The maxima appear at the midlatitude zones between about 10 and 60 degrees geomagnetic latitude. Also seen is a striking reduction in *Es* occurrence inside a narrow zone of about 3–5 degrees in latitude centred at the magnetic equator, which prevails clearly during all seasons. This is in good agreement with the windshear theory which inhibits layer formation at small dip angles because of the unique magnetic filed geometry at the equator and the need for the plasma to remain neutral, as discussed before in the theoretical part of this paper. Also there are deep minima located at high latitudes where dip angles are larger than 70 to 80 degrees, which again agrees with the windshear theory, as discussed earlier in 2.1. Also, as pointed by Arras et al. (2008), *Es* occurrence increases at midlatitudes during summer solstices with two noticeable exceptions, that is, two deep minima: one over the south Atlantic anomaly in the southern hemisphere and the other over northern America in the northern hemisphere, that is, at the two midlatitude regions of the globe where the horizontal magnetic field, *H*, is strongly reduced. This points to a decisive role of *H* in *Es* formation through the Lorentz force that drives the ion-convergence in the zonal windshear mechanism.

To strengthen the importance of this last finding, Fig. 29.10 is provided that shows the global distribution of the horizontal magnetic field intensity *H,* computed from IGRF (international geomagnetic reference field) at 105 km altitude. A qualitative comparison of Figs. 29.9 and 29.10 shows that the horizontal magnetic field is the main reason behind the world-wide distribution of sporadic *E* occurrence at midlatitude, a result that is in good agreement with windshear theory. Since *H* is entering only in the zonal windshear mechanism, the evidence here favors the zonal windshear process as the main player in *Es* forming relative to the meridional windshear. This is logical

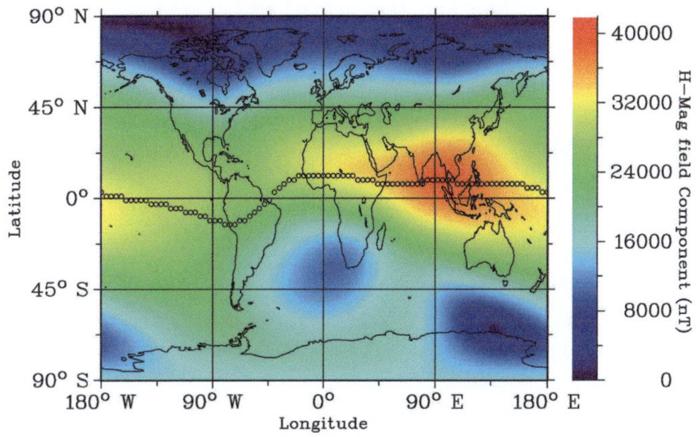

Fig. 29.10 Global map of the horizontal magnetic field component *H,* computed from the international geomagnetic reference (magnetic) field (IGRF) model. Comparison with Fig. 29.9, shows that the horizontal magnetic field component *H* is the key agent responsible for the global sporadic *E* occurrence distribution. This supports the windshear theory and the dominant role played in *Es* formation by the vertical shears in the horizontal zonal wind

to expect because the zonal windshear mechanism operates more effectively at lower *E* region heights where most of sporadic *E* layers are known to be situated.

29.4 Summary and Concluding Comments

The present paper is a tutorial review that provides a comprehensive update of our present basic knowledge on, and physical understanding of, midlatitude sporadic *E* layers (*Es*). It starts with a description of the *Es* windshear theory basics placing the emphasis on the physical picture defined by the driving forces and ion-convergence mechanisms of layer forming at different *E* region altitudes. Next, key observations are selected which are presented and discussed, showing that sporadic *E* layer formation and altitude descent are controlled by the vertical wind shears of atmospheric waves in the lower thermosphere. Although there is a number of parameters involved in sporadic *E* layer occurrence and intensity, which all may affect the overall *Es* formation and dynamics, the effects of the diurnal and semidiurnal tides though their vertical wind shears, remain prominent and are of fundamental importance.

Provided there is an abundance of metallic ions and atoms in the lower thermosphere, the close dependence of sporadic *E* on tidal wind shears, as the Arecibo incoherent scatter radar confirms, show that *Es* is a variable but a non-sporadic phenomenon. The present review places an emphasis on the "non-sporadic" character of sporadic *E* in order to draw attention to a fact that has so far been overlooked. That is, the non-sporadic character of *Es* is important because it implies that the physics of *Es* can be integrated in the existing atmosphere-ionosphere coupling models. In addition, the present paper discusses recent observations which provided long-waited explanations to problems associated with the seasonal dependence and global distribution of sporadic *E* layers. These findings identified the decisive contributions in sporadic *E* occurrence and intensity of the prominent seasonal dependence of meteoric deposition, and the global variability in the Earth's horizontal magnetic field component. These, along with the diurnal and semidiurnal tides, are the key agents that control the variability and dynamics of sporadic *E*.

The present paper is far from a complete review of a rather diversified and long-studied subject. A complete treatise of sporadic E would have required dealing with additional *Es*-related processes and properties. Other than a brief reference to additional phenomena, the present paper does not deal with: (1) short term and/or quasi-periodic variations in sporadic *E* caused, possibly and partly, by gravity waves, and wind shear Kelvin-Helmholtz type instabilities, (2) short- and medium-scale electrostatic irregularities and plasma instabilities, and (3) electrodynamics and sporadic E – spread F coupling processes. All these are important topics of active and undergoing research. Also, other than a brief mention, the present paper did not deal with long-term variability in Es caused by planetary waves. The latter, which is a relatively new topic, is hoped to be dealt with in a separate publication.

Acknowledgements I wish to thank Dora Pancheva, Chris Meek, Sergei Shalimov, Qihou Zhou, Nikos Christakis, Alain Bourdillon, and Glenn Hussey, with whom I worked jointly the last several years in researching midlatitude sporadic E layers. Also wish to express my gratitude to Christina Arras for kindly providing Fig. 29.9 of this chapter. ELKE, University of Crete provided support for this work through grant 3116.

References

Arras C, Wickert J, Beyerle G, Heise S, Schmidt T, Jacobi C (2008) A global climatology of ionospheric irregularities derived from GPS radio occultation. Geophys Res Lett 35:L14809. doi:10.1029/2008GL034158

Axford WI (1963) The formation and vertical movement of dense ionized layers in the ionosphere. J Geophys Res 68:769

Bishop RL, Earle GD (2003) Metallic ion transport associated with midlatitude intermediate layer development. J Geophys Res 108:1019

Chapman S, Lindzen RS (1970) Atmospheric tides. D. Reidel, Hingham, MA

Chimonas G, Axford WI (1968) Vertical movement of temperate zone sporadic E layer. J Geophys Res 73:111

Christakis N, Haldoupis C, Zhou Q, Meek C (2009) Seasonal variability and descent of mid-latitude sporadic E layers at Arecibo. Ann Geophys 27:923–931

Djuth FT, Sulzer MP, Gonzales SA, Mathews JD, Elder JH (2004) A continuum of gravity waves in the Arecibo thermosphere. Geophys Res Let 31:L16801. doi:10.1029/2003GL019376

Forbes JM, Zhang X, Palo S, Russell J, Mertens CJ, Mlynczak M (2008) Tidal variability in the ionospheric dynamo region. J Geophys Res 113:A02310. doi:10.1929/2007JA012737

Haldoupis C, Meek C, Christakis N, Pancheva D, Bourdillon A (2006) Ionogram height-time intensity observations of descending sporadic E layers at mid-latitude. J Atmos Solar-Terr Phys 68:539

Haldoupis C, Pancheva D (2002) Planetary waves and midlatitude sporadic E layers: strong experimental evidence for a close relationship. J Geophys Res 107. doi:10.1029/2001JA000212

Haldoupis C, Pancheva D (2006) Terdiurnal tidelike variability in sporadic E layers. J Geophys Res 111:A07303. doi:10.1029/2005JA011522

Haldoupis C, Pancheva D, Mitchell NJ (2004) A study of tidal and planetary wave periodicities present in midlatitude sporadic E layers. J Geophys Res 109:A02302. doi:10.1029/2003JA010253

Haldoupis C, Pancheva D, Singer W, Meek C, MacDougall J (2007) An explanation for the seasonal dependence of midlatitude sporadic E layers. J Geophys Res 112:A06315. doi:10.1029/2007JA012322

Harper RM (1997) Tidal winds in the 100- to 200-km region at Arecibo. J Geophys Res 82:3243

Hocke K, Tsuda T (2001) Gravity waves and ionospheric irregularities over tropical convection zones observed by GPS/MET radio occultation. Geophys Res Lett 28: 2815–2818

Hocke K, Igarashi K, Nakamura M, Wilkinson P, Wu J, Pavelyev A, Wickert J (2001) Global sounding of sporadic E layers by the GPS/MET radio occultation experiment. J Atmos Solar-Terr Phys 63:1973–1980

Hysell DL, Yamamoto M, Fukao S (2000) Imaging radar observations and theory of type I and type II quasiperiodic echoes. J Geophys Res 107:1360

Janches D, Palo SE, Lau EM, Avery SK, Avery JP, de la Pena S, Makarov NA (2004) Diurnal and seasonal variability of the meteoric flux at the South Pole measured with radars. Geophys Res Lett 31:L20807. doi:10.1029/2004GL021104

Kelley MC (2009) The earth's ionosphere: plasma physics and electrodynamics, 2nd edn. Academic, San Diego, CA, p 281

Larsen MF (2000) A shear instability seeding mechanism for quasiperiodic echoes. J Geophys Res 105(24):931

Larsen MF (2002) Winds and shears in the mesosphere and lower thermosphere: Results from four decades of chemical release wind measurements. Geophys Res Lett 107:1215. doi:10.1029/2001JA000218

Larsen MF, Liu AZ, Gardner CS, Kelley MC, Collins S, Friedman J, Hecht JH (2004) Observations of overturning in the upper mesosphere and lower thermosphere. J Geophys Res 109:D02S04. doi:10.1029/2002JD003076

Lau EM, Avery SK, Avery JP, Janches D Palo SE, Schafer R, Makarov NA (2006) Statistical characterization of meteor trail distribution at the south pole as seen by a VHF interferometric meteor radar. Radio Sci 41:RS4007. doi:10.1029/2005RS003247

MacDougall JW, Plane JM, Jayachandran PT (2000) Polar cap Sporadic E: part 2, modeling. J Atmos Solar-Terr Phys 62:1169–1176

Mathews JD (1998) Sporadic E: current views and recent progress. J Atmos Solar-Terr Phys 60:413

Pancheva D, Haldoupis C, Meek CE, Manson AH, Mitchell NJ (2003) Evidence of a role for modulated atmospheric tides in the dependence of sporadic E on planetary waves. J Geophys Res 108. doi:10.1029/2002JA009788

Singer W, von Zahn U, Weiss J (2004) Diurnal and annual variations of meteor rates at the arctic circle. Atmos Chem Phys 4:1355–1363

Whitehead JD (1961) The formation of the sporadic E layer in the temperate zones. J Atmos Solar-Terr Phys 20:49

Whitehead JD (1989) Recent work on midlatitude and equatorial sporadic E. J Atmos Solar-Terr Phys 51:401

Wu DL (2006) Small-scale fluctuations and scintillations in high-resolution GPS/CHAMP SNR and phase data. J Atmos Solar-Terr Phys 68:999–1017

Wu DL, Ao CO, Hajj GA, de la Torre Juarez M, Mannucci J (2005) Sporadic E morphology from GPS-CHAMP radio occultation. J Geophys Res 110:A01306. doi:10.1029/2004JA010701

Chapter 30

Long-Term Trends in the Upper Atmosphere – Recent Progress

Jan Laštovička

Abstract Anthropogenic emissions of greenhouse gases influence the atmosphere at nearly all altitudes between ground and space, thus affecting not only life on the surface, but also the space-based technological systems on which we increasingly rely. Long-term changes and trends in the upper atmosphere (mesosphere, thermosphere and ionosphere) are affected also by other drivers, stratospheric ozone depletion, changes in atmospheric wave forcing, changes in water vapour concentration, long-term changes of geomagnetic activity, and secular changes of Earth's magnetic field. The global trend scenario in the upper atmosphere is presented. It forms a consistent pattern of global change at heights above 50 km. The upper atmosphere is generally cooling and contracting, and related changes in chemical composition (minor components) are affecting the ionosphere. A significant progress reached in the last few years in the three areas, which did not fit the global scenario of trends in the upper atmosphere, is described in more detail.

30.1 Introduction

The upper atmosphere is located above an altitude of about 50 km and consists of the mesosphere, thermosphere and ionosphere, the ionosphere being ionized part embedded in the mesosphere and particularly thermosphere. This region is affected by the solar radiation variability and space weather processes like geomagnetic storms on one hand, and by neutral atmospheric processes and atmospheric waves coming from below on the other hand. Neither of these natural drivers is stable and, moreover, the upper atmosphere is affected also by anthropogenic phenomena and processes. Therefore it is not surprising that the upper atmosphere is changing on both short- and long-term time scales, most of these changes being quasi-periodic variability but long-term trends also do occur.

The main driver of long-term trends in the atmosphere as a whole is now the increasing concentration of greenhouse gases, particularly of CO_2. In the troposphere this increase results in the well-known greenhouse warming. However in the upper atmosphere the infrared radiative cooling by CO_2 dominates over the effect of CO_2 trapping of the outgoing longwave radiation, because the CO_2 layer in the upper atmosphere is optically thin for the outgoing longwave radiation and is also optically thin for the infrared radiation emitted by CO_2. Thus the increasing concentration of greenhouse gases results in cooling of the upper atmosphere. The cooling is expected to result in thermal contraction of the upper atmosphere and we may expect a significant decline in thermospheric density at fixed heights as well as a downward displacement of ionospheric layers; this was supported already by the first model studies by Roble and Dickinson (1989), Rishbeth (1990) and Rishbeth and Roble (1992), which opened this area of research, and later supported by observational studies. However, it is necessary to mention that for some parameters the observational results of various authors are controversial and some of them differ from or are even opposite to model expectations.

Further on we shall deal with one important feature of the long-term trend in the upper atmosphere, which

J. Laštovička (✉)
Institute of Atmospheric Physics ASCR, 14131 Prague, Czech Republic
e-mail: jla@ufa.cas.cz

M.A. Abdu, D. Pancheva (eds.), A. Bhattacharyya (Coed.), *Aeronomy of the Earth's Atmosphere and Ionosphere*, IAGA Special Sopron Book Series 2, DOI 10.1007/978-94-007-0326-1_30, © Springer Science+Business Media B.V. 2011

is its long-term non-stability due to changing role of various trend drivers. Then a global scenario of trends will briefly be introduced and a recent progress in areas of problems and contradictions will be summarized.

30.2 What Causes Upper Atmospheric Trends?

Which agents are the main drivers of the observed long-term trends? The main driver of the observed long-term trends in the upper atmosphere is at present considered to be the increasing concentration of greenhouse gases in the atmosphere, which affects the whole atmosphere from surface up to heights of several hundred kilometres (e.g., Laštovička et al., 2006a), both in the neutral and ionized component. However, there are also other drivers of long-term changes and trends in the upper atmosphere.

In the thermosphere and in the ionized component, the ionosphere, several other drivers can play a role in long-term trends. The solar activity in terms of variability of the solar radiation absorbed in the thermosphere and ionizing the ionosphere is rather decreasing in the second half of the 20th century, and particularly in the beginning of the 21st century, which is a tendency opposite to what is requested to explain the observed ionospheric trends in the E and F1 regions. Moreover, the effect of solar activity in terms of solar cycle is carefully removed and/or very reduced when long-term trends are computed both in the ionosphere and thermosphere (different corrections to solar activity is one of the sources of differences between different trend results in the F2-region parameters, foF2 and hmF2). Thus the solar activity itself has no direct effect on the observed trends. However, it is necessary to mention that trends may be quantitatively different under solar activity maximum and minimum conditions, which is the case for thermospheric density. Whether it is the case for some other upper atmospheric and ionospheric variables is one of topics of the new SCOSTEP project CAWSES-II. This may be considered as an indirect effect of solar activity on trends in the thermosphere and ionosphere.

A potentially important driver is geomagnetic activity and its long-term changes. An increase of geomagnetic activity throughout the twentieth century has been reported (Clilverd et al., 1998; Mursula and Martini, 2006), even though due to some calibration/instrumental problems in the 1950s the increase is probably smaller than thought earlier (e.g., Martini and Mursula, 2008). However, at present the geomagnetic activity is no more increasing. The geomagnetic activity changes might be even dominant in trends in foF2. This is discussed in more detail later. Laštovička (2005) found that in the 20th century the role of geomagnetic activity in the observed long-term trends/changes in the atmosphere-ionosphere system was decreasing from its beginning towards its end. Mikhailov and de la Morena (2003) found that trends in foE were controlled by geomagnetic activity before about 1970, but not in more recent years. The higher we go in the ionosphere, the later the geomagnetic control is lost, as shown in Fig. 30.1 for two European stations, Slough/Chilton (England) and Rome (Italy). Decreasing tendency of critical frequencies foE, foF1 and foF2 (i.e. maximum electron densities in the ionospheric E, F1 and F2 layers) with increasing geomagnetic activity corresponds to the predominant effect of geomagnetic storms on the ionosphere and, therefore, means the geomagnetic activity control. This is lost around 1970 for foE, in the early 1990s for foF1 and maybe near 2000 for foF2 (more data are need to confirm it).

Another regionally important factor is the secular change of the main magnetic field of the Earth, which forces the International Association of Geomagnetism and Aeronomy (IAGA) to issue each 5 years a new reference map of the surface magnetic field. This effect is insignificant in most regions, but in the equatorial to mid-latitude southern Atlantic Ocean and in most of the South America it seems to play a substantial role in trends in hmF2 and foF2 after model calculations by Cnossen and Richmond (2008). A physical analysis by Elias (2009) confirms important regional role of secular changes of the Earth's magnetic field in long-term trends in the F2 region; lowering and lifting of the F2-region peak is induced together with changes in foF2 depending on season, time and location. The effect is dominated by changes of inclination and to some extent also declination of the Earth's magnetic field, but changes of field strength can hardly be fully neglected, as well.

Another possible factor may be a long-term decrease of thermospheric atomic oxygen concentration, as hypothesised by Danilov (2005). The most probable origin of this negative trend is the

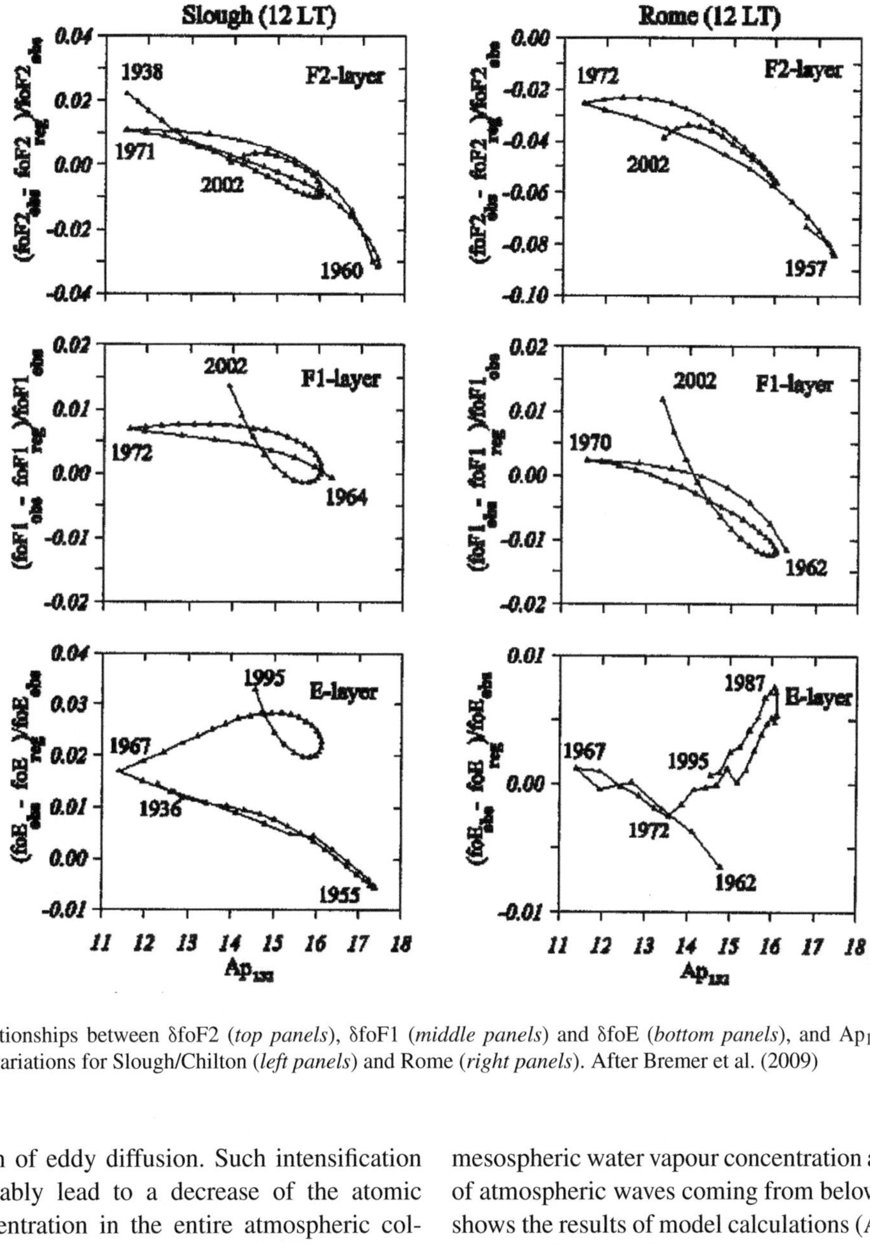

Fig. 30.1 Relationships between δfoF2 (*top panels*), δfoF1 (*middle panels*) and δfoE (*bottom panels*), and Ap_{132} (132-months smoothed Ap) variations for Slough/Chilton (*left panels*) and Rome (*right panels*). After Bremer et al. (2009)

intensification of eddy diffusion. Such intensification should inevitably lead to a decrease of the atomic oxygen concentration in the entire atmospheric column above the E region (Danilov, 2005). However, this is not generally accepted as an important mechanism of global change in the upper atmosphere. Other until now unknown potential sources of long-term trends in the thermosphere and ionosphere cannot be excluded, but if they exist, their role is probably small or negligible.

At lower heights, in the mesosphere and lower thermosphere (MLT region), the most important additional driver seems to be the stratospheric ozone depletion and some role may be played by changes in mesospheric water vapour concentration and in activity of atmospheric waves coming from below. Figure 30.2 shows the results of model calculations (Akmaev et al., 2006) of impact of the decreasing ozone and increasing water vapour concentrations on trends in atmospheric density. Maximum effect of the stratospheric ozone depletion, which amplifies greenhouse thermal contraction of the upper atmosphere, is simulated at 110 km, i.e. in the lower thermosphere, and it seems to be detectable well into the thermosphere, at least up to 200 km. Water vapour also seems to enhance the contraction of the atmosphere but its role is less important. This model probably somewhat overestimates trends in atmospheric density caused by these two drivers but

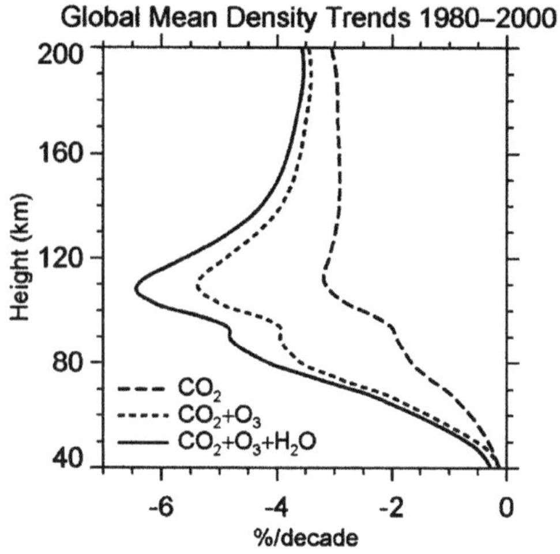

Fig. 30.2 Trends in global-mean atmospheric density at 40–200 km caused by changes of CO_2, ozone and water vapour concentrations over 1980–2000 according to model calculations by Akmaev et al. (2006)

the existence of some influence of these two drivers is undisputable. Unfortunately, there is no observational information about trends in ozone concentration in the mesosphere and mesopause region, which could affect trends in various quantities in the mesosphere and lower thermosphere region, except for the analysis of ozone data from UARS/HALOE satellite observations, which provides no detectable trend for sunrise ozone but a strong negative trend for sunset ozone in a narrow region near 80 km, in anticorrelation with the water vapour trends (Marsh et al., 2003). However, due to the length of the analysed interval and uncertainty in the present-day water vapour trend, the mesospheric ozone trend remains uncertain. Further support for the role of stratospheric ozone in long-term trends in the MLT region is provided by ionospheric data analyses. An analysis of long-term variations of the low frequency reflection height near 81 km and the total ozone content from Arosa (Switzerland) revealed similarity in trends in the sense that in the period of the midlatitude ozone decrease of 1980–1995 the trend in reflection heights was steeper (Bremer and Peters, 2008). Bremer (2008) investigated longitudinal variation of deviations of foE trends and total ozone trends from the zonal mean trends (ionosondes and ERA40 in 30°N–75°N). He found a clear anticorrelation between deviations of foE and total ozone trends both for winter and summer with slight phase shift, and more pronounced effect in winter. On the other hand, at F1 region heights in foF1 no relation between trends in foF1 and total ozone has been observed (Bremer, 2008).

Atmospheric waves coming from below appear to be the main agent responsible for coupling of the lower atmosphere to the upper atmosphere. Very little is known about their long-term changes and trends. However, their trends may partly be caused by changes in the source region, i.e. in the troposphere and, therefore, they may differ from the trends in the upper atmosphere, which are mostly excited in situ. Portnyagin et al. (2006) and Merzlyakov et al. (2009) showed that the prevailing wind and semidiurnal tide in the mesopause region on the Northern Hemisphere as well as in Antarctica changed the trend (mostly to opposite one) around 1990. Changing activity of gravity waves was suggested as possible explanation (Laštovička et al., 2010) but more data are needed to test this hypothesis.

Greenhouse gas concentration is steadily and more or less homogeneously increasing. However, other drivers change their behaviour both in time and space. Stratospheric ozone concentration after a period of depletion reversed trend and began to recover; its changes are large at high latitudes but almost negligible at low latitudes. Geomagnetic activity is no more increasing and its effect is more important at high than at low latitudes. Secular changes of the main magnetic field of the Earth are very regional phenomenon introducing in various regions even opposite effects on trends. *Therefore we can hardly expect stable, non-changing trends and stable role of various factors responsible for trends – they will be changing both with time and space (location).*

What is to be understood under the term trend? Most authors understand the term "trend" as a "long-term trend", i.e. as a long-term tendency of essentially monotonic linear/quasi-linear change, either an increase or decrease of the values of the studied variable with time. Strictly speaking, trends are often not quite linear. However, in many cases the linear trend approximation is sufficient, and in other cases we use the linear approximation for easier comparison with trends in other parameters. Since the relative role of some trend drivers and in some cases even sign of their effect change with time (e.g. ozone), an overall linear trend cannot be stable in time and in reality a quasi-stable trend should begin and end within a certain period of time. Such behaviour may be characterized by a sophisticated mathematical curve, but may often be approximated by the so-called piecewise contiguous linear trend model (Weatherhead et al., 2000; Reinsel et al., 2005), wherein substantially different linear trends are fit to the data within different time intervals, the latter also being model parameters. The upper atmospheric and ionospheric linear trends should be understood in such a way, even though for some parameters the period of quasi-stable trend may be identical with or longer than the period of available data for trend studies, i.e. such trends appear to be monotonic.

It is also possible to use a different and more mathematical approach. A true linear trend is located at infinite period and is not satisfactorily taken into account in a spectral decomposition. Assuming the trend is not a small temporal part of a major very long period cycle or due to some temporally-coincident stochastic change, then it is possible to decompose temporally and spectrally the time series to separate the oscillatory and true trend elements using, for instance, the wavelet transform technique. However, practical application of such an approach is limited and has to be made with care, as the trends have a non-infinite period limited by the length of the dataset or, which is more important, the length of the interval of a stable linear trend is limited and after some time the trend changes due to change of behaviour of trend drivers, as it is shown above.

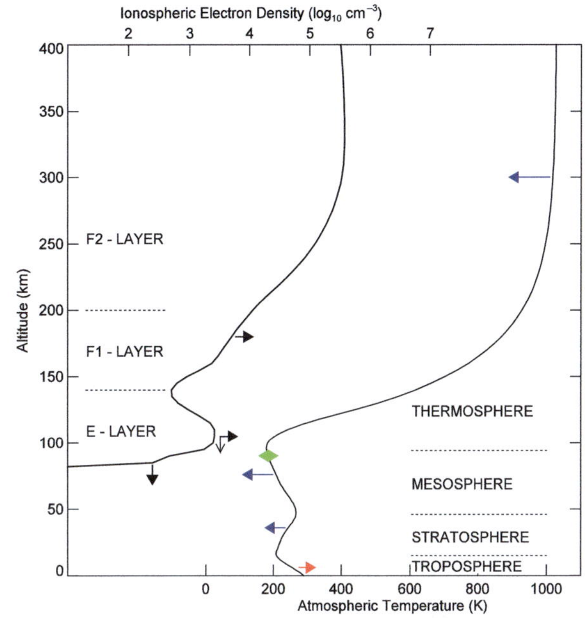

Fig. 30.3 Scenario of trends in the Earth's atmosphere. The atmospheric layers are defined by the temperature profile. The ionospheric layers are defined by the electron density profile (midnight at equator). *Arrows* indicate the direction of change. *Red* – warming; *blue* – cooling; *green* – no change of temperature; *black* – changes in electron density (*horizontal*) and heights of ionospheric layers (*vertical*). After Laštovička et al. (2008a)

30.3 Global Scenario of Trends

The first global pattern or rather scenario of long-term trends in the upper atmosphere and ionosphere has been introduced by Laštovička et al. (2006a, 2008a). The scenario (Fig. 30.3) is formed by trends in mesospheric temperature (mesospheric cooling and no trend in mesopause region), electron concentration in the lower ionosphere below 100 km (increase at fixed heights), maximum electron concentration (slight increase) and height of this maximum (statistically rather insignificant decrease due to coarse resolution of height measurements by ionosondes) in the E region, electron concentration in the F1-region maximum (slight increase), thermospheric neutral density (moderate decrease) and F2-region ion temperature (some decrease – rather preliminary result – not shown in Fig. 30.3). All these trends are qualitatively mutually consistent and qualitatively agree with model simulations of consequences of the enhanced greenhouse effect, and with the hypothesis of global cooling and contracting of the upper atmosphere.

The above scenario together with information about long-term trends in the stratosphere and troposphere demonstrate that anthropogenic emissions of greenhouse gases influence the atmosphere at nearly all altitudes between ground and space, thus affecting not only life on the surface, but also the space-based technological systems on which we increasingly rely. Life on Earth is more directly affected by climate change near the surface than in the upper atmosphere, but as the story of the Earth's ozone layer illustrates, changes at higher levels of the atmosphere may be important as well.

As regards trends in mesospheric temperatures, Beig et al. (2003) provided a first comprehensive review of observational and model results for temperature trends in the mesosphere (50–80 km – cooling by about 2–3 K/decade) and mesopause region (80–100 km – no significant trend). The comparison of the results obtained by different observations distributed over several decades was complicated. Nevertheless, there were a number of occasions where the temperature trend results indicated consistency and some of the differences were explainable. Beig (2006), Kubicky et al. (2008), Remsberg (2009) and others made the results more accurate and removed and/or explained some discrepancies and thus strengthened the conclusions of Beig et al. (2003). A lack of any significant temperature trend near the mesopause has been confirmed by model simulations (e.g., Garcia et al., 2007).

Trends in the lower ionosphere in older data seem to be fairly known and understood (Laštovička and Bremer, 2004) but it is well possible that the change of ozone trends in the mid-1990s and/or the second half of the 1990s affected and modified trends in the lower ionosphere. Unfortunately, there are little data to test possible change of trends.

Bremer (2008) investigated global trends in foE based on trends from 71 individual ionosonde stations worldwide. Positive trends dominate but number of stations providing negative trends is larger than negligible. The average trend is $+0.013\pm 0.005$ MHz/decade. Trends slightly weaken with increasing latitude; their longitudinal dependence seems to be affected, if not determined, by longitudinal dependence of ozone trends. Positive trends in foE are supported also by model calculations of Qian et al. (2008, 2009). Differences in trends obtained from data of individual stations, which do not seem to be a systematic function of any parameter, probably reflect long-term data homogeneity problems. Bremer (2008) also studied trends in hmE, height of the E-region maximum. The global trend is negative, -0.29 ± 0.20 km/decade, as an expected consequence of mesospheric cooling and, thus, of thermal shrinking. There is rather large scatter of trends obtained from data of individual stations. This makes the trend somewhat questionable, mainly due to data problems (poor height resolution of historical measurements).

Bremer (2008) analyzed foF1 in the same way as foE and obtained from global ionosonde network a weak positive trend, which is only slightly stronger than that for foE and reaches a value of 0.019 ± 0.011 MHz/year. However, there is one significant difference between trends in foE and foF1 – no influence of ozone depletion on trends in foF1 was found contrary to that in foE trends.

Trends in the thermospheric neutral density are of high practical importance as they can change orbital lifetime of both satellites and space debris. Emmert et al. (2008) analyzed data on satellite drag of about 5000 space objects, thus their results are really representative. They essentially confirmed previous results with only slight quantitative corrections. Thermospheric density trends are clearly negative, for 400 km typically a decrease by about 1.5–2%/decade for high and medium solar activity, which becomes much stronger for low solar activity, up to 5–7%/decade at solar minimum. The reason is the increasing role of CO_2 compared with NO in radiative cooling of the thermosphere with decreasing solar activity. While the CO_2 concentration is not much dependent on solar cycle, the NO concentration decreases with decreasing solar activity. The NO concentration has probably little trend (if any) at these heights, while concentration of CO_2 is increasing. Thermospheric density trends at 200–600 km are similar; they become slightly stronger with increasing height.

The first paper on long-term trends in F-region ion temperatures has been published by Holt and Zhang (2008) based on long-term data on ion temperatures at 350–450 km as measured by incoherent scatter radar above Millstone Hill. The trend is quite large, a cooling by 47 K/decade, even though it is substantially less than the solar cycle effect. Scatter of data is also quite large but the cooling trend is large enough to make the cooling itself reliable. The negative trends in thermospheric ion temperatures and thermospheric neutral density indicate a negative trend in neutral temperature, as well.

All the above parameters form a consistent scenario of trends in the upper atmosphere. However, there are three areas, characterized by key words F2 region, MLT region dynamics, mesospheric water vapour, which were found to be not fitting the above global scenario of long-term trends in the upper atmosphere (Laštovička et al., 2006a, 2008a). The reasons are either mutually controversial results, or unexpected changes of trends, or missing observational information. Recent progress in solving these open questions will be summarized in the next Sections.

30.4 F2 Region of the Ionosphere

There are large discrepancies and controversies between the results of different authors and different methods as for trends in the F2 region parameters foF2, which reflects peak electron density in the F2 region, and hmF2, height of this peak. Laštovička et al. (2006b) compared the results of different methods applied to a high-quality test data set (1976–1996, two comparable solar cycles) from Juliusruh (northern Germany). It was found that the correction for the solar cycle effect with F10.7 or E10.7 was evidently better than that with R or R12. Various regression-based methods provided comparable weak negative trends in foF2. A wavelet-based method as used was found erroneous (Laštovička et al., 2008b). Also a neural network-based method (Yue et al., 2006) gave results comparable with regression-based results (Laštovička et al., 2008b). Only the foF2 trends derived by Mikhailov's method, which uses a special way of eliminating geomagnetic activity effects, are substantially different from other results; however without his special elimination of the geomagnetic activity effects, the results are consistent with others (Laštovička et al., 2008b). Thus some discrepancies were caused by different corrections for solar (and geomagnetic) activity, others by different methods applied, but it does not explain problems in full extent. An important source of many discrepancies in trend results among various authors seems to be different data sets, quality of data, and different analyzed periods. With trends in hmF2 there is one more problem with calculating hmF2 from M(3000)F2 (direct measurements by incoherent scatters or deduced from digisonde electron density profiles do not provide sufficiently long and continuous data sets), because various formulas used for re-calculating are applicable only under some conditions, and under specific conditions different formulas are the best (McNamara, 2008).

Bencze (2009) introduced a new factor into hmF2 trend investigations. He claims that at stations located relatively nearby seashore negative trends in hmF2 are observed, whereas well-inland stations reveal more often positive trends, and that this distribution of trends is affected essentially by non-migrating tidal winds.

Another controversy concerns origin of trends in foF2 and hmF2. The two main candidates are increasing concentration of greenhouse gases and long-term changes of geomagnetic activity (the role of secular changes of the Earth's magnetic field was already discussed). In the past it was not easy to distinguish between them, as their effects on trends were in the same direction over most of the 20th century (but no more at present).

Based on analyses of foF2 and hmF2 long-term variations of Eurasian ionosonde stations, the so-called geomagnetic control concept has been developed by Mikhailov (2002). According to this concept the observed long-term variations of electron concentration are mainly controlled by the geomagnetic activity long-term variations (11-year running mean Ap indices Ap_{132} are used). The revealed morphological pattern can basically be explained by the geomagnetic storm effect mechanisms. The long-term trend in the ionosphere might be of geomagnetic origin due to an increase of geomagnetic activity throughout the 20th century (Clilverd et al., 1998; Martini and Mursula, 2008). However, this geomagnetic control was valid only in the past as Fig. 30.1 shows for foF2, where the geomagnetic control was probably lost near 2000. In hmF2 the geomagnetic control was lost earlier, in the first half of the 1970s as it follows from the results of Mikhailov (2006).

Model calculations of CO_2 doubling effect on the ionosphere (Fig. 30.4) by Qian et al. (2008) qualitatively explain how such a different behaviour of

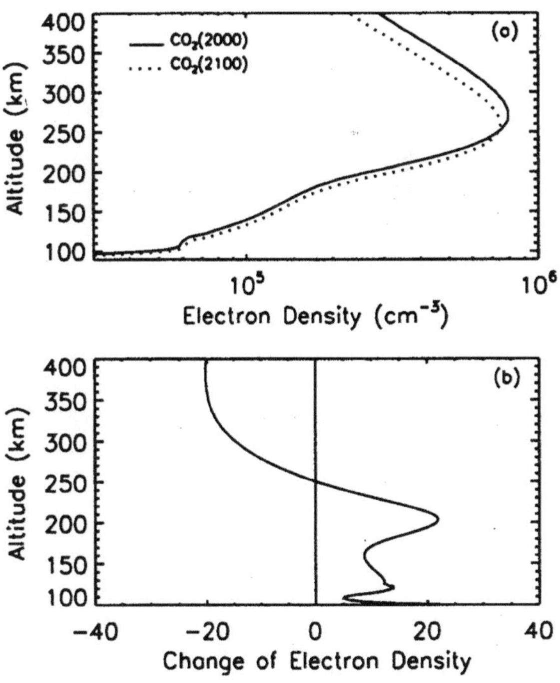

Fig. 30.4 Electron density profiles for the base case (2000 – *solid line*) and the doubled CO_2 case (2100 – *dotted line*), under medium solar activity conditions (F10.7 = 150), *top panel*. Percentage change of electron density as a consequence of doubling CO_2 concentration, *bottom panel*. After Qian et al. (2008)

foF2 and hmF2 is possible. These calculations confirm observational findings (Fig. 30.3) of increasing electron densities at fixed heights in the lower ionosphere, increasing foE and foF1, and decreasing hmE. They also show evident decrease of hmF2, while decrease of foF2 is small (this all agrees with the best quality observational results), because the cross point between increasing and decreasing electron densities is only a little below the height of maximum electron density in the F2 region. Qian et al. (2009) demonstrated that these results are typical for daytime, at night the pattern might be different; trends in F2 region parameters reveal to some extent latitudinal, longitudinal, diurnal, seasonal and solar cycle variations of trends. Model results allow simultaneous greenhouse control of trend in hmF2 (strong greenhouse influence) and geomagnetic control of trend in foF2 (weak greenhouse influence); change in foF2 trend around 2000 is associated with disappearance of positive trend in geomagnetic activity. *Trends in the F2-region are becoming to be controlled by the greenhouse gas increase and, thus, consistent with the scenario of trends in the upper atmosphere (Fig. 30.3).*

30.5 Mesosphere and Lower Thermosphere (MLT) Dynamics

Routine wind measurements in the MLT region at heights 80–100 km have been carried out for several decades, particularly at several stations at northern higher middle latitudes. The most prominent components of the mid-latitude MLT winds are the prevailing (mean) winds and the semidiurnal tidal winds.

Figure 30.5 shows time series of the annual mean zonal and meridional prevailing winds over Obninsk (55°N, 37°E) and Collm (52°N, 15°E). In spite of differences in the measurement techniques and strong year-to-year wind variations, the general tendencies in the climatic MLT wind variations at both stations are similar. Before 1990 both the annual prevailing zonal and meridional wind weakened, while after 1990 zonal wind strengthened and the negative trend in meridional wind levelled off. Shorter-period wind data series from some other stations confirm that this change of trend is characteristic for the whole Northern Hemisphere mid-latitude belt.

The change of trends around 1990 occurs not only in the MLT prevailing wind, it occurs also in the semidiurnal tide. Moreover it occurs also in the lower stratospheric winds at 100 hPa at 50°N –70°N and in the occurrence frequency of small laminae in ozone profiles in Europe near 50°N (Laštovička et al., 2010), and in the stationary planetary wave with zonal wavenumber one near 50°N (Jacobi et al., 2009). This indicates possibility of changes in trends in dynamics of the whole northern midlatitude middle atmosphere.

An important role in dynamics and energy deposition in the MLT region is played by atmospheric waves. Sources and atmospheric filtering of these waves might be affected by anthropogenic changes in the atmosphere. The MLT dynamics is forced through wave coupling, so trends in the mean circulation may be indicative of possible long-term trends in wave activity. The MLT wave activity is excited by wave sources generally located predominantly in the troposphere and partly stratosphere, therefore its trends might differ from trends of "in situ" mesospheric parameters. The wave propagation conditions in the stratosphere/mesosphere and various interaction processes in the MLT region contribute substantially to resulting MLT wave activity. Usually, wave activity is divided into three main categories: gravity waves (periods of tens of minutes to hours), tides with periods of 24 h plus harmonics, and planetary waves of periods of about 2–30 days.

A negative trend in the semidiurnal tidal winds in the MLT region seems to cease after the mid 1980s or 1990 (e.g., Portnyagin et al., 2006), or may even reverse. The most recent results from Collm, however, indicate rather negative trend (weakening) of the semidiurnal tide (with large seasonal variation of trend) over 1979–2007. More observations are needed.

Information on trends in planetary wave activity as summarized by Laštovička et al. (2008a) does not provide clear pattern. The planetary wave activity in the MLT region seems rather to increase, even though this increase appears to be intermittent. Jacobi et al. (2008) report trends in planetary wave activity, which are highly variable with period range and season, being generally positive in zonal wind and weaker and negative in meridional wind, and in total wind being in average slightly positive. They observed some trends only for quasi 5-day waves and the overall (2–30 days) wave activity, while for other period ranges the trends

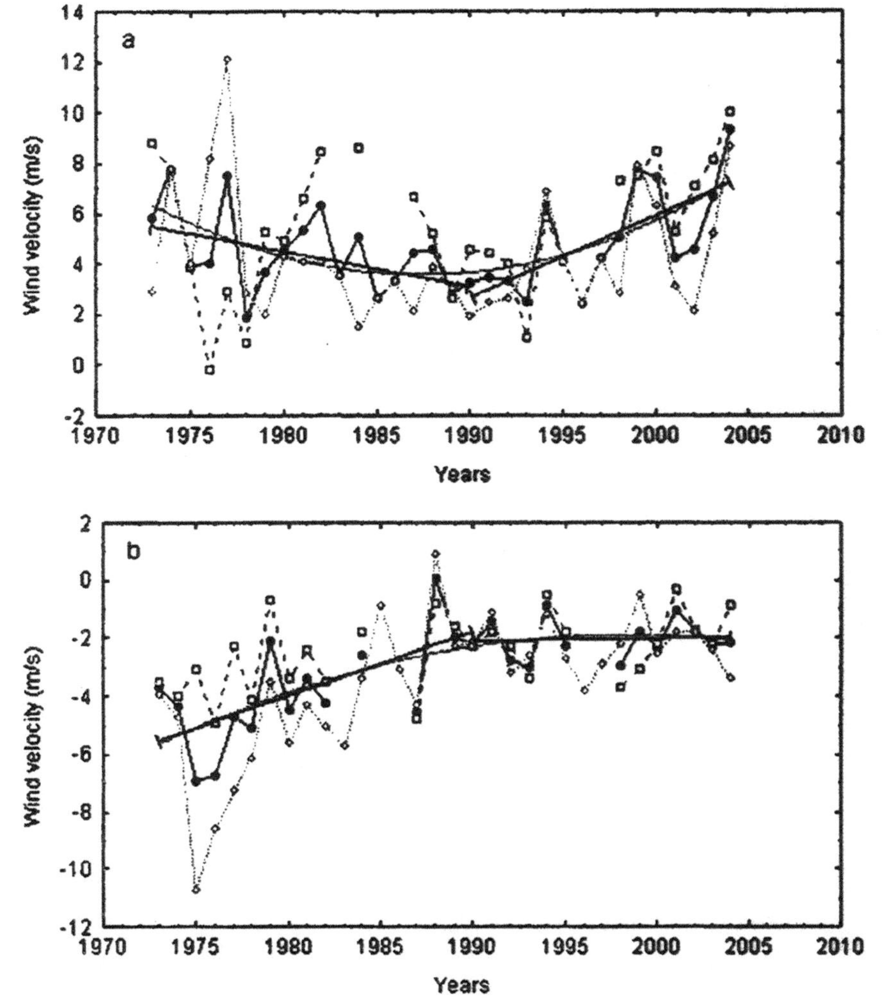

Fig. 30.5 MLT annual mean prevailing winds over Obninsk (55°N, 37°E) and Collm (52°N, 15°E), *empty squares* (*dashed line*) – Obninsk, *empty cubes* (*dotted line*) – Collm, *full cubes* (*solid line*) – mean values; trend lines are included; (**a**) – zonal component, (**b**) – meridional component (Laštovička et al., 2008a)

are weak if any. The obtained trends are statistically insignificant except for peak trends in summer for the quasi 5-day (3–7 days) oscillations.

Results of gravity wave analyses from ground-based systems can be strongly dependent on system characteristics. The wave periods seen in wind data measured by MF radars (>10 min), meteor radars (>1 h) and LF drifts (0.7–3 h) are very different and results are not simply comparable. Indirect evidence was rather in favour of some trends in gravity wave activity, but wind measurements themselves did not reveal a trend, as summarized by Laštovička et al. (2008a). Information on trends in the MLT region gravity wave activity is still very limited. Potential mechanisms of trends in gravity wave activity may be related either to changes in middle atmosphere filtering, or to changes in tropospheric/stratospheric sources like deep tropical convection or changes of storm tracks at midlatitudes; the latter may result in regionally/locally different trends in gravity wave activity even as to sign of trends.

30.6 Mesospheric Water Vapour and Related Phenomena

According to regular frost point hygrometer balloon-borne measurements, the water vapour content in the lower stratosphere was increasing by ~1%/year over last 40–45 years of the 20th century (Oltmans et al., 2000), which was recently corrected to 0.6–0.7%/year by Scherer et al. (2008).

Measurements of water vapour in the mesosphere have been available since the early 1990's from HALOE and from the WVMS instruments. HALOE measurements in the upper mesosphere show that after almost a full solar cycle there has been an increase in mesospheric water (Nedoluha et al., 2003); however, this was not the case for more recent years since about 1997. More measurements are needed. Similar to trends in dynamics, trends in water vapour are largely determined by trends in tropospheric sources and conditions of upward propagation and, therefore, can hardly be directly comparable with other trends in the mesosphere region.

The highest atmospheric clouds, noctilucent clouds (NLC) appear in extremely cold summer polar mesopause region at heights of about 82–85 km. Their appearance is controlled by temperature and water vapour content. NLCs observed from above by satellites are called polar mesospheric clouds (PMC). Long-term trends in the occurrence frequency and brightness of NLCs observed from ground at latitudes below about 65°N in Europe do not display a detectable trend in the NLCs according to Kirkwood et al. (2008). Satellite observations cover a larger geographical area and the highest occurrence frequency of PMCs is around 80°N. The satellite observations indicate an increase in PMC occurrence frequency and brightness. Shettle et al. (2009) analyzed 28 years of satellite PMC observations for latitudinal bands 50°–64° (NLC observation latitudes), 64°–74° and 74°–82°. They observed statistically significant increase only in 74°–82°, whereas in 50°–64° it was half in magnitude and statistically insignificant. At NLC observational latitudes the PMC trend of Shettle et al. (2009) is +9.9%/decade, the NLC trend of Kirkwood et al. (2008) is +4.4%/decade for moderate and bright NLCs and ~14%/decade for all NLCs, all trends being statistically insignificant. This means that the *PMC and NLC trends are comparable and do not differ within accuracy of their determination* in spite of partly different analyzed periods.

What is the cause of PMC trends? There is no detectable temperature trend in the mesopause region NLC/PMC heights. Water vapour trends in the mesopause region are uncertain. Further investigations are necessary.

30.7 Conclusions

The trend scenario forms a consistent pattern of global change in the upper atmosphere at heights above 50 km, as illustrated in Fig. 30.3. The upper atmosphere is generally cooling and contracting, and related changes in chemical composition (minor components) are affecting the ionosphere. The dominant driver of long-term trends in the last 3–4 decades is the increasing greenhouse forcing; some role may be played by anthropogenic changes of the ozone layer (in the lower part), long-term increase of geomagnetic activity throughout the 20th century (in the upper part) and other mentioned factors. As the role of various trend drivers changes with time and location, we can hardly expect stable, non-changing trends and stable role of various factors responsible for trends – they will be changing both with time and space (location).

In the last few years a significant progress was reached in the three areas, which did not fit the global scenario of trends in the upper atmosphere. Discrepancies in the F2 region trends have mostly been explained/removed and the F2 region is "on the track" to fit the global scenario. As for mesospheric water vapour, the discrepancy between NLC and PMC measurements seems to be removed. However it is not clear what is the origin of PMC trends. Change of trends in the MLT region winds appears to coincide with change of trends of some stratospheric parameters, which supports its origin "from below", probably via change sin atmospheric waves. However, there was almost no progress in the area of trends in atmospheric wave activity, which remains the key open question of trends in the upper atmosphere. A new question is emerging – would it be possible to join stratospheric and upper atmospheric trends into one scenario? These two questions appear to be the two principal questions in investigating long-term trends in the upper atmosphere.

Acknowledgement This work was supported by the Grant Agency of the Czech Republic through grant P209/10/1792.

References

Akmaev RA, Fomichev VI, Zhu X (2006) Impact of middle-atmospheric composition changes on greenhouse cooling in the upper atmosphere. J Atmos Solar-Terr Phys 68: 1879–1889

Beig G (2006) Trends in the mesopause region temperature and our present understanding – an update. Phys Chem Earth 30:3–9

Beig G, Keckhut P, Lowe RP et al (2003) Review of mesospheric temperature trends. Rev Geophys 41:1015. doi:10.1029/2002RG000121

Bencze P (2009) Geographical distribution of long-term changes in the height of the maximum electron density of the F region: a nonmigrating tidal effect? J Geophys Res 111:A06304. doi:10.1029/2008JA013492

Bremer J (2008) Long-term trends in the ionospheric E, F1 regions. Ann Geophys 26:1189–1197

Bremer J, Peters D (2008) Influence of stratospheric ozone changes on long-term trends in the meso- and lower thermosphere. J Atmos Solar-Terr Phys 70:1473–1481

Bremer J, Laštovička J, Mikhailov AV, Altadill D, Bencze P, Burešová D, De Franceschi G, Jacobi C, Kouris S, Perrone L, Turunen E (2009) Climate of the upper atmosphere. Ann Geophys 52:273–279

Clilverd MA, Clark TGC, Clarke E, Rishbeth H (1998) Increased magnetic storm activity from 1868 to 1995. J Atmos Solar-Terr Phys 60:1047–1056

Cnossen I, Richmond, AD (2008) Modelling the effect of changes in the earth's magnetic field from 1957 to 1997 on the ionospheric hmF2 and foF2 parameters. J Atmos Solar-Terr Phys 70:1512–1524

Danilov AD (2005) Long-term trends in F2-layer parameters and their relation to other trends. Adv Space Res 35:1405–1410

Elias AG (2009) Trends in the F2 ionospheric layer due to long-term variations in the Earth's magnetic field. J Atmos Solar-Terr Phys 71:1602–1609

Emmert JT, Picone JM, Meier RR (2008) Thermospheric global average density trends 1967–2007, derived from orbits of 5000 near-Earth objects. Geophys Res Lett 35:L05101. doi:10.1029/2007GL032809

Garcia RR, Marsh DR, Kinnison DE, Boville BA, Sassi F (2007) Simulation of secular trends in the middle atmosphere, 1950–2003. J Geophys Res 112:D09301. doi:10.1029/2006JD007485

Holt JM, Zhang S-R (2008) Long-term temperature trends in the ionosphere above Millstone Hill. Geophys Res Lett 35:L05813. doi:10.1029/2007GL031148

Jacobi C, Hoffmann P, Kürschner D (2008) Trends in MLT region winds and planetary waves, Collm (52°N, 15°E). Ann Geophys 26:1221–1232

Jacobi C, Hoffmann P, Liu RQ, Križan P, Laštovička J, Merzlyakov EG, Solovjova TV, Portnyagin YuI (2009) Midlatitude mesopause region winds and wave sand comparison with stratospheric variability. J Atmos Solar-Terr Phys 71:1540–1546

Kirkwood S, Dalin P, Réchou A (2008) Noctilucent clouds observed from the UK and Denmark – trends and variations over 43 years. Ann Geophys 26:1243–1254

Kubicky A, Keckhut P, Chanin M-L, Golitsyn GS, Lysenko E (2008) Temperature trends in the middle atmosphere as seen by historical Russian rocket launches: part II, Heiss Island (80.6°N, 58°E). J Atmos Solar-Terr Phys 70: 145–155

Laštovička J, Bremer J (2004) An overview of long-term trends in the lower ionosphere below 120 km. Surv Geophys 25:69–99

Laštovička J (2005) On the role of solar and geomagnetic activity in long-term trends in the atmosphere-ionosphere system. J Atmos Solar-Terr Phys 67:83–92

Laštovička J, Akmaev RA, Beig G, Bremer J, Emmert JT (2006a) Global change in the upper atmosphere. Science 314:1253–1254

Laštovička J, Akmaev RA, Beig G, Bremer J, Emmert JT, Jacobi C, Jarvis MJ, Nedoluha G, Portnyagin Y, Ulich T (2008a) Emerging pattern of global change in the upper atmosphere and ionosphere. Ann Geophys 26:1255–1268

Laštovička J, Križan P, Kozubek M (2010) Long-term trends in the middle atmosphere dynamics at northern middle latitudes – one regime or two different regimes? Atmos Chem Phys Discuss 10:2633–2668

Laštovička J, Mikhailov AV, Ulich Th, Bremer J, Elias AG, Ortiz de Adler N, Jara V, Abarca del Rio R, Foppiano AJ, Ovalle, E, Danilov, AD (2006b) Long-term trends in foF2: a comparison of various methods. J Atmos Solar-Terr Phys 68:1854–1870

Laštovička J, Yue X, Wan, W (2008b) Long-term trends in foF2: their estimating and origin. Ann Geophys 26: 593–598

Marsh D, Smith A, Woble E (2003) Mesospheric ozone response to changes in water vapour. J Geophys Res 108:4109. doi:10.1029/2002JD002705

Martini D, Mursula K (2008) Centennial geomagnetic activity studied by a new, reliable long-term index. J Atmos Solar-Terr Phys 70:1074–1087

McNamara LF (2008) Accuracy of models of hmF2 used for long-term trend analyses. Radio Sci 43:RS2002. doi:10.1029/2007RS003740

Merzlyakov EG, Jacobi Ch., Portnyagin YuI, Solovjova TV (2009) Structural changes in trend parameters of the MLT winds based on wind measurements at Obninsk (55°N, 37°E) and Collm (52°N, 15°E). J Atmos Solar-Terr Phys 71:1547–1557

Mikhailov AV (2002) The geomagnetic control concept of the F2-layer parameter long-term trends. Phys Chem Earth 27:595–606

Mikhailov AV (2006) Ionospheric long-term trends: can the geomagnetic control and the greenhouse hypothesis be reconciled? Ann Geophys 24:2533–2541

Mikhailov AV, de la Morena BA (2003) Long-term trends of foE and geomagnetic activity variations. Ann Geophys 21: 751–760

Mursula K, Martini D (2006) Centennial increase in geomagnetic activity: Latitudinal difference and global estimates. J Geophys Res 111:A08209. doi:10.1029/2005JA011549

Nedoluha GE, Bevilacqua RM, Gomez RM, Hicks BC, Russell JM III, Connor BJ (2003) An evaluation of trends in middle atmospheric water vapor as measured by HALOE, WVMS, and POAM. J Geophys Res 108:4391. doi:10.1029/2002JD003332

Oltmans SJ, Vömel H, Hofmann DJ, Rosenlof KH, Kley D (2000) The increase in stratospheric water vapor from balloonborne, frostpoint hygrometer measurements at Washington, D.C., and Boulder, Colorado. Geophys Res Lett 27:3453–3456

Portnyagin YuI, Merzlyakov EG, Solovjova TV, Jacobi Ch, Kürschner D, Manson A, Meek C (2006) Long-term trends and year-to-year variability of mid-latitude mesosphere/lower thermosphere winds. J Atmos Solar-Terr Phys 68:1890–1901

Qian L, Burns AG, Solomon SC, Roble RG (2009) The effect of carbon dioxide cooling on trends in the F2 layer ionosphere. J Atmos Solar-Terr Phys 71:1592–1601

Qian L, Solomon SC, Roble RG, Kane TJ (2008) Model simulations of global change in the ionosphere. Geophys Res Lett 35:L07811. doi:10.1029/2007GL033156

Reinsel GC, Miller AJ, Weatherhead EC, Flynn LE, Nagatani RM, Tiao GC, Wuebbles DJ (2005) Trend analysis of total ozone data for turnaround and dynamical contributions. J Geophys Res 110:D16306. doi:10.1029/2004JD004662

Remsberg EE (2009) Trends and solar cycle effects in temperatures versus altitude from the halogen occultation experiment for the mesosphere and upper stratosphere. J Geophys Res 114:D12303. doi:10.1029/2009JD011897

Rishbeth H (1990) A greenhouse effect in the ionosphere? Planet Space Sci 38:945–948

Rishbeth H, Roble RG (1992) Cooling of the upper atmosphere by enhanced greenhouse gases – modelling of thermospheric and ionospheric effects. Planet Space Sci 40: 1011–1026

Roble RG, Dickinson RE (1989) How will changes in carbon dioxide and methane modify the mean structure of the mesosphere and lower thermosphere? Geophys Res Lett16: 1441–1444

Scherer M, Vömel H, Fueglistaler S, Oltmans SJ, Staehelin J (2008) Trends and variability of midlatitude stratospheric water vapour deduced from the re-evaluated Boulder balloon series and HALOE. Atmos Chem Phys 8:1391–1402

Shettle PE, DeLand MT, Thomas GE, Olivero JJ (2009) Long term variations in the frequency of polar mesospheric clouds in the Northern Hemisphere from SBUV. Geophys Res Lett 36:L02803. doi:10.1029/2008GL036048

Weatherhead EC, Reinsel GC, Tiao GC et al (2000) Detecting the recovery of total column ozone. J Geophys Res 105:22201–22210

Yue X, Wan W, Liu L, Ning B, Zhao B (2006) Applying artificial neural network to derive long-term foF2 trends in Asia/Pacific sector from ionosonde observations. J Geophys Res 111:D22307. doi:10.1029/2005JA011577

Chapter 31

Parameters of the Ionospheric F2 Layer as a Source of Information on Trends in Thermospheric Dynamics

Alexey D. Danilov

Abstract The problem of determination of trends in thermospheric dynamics parameters (horizontal winds) on the basis of analysis of trends in various combinations of ionospheric F2-layer parameters is formulated. The previous attempts of the author in this direction are briefly described. It is shown that all studied parameters lead to the same result: after the "boundary date" (approximately 1980) a systematic change in these parameters (a long-term trend) is observed, this fact manifesting changes in the dynamical regime of the thermosphere because of cooling and contraction of the entire middle and upper atmosphere. The results of a search for trends in the hmF2 height for the moment (T(ss) + 2h) are described. These trends are found higher than the hmF2 trends obtained earlier by various authors analyzing the hmF2 behavior at fixed moments of local time.

31.1 Introduction

The problem of long-term trends in parameters of the middle and upper atmosphere attracts attention of researchers already during two decades. The cause of such an attention is evident: reliably detected changes in the composition and thermal regime of the troposphere and stratosphere (increase in the amount of greenhouse gases, decrease in the ozone amount in the stratosphere) should be manifested in the structure and time behavior of the atmospheric layers located above (mesosphere and thermosphere). The author has already numerously considered this theme (see, for example, Danilov, 2002, 2006a), so there is no need here to describe in detail all the aspects of this complicated problem. We note only that recently a concept of cooling and contraction of the entire thickness of the atmosphere from the stratosphere to thermosphere due to the increase in the greenhouse gases amount (first of all, CO_2) has been formulated (Laštovička et al., 2008). This concept is based on the group of parameters of the middle and upper atmosphere trends of which were detected fairly reliably. It is important that trends of all these parameters are qualitatively (and partly quantitatively) described by current theoretical models taking into account real current change in the greenhouse gas amount (for references see Laštovička et al., 2008). In his recent paper, Laštovička (Chapter 30, this volume) confirmed the principal statements of the above mentioned concept and presented a series of new results.

At the same time, Laštovička et al. (2008) did not include into the list of parameters trends of which are known and mutually agree in the scope of the above indicated concept parameters of the ionospheric F2 layer (foF2 and hmF2) and dynamical processes in the thermosphere and mesosphere. The problem of determination of trends in the critical frequency and height of the F2 layer has been numerously considered (see, for example, review papers by Danilov (2006a) and Laštovička et al. (2006)), but no consistent picture of the foF2 and hmF2 trends was obtained. As for trends in the thermospheric dynamics (horizontal thermospheric winds), the experimental data on this problem are scarce and controversial (see, for example, Bremer et al. (1997), Jacobi et al. (2001, 2005), Merzlyakov

A.D. Danilov (✉)
Institute of Applied Geophysics, 129128, Moscow, Russia
e-mail: adanilov99@mail.ru

M.A. Abdu, D. Pancheva (eds.), A. Bhattacharyya (Coed.), *Aeronomy of the Earth's Atmosphere and Ionosphere*, IAGA Special Sopron Book Series 2, DOI 10.1007/978-94-007-0326-1_31, © Springer Science+Business Media B.V. 2011

et al. (2008), and Portnyagin et al., (2006)). At the same time, from the most general consideration, it is evident that the above indicated cooling and contraction of the middle and upper atmosphere should lead to a redistribution of factors generating winds in the mesosphere and thermosphere and, therefore, to substantial changes in the winds themselves.

As far as one can hardly expect in the nearest years reliable data on trends in the thermospheric dynamics because of the absence of corresponding long-period observations of winds, the author (Danilov, 2006b, 2009a–c; Danilov and Vanina-Dart, 2009) undertaken a series of attempts to estimate (at least qualitatively) possible trends in the thermospheric winds on the basis of the analysis of the data on F2-layer parameters, foF2 and hmF2. Such approach seems reasonable because of two reasons. First, the F2 layer is very sensitive to changes in thermospheric winds. Second, there are observations of the F2 layer by the vertical sounding method during a few decades at the vast global network of stations, whereas the measurements of thermospheric winds are limited both in time and space.

The aim of this chapter is a brief description of the results already published by the author on the trends in thermospheric dynamics from the ionospheric vertical sounding data and also presentation of new results aimed to the same direction. The author hope to attract by this paper attention of specialists who have access to the vertical sounding data over a few decades to the problem in question and encourage them to check, specify, and continue the obtained results using more complete (and especially covering the recent decade) sets of ionospheric data.

31.2 Variations in the Correlation Coefficient Between foF2(Night) and foF2(Day)

Vanina and Danilov (2006) were the first to find that there is a significant negative correlation between the values of foF2 taken for the nighttime (0200 LT) and daytime (1400 LT) moments of the same day. The coefficient of this correlation was designated as *R*(foF2).

In order to analyze the behavior of any ionospheric parameter over time at scales of a few decades, one has to reject the solar activity influence. Looking for trends of all the parameters considered in this chapter, we used for this aim a regression dependence relative to the annual mean value of the solar index *F*(10.7). Since it was found that one could expect some trend after approximately 1980, we used for drawing this dependence only the data before 1980. Using these data, we were drawing a regression curve and for each point were finding the deviation from it: ΔR(foF2) = *R*(foF2)(obs)–*R*(foF2)(reg), where *R*(foF2)(obs) and *R*(foF2)(reg) are the values of *R*(foF2) really obtained and corresponding to the regression line, respectively. Examples of the dependence of analyzed ionospheric parameters on the *F*(10.7) index will be presented below. Further analysis of all long-term trends was performed for the parameters similar to ΔR(foF2), that is, free of solar activity influence.

The minimum of the negative correlation coefficient *R*(foF2) falls on spring and in some cases reaches 0.8–0.9 by absolute value. The analysis of the behavior with time of the *R*(foF2)(min) value at a series of vertical sounding stations showed that after approximately 1980 (this moment were called a "boundary date") a systematic increase in the absolute value of *R*(foF2)(min) is observed (Danilov and Vanina-Dart, 2007).

The author have considered possible processes leading to the above indicated systematic changes in *R*(foF2)(min). Danilov (2006b) and Danilov and Vanina-Dart (2007) obtained that the increase in *R*(foF2)(min) is related to intensification with time after the "boundary date" of the poleward horizontal wind in the thermosphere or to an increase in its variability.

31.3 Variations in the foF2(Night)/foF2(Day) Ratio

In a series of publications of the author (Danilov, 2008a, b, 2009b), trends in the ratio of the critical frequencies foF2 taken for the nighttime (0200 LT) and daytime (1400 LT) moments of the same day were considered. The main idea of such approach is that the daytime values of foF2 almost are not sensitive to dynamical processes and are governed mainly by the photochemical processes balance, whereas at night (because of the absence of ionization by direct solar radiation) the impact of dynamical processes (first of all, the horizontal wind leading to the vertical shift of the F2 layer) could be quite important.

The analysis of the foF2(02)/foF2(14) value was performed for the days with Ap < 30. This corresponds to the major part of days except rather strong geomagnetic disturbances. For details see Danilov (2009a). Two periods of the year were considered: January–February and June–July. The rejection of the dependence of foF2(02)/foF2(14) on solar activity was performed by the method described above.

The first publications on the analysis of the behavior of the foF2(02)/foF2(14) ratio (Danilov and Vanina-Dart, 2007; Danilov, 2008b) showed that after approximately 1980 (a "boundary date") a systematic behavior with time of this ratio (long-term trend) is observed and the sign of this trend could be both positive and negative. An example of the foF2(02)/foF2(14) dependence on time is shown in Fig. 31.1. The determination coefficient R^2 shows that according to Fisher F-test, the statistical significance of the obtained dependence is about 99%. More examples of the change of foF2(02)/foF2(14) with time could be found in the above publications. It was found that the above indicated changes after the "boundary date" are statistically significant with a high level of significance.

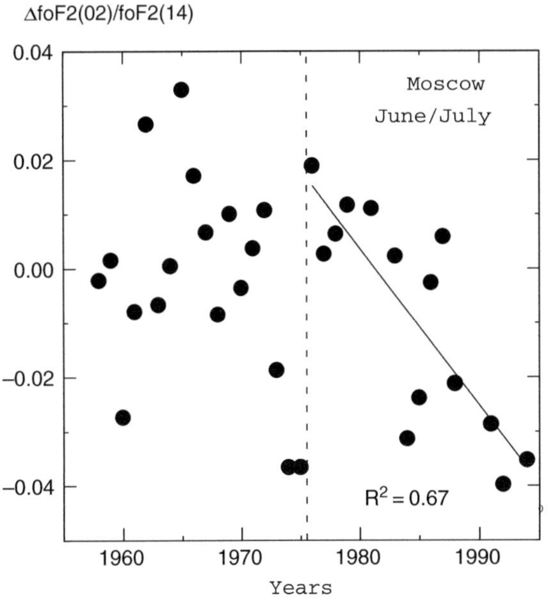

Fig. 31.1 Dependence of foF(02)/foF(14) on time for Moscow station

More ionospheric stations of the global network were analyzed (Danilov, 2009b) and it was found that the sign of the foF2(02)/foF2(14) trend is governed by the sign of the magnetic declination *D*. This made it possible to conclude that the above indicated trend is related to the trend in the horizontal zonal wind in the thermosphere: the eastward wind is intensified with time.

31.4 Scatter of hmF2 Values

As far as the layer maximum height hmF2 serves as a main intermediate link in the mechanism of the influence of the wind regime changes on the electron concentration in the F2 layer, one would expect that it should react to changes in the wind regime of the thermosphere. It has been already indicated above that the author (Danilov, 2008a, b; Danilov and Vanina-Dart, 2007) obtained that after some "boundary date" (1980±2–3 years) there occur systematic changes in the foF2(night)/ foF2(day) and *R*(foF2) parameters till the end of the 1990s (up to the end of the data series analyzed). If it is so, the value of hmF2 should in a different way behave in the first years of the time interval considered (the end of the 1950s–beginning of the 1960s) and at the end of this interval (the 1990s).

Danilov (2009c) made an attempt to check this statement. To do that variations from the 1960s to the 1990s of the degree of hmF2 variability, the standard deviation *S*(hmF2), were considered.

The monthly median hmF2 data at each considered station available in the databanks of international centers for geophysical data were used. Danilov (2009c) rejected the solar activity effect in the following way. For the chosen dataset, a dependence of hmF2 on the *F*(10.7) solar activity index was drawn. Then a standard deviation of the hmF2 value from this dependence *S*(hmF2) was considered. If we take years in various intervals of the entire period analyzed, we could find whether or not the value of *S*(hmF2) was changing with time during the recent decades.

An example of such analysis is shown in Fig. 31.2 for Ashkhabad station. For this station, the entire period 1957–1999 for which the data on hmF2 were available was split into two approximately equal parts: 1957–1979 and 1980–1999.

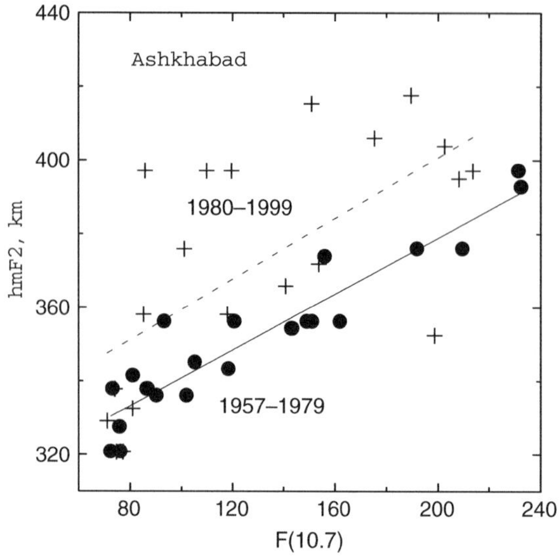

Fig. 31.2 The hmF2 dependence on *F*(10.7) for two periods for Ashkhabad station

Points and crosses in Fig. 31.2 present real values of hmF2 medians for March and 0200 LT time moment for the first and second indicated above periods, respectively. Solid and dashed lines present linear approximations of the hmF2 dependence on *F*(10.7) for points and crosses, respectively. The value of the standard deviation itself is calculated for all points relative to the solid line and for all crosses relative to the dashed line.

Even without calculation of the *S*(hmF2) values, at a sight at Fig. 31.2 one can see that the character of the scatter of the points relative to the approximating line strongly differs for 1957–1979 and 1980–1999. In the latter case this scatter is much stronger than in the former one. The *S*(hmF2) values confirm quantitatively the picture seen by eyes: the values of *S*(hmF2) for Ashkhabad are 7.3 and 25 km for the earlier and later periods, respectively.

Danilov (2009c) showed that for all the midlatitude stations studied such a dependence is confirmed: the *S*(hmF2) value for the period after 1980 is substantially higher than for the period before 1980.

The second approach to the analysis of the same problem was the following (Danilov, 2009c). Out of the same series of medians for 0200 LT and March, periods 21 years long were chosen in sequence. Using the same method, the value of *S*(hmF2) was calculated for them and referred to the middle of the interval. In such way, a continuous series of *S*(hmF2) values showing a gradual change in *S*(hmF2) during the entire period analyzed was obtained.

Variations of the *S*(hmF2) values calculated by this method for three midlatitude stations (Moscow, Ashkhabad, and Poitiers) are shown in Fig. 31.3. One can distinctly see that, in spite of some scatter of the points, on the whole all three stations provide absolutely similar picture: from 1967 (the 1957–1977 interval) to 1989 (the 1979–1999 interval) the value of *S*(hmF2) grows continuously. At that, both the growth amplitude and the slope of the *S*(hmF2) dependence on time are approximately the same for all three stations. Naturally, this fact increases substantially the statistical significance of the results obtained.

All the values of *S*(hmF2) presented above corresponded to March and the 0200 LT moment. This moment was chosen because of the following reasons. As it has been mentioned above, we are interested, first of all, in changes in the hmF2 at night, because nighttime values of the electron concentration are more sensitive to changes in the height of the layer. In all studies of the behavior of the foF2(night)/foF2(day) and *R*(foF2) parameters described above, this very moment 0200 LT was taken as "night" because it is the most characteristic for nighttime conditions.

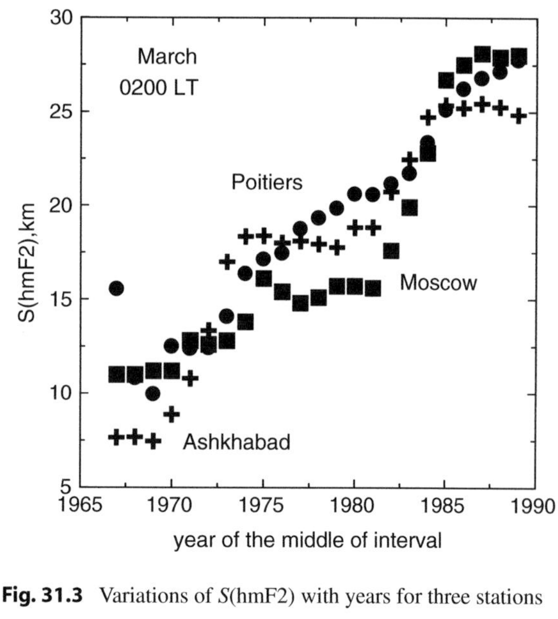

Fig. 31.3 Variations of *S*(hmF2) with years for three stations

As for the choice of the month, the analysis of the seasonal behavior of the *S*(hmF2) showed (Danilov, 2009c) that the effect of *S*(hmF2) increase from the earlier period to the later one is observed during the entire year, being most distinctly pronounced during the spring.

Danilov (2009) considered the question whether the detected effect depends on choosing days with different geomagnetic activity. Several limitations on geomagnetic activity (Ap < 8, 12, 16, 20 and 30) were taken and no substantial differences in the behavior of *S*(hmF2) at introduction of limitations on Ap were found. The latter fact manifests independence of the detected effect on geomagnetic activity.

Danilov (2009c) analyzed the effect in question only for midlatitude stations ($30° < \varphi < 60°$). At high latitudes during geomagnetic disturbances winds in the thermosphere are generated, completely changing the entire picture of the thermospheric circulation. The same is true for low latitudes: specific processes (first of all, variations in electric fields) take part in formation of the F2 layer and are able to distort completely the picture of the hmF2 height behavior.

Thus, the results obtained by Danilov (2009c) show that the scatter of the hmF2 values characterized by the *S*(hmF2) parameter grew substantially during the period from the 1950s to 1990s. This result is very important for the concept developed by the author (see above) that the observed trends of the foF2(night)/foF2(day) and *R*(foF2) parameters are caused by changes during the recent decades in the system of thermospheric winds. At that, the above indicated changes could be both one-side directed (for example, a systematic intensification of the meridional wind to the pole) or having oscillating character (for example, more frequent changes in the direction of zonal or meridional wind to the opposite one). Such changes in parameters of the global circulation could occur because of changes in the temperature regime of the thermosphere and middle atmosphere happening due to the increase of greenhouse gases amount in the atmosphere (see Chapter 30, this volume; Laštovička et al., 2008).

One should note that the detected strong increase in the scatter of the hmF2 values to the 1990s could significantly contribute to the difficulties in attempts to reveal long-term trends in hmF2 over the 1950s–1990s period (see, for example, Bremer, 1998; Marin et al., 2001; Ulich and Turunen, 1997).

31.5 Behavior of foF2 After Sunset

As far as the role of the dynamics is different at different moments of the day, we undertook an attempt to analyze variation in the foF2 ratio not for two fixed moments of local time (as it has been done in the previous papers, see above) but for 1400 LT and one of the moments after sunset. The first moment corresponded to typical daytime conditions. The second moment (*T*(ss)+2*h*) was chosen on the basis of the following considerations. According to Rishbeth (1998), the input into foF2 of the vertical drift caused by wind should be maximal namely in the after-sunset period, 1–2 h after the sunset.

The ratio of the foF2(ss+2) value to the foF2(14) value was considered. In the same way as in the previous cases, to reject solar activity effects a dependence of the foF2(ss+2)/ foF2(14) value on the solar activity index *F*(10.7) was drawn. An example of such dependence for Chung-Li station is presented in Fig. 31.4

The standard deviation *S*(foF2) of the foF2(ss+2)/ foF2(14) values from the model given by the approximation of the type of Fig. 31.4 for the running mean 11-year intervals (1958–1968, 1959–1969, etc.) was calculated, the obtained values being referred to the middle of the interval (1963, 1964, etc.).

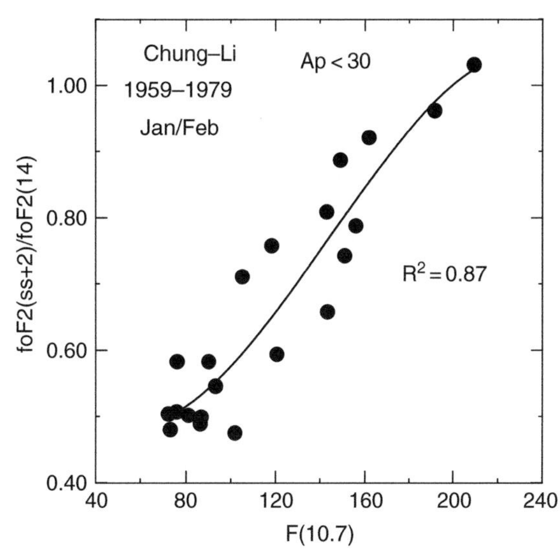

Fig. 31.4 Dependence of the foF2(ss+2)/foF2(14) ratio on solar activity index *F*(10.7)

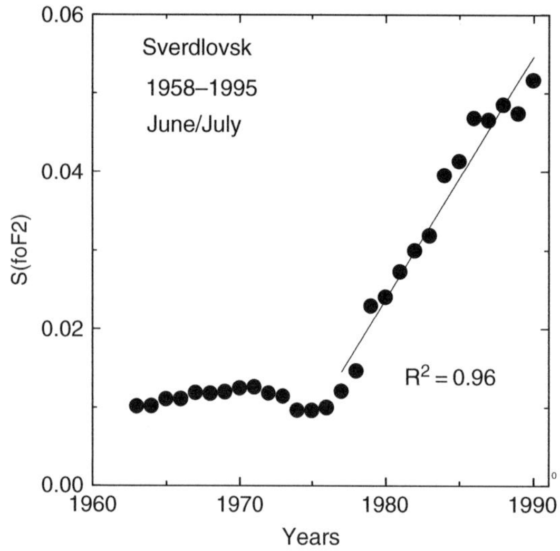

Fig. 31.5 Variations in *S*(foF2) with time

An example of variations in the *S*(foF2) value with time for June–July at Sverdlovsk station are shown in Fig. 31.5. Figure 31.5 presents an example of a picture close to an "ideal" one. At the first time stage (before the "boundary date"), the *S*(foF2) value is small and almost constant (there is no pronounced behavior with time). After the "boundary date", a well pronounced increase in *S*(foF2) with time (a positive trend) is observed. Figure 31.5 shows that the "boundary date" coincides with the average value (1980) obtained in the earlier studies of the author.

Not at all stations the picture looks such ideal as in Fig. 31.5, however on the average the values of *S*(foF2) for the period after 1980 are substantially (and statistically significantly) higher than for the period before 1980. For details of the *S*(foF2) statistics at various stations see Danilov and Vanina-Dart (2010).

Now we consider the conclusions following from the above presented fact. First, it is confirmed that around 1980 (± 2–3 years) there happens some change in the regime of the ionosphere: after this "boundary date" trends in its parameters are observed.

Second, due to the moments taken for the comparison of foF2 (see above), it is natural to assume that the scatter of *S*(foF2) manifests the behavior of the vertical drift induced by the horizontal winds in the thermosphere. The obtained result confirms the conclusions obtained above on the basis of the analysis of variability in the F2-layer height (see the previous section). Both results mean that after the "boundary date" the wind regime in the thermosphere changes systematically. This change could, in principle, occur in several ways. A systematic increase in the wind velocity is possible. Also possible is its systematic decrease. An increase in the wind variability (the number of reversals from the given direction to the opposite one) is possible. In all these cases, both *S*(foF2) and *S*(hmF2) values should increase.

31.6 Behavior of hmF2 After Sunset

The results of two previous sections directly indicate to changes in the thermospheric wind regime after the "boundary date". As far as the wind regime influences foF2 via the induced drift and the corresponding change in the layer height, it is natural to expect changes in the hmF2 value itself.

We performed an analysis of the hmF2 values for the same moment (*T*(ss)+2 *h*), which was considered in the previous section The absolute values of hmF2 calculated from the foF2 and M3000 values available in databanks by the Shimazaki (1955) method were considered. We succeeded in analyzing only a few midlatitude stations (high-latitude and near-equatorial stations were not considered due to the reasons presented above), because the data on M3000 are poorly presented in the international databanks.

The analysis was performed by the same method as for other parameters of the F2 layer described above. The dependence of the hmF2(ss+2) on *F*(10.7) was drawn and the time behavior of the smoothed over the 11-year periods values of ΔhmF2(ss+2) for two periods of the year (January–February and June–July) was analyzed.

Examples of the obtained dependencies on time of the ΔhmF2(ss+2) values are shown in Figs. 31.6 and 31.7. One can see that, as in the case of other parameters considered above, a well pronounced systematic behavior (trend) with time after the "boundary date" is observed. The result of consideration of this trend (the slope of the approximating line *k* after 1980) for January–February for all the stations analyzed is shown in Table 31.1 Similar results are obtained for June–July.

One can easily see that the method used is similar to the method numerously used for looking for

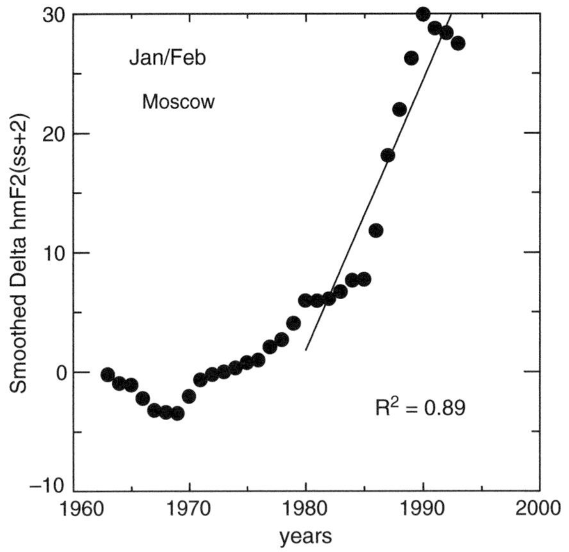

Fig. 31.6 Behavior of ΔhmF2(ss+2) with time for Moscow station

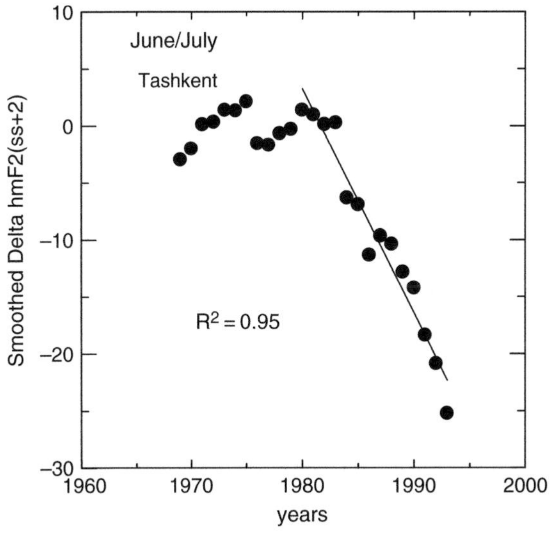

Fig. 31.7 Behavior of ΔhmF2(ss+2) with time for Tashkent station

trends in hmF2 (see, e.g., Bremer, 1998). The difference (but very important!) is only in the fact that earlier the values of hmF2 were taken for fixed moment of LT and the data were averaged over the day and months of the year. In this case we consider not the formal time moment (about which the Nature knows nothing), but the moment related to particular physical process:

Table 31.1 Results for various stations

Station	*k*	R^2
Ashkhabad	–2.3	0.70
Boulder	–1.0	0.56
Chang-Li	+0.4	0.55
Hobart	+1.1	0.83
Irkutsk	+0.7	0.98
Juliusruh	–1.6	0.74
Kaliningrad	–0.7	0.85
Khabarovsk	+0.5	0.52
Moscow	+2.3	0.89
Slough	+0.5	0.84
Sverdlovsk	+1.3	0.97
Tashkent	+0.3	0.66
Tomsk	+0.9	0.87
Wakkanai	–0.6	0.94

sunset. Moreover, the summer and winter data are considered separately. One can see in Figs. 31.6 and 31.7 and Table 31.1 that in all cases a well pronounced and statistically significant trend in hmF2 after the "boundary date" is observed. One can also see that both positive and negative trends are observed.

No regularity in the geographic distribution of positive and negative trends is found. For 11 stations out of 14, the signs of the trends in January–February and June–July are the same. In summer the numbers of positive and negative trends are equal (7 and 7, respectively). In winter positive trends slightly prevail (9 out of 14).

31.7 Discussion

Thus, the analysis of time variations of several parameters based on characteristics of the F2 ionospheric layer (*R*(foF2), foF2(night)/foF2(day), *S*(hmF2), *S*(foF2), and hmF2(ss+2)) shows that after some "boundary date" (1980 ± 2–3 years) a systematic change (a trend) in this parameters with time is observed.

All the parameters were chosen in such a way that according to the current ideas on the physics of the F2 layer these parameters should be sensitive to changes in the vertical drift which at middle latitudes is induced by the horizontal thermospheric winds. The fact that all the considered parameters demonstrate statistically significant trends after the "boundary date" manifests a systematic change in the wind regime of the thermosphere after this date. The analysis of various characteristics makes it possible to obtain various information on possible trends in the thermospheric winds. Some data indicate to an increase in the poleward meridional wind, other data show an increase in the eastward zonal wind. The data on *S*(hmF2) and *S*(foF2) make is possible to state only that after approximately 1980 the variability of the horizontal wind increased sharply. The data on hmF2(ss+2) show that the character (and sign) of trends in thermospheric winds could be different at different geographic regions.

The question is inevitable, why systematic changes of all the analyzed parameters begin at about 1980. We have to confess that we have yet no answer to this question. However one should note that in various publications indications to changes in ionospheric and thermospheric parameters beginning from the 1970s are met. Emmert et al. (2001) found systematic changes in the thermospheric wind parameters during the previous decades on the basis of measurements by the WINDII device. Portnyagin et al. (2006) found changes in the prevailing wind in the 1980s. Mikhailov (2006) found that after approximately 1972 (unlike in the 1940s–1960s) the changes in hmF2 with time could not be described by the long-term variations in geomagnetic activity. Bremer (2006) found changes in the trends in the ionospheric F1 and E layers around 1979.

We consider here in more detail the results of the analysis of hmF2(ss+2) described in the previous section. One of the results is that the obtained trends in hmF2 are of opposite sign for different ionospheric stations. At a first sight, an assumption is inevitable that in different geographic regions changes in the winds could occur in the opposite direction. However the changes in the zonal and meridional wind could occur in a different degree at various places. The influence of these changes on the vertical drift velocity is also different and depends on the magnetic declination *D* and inclination I at the given geographic point. Therefore, it is evident that effects of drift of opposite sign at different points (at different stations) could appear at changes in both zonal and meridional wind in the same direction at all stations. In other words, different signs of the hmF2(ss+2) trends do not require obligatory assumption on a "spot-like" character of trends in each component of the horizontal thermospheric wind. The "spot-like" structure of the hmF2(ss+2) trends could manifest a different relation between the input of trends in the meridional and zonal winds at points with different values of *D* and *I*.

The fact that trends in hmF2(ss+2) have opposite signs for different stations and even in different seasons makes it possible to understand why at the approaches applied earlier (Bremer, 1998; Danilov, 2006a; Ulich and Turunen, 1997; Marin et al., 2001) no clear and reliable picture of hmF2 trends was obtained.

Nevertheless, it is reasonable to compare the hmF2(ss+2) trends obtained in this chapter with the hmF2 trends obtained earlier by Bremer (1998) and Danilov (2006a). The comparison shows that the trends obtained in this chapter are systematically higher than the trends obtained using fixed LT moments. This is one more argument in favor of the statement that trends in hmF2 are caused by changes in dynamical processes, so it is correct to look for them not at fixed LT moments but at moments related to the natural process: sunset.

It is worth paying attention to the fact that the sign of the trends obtained in this chapter essentially coincide with the trends obtained by Bremer (1998) and Danilov (2006a) earlier.

31.8 Conclusions

The analysis of long-term changes (trends) in a series of parameters of the ionospheric F2 layer shows that in the behavior with time of all these parameters a sharp change in the period around 1980 (called by the author a "boundary date") is observed. Those parameters are considered which according to the current ideas should be sensitive to changes in the dynamical regime (horizontal winds) in the thermosphere.

The obtained result shows that the dynamical regime of the thermosphere is also changing (which seems to be inevitable) with the change in the general state of the atmosphere: with cooling and contraction of the middle and upper atmosphere due to the increase in the greenhouse gases amount (Laštovička et al., 2008). At the same time, the results obtained in the earlier publications by the author and in this paper show how the changes in ionospheric parameters (which are available for much longer period than the measurements of thermospheric wind) could be used for the analysis of the general situation with trends in the thermospheric dynamics and in its particular characteristics: velocities of the meridional and zonal wind.

The fact is found in this paper that trends in the height of the F2 layer are better pronounced and of higher magnitude if they are derived not for fixed moments of local time (as it has been done before) but for a moment connected to the natural process: sunset. This fact explains the cause of failures with searches for the hmF2 trends in a series of earlier publications of various authors. Moreover, the above indicated fact shows that (together with possible input from the thermosphere cooling and contraction) the hmF2 trends should be related to trends in the wind regime of the thermosphere. Apparently, it is senseless to look for their relation to aeronomical parameters (the atomic oxygen concentration, recombination coefficient, etc.) only. The wind influence on the hmF2 trends makes it possible to understand different signs of trends for different stations obtained in this paper and by all the researchers having tried to derive the hmF2 trends earlier. The difference in the sign of the hmF2 trends is related, evidently, to the different input of trends in the zonal and meridional wind at points with different values of the magnetic declination D and inclination I.

Acknowledgments The work was supported by the Russian Foundation for Basic Research (project No. 07-05-00095).

References

Bremer J (1998) Trends in the ionospheric E and F regions over Europe. Ann Geophys 16:986–996

Bremer J (2006) Long-term trends in ionospheric E and F1 regions. Paper presented at the 4th workshop on long-term changes and trends in the atmosphere, Sodankyla (Finland), 2–8 September 2006

Bremer J, Schminder R, Greisiger KM, Hoffmann P, Kurschner D, Singer W (1997) Solar cycle dependence and long-term trends in the wind field of the mesosphere/lower thermosphere. J Atmos Solar-Terr Phys 59:497–509

Danilov AD (2002) Overview of the trends in the ionospheric E and F2 regions. Phys Chem Earth (C) 20: 579–588

Danilov AD (2006a) Progress in studies of the trends in the ionosphere F region. Phys Chem Earth (C) 31(1–3): 34–40

Danilov AD (2006b) Relation between daytime and night-time values of the critical frequency foF2. Int J Geomagn Aeron 6(3). doi: 10.1029/2005GI000129

Danilov AD (2008a) Long-term trends in the relation between daytime and nighttime values of foF2. Ann Geophys 26(5):1199–1206

Danilov AD (2008b) Time and spatial variations in the ratio of nighttime and daytime critical frequencies of the F2 layer. J Atmos Solar-Terr Phys 70(8–9):1201–1212

Danilov AD (2009a) Critical frequencies foF2 as an indicator of trends in thermospheric dynamics. J Atmos Solar-Terr Phys 71(13):1430–1440

Danilov AD (2009b) Time and spatial variations of the foF2(night)/foF2(day) values. Adv Space Res 43(11): 1786–1793

Danilov AD (2009c) Scatter of hmF2 values as an indicator of trends in thermospheric dynamics. J Atmos Solar-Terr Phys 71(14–15):1586–1591

Danilov AD, Vanina-Dart LB (2007) Long-term trends in the ratio of daytime and nighttime values of foF2. Geomagn Aeron (in Russian) 47(2):236–241

Danilov AD, Vanina-Dart LB (2009) Scatter of hmF2 values as an indicator of trends in thermospheric dynamics. Geomagn Aeron (in Russian) 49(1):58–62

Danilov AD, Vanina-Dart LB (2010) Comparison of foF2 values in the daytime and after the sunset. Geomagn Aeron (in Russian) 50(1):61–66

Emmert JT, Fejer BG, Fesen CD, Sheperd GG, Solheim BH (2001) Climatology of middle- and low-latitude daytime F-region disturbance in neutral wind measured by wind imaging interferometer (WINDII). J Geophys Res 106(24): 701–712

Jacobi Ch, Lange M, Kurschner D, Manson AH, Meek CE (2001) A long-term comparison of Saskatoon MF radar and Collm LFD mesosphere-lower thermosphere wind measurements. Phys Chem Earth 26:419–424

Jacobi Ch, Portnyagin YuI, Merzlyakov EG, Solovjova TV, Makarov NA, Kurschner D (2005) A long-term comparison of mesopause region wind measurements over Eastern and Central Europe. J Atmos Solar-Terr Phys 67(3):229–240

Laštovička J et al (2006) Long-term trends in foF2: a comparison of various methods. J Atmos Solar-Terr Phys 68(17):1854–1870

Laštovička J et al (2008) Emerging pattern of global change in the upper atmosphere and ionosphere. Ann Geophys 26(5):1255–1268

Marin D, Mikhailov AV, de la Morena BA, Herraiz M (2001) Long-term hmF2 trends in the Eurasian longitudinal sector on the ground-based ionospheric observations. Ann Geophys 19(7):761–772

Merzlyakov E, Jacobi Ch, Solovjova TV, Portnyagin YuI (2008) Structural changes in trends of MLT winds at middle and high latitudes. Paper presented at 5th IAGA/ICMA/CAWSES workshop "long-term changes and trends in the atmosphere", St. Petersburg (Russia), 9–3 September, 2008

Mikhailov AV (2006) Ionospheric long-term trends: if the geomagnetic control and the greenhouse hypotheses can be reconciled? Ann Geophys 24(10):2533–2541

Portnyagin YuI, Merzlyakov EG, Sokolova TV, Jacobi Ch, Kurschner D, Manson A, Meek C (2006) Long-term trends and year-to-year variability of midlatitude mesosphere/lower thermosphere winds. J Atmos Solar-Terr Phys 68(17): 1890–1901

Rishbeth H (1998) How the thermospheric circulation affects the ionospheric F2 layer. J Atmos Solar-Terr Phys 60(14): 1385–1402

Shimazaki T (1995) World wide daily variations in the height of the maximum electron density in the ionospheric F2 layer. J Radio Res Labs (Jpn) 2(7):85–97

Ulich T, Turunen E (1997) Evidence for long-term cooling of the upper atmosphere in ionospheric data. Geophys Res Lett 24(9):1103–1106

Vanina LB, Danilov AD (2006) Relation between daytime and nighttime values of the F2-layer critical frequency. Geomagn Aeron (in Russian) 46(2):219–224

Part V
Ionosphere–Thermosphere Disturbances and Modelling

Chapter 32

Storm-Time Response of the Thermosphere–Ionosphere System

Timothy J. Fuller-Rowell

Abstract During a geomagnetic storm, the magnetospheric energy injected into the upper atmosphere increases by at least an order of magnitude, and during these times far exceeds the solar EUV and UV energy input. The energy is initially deposited towards higher latitudes where it heats and expands the thermosphere, increasing temperature and neutral density. Ionospheric plasma at high latitudes accelerates in response to the magnetospheric forcing, and through collisions can drive neutral winds in excess of 1 km/s. Large scale gravity waves are launched equatorward preceding a change in global circulation. Upwelling at high latitude and equatorward winds transport molecular rich neutral gas towards mid and low latitudes, particularly in the summer hemisphere, where it speeds up recombination and depletes the ionosphere. Additional electrodynamic processes, such as prompt penetration and disturbance dynamo electric fields, accompany the dynamic response to storms and can cause a huge redistribution and increase of ionospheric plasma. The papers following this one will elucidate many of the details in the storm-time response and provide a broader perspective.

32.1 Introduction

This chapter will review the upper atmosphere response to a geomagnetic storm. A storm is driven by the arrival at Earth of a cloud of plasma from the Sun, such as a coronal mass ejection (CME). The magnetosphere filters the energy from the solar wind and interplanetary magnetic field, and modulates the electrodynamic and particle energy and momentum that is deposited into the upper atmosphere. The sequence of neutral temperature, winds, composition, plasma, and electrodynamic response will be traced through the history of an event.

We will begin this section by quantifying the energy sources during a typical geomagnetic storm. We will then proceed with a review of our understanding of the response of the upper atmosphere to these types of injections of energy, and describe the range of physical processes that follow. Understanding the neutral dynamics, composition, temperature, mass density, plasma, and electrodynamic response of the upper atmosphere to geomagnetic storms is a significant challenge, and is particularly relevant due to the impact of space weather on operational systems.

32.2 High Latitude Energy Injection

The magnitude of the magnetospheric energy source, absorbed by the atmosphere, can be surprisingly well quantified, not because we can estimate the magnitude of the energy source itself, but because we can measure the atmospheric response to a storm. The energy injection causes a thermal expansion and an increase in atmospheric density at low Earth orbiting (LEO) satellite altitudes. For instance, the accelerometer onboard the CHAMP satellite measures the neutral mass density at around 400 km (see, for example, Liu and Lühr, 2005). The increase in density is a good measure of the thermal expansion and hence the

T.J. Fuller-Rowell (✉)
Cooperative Institute for Research in Environmental Sciences, University of Colorado, Boulder 80309, CO, USA; Space Weather Prediction Center, National Oceanic and Atmospheric Administration, Boulder, CO 80305, USA
e-mail: tim.fuller-rowell@noaa.gov

M.A. Abdu, D. Pancheva (eds.), A. Bhattacharyya (Coed.), *Aeronomy of the Earth's Atmosphere and Ionosphere*,
IAGA Special Sopron Book Series 2, DOI 10.1007/978-94-007-0326-1_32,

energy injection. The upper panel of Fig. 32.1, from M. Fedrizzi (private communication, 2010) shows a comparison of CHAMP satellite orbit-averaged neutral mass density with the results of a physical model numerical simulation for the first 15 days of January 2005. The black and red lines are the orbit averaged density for CHAMP and the model, respectively.

The model is able to follow the density response and recovery with quite high fidelity. The agreement enables the model to be used to estimate the rate of energy influx from magnetospheric sources. This example is by no means an extreme period; many larger events have been observed that could be twice the magnitude of the event seen on January 7/8th.

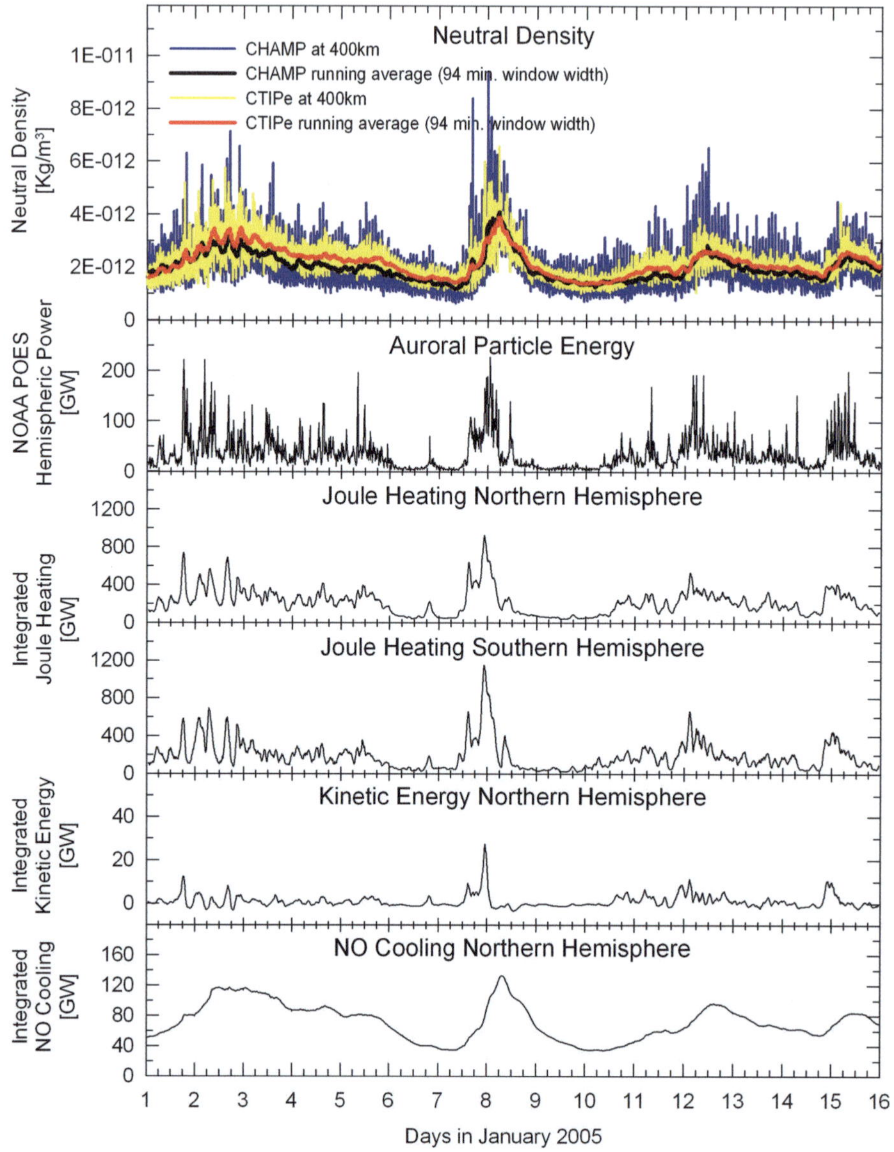

Fig. 32.1 The *upper panel* shows a comparison of CHAMP neutral density measurements at 400 km altitude with a numerical simulation for a stormy period in January 2005 (Courtesy of M. Fedrizzi, 2010). The *lower panels* show, from *top* to *bottom*, estimates of global auroral power, Joule heating in the Northern and Southern hemispheres, kinetic energy deposition, and nitric oxide infrared cooling rates

Figure 32.1 also shows, in the lower panels, the corresponding energy injection that was used in the model, as well as the NO cooling rates. The second panel shows the global auroral particle energy. On a quiet day, auroral precipitation is 10–20 GW, and during the storm on the 7/8th, it rises to about 200 GW in each hemisphere. The majority of the energy input, however, comes in the form of Joule or frictional heating, and at times exceeds 1000 GW in both the Northern and Southern hemispheres, shown in the third and fourth panels, respectively. The total power input therefore reaches 2–3 TW, which even for this fairly typical storm event, is many times the power input from solar radiation. In addition to Joule heating, kinetic energy is injected by the action of the $\mathbf{J}\times\mathbf{B}$ force, also known as ion drag. The fifth panel shows the magnitude of the global kinetic energy deposition during this event, which is typically 10–15% of the total Joule heating. The kinetic energy will itself dissipate in time, over a period of hours, as viscosity and ion drag convert the kinetic energy into internal thermal energy of the bulky neutral gas.

The numerical simulation is from a coupled model of the thermosphere, ionosphere, plasmasphere, and electrodynamics (CTIPe; Millward et al., 1996; Fuller-Rowell et al., 1996a). The magnetospheric energy input are defined by using the Weimer (2005) high latitude convection electric field model, which is driven by the solar wind and interplanetary magnetic field (IMF) parameters, and the auroral precipitation patterns derived from the TIROS/NOAA satellite observations (Evans et al., 1988). In the same way that the increase in atmospheric density is driven by Joule and auroral particle heating, the rapid recovery of neutral density is a consequence of heat loss processes. One process is the vertical heat conduction downward to the colder lower thermosphere, but the more important time-varying process is the infrared radiative cooling by nitric oxide (NO). The sixth panel of Fig. 32.1 shows the time history of the global NO radiative cooling, which can vary by an order of magnitude. The increase comes from two effects: the first is that storm auroral precipitation increases NO production from dissociation and ionization of molecular nitrogen; the second is that O atom vibrational relaxation of excited NO is temperature dependent. The CTIPe simulation in Fig. 32.1 uses the time-dependent estimates of NO from the Marsh et al. (2004) empirical model based on SNOE data.

32.3 Auroral Heating and Ionization

Auroral precipitation contributes only a modest 10–20% of the energy influx during a geomagnetic storm, yet it is the auroral displays that are the visual manifestation of geomagnetic storms. Although a modest energy source, the auroral ionization provides the catalyst for the electromagnetic energy transfer. Auroral precipitation is most effective in increasing the *E*-region plasma densities, which produces a peak of the Pedersen conductivity profile around 125 km altitude. A doubling of the plasma density doubles both the conductivity and Joule heating dissipation from the auroral currents, a contribution to the energy budget significantly larger than the particle energy itself.

32.4 Storm Dynamics at High Latitudes

At high latitude, ionospheric plasma responds directly to the strong magnetospherically imposed electric fields, which drive ion drifts of many hundreds, if not thousands, of meters per second. Figure 32.2 shows the response of the upper atmosphere in the Southern hemisphere shortly after impact on the magnetosphere of a coronal mass ejection. On the left side, the vectors represent plasma drift velocity in the upper thermosphere, where the ion drift motion is close to the $\mathbf{E}\times\mathbf{B}$ drift velocity, since the collisions with the thermosphere are relatively infrequent. The ion drift exceeds 1500 m/s over wide areas. The color contours represent plasma density near the *F*-region peak, close to 300 km altitude. The high latitude plasma densities are actually depleted by about 30% in spite of the increase in auroral ionization. The cause of the depletion is the change in the neutral composition resulting from the Joule heating, which will be expanded upon later. Increased up-welling enhances the molecular oxygen and nitrogen neutral species, compared to atomic oxygen, so there is a significant increase in ionospheric loss rates.

On the right side of Fig. 32.2, the neutral wind and temperature response over the same region are shown. Peak winds during quiet a case (not shown) are around 300 m/s. With the imposition of more intense magnetospheric convection, even with the infrequent collisions at this altitude, there is a sufficient momentum source

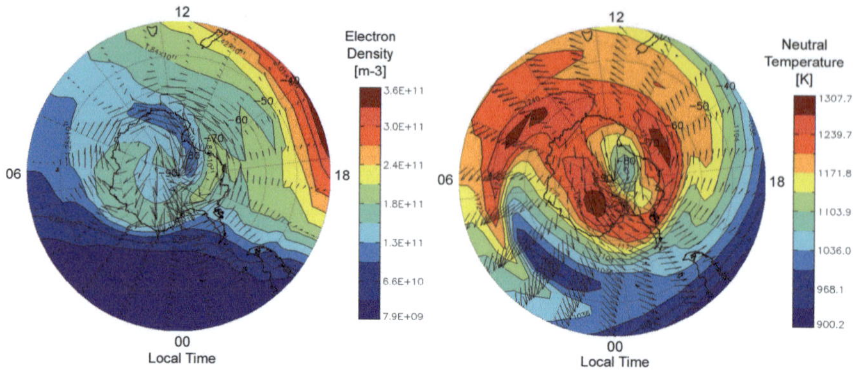

Fig. 32.2 Simulated response of the *F*-region plasma drifts and densities (*left*) and neutral winds and temperature (*right*) at the peak of the storm event at 1:30 UT on January 8, 2005, in the Southern hemisphere. Both represent the response in the upper thermosphere at about 300 km altitude. Peak plasma drift vectors are ~1500 m/s on the *left*, and peak neutral wind vectors are ~750 m/s on the *right*

to accelerate the medium close to 800 m/s after 6 h (see Killeen et al., 1984, 1988; for observations of these winds). For this Southern hemisphere view, the strong dusk sector neutral wind vortex is anticlockwise, or anticyclonic due to the inertial resonance between the ion and neutral convection (Fuller-Rowell et al., 1984; Fuller-Rowell, 1995). Overall, the dramatic increase in Joule heating raises the average temperature at high latitude by several hundred Kelvin in 6 h. The weaker cyclonic dawn vortex, which does not resonate with the plasma motion, is always divergent and tends to form a cool, low-density core, superimposed on the overall increase from Joule heating. The dusk cell in the quiet case is not strong enough to be divergent; in this more intense case it is divergent, and so also tends to form a cold, low-density region at its core (Crowley et al., 1996), but it is still not as well developed as the dawn cell. The numerical simulation results depicted in Fig. 32.2 are taken from a CTIPe simulation.

The right side of Fig. 32.3 shows the wind and temperature in the lower thermosphere, also for the Southern hemisphere at 140 km altitude, at the same storm time as Fig. 32.2. The pattern is similar to that at 300 km shown on the right side of Fig. 32.2, but reduced in magnitude. At 140 km altitude, collisions between the neutrals and ions are more frequent, but the inertia of the more massive neutral atmosphere more than outweighs the increased collisional force. The peak winds at this altitude are about half those at 300 km, and the balance between the main forces of inertial, Coriolis, ion drag, and pressure gradients is different (see Kwak and Richmond, 2007). Forty kilometers lower in altitude, the impact of the ion motion would be barely discernable.

One hundred and forty kilometers is above the altitude where eddy diffusion tends to mix and damp the winds, and molecular diffusion is still slow in the dense lower thermosphere. The momentum in the neutral

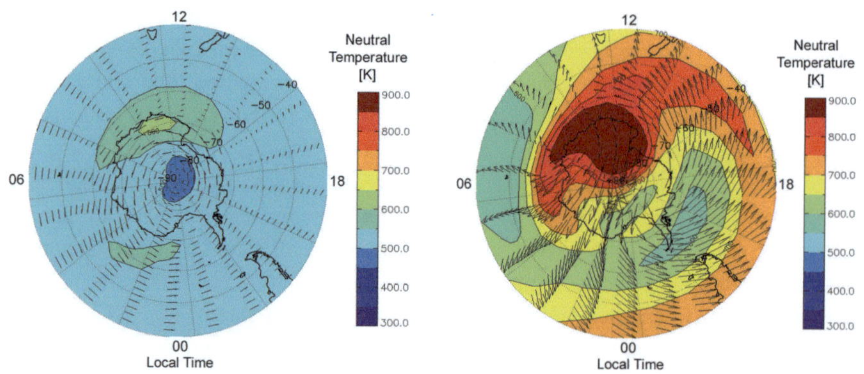

Fig. 32.3 Neutral winds in the lower thermosphere at around 140 km altitude at the peak of the storm at 1:30 UT on January 8, 2005, in the Southern hemisphere (*right*), and at the same UT as the quiet day preceding the storm (*left*). Winds in the lower thermosphere are about half those at 300 km, and are much slower to recover or dissipate. Peak neutral wind vectors are ~120 m/s for the quiet case, on the *left*, and ~350 m/s for the active case, on the *right*

winds is therefore slow to dissipate; the winds remain elevated for several hours, at speeds exceeding the ion drifts driven by the imposed, but waning, electric field—the so-called flywheel effect (Richmond, 1995; Richmond and Roble, 1987). The plasma densities recombine rapidly in this dense chemically controlled region as auroral precipitation returns to its quiet-time levels. The elevated winds and the background conductivity produce a dynamo action, which drives electromagnetic energy, or Poynting flux, upward from the lower thermosphere to the magnetosphere. The total power is fairly small compared with the downward flux during the solar-wind driven phase; the impact on the magnetosphere is likely to be fairly modest.

32.5 Global Dynamics, Temperature, and Density Response

The dynamic changes during a storm provide the conduit for many of the physical processes that ensue in the upper atmosphere. For instance, the increased meridional winds at mid latitudes push plasma parallel to the magnetic field to regions of different neutral composition. The global circulation also carries molecular rich air from the lower thermosphere upwards and equatorwards, changing the ratio of atomic and molecular neutral species, and changing loss rates for the ionosphere. The physical processes behind the composition change are similar to those behind the interhemispheric seasonal circulation driving the global composition structure. The storm wind system also drives the "disturbance dynamo," which through plasma transport modifies the strength and location of the equatorial ionization anomaly peaks. These electrodynamic processes will be elaborated upon by Maruyama et al. (Chapter 35, this volume).

During geomagnetic storms, the global dynamics of the upper atmosphere changes dramatically (Buonsanto, 1999; Fuller-Rowell et al., 1994, 1997; Fejer et al., 2002; Emmert et al., 2001, 2002). The response is complex even during the simplest of events. The thermosphere, although thought of as a sluggish medium, can respond quite quickly (in tens of minutes) and can support high-speed, large-scale gravity waves that propagate globally, initiated by impulsive forcing at high latitudes (Richmond and Matsushita, 1975). The large-scale waves have typical wavelengths of 1000 km or more and phase propagation speeds ranging from 400 to 1000 m/s (Hunsucker, 1982; Shiokawa et al., 2002). Gravity waves propagate at close to sound speeds, so waves launched by auroral heating can reach mid latitudes in an hour, and can reach the equator and penetrate into the opposite hemisphere within 3 h. Waves launched from both hemispheres interact to form a quite complex wave train (Shiokawa et al., 2002; Lu et al., 2008), even for the simplest forcing time histories. Real events with complex time histories are more difficult to unravel.

Large-scale gravity waves provide the mechanism for transmitting changes in pressure gradients around the globe. A new global circulation can therefore be imposed on the same time-scale as gravity wave propagation; it does not rely on, nor require, the bulk physical transport of mass by the wind field, which is typically much slower at mid latitude, 100–200 m/s.

Figure 32.4 shows the change in neutral wind at mid and low latitudes at 250 km, 3 h into a numerical simulation of a step-function increase in high-latitude forcing in the auroral oval (65–75° geomagnetic latitude). The wind response is shown within 50° latitude of the geographic equator, to allow for a scale that clearly shows the mid- and low-latitude dynamic response. Whereas at auroral latitude the peak neutral winds would be approaching 1000 m/s, at mid and low latitudes the winds are much more modest, with 100–200 m/s wind surges above the background circulation. At this time, 3 h into the simulation, the disturbance winds have reached the equator and are beginning to penetrate the other hemisphere and interact with the opposing wave front from the other pole. The arrival of the wave front at the geographic equator within 3 h indicates a propagation speed of about 700 m/s, in this case. The dependence on longitude, or local time, is quite prominent with the strongest intensity of propagation in the 100–200° longitude sector, which for the Universal Time of the image (15 UT) is on the nightside. The peak response appears to be more dependent on day or night difference, rather than the longitude sector of the magnetic pole. Stronger nightside wave propagation can be due to reduced ion drag (Fuller-Rowell et al., 1994).

A vertical cut through the thermosphere would reveal a tilted wave front with wave propagating more slowly at the lower altitudes (Richmond and Matsushita, 1975). Two hours later in the numerical simulation, the wave surges penetrate the opposite

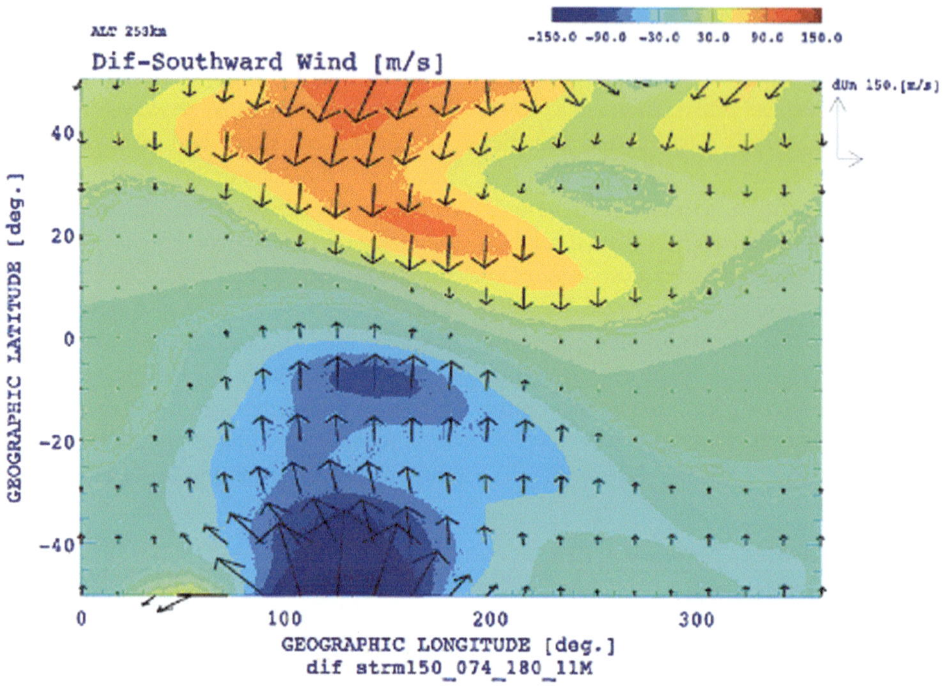

Fig. 32.4 Simulation of the response of the neutral winds at mid and low latitudes at 250 km altitude, shortly after a sudden increase in high-latitude Joule heating. The region within 50° of the geographic equator is shown at 15 UT, 3 h after the increase in high latitude magnetospheric forcing, equivalent to a $K_p \sim 7$. Wind surges of ~150 m/s are produced, mainly on the nightside. 0° longitude on the left is at 15 LT, 135° longitude, near the peak of the wind surge in both hemispheres, corresponds to midnight LT

hemisphere and drive poleward winds at mid latitude, at a time when the high-latitude forcing is still at its strongest. The complex wave train of equatorward and poleward winds during geomagnetic disturbances is a typical characteristic of neutral wind observations (Shiokawa et al., 2002), and will have a signature in many of the neutral and plasma parameters.

Observations and model simulations reveal a "sloshing" of winds between hemispheres in response to the high latitude heating during a storm. The net integrated wind effect is for an increase in the global circulation from pole to equator in both hemispheres (Roble, 1977). The change in circulation transports all neutral parameters including temperature, density, and species composition. The neutral composition changes and their impact on the ionosphere will be dealt with below.

As described in Fig. 32.4, the high latitude magnetospheric convection is the main source of the energy during a storm. The temperature response to a major storm can raise temperatures in the upper thermosphere by 500–1000 K, more than doubling the local internal energy of the system. The increase in temperature also drives a thermal expansion, which is seen from observations of the height of the F-region peak (Fedrizzi et al., 2008), and increases in mass density at fixed height (see Fig. 32.1). The increase in density at the source can subsequently be transported by the horizontal wind field, both from the background wind field and by increases in the equatorward circulation due to the storm. The prevailing quiet-time circulation is from summer to winter so the storm circulation assists the equatorward transport by the prevailing wind in the summer hemisphere, and tries to compete in winter (see for example Forbes, 2007; Forbes et al., 2005; and Sutton et al., 2005). They studied the evolution of atmospheric density at a constant height from a fixed local time sector from the CHAMP satellite. The data clearly show the penetration of regions of increased neutral density from the summer polar regions, and the lack of such penetration in the winter hemisphere.

32.6 Neutral Composition Response

The change in the global circulation induces upwelling at high latitudes and transport of molecular rich air (O_2, N_2) from the mid and lower thermosphere, upward and equatorward (Rishbeth et al., 1987; Prölss, 1997; Burns et al., 1991). The circulation during prolonged storms can transport neutral composition to low latitudes, which has been observed by space-based composition measurements (Crowley and Meier, 2008). The same can happen during solstice for even quite modest storms due to the additional transport by the prevailing summer to winter circulation (Fuller-Rowell et al., 1996b).

The region of increased mean molecular mass during a storm has been termed a "composition bulge," which is distinct from the background seasonal/latitudinal structure. The bulge can be transported by the background and storm-induced wind fields. The seasonal dependence in the transport of the composition bulge is depicted in Fig. 32.5. The equinox case is shown on the left; summer in the middle; and winter on the right. Each panel shows a snap-shot from simulations at the three seasons of the storm-time change in mean molecular mass in the Northern hemisphere from 10° latitude to the pole, at about 300 km altitude, 6 h after a 12-h storm. The figure is from numerical simulations using CTIPe in response to a substantial storm-like increase in high-latitude Joule heating. The composition change is an integrated effect of upwelling over the storm period, which gradually decays as transport by molecular diffusion tries to restore equilibrium. The diffusion timescale, however, can be long, typically a day or two. For reference, a change in mean mass from 19 to 22 atomic mass units (amu) is equivalent to a change in the proportion of molecular nitrogen from 25 to 50%. In the middle panel, in summer, the bulge of increased mean molecular mass (which is also equivalent to a decrease in the height-integrated ratio of O/N_2) has been transported by the wind field to low latitudes. In the right panel, for the northern winter solstice, the composition bulge has been constrained to high latitudes. The left panel for equinox is the intermediate case.

In summer, the storm-induced circulation augments the normal seasonal circulation from summer to winter, so the composition disturbance can very easily be transported to mid and low latitudes. In winter, the storm circulation competes with the seasonal flow, so the composition is constrained to high latitudes. Stronger storms can drive an equatorward circulation that can overpower the seasonal circulation, but it still has to compete with the opposing forcing of winds from the opposite hemisphere. The seasonal effect on meridional transport can also be modulated by hemispheric asymmetries in the strength of the magnetospheric sources themselves. This can arise either from asymmetries induced by the tilt of the magnetosphere with respect to the solar wind, or from different ionospheric conductivities in the polar region from solar illumination. Joule heating in the North and South polar regions are very likely to be different, and to be further modulated by UT.

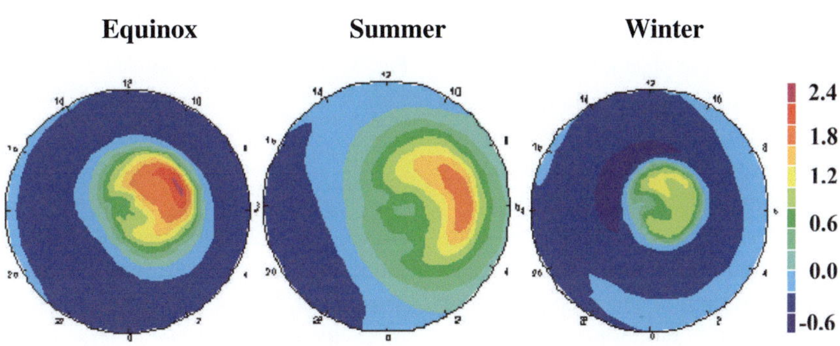

Fig. 32.5 Numerical simulation of the equatorial extent of the "composition bulge" at 1200 UT, for equivalent storms in the Northern hemisphere for equinox (*left panel*), summer (*middle panel*), and winter (*right panel*). The seasonal circulation assists the transport to low latitudes in the summer hemisphere and inhibits the transport in winter. The region shown is from 10° latitude to the pole

The composition bulge theory also predicts a diurnal modulation of the storm-time composition change. In the same way that the seasonal circulation transports the bulge, the normal background diurnal variation of the upper thermosphere wind at mid latitude can influence the bulge. On the dayside, upper thermosphere winds are poleward, responding to the dayside solar-generated increase in temperature and pressure, so limiting the equatorward transport. Similarly, on the nightside, the diurnal equatorward winds assist the flow to mid and low latitudes. As the globe rotates the bulge rotates with it, and the transport is diurnally modulated by this alternating dayside poleward and nightside equatorward wind field.

32.7 Positive and Negative Ionospheric Storms

It is well known that changes in neutral composition can impact the ionosphere by changing the ion loss rate. A decrease in the O/N_2 ratio can cause substantial decreases in plasma density (Strickland et al., 2001), often referred to as a "negative phase" ionospheric storm (Prölss, 1997; Rodger et al., 1989). In spite of the complexities in the observed response of the ionosphere to a geomagnetic storm, systematic features are apparent. One of the breakthroughs in understanding the storm-time ionosphere came from analysis of extensive ionospheric observations, and from interpretation of the data by physical models. Figure 32.6, from Rodger et al. (1989), shows that the storm-time response of the ionosphere reveals both seasonal and local-time (LT) dependencies. The figure shows the diurnal variation of the natural logarithm of the ratio of the storm to quiet peak F-region plasma density, NmF_2. For reference, a decrease of 0.5 indicates a decrease in the plasma density by 40%. Rodger et al. demonstrated that at this southern magnetic mid-latitude station of Argentine Islands, a consistent local time signature in the ratio of disturbed to quiet NmF_2 existed throughout the year, with a minimum in the morning hours around 06 LT and a maximum in the evening around 18 LT. The local-time "AC" variation was superimposed on a "DC" shift of the mean level that varied with season, being most positive in winter (May–July) and most negative in summer (October–February). The data support the widely held belief that "positive phases" of storms (increases in electron density) are more likely in winter mid latitudes, and "negative phases" of storms (decreases in electron density) are more likely in summer. Field and Rishbeth (1997) showed that these same characteristics are true for other longitude sectors. Rodger et al. (1989) stressed the point that individual storms show large deviations from the average behavior.

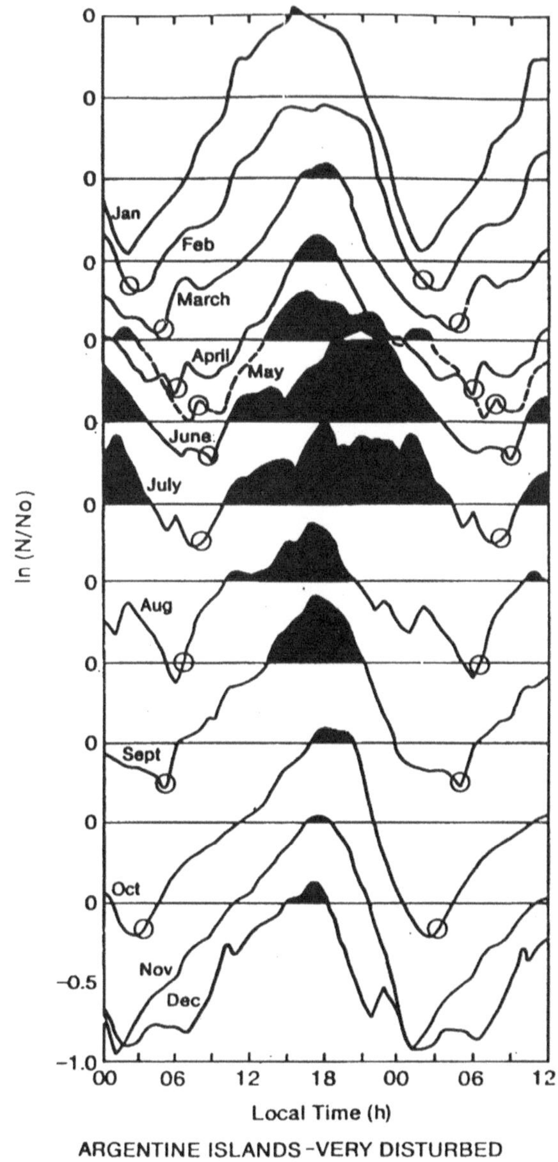

Fig. 32.6 The average seasonal and local time variations in ln(N/N_0), at Argentine Islands (65°S) for 1971–1981 (Courtesy of A.S. Rodger)

The response is entirely consistent with numerical simulations and the discussion above on storm-time changes in neutral composition. Prölss (1997) reviewed the evidence that negative storm effects are due to regions in which the neutral gas composition is changed—the ratio of molecular gas concentration (N_2+O_2) to the atomic oxygen concentration is increased. It is shown above that such a region, which Fuller-Rowell et al. (1994) called a "composition bulge" because it represents a region of increased mean mass, is originally produced through heating, upwelling, and transport of air by the magnetospheric energy inputs at auroral latitudes. The likely cause of the LT variation in the ionosphere, therefore, is simply the oscillation in latitude of the composition bulge in response to the diurnally varying winds (Fuller-Rowell et al., 1994). Skoblin and Förster (1993) also showed a case where steep gradients in thermospheric composition could be advected by meridional wind, and cause a rapid change in plasma content.

Similarly, the seasonal variations in storm-time mid-latitude plasma density is a consequence of the summer/winter difference in the global circulation, and the degree to which composition changes can be transported to mid latitudes. The seasonal migration of the bulge is superimposed on the diurnal oscillation.

32.8 Structure in the Plasma Response

Not too many years ago, storm-time ionospheric changes were typically characterized purely as "positive" and "negative" phases, as illustrated above. This characterization was appropriate for the interpretation of a limited number of point or local measurements from the few ionosonde stations scattered around the world and the still fewer number of incoherent scatter radar facilities. The conventional wisdom was that the negative phase at mid latitude was a consequence of neutral composition change (a basic concept that is still largely accepted). At the same time, the cause of the large-scale positive phase was still very much open to question, but was thought to be mainly due to winds or decreases in mean mass (a tenet now challenged). The wind effect is the concept whereby the equatorward winds at mid latitudes, in the presence of an inclined dipole-like magnetic field, tends to raise the plasma to regions where there are fewer heavy molecular neutral species, and hence experience reduced loss rates. The composition effect is the reverse of the argument for the negative phase: downwelling at mid or low latitudes, i.e., a closing of the global circulation, or Hadley cell, increases the O/N_2 ratio and drives a positive phase. This basic interpretation of observations is described in the reviews by Prölss (1997), Fuller-Rowell et al. (1997), and Buonsanto (1999).

Three significant events have altered this perspective: (1) "Mapping" of the plasma response to storms is now possible due to the explosion of the number of ground-based, dual-frequency Global Positioning System (GPS) observations of total electron content (TEC) (e.g., Foster et al., 2002; Coster et al., 2003); (2) The global mapping of neutral composition change from the GUVI instrument on the TIMED satellite (Paxton et al., 1999); and (3) The spectacular images of the plasmasphere from the IMAGE spacecraft (Goldstein et al., 2003). Figure 32.7, from Foster et al. (private communication, 2009), shows large increases in TEC at mid latitudes, including features described as TEC plasma "bulges" and "SEDs" (storm-enhanced densities). These features are in addition to the formation of the equatorial ionization anomaly (EIA), which is the most prominent ionospheric feature during quiet geomagnetic activity. The new ground-based TEC "imaging" capability can follow changes in the EIA as the storm-time electrodynamics evolves, but can also reveal these new features, raising new science questions, and stimulating the introduction of a whole new vocabulary for ionospheric phenomena.

A dramatic increase in TEC during a storm was shown by Mannucci et al. (2005); see Fig. 32.8. They used upward-looking GPS data from the CHAMP satellite at ~400 km altitude. The data show a huge increase in over-the-satellite electron content (OSEC), which increased at mid latitudes from about 50 to over 300 TEC units (1 TEC unit $\equiv 10^{16}$ electrons m^{-2}). At least two mechanisms have been proposed for the increase. The first, from Foster et al. (private communication, 2009), suggests that the storm-time electrodynamics at low latitudes transport plasma from the equatorial ionization anomaly towards mid latitudes. The mechanism relies on the idea that strong polarization electric fields are established as the *E* region moves into darkness at either end of an ionospheric flux tube. The process is also strongly influenced by the distortion of the Earth's magnetic field in the American longitude sector.

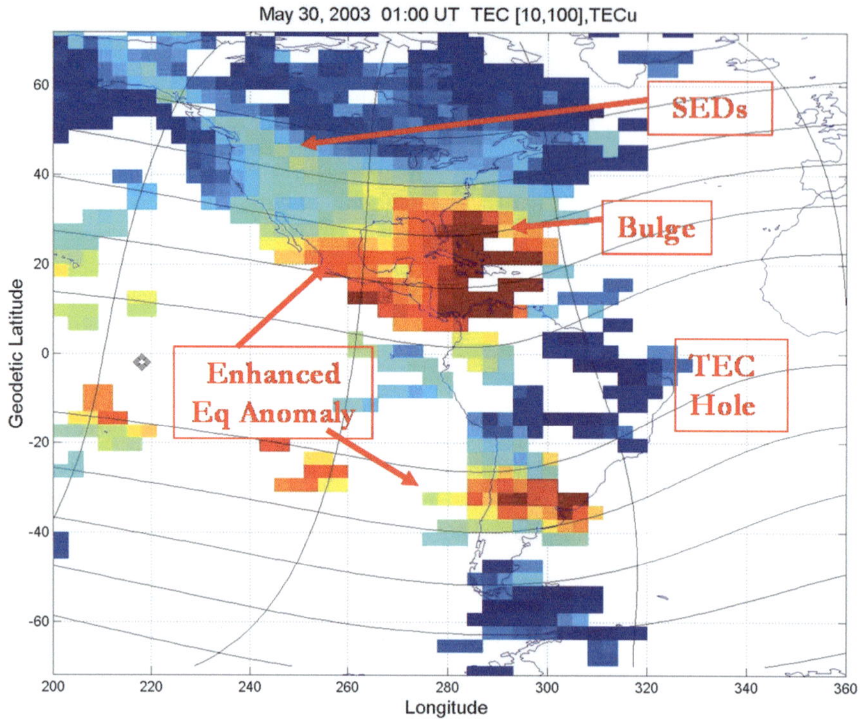

Fig. 32.7 Illustration of the large enhancement (bulge) in TEC at mid latitudes during a geomagnetic storm and showing the plume of plasma (SED) connecting the bulge to the high latitudes (Courtesy of J.C. Foster). The figure covers longitudes from 200° to 360°, which for 1 UT of this figure, covers the range of LT from 14:20 h to 01:00 h

Fig. 32.8 Order of magnitude increases in over-the-satellite electron content (OSEC) above 400 km during the Halloween storm as measured by the CHAMP satellite (Courtesy of A.J. Mannucci)

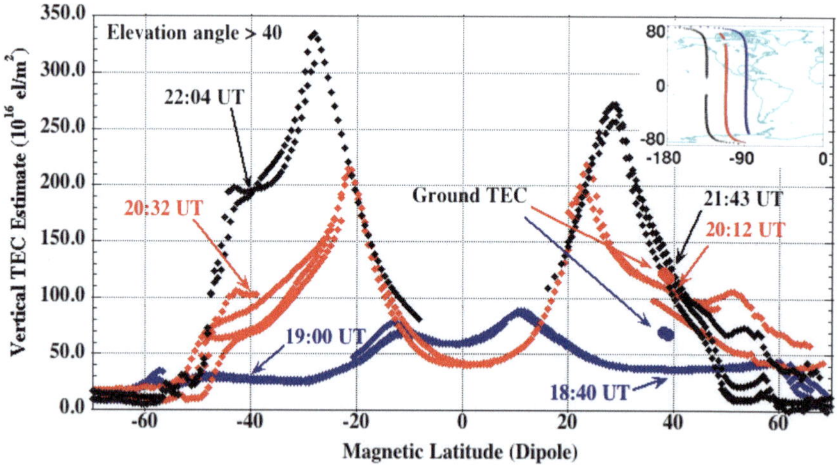

The second mechanism suggests that the build up of plasma at mid latitudes is simply a consequence of the expansion of the high latitude magnetospheric convection (Heelis et al., 2009). In the mid-latitude dusk sector, the expanded two-cell pattern of high latitudes would produce an electric field that is directed poleward and eastward. The plasma drift from the poleward field would tend to stagnate the plasma in local time, holding it in sunlight for longer and allowing plasma densities to build up from solar ionization. At the same time, because of the more inclined magnetic field at mid latitudes, the drift induced by the eastward electric field tends to raise the height of the ionosphere to regions of reduced neutral molecular species, and hence reduced ion recombination rates, which also tend to store plasma.

Lei et al. (2008) showed an increase of mid-latitude TEC during a storm from numerical simulations of a coupled magnetosphere-ionosphere-thermosphere model (CMIT). In their simulations, the increase appeared to arise from an increase in the equatorial eastward electric field followed by an increase in the strength of the equatorial ionization anomaly.

Whatever the mechanism, it is now clear that electrodynamics plays a much more important role in understanding the storm-time increases in plasma density at mid latitudes. Before the "imaging" techniques became widely used, observing the nature of these more localized (in local time) regions of increase, and understanding the mechanisms responsible for their formation, was elusive.

The spatially resolved feature in Fig. 32.7, now referred to as the storm-enhanced density (SED; Foster and Rideout, 2005), was originally called the "dusk effect" by Mendillo et al. (1970) when discovered in earlier observations. SED is a plume of increased plasma density that appears to emanate from the mid-latitude plasma bulge. SEDs are likely to be the ionospheric counterpart of the plasmaspheric plumes seen by the IMAGE satellite; see for example Goldstein et al. (2005). Foster et al. (2002) traced the expected field line geometry between the two features, which indicated the two structures are the respective plasma signatures at either end of the flux tubes responding to the same electric field.

The SED feature is likely to be a consequence of plasma transport by a fairly narrow mid-latitude electric field structure, now referred to as the sub-auroral polarization stream (SAPS). SAPS result from a polarization of inner magnetosphere plasma in the dusk sector during storms (Sazykin et al., 2005). It tends to form in regions of low E-region conductivity, which inhibits the closure of magnetosphere currents and leads to strong polarizing electric fields, often exceeding tens of millivolts per meter. The argument is that the fast convection from the SAPS picks up plasma from the bulge and transports it westward and poleward. This plasma can subsequently be carried over the pole, in the traditional way that a tongue of ionization forms, as dayside plasma is picked up by the "normal" magnetospheric convection and transported into the polar cap.

The apparent dichotomy is that the SAPS requires low conductivity, but is in the vicinity of the high F-region plasma densities associated with the SED. The SAPS transport theory for creating a SED is further complicated by the fact that when high plasma velocities flow through the more sluggish neutral medium, it tends to heat the plasma and increase loss rates (Schunk et al., 1975). The high velocity plasma can therefore also be associated with a trough, in direct contradiction to the appearance of the SED. The details of the feature have yet to be simulated realistically in a physical model in order to be able to analyze and test the theories, and to understand the balance between the various physical processes.

32.9 Storm-Time Electrodynamics

SAPS is an example of an electrodynamic response to a geomagnetic storm. Although a distinct mechanism, it is often combined with the general high-latitude magnetospheric convection driven by the solar wind dynamo, so is part of the high latitude source, at least as far as the thermosphere–ionosphere system is concerned. This of course is an oversimplification, but it is sometimes difficult to separate the SAPS from the main solar-wind driven dynamo when the patterns are complex. The main part of this section, however, addresses the other two sources of storm-time electrodynamics: (a) the "disturbance dynamo" (DD), driven by the storm-time neutral winds; and (b) the "prompt penetration" (PP) electric field, driven by the leakage of high latitude convection electric fields to low latitudes. Understanding the balance, separation, and interaction between this prompt penetration (PP) and the disturbance dynamo (DD) fields still remains a challenge (see review by Fuller-Rowell et al., 2008). Other papers in this section will address this question further.

Electrodynamics at low latitudes is important because of the configuration of the magnetic field. At low latitude the structure of the ionosphere is strongly controlled by electrodynamics (see review by Richmond in Chapter 13, this volume). During quiet times, the electric fields are driven by a combination of the E- and F-region dynamo processes (Fesen et al., 2000; Millward et al., 2001; Heelis, 2004). The net result at the magnetic equator is eastward electric fields, or upward plasma drift, during the day, downward drift at night, and a pre-reversal enhancement (PRE) just after sunset. The upward plasma transport induced by the electrodynamics on the dayside, generates the equatorial ionization anomaly (EIA). A strong PRE after sunset can also raise the ionosphere giving rise to the Rayleigh-Taylor instability, producing conditions conducive to the generation of ionospheric irregularities. The latter are notoriously difficult to predict on a day-to-day basis.

During geomagnetic storms, the dynamo electric fields are altered because the normal quiet-day thermospheric neutral winds are disrupted. Blanc and Richmond (1980) were the first to describe the characteristics of the storm-time disturbance dynamo, and their results are strongly supported by observations (Scherliess and Fejer, 1997; Fejer and Emmert, 2003). The Blanc and Richmond theory relies on the build up of zonal winds at mid-latitude under the action of the Coriolis force, in response to the increased equatorward winds. The meridional winds are forced by high latitude heating. The dynamo action of the zonal winds drives an equatorward Pedersen current. Positive charge builds up at the equator producing a poleward-directed electric field, which balances the wind driven equatorward current. The poleward electric field subsequently drives an eastward Hall current, which cause positive charge build up at the dusk terminator and negative charge build up at dawn. The zonal electric field driven by the disturbance dynamo opposes the normal dayside eastward, and nightside westward, quiet-time dynamo electric field and magnetic perturbations. The disturbance dynamo therefore act as a reverse Sq current vortex, reducing or even reversing the eastward electric field on the dayside, and reducing or reversing the normal westward electric field on the nightside.

Figure 32.9, from N. Maruyama (private communication, 2009), shows the simulation of the disturbance dynamo at the magnetic equator for a real event, during the storm in November 2004. Four days of simulation are presented beginning on November 6th. The storm had two main driven phases, when the interplanetary magnetic field B_Z component was directed strongly southward. The first occurred on November 7th, shown by the orange line, and the second on November 9th, shown in blue. The dotted curve shows the normal quiet-day variation from the numerical simulation. On November 7th the normal downward drift is reversed to strongly upward in the post-midnight sector. On November 8th, shown by the green line, the upward drift begins to abate in this sector, as the forcing from the first pulse begins to decline. At the same time, the strength of the pre-reversal enhancement is reduced. On the 9th (blue line), the second pulse in the solar wind further reduces the pre-reversal enhancement and a second surge in the upward drift on the nightside is produced. The basic theory presented by Blanc and Richmond for the nightside is consistent with the CTIPe simulation results, but the three-dimensional model contains significantly more local time structure. In particular, rather than being a uniform reduction in the downward plasma drift on the nightside, the response is much more localized in local time, and even reverses the direction of the drift to upward in the post-midnight or predawn sector.

It is interesting to note that one of the documented storm responses is that irregularities, which are normally associated with post-sunset enhancement in vertical plasma drift during quiet times, often appear post-midnight or pre-dawn during a storm. This is consistent with the results of the numerical simulation. The height of the ionosphere will be raised in the pre-dawn sector by the upward drift leading to conditions that are ripe for the initiation of plasma bubbles, or irregularities, from the Rayleigh-Taylor instability mechanism. These irregularities cause scintillation in ground-to-satellite radio signals at a range of wavelengths, including UHF and GPS L-band frequencies (Basu et al., 1996; Fejer and Kelley, 1980; Fejer et al., 1999; Groves et al., 1997).

The other significant feature in the simulations is the apparent reduction in the magnitude of the pre-reversal enhancement. Simulations of real events (Maruyama et al., 2005, 2007) show that this nightside response is fairly typical, and is in reasonable agreement with some of the observations of storm time response seen by the Jicamarca incoherent scatter radar facility on the magnetic equator in Peru (Fejer and Scherliess, 1997). The response on the dayside is more uniform with a tendency for reduced upward drift in response to the storm, which is consistent with the Blanc and Richmond (1980) theory.

The Blanc and Richmond theory predicts that the disturbance dynamo is slow to develop, due to the gradual build up of the zonal winds, and also slow to abate. An additional mechanism was

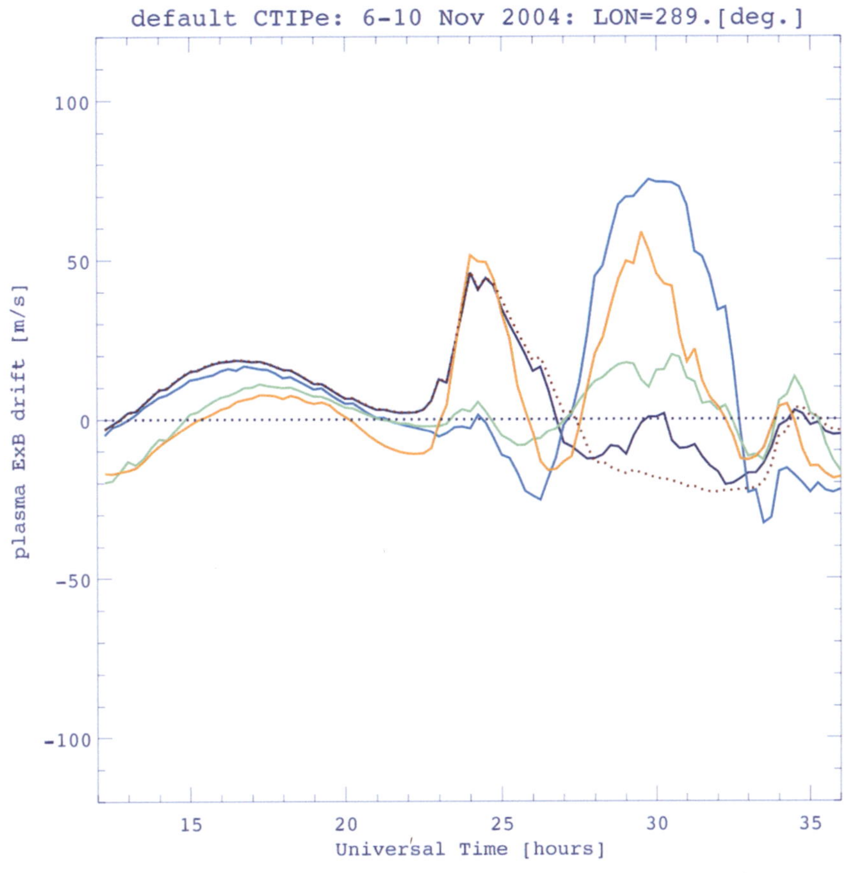

Fig. 32.9 Simulation of a vertical plasma drift at the magnetic equator during a real storm covering the period from November 6 to 9, 2004 (Courtesy of N. Maruyama). Four days of simulation are presented beginning on November 6th (*purple line*). The storm had two main driven phases, when the interplanetary magnetic field B_Z component was directed strongly southward. The first occurred on November 7th (*orange line*), and the second on November 9th (*blue line*). The *dotted curve* shows the normal quiet-day variation from the numerical simulation. On November 7th, the normal downward drift is reversed to strongly upward in the post-midnight sector. On November 8th, (*green line*), the upward drift begins to abate in this sector, as the forcing from the first pulse begins to decline. At the same time, the strength of the pre-reversal enhancement is reduced. On November 9th (*blue line*), the second pulse in the solar wind further reduces the pre-reversal enhancement and a second surge in the upward drift on the nightside is produced

mentioned by Blanc and Richmond, and was explored by Fuller-Rowell et al. (2002) in numerical simulations. The new mechanism appears to provide a means of generating a disturbance dynamo response about an hour or two after the onset of a geomagnetic storm. This second mechanism is driven by the meridional wind surges that respond within an hour or two of the high latitude heating (see Fig. 32.4). The mechanisms for the rapid disturbance dynamo onset is a combination of two effects. The first follows the Blanc and Richmond (1980) theory. The meridional wind surges in the geographic frame, have components in both the meridional and zonal magnetic frame. The zonal component produces the same response as the Blanc and Richmond theory, except that it does not require the slow build up of the zonal wind via the Coriolis force. The second effect arises from a direct effect of the meridional wind at mid-latitudes. The dynamo equations in Blanc and Richmond show that an equatorward wind in the magnetic frame drives an eastward directed zonal Pedersen current at midlatitudes. In both cases, the electrodynamic response is to the wind surge, which drives the dynamo at midlatitudes within 1–2 h of storm onset at high latitudes, and is experienced at the equator on the same timescale. Note that the equatorial electrodynamics responds to this rapid disturbance dynamo effect before the wind surge itself has reached the equator.

In addition to the dynamo fields, prompt penetration electric fields are also a major source of disruption of the low-latitude ionosphere during geomagnetic storms. When the high-latitude magnetospheric convection increases, usually associated with a southward turning of the interplanetary magnetic field (IMF), the high-latitude electric fields are unshielded as the magnetospheric plasma begins to respond. As a result, the electric fields can penetrate directly to the equator (Kelley et al., 1979; Spiro et al., 1988; Fejer et al., 1990).

The observed electrodynamic response can therefore be a complex combination of prompt penetration and disturbance dynamo effects. The empirical model of Fejer and Scherliess (1997) attempts to separate the prompt penetration and disturbance dynamo electric fields using the time history of the AE geomagnetic index. The results from the Rice magnetospheric convection model (RCM; Sazykin et al., 2005) are in good agreement with the empirical penetration electric field model (Fejer and Scherliess, 1997). The magnitude of the penetration fields can be significantly larger than dynamo fields, but their duration tends to be shorter, typically lasting less than an hour. There has been speculation recently that penetration electric fields can remain unshielded for several hours (Huang et al., 2005). Care has to be taken, however, not to confuse the penetration fields with the rapid onset dynamo fields, described above, that can appear within 2 h of an event and before wave surges have reached the equator (Fuller-Rowell et al., 2002). Whether of short or long duration, penetration fields can be intense and can cause significant redistribution of plasma at low latitudes.

Penetration electric fields are also generated in response to a decrease in high-latitude convection with opposite sign with an increase in convection. The decrease is often associated with a northward turning of the IMF. Even though penetration and dynamo fields can become confused in this over-shielded case, the Fejer and Scherliess (1997) model is able to separate the two effects. If the solar wind is highly variable, a series of prompt penetration electric fields will be initiated at low latitudes. The penetration field will interfere constructively or destructively with the dynamo component depending on the particular time history of the solar wind.

Satellite observations of the dramatic changes that can occur at low latitude in response to geomagnetic storms were presented by Basu et al. (2001). They showed data from the DMSP polar orbiting satellite at 850 km altitude during the Bastille Day storm in July 2000 indicating that a wide swath of plasma had disappeared over tens of degrees in latitude. A similar event occurred during the March 1989 storm (Greenspan et al., 1991) when vertical drift measurements exceeded 100 m/s. Modeling this event using the observed drifts raised the F-region ionosphere to over 800 km (Batista et al., 1991).

This huge restructuring of the ionosphere at low latitudes, as depicted in the DMSP observations, has never been successfully modeled. Numerical simulations are able to produce similar effects at 400 km altitude from the action of the various dynamo processes but this in no way compares to the extreme changes that are observed. The most likely scenario is that the dynamo and prompt penetration fields are acting together, due to the particular time history of the solar wind drivers. Coupled physics-based models of the inner magnetosphere and thermosphere/ionosphere are currently being developed to simulate the effects of both dynamo and penetration electric fields during storms, and their interaction (Maruyama et al., 2005, 2007; Huba et al., 2005). Maruyama et al. (Chapter 35, this volume) elaborates further on the non-linear interaction between "prompt penetration" and "disturbance dynamo" electric fields.

32.10 Conclusions

This chapter has provided a broad overview of the many physical processes that operate in the upper atmosphere during a geomagnetic storm. Subsequent papers will elaborate on many of the details. It is interesting to note that changes in thermospheric winds during a storm can be a conduit for many of the global ionospheric and thermospheric storm-time changes. For instance, the winds transport the composition changes that are responsible for much of the large scale decreases in plasma density during a storm, and the winds drive at least one component of the electrodynamic changes that are clearly involved in the processes leading to some of the large increases in plasma density at midlatitude. The dynamical and physical processes involved in the response of the thermosphere to a geomagnetic storm are reasonably well understood, and numerical simulations have illustrated how the complexity of the response is influenced by interactions among dynamics, composition, and electrodynamics. The energy input to the upper atmosphere during a large geomagnetic storm completely overwhelms the solar radiative input. The energy is initially deposited towards higher latitudes where it heats and expands the thermosphere, increasing temperature and neutral density. The impact of the energy injection, however, is felt by the global system very quickly, firstly in seconds through prompt penetration of the high latitude convection, and shortly after by the large-scale gravity waves that precede the change in global circulation, composition change, and dynamo effects.

References

Basu S, Basu S, Groves KM, Yeh H-C, Su S-Y, Rich FJ, Sultan PJ, Keskinen MJ (2001) Response of the equatorial ionosphere in the South Atlantic region to the great magnetic storm of July 15, 2000. Geophys Res Lett 28(18):3577–3580

Basu S, Kudeki E, Basu S, Valladares CE, Weber EJ, Zengingonul HP, Bhattacharyya S, Sheehan R, Meriwether JW, Biondi MA, Kuenzler H, Espinoza J (1996) Scintillations, plasma drifts, and neutral winds in the equatorial ionosphere after sunset. J Geophys Res 101(A12):26795–26810

Batista IS, de Paula ER, Abdu MA, Trivedi NB, Greenspan ME (1991) Ionospheric effects of the March 13, 1989, magnetic storm at low and equatorial latitudes. J Geophys Res 96(A8):13943–13952

Blanc M, Richmond AD (1980) The ionospheric disturbance dynamo. J Geophys Res 85:1669–1686

Buonsanto MJ (1999) Ionospheric storms – A review. Space Sci Rev 88:563–601

Burns AG, Killeen TL, Roble RG (1991) A theoretical study of thermospheric composition perturbations during an impulsive geomagnetic storm. J Geophys Res 96(A8): 14153–14167

Coster AJ, Foster J, Erikson P (2003) Monitoring the ionosphere with GPS. Space Weather, GPS World 14(5):42–49

Crowley G, Meier RR (2008) Disturbed O/N_2 ratios and their transport to mid and low latitudes. In: Kintner PM, Coster AJ, Fuller-Rowell TJ, Mannucci AJ, Mendillo M, Heelis R (eds) Midlatitde ionospheric dynamics and disturbances. Geophysical monograph series, vol 181. American Geophysical Union, Washington, DC, pp 221–234

Crowley G, Schoendorf J, Roble R, Marcos F (1996) Cellular structures in the high-latitude thermosphere. J Geophys Res 101(A1):211–223

Emmert JT, Fejer BG, Fesen CG, Shepherd GG, Solheim BH (2001) Climatology of middle- and low-latitude daytime *F*- region disturbance neutral winds measured by Wind Imaging Interferometer (WINDII). J Geophys Res 106(A11): 24701–24712

Emmert JT, Fejer BG, Shepherd GG, Solheim BH (2002) Altitude dependence of middle and low-latitude daytime thermospheric disturbance winds measured by WINDII. J Geophys Res. 107(A12):1453

Evans DS, Fuller-Rowell TJ, Maeda S, Foster J (1988) Specification of the heat input to the thermosphere from magnetospheric processes using TIROS/NOAA auroral particle observations. Adv Astronautical Sci 65:1649–1668

Fedrizzi M, Fuller-Rowell TJ, Codrescu MV (2011) Physics-based modeling of upper atmosphere neutral density. Space Weather (submitted)

Fedrizzi M, Fuller-Rowell TJ, Maruyama N, Codrescu MV, Khalsa H (2008) Global modeling of storm-time thermospheric dynamics and electrodynamics. In: Kintner PM, Coster AJ, Fuller-Rowell TJ, Mannucci AJ, Mendillo M, Heelis R (eds) Midlatitde ionospheric dynamics and disturbances. Geophysical monograph series, vol 181. American Geophysical Union, Washington, DC, pp 187–200

Fejer BG, Emmert JT (2003) Low-latitude ionospheric disturbance electric field effects during the recovery phase of the 19–21 October 1998 magnetic storm. J Geophys Res 108(A12):1454

Fejer BG, Emmert JT, Sipler DP (2002) Climatology and storm time dependence of nighttime thermospheric neutral winds over Millstone Hill. J Geophys Res. doi:10.1029/2001JA000300

Fejer BG, Kelley MC (1980) Ionospheric irregularities. Rev Geophys Space Phys 18:401

Fejer BG, Kelley MC, Senior C, de la Beaujardière O, Holt JA, Tepley CA, Burnside R, Abdu MA, Sobral JHA, Woodman RF, Kamide Y, Lepping R (1990) Low- and mid-latitude ionospheric electric fields during the January 1984 GISMOS campaign. J Geophys Res 95(A3):2367–2378

Fejer BG, Scherliess L (1997) Empirical models of storm time equatorial zonal electric fields. J Geophys Res 102(A11):24047–24056

Fejer BG, Scherliess L, de Paula ER (1999) Effects of the vertical plasma drift velocity on the generation and evolution of equatorial spread *F*. J Geophys Res 104(A9):19859–19870

Fesen CG, Crowley G, Roble RG, Richmond AD, Fejer BG (2000) Simulations of the pre-reversal enhancement in the low latitude vertical ion drifts. Geophys Res Lett 27: 1851–1854

Field P, Rishbeth H (1997) The response of the ionospheric F_2 layer to geomagnetic activity: An analysis of worldwide data. J Atmos Solar-Terr Phys 59:163–180

Forbes JM (2007) Dynamics of the thermosphere. J Met Soc Jpn 85B:193–213

Forbes JM, Lu G, Bruinsma S, Nerem S, Zhang X (2005) Thermosphere density variations due to the 15–24 April 2002 solar events from CHAMP/STAR accelerometer measurements. J Geophys Res. 107(A5):1052

Foster JC, Erickson PJ, Coster AJ, Goldstein J, Rich FJ (2002) Ionospheric signatures of plasmaspheric tails. Geophys Res Lett. 29(13):1623

Foster JC, Rideout W (2005) Midlatitude TEC enhancements during the October 2003 superstorm. Geophys Res Lett. 32:L12504

Fuller-Rowell TJ (1995) Dynamics of the lower thermosphere. In: Johnson RM, Killeen TL (eds) The upper mesosphere and lower thermosphere: a review of experiment and

theory. Geophysical monograph series, vol 87. American Geophysical Union, Washington, DC, p 23
Fuller-Rowell TJ, Codrescu MV, Moffett RJ, Quegan S (1994) Response of the thermosphere and ionosphere to geomagnetic storms. J Geophys Res 99(A3):3893–3914
Fuller-Rowell TJ, Codrescu MV, Risbeth H, Moffett RJ, Quegan S (1996b) On the seasonal response of the thermosphere and ionosphere to geomagnetic storms. J Geophys Res 101(A2):2343–2354
Fuller-Rowell TJ, Codrescu MV, Roble RG, Richmond AD (1997) How does the thermosphere and ionosphere react to a geomagnetic storm? In: Bruce BT, Gonzalez WD, Kamide Y, Arballo JK (eds) Magnetic storms. Geophysical monograph series, vol 98. American Geophysical Union, Washington, DC, pp 203–225
Fuller-Rowell TJ, Millward GH, Richmond AD, Codrescu MV (2002) Storm-time changes in the upper atmosphere at low latitudes. J Atmos Solar-Terr Phys 64:1383–1391
Fuller-Rowell TJ, Rees D (1984) Interpretation of an anticipated long-lived vortex in the lower thermosphere following simulation of an isolated substorm. Planet Space Sci 32:69–85
Fuller-Rowell TJ, Rees D, Quegan S, Moffett RJ, Codrescu MV, Millward GH (1996a) A coupled thermosphere ionosphere model (CTIM). In: Schunk RW (ed) Handbook of ionospheric models. STEP Report. Utah State University, Logan, pp 217–238
Fuller-Rowell TJ, Richmond AD, Maruyama N (2008) Global modeling of storm-time thermospheric dynamics and electrodynamics. In: Kintner PM, Coster AJ, Fuller-Rowell TJ, Mannucci AJ, Mendillo M, Heelis R (eds) Midlatitde ionospheric dynamics and disturbances. Geophysical monograph series, vol 181. American Geophysical Union, Washington, DC, pp 187–200
Goldstein J, Burch JL, Sandel BR, Mende SB, Brandt PC, Hairston MR (2005) Coupled response of the inner magnetosphere and ionosphere on 17 April 2002. J Geophys Res. 110:A03205
Goldstein J, Sandel BR, Hairston MR, Reiff PH (2003) Control of plasmaspheric dynamics by both convection and subauroral polarization stream. Geophys Res Lett. 30(24):2243
Greenspan ME, Rasmussen CE, Burke WJ, Abdu MA (1991) Equatorial density depletions observed at 840 km during the great magnetic storm of March 1989. J Geophys Res 96(A8):13931–13942
Groves KM, Basu S, Weber EJ, Smitham M, Kuenzler H, Valladares CE, Sheehan R, MacKenzie E, Secan JA, Ning P, McNeill WJ, Moonan DW, Kendra MJ (1997) Equatorial scintillation and systems support. Radio Sci 32(5): 2047–2064
Heelis RA (2004) Electrodynamics in the low and middle latitude ionosphere: A tutorial. J Atmos Solar-Terr Phys 66:825–838
Heelis RA, Sojka JJ, David M, Schunk RW (2009) Storm time density enhancements in the middle-latitude dayside ionosphere. J Geophys Res. 140:A03315
Huang C-S, Foster JC, Kelley MC (2005) Long-duration penetration of the interplanetary electric field to the low-latitude ionosphere during the main phase of magnetic storms. J Geophys Res 110:A11309
Huba JD, Joyce G, Sazykin S, Wolf R, Spiro R (2005) Simulation study of penetration electric field effects on the low- to mid-latitude ionosphere. Geophys Res Lett. 32:L23101
Hunsucker RD (1982) Atmospheric gravity waves generated in the high-latitude ionosphere: a review. Rev Geophys 20: 293–315
Kelley MC, Fejer BG, Gonzales CA (1979) An explanation of anomalous ionospheric electric fields associated with a northward turning of the interplanetary magnetic field. Geophys Res Lett 6:301
Killeen TL, Craven JD, Frank LA, Ponthieu J-J, Spencer NW, Heelis RA, Brace LH, Roble RG, Hays PB, Carignan GR (1988) On the relationship between dynamics of the polar thermosphere and morphology of the aurora: Global-scale observations from Dynamics Explorers 1 and 2. J Geophys Res 93(A4):2675–2692
Killeen TL, Hays PB, Carignan GR, Heelis RA, Hanson WB, Spencer NW, Brace LH (1984) Ion-neutral coupling in the high latitude *F* region: Evaluation of ion-neutral heating terms from the Dynamics Explorer 2. J Geophys Res 89:7495–7509
Kwak Y-S, Richmond AD (2007) An analysis of the momentum forcing in the high-latitude lower thermosphere. J Geophys Res. 112:A01306
Lei J, Wang W, Burns AG, Solomon SC, Richmond AD, Wiltberger M, Goncharenko LP, Coster AJ, Reinisch BW (2008) Observations and simulations of the ionospheric and thermospheric response to the December 2006 geomagnetic storm: Initial phase. J Geophys Res. 113:A01314
Liu H, Lühr H (2005) Strong disturbance of the upper thermospheric density due to magnetic storms: CHAMP observations. J Geophys Res. 110:A09529
Lu G, Goncharenko LP, Richmond AD, Roble RG, Aponte N (2008) A dayside ionospheric positive storm phase driven by neutral winds. J Geophys Res. 113:A08304
Mannucci AJ, Tsurutani BT, Iijima BA, Komjathy A, Saito A, Gonzalez WD, Guarnieri FL, Kozyra JU, Skoug R (2005) Dayside global ionospheric response to the major interplanetary events of October 29–30, 2003 "Halloween Storms." Geophys Res Lett. 32:L12S02
Marsh DR, Solomon SC, Reynolds AE (2004) Empirical model of nitric oxide in the lower thermosphere. J Geophys Res. 109:A07301
Maruyama N, Richmond AD, Fuller-Rowell TJ, Codrescu MV, Sazykin S, Toffoletto FR, Spiro RW, Millward GH (2005) Interaction between direct penetration and disturbance dynamo electric fields in the storm-time ionosphere. Geophys Res Lett. 32:L17105
Maruyama N, Sazykin S, Spiro RW, Fejer BG, Wolf R, Anderson DN, Anghel A, Toffoletto FR, Fuller-Rowell TJ, Codrescu MV, Richmond AD, Millward GH (2007) Modeling storm-time electrodynamics of the low-latitude ionosphere-thermosphere system: Can long lasting disturbance electric fields be accounted for? J Atmos Solar-Terr Phys 69:1182–1199
Mendillo M, Papagiannis MD, Klobuchar JA (1970) Ionospheric storms at midlatitudes. Radio Sci 5:895–898
Millward GH, Moffett RJ, Quegan S, Fuller-Rowell TJ (1996) A coupled thermosphere ionosphere plasmasphere model (CTIP). In: Schunk RW (ed) Handbook of ionospheric models. STEP Report. Utah State University, Logan, pp 239–279

Millward GH, Müller-Wodarg ICF, Aylward AD, Fuller-Rowell TJ, Richmond AD, Moffett RJ (2001) An investigation into the influence of tidal forcing on *F* region equatorial vertical ion drift using a global ionosphere-thermosphere model with coupled electrodynamics. J Geophys Res 106: 24733–24744

Paxton LJ, Christensen AB, Humm DC, Ogorzalek BS, Pardoe CT, Morrison D, Weiss MB, Crain W, Lew PH, Mabry DJ, Goldstein JO, Gary SA, Persons DF, Harold MJ, Alvarez EB, Ercol CJ, Strickland DJ, Meng C-I (1999) Global ultraviolet imager (GUVI): Measuring composition and energy inputs for the NASA Thermosphere Ionosphere Mesosphere Energetics and Dynamics (TIMED) mission. In: Larer MA (ed) SPIE optical spectroscopic techniques and instrumentation for atmospheric and space research III, vol 3756. Society of Photo-Optical Instrumentation Engineers, Bellingham, WA, pp 265–276

Prölss GW (1997) Magnetic storm associated perturbations of the upper atmosphere. In: Bruce BT, Gonzalez WD, Kamide Y, Arballo JK (eds) Magnetic storms. Geophysical monograph series, vol 98. American Geophysical Union, Washington, DC, pp 227–241

Richmond AD (1995) Ionospheric electrodynamics. In: Volland H (ed) Handbook of atmospheric electrodynamics, vol II. CRC Press, Boca Raton, FL, pp 249–290

Richmond AD, Matsushita S (1975) Thermospheric response to a magnetic substorm. J Geophys Res 80(19): 2839–2850

Richmond AD, Roble RG (1987) Electrodynamic effects of thermospheric winds from the NCAR thermospheric general circulation model. J Geophys Res 92:12365–12376

Rishbeth H, Fuller-Rowell TJ, Rees D (1987) Diffusive equilibrium and vertical motion in the thermosphere during a severe magnetic storm: A computational study. Planet Space Sci 35:1157–1165

Roble RG (1977) The upper atmosphere and magnetosphere. National Academy of Science, Washington, DC

Rodger AS, Wrenn GL, Rishbeth H (1989) Geomagnetic storms in the Antarctic *F* region, II, Physical interpretation. J Atmos Solar-Terr Phys 51:851–866

Sazykin S, Spiro RW, Wolf RA, Toffoletto FR, Tsyganenko N, Goldstein J, Hairston M (2005) Modeling inner magnetospheric electric fields: Latest self-consistent results. In: Pulkkinen TI, Tsyganenko NA, Friedel RHW (eds) The inner magnetosphere: physics and modeling. Geophysical monograph series, vol 115. American Geophysical Union, Washington, DC, pp 263–269

Scherliess L, Fejer BG (1997) Storm time dependence of equatorial disturbance dynamo zonal electric fields. J Geophys Res 102(A11):24037–24046

Schunk R, Raitt W, Banks P (1975) Effect of electric fields on the daytime high-latitude *E* and *F* regions. J Geophys Res 80(22):3121–3130

Shiokawa K, Otsuka Y, Ogawa T, Balan N, Igarashi K, Ridley AJ, Knipp DJ, Saito A, Yumoto K (2002) A large-scale traveling ionospheric disturbance during the magnetic storm of 15 September 1999. J Geophys Res. 107(A6):1088

Skoblin MG, Förster M (1993) An alternative explanation of ionization depletions in the winter night-time storm perturbed F_2 layer. Ann Geophys 11:1026–1032

Spiro RW, Wolf RA, Fejer BG (1988) Penetration of high-latitude-electric-field effects to low latitudes during SUNDIAL 1984. Ann Geophys 6:39–50

Strickland DJ, Daniell RE, Craven JD (2001) Negative ionospheric storm coincident with DE 1-observed thermospheric disturbance on October 14, 1981. J Geophys Res 106(A10):21049–21062

Sutton EK, Forbes JM, Nerem RS (2005) Global thermospheric neutral density and wind response to the severe 2003 geomagnetic storms from CHAMP accelerometer data. J Geophys Res. 110:A09540

Weimer DR (2005) Improved ionospheric electrodynamic models and application to calculating Joule heating rates. J Geophys Res. 110:A05306

Chapter 33

Ionosphere Data Assimilation: Problems Associated with Missing Physics

R.W. Schunk, L. Scherliess, and D.C. Thompson

Abstract Physics-based data assimilation models can be used for a wide range of applications in space physics, but as with all models the data assimilation models have limitations. The limitations can be associated with the data, the assimilation technique, or the background physics-based model. Here, we focused on the ionosphere and on elucidating the problems associated with missing physics in the background ionosphere model. The study was conducted with the Global Assimilation of Ionospheric Measurements-Gauss-Markov (GAIM-GM) physics-based data assimilation model. Simulations relevant to the low and middle latitude ionosphere were conducted in order to show how missing physics in the background ionosphere model affects the reconstructions. The low-latitude simulation involved the presence of equatorial plasma bubbles and a background physics-based ionosphere model that does not self-consistently describe bubbles. This problem, coupled with insufficient data, led to a Gauss-Markov reconstruction that contained a broad region of relatively low nighttime Total Electron Density (TEC) values instead of the individual plasma bubbles. The implications of plasma bubbles for reconstructions with the GAIM-FP (Full Physics) data assimilation model, where the electric fields and neutral winds are determined self-consistently, are noted. The mid-latitude simulation involved a Gauss-Markov ionosphere reconstruction for a geomagnetic storm where a Storm Enhanced Density (SED) appeared across the United States. Again, the background physics-based model (Ionosphere Forecast Model) could not produce the SED feature because this model does not take account of penetrating electric fields. Nevertheless, in this case there were sufficient data to overcome the deficiency in the background ionosphere model and the Gauss-Markov data assimilation reconstruction successfully described the SED feature and surrounding ionosphere.

R.W. Schunk (✉)
Center for Atmospheric and Space Sciences, Utah State University, Logan, UT 84322-4405, USA
e-mail: robert.schunk@usu.edu

33.1 Introduction

The ionosphere has been extensively studied for more than 60 years with a variety of measurement techniques and modeling approaches (cf. Schunk and Nagy, 2009), and it is now clear that the ionosphere is highly dynamic and exhibits a significant amount of spatial structure. Over the years, numerous modeling schemes have been used to describe both ionosphere climatology and weather, including empirical models, parameterized and analytical models, stand-alone global and regional numerical models, nested grid numerical models, data-driven numerical models, and global numerical models that couple to different regions (thermosphere, plasmasphere, polar wind, magnetosphere, and lower atmosphere). Recently, however, ionosphere data assimilation models have gained prominence because of their ability to provide both reliable reconstructions of the global ionosphere and the self-consistent ionospheric drivers (Schunk et al., 2004a; Scherliess et al., 2004; Bust and Cowley, 2007; Komjathy et al., 2010). The data assimilation models can be used for both science and applications, but as with all models the data assimilation

M.A. Abdu, D. Pancheva (eds.), A. Bhattacharyya (Coed.), *Aeronomy of the Earth's Atmosphere and Ionosphere*, IAGA Special Sopron Book Series 2, DOI 10.1007/978-94-007-0326-1_33, © Springer Science+Business Media B.V. 2011

models have limitations. These can be associated with the data, the assimilation technique, or the physics-based model. Here, we focus on elucidating the problems associated with missing physics in the ionosphere model coupled with insufficient data. Two examples are presented that pertain to the middle and low latitude ionosphere.

33.2 Data Assimilation Models

At Utah State University we developed two data assimilation models for the ionosphere as part of a program called Global Assimilation of Ionospheric Measurements (GAIM). The Gauss-Markov model (GAIM-GM) uses a physics-based model of the ionosphere and a Kalman filter as a basis for assimilating a diverse set of real-time (or near real-time) measurements (Schunk et al., 2004a, b, 2005a, b; Scherliess et al., 2004, 2006, 2009, 2010). The physics-based model is the Ionosphere Forecast Model (IFM), which is global and covers the E-region, F-region, and topside from 90 to 1400 km (Schunk et al., 1997). It takes account of five ion species (NO^+, $O_2{}^+$, $N_2{}^+$, O^+, H^+). However, the main output of the model is a 3-dimensional electron density distribution at user specified times. In GAIM-GM the ionosphere densities obtained from the IFM constitute the background ionosphere density field on which perturbations are superimposed based on the available data and their errors. The GAIM-GM model assimilates bottom-side N_e profiles from a variable number of ionosondes, slant TEC from a variable number of ground-based GPS/TEC stations, in situ N_e from the DMSP satellites, and line-of-sight UV emissions measured by satellites. Quality control algorithms for all of the data types are provided as an integral part of the model. The GAIM-GM model takes account of latent data (up to 3 h) and provides a first attempt at a 24-h forecast. However, realistic ionosphere forecasts require accurate forecasts of the solar and magnetosphere drivers, which are currently not possible. The model can be run in a global, regional, or nested grid mode. The primary output from GAIM-GM is a time-dependent, 3-dimensional, global, electron density distribution. Validation studies have been conducted by USU (Scherliess et al., 2006; Sojka et al., 2007; Thompson et al., 2006, 2009; Zhu et al., 2006) and by independent scientists (Decker and McNamara, 2007; McNamara et al., 2007, 2008; McDonald et al., 2006) and the model has been shown to provide reliable ionosphere reconstructions. In addition, the model has been used both to study ionosphere decay rates and to drive a global thermosphere model (Jee et al., 2007, 2008).

The Full Physics model (GAIM-FP) uses a physics-based model of the ionosphere–plasmasphere system and a Kalman filter as a basis for assimilating a diverse set of real-time (or near real-time) measurements (Schunk et al., 2004b; Scherliess et al., 2004, 2009, 2010). The physics-based model is the Ionosphere–Plasmasphere Model (IPM), which is global and covers the E-region, F-region, topside ionosphere and plasmasphere from 90 to 30,000 km. It takes account of six ion species (NO^+, $O_2{}^+$, $N_2{}^+$, O^+, H^+, He^+). The GAIM-FP model assimilates data via an ensemble Kalman filter and typically 30 ensemble members are used. The primary output of the GAIM-FP model is a time-dependent, 3-dimensional, plasma density distribution at user specified times. However, the model also provides quantitative estimates of the accuracy of the reconstructed plasma densities as well as the self-consistent drivers of the ionosphere–plasmasphere system (e.g., neutral winds and electric fields). In addition, auxiliary parameters are provided, including N_mE, h_mE, N_mF_2, h_mF_2, slant and vertical TEC. The GAIM-FP model assimilates bottom-side N_e profiles from a variable number of ionosondes, slant TEC from a variable number of ground-based GPS/TEC stations, in situ N_e from the DMSP satellites, occultation data, and line-of-sight UV emissions measured by satellites. The model can be run in a global, regional, or local mode. Forecasting is accomplished by running the IPM forward in time from a given GAIM-FP specification using predictions for the IPM inputs.

33.3 Problems with Missing Physics

There are several problems that need to be addressed in developing data assimilation models and they include problems with the data, the assimilation technique, and the physics-based model. With regard to the data, the important questions are as follows: (1) Are sufficient data available? (2) Are different data types available? (3) What is the quality of the data? and (4) Are error estimates available? An issue with the Kalman filter technique is that a rigorous application of the technique is not tractable, and therefore, approximations are needed. For the GAIM data assimilation models, we have used a reduced state (Scherliess et al., 2004), a Gauss-Markov (Scherliess et al., 2006), and an ensemble Kalman filter approximation (Scherliess et al., 2009, 2010). Here, the focus is on the physics-based model, and in particular, on the problem associated with missing physics. Two examples are given; one in which there are insufficient data to overcome the model deficiency and the other where sufficient data exist to overcome the model deficiency.

The first example involves the use of the GAIM-GM model, which is based on a global ionosphere model (IFM) that does not include the physics needed to allow for the self-consistent excitation of equatorial plasma bubbles. Plasma bubbles are common in the low-latitude ionosphere at night, particularly at solar maximum (Schunk and Nagy, 2009). The bubbles form on the bottom-side of the F-layer as a result of the Rayleigh-Taylor instability and then drift upwards and to the east. As a bubble drifts upward and eastward, the entire north-south extent of the individual flux tubes in the bubble becomes depleted, and the bubble takes the form of a vertically elongated wedge of depleted plasma that can be hundreds of km wide. The east-west extent of a disturbed region can be several thousand km, with the horizontal distance between separate bubbles being tens to hundreds of km. The density depletion in the bubbles is typically about a factor of 10, but factors of 100 and 1000 are also fairly common.

To determine the effect of equatorial plasma bubbles on a data assimilation reconstruction of the ionosphere, we used the GAIM-GM model with synthetic data (empirically generated data) so that the bubble effect could be clearly established. First, the ionosphere–plasmasphere system was simulated with the IPM for day 82 in 2004, which yielded global electron densities as a function of time between 60° south and 60° north geomagnetic latitudes and at altitudes from 90 to 20,000 km. Next, at a selected time, four plasma bubbles were imposed on the nighttime IPM N_e distribution and the bubbles were allowed to co-rotate with the earth. The plasma depletions in the bubbles followed the magnetic field lines and the depletions extended from low altitudes to a specified apex altitude. The depleted regions took the form of vertically elongated wedges of depleted plasma, as observed (i.e., upside-down watermelon slices with a given height (apex altitude), thickness (zonal extent) and separation). Figure 33.1 (top panel) shows a snapshot of the IPM Total Electron Density (TEC) distribution with the imposed bubbles at 03:00 LT in the American sector. The four bubbles are similar, with depletions of a factor of 10, zonal widths of 5°, apex altitudes of 1500 km, and separations that are approximately equal to the bubble widths at their northern most extent.

Using the time-dependent IPM N_e distribution that was modified by the four bubbles, synthetic slant GPS/TEC measurements were generated from the real locations of the GPS satellites and from the real locations of the GPS receivers that were available in 2004. For illustrative purposes, the synthetic slant TEC was converted to vertical TEC with an angle factor and plotted at 350 km, and the results at 03:00 LT are shown in Fig. 33.1 (middle panel). The synthetic slant TEC data were assimilated into the GAIM-GM model for a 24-h period on day 82. During the daytime, when there were no bubbles, the Gauss-Markov model successfully reconstructed the ionosphere with the synthetic data that were available (not shown). At night, however, the presence of equatorial plasma bubbles, which are not described by the background physics-based model (IFM), led to an incorrect reconstruction of the ionosphere (Fig. 33.1, bottom panel). With the Gauss-Markov model, an overall low-density region was obtained. If the Full Physics data assimilation model had been used, the same general N_e distribution would have been obtained, but the self-consistent electric fields and neutral winds deduced by the model would be wrong.

Missing physics in the background physics-based model may not adversely affect the ionosphere reconstruction if there are sufficient data. Figure 33.2 shows such a case. The figure is a snapshot from a regional run of the Gauss-Markov data assimilation model for the large geomagnetic storm that occurred on November 20–21, 2003. Ground GPS/TEC measurements from 332 GPS receivers covering the continental US and Canada were used in the Gauss-Markov model. More than 2000 slant TEC values were assimilated every 15 min. Bottom-side electron density profiles from the ionosondes at Dyess and Eglin Air Force bases were also assimilated. Figure 33.2 shows a snapshot from the storm simulation at 20 UT on day 324. The top panel shows a snapshot of the IFM output (no data assimilation), the middle panel shows the

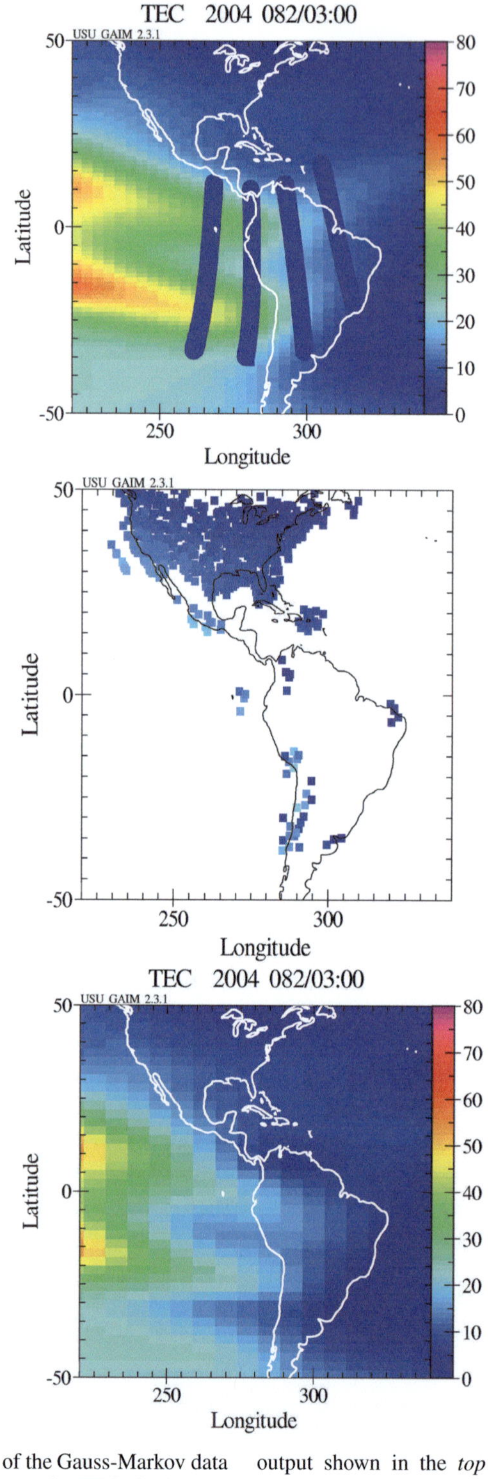

Fig. 33.1 Output from a regional run of the Gauss-Markov data assimilation model. The *top panel* shows the TEC distribution obtained from the Ionosphere–Plasmasphere Model (IPM) with four equatorial plasma bubbles superimposed. The *middle panel* shows the synthetic TEC data obtained from the IPM-bubble output shown in the *top panel* (slant TEC converted to the vertical and plotted at 350 km). The *bottom panel* shows the reconstructed TEC distribution when the synthetic slant TEC data are assimilated into the Gauss-Markov data assimilation model. The results are for day 82 in 2004 at 03:00 LT

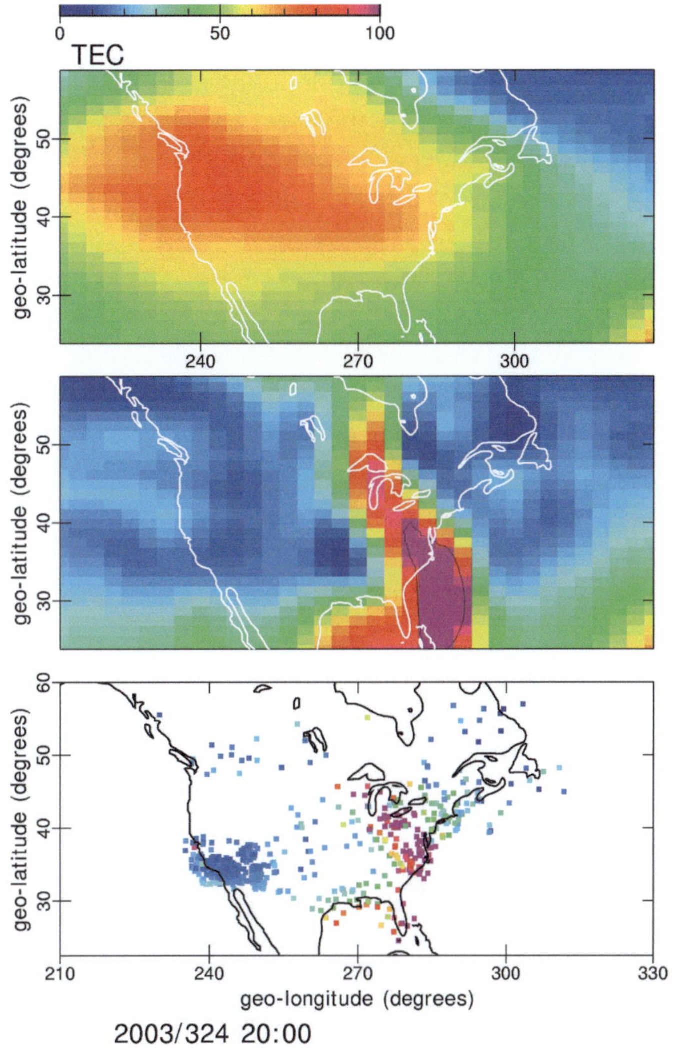

Fig. 33.2 The output from a regional run of the Gauss-Markov data assimilation model for the geomagnetic storm that occurred on November 20–21, 2003. The *top panel* shows the TEC distribution obtained from the physics-based IFM (no data assimilation), the *bottom panel* shows the measured TEC at the 350 km pierce point (slant TEC converted to the vertical), and the *middle panel* shows the Gauss-Markov reconstruction, where the slant TECs are assimilated into the IFM. The snapshot is for day 324 in 2003 at 2000 UT

Gauss-Markov TEC reconstruction, and the bottom panel shows the measured TEC plotted at 350 km (slant TEC converted to the vertical). Note that the region of Storm Enhanced Density (SED), which extends from Florida to the Great Lakes, is clearly evident in the data and is reproduced by the Gauss-Markov data assimilation model.

However, the background physics-based model (IFM) does not display the SED feature and yields TEC values outside of the SED region that are too high. This erroneous output is a result of missing physics in the IFM. Specifically, the SED is a result of high-latitude electric fields that penetrate to the middle-low latitude region (Foster et al., 2005; Heelis et al., 2009) and these penetrating electric fields are not included in the IFM. Likewise, the neutral atmosphere model used in the IFM (MSIS; Hedin, 1987) does not properly describe the large O/N_2 depletions that occur in major magnetic storms, and this is why the TECs obtained from the IFM are too large. Nevertheless, the data available for the storm period were sufficient to overcome the shortcomings in the IFM, and as a consequence, the Gauss-Markov model was able to correctly reproduce the regional, 3-dimensional, electron density distribution.

Acknowledgments This research was supported by Office of Naval Research grant N00014-09-1-0292 to Utah State University.

References

Bust GS, Crowley G (2007) Tracking of polar cap ionospheric patches using data assimilation. J Geophys Res 112:A05307. doi:10.1029/2005JA011597

Decker DT, McNamara LF (2007) Validation of ionospheric weather predicted by global assimilation of ionospheric measurements (GAIM). Radio Sci 42:RS4017. doi:10.1029/2007RS003632

Foster JC et al (2005) Multiradar observations of the polar tongue of ionization. J Geophys Res 110:A09S31. doi:10.1029/2004JA010928

Hedin AE (1987) MSIS-86 Thermospheric Model. J Geophys Res 92:4649–4663

Heelis RA, Sojka JJ, David M, Schunk RW (2009) Storm time density enhancements in the middle-latitude dayside ionosphere. J Geophys Res 114(A3):A03315

Jee G, Burns AG, Wang W, Solomon SC, Schunk RW, Scherliess L, Thompson DC, Sojka JJ, Zhu L (2007) Duration of an ionospheric data assimilation initialization of a coupled thermosphere-ionosphere model. Space Weather 5:S01004. doi:10.1029/2006SW000250

Jee G, Burns AG, Wang W, Solomon SC, Schunk RW, Scherliess L, Thompson DC, Sojka JJ, Zhu L (2008) Driving the TING model with GAIM electron densities: ionospheric effects on the thermosphere. J Geophys Res 113:A03305. doi:10.1029/2007JA012580

Komjathy A, Wilson B, Pi X, Akopian V, Dumett M, Iijima B, Verkhoglyadova O, Mannucci AJ (2010) JPL/USC GAIM: On the impact of using COSMIC and ground-based GPS measurements to estimate ionospheric parameters. J Geophys Res 115:A02307. doi:10.1029/2009JA014420

McDonald SE, Su Basu, Basu S, Groves KM, Valladares CE, Scherliess L, Thompson DC, Schunk RW, Sojka JJ, Zhu L (2006) Extreme longitudinal variability of plasma structuring in the equatorial ionosphere on a magnetically quiet equinoctial day. Radio Sci 41:RS6S24. doi: 10.1029/2005RS003366

McNamara LF, Baker CR, Decker DT (2008) Accuracy of USU-GAIM specifications of foF2 and M(3000)F2 for a world-wide distribution of ionosonde locations. Radio Sci 43:RS1011. doi:10.1029/2007RS003754

McNamara LF, Decker DT, Welsh J, Cole DG (2007) Validation of the USU GAIM model predictions of the maximum usable frequency for a 3000 km circuit. Radio Sci 42:RS3015. doi:10.1029/2006RS003589

Scherliess L, Schunk RW, Sojka JJ, Thompson DC (2004) Development of a physics-based reduced state Kalman filter for the ionosphere. Radio Sci 39:RS1S04. doi:10.1029/2002RS002797

Scherliess L, Schunk RW, Sojka JJ, Thompson DC, Zhu L (2006) The USU GAIM Gauss-Markov Kalman filter model of the ionosphere: model description and validation. J Geophys Res 111:A11315. doi:10.1029/2006JA011712

Scherliess L, Thompson DC, Schunk RW (2009) Ionospheric dynamics and drivers obtained from a physics-based data assimilation model. Radio Sci 44:RS0A32. doi:10.1029/2008RS004068

Scherliess L, Thompson DC, Schunk RW (2010) Data assimilation models: a 'new' tool for ionospheric science and applications. In: Liu W, Fujimoto M (eds) The dynamic magnetosphere, IAGA. Springer, NY (in press)

Schunk RW, Nagy AF (2009) Ionospheres, 2nd edn. Cambridge University Press, Cambridge

Schunk RW, Sojka JJ, Eccles JV (1997) Expanded capabilities for the ionospheric forecast model, Final Report, AFRL-VS-HA-TR-98-0001, 1–142

Schunk RW et al (2004a) Global assimilation of ionospheric measurements (GAIM). Radio Sci 39:RS1S02. doi:10.1029/2002RS002794

Schunk RW et al (2004b) USU Global ionospheric data assimilation models. Proc SPIE 5548:327–336. doi:10.1117/12.562448

Schunk RW et al (2005a) An operational data assimilation model of the global ionosphere. In: Goodman JM (ed) Proceedings of 2005 ionospheric effects symposium. JMG Associates Ltd, Alexandria, VA, pp 512–518

Schunk RW, Scherliess L, Sojka JJ, Thompson DC, Zhu L (2005b) Ionospheric weather forecasting on the horizon. Space Weather 3:S08007. doi:10.1029/2004SW000138

Sojka JJ, Thompson DC, Scherliess L, Schunk RW (2007) Assessing models for ionospheric weather specification over Australia during the 2004 CAWSES campaign. J Geophys Res 112:A09306. doi:10.1029/2006JA012048

Thompson DC, Scherliess L, Sojka JJ, Schunk RW (2006) The Utah State University Gauss-Markov Kalman filter in the ionosphere: the effect of slant TEC and electron density profile data on model fidelity. J Atmos Solar-Terr Phys 68:947–958

Thompson DC, Scherliess L, Sojka JJ, Schunk RW (2009) Plasmasphere and upper ionosphere contributions and corrections during the assimilation of GPS slant TEC. Radio Sci 44:RS0A02. doi:10.1029/2008RS004016

Zhu L et al (2006) Validation study of the ionosphere forecast model (IFM) using the TOPEX total electron content measurements. Radio Sci 41:RS5S11. doi:10.1029/2005RS003336

Chapter 34

Penetration of Magnetospheric Electric Fields to the Low Latitude Ionosphere During Storm/Substorms

Takashi Kikuchi, Kumiko K. Hashimoto, Atsuki Shinbori, Yuji Tsuji, and Shin-Ichi Watari

Abstract Penetration of the magnetospheric electric fields to low latitude ionosphere is examined using magnetometer data from high latitude to the dip equator during substorms and geomagnetic storms. To detect the penetration electric fields, we analyzed magnetic disturbances at the dip equator, subtracted by those at low latitude (DP2). During substorm growth phase, the DP2 currents are enhanced by the dawn-to-dusk convection electric field, which are supplied by the Region-1 field-aligned currents (R1 FACs) via the mid and low latitude ionosphere. On the other hand, the DP2 currents decrease significantly during the substorm expansion, superposed by reversed currents flowing from the R2 FACs. In particular, when the IMF turns northward during the substorm, the DP2 currents change to the counterelectrojet (CEJ), i.e., overshielding currents at the dip equator. During the storm main phase, the DP2 currents are very much intensified by the enhanced convection electric field. However, the DP2 currents change to the CEJ at the beginning of the storm recovery phase. The overshielding electric field as well as the convection electric field causes dramatic changes in the low latitude ionosphere.

34.1 Convection Electric Field and Global DP2 Currents

The magnetospheric convection is generated by the reconnection between the interplanetary magnetic field (IMF) and the Earth's magnetic field at the magnetopause (Dungey, 1961). Global MHD simulations have provided detailed structures of a dynamo generating the Region 1 field-aligned currents (R1 FACs), which is located around the cusp where the solar wind energy is converted to the thermal energy of high pressure plasma and generates the convection electric field and the R1 FACs (Tanaka, 1995, 2007; Siscoe et al., 2000). The R1 FACs flow into/out from the polar ionosphere in the morning/afternoon sector (Iijima and Potemra, 1976, 1978).

The convection electric field causes DP2 magnetic fluctuations at high latitude and dip equator (Nishida et al., 1966; Nishida, 1968a, b; Kikuchi et al., 1996; Kobea et al., 1998). The DP2 fluctuations were coherent with variations in the southward IMF, suggesting direct penetration of the magnetospheric electric field to the dip equator (Nishida, 1968b). Kikuchi et al. (1996), as shown in Fig. 34.1, demonstrated that the DP2 fluctuations observed at afternoon auroral latitude (Kilpisjarvi) during the growth phase of a substorm (1200–1400 UT) are coherent with those at mid latitude (Nurmijarvi) and dip equator (Mokolo) with a correlation coefficient of 0.9 and no time shift greater than 25 s. During the DP2 fluctuation event, EISCAT measured the convection electric field and height-integrated Hall and Pedersen conductivities. The Hall current calculated from these parameters well agrees with the DP2 fluctuations at Kilpisjarvi located near EISCAT. The high coherency between high latitude and equator implies that the convection electric field was transmitted to the equator near-instantaneously, driving the Pedersen current amplified by the Cowling effect at the dayside dip equator (Hirono, 1952; Baker and Martyn, 1953). As a result, the DP2 decreases with decreasing latitude, but increases at the dip equator

T. Kikuchi (✉)
Solar-Terrestrial Environment Laboratory, Nagoya University, 464-8601 Nagoya, Aichi, Japan
e-mail: kikuchi@stelab.nagoya-u.ac.jp

M.A. Abdu, D. Pancheva (eds.), A. Bhattacharyya (Coed.), *Aeronomy of the Earth's Atmosphere and Ionosphere*,
IAGA Special Sopron Book Series 2, DOI 10.1007/978-94-007-0326-1_34,

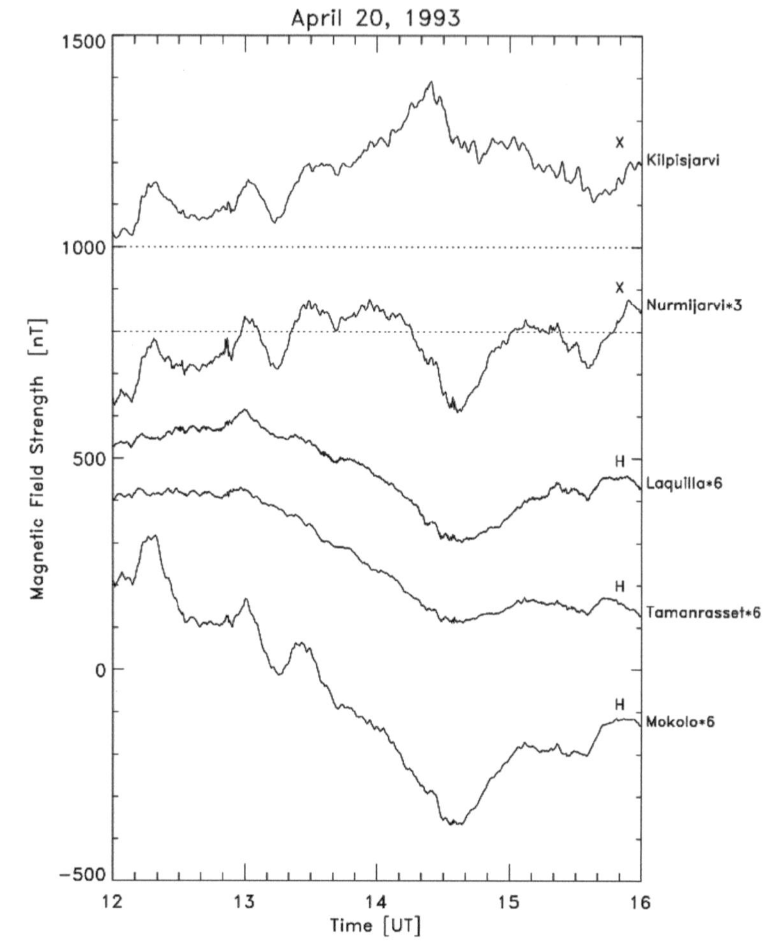

Fig. 34.1 The *X*-component at the IMAGE stations, Kilpisjarvi (65.76° CGM Lat), Nurmijarvi (56.79° CGM Lat), and the H-component at Laquilla (36.19° CGM Lat), Tamanrasset (5.86° CGM Lat) and Mokolo (−1.5° Dip Lat) for the time interval of 12–16 U T on April 20, 1993. The dotted lines for the IMAGE stations indicate the quiet nighttime level. The data are multiplied by a factor of 3 for Nurmijarvi and 6 for the mid to equatorial latitudes for the sake of a better comparison of the wave forms (figure 2 of Kikuchi et al. (1996), reproduced by permission of American Geophysical Union)

with an equatorial to low latitude amplitude ratio of 4 (Nishida, 1968a). The DP2 is different from the DP1 associated with the auroral electrojet intensified during the substorm expansion phase (Nishida, 1968a). The equatorial enhancement of geomagnetic perturbations is an important feature of the equatorward extension of ionospheric currents driven by the potential electric field in the polar ionosphere. Thus, the equatorial DP2 is connected with the R1 FACs via the Pedersen currents at mid and low latitude (Kikuchi et al., 1996).

The diurnal magnetic variation at the geomagnetic equator is often depressed substantially during disturbed periods (Matsushita and Balsley, 1972; Onwumechilli et al., 1973; Kikuchi et al., 1996). Matsushita and Balsley (1972) critically discussed that the DP2 fluctuations should be measured negatively from the quiettime level. However, the DP2 should be measured positively from the smoothed diurnal variation, because of the high coherency between the high latitude and equator with time shift of 2 min (Nishida, 1968b) and 25 s (Kikuchi et al., 1996). The depression of the diurnal variation may be caused by a westward electric field due to the disturbance dynamo (Blanc and Richmond, 1980).

34.2 Overshielding Electric Field and CEJ

The enhanced convection electric field drives an earthward motion of plasma in the plasmasheet, generating a partial ring current and the Region-2 field-aligned currents (R2 FACs) in the inner magnetosphere. The R2 FACs build up an electric field in the ionosphere with an opposite direction to that of the convection electric field at mid and low latitudes (Vasyliunas, 1972; Jaggi and Wolf, 1973; Southwood, 1977; Senior and Blanc, 1984). The R2 FACs provide a northward electric field at auroral latitude and southward electric field at mid latitude in the afternoon sector. As a result, the eastward auroral electrojet intensifies, but the eastward current reduces at subauroral-mid latitudes. In other words, the electric field associated with the R2 FACs tends to cancel the convection electric field, i.e., shielding works at lower latitude. The time constant of the shielding has been estimated as 17–20 min from the magnetometer observations (Somayajulu et al., 1987; Kikuchi et al., 2000) and 20–30 min from the theoretical calculations (Senior and Blanc, 1984; Peymirat et al., 2000).

When the convection electric field reduces abruptly because of the northward turning of the IMF, the electric field reverses its direction at mid-equatorial latitudes. The reversal of the penetrated electric field was observed by the Jicamarca incoherent scatter radar at the equator, which was identified as the overshielding electric field (Kelley et al., 1979; Gonzales et al., 1979; Fejer et al., 1979). The reversed current at the equator appears as the counter-electrojet (CEJ) (Rastogi and Patel, 1975; Rastogi, 1977, 1997; Kobea et al., 2000; Kikuchi et al., 2000, 2003). The equatorial CEJ is connected with the R2 FACs via the auroral ionosphere.

34.3 DP2 Currents and CEJ During Substorms

The substorm growth phase is initiated by the southward turning of the IMF, which causes DP2 currents in the ionosphere (Iijima and Nagata, 1972; Nishida and Kamide, 1983; Kamide et al., 1996). Kikuchi et al. (1996, 2000) demonstrated that the eastward ionospheric currents increased at auroral-mid latitudes in the afternoon sector, coherently with an enhancement of the EEJ during the substorm growth phase (1200–1400 UT, Fig. 34.1). However, the coherency between the auroral (Kilpisjarvi) and mid latitude (Nurmijarvi) broke down during the substorm expansion phase (1400–1510 UT, Fig. 34.1). Figure 34.2 shows the height-integrated Hall (solid curve in the upper panel) and Pedersen (dotted curve) conductivities, northward (solid curve in the middle panel) and eastward (dotted curve) electric fields measured by EISCAT, and the *X*-component magnetic field at Nurmijarvi measured from the quiettime diurnal variation during the growth phase (DP2 in the upper panel) and expansion phase of the substorm (Kikuchi et al., 2000). Under an assumption that the convection electric field at EISCAT penetrated to Nurmijarvi with geometrical attenuation (factor = 0.4) and the R2 FACs developed between the two latitudes and increased/decreased the electric field at the auroral/mid latitude, the convection and shielding electric fields are derived from these parameters. Figure 34.3 (upper panel) shows the convection electric field, E_1 (solid line) and shielding electric field, E_2 (dashed line) at auroral latitude (EISCAT). The shielding electric field is in the same direction as the convection electric field, intensifying the electric field at auroral latitude. Figure 34.3 (lower panel) shows that the convection electric field decreases to be $E_1 \times 0.4$ at mid latitude (Nurmijarvi) because of the geometrical attenuation, while the shielding electric field is $-E_2$ with an opposite direction to the convection electric field. During the growth phase before 14 UT, the convection electric field is dominant, but the shielding electric field increased significantly during the expansion phase after 1400 UT. Thus, the shielding electric field reduces the electric field at mid latitude, causing the overshielding when E_2 is greater than $E_1 \times 0.4$ at around 1430 UT (Fig. 34.3, lower panel). As a result, the equatorial CEJ occurred, causing an equatorial enhancement of the negative bay at the dayside dip equator, Mokolo (Fig. 34.1). It is to be noted that the shielding electric field, i.e., the R2 FACs started to increase at around the onset of the substorm and continued growing during the expansion phase. When the IMF turned northward during the expansion phase, the well-grown R2 FACs cause a large amplitude CEJ (Kikuchi et al., 2003). Figure 34.4 shows a schematic diagram of the current circuit between the inner magnetosphere and the ionosphere at the dayside dip equator during the period of overshielding in substorm expansion phase. The R2 FACs and the equatorial CEJ are connected by the Pedersen current at mid and low latitudes.

34.4 DP2 Currents and CEJ During Geomagnetic Storms

During the storm main phase, intense dawn-to-dusk electric field is generated deep inside the magnetosphere, which would cause equatorward shift

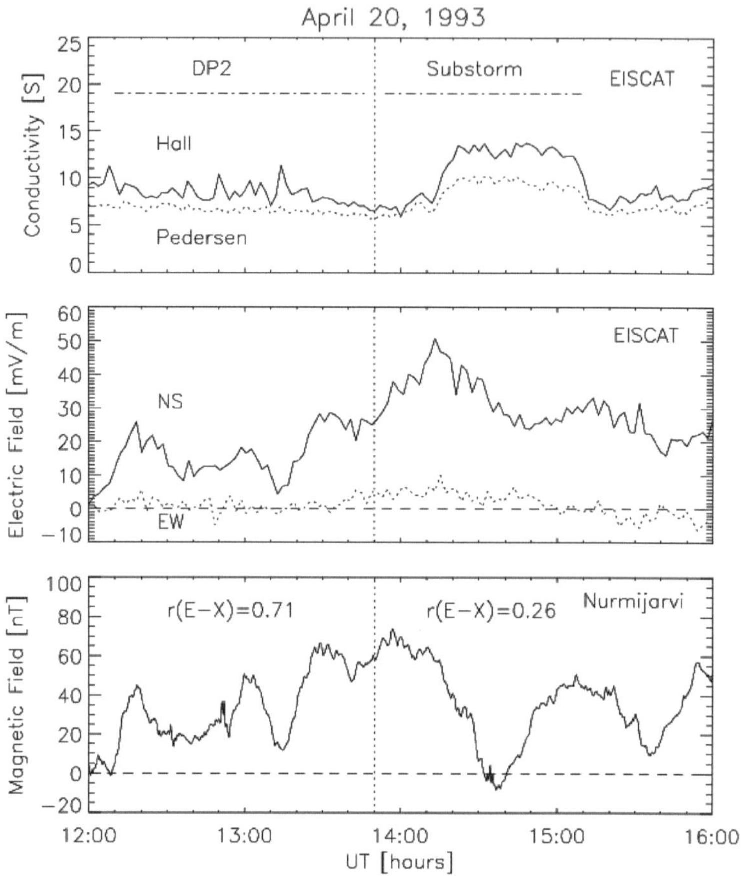

Fig. 34.2 The Hall and Pedersen conductivities (*upper panel*), NS and EW components of the electric field (*middle panel*) observed by EISCAT (66.16° CGM Lat), and the *X* component magnetic field at Nurmijarvi measured from the quiettime level. The correlation coefficients, *r*(*E*–*X*), are calculated between the NS electric field at EISCAT and *X*-component at Nurmijarvi during the growth phase (DP2) and expansion phase of the substorm. (figure 6 of Kikuchi et al. (2000), reproduced by permission of American Geophysical Union)

of auroral electrojet to mid latitude (Feldstein et al., 1997). The stormtime electric field penetrated to the inner magnetosphere as observed by CRRES and Akebono satellites at $L = 2$–6 (Wygant et al., 1998; Burke et al., 1998; Shinbori et al., 2005), driving the storm ring current. Shinbori et al. (2005) demonstrated temporal variations of the convection electric field at a distance of 2–5 Re during the main phase of the major storm on March 13, 1989, with maximum magnitude of 46 mV/m.

Wilson et al. (2001) demonstrated that the DP2 currents developed at mid latitude during a geomagnetic storm, when a significant electric field was detected by CRRES inside the ring current. They suggested that the electric field associated with the DP2 currents might have contributed to the development of the storm ring current. The convection electric field further penetrates to the equatorial ionosphere, driving the DP2 currents and intensifying the equatorial EEJ (Tsurutani et al., 2004; Huang et al., 2005; Kikuchi et al., 2008a, b).

Kikuchi et al. (2008a) analyzed a geomagnetic storm recorded at the dayside dip equator, Yap (YAP, –0.3° GM Lat), off equator, Guam (GAM, 4.89° GM Lat) and low latitude, Okinawa (OKI, 14.47° GM Lat) as shown in Fig. 34.5. The temporal variations at OKI indicate those of the ring current, which developed immediately after the storm sudden commencement (SC) because of the strong southward IMF. The storm at YAP, on the other hand, increased considerably for more than 1 h during the main phase, while decreased considerably during the recovery phase. As a result, the geomagnetic storm was enhanced at the equator with an equatorial to low latitude amplitude ratio of 2.7. Rastogi (2004) demonstrated that the magnitude of the geomagnetic storm was significantly enhanced at the dayside dip equator, which was attributed to the CEJ caused by the northward turning of the IMF. Kikuchi et al. (2008a) pointed out that the geomagnetic storm was enhanced by a combined effect of the DP2 currents during the main phase and the CEJ during the recovery phase.

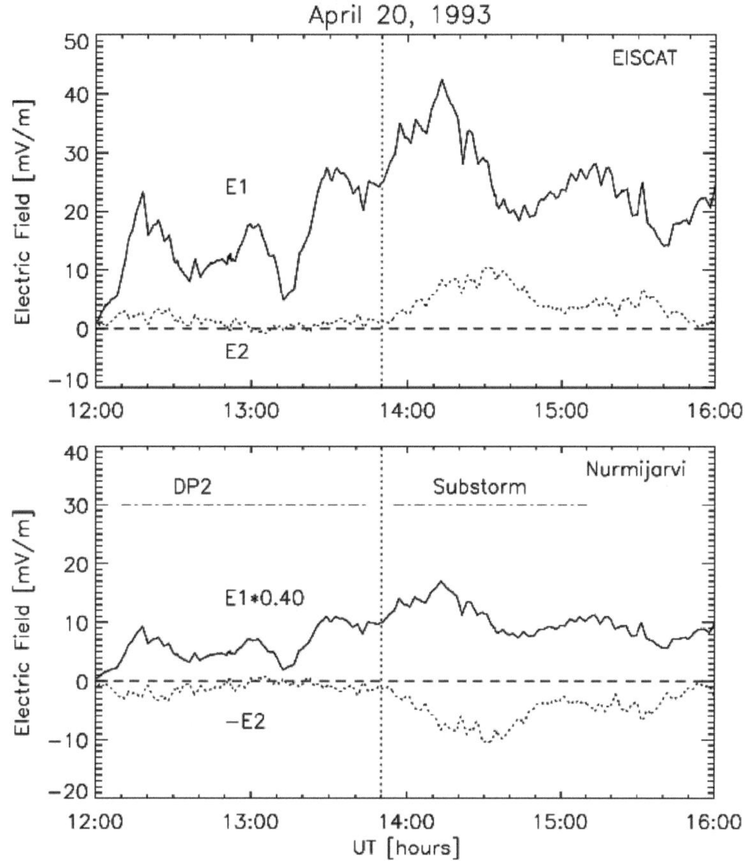

Fig. 34.3 Estimated electric fields associated with the *R*1 and *R*2 FACs at EISCAT (*E*1, *E*2) and at Nurmijärvi (*E*1 × 0.4, –*E*2) during the growth (DP2) and expansion phases of the substorm, deduced from the electric field measured by EISCAT and the *X*-component magnetic field at Nurmijärvi (Fig. 34.2). (figure 8 of Kikuchi et al. (2000), reproduced by permission of American Geophysical Union)

Difference between YAP and OKI provides us with DP2/CEJ during the storm, which is shown in the bottom panel of Fig. 34.6 together with contours of westward auroral electrojet derived from the IMAGE magnetometer data at the dawn sector (upper panel). The DP2 currents are remarkably enhanced during the main phase (02–04 UT), but it changes to CEJ during the recovery phase (04–07 UT). The auroral electrojet developed during the main phase at mid latitude (57° CGM Lat), but it shifted poleward rapidly to the normal auroral latitude (67° CGM Lat) at the beginning of the recovery phase. Huang et al. (2005) pointed out that the convection electric field continued to penetrate to low latitude for many hours during the whole period of storm main phase. This tendency is observed in Fig. 34.6, but the DP2 started to decrease at 0250 UT, indicating that the shielding became effective in late main phase (Kikuchi et al., 2008a). The AEJ was intensified over 0240–0340 UT with a peak at 0310 UT, when the equatorial DP2 was decreasing its intensity. The distinct temporal behavior at the two latitude regions implies development of the R2 FACs below 55° CGM Lat, which intensifies/reduces the electric field at auroral/lower latitude. Because of the developing R2 FACs, overshielding occurred at the beginning of storm recovery phase, when the convection electric field decreased because of a decrease in the southward IMF (Kikuchi et al., 2008a). The overshielding electric field responsible for the CEJ would be transmitted to the inner magnetosphere, where the ring current must have ceased developing because of the reversed electric field. The reversed electric field was observed by CRRES during the storm recovery phase (Wygant et al., 1998). These observations imply that the electric field associated with the DP2 currents contributes to the ring current development, while the overshielding electric field may contribute to the decay of the ring current.

The stormtime current circuits are composed of the R1 FAC-DP2 during the main phase and the R2 FAC-CEJ during the recovery phase, which are similar to those of the substorm growth and expansion

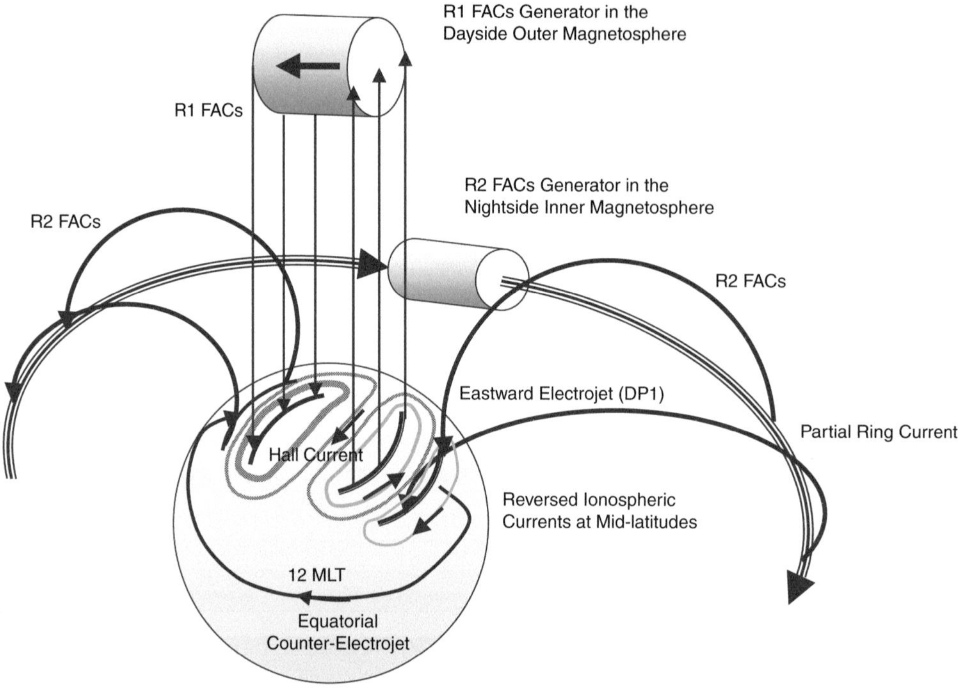

Fig. 34.4 Schematic diagram of the current system in the magnetosphere and ionosphere during the substorm. The $R1$ and $R2$ FACs flow into the equatorial ionosphere through the polar ionosphere, resulting in the equatorial DP2 during the growth phase, while the equatorial CEJ appears during the expansion phase. The diagram is pertinent to the equatorial CEJ. (figure 11 of Kikuchi et al. (2003), reproduced by permission of American Geophysical Union)

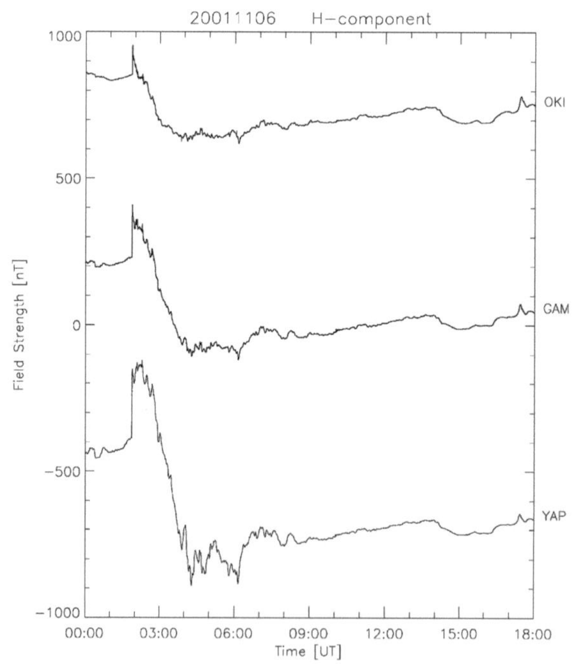

Fig. 34.5 H-component magnetic fields recorded at low latitude, Okinawa (OKI, 14.47° GML), and near the dip equator, Guam (GAM, 4.89° GM Lat), and Yap (YAP, –0.3° GM Lat) during the storm on November 6, 2001. The stations were located at around noon when the storm started (MLT= UT+9). (figure 6 of Kikuchi et al. (2008a), reproduced by permission of American Geophysical Union)

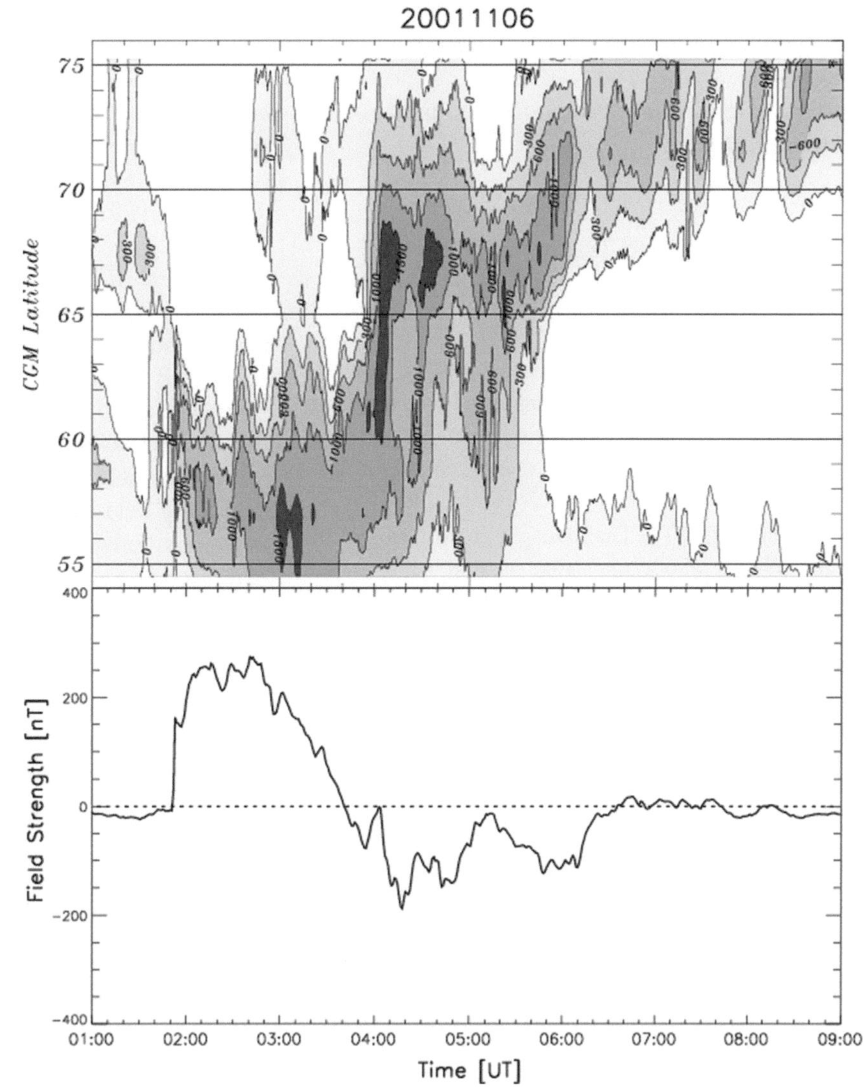

Fig. 34.6 (*Upper panel*): Contour map of the magnetic disturbances caused by the westward electrojet at auroral – mid latitudes in the dawn sector derived from the IMAGE magnetometer array data. (*Lower panel*): Magnetic deflections caused by storm-time ionospheric currents at YAP obtained by subtracting magnetic disturbances at OKI after eliminating quiettime diurnal variations on November 4, 2001. The magnetic disturbance at OKI is used to eliminate effects of the magnetospheric currents. (figure 8 of Kikuchi et al. (2008a), reproduced by permission of American Geophysical Union)

phases, respectively. Kikuchi et al. (2008b) confirmed the current circuits of the main and recovery phases, by analyzing two other storm events with different conditions of the IMF. The rapid poleward shift of the auroral electrojet was commonly observed in the three events, which must be related to reconfiguration of the magnetosphere during substorms. The role of the substorm in the evolution of the geomagnetic storm remains a crucial issue in the storm-substorm relationship.

It should be noted that the CEJ was observed even during the storm main phase. Fejer et al. (2007) indicated that the electric field at the equator was opposite in direction to that of the convection electric field on both the day- and night-sides for the storm event on November 10, 2004. On the other hand, the electric field was the normal convection electric field during the storm main phase on November 9, 2004. The disturbance dynamo caused by the thermospheric wind (Blanc and Richmond, 1980) must have been activated by the first storm, of which electric field overcame the convection electric field in main phase of the second storm. It should be reminded that the disturbance dynamo starts to work with a time lag of several hours from the beginning of storm and continues to work for a long time, say, 10 h (Fejer and Scherliess, 1997). The overshielding electric field is caused by the R2 FACs flowing into the auroral ionosphere, while the disturbance dynamo works in the mid latitude thermosphere. Latitude and local time distribution of the ionospheric electric field will enable us to distinguish the overshielding electric field from that of the disturbance dynamo.

34.5 Electric Field Transmission Mechanism

The near-instantaneous transmission of the convection electric field to the equator was first suggested by Araki (1977), based on his finding of the simultaneous onset of the preliminary reverse impulse (PRI) of SCs at high latitude and dip equator with a temporal resolution of 10 s. The ionospheric current pattern of the PRI is about the same as that of the DP2 (Nagata and Abe, 1955; Nishida et al., 1966), driven by a dusk-to-dawn electric field. The PRI electric field was directly observed by HF Doppler measurements as a downward/upward motion of low latitude ionosphere on the day/nightside (Kikuchi et al., 1985). The PRI electric field is a potential electric field same as the DP2, propagating near-instantaneously to low latitude within 10 s (Kikuchi, 1986). The simultaneity within 10 s is explained by means of the TM_0 mode wave propagating at the speed of light in the Earth-ionosphere waveguide (Kikuchi et al., 1978; Kikuchi and Araki, 1979).

Figure 34.7 shows a schematic diagram of the three-layered Earth-ionosphere waveguide where a downward FAC is given at one end of the waveguide (Kikuchi, 2005). The electric field carried along the FAC is subject to reflection at the ionosphere. However, a fraction of the electric field is transmitted into the Earth-ionosphere waveguide, providing a vertical electric field, E_z below the ionosphere. The vertical electric field excites the TM_0 mode wave, which propagates horizontally (x-direction) at the speed of light, accompanying a magnetic field, H_y transverse to the propagation plane. Thus, the Poynting flux, S_x composed of E_z and H_y is transmitted to low latitude (x-direction). The TM_0 mode wave accompanies electric currents in the ionosphere and at the surface of the ground, which are connected by the displacement

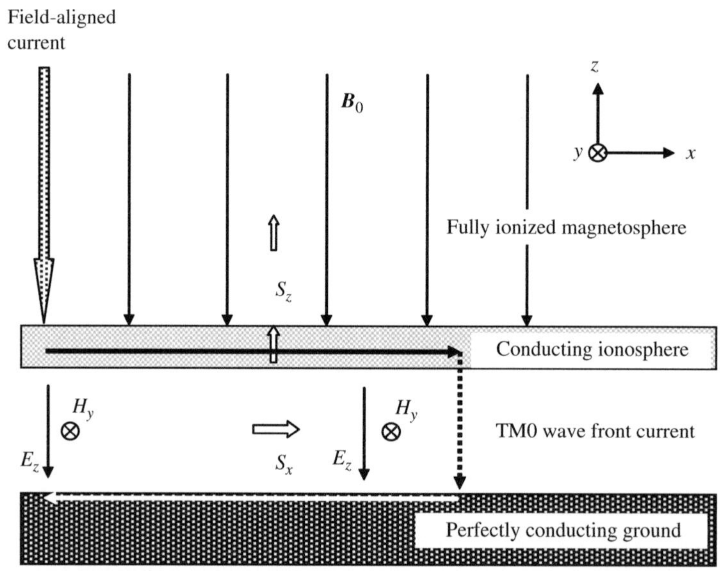

Fig. 34.7 Three-layered Earth-ionosphere waveguide composed of the magnetosphere, ionosphere and neutral atmosphere terminated by the perfectly conducting ground. The TM_0 mode wave propagates in x-direction in the waveguide, characterized by the magnetic field, H_y, transverse to the propagation plane (x–z plane) and vertical electric field, E_z, in the plane. The TM_0 mode wave carries electric currents in the ionosphere and on the ground, connected by the displacement current at the wave front. The Poynting flux, S_x, composed of E_z and H_y transmits horizontally to low latitude (x-direction). A fraction of energy, S_z, escapes vertically into the ionosphere and the magnetosphere, which will cause quick response of the low latitude ionosphere and prompt development of the electric field in the inner magnetosphere

current at the wave front. The TM_0 mode wave is applied not only to the short-term electric field of the PRI, but also to long lasting electric fields of the DP2 and storm/substorms, because the TM_0 mode wave has no lower cutoff frequency (Kikuchi and Araki, 1979).

A fraction of the Poynting flux is transmitted from the ionosphere to the inner magnetosphere as shown in Fig. 34.7, causing attenuation of the TM_0 mode wave. However, the attenuation is much less than that of the geometrical attenuation due to the finite size of the polar cap electric field even under the nighttime ionospheric condition (Kikuchi and Araki, 1979; Kikuchi, 2005). Under a daytime ionospheric condition with high ionospheric conductance to Alfven conductance ratio (>10), the electric field associated with the ionospheric current is transmitted by the Alfven wave upward along the field lines with no appreciable attenuation. The polarization current at the wave front of the Alfven wave exerts a $\boldsymbol{J} \times \boldsymbol{B}$ force on plasma in the F-region ionosphere and in the inner magnetosphere. This results in the coherent variations in the ground magnetic field and in the F-region ionosphere at the geomagnetic equator, as observed by the HF Doppler measurements (Abdu et al., 1998) and by the Jicamarca incoherent scatter radar (Kikuchi et al., 2003). The upward transmission of the Poynting flux would cause the quick development of the electric field in the inner magnetosphere (Nishimura et al., 2009), ring current (Hashimoto et al., 2002), plasmasheet thinning (Hashimoto and Kikuchi, 2005), and reconfiguration of the plasmapause (Murakami et al., 2007).

34.6 Conclusion

1. The convection electric field is transmitted by the Alfven wave from the dynamo in the outer magnetosphere to the polar ionosphere, accompanying the Region-1 field-aligned currents and driving the DP2 currents composed of ionospheric Hall currents at high latitude and the Pedersen currents amplified by the Cowling effect at the dip equator.
2. The convection electric field is transmitted to low latitude near-instantaneously, resulting in high coherency of the DP2 fluctuations and simultaneous enhancement of the equatorial electrojet during storm/substorms.
3. The electric field associated with the DP2 currents is transmitted into the F-region ionosphere and into the inner magnetosphere, causing quick response of the low latitude ionosphere and ring current development during storm/substorms.
4. The overshielding electric field due to the Region-2 field-aligned currents causes reversal of the ionospheric currents at mid latitude and the counterelectrojet at the dip equator during substorm expansion phase and storm recovery phase.
5. The TM_0 mode wave propagating at the speed of light in the Earth-ionosphere waveguide carries the polar electric field to low latitude near-instantaneously and further into the inner magnetosphere with a time delay no greater than the propagation time of the Alfven wave.

References

Abdu MA, Sastri JH, Luhr H, Tachihara H, Kitamura T, Trivedi NB, Sobral JHA (1988) DP 2 electric field fluctuations in the dusk-time dip equatorial ionosphere. Geophys Res Lett 25:9. doi:10.1029/98GL01096

Araki T (1977) Global structure of geomagnetic sudden commencements. Planet Space Sci 25:373–384

Baker WG, Martyn DF (1953) Electric currents in the ionosphere I. The conductivity. Phil Trans R Soc Lond Ser A 246:281–294

Blanc M, Richmond AD (1980) The ionospheric disturbance dynamo. J Geophys Res 85:1669–1686

Burke WJ, Maynard NC, Hagan MP, Wolf RA, Wilson GR, Gentile LC, Gussenhoven MS, Huang CY, Garner TW, Rich FJ (1998) Electrodynamics of the inner magnetosphere observed in the dusk sector by CRRES and DMSP during the magnetic storm of June 4–6, 1991. J Geophys Res 103(A12):29,339–29,418

Dungey JW (1961) Interplanetary magnetic field and the auroral zones. Phys Rev Lett 6:47

Fejer BG, Gonzales CA, Farley DT, Kelley MC (1979) Equatorial electric fields during magnetically disturbed conditions 1. The effect of the interplanetary magnetic field. J Geophys Res 84:5797–5802

Fejer BG, Jensen JW, Kikuchi T, Abdu MA, Chau JL (2007) Equatorial ionospheric electric fields during the November 2004 magnetic storm. J Geophys Res 112:A10304. doi:10.1029/2007JA012376

Fejer BG, Scherliess L (1997) Empirical models of storm time equatorial zonal electric fields. J Geophys Res 102(A11):24,047–24,056

Feldstein YI, Grafe A, Gromova LI, Popov VA (1997) Auroral electrojets during geomagnetic storms. J Geophys Res 102:14223–14235

Gonzales CA, Kelley MC, Fejer BG, Vickrey JF, Woodman RF (1979) Equatorial electric fields during magnetically disturbed conditions 2. Implications of simultaneous auroral and equatorial measurements. J Geophys Res 84:5803–5812

Hashimoto KK, Kikuchi T (2005) Quick response of the near-earth Magnetotail to changes in the interplanetary magnetic field. In: Pulkkinen TI, Tsyganenko NA, Friedel RHW (eds) The inner magnetosphere: physics and modeling. Geophysical monograph series, vol 155. AGU, Washington, DC, pp 47–53

Hashimoto KK, Kikuchi T, Ebihara Y (2002) Response of the magnetospheric convection to sudden interplanetary magnetic field changes as deduced from the evolution of partial ring currents. J Geophys Res 107(A11):1337. doi:10.1029/2001JA009228

Hirono M (1952) A theory of diurnal magnetic variations in equatorial regions and conductivity of the ionosphere E region. J Geomag Geoelectr Kyoto 4:7–21

Huang C-S, Foster JC, Kelley MC (2005) Long-duration penetration of the interplanetary electric field to the low-latitude ionosphere during the main phase of magnetic storms. J Geophys Res 110:A11309. doi:10.1029/2005JA011202

Iijima T, Nagata T (1972) Signatures for substorm development of the growth phase and expansion phase. Planet Space Sci 20:1095–1112

Iijima T, Potemra T (1976) The amplitude distribution of field-aligned currents at northern high latitudes observed by Triad. J Geophys Res 81:13. doi:10.1029/JA081i013p02165

Iijima T, Potemra T (1978) Large-scale characteristics of field-aligned currents associated with substorms. J Geophys Res 83:A2. doi:10.1029/JA083iA02p00599

Jaggi R, Wolf R (1973) Self-consistent calculation of the motion of a sheet of ions in the magnetosphere. J Geophys Res 78(16):2852–2866

Kamide Y, Sun W, Akasofu S-I (1996) The average ionospheric electrodynamics for the different substorm phases. J Geophys Res 101:A1. doi:10.1029/95JA02990

Kelley MC, Fejer BG, Gonzales CA (1979) An explanation for anomalous equatorial ionospheric electric fields associated with a northward turning of the interplanetary magnetic field. Geophys Res Lett 6:301–304

Kikuchi T (1986) Evidence of transmission of polar electric fields to the low latitude at times of geomagnetic sudden commencements. J Geophys Res 91:3101–3105

Kikuchi T (2005) Transmission line model for driving plasma convection in the inner magnetosphere. In: Pulkkinen TI, Tsyganenko NA, Friedel RHW (eds) The inner magnetosphere: physics and modeling. Geophysical monograph series, vol 155. AGU, Washington, DC, pp 173–179

Kikuchi T, Araki T (1979) Horizontal transmission of the polar electric field to the equator. J Atmos Terr Phys 41:927–936

Kikuchi T, Araki T, Maeda H, Maekawa K (1978) Transmission of polar electric fields to the equator. Nature 273:650–651

Kikuchi T, Hashimoto KK, Kitamura T-I, Tachihara H, Fejer B (2003) Equatorial counterelectrojets during substorms. J Geophys Res 108(A11):1406. doi:10.1029/2003JA009915

Kikuchi T, Hashimoto KK, Nozaki K (2008a) Penetration of magnetospheric electric fields to the equator during a geomagnetic storm. J Geophys Res 113:A06214. doi:10.1029/2007JA012628

Kikuchi T, Hashimoto KK, Nozaki K (2008b) Storm phase dependence of penetration of magnetospheric electric fields to mid and low latitudes. In: Kintner PM Jr, Coster AJ, Fuller-Rowell T, Mannucci AJ, Mendillo M, Heelis R (eds) Midlatitude ionospheric dynamics and disturbances. Geophysical Monograph Series, vol 181. AGU, Washington, DC, pp 145–155

Kikuchi T, Ishimine T, Sugiuchi H (1985) Local time distribution of HF Doppler frequency deviations associated with storm sudden commencements. J Geophys Res 90:4389–4393

Kikuchi T, Luehr H, Schlegel K, Tachihara H, Shinohara M, Kitamura T–I (2000) Penetration of auroral electric fields to the equator during a substorm. J Geophys Res 105: 23251–23261

Kikuchi T, Lühr H, Kitamura T, Saka O, Schlegel K (1996) Direct penetration of the polar electric field to the equator during a DP2 event as detected by the auroral and equatorial magnetometer chains and the EISCAT radar. J Geophys Res 101:17161–17173

Kobea AT, Amory-Mazaudier C, Do JM, Luehr H, Hougninou E, Vassal J, Blanc E, Curto JJ (1998) Equatorial electrojet as part of the global circuit: a case-study from the IEEY. Ann Geophys 16:698–710

Kobea AT, Richmond AD, Emery BA, Peymirat C, Luehr H, Moretto T, Hairston M, Amory-Mazaudier C (2000) Electrodynamic coupling of high and low latitudes: observations on May 27, 1993. J Geophys Res 105(A10): 22979–22989

Matsushita S, Balsley BB (1972) A question of DP2 magnetic fluctuations. Planet Space Sci 20:1259–1267

Murakami G, Hirai M, Yoshikawa I (2007) The plasmapause response to the southward turning of the IMF derived from sequential EUVimages. J Geophys Res 112:A06217. doi:10.1029/2006JA012174

Nagata T, Abe S (1955) Notes on the distribution of SC* in high latitudes. Rept Ionosph Res Jpn 9:39–44

Nishida A (1968a) Geomagnetic Dp2 fluctuations and associated magnetospheric phenomena. J Geophys Res 73:1795–1803

Nishida A (1968b) Coherence of geomagnetic DP2 magnetic fluctuations with interplanetary magnetic variations. J Geophys Res 73:5549–5559

Nishida A, Iwasaki N, Nagata T (1966) The origin of fluctuations in the equatorial electrojet: a new type of geomagnetic variation. Ann Geophys 22:478–484

Nishida A, Kamide Y (1983) Magnetospheric processes preceding the onset of an isolated substorm: a case study of the March 31, 1978, Substorm. J Geophys Res 88(A9): 7005–7014

Nishimura Y, Kikuchi T, Wygant J, Shinbori A, Ono T, Matsuoka A, Nagatsuma T, Brautigam D (2009) Response of convection electric fields in the magnetosphere to IMF orientation change. J Geophys Res 114:A09206. doi:10.1029/2009JA014277

Peymirat C, Richmond AD, Kobea AT (2000) Electrodynamic coupling of high and low latitudes: simulations of shielding/overshielding effects. J Geophys Res 105(A10): 22991–23003

Rastogi RG (1977) Geomagnetic storms and electric fields in the equatorial ionosphere. Nature 268:422–424

Rastogi RG (1997) Midday reversal of equatorial ionospheric electric field. Ann Geophys 15:1309–1315

Rastogi RG, Patel VL (1975) Effect of interplanetary magnetic field on ionosphere over the magnetic equator. Proc Indian Acad Sci 82:121–141

Senior C, Blanc M (1984) On the control of magnetospheric convection by the spatial distribution of ionospheric conductivities. J Geophys Res 89:261–284

Shinbori A, Nishimura Y, Ono T, Iizima M, Kumamoto A, Oya H (2005) Electrodynamics in the duskside inner magnetosphere and plasmasphere during a super magnetic storm on March 13–15, 1989. Earth Planets Space 57:643–659

Siscoe GL, Crooker NU, Erickson GM, Sonnerup BUO, Siebert KD, Weimer DR, White WW, Maynard NC (2000) Global geometry of magnetospheric currents inferred from MHD simulations. In: Ohtani S, Fujii R, Hesse M, Lysak RL (eds) Magnetospheric current systems. Geophysical monograph series, vol 118. AGU, Washington, DC, pp 41–52

Somayajulu VV, Reddy CA, Viswanathan KS (1987) Penetration of magnetospheric convective electric field to the equatorial ionosphere during the substorm of March 22, 1979. Geophys Res Lett 14:876–879

Southwood DJ (1977) The role of hot plasma in magnetospheric convection. J Geophys Res 82:5512–5520

Tanaka T (1995) Generation mechanisms for magnetosphere-ionosphere current systems deduced from a three-dimensional MHD simulation of the solar wind-magnetosphere-ionosphere coupling processes. J Geophys Res 100:A7. doi:10.1029/95JA00419

Tanaka T (2007) Magnetosphere–ionosphere convection as a compound system. Space Sci Rev. doi:10.1007/s11214-007-9168-4

Tsurutani B et al (2004) Global dayside ionospheric uplift and enhancement associated with interplanetary electric fields. J Geophys Res 109:A08302. doi:10.1029/2003JA010342

Vasyliunas VM (1972) The interrelationship of magnetospheric processes. In: McCormac BM (ed) Earth's magnetospheric processes. Reidel, Norwell, Massachusetts, pp 29–38

Wilson, GR, Burke WJ, Maynard NC, Huang CY, Singer HJ (2001) Global electrodynamics observed during the initial and main phases of the July 1991 magnetic storm. J Geophys Res 106(A11):24517–24539

Wygant J, Rowland D, Singer HJ, Temerin M, Mozer F, Hudson MK (1998) Experimental evidence on the role of the large spatial scale electric field in creating the ring current. J Geophys Res 103(A12):29527–29544. doi:10.1029/98JA01436

Chapter 35

Modeling the Storm Time Electrodynamics

Naomi Maruyama, Timothy J. Fuller-Rowell, Mihail V. Codrescu, David Anderson, Arthur D. Richmond, Astrid Maute, Stanislav Sazykin, Frank R. Toffoletto, Robert W. Spiro, Richard A. Wolf, and George H. Millward

Abstract A model that electrodynamically couples inner magnetosphere, ionosphere, plasmasphere, thermosphere, and electrodynamics has been developed and is used to separate sources of the storm time electric fields between the magnetospheric, ionospheric, and thermospheric processes and to investigate their nonlinear interactions. The two sources of the electric-field disturbances, prompt penetration (PP) and disturbance dynamo (DD), have been identified in the coupled model results. Furthermore, the results suggest that the sources of variability in storm time electric fields are associated with the nonlinear interaction between the PP and DD, such that the response depends on the preconditioning of the coupled system. The preconditioning in this study is caused by the fact that the magnetosphere, ionosphere, and thermosphere respond to external forcing as a coupled system. The results clearly demonstrate the need for a fully coupled model of magnetosphere–ionosphere–thermosphere, in order to determine the preconditioning effect.

35.1 Introduction

Ionospheric currents and electric fields at middle and low latitudes are significantly disturbed due to the magnetosphere–ionosphere–thermosphere interactions during geomagnetic storms. The basic mechanisms of generating the storm time electric fields have been recognized for many years: prompt penetration (PP) and disturbance dynamo (DD). The magnetometer measurements by work of Nishida (1968) reported that the solar wind-induced electric fields could penetrate deep into the inner magnetosphere, and even further down to the magnetic equator. The relationship between rapid changes in Interplanetary Magnetic Field (IMF) Bz and nearly simultaneous short-lived electric field in the ionosphere has been studied for many years (e.g., Kelley et al., 1979). The Region-2 field-aligned currents (FACs) respond to an enhancement of the polar cap potential (PCP) and tend to generate electric fields that cancel the polar cap electric field, a process known as shielding (e.g., Jaggi and Wolf, 1973). The PP timescale is defined from the enhancement of the PCP until the time when the shielding has established, ranging from 5 to 300 min. Conversely, overshielding happens when the IMF Bz turns northward (e.g., Kelley et al., 1979). It produces the reversed electric field at the magnetic equator, with the timescale of 10–60 min (e.g., Spiro et al., 1988; Peymirat et al., 2000). Shielding efficiency is a key to determining the PP timescale. The shielding time constant depends on plasmasheet (PS) conditions and ionospheric conductivities (e.g., Jaggi and Wolf, 1973; Wolf et al., 1982; Senior and Blanc, 1984; Spiro et al., 1988; Peymirat et al., 2000; Garner, 2003). In addition, magnetospheric reconfiguration effect has also been suggested to impact the PP process (Fejer et al., 1990; Wolf et al., 2007), as well as the PCP changes. The reconfiguration of the magnetospheric magnetic field in a period of dramatic geomagnetic quieting consequently results in tailward motion of the nightside shielding layer caused by a dusk-to-dawn (eastward) potential electric field

N. Maruyama (✉)
Cooperative Institute for Research in Environmental Sciences, Space Weather Prediction Center, University of Colorado, Boulder, CO, USA; Space Weather Prediction Center, National Oceanic and Atmospheric Administration, Boulder, CO, USA
e-mail: Naomi.maruyama@noaa.gov

M.A. Abdu, D. Pancheva (eds.), A. Bhattacharyya (Coed.), *Aeronomy of the Earth's Atmosphere and Ionosphere*, IAGA Special Sopron Book Series 2, DOI 10.1007/978-94-007-0326-1_35,

near the inner edge of the PS. The eastward potential field penetrates to the ionosphere.

On the other hand, the original DD mechanism suggested by Blanc and Richmond (1980) has timescales of about 6 h and longer. The westward wind is built up, through conservation of angular momentum, by the initial equatorward motion driven by an enhanced energy deposition into the high latitude ionosphere and thermosphere. An additional, relatively fast timescale of 2–3 h was reported by Scherliess and Fejer (1997), and the mechanism is attributed to an equatorward surge of the neutral wind (Fuller-Rowell et al., 2002, 2008). The response observed after 24 h is attributed to a combined effect of neutral wind and ionospheric conductivity caused by the neutral composition effect (Scherliess and Fejer, 1997).

Recent observations during super storms have suggested that magnitudes and timescales of the disturbance electric fields can be highly variable (Fejer et al., 2007). In spite of our knowledge of the basic mechanisms, we still cannot fully interpret the individual observations and identify the relative contribution of the two sources, which is essential for prediction. One of the challenges in understanding the response of the storm time electric fields is our insufficient understanding of the storm time electrodynamics, including the magnetosphere–ionosphere coupling, its consequences on the ionosphere–thermosphere, and the feedback between the magnetosphere and ionosphere–thermosphere. An attempt to separate out the two disturbance sources in the observations has been made by considering the different timescales. Spiro et al. (1988) suggested the fossil wind mechanism to explain the extremely long overshielding timescale estimated from the observation, while Fejer et al. (1990) suggested the magnetospheric reconfiguration effect. Fejer and Scherliess (1997) developed an empirical model to separate out the two processes based on the previous knowledge about the different timescales. Richmond et al. (2003) examined the fossil wind concept using Magnetosphere-Thermosphere-Ionosphere-Electrodynamics General Circulation Model (MTIEGCM). Maruyama et al. (2005) suggested a possible interaction and feedback between PP and DD. The comparison between the model results and observations indicated that the interaction seems to become stronger more significant for super storms (Maruyama et al., 2007).

Physics-based modeling has been a useful approach to combine the two different processes, with an attempt to predict equatorial electric field. Forbes and Harel (1989) incorporated analytical neutral winds into Rice Convection Model (RCM). Peymirat et al. (1998) has developed MTIEGCM. Huba et al. (2005) has coupled SAMI3 and RCM. Coupled Magnetosphere–Ionosphere–Thermosphere (CMIT) model (Wang et al., 2004; Wiltberger et al., 2004) has been developed by coupling Lyon–Fedder–Mobary (LFM) global MHD model and Thermosphere–Ionosphere Nested Grid (TING) model. Maruyama et al. (2005, 2007) combined the separate electric field components from the RCM as the source of PP and the Coupled Thermosphere Ionosphere Plasmasphere Electrodynamics (CTIPe) model, which simulates DD to interpret the observed equatorial electric field.

The purpose of this study is to address the following problems: (1) identifying/separating out the dominant drivers of the storm time electric field and their timescales, including separating the contribution from magnetospheric (penetration), ionospheric (conductance), and thermospheric (neutral wind dynamo) sources; (2) exploring possible feedback processes between PP and DD and evaluate their nonlinear interactions that can explain the variability of the storm time electric fields. In order to address the questions, a self-consistently coupled model of the inner magnetosphere, ionosphere, plasmasphere, thermosphere and electrodynamics system has been developed to take into account all the physical processes required to address the possible feedback processes between PP and DD.

In this chapter, we show that the two sources of the storm time electric field: PP and DD have been identified in the results from the coupled model. Furthermore, our model results demonstrate that the sources of variabilities in the storm time electric fields are associated with the nonlinear interaction between PP and DD.

35.2 Coupled Model Descriptions

35.2.1 The Coupled RCM–CTIPe Model

The model describes the electrodynamic coupling and interactions and feedback between the inner magnetosphere and the thermosphere–ionosphere–plasmasphere system and, therefore, is a unique and adequate tool for our study. The coupled model is composed of three well developed and tested physical models: (1) the Coupled Thermosphere-Ionosphere-Plasmasphere-Electrodynamics (CTIPe) model (Millward et al., 2001; Fuller-Rowell et al., 2002); (2) the Rice Convection Model (RCM) (Wolf et al., 1983; Toffoletto et al., 2003); (3) the Global electrodynamic solver based on the National Center for Atmospheric Research Thermosphere–Ionosphere Electrodynamics GCM (NCAR-TIEGCM) (Richmond et al., 1992). The core part of the first two models has not been modified associated with coupling except that the potential solvers in both models have been replaced by the new global electrodynamic solver. Self-consistency in the coupling between RCM and CTIPe has been accomplished by using a common electrodynamic solver so that the same global electric fields are used in both models. In this section, we focus on briefly describing the electrodynamic solver and its interfaces with RCM and CTIPe.

The global potential solver solves the electrostatic electrodynamic equations, in a realistic magnetic field through the use of Magnetic Apex coordinates, under the assumption that the field lines are equipotential (Richmond, 1995). At low and middle latitudes, hemispheric symmetry of the electric potential is assumed and therefore the electrodynamic equation is only solved in one hemisphere, satisfying the assumption of equipotential field lines. The solver takes into account several inputs from both RCM and CTIPe, in order to accomplish the electrodynamic coupling and accurately describe the storm time electric field.

The distribution of high-latitude FACs is the primary parameter of the RCM used in the potential solver at every time step of RCM (typically 1–5 s). This FAC distribution contains the information about the inner magnetosphere concerning under- or over-shielding and provides the necessary timescales for the PP field and the shielding process from the reconfiguration of magnetospheric plasma in response to given electric and magnetic fields.

The global ionospheric conductivity and neutral wind are calculated from CTIPe with a cadence of 1 min, and both are used in the potential solver. The time-dependent distributions of these two parameters are crucial in producing the DD field. Furthermore, it is important to ensure that the PP process responds to storm time changes in ionospheric conductivity and neutral winds.

The potential needs to be specified at the high latitude boundary of the RCM computational domain; poleward of this boundary the RCM FACs are not available to solve the potential self-consistently. In this study, we specify the boundary potential using the analytical model of Heelis et al. (1982) driven by the PCP. The equatorward and poleward movement of the boundary is prescribed using a time varying magnetospheric magnetic field model (e.g., Tsyganenko et al., 2003). The potential solver has been modified to accommodate the temporal and spatial variations of the boundary. Furthermore, the latitude grid spacing we use is finer than that used in the standard TIEGCM, in order to capture the full structure of the FACs and to ensure numerical stability.

35.2.2 Model Input Parameters

A generic double storm was investigated in order to investigate possible feedback processes between PP and DD. In the first storm (Storm A) the PCP increases at 3UT from 25 to 150 kV over a period of 20 min, remains at this level for 6 h, and returns to the background level (25 kV) over another 20 min. Storm B begins at 15UT, 12 h after the start of Storm A with the same intensity and duration. The auroral precipitation level is based on Fuller-Rowell and Evans (1987). The solar activity index during this interval was set to $F_{10.7} = 150 \times 10^{-22}$ W m^{-2} Hz^{-1}. A spring equinox condition was selected. The description of the upward propagating tides at the lower boundary of the thermosphere model is given by Millward et al. (2001), as was also used in Maruyama et al. (2005, 2007). RCM input parameters including PS boundary conditions and magnetospheric magnetic fields were fixed in time to simplify the problem in the complex coupled system. The PS density and temperature at the RCM high-latitude boundary are set to the values of 0.4 cm^{-3} and 5000 eV, respectively, to describe statistically averaged values for the PS moments at ~13 Re at midnight (Huang and Frank, 1986, 1994).

35.3 Results and Discussion

35.3.1 Identification of the Two Sources of the Storm Time Electric Fields

We have been able to identify the two sources of the disturbed electric fields in the coupled model results.

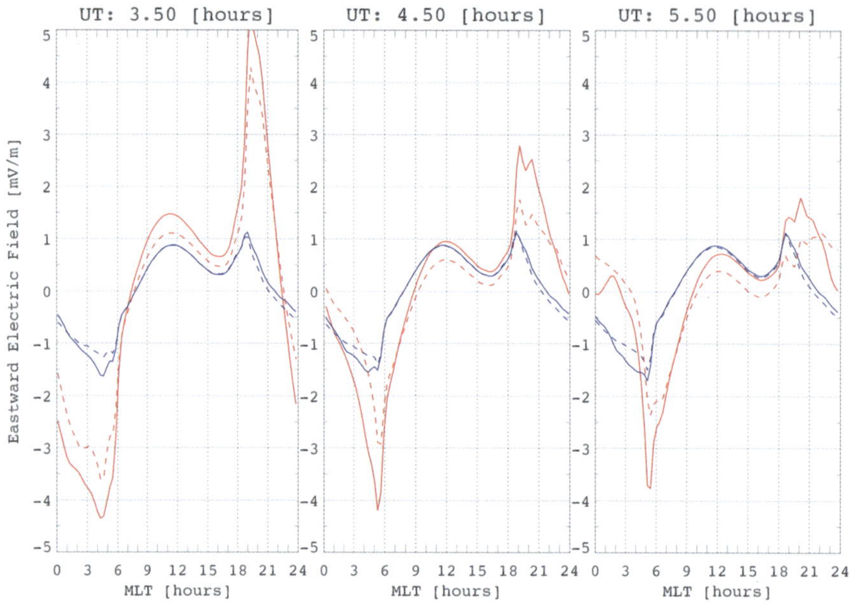

Fig. 35.1 Eastward electric field [mV/m] at the magnetic equator as a function of MLT at various times: (**a**) 3.5UT, 0.5 h from commencement of Storm A; (**b**) 4.5UT, 1.5 h after Storm A commencement; (**c**) 5.5UT, 2.5 h after Storm A commencement. The *red solid curves* show the storm simulation, whereas the *blue solid lines* show the quiet time simulation in which the PCP value is kept constant at 25 kV for the entire simulation period. The corresponding field for Storm B starting at 15UT is plotted by the *red dashed curves* for comparison. The *blue dashed curves* show the quiet time reference for Storm B

Figure 35.1 shows examples of the eastward electric field at the magnetic equator obtained from the coupled model results. PP features are dominant in Fig. 35.1a, where the response of the electric field to the generic storm is shown by the red solid line, 0.5 h after the PCP enhancement of Storm A (3.5UT). A reference quiet simulation is shown by the blue solid line, for which the PCP was held constant at 25 kV. The PP features can be identified as an enhancement of the eastward electric field during the day and in the evening, whereas during the night, the PP effect causes an enhancement of the westward field. The PP features are qualitatively consistent with those from previous studies (e.g., Spiro et al., 1988). In Fig. 35.1b, 1.5 h after the PCP enhancement (4.5UT, red solid line), the PP effect is decaying as the shielding becomes established. The timescale of the PP response is, in general, consistent with previous studies (e.g., Spiro et al., 1988; Sazykin, 2000). The PP effect tends to persist longer near sunrise and sunset since the enhancement of the PP effect near the terminators mostly due to the conductivity gradient between day and night. On the other hand, the DD features become dominant in Fig. 35.1c, 2.5 h after the PCP enhancement (5.5UT, red solid line). The DD features can be identified as a reduction of the daytime eastward field and development of a more-eastward field in the central nighttime hours. The qualitative features and the timescale of the rapid DD response, which was around 2.5 h in this study, are roughly consistent with previous studies (e.g., Fuller-Rowell et al., 2002; Maruyama et al., 2005, 2007). The PP effect causes positive potential at dawn and negative at dusk at the equator, the same as at high latitude. In contrast, in Fig. 35.1c, the nighttime disturbed neutral winds produce an equatorward dynamo current. The current tends to charge the low latitude ionosphere positively around midnight, which can lead to a reduction or reversal of the quiet time field. The features are consistent with Blanc and Richmond (1980) and Richmond et al. (2003). The balance between the two processes determines the net equatorial electric field.

The response of the eastward electric field to Storm B (red dashed lines) basically shows the same qualitative features as those of Storm A (red solid lines), but the magnitudes of the disturbed electric field are smaller in Fig. 35.1a and b. Furthermore, in Fig. 35.1c, development of the more-eastward field in the central nighttime hours becomes much more prominent. Note that difference between the blue solid and dashed lines reflects the longitudinal differences in the quiet time electric field, which is much smaller than the difference of the storm time fields between Storms A (red solid lines) and B (red dashed lines), respectively. In the next section, the difference of the disturbed fields between Storms A and B is examined in further detail.

35.3.2 Nonlinear Interactions Between PP and DD

Figure 35.2 shows a comparison of the disturbed eastward electric fields at 11 MLT between Storm A, starting at 3UT (Fig. 35.2a), and Storm B, starting at 15UT (Fig. 35.2b). The comparison clearly demonstrates that the storm time fields give different responses to the same magnitude and duration of the PCP enhancements depending on the previous conditions of the system. The differences in the disturbed fields between Storms A and B are summarized as follows: (1) the magnitude of the disturbed field at Storm B is smaller than that of Storm A; (2) the disturbed field at Storm B decays faster than that of Storm A. The values of the disturbed fields go negative at low latitudes at Storm B (Fig. 35.2b) after 45 min; (3) the latitude profiles of the disturbed fields at Storm B are modified from those of Storm A. In other words, Storm A shows a gradual profiles from mid to low latitude, whereas Storm B displays steeper profiles at mid latitude and a relatively flat profiles toward low latitude. The results from the coupled model suggest that the differences between Storms A and B are attributed to a nonlinear interaction between PP and DD. In the following sections, we attempt to address what physical processes are responsible for the differences of the disturbed fields between Storms A and B.

35.3.3 Preconditioning of the Inner Magnetosphere

Our results demonstrate that the magnetospheric conditions are significantly different between Storm A (Fig. 35.3a) and Storm B (Fig. 35.3b). The plasma pressure (P) in the inner magnetosphere becomes larger and is distributed more earthward in Storm B since the ring currents have been injected by the strong convection electric field during Storm A that occurred 12 h earlier. The applied cross-tail electric field moves the PS edge sunward, causing a partial westward ring across nightside. The cross-tail dawn to dusk electric field charges the dusk side of PS edge positively and the dawn side negatively, both in the magnetospheric equatorial plane and in the ionosphere. Furthermore, the ring current pressure distribution has become more symmetrical in Storm B. Toffoletto et al. (2003) pointed out three mechanisms to symmetrize the pressure distribution: (1) lack of fresh injection, (2) trapping of already injected particles on closed trajectories as the convection strength decreases, and (3) charge-exchange loss with a characteristic lifetime of many hours. Furthermore, the second-from-left plots in Fig. 35.3a, b show a form of the adiabatic pressure invariant, $PV^{5/3}$, extending to lower L values in Storm B, where P is thermodynamic pressure and V is the flux tube volume (as described in equation (5) of Toffoletto et al. (2003)). Shielding efficiency in the

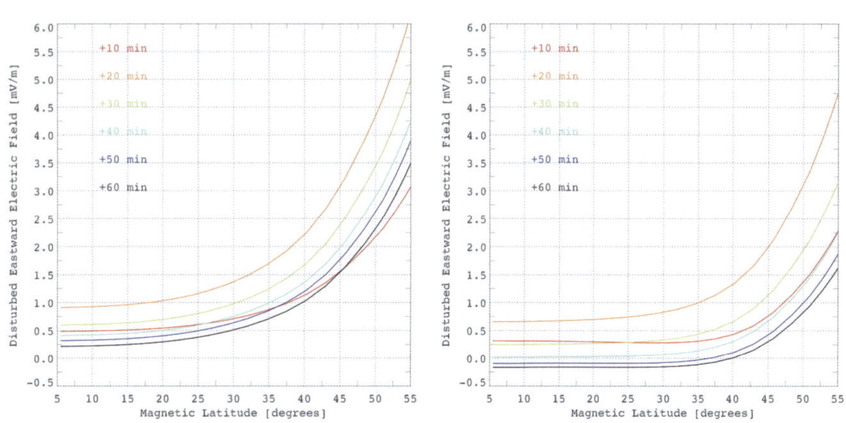

Fig. 35.2 Comparison of the temporal evolutions (every 10 min) of the latitude profiles of the disturbed eastward electric fields [mV/m] at 11 MLT in the first 60 min between Storm A starting at 3UT (**a**); and Storm B starting at 15UT (**b**)

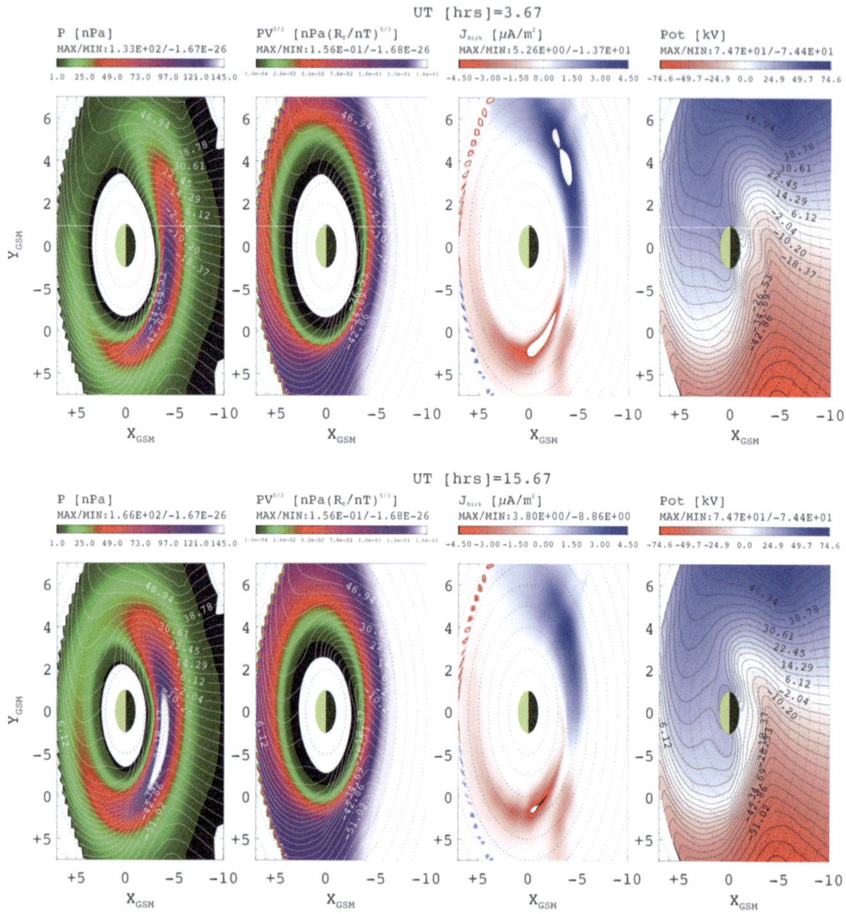

Fig. 35.3 Comparison of the RCM parameters in the GSM X–Y (equatorial) plane between Storm A at 3.67UT (**a**); and Storm B at 15.67UT (**b**). *Color contour plots* show plasma sheet particle pressure (P) [nPa], $PV^{5/3}$ [nPa(R_E/nT)$^{5/3}$], Birkeland current into the ionosphere (positive earthward, both hemispheres together) (J_{BIRK}) [μA/m^2], and ionospheric electrostatic potential in the rotating frame mapped out to the magnetospheric equatorial plane (Pot) [kV]. *Contour lines* show the same electrostatic potential. The Sun is to the left

RCM increases with increasing $PV^{5/3}$ (Wolf et al., 2007). Accordingly, the field-aligned (Birkeland) currents (J_{BIRK}) calculated from the pressure gradient via the so-called Vasyliunas equation (Vasyliunas, 1970) are distributed more earthward in Storm B, due to the fact that $PV^{5/3}$ is directly related to the Birkeland currents (e.g., Garner, 2003). The currents play a crucial role in the electrodynamic coupling between the magnetosphere and ionosphere: the Region-2 Birkeland currents flow upward from the dawn side ionosphere near the inner edge of the PS and downward into the dusk side. They tend to shield the near-Earth region from the dawn–dusk convection field by creating an opposing dusk–dawn polarization electric field in the inner magnetosphere. The electric potential distribution (without the corotation potential) shown in the right plots of Fig. 35.3 (Pot) indicates a stronger shielding generated in Storm B, especially pronounced from the dusk to midnight sector, whereas in Storm A, a weak asymmetric ring current is unable to provide efficient shielding of the low latitude ionosphere. As a consequence, the response of the daytime electric field becomes smaller in Storm B. Our results suggest that the preceding period of the enhanced convection electric field provides a stronger and quicker shielding on the dayside at lower L values in Storm B even though the magnitude and duration of the PCP enhancements are identical between Storms A and B.

Note that both the PS parameters and the magnetospheric configuration were fixed in time for these modeled storms. Although we expect that results would be modified if the PS parameters and magnetospheric configuration vary with time as both have a significant impact on the shielding efficiency (e.g., Spiro et al., 1988; Fejer et al., 1990), such investigation is beyond the scope of this study. Furthermore, our study focuses on the PP response of the distributed electric field in the undershielding condition when the convection field suddenly increases due to a southward turning of the IMF. It would make an interesting study to evaluate the impact of the previous storm activity on the overshielding condition when the convection field suddenly decreases due to a northward turning of the IMF.

35.3.4 Residual Disturbance Dynamo Effect

The preceding period of the enhanced convection electric field disturbs not only the inner magnetosphere but also the ionosphere and thermosphere. Huang et al. (2005) showed that the DD effects take days to die away completely as the disturbance winds tend to persist after the geomagnetic activity ceases. Our results suggest that the differences of the disturbed electric fields between Storms A and B are caused by the residual DD effect of Storm A as well as the preconditioning of the inner magnetosphere as explained in the previous section. In this section, we discuss another important role played by the redistribution of the global ionospheric conductivity as well as the disturbance winds in determining the DD process, since it is the conductivity-weighted winds that largely drive the ionospheric wind dynamo.

We found a significant modification mainly in the Pedersen conductivity distribution in Storm B (Fig. 35.4a) as compared to that of Storm A (Fig. 35.4b), whereas the Hall conductivity does not seem to change very much (not shown here). This indicates that the modification to the ionospheric conductivity is mainly caused by the transport of ionospheric plasma in the F-region. Note that the oval-shaped area around 60° magnetic latitude, saturated with white color, is associated with the auroral conductance, and here we mainly focus on features equatorward of the auroral oval. Strong electric fields generated during magnetic storms, such as including the PP and Sub-Auroral Polarization Stream (SAPS) electric fields, tend to reduce the ionospheric electron density above ~160 km. Stronger electric fields generate more frictional heating due to relative motions of charged particles to neutral species, resulting in an increase in the ion temperature. The elevated ion temperature, in turn, accelerates the recombination chemical reactions between atomic oxygen ions with molecular neutral species ($N_2 + O^+ \rightarrow NO^+ + N$; $O_2 + O^+ \rightarrow O_2^+ + O$), whose chemical reaction rates depend on the ion and neutral temperatures and, relative drift velocities between ionized and neutral species (e.g., Torr and Torr, 1979; Millward et al., 1996). On the other hand, the storm time neutral winds driven by the Joule heating deposited into the high-latitude upper atmosphere and ion-drag acceleration lifts the ionospheric plasmas up to a higher altitude where the recombination is slower, resulting in a modification to height profiles of the ionospheric conductivity. Note that the thermal

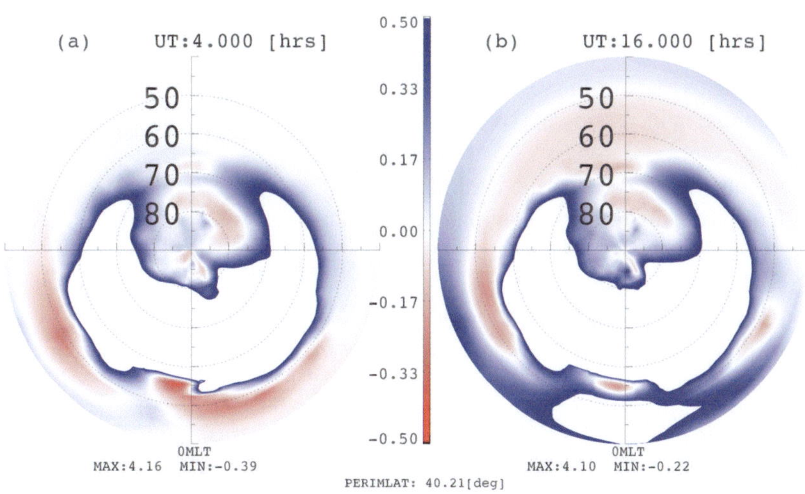

Fig. 35.4 Comparison of the response of the ionospheric Pedersen conductance (sum of the two hemispheres) between Storm A at 4UT (**a**) and Storm B at 16UT (**b**). The differences between the storm and quiet values normalized by the quiet time values are plotted

expansion process in the storm time upper atmosphere itself can also contribute to lifting the ionospheric layer to higher altitude as well. On the nightside, at latitudes around 50 (add a symbol of degree) and lower, there are differences between the two storms not only in magnitude but also in sign. During Storm A, the ionospheric conductivity generally decreases (red) as compared to that of the quiet time as shown in Fig. 35.4a due to the effect of the storm time electric field, whereas in Storm B, the conductivity increases (blue) as compared to that of the quiet time as shown in Fig. 35.4b due to the disturbed neutral wind effect.

Storm time changes of both conductivity and wind are responsible for the residual DD effect. Our results suggest that the residual DD effect as well as the preconditioning of the inner magnetosphere produce stronger and quicker shielding. Our simulations support the prediction by Richmond et al. (2003) that storm time neutral winds tend to enhance steady-state shielding, by reducing the net electric field that penetrates to middle and low latitudes. We find that the disturbed electric fields at Storm B in the simulation with the neutral wind show smaller magnitudes and faster decay than those without the neutral wind (results not shown). On the other hand, the modified latitudinal profiles are mainly caused by the residual DD effect, including the fact that the values of the disturbed fields go negative at low latitudes in after 45 min Fig. 35.2b. A more detailed quantification of the relative importance between the ionospheric conductivity and neutral wind in determining the residual DD effect is beyond the scope of this study. An interesting question remains to be solved as to whether the preconditioned magnetosphere or the residual DD effect plays a dominant role in increasing the shielding efficiency.

In this chapter, we mainly focused on the qualitative evaluation of the impact of the preconditioning of the coupled system in just a single, idealized storm case. More comprehensive studies with a quantitative approach would be required including various intensities and durations of the PCP enhancement as well as the temporal variations of the PS parameters and magnetospheric configuration and real storm events, until we would be able to draw a more generalized conclusion in evaluating the feedback effect of the ionospheric conductivity and neutral wind on the PP and shielding processes.

35.4 Conclusions

We have developed a model that electrodynamically couples inner magnetosphere, ionosphere, plasmasphere, thermosphere, and electrodynamics in a self-consistent manner, in order to separate out sources of the storm time electric fields between the magnetospheric, ionospheric, and thermospheric processes and to evaluate their nonlinear interactions associated with changes in magnetospheric plasma density, ionospheric conductivity, and neutral wind. The two sources of the storm time electric field, prompt penetration (PP) and disturbance dynamo (DD), have been identified in the coupled model results, as well as the realistic inner magnetosphere shielding process. Furthermore, our results demonstrate that the sources of variability in storm time electric fields are associated with the nonlinear interaction between the PP and DD, such that the response depends on the preconditioning of the coupled system. The preconditioning of the system in this study is caused as a consequence of the response of the magnetosphere, ionosphere, and thermosphere to external forcing. The results clearly demonstrate the need for a fully coupled magnetosphere–ionosphere–thermosphere model, in order to determine the preconditioning effect. Our study suggests that an appreciation and quantitative evaluation of the variabilities associated with the interaction are vital in interpreting the observations.

Acknowledgments NM was supported by the National Science Foundation under Agreement Number ATM0720406, and NASA GSFC: LWS TRT NNX06AC68G and Guest Investigator C/NOFS program NNX09AN58G. AR and AM were supported in part by the NASA LWS program. The National Center for Atmospheric Research is sponsored by the National Science Foundation.

References

Blanc M, Richmond AD (1980) The ionospheric disturbance dynamo. J Geophys Res 85:1669–1686

Fejer BG, Jensen JW, Kikuchi T, Abdu MA, Chau JL (2007) Equatorial ionospheric electric fields during the November 2004 magnetic storm. J Geophys Res 112:A10304. doi:10.1029/2007JA012376

Fejer BG, Scherliess L (1997) Empirical models of storm time equatorial electric fields. J Geophys Res 102:24,047

Fejer BG, Spiro RW, Wolf RA, Foster JC (1990) Latitudinal variation of perturbation electric fields during magnetically

disturbed periods: 1986 SUNDIAL observations and model results. Ann Geophys 8:441
Forbes JM, Harel M (1989) Magnetosphere-thermosphere coupling: an experiment in interactive modeling. J Geophys Res 94:2631–2644
Fuller-Rowell TJ, Evans DS (1987) Height-integrated Pedersen and Hall conductivity patterns inferred from the TIROS/NOAA satellite data. J Geophys Res 92:7606–7618
Fuller-Rowell TJ, Millward GH, Richmond AD, Codrescu MV (2002) Stormtime changes in the upper atmosphere at low latitudes. J Atmos Solar-Terr Phys 64: 1383
Fuller-Rowell TJ, Richmond AD, Maruyama N (2008) Global modeling of storm-time thermospheric dynamics and electrodynamics. In: Kintner PM Jr, Coster AJ, Fuller-Rowell T, Mannucci AJ, Mendillo M, Heelis R (eds) Midlatitude ionospheric dynamics and disturbances. Geophysical monograph series, vol 181, 10.1029/181GM18. American Geophysical Union, Washington, DC 187–200
Garner TW (2003) Numerical experiments on the inner magnetospheric electric field. J Geophys Res 108(A10): 1373. doi:10.1029/2003JA010039
Heelis RA, Lowell JK, Spiro RW (1982) A model of the high-latitude ionospheric convection pattern. J Geophys Res 87:6339–6345
Huang CY, Frank LA (1986) A statistical study of the central plasma sheet: implications for substorm models. Geophys Res Lett 13:652–655
Huang CY, Frank LA (1994) A statistical survey of the central plasma sheet. J Geophys Res 99:83
Huang C-M, Richmond AD, Chen M-Q (2005) Theoretical effects of geomagnetic activity on low-latitude ionospheric electric fields. J Geophys Res 110:A05312. doi:10.1029/2004JA010994
Huba JD, Joyce G, Sazykin S, Wolf RA, Spiro RW (2005) Simulation study of penetration electric field effects on the low- to mid-latitude ionosphere. Geophys Res Lett 32(L23101). doi:10.1029/2005GL024162
Jaggi RK, Wolf RA (1973) Self-consistent calculation of the motion of a sheet of ions in the magnetosphere. J Geophys Res 78(16):2852–2866
Kelley MC, Fejer BG, Gonzales CA (1979) An explanation for anomalous ionospheric electric fields associated with a northward turning of the interplanetary magnetic field. Geophys Res Lett 6:301
Maruyama N, Richmond AD, Fuller-Rowell TJ, Codrescu MV, Sazykin S, Toffoletto FR, Spiro RW, Millward GH (2005) Interaction between direct penetration and disturbance dynamo electric fields in the storm-time equatorial ionosphere. Geophys Res Lett doi:10.1029/2005GL 023763
Maruyama N, Sazykin S, Spiro RW, Anderson D, Anghel A, Wolf RA, Toffoletto FR, Fuller-Rowell TJ, Codrescu MV, Richmond AD, Millward GH (2007) Modeling stormtime electrodynamics of the low latitude ionosphere thermosphere system: can long lasting disturbance electric fields be accounted for? J Atmos Solar-Terr Phys. doi:10.1016/j.jastp.2006.08.020
Millward GH, Moffett RJ, Quegan S, Fuller-Rowell TJ (1996) A coupled thermosphere-ionosphere-plasmasphere model (CTIP). In: Schunk RW (ed) STEP hand book. Utah State University, Logan, UT, pp 239–279
Millward GH, Muller-Wodarg ICF, Aylward AD, Fuller-Rowell TJ, Richmond AD, Moffett RJ (2001) An investigation into the influence of tidal forcing on F region equatorial vertical ion drift using a global ionosphere-thermosphere model with coupled electrodynamics. J Geophys Res 106: 24,733–24,744
Nishida A (1968) Coherence of geomagnetic DP2 fluctuations with interplanetary magnetic variations. J Geophys Res 75:5549
Peymirat C, Richmond AD, Emery BA, Roble RG (1998) A magnetosphere-thermosphere-ionosphere-electrodynamics general-circulation model. J Geophys Res 103:17,467–17,477
Peymirat C, Richmond AD, Kobea AT (2000) Electrodynamic coupling of high and low latitudes: Simulations of shielding/overshielding effects. J Geophys Res 105: 22,991
Richmond AD (1995) Ionospheric electrodynamics using magnetic APEX coordinates. J Geomag Geoelect 47:191–212
Richmond AD, Peymirat C, Roble RG (2003) Long-lasting disturbances in the equatorial ionospheric electric field simulated with a coupled magnetosphere-ionosphere-thermosphere model. J Geophys Res 108(A3). doi:10.1029/2002JA009758
Richmond AD, Ridley EC, Roble RG (1992) A thermosphere/ionosphere general circulation model with coupled electrodynamics. Geophys Res Lett 19:601–604
Sazykin S (2000) Theoretical studies of penetration of magnetospheric electric fields to the ionosphere. Ph.D. thesis, Utah State University, Logan, UT
Scherliess L, Fejer BG (1997) Storm time dependence of equatorial disturbance dynamo zonal electric fields. J Geophys Res 102(A12):24,037
Senior C, Blanc M (1984) On the control of magnetospheric convection by the spatial distribution of ionospheric conductivities. J Geophys Res 89:261
Spiro RW, Wolf RA, Fejer BG (1988) Penetration of high-latitude electric-field effects to low latitudes during sundial 1984. Ann Geophys 6:39
Toffoletto FR, Sazykin S, Spiro RW, Wolf RA (2003) Inner magnetospheric modeling with the rice convection model. Space Sci Rev 107:175–196
Torr DG, Torr MR (1979) Chemistry of the thermosphere and ionosphere. J Atmos Solar-Terr Phys 41:797–839
Tsyganenko NA, Singer HJ, Kasper JC (2003) Storm-time distortion of the inner magnetosphere: how severe can it get? J Geophys Res 108(A5). doi:10.1029/2002JA009808
Vasyliunas VM (1970) Mathematical models of magnetospheric convection and its coupling to the ionosphere. In: McCormac BM (ed) Particles and fields in the magnetosphere. D. Reidel, Hingahm, MA, pp 60–71
Wang W, Wiltberger M, Burns AG, Solomon SC, Killeen TL, Maruyama N, Lyon JG (2004) Initial results from the coupled magnetosphere-ionosphere-thermosphere model: thermosphere-ionosphere responses. J Atmos Solar-Terr Phys 66:1425. doi:10.1016/j.jastp.2004.04.008
Wiltberger M, Wang W, Burns AG, Solomon SC, Lyon JG, Goodrich CC (2004) Initial results from the coupled

magnetosphere-ionosphere-thermosphere model: magnetospheric and ionospheric responses. J Atmos Solar-Terr Phys 66:1411. doi:10.1016/j.jastp.2004.04.026
Wolf RA (1983) The quasi-static (slow-flow) region of the magnetosphere. In: Carovillano RL, Forbes JM (eds) Solar-terrestrial physics: principles and theoretical foundations. D. Reidel, Hingahm, MA, pp 303–368
Wolf RA, Spiro RW, Sazykin S, Toffoletto FR (2007) How the earth's inner magnetosphere works: an evolving picture. J Atmos Solar-Terr Phys. doi:10.1016/j.jastp.2006.07.026
Wolf RA, Spiro RW, Voigt G-H, Reiff PH, Chen CK, Harel M (1982) Computer simulation of inner magnetospheric dynamics for the magnetic storm of July 29, 1977. J Geophys Res 87:5949–5962

Chapter 36

A Physical Mechanism of Positive Ionospheric Storms

Nanan Balan and Graham J. Bailey

Abstract A physical mechanism of the positive ionospheric storms at low and mid latitudes reported recently is reviewed, and comapred with the positive ionospheric storms observed during a super storm. In addition, the possible variations of the mechanism with the strength of the equatorward winds, intensity of prompt penetration electric field (PPEF), local time and season are discussed. According to the mechanism, the mechanical effects of the equatorward wind (1) reduce (or stop) the downward diffusion of plasma along the geomagnetic field lines, (2) raise the ionosphere to high altitudes of reduced chemical loss, and hence (3) accumulate the plasma at altitudes near and above the ionospheric peak centered at ±15°–30° magnetic latitudes. The daytime eastward PPEF, if occurs, also shifts the EIA crests to higher than normal latitudes. The positive ionospheric storms are most likely in the longitudes of morning-noon onset of the geomagnetic storms. The mechanism agrees with the positive ionospheric storms observed during the super storm of 07–08 November 2004.

36.1 Introduction

Following geomagnetic storms the ionospheric density often increases very much above its average quiet-time level, which are known as positive ionospheric storms. It is important to monitor the occurrence of the positive storms and understand their physical mechanisms because these storms can cause serious problems in satellite navigation and telecommunication. Thanks to the works of a number of scientists it is known that ionospheric electric fields and thermospheric neutral winds (and waves) are involved in producing positive ionospheric storms (e.g., Matuura, 1972; Mendillo and Klobuchar, 1975; Reddy et al., 1990; Rishbeth, 1991; Batista et al., 1991; Prolss, 1995; Sastri et al., 2000; Abdu et al., 2006; Maruyama et al., 2007; Zhao et al., 2008; Shiokawa et al., 2008).

Recently Kelley et al. (2004) and Tsurutani et al. (2004) suggested that a strong eastward prompt penetration electric field (PPEF) in the presence of daytime production of ionization can strengthen the equatorial plasma fountain (e.g., Hanson and Moffett, 1966; Balan and Bailey, 1995) to a super plasma fountain, which, in turn, can lead to positive ionospheric storms at mid latitudes. However, modeling studies later showed that a daytime eastward PPEF event on its own is unlikely to produce positive ionospheric storms, and the nature of the ionospheric storms depends on the state of the thermosphere. The studies suggest that an equatorward neutral wind is required to produce positive ionospheric storms (e.g., Werner et al., 1999; Lin et al., 2005; Vijaya Lekshmi et al., 2007; Lu et al., 2008; Balan et al., 2009a). Quite recently, we explained a physical mechanism of how the equatorward wind produces positive ionospheric storms at low and mid latitudes (Balan et al., 2009b).

In this chapter we review the physical mechanism. The mechanism is based on the theoretical modeling and basic principles during the super geomagnetic storm of 08 November 2004. Repeated eastward PPEF

N. Balan (✉)
Control and Systems Engineering, University of Sheffield,
S1 3JD Sheffield, UK
e-mail: B.Nanan@sheffield.ac.uk

M.A. Abdu, D. Pancheva (eds.), A. Bhattacharyya (Coed.), *Aeronomy of the Earth's Atmosphere and Ionosphere*,
IAGA Special Sopron Book Series 2, DOI 10.1007/978-94-007-0326-1_36,

events occurred during the main phase of the storm in Japan-Australian longitudes which were in daytime during the main phase. The ionospheric observations also show the presence of equatorward neutral winds developed during the long-lasting (18 h) initial phase of the super storm.

The geomagnetic storm, and the storm-time electric fields, neutral winds and neutral densities required for the model calculations are described in Section 36.2. The relative effects of the electric fields and neutral winds on producing the positive ionospheric storms are evaluated using the Sheffield University Plasmasphere Ionosphere Model (SUPIM) (Bailey and Balan, 1996) described in Section 36.3. The model results (Section 36.4) reveal that it is the mechanical effects of the storm-time equatorward neutral winds that drive the positive ionospheric storms. Basic principles are then used to explain how the mechanical effects of the winds drive the positive storms (Section 36.5). The mechanism is tested using the storm-time ionospheric data in Section 36.6. The possible variations of the mechanism are discussed in Section 36.7, with conclusions in Section 36.8.

36.2 Geomagnetic Storm and Data

The super geomagnetic storm of 07–08 November 2004 occurred during the passage of a strong CME (or solar storm). As shown in Fig. 36.1 (panel a) the solar storm (at ACE spacecraft) undergoes several dynamic pressure pulses with the pressure reaching up to 60 nP and velocity (panel b) reaching up to 750 km s^{-1}. The super geomagnetic storm (panel d) starts with the sudden commencement (SC) at the arrival of the first solar storm pressure pulse at 03:00 UT on 07 November. The main phase (MP) of the storm starts at 21:00 UT on 07 November and maximum main phase (Dst = −373 nT) occurs at around 06:00 UT on 08 November when Kp (Fig. 36.1, panel e) reaches 9^{-}. IMF Bz (panel c) remains strongly southward during the main phase (21:00–06:00 UT), with large fluctuations at the beginning. That caused repeated occurrence of eastward PPEF during the morning-noon (06:00–15:00 LT) main phase in Japan-Australian longitudes (Fejer et al., 2007). The super storm also has an unusually long (18 h) initial phase (panel d) during which storm-time equatorward neutral wind and hence westward electric fields developed.

36.2.1 Storm-Time Electric Fields

The storm-time equatorial electrojet index EEJ(storm) has been used as a measure of the equatorial east-west electric fields (e.g., Alex et al., 1986). The electrojet index (not shown) is obtained from the horizontal component (H-component) of the geomagnetic field at the equatorial station Yap (9.30°N, 138.5°E, 0.30° dip latitude) and low latitude station Okinawa (26.75°N, 128.22°E, 26.75°N magnetic latitude) in nearly same longitudes. The EEJ(storm) converted to the equivalent vertical $\mathbf{E}\times\mathbf{B}$ drift (Balan et al., 2009b) during the morning-noon MP (06:00–14:00 LT, 21:00–05:00 UT) is shown in Fig. 36.2 (panel a, thick curve). The drift velocity undergoes three strong upward events centered at around 23:30 UT, 01:00 UT and 03:00 UT, and three downward events centered at 22:30 UT, 00:15 UT and 01:30 UT. The upward and downward events last approximately 1–2 h, and correspond respectively to eastward PPEFs and westward electric fields during the morning-noon MP. The drift velocities during the remaining period on the storm day are assumed to be equal, to the quiet-day values (thin curve, panel a), following Fejer et al. (2007).

36.2.2 Storm-Time Neutral Winds

The magnetic meridional wind velocities are obtained from HWM91 (Hedin et al., 1991). The wind velocities given by HWM for the storm conditions are found to be inadequate. The storm-time winds are therefore obtained from those for quiet (Ap = 4) conditions. The quiet-time winds during 06–18 LT are multiplied by −1.5 for latitudes >30°N, −0.75 for latitude 15°N, −0.25 for latitude 7°N and 0 for latitude 0°N; and interpolations are used for the latitudes in between. The resulting winds are used in both hemispheres. A sample latitude variation of the resulting effective magnetic meridional wind velocity at 13.5 LT (end of eastward PPEF) at 350 km altitude is shown in Fig. 36.2 (panel b). The wind is equatorward in both hemispheres.

36.2.3 Neutral Densities

The neutral densities are from MSIS (Hedin, 1987). The quiet-time (Ap = 4) neutral densities are used in the model calculations not to mix the mechanical

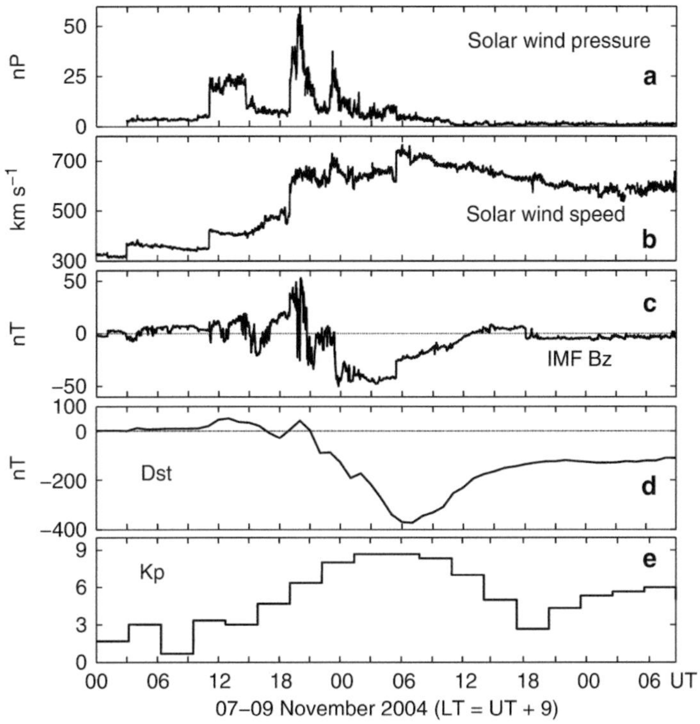

Fig. 36.1 Variations of solar wind dynamic pressure (*panel a*), solar wind velocity (*panel b*), IMF Bz (*panel c*), Dst index (*panel d*) and Kp index (*panel e*) form 00 UT on 07 November to 09 UT 09 November 2004; the solar wind and IMF data measured by ACE spacecraft are shifted by 64 min for the initial velocity of the CME

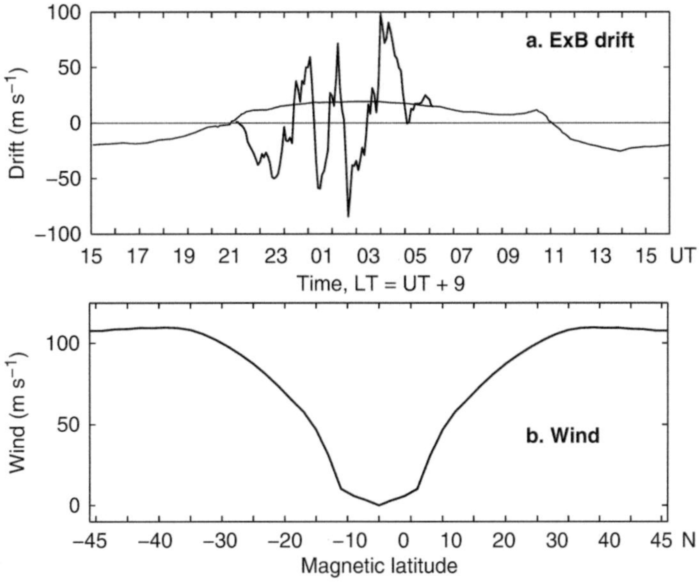

Fig. 36.2 Model inputs. Local time variation of the vertical **E**×**B** drift velocity over the equator (*panel a*), and latitude variations of the magnetic meridional equivalent neutral wind at 350 km height at 13:30 LT for the storm conditions constructed from HWM (*panel b*), only sample variation is shown for simplicity (see text)

effects of the storm-time equatorward (or upwelling) winds with its chemical effect. The chemical effect of the upwelling winds can decrease the $[O]/[N_2]$ ratio at mid latitudes and hence reduce the positive ionospheric storms at these latitudes. However, as discussed in Section 36.5, the chemical effect of the wind becomes less effective once the mechanical effects get the dominance.

36.2.4 Storm-Time Ionospheric Data

The ionospheric data to be used to verify the physical mechanism include electron density (Ne), peak electron density (Nmax), peak height (hmax) and total electron content (GPS-TEC) in Japan-Australian longitudes (120°E–150°E). The Nmax and hmax are obtained from the ionosondes in the Japan-Australian longitudes, and GPS-TEC in the same longitude is obtained using GEONET (Otsuka et al., 2002). The Ne data at ≈400 km height is obtained from the CHAMP satellite (Reigber et al., 2002; Liu et al., 2007).

36.3 Model and Inputs

The physics based model SUPIM (Bailey and Balan, 1996) is used for the calculations. The model solves the coupled time-dependent equations of continuity, momentum and energy for the electrons and ions (O^+, H^+, He^+, N_2^+, NO^+, and O_2^+) along closed eccentric-dipole geomagnetic field lines. For this study, 165 field lines with apex altitude distributed from 150 km to 12000 km are used, with apex intervals of 10 km up to 400 km and 25 km from 400 to 1000 km. The calculations are for the Japan-Australian longitude (135°E) for day number 312 (November 8) under medium solar activity (F10.7 = 105).

The model inputs are the east-west electric field (or vertical **E**×**B** drift), neutral densities and neutral winds (described in Section 36.2). The **E**×**B** drift (Fig. 36.2, panel a) is applied to all field lines and hence to all altitudes and latitudes. The model results are presented (Section 36.4) for one combination of the model inputs (Fig. 36.2). Other combinations of the inputs can also produce similar results. No data-model comparisons are attempted because the measured values of the model inputs are not available.

36.4 Model Results

To evaluate the relative effects of the electrodynamic drift and neutral wind, the model calculations are carried out (1) with drift and no wind, and (2) with drift and wind. In each case, the model calculations are performed at 15-min intervals, with the local time of the calculations coinciding with that of the inputs. The calculations are continued for several consecutive 24-h periods (calculation days) until the day-to-day changes in the model outputs are negligible. The model outputs from the final day's calculations are used for the study.

36.4.1 Effect of Electric Field

Figure 36.3 shows the altitude-latitude variation of the model electron density at 13:30 LT (end of PPEF) obtained from the calculations using the storm-time **E**×**B** drift (Fig. 36.2) with no neutral wind. As shown by Fig. 36.3, the storm-time **E**×**B** drift shifts the equatorial ionization anomaly (EIA) crests to about ±35° latitudes. However, the EIA crests are lower in altitude by about 50 km and its electron density is smaller compared to the EIA obtained using the quiet-time **E**×**B** drift (Balan et al., 2009b). This result (Fig. 36.3) suggests that daytime eastward PPEF on its own can produce only negative effects in electron density at low and mid latitudes though it can shift the EIA crests to higher than normal latitudes. The EIA becomes weak because as the crests shift to higher latitudes they encounter regions where geomagnetic field lines are more inclined and hence plasma gets lost by downward diffusion to low altitudes of heavy chemical loss.

36.4.2 Effects of Neutral Wind

A second set of calculations is carried out using the storm-time **E**×**B** drift and storm-time winds (Fig. 36.2); quiet-time MSIS is also used (as for Fig. 36.3) so that only the mechanical effects of the equatorward winds become effective. As shown by Fig. 36.4 (similar to Fig. 36.3) the equatorward

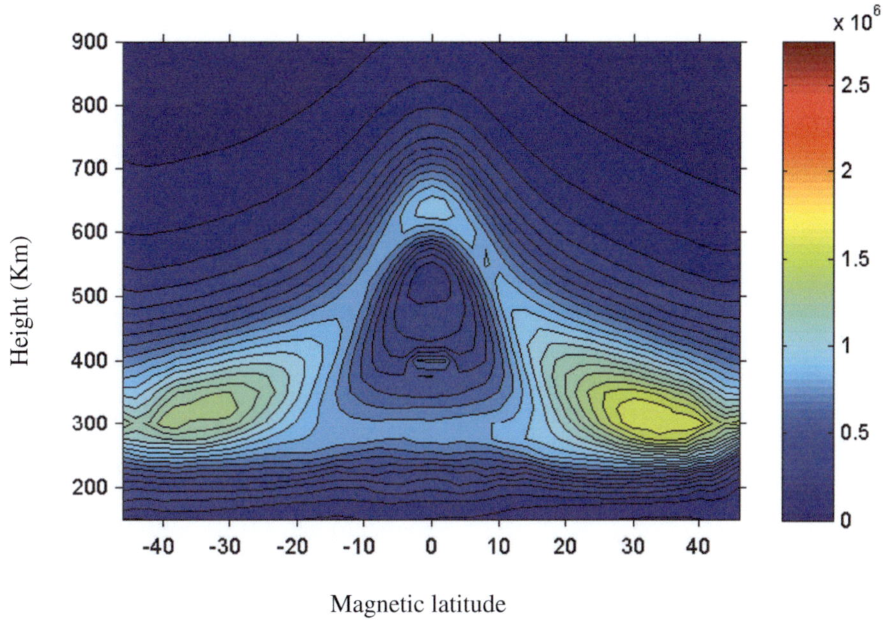

Fig. 36.3 Altitude-latitude variation of the model electron density at 13:30 LT (end of PPEF) obtained from the calculations using storm-time $\mathbf{E}\times\mathbf{B}$ drift, no neutral wind and quiet-time MSIS. Positive latitude is North

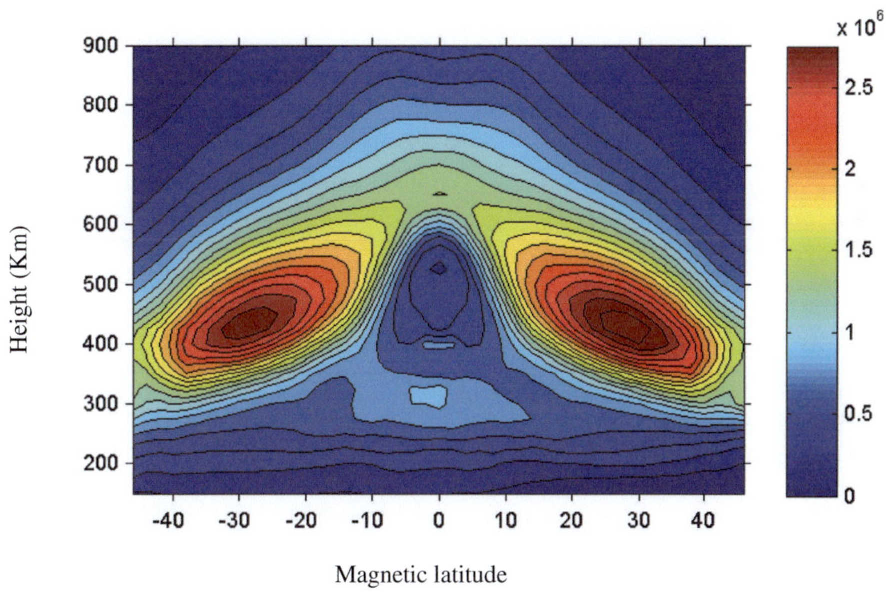

Fig. 36.4 Altitude-latitude variation of the model electron density at 13:30 LT (end of PPEF) obtained from the calculations using storm-time $\mathbf{E}\times\mathbf{B}$ drift, storm-time neutral wind and quiet-time MSIS. Positive latitude is North

winds raise the ionosphere in altitude and increase the electron density at low and mid latitudes, except around the equator. The EIA crests (Fig. 36.4) occur at around $\pm 30°$ latitudes, they raise by over 100 km and their density increases by over 150% (compared to Fig. 36.3). Hence the mechanical effects of the equatorward winds accumulate the plasma at altitudes near and above an elevated ionosphere and produce positive ionospheric storms. How the winds accumulate the plasma are explained further using basic principles in the next section.

36.5 Physical Mechanism

The equatorward winds accumulate the plasma by (1) raising the ionospheric height and (2) reducing (or stopping or reversing) the downward diffusion of plasma (due to gravity and pressure gradient) along the geomagnetic field lines. These mechanical effects are explained using Fig. 36.5. Two latitude variations of the horizontal effective magnetic meridional equatorward wind of velocity U are shown in panel a; solid curve represents typical winds during geomagnetic storms, and dashed curve represents an extreme case; the maximum of U (assumed 100 m s^{-1}) varies with the intensity of the storms (e.g., Rishbeth, 1991; Fuller Rowell et al., 1994; Prolss, 1995). The two sets of solid and dashed curves (1 and 2) in panel b show the latitude variations of the mechanical effects of U.

The equatorward wind of velocity U drives the ionospheric plasma up the field lines with a velocity $U\ cosI$ of which the vertical component is $U\ cosI\ sinI$, with I being dip angle. This velocity raises the ionosphere to altitudes of reduced chemical loss and hence accumulates plasma. Roughly, the ionospheric peak rises vertically by $\Delta h \approx (H_p/D_p)U\ cosI\ sinI$ where H_p is scale height and D_p is diffusion coefficient. This height rise can maximize at (±45°) dip latitudes as shown by the variations of $U\ cosI\ sinI$ in panel b (curves 1). The downward plasma velocity along the field lines due to diffusion is $V_{\parallel} = -(D_p/n)(dn/dh + n/H_p)sinI$ where n is plasma density. The upward wind velocity along the field lines ($=U\ cosI$) reduces (or stops or reverses) the downward diffusion velocity, and hence turns the plasma flux vectors toward the equator (Balan et al., 2009b) and accumulates the plasma. This effect can maximize at low-mid to equatorial latitudes (curves 2, panel b).

The plasma gets accumulated centered at those latitudes where the two mechanical effects (1 and 2) optimize. As shown by the sum of the modulus of $U\ cosI$ and $U\ cosI\ sinI$ (panel c), the two effects can optimize at ≈±30°–50° magnetic dip angles (≈±15°–30° magnetic latitudes) for typical to extreme variations of the equatorward winds; however, the two terms involving $UcosI$ and $U\ cosI\ sinI$ cannot be added directly for vector reasons. The mechanical effects of the equatorward winds can therefore produce positive ionospheric storms by accumulating the plasma at altitudes near and above the ionospheric peak centered at ±15°–30° magnetic latitudes. The eastward PPEF, if occurs, also shifts the EIA crests to higher than normal latitudes.

36.5.1 Local Time Dependence

The occurrence of ionospehric storms depend on the local time of onset of geomagnetic storms because (1) the speed of the equatorward winds and their equatorward extend vary with ionospheric density (or ion-drag) and (2) thermosphere becomes richer and richer in N_2 as time progresses from the onset of the storms. The positive ionospheric storms are most likely in the longitudes of morning-noon onset of the geomagnetic storms; in these longitudes the mechanical effects of the wind together with ionization production can largely exceed the chemical effect of the wind. Once the mechanical effects get the dominance and equatorward winds continue to flow the positive ionospheric storms can exist for long periods because then the plasma is accumulated at higher altitudes where the chemical effect becomes less effective. In the longitudes of afternoon-evening storm onsets, there can be initial positive ionospheric storms followed by negative storms. In the longitudes of nighttime (up to about 03:00 LT) onsets, positive ionospheric storms are unlikely, and there can be large negative ionospheric storms extending to the next one to 3 days (e.g., Rishbeth, 1991; Prolss, 1995).

36.5.2 Seasonal Dependence

The occurrence of ionospheric storms also depends on season through the background thermospheric composition. The background composition is such that the [O]/[N_2] ratio at any thermospheric pressure (or height) level is greater in winter than in summer, with equinox falling in-between. When geomagnetic storms occur in this background, positive ionospheric storms are most probable in winter and least probable in summer.

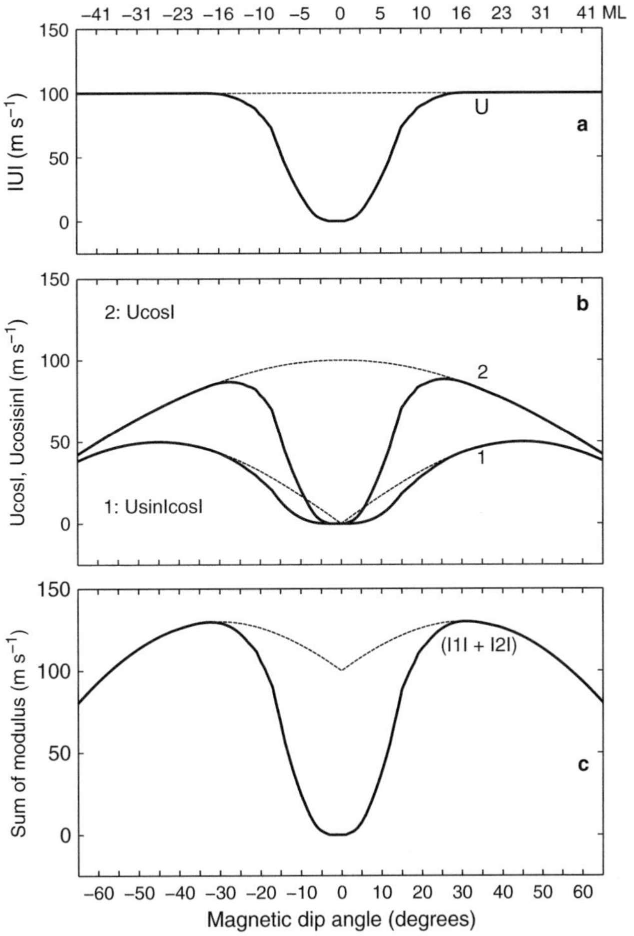

Fig. 36.5 Panel a – Latitude variations of the magnitudes of typical (*solid*) and extreme (*dashed*) magnetic meridional equatorward neutral wind (U); direction of the wind is opposite in north and south. *Panel b* – Latitude variations of (1) *U cosI sinI* involved in the height rise of ionospheric peak ($\approx(H/D)U\ cosI\ sinI$) and (2) the component *U cosI* of *U* upward along the field lines that reduces (or stops) the downward diffusion velocity of plasma ($= -(D_p/n)(dn/dh + n/H_p)sinI$) along the field lines. *Panel c* – Sum of the modulus of (1) and (2)

36.6 Verification of Mechanism

Figure 36.6 (panels a and b) compares the Nmax variations from 00:00 UT on 07 November to 09:00 UT on 09 November 2004 (solid curves) with the seven quiet-day averages prior to the storm (dashed curves). The data shown are from the mid latitude stations Brisbane (153.17°E, 27.5°S; 36.7°S magnetic) in Australia and Kokubunji (139.5°E, 35.7°N; 26.8°N magnetic) in Japan. The Nmax shows strong positive ionospheric storms on 08 November in both hemispheres (panels a and b), starting at sunrise which happens to be at MP onset (21:00 UT, 06:00 LT). The GPS-TEC (panel c) also shows strong positive ionospheric storms starting from the MP onset at sunrise. However, the eastward PPEF events occurred later (from 23:00 LT, compare Figs. 36.2 and 36.6). Figure 36.6 also shows that, unlike Nmax and TEC, hmax starts increasing well before the MP onset. The observations (Fig. 36.6) therefore seem to confirm that the (1) eastward PPEF (Fig. 36.2) is not involved in initiating the positive storms, (2) storm-time equatorward neutral winds raised the mid latitude ionosphere in both hemispheres before sunrise, and (3) by sunrise the wind initiated the positive ionospheric storms in Nmax and TEC.

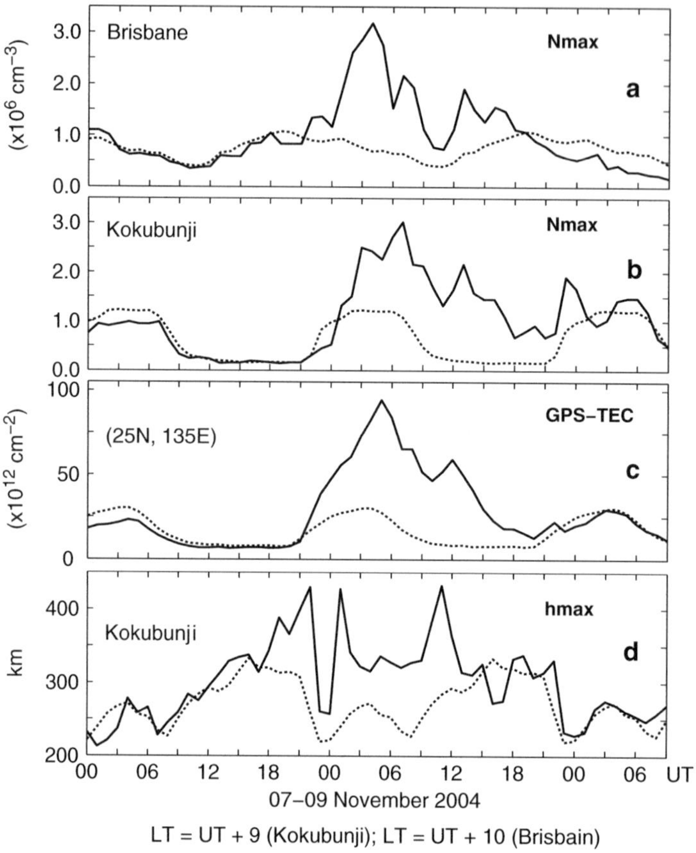

Fig. 36.6 Nmax variations at Brisbane (153.17°E, 27.5°S; 36.7°S magnetic; LT = UT + 10, *panel a*) and Kokubunji (139.5°E, 35.7°N; 26.8°N magnetic; LT = UT + 9, *panel b*), and GPS-TEC close to Kokubunji (*panel c*) from 00 UT on 07 November to 09 UT on 09 November 2004 (*solid curves*); *dashed curves* show the seven quiet day averages prior to the storm. *Panel d* shows the corresponding hmax at Kokubunji

Figure 36.7 compares the latitude variations of the electron density Ne near 400 km height obtained from the CHAMP satellite during the super storm (thick curves 1) with those during a quiet period (06 November, thin curves 2). The variations are for the equatorial crossing times of the satellite before MP onset (panel a, 17:21 UT on 07 November), early MP (panel b, 21:56 UT), during MP (panel c, 02:35 UT on 08 November) and after MP (panel d, 15:34 UT on 08 November). The crossing times in UT (UT1 for storm and UT2 for quiet) are given in the top axes. The corresponding magnetic local times (MLT1 and MLT2) and geographic longitudes (GLOG1 and GLOG2) are noted in the bottom axes. The data during the storm period and quiet period are for nearly same MLTs.

As shown by panel a, the electron density before the MP onset (thick curve) is two to four times greater than that during the corresponding quiet period (thin curve) at mid and low latitudes down to about ±20°. With the progress of time the density increases further and extends to cover the equatorial latitudes also (panel b). These observations (panels a and b), which are before the start of the eastward PPEF events, also indicate that the storm-time equatorward neutral winds initiated the positive ionospheric storms. The satellite pass in panel c is during the third eastward PPEF event (Fig. 36.2, panel a) and close to Japan-Australian longitudes. The satellite detected strong EIA, the crests shift to higher latitudes by about 10° compared to the quiet-time EIA, and the crests become strong at the expense of equatorial ionization as expected from the super plasma fountain in the presence of equatorward neutral winds (Balan et al., 2009a). The crests become sharp and strong because of plasma convergence, with

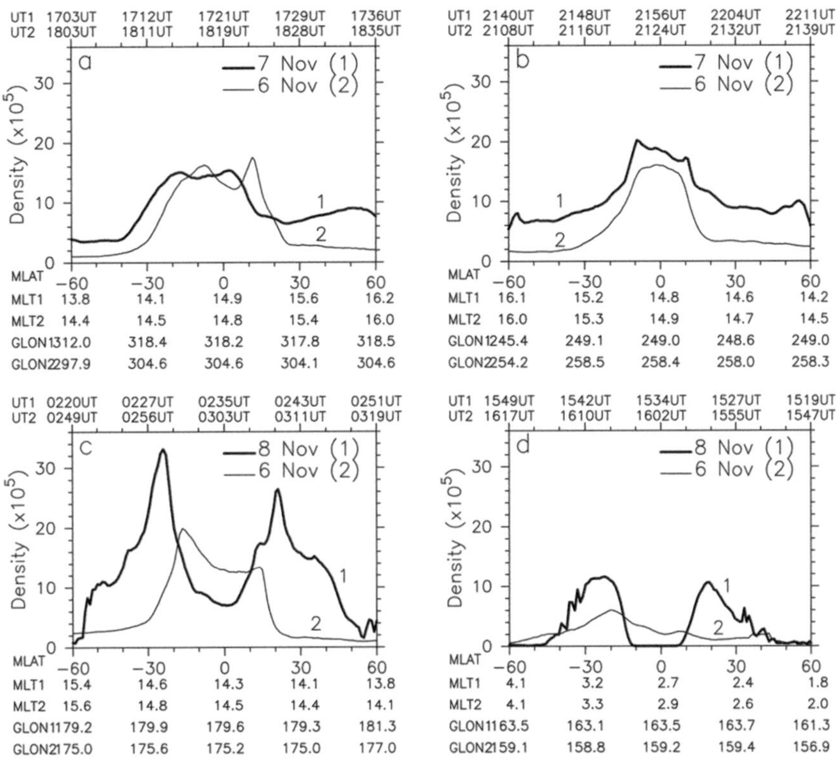

Fig. 36.7 Latitude variations of the electron density Ne near 400 km height obtained from the CHAMP satellite during the super storm (*thick curves 1*) and during quiet periods (06 November, thin *curves 2*). The variations are shown at the equatorial crossing times of the satellite before MP onset (*panel a*, 17:21 UT on 07 November), early MP (*panel b*, 21:56 UT on 07 November), during MP (*panel c*, 02:35 UT on 08 November) and after MP (*panel d*, 15:34 UT on 08 November). The satellite crossing times in UT (UT1 for storm and UT2 for quiet) are given in the top axes. The corresponding magnetic local times (MLT1 and MLT2) and geographic longitudes (GLOG1 and GLOG2) are noted in the *bottom axes*. Positive latitude is North

the super fountain driving the plasma poleward and wind driving the plasma equatorward. As shown by panel d, the nearly symmetric strong EIA continues until after the following midnight. The observations (Figs. 36.6 and 36.7) agree with the physical mechanism.

36.7 Discussion of Physical Mechanism

According to the physical mechanism, equatorward neutral winds are required to produce positive ionospheric storms. The positive storms are expected to start equatorward of the auroral oval and move toward lower latitudes with the equatorward propagation of the winds. If the winds reach low and equatorial latitudes, then the positive ionospheric storms can be strong and centered at $\approx\pm15°$–$30°$ magnetic latitudes. The eastward PPEF, if occurs, also shifts the EIA crests to higher than normal latitudes. The convergence of the plasma then can cause positive ionospheric storms centered at latitudes poleward of the quiet-time EIA crests as has been observed during super geomagnetic storms (e.g., present paper; Tsurutani et al., 2004). However, the equatorward winds alone during super storms are expected to produce stronger positive ionospheric storms extending more to lower latitudes than together with eastward PPEF. The TADs (e.g., Shiokawa et al., 2008) can contribute similar mechanical effects as the equatorward winds. The storm-time changes of the other parameters involved in the height rise and diffusion velocity (Section 36.5) can also affect the ionospheric storms.

If eastward PPEF occurs during daytime before the development of storm-time equatorward winds, then there can be negative effects in electron density at low and mid latitudes because the usual daytime neutral wind is poleward. However, if eastward PPEF occurs in the evening (around the pre-reversal enhancement of the eastward electric field), it may lead to positive storms originating from the equator even before the development of storm-time equatorward winds. That is because, the normal neutral wind usually changes from poleward to equatorward before evening. Weak daytime eastward PPEF may also produce weak positive storms at low latitudes even in the absence of equatorward neutral winds, provided the EIA crests do not shift beyond about 20° magnetic dip angles where plasma diffusion becomes strong.

As shown in Fig. 36.5, both mechanical effects of the equatorward wind are significant at subauroral latitueds. These effects can therefore strengthen the positive ionospheric storms at these latitudes, which are interpreted in terms of the equatorward expansion of convection electric fields (e.g., Foster, 1993; Heelis et al., 2009). The effect of the equatorward wind to reduce diffusion (through $U \cos I$) can be significant to maximum at the equator (panel b, Fig. 36.5) provided U is not zero. Due to this effect, a daytime westward electric field and equatorward winds and /or waves can produce positive ionospheric storms at low latitudes. The chemical effect of the downwelling equatorward wind that increases the $[O] / [N_2]$ ratio at low latitudes (e.g., Roble et al., 1982) can also contribute to the positive ionospheric storms at low latitudes (e.g., Burns et al., 1995; Vijaya Lekshmi et al., 2007).

36.8 Conclusions

A physical mechanism of positive ionospheric storms at low and mid latitudes reported recently has been reviewed. According to the mechanism, an equatorward neutral wind is required. The mechanical effects of the wind reduce (or stop) the downward diffusion of plasma along the geomagnetic field lines, raise the ionosphere to high altitudes of reduced chemical loss, and hence produce positive ionospheric storms centered at $\approx\pm 15°$–$30°$ magnetic latitudes. The daytime eastward PPEF, if occurs, also shifts the EIA crests to higher than normal latitudes. The positive ionospheric storms are most likely in the longitude of the morning-noon onset of geomagnetic storms. The physical mechanism agrees with the ionospheric storms observed in Japan-Australian longitudes during the super storm of 07–08 November 2004. The possible variations of the physical mechanism have been discussed.

Acknowledgments We thank NICT (Tokyo) for the Yap and Okinawa magnetometer and ionosonde data, Y. Otsuka of STEL (Nagoya University) for the GPS-TEC data, H. Liu of RISH (Kyoto University) for plotting the CHAMP data, and ACE team for the solar wind and IMF data. The CHAMP mission is supported by the German Aerospace Center (DLR) in operation, and by the Federal Ministry of Education and Research (BMBF) in data processing.

References

Abdu MA, de Souza JR, Sobral JHA, Batista IS (2006) Magnetic storm associates disturbance dynamo effects over low and equatorial latitude ionosphere. In: Tsurutani B (eds) Recurrent magnetic storms: corotating solar wind streams. Geophysical monograph series, vol 167, pp 283–304. Copyright 2006 by the AGU, 10.1029/167GM22

Alex S, Patil A, Rastogi RG (1986) Equatorial counter electrojet – Solution of some dilemma. Indian J Radio Space Phys 15: 114–118

Bailey GJ, Balan N (1996) A low-latitude ionosphere-plasmasphere model. In: Schunk RW (ed) STEP Handbook. Utah State University, Logan, UT 84322–4405, p 173

Balan N, Bailey GJ (1995) Equatorial plasma fountain and its effects: possibility of an additional layer. J Geophys Res 100(11): 21421

Balan N, Shiokawa K, Otsuka Y, Watanabe S, Bailey GJ (2009a) Super plasma fountain and equatorial ionization anomaly during penetration electric field. J Geophys Res 114: A03310. doi:10.1029/2008JA013768

Balan N, Shiokawa K, Otsuka Y, Kikuchi T, Vijaya Lekshmi D, Kawamura S, Yamamoto M, Bailey GJ (2009b) A physical mechanism of positive ionospheric storms at low and mid latitudes. J Geophys Res 115: A02304. doi:10.1029/2009JA014515

Batista IS, de Paula E, Abdu MA, Trivedi N, Greenspan M (1991) Ionospheric effects of the March 13, 1989, magnetic storm at low and equatorial latitudes. J Geophys. Res 96(A8): 13943

Burns AG, Killeen TL, Deng W, Carignan GR, Roble RG (1995) Geomagnetic storm effects in the low- to middle-latitude upper thermosphere. J Geophys. Res 100(14):673

Fejer BG, Jensen JW, Kikuchi T, Abdu MA, Chau JL (2007) Equatorial ionospheric electric fields during the November 2004 magnetic storm. J Geophys Res 112: A10304. doi:10.1029/2007JA012376

Foster JC (1993) Storm-time plasma transport at middle and high latitudes. J Geophys Res 98: 1675–1689

Fuller-Rowell TJ, Codrescu MV, Moffett RJ, Quegan S (1994) Response of the thermosphere and ionosphere to geomagnetic storms. J Geophys Res 99: 3893

Hanson WB, Moffett RJ (1996) Ionization transport effects in the equatorial *F* region. J Geophys Res 71: 5559

Hedin AE (1987) MSIS-86 thermospheric model. J Geophys Res 92: 4649

Hedin AE et al (1991) Revised global model of thermosphere winds using satellite and ground-based observations. J Geophys Res 96: 7657

Heelis RA, Sojka JJ, David M, Schunk RW (2009) Storm time density enhancements in the middle-latitude dayside ionosphere. J Geophys Res 114:A03315. doi:10.1029/2008JA013690

Kelley MC, Vlasov MN, Foster JC, Coster AJ (2004) A quantitative explanation for the phenomenon known as storm-enhanced density. Geophys Res Lett 31: L19809. doi:10.1029/2004GL020875

Kikuchi T, Araki T, Maeda H, Maekawa K (1978) Transmission of polar electric fields to the equator. Nature 273: 650–651

Lin CH, Richmond AD, Heelis RA, Bailey GJ, Lu G, Liu JY, Yeh HC, Su SY (2005) Theoretical study of the low and mid latitude ionospheric electron density enhancement during the October 2003 storm: relative importance of the neutral wind and the electric field. J Geophys Res 110: A12312. doi:10.1029/2005JA011304

Liu H, Stolle C, Forster M, Watanabe S (2007) Solar activity dependence of the electron density in the equatorial anomaly regions observed by CHAMP. J Geophys Res 112: A11311. doi:10.1020/2007JA012616

Lu G, Goncharenko LP, Richmond AD, Roble RG, Aponte N (2008) A dayside ionospheric positive storm phase driven by neutral winds. J Geophys Res 113: A08304. doi:10.1029/2007JA012895

Maruyama T, Nakamura M (2007) Conditions for intense ionospheric storms expanding to lower mid latitudes. J Geophys Res 112: A05310. doi:10.1029/2006JA012226

Matuura N (1972) Theoretical models of ionospheric storms. Space Sci Reviews 13: 124–189

Mendillo M, Klobuchar J (1975) Investigations of the ionospheric F region using multistation total electron content observations. J Geophys Res 80(4): 643–650

Otsuka Y et al (2002) A new technique for mapping of total electron content using GPS networks in Japan. Earth Planets Space 54: 63–70

Prolss GW (1995) Ionospheric F region storms. In: Volland (ed) Handbook of atmospheric electrodynamics. CRC Press, Boca Raton, FL, pp 195–248

Reddy CA, Fukao S, Takami T, Yamamoto M, Tsuda T, Nakamura T, Kato S (1990) A MU radar-based study of mid-latitude F region response to a geomagnetic disturbance. J Geophys. Res 95: 21077

Reigber C, Luhr H, Schwintzer P (2002) CHAMP mission status. Adv Space Res 30: 129–134

Rishbeth H (1991) F-region storms and thermospheric dynamics. J Geomagn Geoelectr 43: 513

Roble RG, Dickinson RE, Ridley EC (1982) Global circulation and temperature structures of thermosphere with high-latitude plasma convection. J Geophys Res 87: 1599

Sastri J, Jyoti N, Somayajulu V, Chandra H, Devasia C (2000) Ionospheric storm of early November 1993 in the Indian equatorial region. J Geophys Res 105(A8): 18443

Shiokawa K, Tsugawa T, Otsuka Y, Ogawa T, Lu G, Saito A, Yamamoto M (2008) Optical and radio observations and AMIE/TIEGCM modeling of nighttime traveling ionospheric disturbances a storms. In: AGU, Washington, DC monograph on mid-latitude ionospheric dynamics and disturbances. pp 271–281

Tsurutani B et al (2004) Global ionospheric uplift and enhancement associated with interplanetary electric fields. J Geophys Res 109: A08302. doi:10.1029/2003JA010342

Vijaya Lekshmi D, Balan N, Vaidyan VK, Alleyne H, Bailey GJ (2007) Response of the ionosphere to super storms. J Adv Space Res. doi:/10.1016/j.asr.2007.08.029

Werner S, Bauske R, Prolss GW (1999) On the origin of positive ionospheric storms. Adv Space Res 24: 1485–1489

Zhao B et al (2008) Ionosphere disturbance observed throughout Southeast Asis of the superstorm of 20–22 November 2003. J Geophys Res 113: A00A04. doi:10.1029/2008JA013054

Index

M.A. Abdu, D. Pancheva (eds.), A. Bhattacharyya (Coed.), *Aeronomy of the Earth's Atmosphere and Ionosphere*, IAGA Special Sopron Book Series 2, DOI 10.1007/978-94-007-0326-1, © Springer Science+Business Media B.V. 2011

MIX
Papier aus verantwortungsvollen Quellen
Paper from responsible sources
FSC® C105338

If you have any concerns about our products,
you can contact us on
ProductSafety@springernature.com

In case Publisher is established outside the EU,
the EU authorized representative is:
Springer Nature Customer Service Center GmbH
Europaplatz 3, 69115 Heidelberg, Germany

Printed by Libri Plureos GmbH
in Hamburg, Germany